Philippe Rahm architectes

PHILIPPE RAHM

CLIMATIC ARCHITECTURE

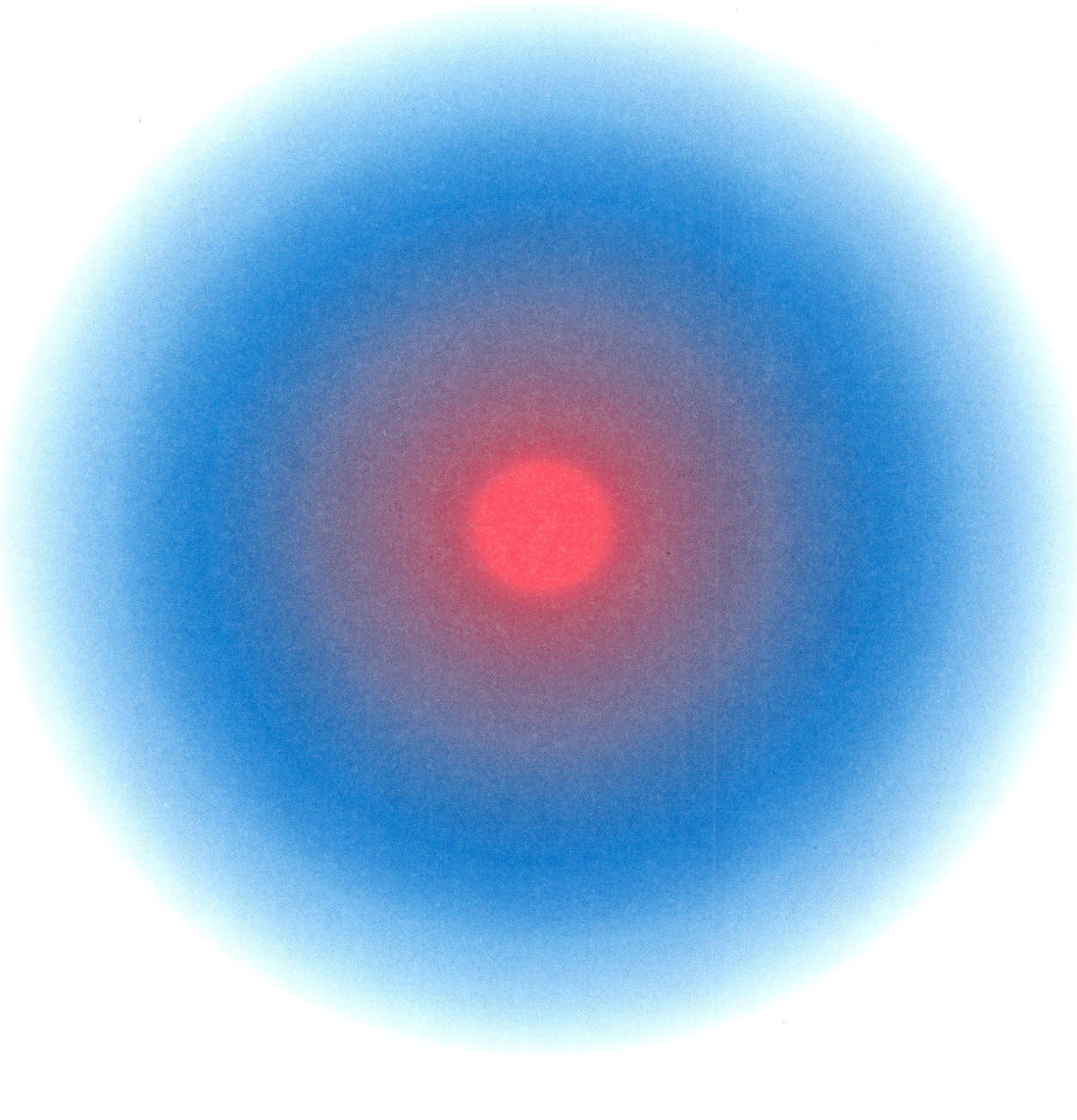

Philippe Rahm architectes

PHILIPPE RAHM

This book is published by Actar Publishers with the help of the École Nationale Supérieure de Versailles, France, committed through research and teaching to promoting architecture engaged in the fight against global warming by reducing CO2 emissions. It has also received the support of Istituto Svizzero, the Graham Foundation, Swissnex, MAAT and the SFAI.

CLIMATIC ARCHITECTURE

This book was written by
the Swiss architect Philippe Rahm
It is about climate and architecture.

This book is also an architect's monograph, a manifesto for a **climatic** architecture to face global warming, an encyclopedia on **meteorological** phenomena applied in architecture, a theoretical and practical treaty of the art of constructing **atmospheres**.

It presents work conducted by Paris-based architecture firm Philippe Rahm architectes active in the field of architecture, urban planning and landscape since 2005

ACTAR

TABLE OF CONTENTS

Introduction

CLIMATIC ARCHITECTURE

Architecture & Climate 11
Architecture & Physiology 16
Meteorological Architecture 18
Climatic Urban Planning 22
Atmospheric Decoration 25

Chapter I.

METEOROLOGICAL COMPOSITIONS

Conduction 30
Convection 56
Emissivity 84
Effusivity 98
Radiation 114
Evaporation 136
Pressure 152

Chapter II.

A SENTIMENTAL METEOROLOGY

Radiation 168
Inertia 170
Altitude 172
Diffusion 174
Afterglow 175
Sunstroke 177
Oblique 178
Obfuscation 179
Condensation 180
Cooling 181
Greenhouse Effect 182
Conduction 183
Rotation 185
Pressure 186
Anaerobic 187
Evaporation 188
Emanation 189
Isolation 190
Acclimated 191

TABLE OF CONTENTS

Chapter III.

BUILT ATMOSPHERES

Climatorium 196
Meteorological Urbanism 212
Light Gradient 220
The Attenuated Exterior 224
Melatonin Café 230
Luminance 234
Gradiation of Interiority 240
Meteorological Garden 248
Structural Densification 262
The Convective Cafeteria 266
Gradiation of Privacy 270
White Geology 278
Spectral Light 280
Evaporated Rooms 284

Chapter IV.

ATMOSPHERIC FRONTS

With Alain Robbe-Grillet 292
With Piero Macola 294

Chapter V.

CLIMATE RESEARCH

Research into Air 302
Research into Light 306
Research into Heat 314

Chapter VI.

WEATHER REPORTS

Meteorological Architecture 332
Climatic Urbanism 341

INDEX

List of Projects 356
Acknowledgements 358
Biography 359

Philippe Rahm architectes

INTRODUCTION

CLIMATIC ARCHITECTURE

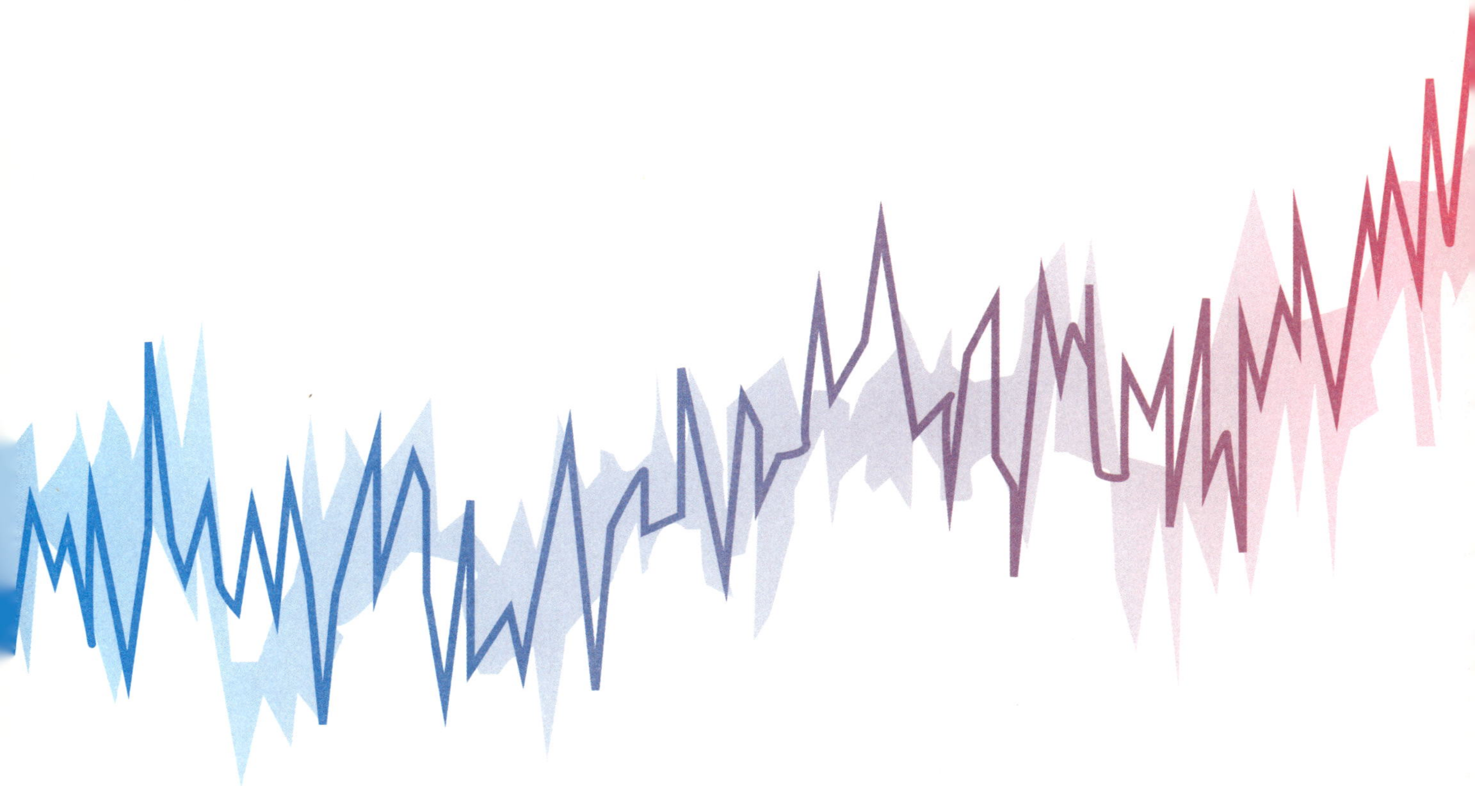

Simulated Annual Global Mean Surface Temperatures
Source: GIEC

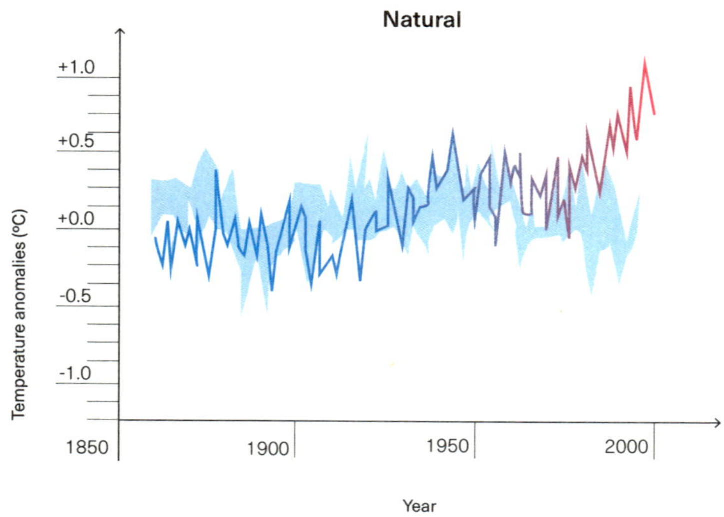

Global warming is caused by greenhouse gas emissions {Fig. 0-01}
Computer simulations show that the addition of human greenhouse gas emissions to the terrestrial model leads to a simulated global temperature increase similar to the values actually measured. Globally, 39% of CO2 emissions responsible for global warming come from the building sector. Architects are therefore at the forefront in the fight against global warming.

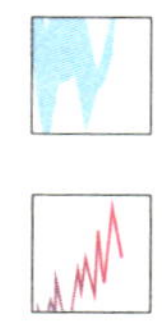

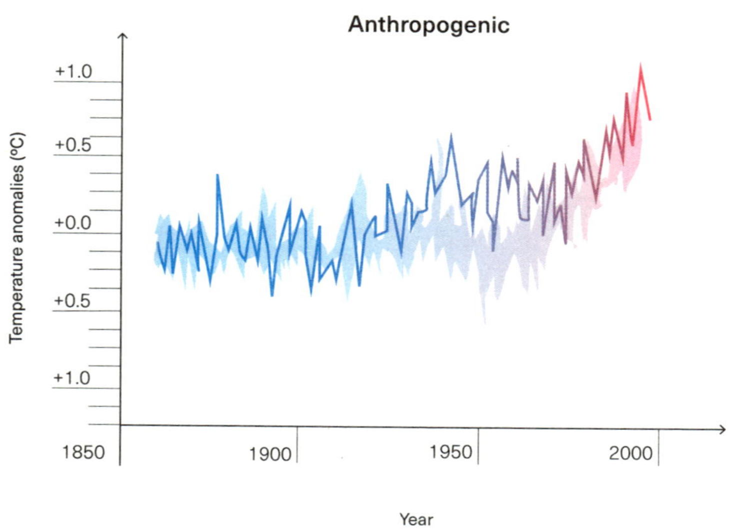

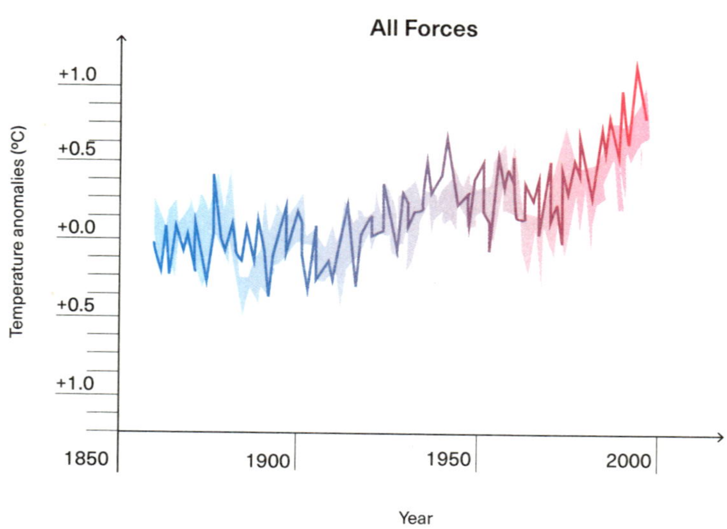

I. Architecture & Climate

How Global Warming is Changing Architecture

Global Mean Surface Temperature
Source: NASA figure adapted from Goddard Institute for Space Studies

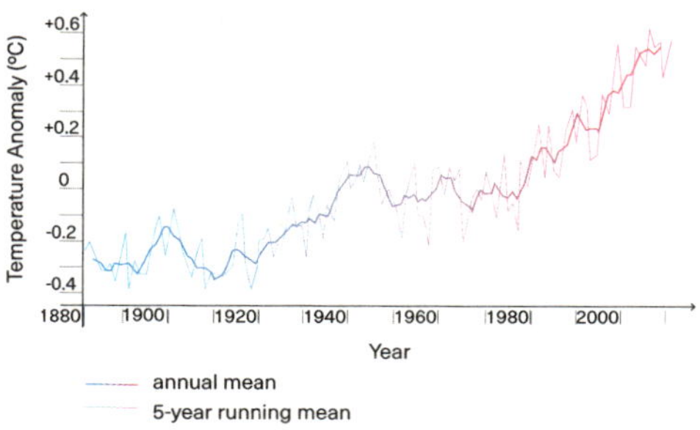

Increase of the concentration of carbon dioxide in the atmosphere
Carbon Dioxide ppm (parts per million)
Source: NASA graphs by Robert Simmon based on data from the NOAA Paleoclimatology and Earth System Researche Laboratory

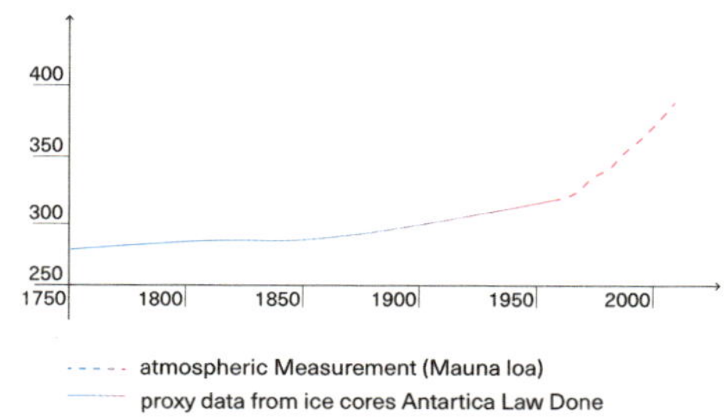

Global warming {Fig. 0-02}

The global temperature has been rising since 1812 and reaches +1.2 °Celsius in 2023

Increase in atmospheric CO2 concentrations {Fig. 0-03}

Since 1832, the concentration of CO2 in the Earth's atmosphere has increased hand in hand with temperature. It reaches 417 ppm in 2023 while it was 284 ppm in 1832

No Global Warming

The longwave radiations emitted by Earth during the night can pass throught the atmosphere, and their heat is lost in space.

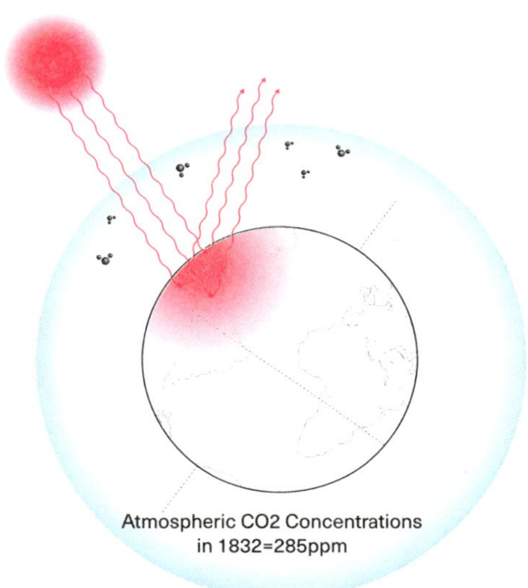

Global Warming

CO2 emitted by humans blocks the longwaves radiations from the Earth to the space, leading to global warming

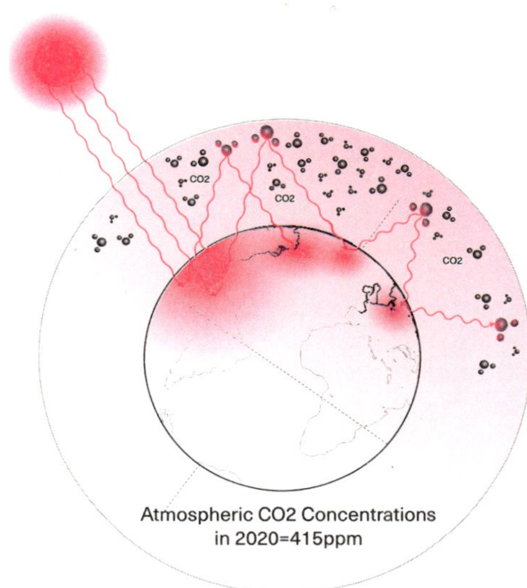

Low greenhouse gas concentration in the atmosphere (1832) {Fig. 0-04}

In 1832, the concentration of CO2 in the atmosphere was low and allowed long-wavelength infrared rays to leave the earth through the atmosphere to the universe.

Significant concentration of greenhouse gases in the atmosphere (currently) {Fig. 0-05}

Nowadays, the concentration of CO2 in the atmosphere is high, which prevents the infrared rays emitted by the Earth from leaving the atmosphere, thus increasing the temperature of the atmosphere and then of the Earth surface.

"Thus art and industry find remedy for the drawbacks of the environment, and in each region a pleasant temperature is assured for the dwellings by exposure adapted to their position on Earth." »

Vitruvius, De Architectura, 20 B.C.

The raison d'être of architecture is climatic: because humans must maintain a constant body temperature of 37°C, and because they need to constantly discharge a little heat through the skin, the environment surrounding their body must be at a slightly cooler temperature than that of their body, between 20°C and 28°C. Below that, it starts to get too cold, while above, it starts to get too hot. This environment, between 20 and 28°C, if no more found naturally outside, in nature, in the ambient air temperature, must be built artificially, shaping an atmosphere like a bubble which will encapsulate air which it will be able to raise to the right temperature - by burning wood initially, then with radiators -, or lower it, thanks to the shade of the roof initially, then thanks to air-conditioning.

Architecture exists thanks to the fact that nature rarely provides the necessary physiological temperature conditions, those of an ideal spring as established by mythology in the image of the Garden of Eden, of the lost paradise or perhaps in the African cradle of humanity, the Olduvai Gorge. In most cases, due to the seasons, and more so with the dispersal of the human species throughout the planet, moving away from the tropics, the human being is confronted with physiologically unbearable climates and seasons, with temperatures that are too low or too high, where they cannot survive. So, they must artificially reconstruct the climate of paradise or the Ethiopian highlands, reconstruct air whose temperature will be between 20°C and 28°C. Architecture is thus the art of building a climatic environment close to the human being's homeothermal condition. It is to construct a temperate temperature anywhere in the world and any time in the year. The purpose of architecture is to create a habitable climate around the human body. The raison d'être of architecture is to build physiologically humanized atmospheres. The purpose of architecture is climatic: to recreate year-round springtime atmospheres in our domestic spaces.

The conditions of architectural production today are also climatic. Indeed, in 2019, 38% of CO_2 emissions causing global warming came from the construction sector. ¾ of this CO_2 is emitted by buildings' heating or air-conditioning and ¼ is emitted in the manufacture of materials (the carbon footprint) and the building process. Architects are thus on the front line in the fight against global warming, by reducing the energy consumed to heat or air-condition indoor spaces, by thermally insulating them from the uninhabitable, icy or hot wild outdoor atmosphere.

If the raison d'être of architecture is climatic, if the ends are climatic, if the conditions today are also climatic, what about the means ? Should they remain geometric, analogical, metaphorical, functionalistic, programmatic like those of 20th century architecture and urban planning? Or could they, in turn, become climatic? Could the tools of architectural compositions become meteorological? If so, then convection is the new symmetry, conduction is the new alignment, emissivity is the new decoration. Meteorological phenomena are, therefore, the new means of composing, designing spaces, structuring cities and arranging interiors. When we talk about meteorological architecture, Climatic urbanism, atmospheric decoration, no metaphor is meant. It is literally from the natural sciences, from the principles of thermodynamics that we draw the means for architecture, urban planning, landscape and interior design.

Before the twentieth century, urban planning and architecture were traditionally based on climate to improve the comfort and health of inhabitants. Roman or Renaissance architecture erected domes to evacuate warm air, aligned windows and doors to let the wind pass and cool off in summer. The Middle Ages hung tapestries against the walls to protect them from the cold. The urban planning of ancient Rome was construed around wind directions for protection in winter or to take advantage of it to cool off in summer; around exposure to the sun, either to hide from it in summer or to welcome it in winter, on the humidity of the air and temperature variations. These fundamental principles of urban planning were ignored during the twentieth century thanks to the vast consumption of fossil fuels by heating and air-conditioning systems, pumps, motors or refrigerators, which today cause air pollution, heat islands, and more generally, the greenhouse effect and global warming. Combating heat waves and climate change is forcing architects and town planners to respond effectively to contemporary climate challenges. It forces them to rethink their urban patterns, the form of buildings, the design of spaces according to these new physical realities. Working with the albedo, with evaporation, with convective air movements are all renewed design strategies to rethink the design of streets, the choice of materials and shapes of buildings in an understanding of the local climate context and energy resources. Faced with the climate challenge of the 21st century, Climatic architecture proposes to anchor urban planning and architecture on its intrinsic climatic qualities, where the convective, the conductive, the emissive or the effusive replace the geometric, the symbolic, the analog and the metaphorical.

The Irish scientist John Tyndall is credited with discovering in 1859 that CO_2 is opaque in certain wavelengths of infrared radiation, such as those emitted by Earth into space and are thus found trapped in the Earth's atmosphere, causing global warming. The CO_2 emitted by humans was formerly, prior to the 19th century, integrated into the general carbon cycle. We burned wood, so we emitted CO_2, but it was quickly recovered by the trees growing, which harnessed it to make use of it, giving off oxygen, which allowed us humans to breathe.

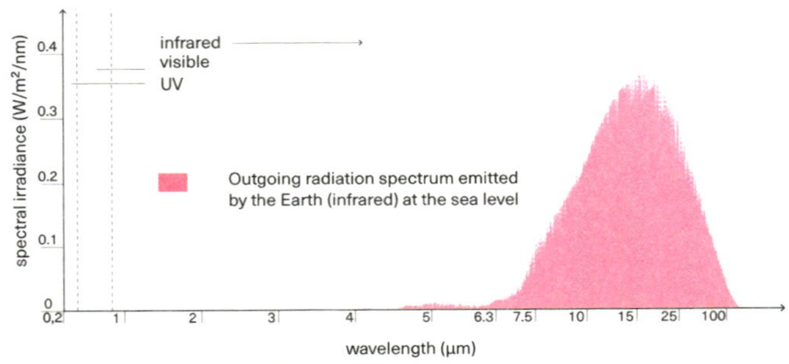

Source: "The greenhouse effect and carbon dioxide" written by Wenyi Zhong and Joanna D.Haigh in Weather - April 2013, Vol. 68, No.4

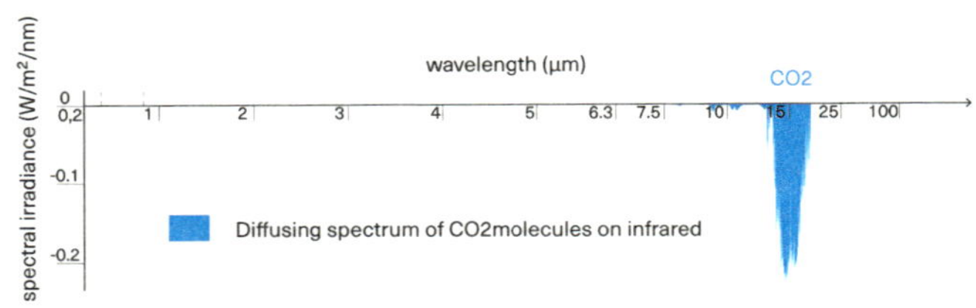

Source: "The greenhouse effect and carbon dioxide" written by Wenyi Zhong and Joanna D.Haigh in Weather - April 2013, Vol. 68, No.4

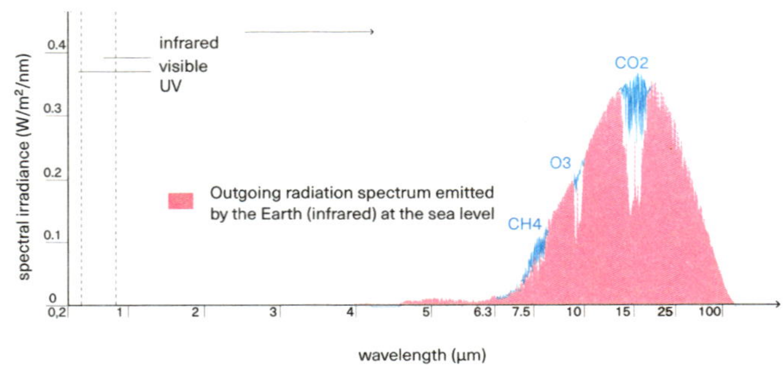

Source: "The greenhouse effect and carbon dioxide" written by Wenyi Zhong and Joanna D.Haigh in Weather - April 2013, Vol. 68, No.4

But the discovery of fossil carbon, that is, of all these forms of energy hidden in the subsoil of our planet, which had for a long time been removed from the atmosphere, natural gas, coal, oil, has disturbed the balance of the ecological cycle of carbon. As of the nineteenth century, with the industrial revolution, these fossil fuels began to be burned and since there were not enough trees to absorb this new CO_2 load, it remained in the air, increasingly blocking the infrared rays emitted by the earth, causing the inevitable warming of the planet. Since the ecological, physiological, social and political consequences of this global warming have begun to be felt in recent years, the fight against CO_2 emissions has become a priority. Here is where architecture is called upon, as the construction sector was responsible in 2019 for 38-39% of global emissions, far higher than air transport, for example, at 2%. Of this 39%, 11% concerns building materials and building construction (the latter figure is tending to increase as the energy efficiency of buildings is improved by better thermal insulation following the thermal regulations imposed by governments). The remaining 28% of global CO_2 emitted by the building sector relates to the operation of buildings during their lifetime, for heating, air-conditioning (growing annually and now accounts for 5% of CO_2 emissions), or lighting. The priority is, therefore, to reduce this figure of 28%, and this is what occurs immediately when thermally insulating buildings to a thickness of at least 20 cm.

Electromagnetic spectrum of the radiation emitted by the Earth to the universe {Fig. 0-06}

Solar radiation reaching the Earth's surface is converted into heat. These hot terrestrial surfaces will in turn emit rays of different wavelengths toward the universe, especially in the infrared.

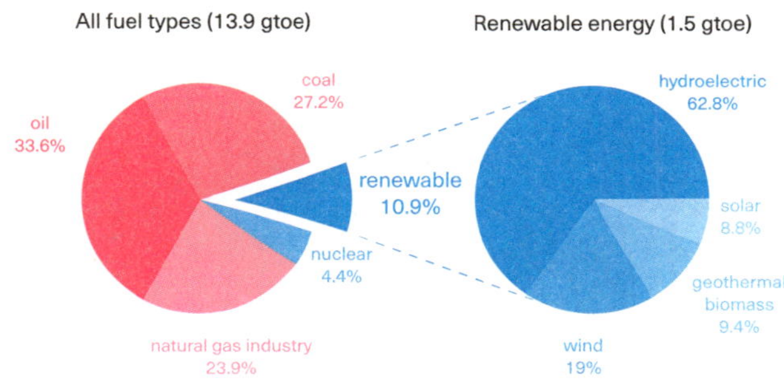

Source: 1965-2018 BP Statical review of World Energy

Global energy consumption {Fig. 0-09}

As the amount of fossil fuels consumed is very high (85%), it will take several years to completely stop using this type of energy resource in favor of renewable energies. This necessary transition from fossil fuels to renewable energies is called the "Energy Transition". In the meantime, it imposes a reduction in energy consumption in order to reduce CO_2 emissions that cause global warming.

CO_2 absorption spectrum {Fig. 0-07}

CO_2 is generally transparent for electromagnetic waves (especially in the visible spectrum of light), with the exception of infrared waves around 15 micrometers for which the molecules are opaque.

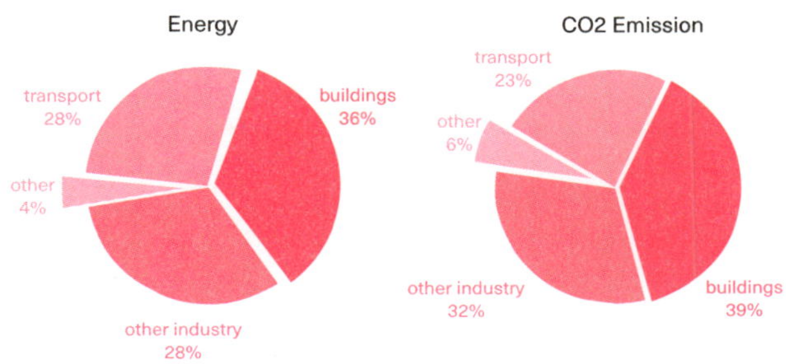

Source: Global Alliance for Building and Construction, 2018 Global Status Report. Towards a zero-emission, efficient and resilient buildings and construction sector p.11

The building sector emits 39% of the CO_2 that causes global warming {Fig. 0-10}

The high amount of CO_2 emitted by the building sector shows that priority must be given to acting on architecture to combat global warming by reducing expenditure on non-renewable energy.

Spectrum of terrestrial infrared emissions emitted by CO_2 {Fig. 0-08}

CO_2 partially blocks infrared radiation emitted from the Earth's surface into space, creating a greenhouse effect that causes global warming.

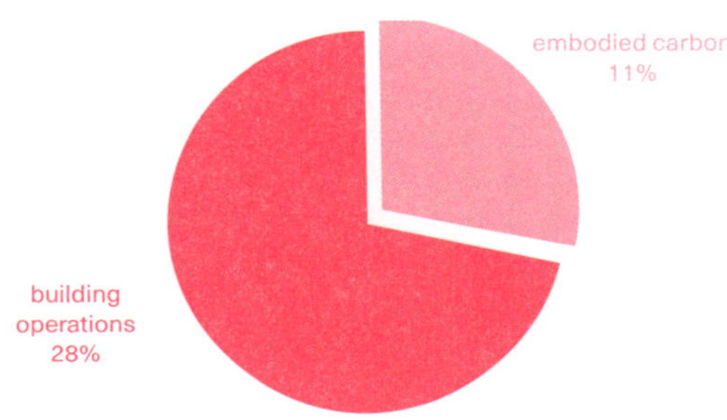

Source: UN Environnemental Status Report 2017, EIA International Energy Outlook 2017

Heating is the main cause of global warming {Fig. 0-11}

Almost 3/4 of CO_2 emissions from the building industry occur during the use of buildings and are mainly due to heating and air-conditioning. The remaining quarter of emissions are due to the production of materials and construction.

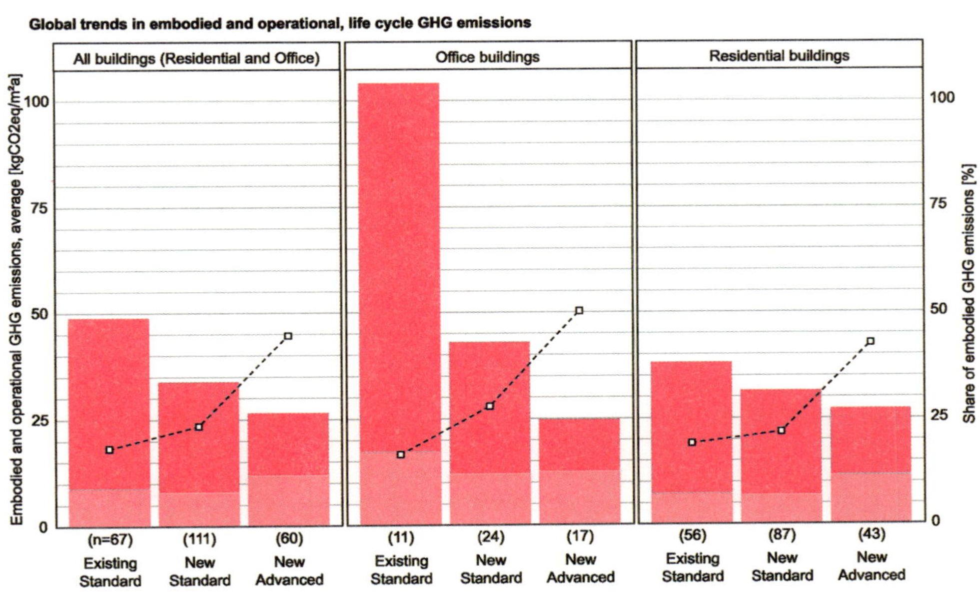

Source: Martin Röck, Embodied GHG emissions of buildings, Applied Energy, Vol. 258, 15 January 2020.

Ratio between the carbon footprint of materials and the CO_2 emitted by building techniques {Fig. 0-12}

Since 2000, thermal regulations and energy optimization programs such as Passivhaus or Minergie have greatly improved buildings' thermal insulation, reducing buildings' energy losses during their functioning. Thus, today, the carbon footprint of building materials is increasing, now accounting for half of the sources of CO_2 emissions.

Building operations
Embodied carbon

Global Warming
Anthropic emission of CO2 by buildings represents more than 39% of the Greenhouse gases emission responsible of Global Warming

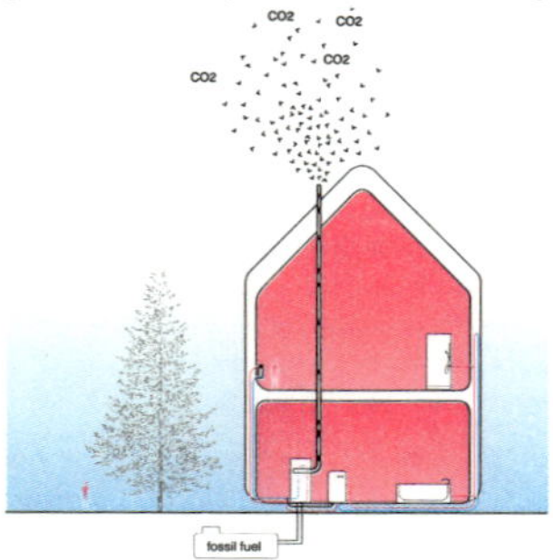

No Global Warming
Use of renewable energies doesn't emit greenhouse gas. Renewable energies don't induce global warming.

Principle of a house emitting greenhouse gas {Fig. 0-13}

The lack of thermal insulation and the absence of controlled ventilation requires a huge demand for fossil energy (gas, oil, coal, fuel oil, etc.) to heat indoor spaces in winter (or cool them with air-conditioning in summer). By burning carbonaceous energy underground in boilers to produce the hot water that will flow through the radiators will emit CO2 during the combustion of these fossil fuels, which will add to global warming.

Principle of a renewable energy house {Fig. 0-14}

If the energy sources are not fossil-based, the house will no longer emit greenhouse gases and will no longer play a role in global warming. Energy consumption will then no longer be a concern, as it will be carbon-neutral and therefore have no impact on the climate. Achieving this carbon neutrality is the goal of the energy transition. But until all energy sources are decarbonized, energy consumption must be reduced to decrease CO2 emissions.

Passivehaus / Minergie
Building technics for decreasing the consumption of energy
-Continuos thermal insulation 25cm thick
-Continuous air proof membrane
-Heat recovery ventilation

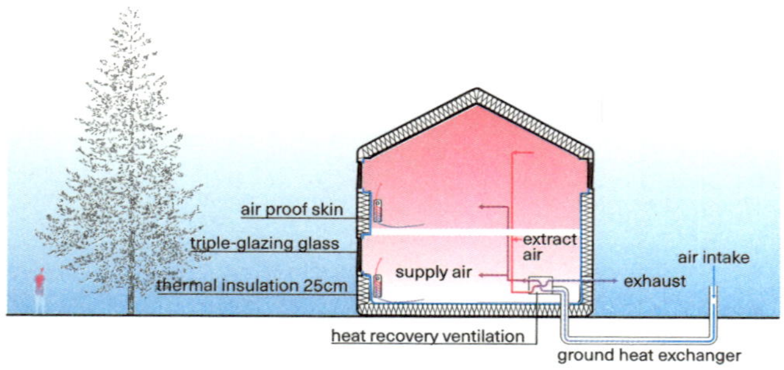

Minergie / Passivhaus {Fig. 0-15}

While waiting to be powered solely by renewable energy sources, i.e., during the energy transition, it is imperative to limit energy expenditure to the extent possible in order to reduce CO2 emissions since they still rely 85% on fossil fuels. To do so, it is necessary to reinforce buildings' thermal insulation to limit heat losses by thermal conduction between the interior and the exterior of the building via the building envelope (walls, roof, floor, windows), and to control the quantity of fresh air brought into the dwelling through ventilation, since each liter of fresh air brought into the dwelling will require an expenditure of heat in order to raise its temperature from that of the cooler exterior to that of the warmer interior, at about 21°C.

Dual-flow ventilation enables preheating the incoming fresh air free of charge by the heat of the outgoing stale air in a heat exchanger which brings the incoming air flow into thermal contact with the outgoing air flow without mixing them chemically, the heat exchange taking place via aluminum or copper plates, which are highly thermally conductive but airtight. This system means that it is not necessary to release the heat of the stale hot air to the outside; it is recovered it in order to save the use of the boiler, and thus reduce CO2 emissions. It is also necessary to ensure the airtightness of buildings, so that there is no air leakage, allowing hot air to escape to the outside, a pure waste of energy. Source: *Minergie*

Thermal insulation enables reducing the costs of heating, which globally, still depends 85% on fossil fuels. For as long as we have not completely abandoned fossil fuels, a responsible building in the fight against global warming is first and foremost a building that reduces the energy consumption for its heating, ventilation and air-conditioning. It is in this scenario that meteorological architecture is deployed, in the energy-related and thermal part of the building, taking this constraint which has become major as the efficiency force of the architectural form. It is then (or at the same time) necessary to reduce construction's carbon footprint, prioritizing the re-use and transformation of existing buildings, the use of materials requiring little energy for their manufacture, transport, implementation, taking into account the sustainability of the material, its maintenance over time, its dismantling, recycling, and its disposal. Thus, materials with a low carbon footprint, such as wood, stone and earth, are increasingly used to the detriment of materials whose production requires high temperatures, such as aluminium (2,300°C) or concrete (1,450°C). The latter, for example, is composed of cement obtained from limestone rock heated to more than 1,450°C. This temperature is difficult to achieve by burning wood whose calorific value (5.5 kWh/kg) is almost half that of coal (8.9 kWh/kg) or fuel oil (11.2 kWh/kg), which makes the production of concrete responsible for the use of non-renewable fossil fuels (rather than renewable energy such as wood), and for significant CO2 emissions as a result (8% of global CO2 emissions). The chemical transformation of limestone into cement also emits high levels of CO2 in itself. Meteorological architecture proposes taking these energy issues, which are directly responsible for 28% of global warming, as the object and means of the architectural project, capable of advising in the choice of form, materials and functions, keeping the primary objective of architecture in mind: to artificially constitute habitable climates in naturally uninhabitable climates.

Faced with the problem of global warming, by gaining awareness of the urgency of the issue of thermal performance, heating, ventilation, lighting, as the cause, subject, objective and principal responsibility of architecture, by accepting that it is no longer secondary as it was in the twentieth century, but the main issue, it is then necessary to agree to renew the formal, material and functional infrastructural principles of architecture, and to discover new social, political, ecological and aesthetic values.

Global warming reminds us that architecture is fundamentally the art of building climates: houses are to be less hot when it is too hot, to take the cover of shade when it is too sunny, to get warmer when it is too cold, to take cover when it is raining. In the fight against today's global warming, could we therefore change the methods of designing the architectural and urban project, and make climate and meteorology the tools of architectural and urban design? Can we base a section on a convective flow? Establish a plan based on gradations of light insolation? Choose a material according to its thermal effusivity value, design a façade according to its thermal emissivity? So what is a Climatic city? What is meteorological architecture? How can we rethink interior design as a response to thermal challenges?

Today we are at a turning point where we must reinvent the modalities of architecture and urban planning in order to cope with the upheaval of global warming and its accompanying chemical pollution (PM 2.5). Modern and post-modern design patterns of the 20th century are giving way to climatic and ecological designs. Thus, there is a need to recover the climatic, energy- and health-related fundamentals of architecture in order to overcome our dependence on fossil fuels, reduce CO2 emissions and combat heat waves and pollution.

Today, the climate has become the main object and subject of architects and it is the engine driving profound changes in their practices in order to integrate this new objective of safeguarding a temperate terrestrial climate. But at the same time that climate and protecting the climate become the goals of architecture, the climate and its physical, chemical and biological phenomena become the resources and tools of architecture. Thus, the meteorological vocabulary used to describe atmospheric phenomena (convection, pressure, depressions, temperature, heat, relative humidity, and glare, for example) becomes the language of architecture, and constitutes the tools of the project. Climatic architecture seeks to integrate architecture's climatic mission not only as the cause and purpose of contemporary architecture but also as a process. Architecture as a meteorology then opens up to other dimensions and definitions of space: on the scale of the building, it explores atmospheric qualities (temperature, air pressure, vapor, light, pollution, radiation, etc.) as well as physical and chemical phenomena related to the building's new climatic techniques such as ventilation, heating, air-conditioning, insulation, lighting. On a microscopic scale, it explores new realms of perception (cutaneous, thermal, hormonal, digestible, breathable), encompassing the biological and chemical relations between space and humans, dealing with the invisible qualities of the environment such as air, light, humidity, electromagnetic waves, the chemical and organic quality of the air, pollution. On an urban scale, climatic urban planning responds with real measures (by convection, evaporation, radiation, glare, albedo, etc.) to environmental crises that have been accentuated and multiplied by global warming, mainly the phenomena of urban heat islands, fine particle air pollution, and flooding. Finally, concerning interior design, the new thermal regulations for insulation or sealing against air or water vapor, are renewed with the warmth-related purposes of the decoration of the past, before vast power enabled by fossil energies operating through modern techniques - central heating, air-conditioning, electric lighting-, rendered the main functional reasons for pre-20th century decorative elements obsolete, -carpets, tapestries, curtains, screens, chandeliers, mirrors, gilding, etc. - which were as much a technical feature to improve the thermal and luminous comfort of the interiors of the past.

Since the climate is architecture's raison d'être (because it is too cold outside, it is raining, the sun is too strong), since today the climate determines the choice of its materialization, since the climate re-steers architecture toward its true object, namely the air, its temperature, its humidity, its chemical composition, light and electromagnetic waves, because architecture determines the physical and chemical nature of the space that we inhabit as a climate, because architecture is on the front line to combat global warming, could we not architecturalize the climate (i.e., characterize the emptiness) rather than design the solid (i.e., what surrounds the emptiness)? Could we not work directly on the climatic dimension of architecture using meteorological phenomena as the tools and purpose of design rather than continue to design spaces resorting to geometry or analogy as in the twentieth century, unrelated to today's climate challenges?

Thus, climatic architecture proposes establishing a new architectural language that would be immediately based on the climatic nature of the space, prescribing a kind of new know-how, new means of composition based on the physical behavior of gases, waves and particles, on the various kinds of energy transfer, on convection, conduction, evaporation, pressure, radiation, effusivity and emissivity like so many forces capable of generating forms of space, of defining the choice of materials, the shape of a building or a city.

2000-Watts Society
A 2000 Watts Society: a diminution od 2/3 of our currently consummation of the fossil energies and the use instead of recycling energies

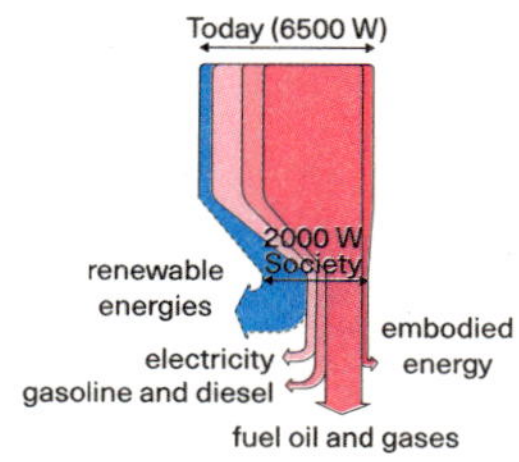

Insulate buildings to reduce their energy consumption {Fig. 0-16}
A building's energy consumption can be divided by 6 or more by providing continuous thermal insulation of 20 cm, airtightness and controlled ventilation. This reduction in energy consumption is necessary for as long as the energy transition is not completed and the building consumes carbon-based energy to operate. (Source: *2000-Watts Society, Zurich, Switzerland)*

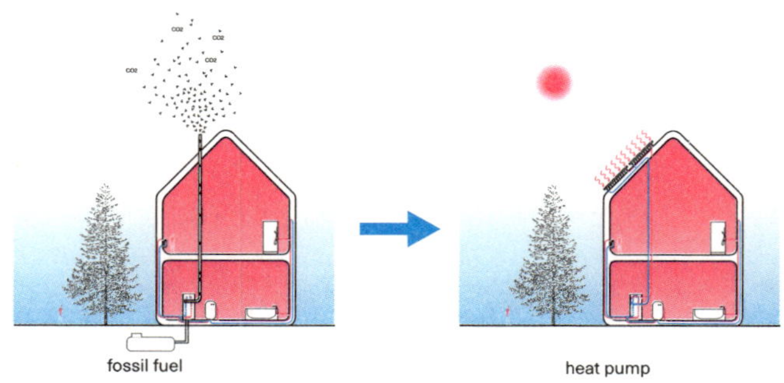

Energy transition {Fig. 0-17}
The energy transition consists of gradually reducing the amount of fossil fuels consumed, by thermally insulating, controlling the amount of fresh air entering the home, while gradually increasing the share of renewable energies until they meet the complete energy demand of the home, which will no longer result in CO2 emissions and therefore stop global warming.

Physical phenomena become the new tools of the architectural and urban project, where climate is the reason behind architecture as an amplified form of thermoregulation, while constituting the objective through the fight against global warming, the effects of urban heat islands, and air pollution. The climate, architecture's raison d'être, is also the goal, while also constituting the means. Architecture is therefore fundamentally climatic, in its objectives, its means, its conditions and its raison d'être. Climatic architecture is based on climatic phenomena that shape space and induce its materiality. Like an encyclopedia or a typological atlas, the physical phenomena underlying architectural and urban compositions, such as conduction, convection, evaporation, pressure, radiation, emissivity or effusivity, like so many chapters, constitute a formal and material catalogue of a climatic architecture. Here, spaces are defined according to the movement of air, the displacements of water vapor or heat, air renewal rates, light intensity and temperature variations, the emissivity or effusivity of the materials: *the form follows the climate.*

II. Architecture & Physiology

Architecture is an Augmented Form of Thermoregulation

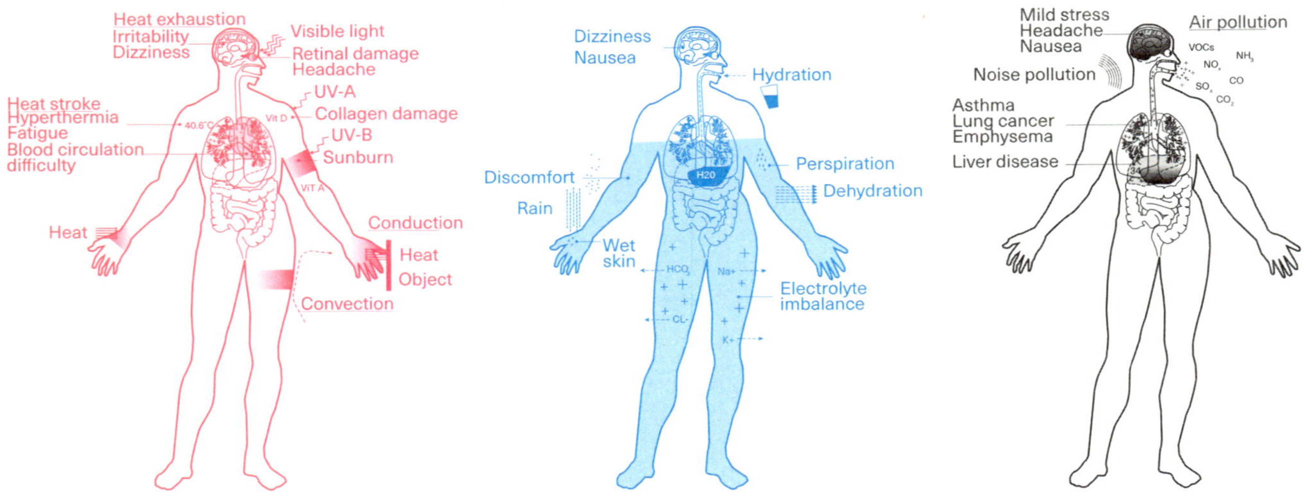

Physiological responses to excess heat, humidity and pollution {Fig. 0-18}

Human beings must keep their core body temperature at 37°C, and to do so must evolve in an external air temperature of between 20°C and 28°C in which they will constantly cool the excess heat they produce.
The external environment is most often incompatible with this physiological requirement; too cold or too hot. To compensate for this, the human body reacts in the first instance naturally through thermal regulation (thermogenesis, thermolysis) and when this no longer suffices, it must use cultural means to mitigate the excesses of the external climate, through clothing and architecture. When the human body is too hot, i.e., when the temperature of the outside air exceeds 28°C, then it activates a series of responses such as vasodilation (the blood vessels dilate to increase the surface area of exchange between the heat of the hotter blood and the colder skin) or perspiration (to enhance this perspiration, you must drink a lot when it is hot). Nevertheless, if the outside air is already overloaded with water vapor, then the sweat can no longer evaporate into the air, leaving it liquid on the body, preventing cooling by evaporation that would have occurred if the sweat could have turned into vapor and therefore lower the temperature of the skin during this physical process. Air pollution is also a major problem, especially in cities, due to the excessive concentration of gases and fine particles that are hazardous to health, particularly carcinogenic PM 2.5.

Because the building industry is responsible for 39% of global CO2 emissions, global warming requires a climatic role of architecture, which involves taking into account energy, thermal and aeraulic issues in the design and choice of architectural forms and materials, in order to initially reduce the consumption of fossil fuels for limiting greenhouse gas emissions. Although this purpose is new, emerging at the end of the twentieth century, brought about by the climate emergency, putting an end to the unbridled use of coal, gas and oil during the twentieth century, the climate issue is actually at the very origin of the appearance of architecture, indeed, its fundamental cause. Architecture's raison d'être is climatic. And if we want to discover the origin of architecture, we must return to our "homeothermal" condition, that is, the need of the human body to maintain a constant temperature of 37°C. Architecture's raison d'être is the billions of enzymes that bring about the biochemical reactions of human metabolism which can only function properly between 35.5° and 37.6°C. Human beings must therefore keep a constant body temperature, regardless of the outside temperature. Hence, the body is equipped with different internal thermoregulatory mechanisms - vasodilation, perspiration, muscle contractions, catecholamine secretions. When these first physiological reactions no longer suffice, it is then necessary to resort to external modes of climatic regulation, involving food, clothing, migration and architecture. Architecture is therefore not a social or a cultural construct, it is one of the physiological means to keep the body temperature at 37° C. Architecture is the construction of climatic shelters, artificially bringing shade when it is too hot, heat when it is too cold, blocking out the rain, the wind and confining ourselves between four walls, beneath a roof and above a floor, a small amount of air, taken from outside, whose temperature can then be more easily modified, by fire, by human heat, by radiator or by air-conditioning. Architecture is one of the solutions for exogenous thermoregulation, along with food and clothing; to lower or increase human body temperature, in addition to endogenous means such as vasodilation, perspiration, muscle contractions and catecholamine secretions for example.
To maintain a constant temperature, various natural or cultural, microscopic or macroscopic, biochemical or meteorological, dietary or urban planning actions are implemented. In its original thermoregulatory mission, architecture appears as a larger form of vasoconstriction or conversely, the burning of brown fat resembles a reduced variant of architecture. From an anthropic point of view, when we say that we are too cold, or conversely, when we are too hot, we evoke causes that are external to us, an inadequate climate at atmospheric level. Then we try to make this outdoor climate habitable and comfortable by correcting it, which is the origin and the intrinsic mission of architecture. *"Fire and groups of men gathered around it to get warm constitute the origin of architecture"* writes Vitruvius, the Roman architect of the 1st century BC. Consequently, architecture can be considered as an augmented form of thermoregulatory mechanisms, an amplified, exogenous and artificial form of thermogenesis or thermolysis.

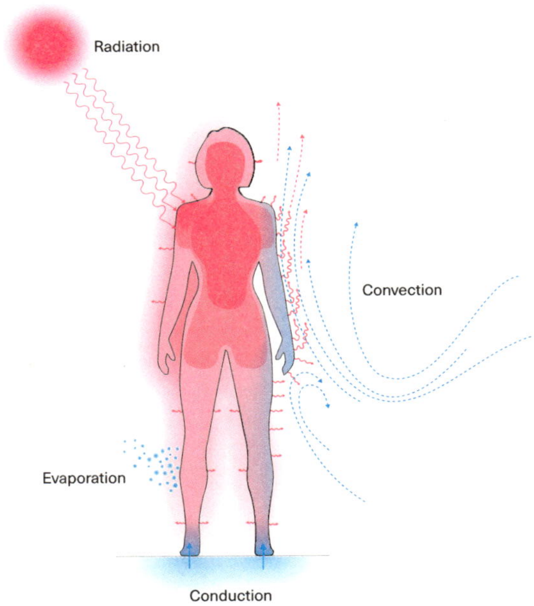

Thermoregulation

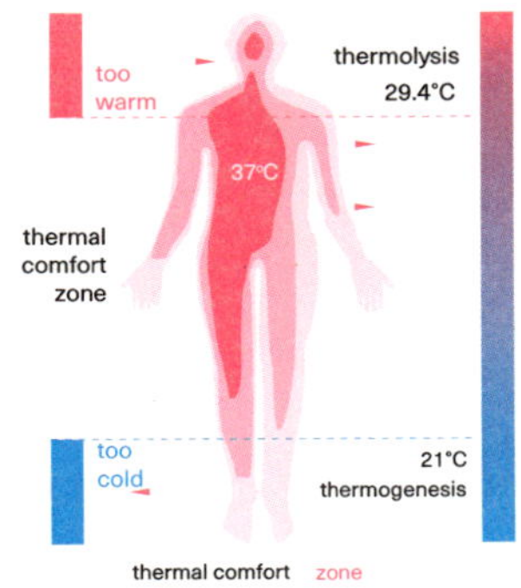

The core body temperature must always be 37 °C. The thermal comfort zone to for the thermoregulation of this temperature is 21°C - 29.4°C. Below 21°C the body begins thermogenesis and above 29.4°C the body begins thermolysis. These thermoregularity processes take energy and put strain on the body. The fundamental role of Architecture is to prevent the necessity of these processes.

Heat exchange between the body and the environment {Fig. 0-19}

The human body constantly exchanges heat with its environment, by radiation, by convection, by conduction, by evaporation at proportions of 35%, 35%, 5% and 25% respectively. If the environment is colder than the skin temperature of the human body (34°C), the body cools down. If it gets warmer, the body warms up. Heat exchange by radiation takes place only through the electromagnetic radiation received by the body (such as that emitted by the sun's rays or by a wall) and that emitted by the body, in the infrared. Exchange by convection (between a solid body and a fluid) occurs between the skin and the air. Exchange by conduction (between two solid bodies) usually takes place with contact between the feet and the ground or the hand and a door handle, for example. Exchange by evaporation is unidirectional, from the body to the air only, and occurs through perspiring and sweating.

Thermoregulation {Fig. 0-20}

To combat the cold, the human body begins to shiver to produce heat by mechanical friction, its blood vessels contract to keep blood away from the cold outside air, and its hairs stand up to brake the wind against the skin to prevent heat loss by convection. To combat heat, the body sweats to produce evaporation on the skin, thus lowering the temperature of the skin, its blood vessels dilate to expose the blood more to the outside air. And when that does not suffice, then humans use clothes and architecture to transform the air around their body by encapsulating a small volume that they can more easily alter artificially. Similarly, if solar radiation is too intense, clothing and architecture create shade, attenuating or eliminating this source of radiant heat.

The origin and raison d'être of architecture in reality can be summed up in this: locally modifying a natural climate so that it adapts to the homeothermal condition of the human being, between 21 and 28° C. In a certain way, architecture is the art of reconstructing, anywhere in the world, the climate perfectly suited to the human physiology of the Olduvai Gorge, the sub-Saharan ancestral home of Homo Erectus/Homo Sapiens. This ideal climate, where it is 26°C all year round, where there is no need to combat cold or heat, where it is possible to live without clothing or housing, is finally the mythical image of paradise, of all paradises, religious, literary, mythological and physiological. It is Ogygia, the paradisiacal island of Calypso, Shangri-La, the land of Cockayne, the Golden Age, the lost paradise, the garden of Eden, so many temperate places, where an eternal spring reigns, not too hot, not too cold, without winter or summer. Architecture's real raison d'être, its origin, is to recreate the eternal spring, a perpetual Eden in a small air bubble contained within four walls, above a floor, beneath a roof and thanks to a fireplace that in cold climates will allow raising the temperature of the air inside the house.

Because what is a house heated in winter in Northern Europe if not the recreation in miniature and indoors of a perpetual spring where it is neither too hot nor too cold? What is a house with air-conditioning in the hot summer heat of the tropics other than the reproduction of the temperate climate of paradise in East Africa? By conduction, convection, radiation, evaporation, the art of architecture is materially to transform an uninhabitable natural climate for the homeothermal body into a habitable climate, between 20°C and 28°C, that of paradise.

Thermogenesis

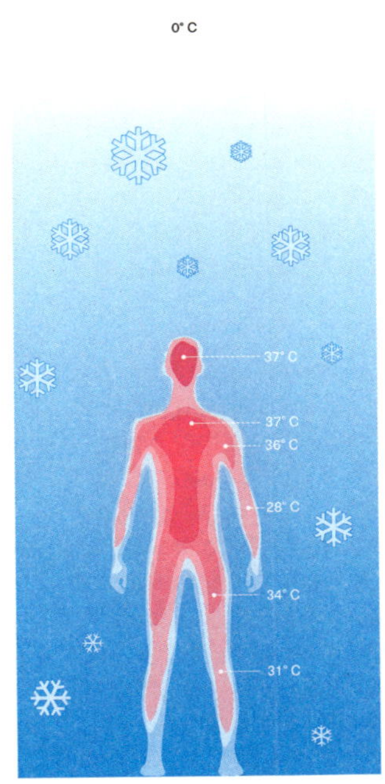

Thermogenic processes :

- Shivering
- Vasoconstriction
- Muscle contraction
-Decreasing blood flow to skin surfaces to minimize heat loss to the air

Thermolysis

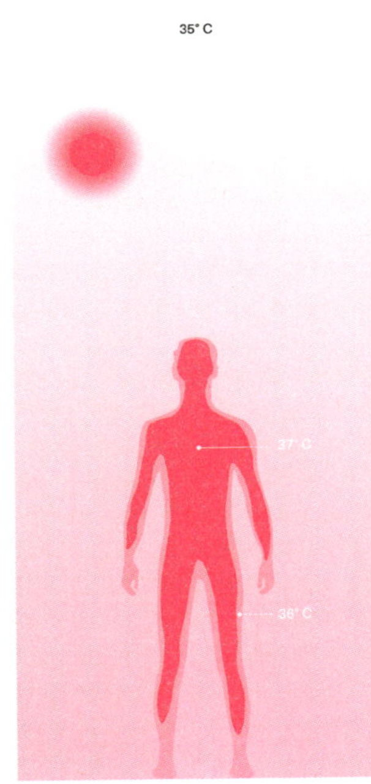

Thermolysis processes :

- Sweating
- Vasodilatation
-Increasing blood flow to skin surfaces to maximize heat transfer to the air

III. Meteorological Architecture

Climatic Value of the Void, Thermal Value of the Solid

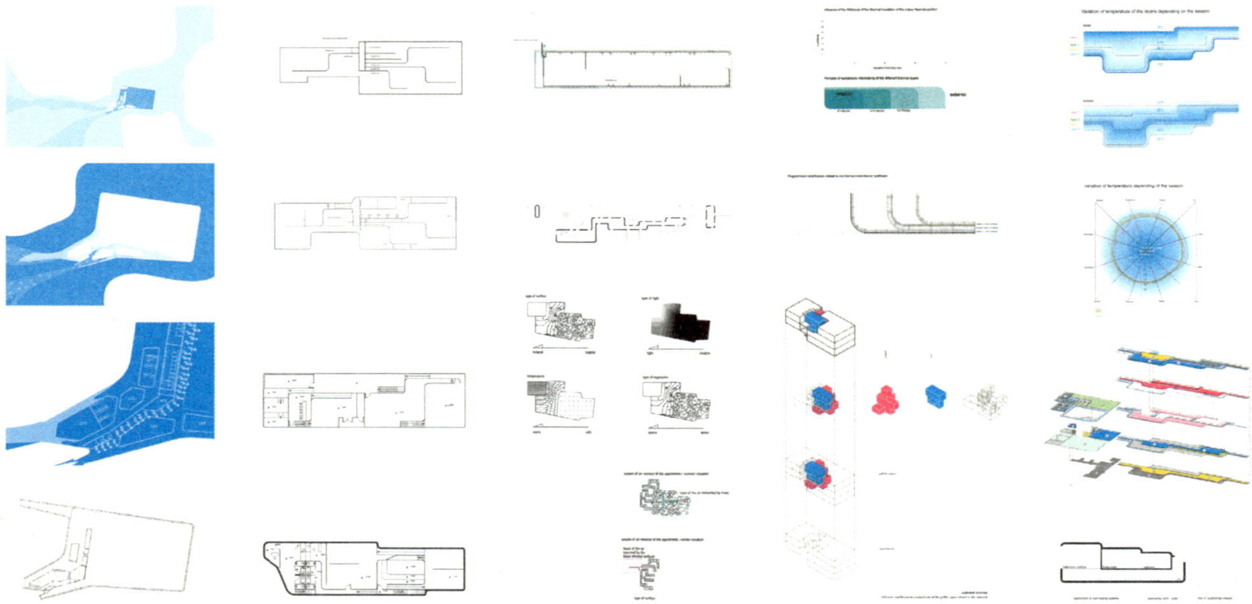

Climatic Typologies {Fig. 0-21}
Projects classified according to their meteorological composition, to the climatic phenomena involved in the design of their form.

The building sector is one of the main contributors to global warming because the combustion of fossil fuels for heating and cooling homes or for manufacturing materials and building is the root cause of nearly 39% of greenhouse gas emissions. Today, the entire industry is mobilized toward sustainable development by advocating for the improved thermal insulation of façades, the use of renewable energies, taking into account the life cycle of materials and their carbon footprint or a more compact form of construction. We can see this; these measures have a very clear goal, which is to fight against climate change by reducing CO_2 emissions. But beyond the goal, beyond these responsible and ecological aims, could climate constitute a new architectural language, that of architecture conceived as meteorology? Could one imagine that climatic phenomena such as convection, conduction and evaporation for example, might become the new tools for architectural composition? Could vapor, heat and light constitute the new building bricks and mortar of today's construction?

Climate change forces us to thoroughly rethink architecture and to shift our interests from a purely visual and functional approach to a more sensitive one concerned rather with the invisible and climatic parameters of space. Gliding from the solid to the empty, from the visible to the invisible, from metric composition to thermal composition, architecture as meteorology opens up other more sensual and more varied dimensions in which boundaries fade and solids evaporate. It is no longer a question of building images and functions, but of opening up climates and interpretations. On a larger scale, meteorological architecture explores the atmospheric and poetic potential of new building techniques such as ventilation, heating, dual-flow air renewal, and insulation. On a microscopic scale, it is in search of new fields of cutaneous, thermal, health-related or hormonal perception. Between the infinitely small of the physiological and the infinitely large of the meteorological, architecture must construct sensual exchanges between the body and space and invent new aesthetics capable of sustainably modifying the way and manner of living today.

Architects regularly remind us that the main subject of their discipline is space. This is in fact how architecture is distinguished from sculpture: we can enter architecture, whereas we must always remain outside sculpture. If sculpture deals with the solid and its forms, architecture must deal with the empty and its atmosphere. But architects have long been unable to define space other than by the solid that surrounded it as they had no real knowledge of space, of this hollow between walls they could neither grasp nor see. But this vacuum gradually gained in thickness, air became heavy with Torricelli and Blaise Pascal in the seventeenth century, it was chemically broken down in the eighteenth century into elementary particles of oxygen and nitrogen with Lavoisier and Daniel Rutherford, it was biologically charged with bacteria with Louis Pasteur in the nineteenth century, it was modulated with non-ionizing and ionizing electromagnetic waves in the twentieth century. If the architects of the 20th century had to limit themselves to working on the solid, they can nowadays work directly on the space whose shape they draw in terms of temperature, brightness or water vapor ratio. Architecture becomes the construction of atmospheres.

Climatic value of the void

The object of architecture is space, that which is removed from the general space of the earth's atmosphere, which is contained between 4 walls, a floor and a ceiling, because it is only by thus confining a certain volume of air that it becomes possible to modify its physical (temperature, humidity, speed, nature of gases) and luminous (visible light, infrared, ionizing or non-ionizing radiation) characteristics in order to make it habitable for humans, while all around the natural space is too hot or too cold, rainy or snowy, too humid or too dry, too sunny or too shady, in other words, uninhabitable.

Humans live in the invisibility of air and not in the visibility of walls. The void we enter and not the solid walls that are inaccessible to us.

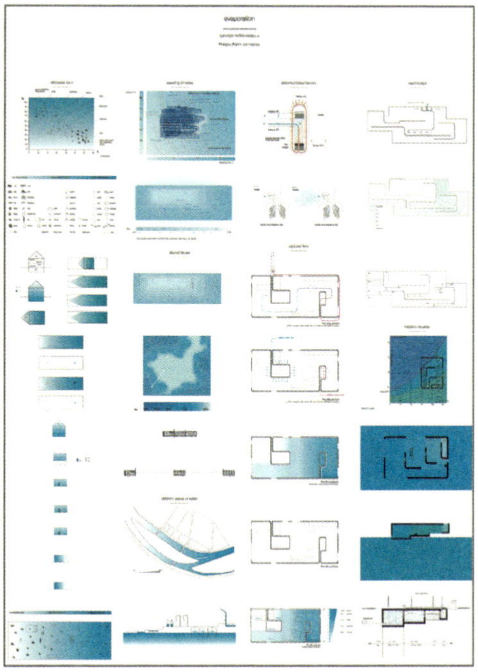

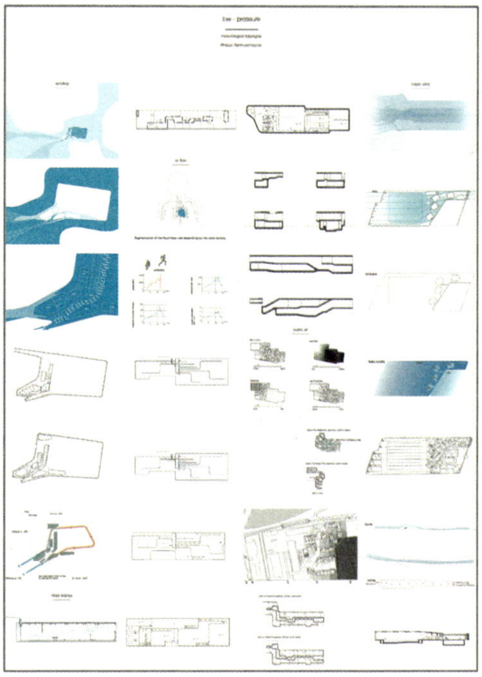

Evaporation {Fig. 0-22}
Architectural projects based on evaporation as a tool of design and formal composition. Water vapor, more or less present or produced by human activities, and its path in space trace architectural forms and arrangements, between drier, fresh air and humid, stale air through the necessary ventilation of interiors.

Pressure {Fig. 0-23}
Architectural projects based on pressure as a tool of design and formal composition. Interiors require air renewal to remove excess water vapor given off by humans and their activities, CO2, indoor pollutants or microbes. The configuration of this passage of air enables conceiving architectural forms as rivers of air whose flow, speed and design can be modeled by the shape. The house reveals its nature as a habitable ventilation duct.

Through art, architecture modifies a portion of the climate, weakens a certain portion of the Earth's atmosphere, and humanizes a natural space.
The purpose of architecture is to provide an artificially habitable climate when the climate is naturally uninhabitable for humankind.
The mission of architecture is to modify the climate's physical parameters to make it habitable for humans.
The whole art of architecture is to develop thermal, hygrometric, chemical and electromagnetic means of measuring and building.

What means can be used to achieve this? Paradoxically, the architect pens the solids (which correspond to the walls and therefore are only of interest to the architect because they contain air) and not the voids (which correspond to spaces and therefore to the architectural object in question). In the days of structuralists and postmodernists, the architect focused only on the visible, while now, interest must focus on the invisible, since it is in the invisible emptiness of air that we live and not in the solidity of the walls.

So we should change the way we design. Design the emptiness rather than the fullness, space rather than walls, represent climate rather than the opacity of the solids, shade emptiness in black and leave solids in white. This revolution in the means of representation, this inversion of the interest in the visible to the invisible, is made possible by the new digital tools to represent the invisible, air, water, vapor; waves and light. First of all, there is software to model the physical behavior of fluids called CFD (Computational Fluid Dynamics) which allow modeling the air and showing its convective movements according to gravity and sources of heat and cold. Air movements can also be modeled according to pressure/ vacuum phenomena organised by ventilation.

Then there are programs that model light, its incidence, its reflection on surfaces and its deployment in space according to the time of day, date and latitude. It is also possible to model and represent the effects of the pipes within the walls, the thermal radiation that depends on the emissivity of the materials. All these new tools for modeling and representation enable approaching the matter from the perspective of space, working directly with emptiness, shaping the invisible, modeling the atmosphere, building climate.

By working on empty space rather than solid boundaries, it becomes clear that patterns of composition are changing. Geometry, morphology, the composition of points, lines and surfaces, geometric transformations; addition, subtraction, inclusion, symmetry, etc. are no longer useful, but a whole new meteorological science is being introduced, in which patterns of architectural composition suddenly emerge in the form of convection, conduction, evaporation, pressure and radiation. Euclidean geometry makes way for meteorology as a pattern of representation with tools to design the space and architectural forms.

The architect must design climates rather than geometric shapes.
Architecture is a meteorological formalization.
Architecture is the art of building climates.

Hot air naturally rises. This phenomenon is known as atmospheric convection. The mass of hot air rises, pushing the cold air that sinks further away. In the house, it is warmer upstairs than at ground level. If we want to save energy, we must take advantage of this natural thermal stratification. You must therefore undress upstairs and dress downstairs. Wash upstairs, sleep downstairs. This is how space takes shape according to meteorological principles: the section is drawn according to meteorology rather than geometry. The aim is to reduce CO2 emissions caused by heating, since 85% of this energy consumed comes from non-renewable sources. To undress upstairs is to take advantage of the wasted heat that leaves the floor to reach the ceiling, and heats the air as it passes.

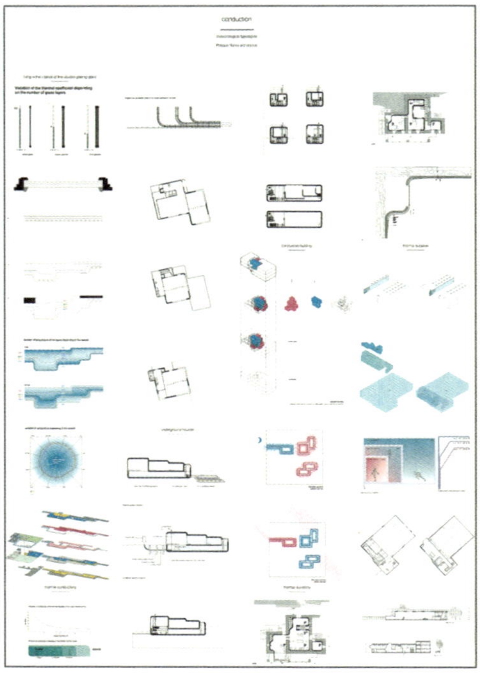

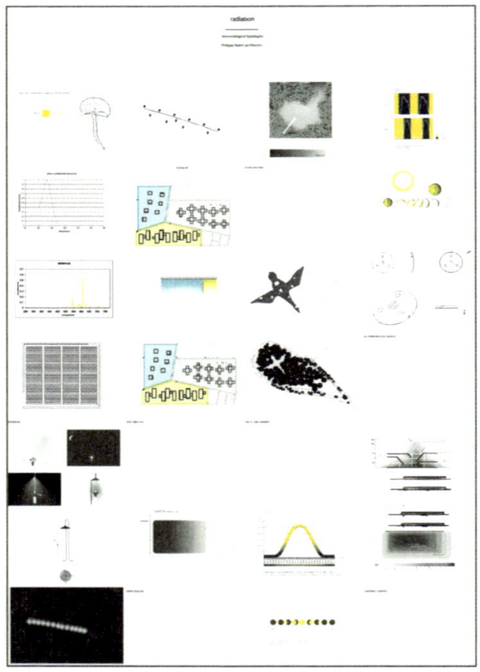

Conduction {Fig. 0-24}
Architectural projects based on conduction as a tool of design and formal composition. The need to thermally insulate buildings implies the addition of layers of wool that can be dissociated to inhabit in-between according to the thermal needs of human programs and activities.

Radiation {Fig. 0-25}
Architectural projects based on radiation as a tool of design and formal composition. The distribution of light creates varying spatial qualities that naturally make up a gradient architecture, a landscape that offers a variety of housing to be appropriated.

Reducing the amount of heating used is the first step toward reducing CO2 emissions. Space is then conceived according to formal meteorological rather than geometric principles, by analyzing the air flows always present in a building to renew oxygen and remove moisture from the air. Thus, it should be agreed to design the building around the air flows, around its speed variations to determine spaces with different uses.
An office occupied by several people who breathe and produce water vapor will necessarily have to be far more ventilated than a library filled with books that do not breathe and a few readers. Ventilation consists of taking outside air and heating it to a comfortable temperature for humans, of about 21°C, which represents considerable energy expenditure and therefore an increase in CO2 emissions. Therefore, to combat global warming, we must try to ventilate as little as possible, or more precisely, we must adapt
the rate of air renewal precisely according to people's activity in the rooms (the more intense the physical activity, the more the air will need to be renewed) and the number of people present. Thus, if it is a question of placing people and books in a room, it is possible to design the space and the building like a river, with faster moving points (like torrents) and slower ones (like lakes), a strong central current and a sidestream, where the current stagnates. By designing the airflow, we can provide work places where the air moves fastest, and reserve areas where the air stagnates for the books. Thus, the air in libraries will not be replaced in vain, thus representing an energy saving. By using fluid mechanics rather than Euclidean geometry, we can solve both space problems and combat global warming, by inventing forms according to meteorology rather than plastic logic.

The final example is that of light. By mapping natural light distribution, it is possible to determine the places that will receive more light on average than the others, and so establish a gradient of light intensities. The workspaces will be positioned in the brightest areas, while the darkest ones will be given over to the storage of food. But beyond this distribution, we can also design the space according to light gradation, using forms that take greater advantage of solar heat gains and will lead to energy savings for both lighting and heating.

These few examples enable explaining how it is possible to design architecture according to the void rather than the solid, how to project a building with space rather than walls.

Here, architecture is based on meteorology rather than geometry, taking advantage of climatic phenomena to reduce energy consumption while discovering new forms of void and spaces, new ways of living.

Thermal value of the solid

Attention should also be paid to the solid, to these barriers erected to contain the air of the house which are its walls, roof and floor. What are their features, what are they made of, what is their influence over the void they encompass?
The postmodernist structuralist architecture dating from a period of the abundant use of fossil fuels in the twentieth century saw them as signs that should be expressed.
The wall said: I am rich if I am made of marble, I am Parisian if I have the color of limestone, I am rock'n'roll if I am black.
In meteorological architecture, the wall no longer says anything but it will be important to know if its chemical composition is not polluting the air by releasing particles or carcinogenic gases such as formaldehyde, if it is not cooling the indoor air thanks to its overly high heat conduction value, if it is not sufficiently insulated, or if it allows cold air to pass in winter, forcing us to burn more fuel oil to heat the house, thus increasing global warming. By analyzing building materials from a meteorological rather than a linguistic standpoint, it is possible to choose their emissivity or effusivity value, to know if by sitting on the floor I will get cool or keep my heat, because if the effusivity of material on the floor is high (such as marble, for example), it will pump out the heat from my body and cool me down, which in winter will force me to turn up my heating and therefore generate more greenhouse gases. We must also take to look at the emissivity of the walls and of the ceiling, since if the emissivity of the materials is high, in winter it will absorb my heat by radiation, forcing me to heat more and contribute to global warming.

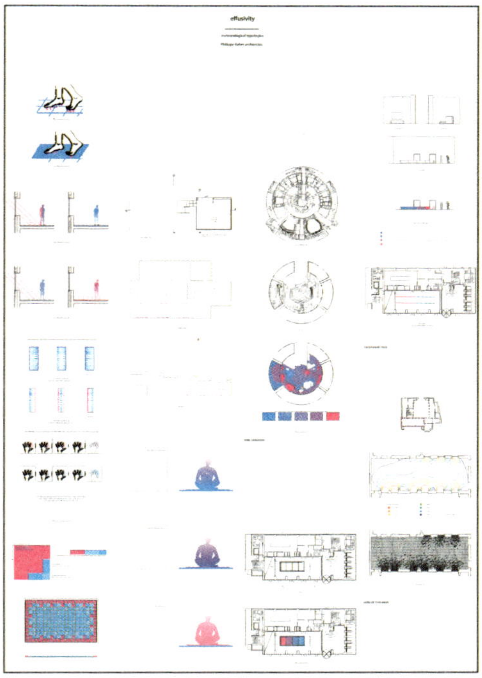

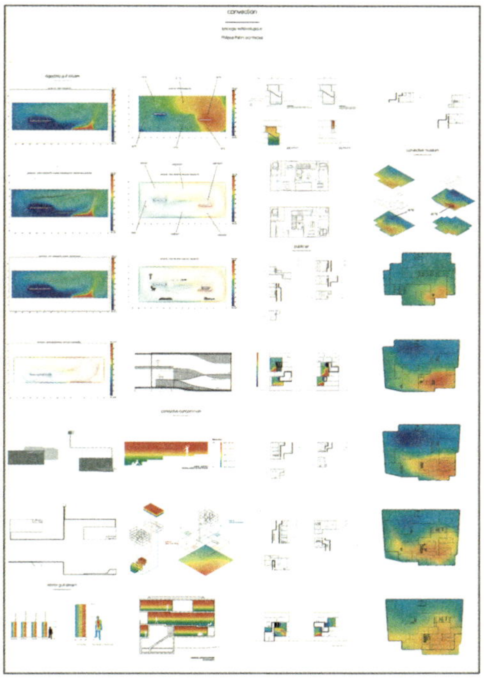

Effusivity {Fig. 0-26}
Architectural projects based on effusivity as a tool of design and formal composition. The choice of building material can be established for thermal reasons: at normal temperature, low effusivity material such as wood or wool will not exchange heat with the air or the skin, the latter remaining warm. Conversely, a high effusivity material such as marble or metal will accelerate heat exchange and cool the air and the skin.

Convection {Fig. 0-27}
Architectural projects based on convection as a tool of design and formal composition. According to Archimedes' law, hot air rises, cold air sinks, naturally. The air in contact with a hot source will rise, creating a movement of air that repels the sinking colder air. Convection can become a tool for architectural formalization, organizing spaces between those that are warmer at the top and colder at the bottom.

Thus, for cold climates and winter operation, we will choose flooring with very low effusivity, made of wood or wool, for example, so that our feet are not cooled by conduction, so that the floor does not drain our heat when in contact with it. Conversely, in warmer climes or in summer, highly effusive flooring (as long as the temperature remains below 32°C and the surface does not receive direct sunlight), such as marble, will be better adapted so that our body can release its excessive heat when it comes into contact with the ground.

In cold climates and in winter, we will choose materials with low emissivity for the walls so as not to lose the heat of our body by radiation and so that, on the contrary, the walls send us back the infrared rays, like a mirror but in invisible wavelengths, to warm us up.

For hot climates and in summer, in indoor air-conditioned environments, the cooled walls will have high emissivity to discharge our excessive heat by radiation. For façades, or for the walls of non-air-conditioned spaces, it will also be necessary to choose materials with high emissivity so that they cool us by radiation provided the air temperature remains lower than that of the skin, that is to say below 34°C, and they are not exposed to sunlight. If the air temperature is above 34°C, a low-emissivity material should be preferable to prevent the material from radiating its heat to our body and warming us by radiation in addition to the air's conductive heating.

Here, the criteria applicable, the values that guide the choice of materials, relate to their climatic properties. Choosing a material for a façade or for an interior partition is no longer a matter of narrative, symbolism or metaphor, but of climate, energy or health.

IV. Climatic Urban Planning

Albedo, Convection, Evaporation

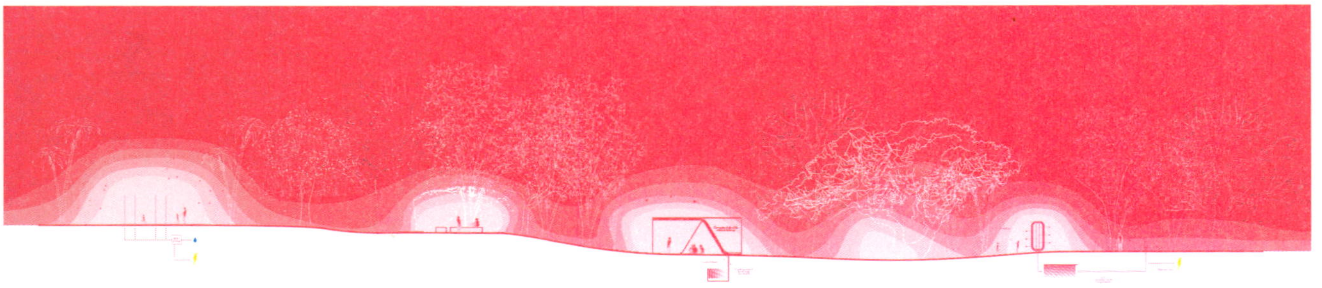

Air temperature cross-section {Fig. 0-28}
Central Park, Taichung, Taiwan.
In the heat represented in red, the white represents the colder areas in the park, cooled by shade, by convective air flows cooled by geothermal energy

Urban planning, that is to say the way in which cities are conceived and designed, the choice of where to build, the orientation of their streets, the shape of islets, the position of the buildings, was traditionally based on health and climate which, moreover, were totally linked to each other. This is what can be seen in the treatises of Vitruvius, the architect of the Roman Empire in the 1st century BC, or Leon Battista Alberti, the Italian Renaissance architect of the 15th century, who recommended analyzing a place's humidity and temperatures before founding a city, relying on exposure to the wind and orientation to the sun to position and draw the layout of streets, squares and buildings, understand the flora and the geology to choose building materials in accordance with the local resources. Choices of exposure to wind and sun, the attention paid to air humidity and temperature, the management of shade and orientation, the passage of the wind, architecture and urban planning are totally intertwined meteorologically and astronomically speaking. The aim sought by this architectural classicism in these analyses of the climate, the measurements of wind directions, of the movement of the sun, is to enable designing the city and its forms in such a way as to ensure the good health of its inhabitants, their thermal survival, so that they are neither too cold nor too hot, in an clearly pre-industrial energy setting, that is to say relatively weak and fragile in the face of climatic hazards, in the face of the seasons, in the face of the heat of summer or the cold of winter and in the face of infectious agents that were not known about and for which there were no cures. Because people died in the cities of yesteryear, of cholera, of the plague, but also of cold.

Vitruvius immediately explains in the first pages of his Ten Books that the architect's knowledge must include medicine: *"The architect should also have knowledge of the study of medicine on account of the questions of climates (in Greek klimata), air, the healthiness and unhealthiness of sites. For without these considerations, the healthiness of a dwelling cannot be assured."* For Vitruvius, the architect's knowledge must then also include astronomy, that is to say knowledge of the movement of the Sun and the Earth in relation to the geography, making it possible to orient cities and buildings according to the path of the sun, to avoid, for example, the excess heat brought by full exposure to the south, both to protect people and to protect foodstuffs: *"From astronomy we find the east, west, south, and north, as well as the theory of the heavens, the equinox, solstice, and courses of the stars. If one has no knowledge of these matters, he will not be able to have any comprehension of the theory of sundials."*

In the Renaissance, Leon Battista Alberti also engages climate and health in architecture, with recommendations concerning architecture where he describes the features of buildings in relation to the wind, the sun, their orientation such as the following: *"Shade and good ventilation will suit summer rooms, and the rays of the sun winter rooms... Winter rooms will be small, with a low ceiling and small windows. Summer rooms, on the contrary, will have a high ceiling, be wide, spacious and open to cool breezes but not to the sun or to the warm air coming from the south. A large amount of air enclosed in a large room is like a large amount of water: it is not easy to heat. "* And also: *"The roof, driving and pushing back the night, the rain and especially the heat of the sun"* or when he breaks down and identifies the different functions of a window: *"Two types of openings: one allows light and wind, the other allows people and things to enter and leave the building."*

There is nothing fundamentally strange in these realistic proposals, which, read in the twenty-first century, come across as being common sense, but they were in fact fully activated by the constant risks of disease that threatened people until the twentieth century, and especially city dwellers, as well as by humans' low availability of energy which, until the coal revolution of the nineteenth century, rendered them extremely fragile and disadvantaged to fight against cold, heat or hunger.

These fundamental climatic causes and means of urban planning were ignored during the twentieth century thanks to the widespread use of fossil fuels by hydraulic pumps, motors, refrigerators, heating and artificial air-conditioning systems, electric lighting to overcome all physical and natural constraints, all considerations concerning the natural climate, the path of the sun and the distribution of the winds. Thanks to coal and antibiotics, we were able to build cities wherever we wanted and buildings at will. Fuel oil or electrically-powered central heating, air-conditioning, artificial lighting, all these techniques that appeared as of the second half of the nineteenth century, enabled inside houses, fighting all climatic problems, cold, heat, humidity, darkness inherent in the natural geographical location of the city or the building. What Vitruvius and Alberti tried to solve in the past by precisely sizing a room, the windows and their orientation, the form of a street, the shape of a building, no longer mattered as of the nineteenth century. All this attention to the natural climate and geography was no longer useful, no longer important in view of the ease, simplicity and efficiency of modern heating,

Air humidity cross-section {Fig. 0-29}
Central Park, Taichung, Taiwan.
In a naturally humid tropical climate represented in blue, the white represents the drier areas in the park, as they are far from water points.

ventilation, air-conditioning, refrigeration and electric lighting techniques that attenuate, compensate for and eradicate all natural excesses of cold, heat, humidity, sun, darkness, lack of water or food. These modern building techniques make it possible to no longer worry about the weather and astronomy, to live henceforth in any climate, any season, any latitude, any altitude, the desert and the mountain be it in the far north or in the burning tropics, at night and in winter, without having to worry about the natural climate, the winds, the sunshine, the temperature and the humidity of the outside air. Thanks to fossil fuels, architecture manufactures within buildings spaces that are perpetually conditioned around 21° Celsius, with constant relative humidity of 50%, which in the few cubic meters of the air of a room infinitely recreate the temperate conditions of a *"perpetual Mediterranean climate"* according to the expression coined by René Dubos in the book *"Mirage of Health"*. Central heating and air-conditioning have allowed humans to colonize and comfortably inhabit any part of the Earth, to equally inhabit the heart of the moist heat of the rainforest and the dark coldness of the icy poles. Because coal and oil are actually the energy of fossilized pleasant prehistoric climates, heat petrified under the earth of an ancient sun, a *"portable climate"* as described by the American philosopher Ralph Waldo Emerson in the mid-nineteenth century: *"Every basket [of coal] is power and civilization. For coal is a portable climate. It carries the heat of the tropics to Labrador and the polar circle; and it is the means of transporting itself whithersoever it is wanted. Watt and Stephenson whispered in the ear of mankind their secret, that a half-ounce of coal will draw two tons a mile, and coal carries coal, by rail and by boat, to make Canada as warm as Calcutta, and with its comforts bring its industrial power.* "

Unfortunately, however, carbon-based energy that was massively combusted during the twentieth century is now causing global warming, air pollution and a collateral effect, the urban heat island as it is called, namely an increase in heat in the city where the air temperature is 5° Celsius higher than in the surrounding rural areas and especially at night when it does not drop, unlike in the countryside. Because the building sector is responsible for 39% of greenhouse gas emissions, because year after year this global warming multiplies heat waves in the cities, because fine particulate pollution PM 2.5, a volatile residue of the combustion of fossil fuels, becomes a public health issue, architects and urban planners must now reintegrate the local climate context and that of renewable energy resources as an engine of urban development and in the design of public spaces to reduce the consumption of fossil energy, reduce the release of fine particles, lower the temperature of cities, in a fight against global warming, against air pollution, against the effects of urban heat islands.

What is required today is to base urban master plans on geography, on the wind, on the movement of the Sun, on the movement of pollutants in order to reduce energy consumption, global warming, urban heat and air pollution. What is needed is to think of a city that is no longer dependent on fossil fuels but on renewable energies, by improving the comfort of public spaces, by building without energy microclimates that are conducive to human life.

The measures to be taken on an urban scale to combat the phenomenon of urban heat islands, air pollution and global warming are the following:

1-Albedo
Increase the albedo of roofs and roads, meaning that all of the city's horizontal surfaces must be as light as possible, or even white.

2-Convective cooling
To avoid excess heat, it is necessary to extend the phenomenon of convective cooling, that is to say, to invent a wind like the breeze which moves the air from a cooler part to a warmer part, sweeping the city, allowing the heat which accumulates and rises in temperature due to a lack of movement of air when it remains stationary to be evacuated.

3-Cooling by evaporation
Evaporation can also be caused by planting trees or activating fountains, that is to say by dissipating liquid form water into the air which will evaporate naturally, turning into gas. The energy necessary for this physical phase change to take place, from the liquid phase to the gaseous phase, will be taken from the temperature of the air, which will decrease.

Every city in the world today also suffers from carcinogenic PM 2.5 fine particle pollution. To limit this pollution, there is an obvious need to act on its cause, that is to say, to limit the emission of fine particles by doing away with diesel engines or by better filtering industrial emissions. Secondarily, these particles can be caught in the branches of trees.

Climatic Urban Planning

Air pollution cross-section {Fig. 0-30}

Central Park, Taichung, Taiwan. In a polluted environment represented in black, the white represents the less polluted areas in the park, because away from roads, or protected by trees with porous trunks to absorb noise or resinous to absorb air pollutants type PM2.5

The species of trees planted to capture fine air particles are above all conifers which, thanks to their resin, catch the fine particles, and some deciduous leaves with a marked, particularly hairy relief.

What we are seeing today is that from now on, urban planning and landscaping plans must integrate the climate issue into their project; or even make it the basis of their design in the fight against urban heat, against air pollution and against global warming. What we see is that once again urban planning is reconnected to the Vitruvian foundations of its discipline, reintegrating the climate, the path of the sun, the passage of the winds. Architecture and urban planning regain their original role: to be *"the art and industry which remedy the inconvenience of the nature of the place, and that in each region a suitable temperature is provided to the dwellings by exposure appropriate to their position on the earth".*

Today, global warming, increasing heat waves, the urban heat island phenomenon, increased flooding, air pollution, are completely disrupting the purpose and means of urban planning and the landscape. These physical, material, climatic, ecological truths and the means put in place to resist them have social as well as aesthetic consequences in the design and use of the city and the landscape, and our job is to transform these new constraints into assets by allowing, by abandoning the urban planning of the twentieth century dominated by the car and visual attractiveness, opening up other more sensual, more egalitarian, more varied ways of life to the diversity of humans and non-humans, with a more ecological and social focus and to promote a healthy environment. The fight against global warming and its catastrophic and everyday effects must be fought on two fronts. Firstly, by acting on the causes, that is to say by reducing CO_2 emissions both in the operation of built structures and by controlling the carbon footprint of building materials and construction, reducing the emissions of pollutants, those emitted by burning fossil fuels, namely PM 10 and PM 2.5. Secondly, by acting on the consequences, by fighting against heat using the wind, the water evaporation, the albedo, shade, through the choice of materials according to their thermal inertia, their thermal emissivity and effusivity.

By fighting floods by thinking of the landscape and its hills as a holding tank to absorb flood waters, by increasing permeable surfaces to absorb and delay rainfall, by thinking about the landscape, parks, gardens as well as squares and streets as places of freshness in summer without forgetting their comfort in winter by creating public spaces sheltered from wind, rain and snow.

In a way, there is also a need to return to pre-modern strategies, before fossil fuels, those described by Vitruvius, Alberti or Palladio, to counter excess heat. Also to return to the vernacular knowledge, the adiabatic refreshment of cities in the Middle East as well as the protections from the cold typical of Swiss villages.

V. Atmospheric Decoration

Thermal Decorative Art: Toward an Anthropocene Style

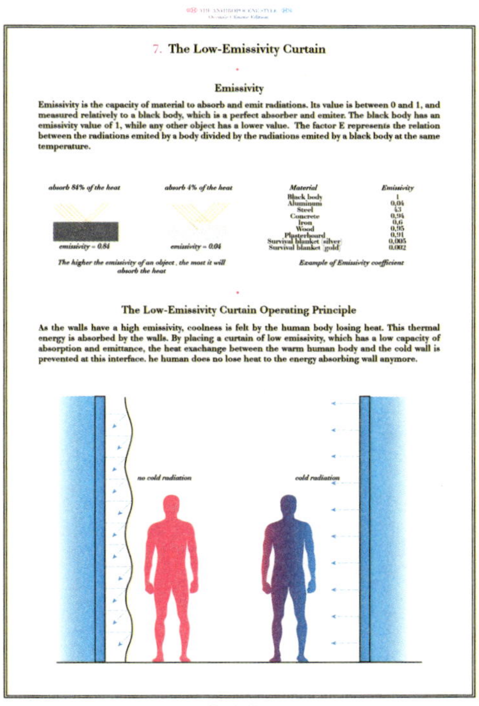

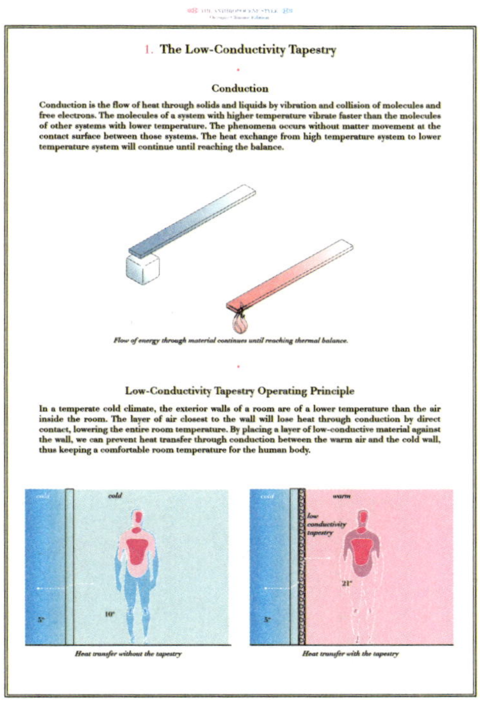

The low-emissivity curtain {Fig. 0-31}

This curtain reflects the infrared emitted by the warmer or colder bodies in front of it. Thus the radiation of the cold wall is reflected on itself, without transferring the cold to the inhabitants while the curtain reflects the infrared emitted by the human body, warming it.

The low-conductivity tapestry {Fig. 0-32}

This tapestry is a modern take on tapestries of the past though using more efficient, thicker wool to reach an extremely low thermal conductivity value, preventing the cold of the wall from being conducted to the air of the room and its inhabitants.

Global warming, which has been taking place since the start of the industrial age, and can be traced back to around 1830, when the first coal-fired plants began to emit greenhouse gases, considerably alters terrestrial ecosystems as well as the physics of the Earth and the atmosphere, climate distribution, terrestrial and marine surfaces, leading to changes in the nature of the planet that are no longer local and superficial but global and geological. Due to the magnitude of these chemical, physical and biological changes that affect both the atmosphere and the lithosphere, some scientists and philosophers have begun to evoke a change of geological era, inventing the term Anthropocene, a new geological era that would succeed the Holocene in the long succession of geological epochs. The causes of this change of epoch would no longer be "natural", volcanic or cosmic, as was the case before, but human, through the release of CO_2 due to the use of fossil fuels, deforestation and other human industrial activities. This change of climate and geological age has serious consequences for human life and for human activities, since it will lead to population migrations in the wake of the disappearance of coastal areas, changes in local economies and in agriculture based on specific climates, and an increase in catastrophic weather events.

The fight against global warming has now become a global priority that major conferences such as COP 21 attempted to legislate in Paris in December 2015. The architecture and construction of buildings is on the front line in the fight against the so-called "white cube". It is nowadays still customary for architects to decry or mock the decorative in the same way as Adolf Loos did and to abandon the subject in favor of designers, interior architects, magazines on decoration and trend books that thrive by continuing to offer curtains and drapes, colors and noble materials, mirrors and a choice of lamps, like a self-referential disciplinary field, most of the time narrative and analogical, often traditionalist and futile.

This divorce between architecture and interior decoration has not been questioned since the early days of modernity and yet it is an unexpected reconciliation that we are witnessing today through the new energy and climate constraints caused by the application of energy-saving measures such as Minergie, Passivhaus or RE2020. Because if the white, minimal neutrality of the Modernist Style was invented and has enjoyed undeniable success until today, it is because the essential and original mission of decorative interior design had been disqualified by the invention of central heating in the late nineteenth century, by the invention of electric lighting in the first half of the twentieth century and by the invention of air-conditioning in the middle of the twentieth century, by the invention and implementation of techniques of air-conditioning, heating, ventilation, lighting, a hundred times more effective than a wall hanging, a curtain or a candle.

Because what was the decorative art of yesteryear if not a series of means to improve the thermal and luminous comfort of cold, dark interiors of olden buildings. The efficiency of central heating and air-conditioning has made us forget the real and practical value of the interior decoration of the olden days, which despite its low efficiency, acted as devices to improve the thermal and luminous comfort of interiors.
In Northern Europe, for example, in the old days to combat the cold by conduction from flooring, thermally insulating carpets and non-conductive wooden floors were implemented, which constituted a blueprint for thermal insulation.
Against the radiant cold of high heat conducting old stone walls, tapestries, woodwork, and drapes were placed to block the cold radiation emitted by the walls. Against the entry of cold air through the imprecise joints between the walls and the old window frames and the draughts generated, these fixed curtains framing the windows were positioned covering the joints of the window frames through which the cold air penetrated as well as by arranging screens.

Atmospheric Decoration

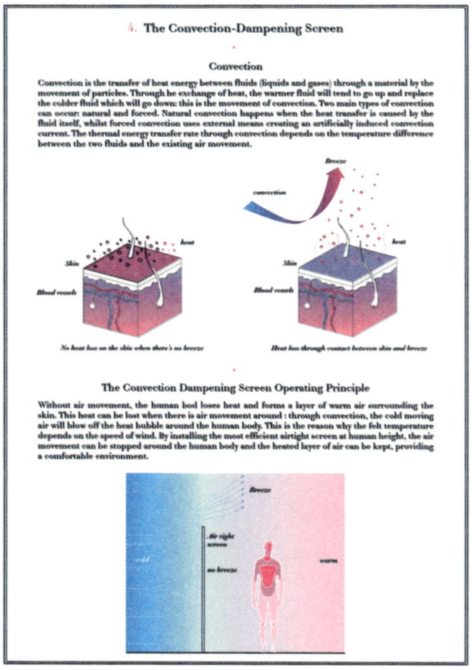

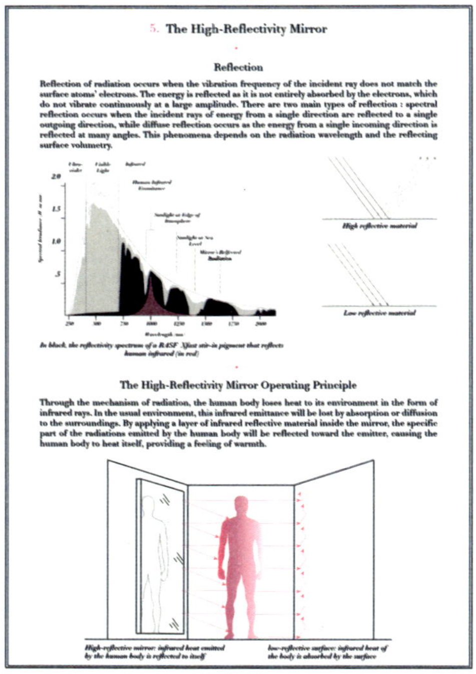

The screen against convection {Fig. 0-33}
This screen made of an extremely airtight material prevents winter air currents from cooling the bodies of the inhabitants by convection.

The infrared mirror {Fig. 0-34}
This mirror, absorbing in the visible and therefore the visually black, reflects in the invisible the wavelengths emitted by the human body in the infrared, thus the human body warms itself.

Against the maximum conduction of old single glazing, heavy velvet curtains were pulled at night to prevent the heat from inside from being sucked out by the cold of the night sky. And to increase the heat contained inside the rooms, the air was heated by radiation and convection by chimneys and then by means of the stoves which constituted so many decorative elements of the interior. In the same way, the tiny amount of daylight allowed in due through the small windows used to be multiplied by mirrors, golds, candlesticks, chandeliers and their myriads of shimmering glassware that diffracted and extended the surface and the quantity of light.

Ease of access to electric energy or gas during the twentieth century, the invention of central heating by radiator at the end of the nineteenth century which uses conductive water to heat the air rather than the air itself, insulating, then the invention of air-conditioning rendered obsolete and depreciated all the decorative art of the interiors of yesteryear, made of woodwork, rugs and tapestries, mirrors and chandeliers and many other decorative elements whose mission was once to increase thermal and luminous comfort, even if only slightly, when heating, air-conditioning or lighting systems were scarcely efficient, before the industrial era. In the 20th century, these olden day decorative elements lost their utility value and became merely superficial decorative elements, without any real function. The stylistic consequence of central heating and air-conditioning allowed Adolf Loos, Le Corbusier or Mies van der Rohe to develop a new stylistic program where the decorative elements of yesteryear, having lost their thermal and luminous legitimacy, can be eliminated in favor of white, empty and neutral minimalism, with no more ornamentation or decorative elements.

What we see today is that empty, white, neutral interiors, most often poorly insulated and with vast amounts of glazing, and the minimalist style that we inherit from modernity, have no consideration for energy, as this problem had been entrusted, in the twentieth century, to the exclusive use of central heating, electric lighting and air-conditioning.

But these modern heating and air-conditioning systems are no longer suitable in our times as the only way to heat or air-condition because they are energy-intensive and often operate by burning fossil fuels such as gas or fuel oil that emit greenhouse CO_2 and contribute to global warming.
The new thermal regulations require a reduction in the energy consumption of buildings, a reduction in heating or air-conditioning, and hence, it is now necessary to combine perfect, total thermal insulation, a vapor barrier, air-tightness, dual-flow ventilation, in order to minimize the energy expended to heat or cool indoor air which are becoming new elements of structural architecture, just as was the case for decorative elements before modern times.

The consequence of the fight against global warming and new thermal regulations is that of reintroducing new elements of interior design, generating a kind of renewed language of interior design, through the installation of insulating wool in walls and floors, the rethinking of ventilation, or the laying of airtight films on walls as new, superficial, decorative elements, once again becoming necessary. The new thermal regulations, therefore, render old-fashioned the modern minimalism of the twentieth century, based on low thermally insulated thin walls and large, single glazed glazing. By abandoning the unique heating and air-conditioning techniques of the 20th century, we must then reappraise the decorative strategies of today's interiors, rethink a decorative style in order to fulfil the new thermal regulations and invent the architectural language of interiors in the Anthropocene era.

CHAPTER I.

METEOROLOGICAL COMPOSITIONS

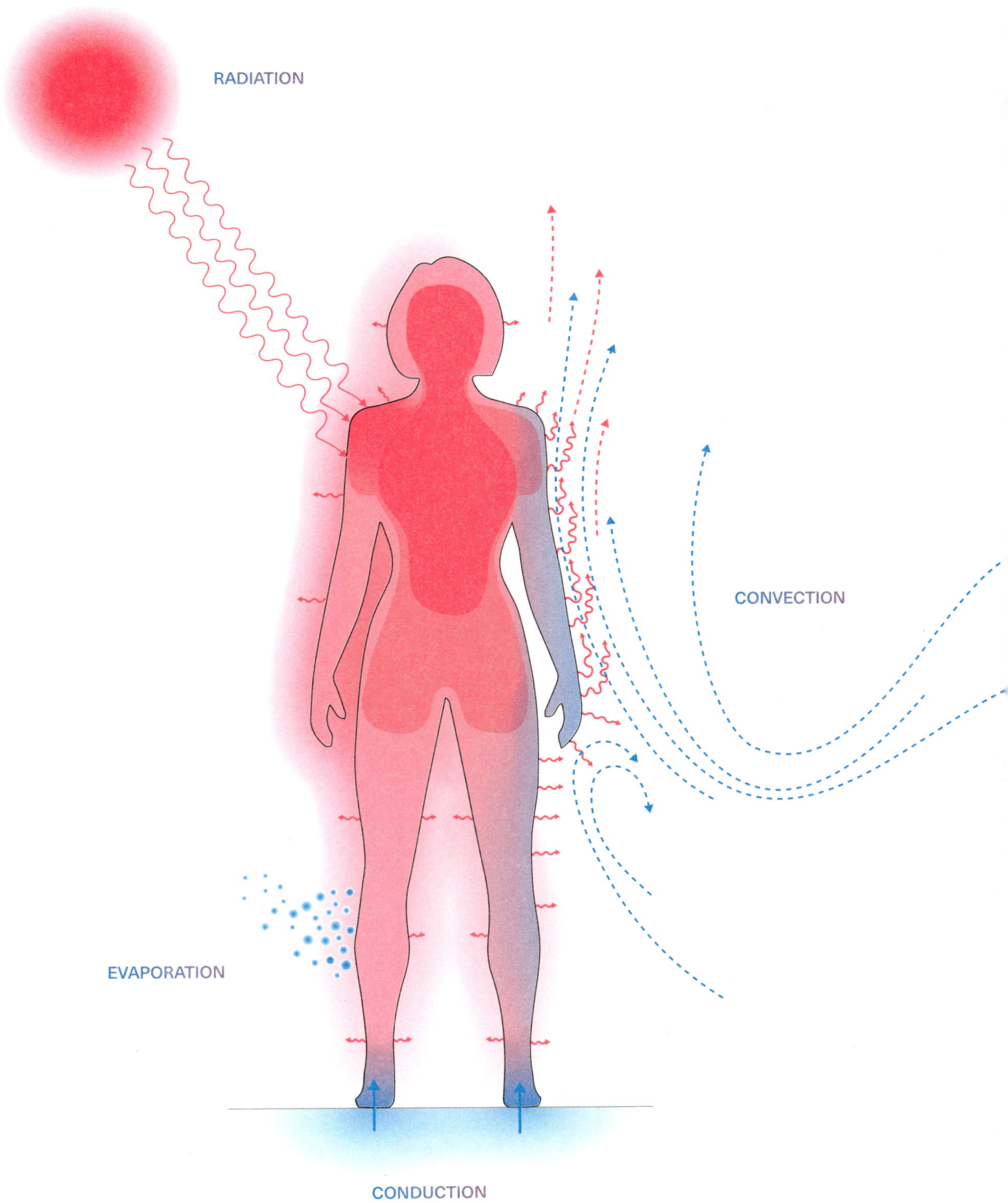

CONDUCTION

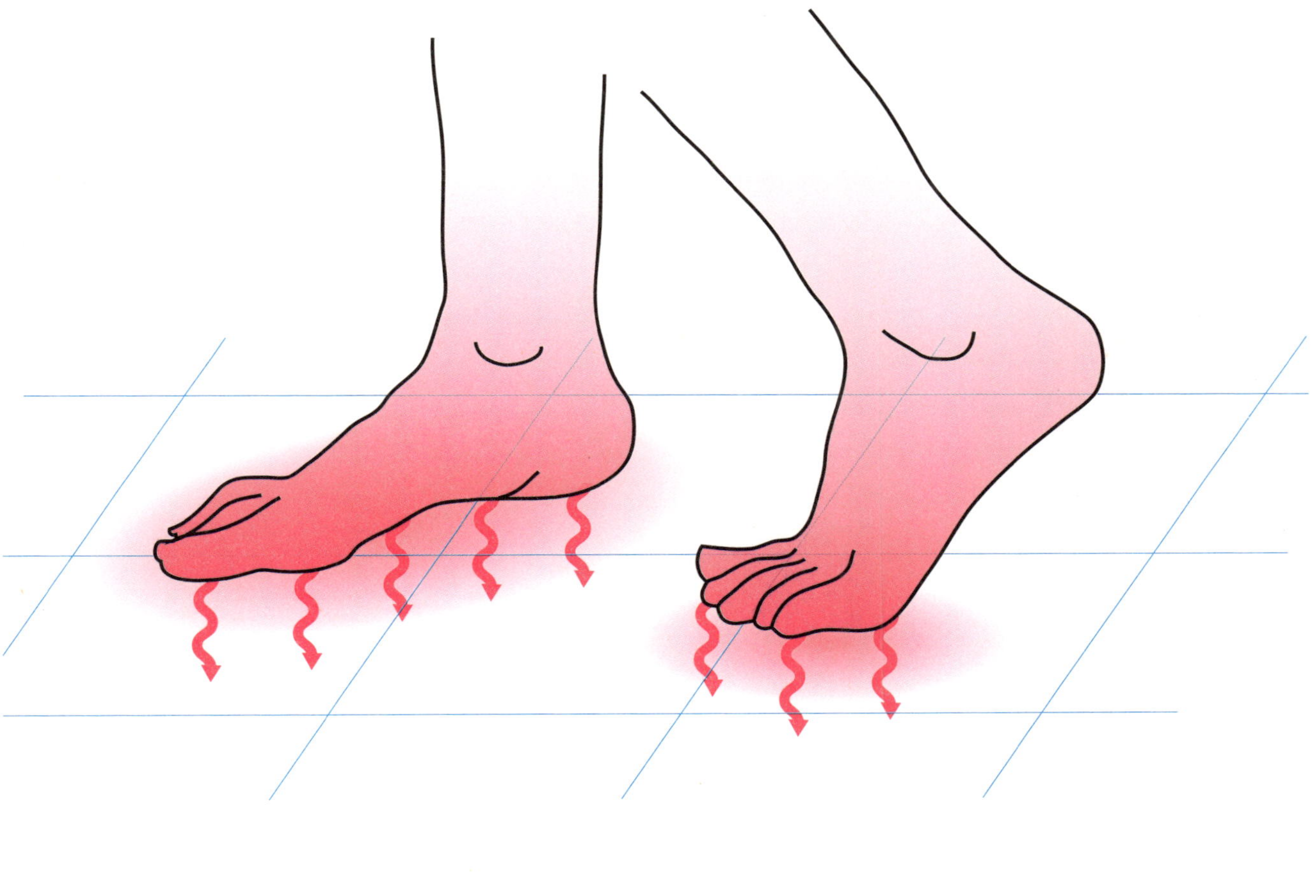

I. Definition

General Properties

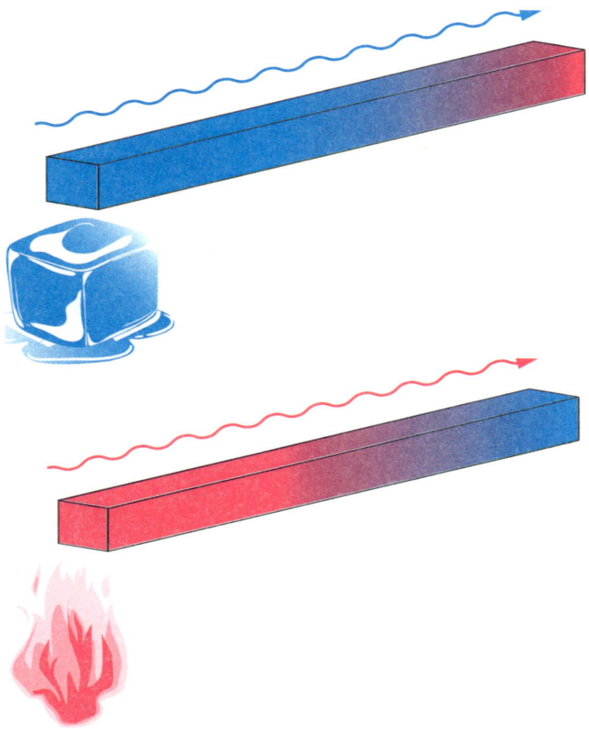

Heat flow from hot to cold {Fig. 1-35}

The process of exchanging heat from one hotter object to another cooler one takes place when solids, liquids or gases come into contact. Microscopic collisions of molecules transfer kinetic energy and potential energy from the ones with the highest temperature, which are more excited, to the cooler molecules, which vibrate more slowly. When a more excited molecule collides with a less excited molecule, it transfers some of its energy to the latter and becomes a little less excited. The role of architecture is to create an interior space, an air bubble separated thermally from the outside air in order more easily to transform its climatic parameters (temperature, humidity, light) to render it habitable. Creating a warm climate in the cold of winter, creating a cool climate during the heat of summer, is the primary goal of architecture.

This separation between the interior and the exterior is carried out by means of a solid layer which must be thermally insulated in order to minimize the necessary energy to heat or cool the interior air. The ability of a material to more or less insulate thermally, to allow more or less cold or heat to pass from one side of the wall to the other, is called thermal conductivity. The more conductive the material, the faster the interior will lose its heat due to the cold coming from the outside.

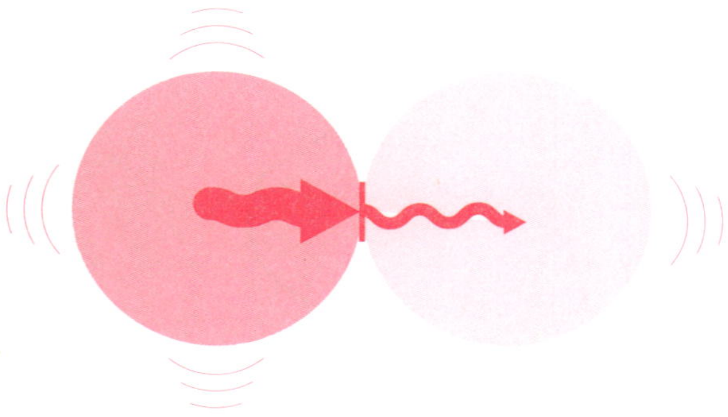

Energy transfer when two molecules collide {Fig. 1-36}

In Buildings

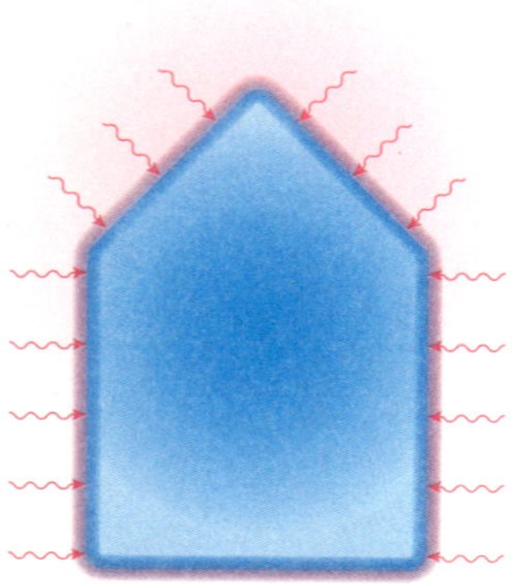

Summer {Fig. 1-37}

Heat from the warmer outdoor environment is transmitted to the interior of the building through the walls. The higher the thermal conductivity of the wall materials, the easier and faster the heat will enter the building, leading to overheating of the interior space.

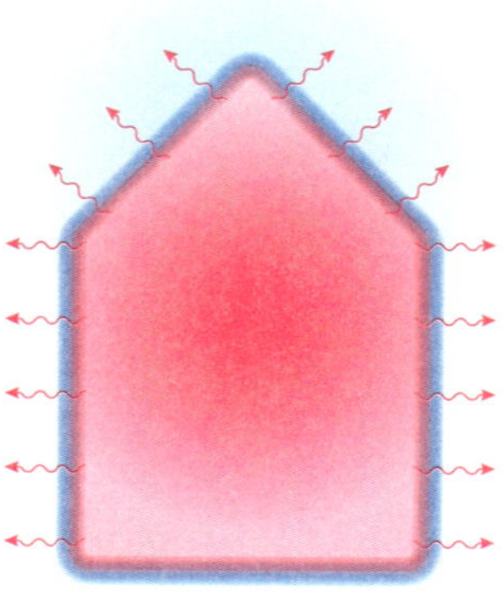

Winter {Fig. 1-38}

Heat from inside the building is transmitted to the cooler outside environment. If the building is poorly thermally insulated, all the heat produced by the radiators inside the building is evacuated into the cold walls and lost, leading to constant cooling of the inside of the building and increased energy consumption to compensate for the heat losses by conduction of the hot air in contact with the cold walls.

II. Properties of Materials

Equation for Heat Flux

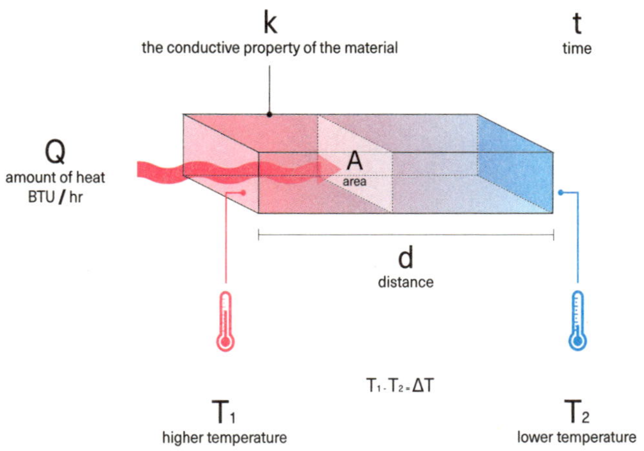

$$\frac{Q}{t} = \frac{kA\Delta T}{d}$$

Diagram of the equation for heat flux {Fig. 1-39}

Heat flux is determined by four factors: the cross-sectional area of the object (A), the thickness of the object (d), the difference between the temperature on both sides of the object (ΔT), and the conductivity of the material (k).

This means that objects with larger surfaces will absorb heat faster than those with a smaller heat-transmitting surface. Thicker objects also need more heat/time to heat than thin ones.

The greater the temperature difference, the faster the heat is transferred, whereas this flux decreases gradually when ΔT approaches 0, or Thermal Equilibrium.

Solids, Liquids and Gases

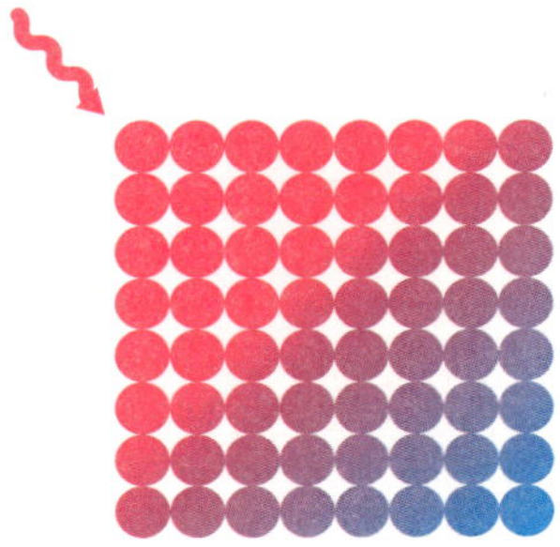

Conduction in liquids {Fig. 1-40}

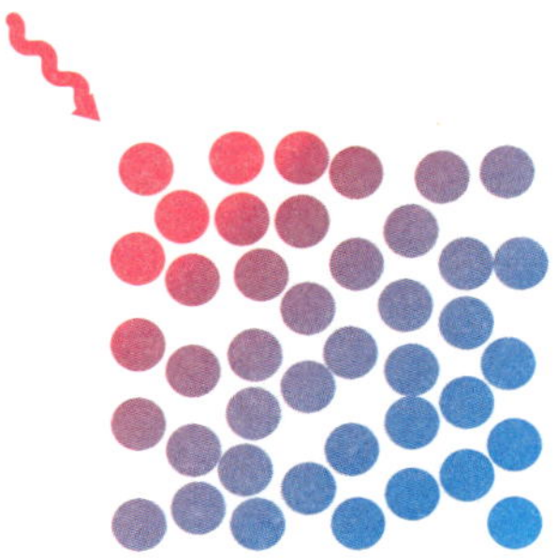

Conduction in solids {Fig. 1-41}

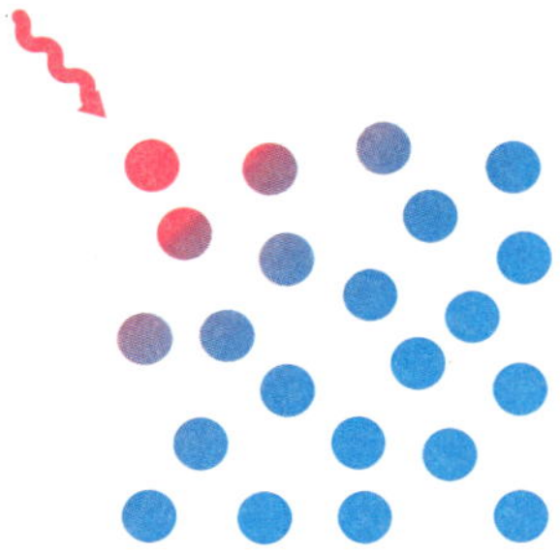

Conduction in gases {Fig. 1-42}

The heat flux depends on the temperature difference between two bodies and their inherent heat conduction properties given by the specific material of which they are made; solids, liquids and gases all have different degrees of conduction. Solids mainly transfer heat through conduction because the molecules are close together, which allows energy transfer through vibration. For liquids and gases, conduction takes place in parallel with other types of heat transfer such as convection or evaporation. Liquids are slightly denser than gases, making transfer possible by conduction, but not as much as for solids. As gases are less dense, their molecules are less likely to collide than in the case of liquids and solids. This is why air is an excellent thermal insulator. However, if two bodies remain in contact, heat exchange by conduction takes place until equilibrium is reached.

K Thermal Conductivity Values

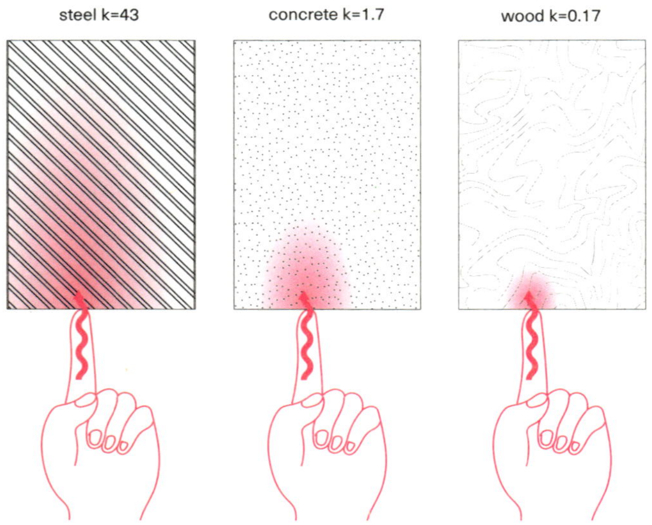

Thermal conductivity in different materials {Fig. 1-43}

The higher the k value (heat transmission coefficient in W/m^2K) of a material, the more heat will pass through this material. The lower the k value, the better the material as a thermal insulator.

Thermal Conductors vs. Thermal Insulators

Some materials are good thermal energy conductors, while others are poor conductors. The latter are therefore good thermal insulators. Good conductors include metals such as steel, aluminum, copper, etc. They have high k values (W m−1 K−1). Good thermal insulators, on the other hand, are materials that contain a large quantity of gases (air for example) such as wool, expanded polystyrene, rockwool or hemp yarn, etc., and have a lower k value. The more heat a material transmits, the higher its k value. Metals generally have a high k value because they transmit heat through moving electrons, while the structure of solids such as wood contains air pockets that work as thermal insulators within the material itself.

Material	Thermal conductivity (k)
Copper	401
Aluminum	205
Steel	43
Concrete	1.7
Brick	1.3
Oak	0.197
Plasterboard	0.5
Wood	0.17
Rockwool	0.045
Glass wool	0.04
Natural wool	0.039
Expanded polystyrene	0.03
Air	0.024

Thermal Conductive Flux over Time

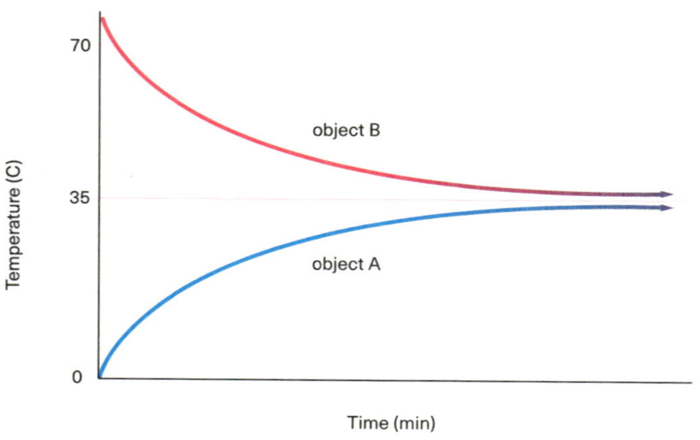

Conductive flux over time {Fig. 1-44}

A hot material will gradually transfer its heat (by decreasing in temperature) to a cooler material (which will increase in temperature) until the two materials have the same temperature, which will cease all energy transfers. This energy transfer will take place more quickly if the materials have a high thermal conductivity (k).

On the contrary, if (one of) the materials (has) have a low k value, then exchange will be slower or almost non-existent. Each material will thus keep its own temperature. Thus, a warm wall protected from the cold exterior by good thermal insulation will retain its heat without sharing it with the outside air.

III. Geographical Scale

Conduction in the Lithosphere and in the Atmosphere

Weather front {Fig. 1-45}

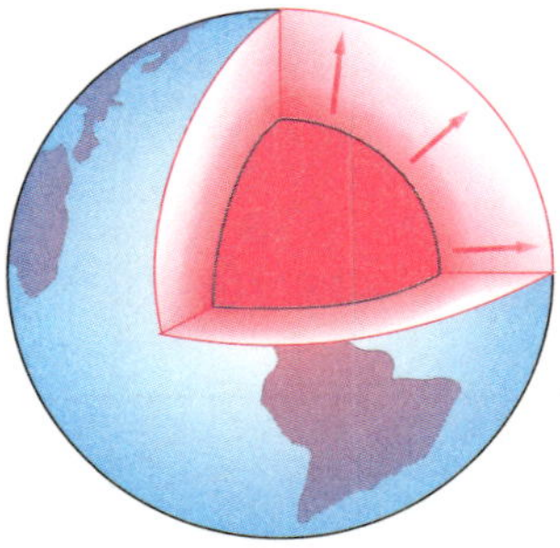

Conduction from the core to the lithosphere {Fig. 1-46}

The conduction phenomenon is less present in the atmosphere than in the lithosphere. In the atmosphere, air masses heated near the Earth's surface rise by convection after receiving energy. Air masses of different temperatures exist in the atmosphere and the level of conduction between these masses is relatively low. As a result, hot and cold air masses move as volumes, and meet at what is called a front. A weather front is the boundary between two air masses of different densities and is mainly the cause of meteorological phenomena due in particular to the condensation of the water vapor contained in the mass of hot air cooling in the front, which causes the formation of clouds, rain, hail, snow, storms, tornadoes, etc.

The Earth constantly produces thermal energy at its core through radioactive decay and residual heat dating back to its formation. This heat is transferred to the lithosphere continuously by conduction from the Earth's core to its surface, creating a thermal gradient that rises as we approach the core. The temperature on the Earth's surface varies according to the temperature of the atmosphere; it cools down at night or in winter and warms up during the day or in summer. Because the lithosphere serves as thermal insulation, digging underground gradually leads to more constant temperatures and at a depth of 5 meters, the temperature remains almost the same all year round, that of the location's annual average temperature.

Geothermal Energy: Heating and Cooling by Conduction with the Earth

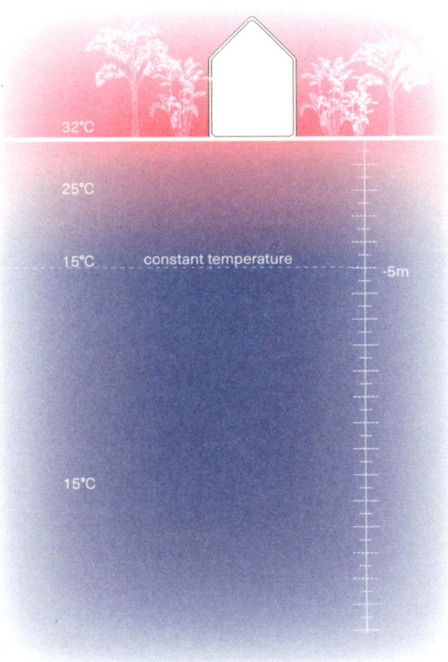

Ground temperature in summer {Fig. 1-47}

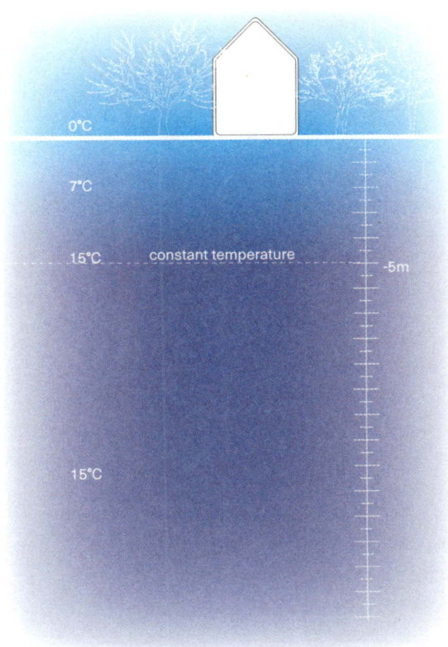

Ground temperature in winter {Fig. 1-48}

Geothermal Warming and Cooling of Air and Water

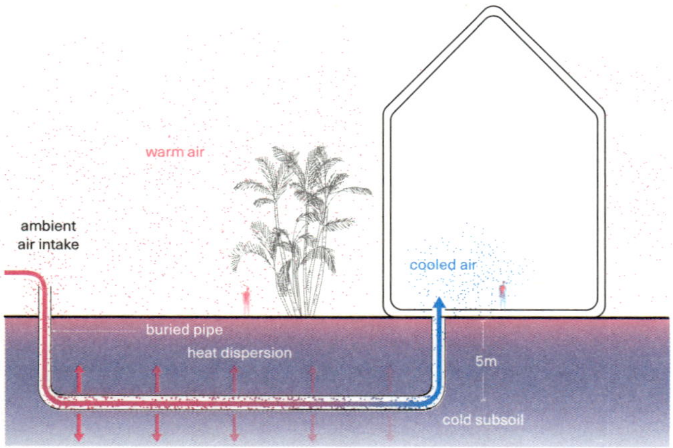

Cooling of air through geothermics in summer {Fig. 1-49}

Air is harnessed from the external ambient atmosphere and forced to pass through underground pipes at a depth of 5m, where the temperature remains constant throughout the year at the average annual temperature of the geographical location, which enables cooling the air in summer, and heating it in winter. This process makes it possible to bring the temperature of fresh air, harnessed outside and introduced into the house for ventilation, closer to the comfort temperature for humans of the indoor air (around 21°C), in order to reduce the energy consumed by heating since we will start from an exterior temperature closer to the temperature required inside the house. The advantage of geothermal energy (through systems called Canadian well, Provencal well or ventiduct) is that it can ventilate the interior of the house by bringing in fresh air during the day in the summer, not the hot air that would enter through the windows.

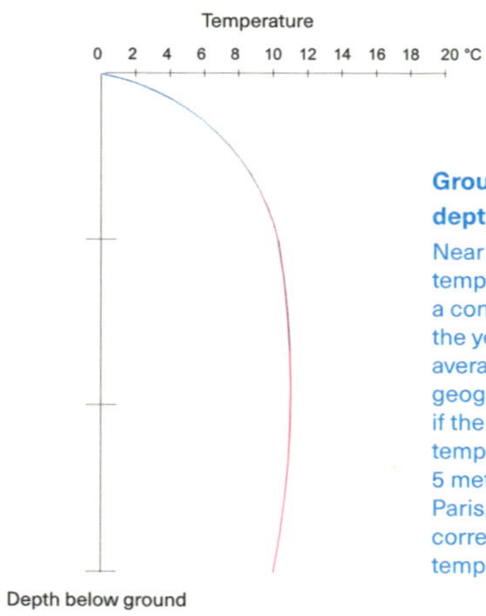

Ground temperature according to depth {Fig. 1-50}

Near the surface, the subsoil temperature at a depth of 5 meters has a constant temperature throughout the year which corresponds to the average annual temperature of its geographical location. Thus, in winter, if the ground surface and the air temperature is 0°C, then, at a depth of 5 meters, it will reach, for example, in Paris, France, a temperature of +11°C corresponding to the annual average temperature in Paris.

In summer, if it is +25°C on the surface, it will still be +11°C at a depth of 5 meters. It is this ability of the earth's depth to be thermally insulated from variations in surface temperature that benefits geothermal systems that allow cooling the house in summer and heating it in winter by passing air or water underground, before introducing it into the house for heating in winter or cooling in summer.

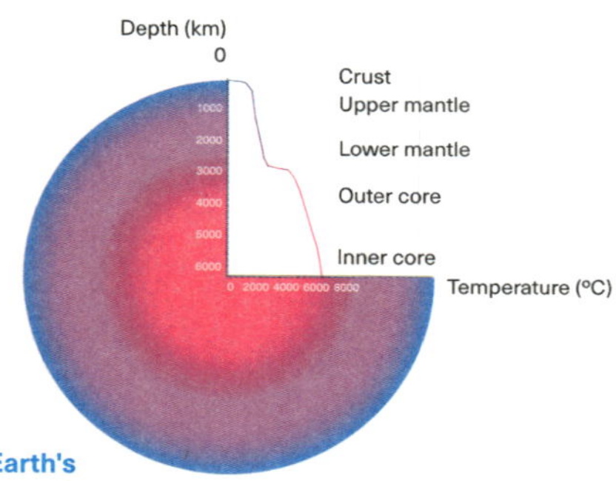

Heat transfer from the Earth's core to the surface {Fig. 1-51}

The temperature of the Earth's core at a depth of 8,000 km is about 5,550°C. As it rises to the surface, it gradually cools reaching an approximately constant temperature (< 30°C) in the last 300 kilometers before the surface. Then, in the last 5 meters, the temperature of the earth will vary according to day and night and the seasons, influenced by thermal conduction by air temperature of the atmosphere above the ground and by the provision or not of direct radiant heat from the Sun.

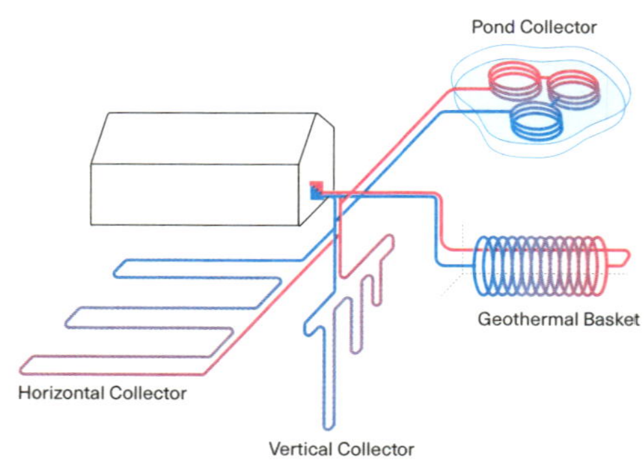

Types of closed network geothermal systems {Fig. 1-52}

The principle of geothermal energy is to circulate air or water underground to raise the temperature in winter or lower it in summer. In winter, for example, cold air is collected outside and as it passes through the subsoil, it is heated by conduction in contact with the constant temperature throughout the year at depth and, therefore, warmer in winter. The air then enters the house, preheated, which makes it possible to reduce the amount of heating spent to bring it to the temperature comfortable for humans of 21°C.

Air and water can be heated geothermally by passing through a long network of buried pipes at a depth of about 5m before entering the house. The fluid thus slowly gains or loses heat until it reaches thermal equilibrium (when it reaches the temperature of the subsoil surrounding the pipes). This fluid is then returned to the house where it passes through a heat exchanger. Horizontal manifolds are used when the floor area is large and it is possible to spread the pipes. Vertical manifolds are used in compact sites. Air-water heat pumps can be installed when there is a body of water nearby which, by thermal inertia, remains, for example, warmer than air in winter and vice versa.

The Earth's surface temperature is more influenced by atmospheric conditions than by the heat transmitted by conduction through the lithosphere. However, at about 5m below the surface, the temperature remains at a constant 13.7°C all year round (annual average temperature for the earth), though this figure varies according to geographical position. The constant temperature can be used to heat or cool the water and air in buildings because it allows them to attain a more comfortable temperature, of 21°C. When the water or air is then heated or cooled to the desired temperature, energy expenditure is lower since the temperature difference between the incoming fresh air and that desired in the house is lower.

Principle of Heat Recovery by Conduction

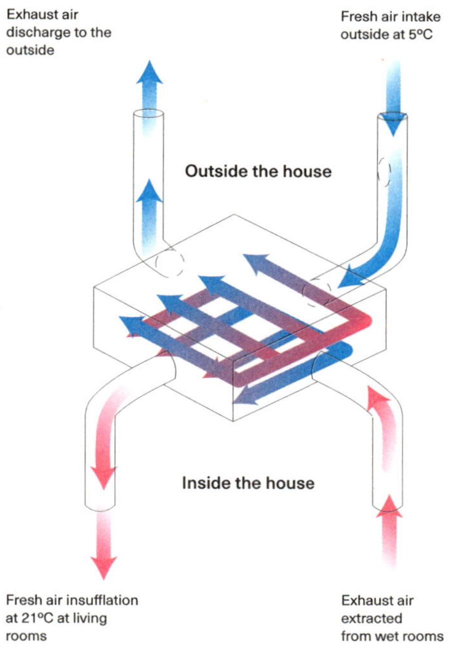

Operating principle of dual-flow ventilation with heat recovery {Fig. 1-53}

In winter, for example, in a heat recovery exchanger the stale air (hot as it has been heated by radiators) which is extracted from the house crosses the fresh air entering the house and preheats it thanks to a cross-ventilation system where, by conduction via a thermally highly conductive material such as aluminum, the heat of the outgoing air is recovered by the incoming air, raising its temperature.

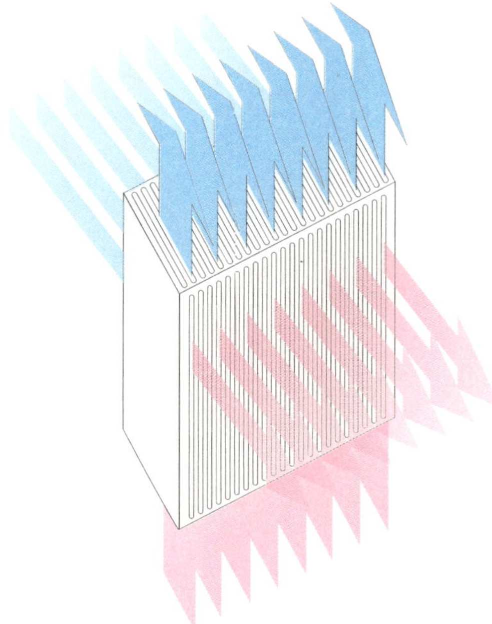

Heat exchange by conduction {Fig. 1-54}

Inside the dual-flow heat exchanger, stale hot air passes through aluminum blades to transfer its heat to the exiting cold air to preheat it before it enters the house. The dual-flow system thermally crosses the outgoing stale air with the incoming fresh air, but without mixing them chemically, keeping the incoming fresh air free from moisture, CO_2, and any indoor pollutants contained in the outgoing stale air.

Heat exchanges and circulation in the building (winter) {Fig. 1-55}

The air heated by geothermal energy passes through the heat exchanger, where it is again heated by the exiting air in the dual-flow exchanger. The fresh air is next introduced into the house, then it will be evacuated later, stale, but will first pass back into the heat exchanger in order, in turn, to heat the incoming cold air. This controlled dual-flow ventilation system with heat recovery allows reducing the energy consumed by radiators since the fresh air entering in winter is already slightly heated by the outgoing air, which thus allows heating it less.

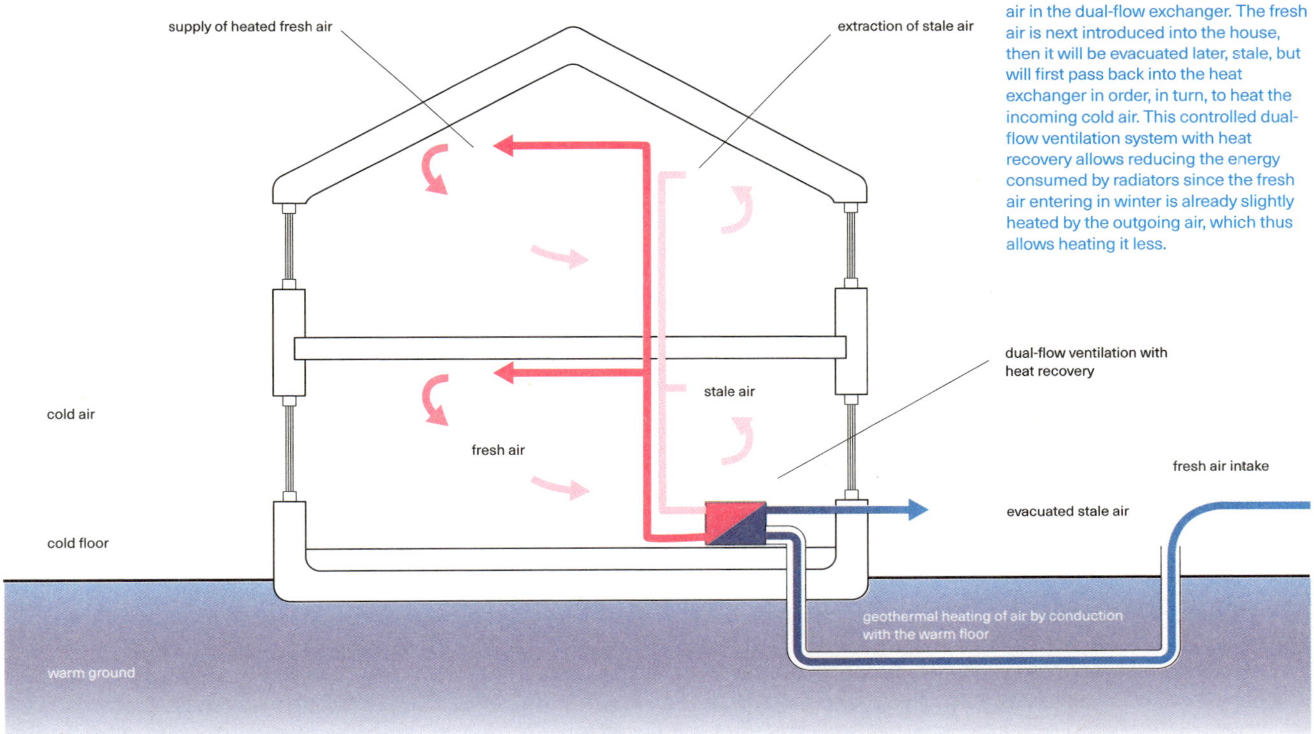

IV. Physiological Scale

Thermoregulation in the Human Body

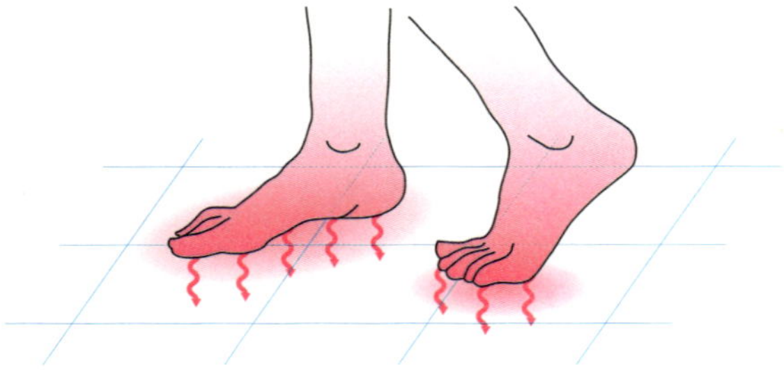

Significant heat loss by conduction with a tiled floor {Fig. 1-56}

Human skin is extremely thermally conductive. In contact with a material that is also conductive (with a significant k value such as metal, tile or stone), the skin will cool by conduction by transferring its heat to the cold ground.

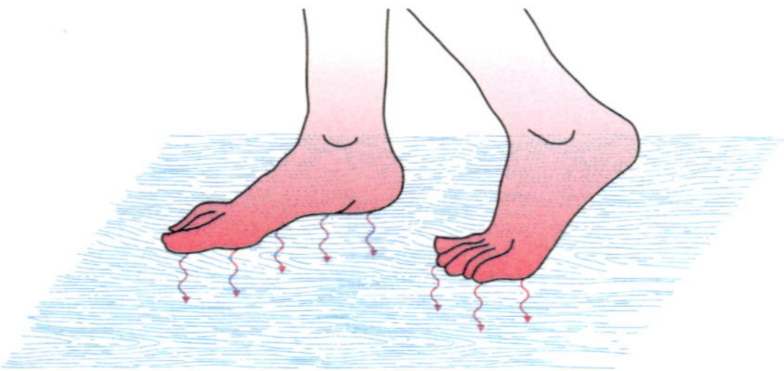

Low heat loss by conduction with a wooden floor {Fig. 1-57}

Conversely, by placing our feet on a surface made of a non-conductive and therefore thermally insulating material (low k value such as wool or wood), the body will not transfer heat to the ground by conduction and will not cool.

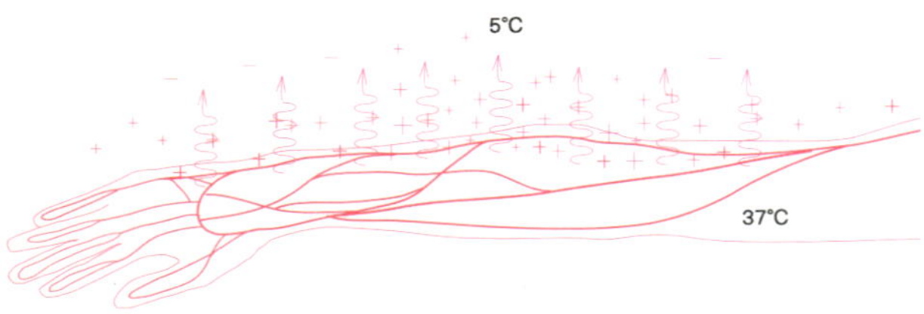

Skin conduction without clothing {Fig. 1-58}

In winter, the skin loses heat by conduction when it comes into contact with cold air. This heat exchange between a solid (the skin) and a gas (the air) is called "convection".

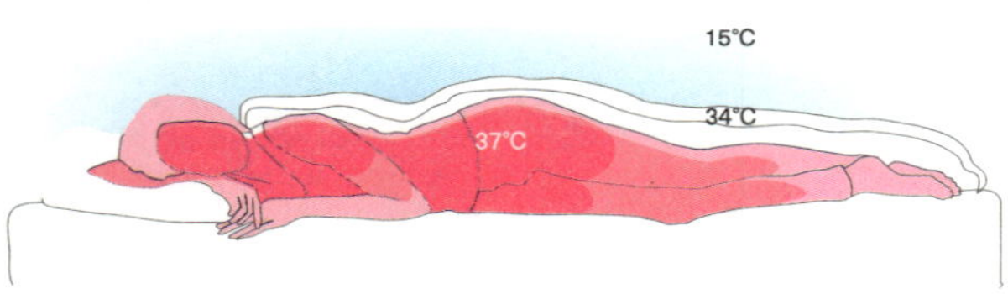

Conduction during sleep with a blanket {Fig. 1-59}

Sleeping with a blanket prevents conduction between the warmer body and the cooler air. The fibers of feathers, wool, cotton surrounding a multitude of air bubbles (the air is a very poor thermal conductor) insulate the body, just like the layer of air trapped under the cover, which is a poor conductor.

The temperature of the human body is about 37°C (98K). However, the body produces excessive heat, and must get rid of it by transferring it to the environment, to the air, for example. To do so, the air must be colder (around 21°C) than the skin temperature which is around 34°C. However, if the air temperature is below 21°C, the body will lose too much heat and conversely if it is above 21°C, the body will start to overheat. The human body loses about 20% of the heat it produces by transferring it to the environment through conduction, including 4% through the feet if they are in contact with a cooler surface, and 15% by convection with the surrounding cold air or by conduction in contact with a scarcely insulating garment cooled by the low temperature of the air. To avoid losing too much heat by conduction, the body must be covered by clothes, shoes, gloves, hats, etc. which in winter are made of thermally insulating materials such as wool, which will limit the loss of heat by conduction between the body and the outside air, preventing the body from cooling too much. Direct heat conduction between the human body and another solid takes place by direct contact, through the hands, the feet.

This is why when we touch a floor tile it feels cold, because the flow of heat between the skin and the material is faster than for other floor materials such as wood or carpets. The use of low heat conduction clothing or blankets is the primary means of thermal resistance because they contain still air which, being a poor heat conductor, will create a barrier limiting heat transfer between the body and the environment. When the air is stationary, it will be warmed by conduction with the body and create a layer of heat that will render the thermal difference between the body and the environment less sharp. The Clo value numerically expresses a garment's thermal resistance [1 Clo = 0.155 m2K/w]. The higher this value, the more insulating the garment will be. Clothing and metabolism activity have a great physical influence on human comfort and can have an impact on the acceptable temperature of a space. Thus, if you do sport or dress well, the air temperature can be lower. The air in sports halls can thus be heated to only 14°C instead of 21°C.

Thermal Insulation

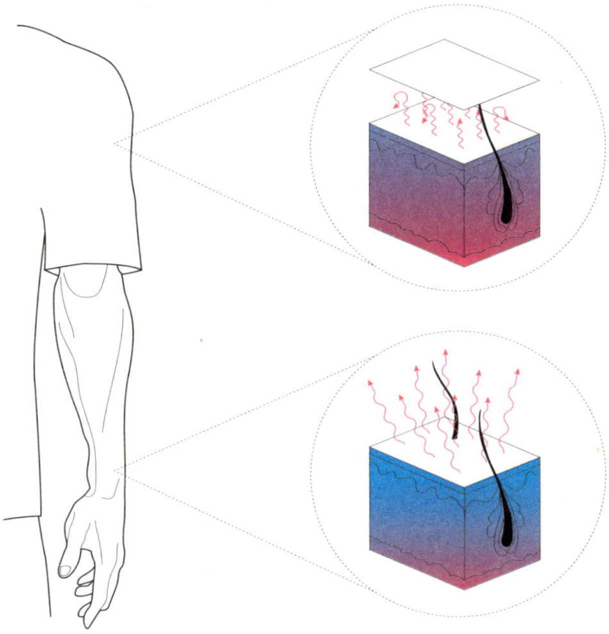

Skin conduction, with and without clothing {Fig. 1-60}

Wearing clothes prevents the body from transferring heat to the atmosphere, keeping the body warm when the air is cold. This is because the layer of air between the skin and the fabric acts as a thermal insulator, limiting conductive exchanges between the skin and the air, while the fabric itself prevents convective exchanges between the skin and the moving air (wind), which cools the skin by preventing the heat from stagnating and accumulating.

Clothing is a form of thermal insulation {Fig. 1-61}

The fibers, and indeed the space between fibers, of wool and indeed of textiles in general, contain tiny bubbles of air which act as an excellent thermal insulator, since air is a very poor conductor. By thickening the textile, or by multiplying the layers, the quantity of air bubbles is increased, proportionally increasing the insulating power of the garment.

At the same time, clothes prevent the skin from coming into direct contact with the wind, which reduces convective cooling. Tight-fitting clothing will also block airflow between fabric and skin, maintaining body heat by preventing convective cooling.

Thermoregulation of the Human Body

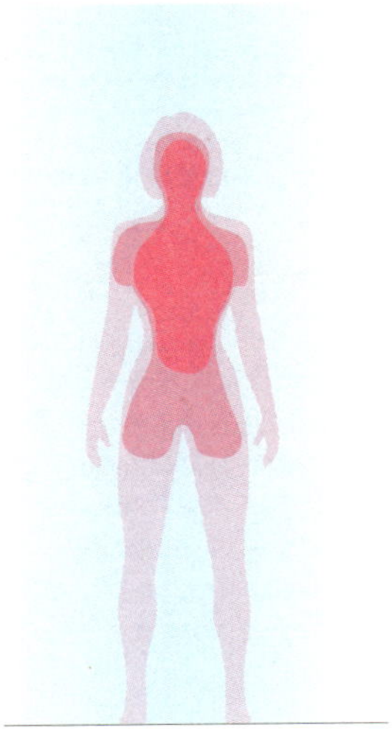

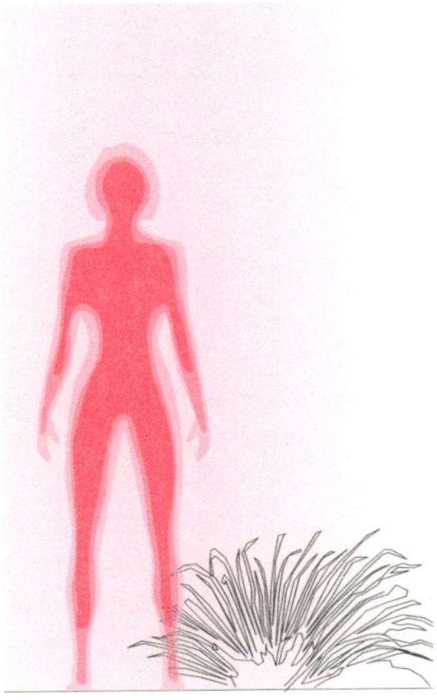

Body temperature at rest in a cold environment {Fig. 1-62}

By conduction, convection, evaporation and radiation, the body loses too much heat to the outside cold environment and reacts through thermogenesis by increasing its internal heat production. This includes vasoconstriction, which reduces the diameter of the veins and thus the surface area of the blood in contact with the exterior, thus limiting heat losses between the cold air and the skin. Thermogenesis also occurs through shivering, which generates heat through the mechanical friction of the skin on itself.

Body temperature during physical activity in a warm environment {Fig. 1-63}

By conduction, convection, evaporation and radiation, the body gains heat and reacts through thermolysis, increasing the body's heat losses. This particularly takes place with vasodilation, when the veins dilate in order to offer a greater surface area in contact with the exterior air via the skin, which enables increasing the exchange surface area and therefore the loss of excess heat, between the blood at 37°C and the outside air which is a little cooler. If the exterior air temperature is higher than 34°C (the same as the temperature of the skin), then the body can no longer get rid of its excess heat by conduction (only sweating can cool down a body hotter than 34°C), which leads to a dangerous rise in body temperature.

V. Architectural Scale

Building Energy Losses

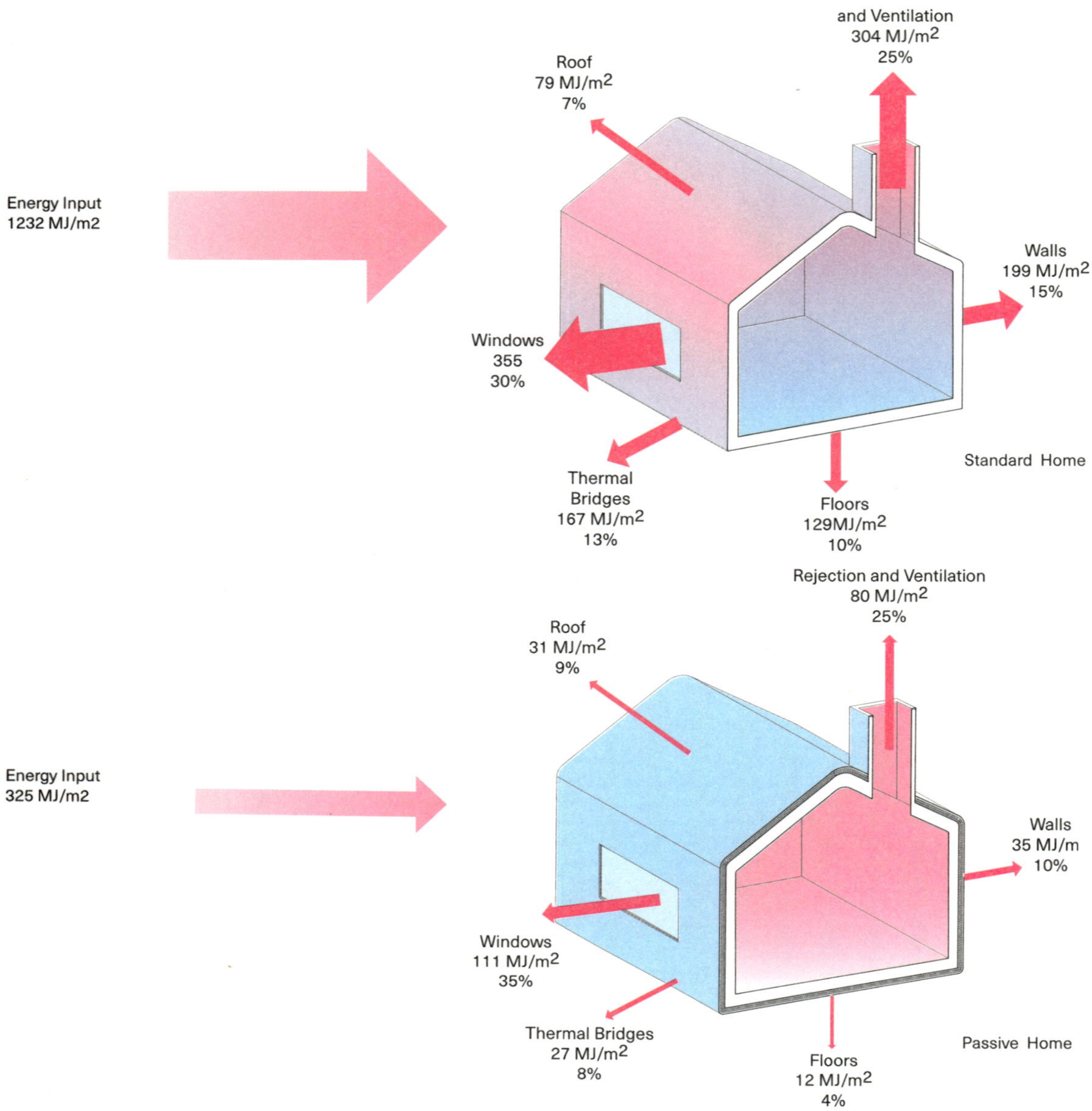

Energy losses in a standard home (above) compared to a passive home (below) {Fig. 1-64}

A standard home consumes more energy than a passive home since a significant amount of heat escapes to the outside environment through conduction through the building envelope. Poorly insulated walls and single-glazed windows are thus highly thermally conductive, constantly cooling the interior of the house thanks to the passage of cold from the outside. A suitably thermally insulated house (25 cm wall insulation, double or triple glazing) loses less heat through the envelope, and therefore consumes less energy for heating. (source: Suisse Énergie / Société à 2000 watts) single-glazed windows are thus highly thermally conductive, constantly cooling the interior of the house thanks to the passage of cold from the outside. A suitably thermally insulated house (25 cm wall insulation, double or triple glazing) loses less heat through the envelope, and therefore consumes less energy for heating. (source: Suisse Énergie / Société à 2000 watts)

A high thermal conduction value of the outer shell of buildings is one of the main causes of heat losses and an increase in energy expenditure to keep the indoor climate at a comfortable temperature. To calculate a building's heat loss, you must check the U-value (heat transmission coefficient) of the different layers that make up the building envelope. U-values are coefficients that (in W/m2K) measure the heat flow passing through each material in relation to its thickness. The lower the U-value, the more the material is considered a good thermal insulator.

The consequence of a highly heat-conducive envelope is that the building needs more energy to operate, which increases its carbon footprint and contributes to global warming. Looking at the diagram above, it becomes clear that the most effective way to reduce energy expenditure is to properly insulate the building. Thus, the U-value of the walls and roof must be less than 0.15 W/m^2K, as recommended by energy efficiency standards such as Passivhaus or Minergie.

Thermal Insulation of Exterior Walls and U-values

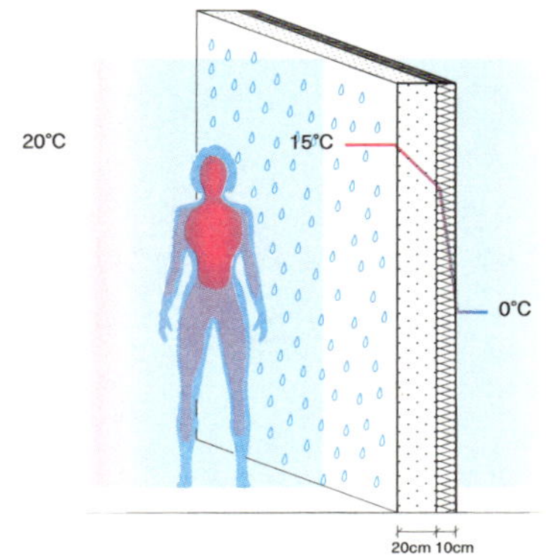

10 cm insulation {Fig. 1-65}

If the temperature difference between the inner surface of the wall and the indoor air is 5°C, as is the case with 10 cm of insulation, the indoor air will cool by conduction and convection in contact with the wall and lead to the overconsumption of energy to compensate, increasing the release of CO_2 and therefore global warming.

There may also be condensation on the wall surface, which can lead to mildew, black spots, and the general degradation of the wall materials. In addition, by radiation, the cold of the wall will cool the inhabitants, leading to considerable discomfort.

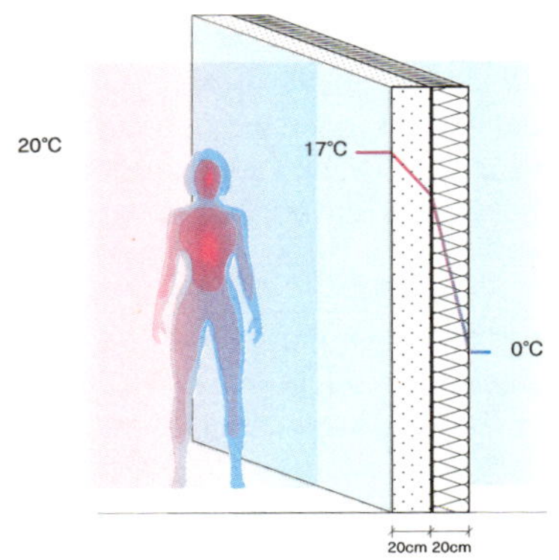

20 cm insulation {Fig. 1-66}

If the temperature difference between the inner surface of the wall and the indoor air is 3°C,as is the case with 20 cm of insulation, there will be no condensation on the wall, but the wall itself will still be a little cold and will still cool the indoor air slightly by conduction and convection, and the human body, by radiation, leading to some discomfort.

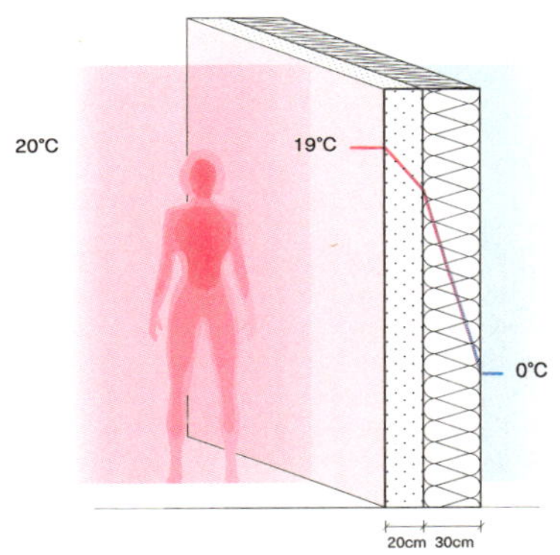

30cm insulation {Fig. 1-67}

If the temperature difference between the inner surface of the wall and the indoor air is 1°C, as is the case with 30cm insulation, the wall will not release cold and the interior will be warmer and more comfortable.

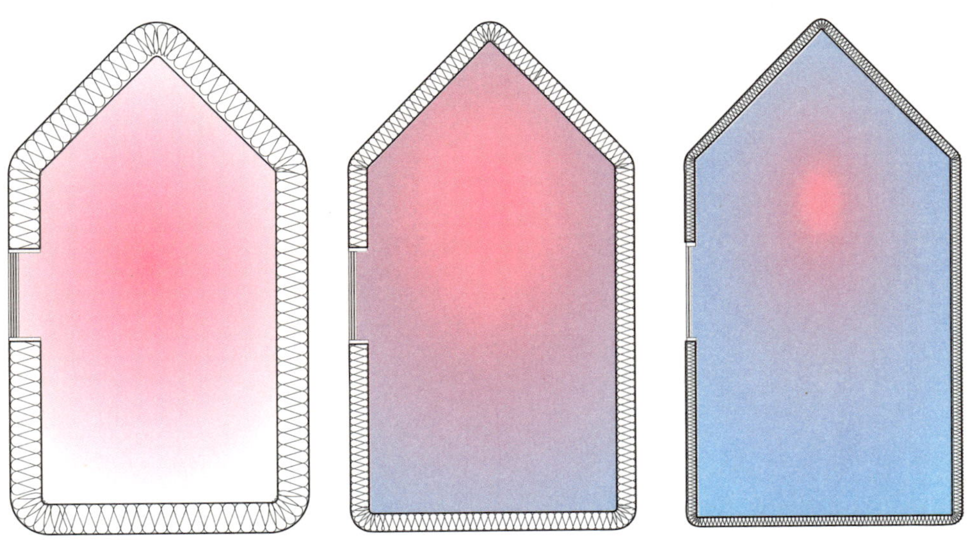

30cm insulation {Fig. 1-68}

U-value: 0.130 W/m^{2}K

20 cm insulation {Fig. 1-69}

U-value: 0.193 W/m^{2}K

10 cm insulation {Fig. 1-70}

U-value: 0.37 at 5 W/m^{2}K

Thermal Insulation Thickness

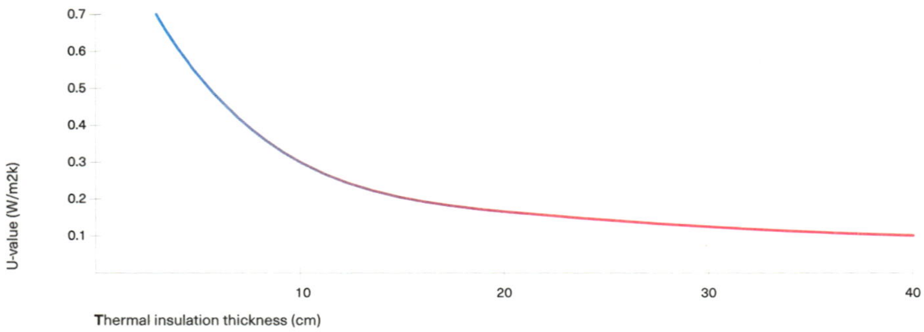

Influence of the thickness of the insulator on insulation performance {Fig. 1-71}

The thicker the thermal insulation, the more it will block heat exchange between one side of the wall and the other. The most effective thickness of insulation is 25 cm; if thicker, the benefit becomes negligible compared to the effort [investment].

In order to minimize the energy spent heating the interior of the house in winter to reduce CO_2 emissions, the building must have a well-insulated, continuous, uninterrupted envelope around the building. Thus, indoor air must be well separated from outdoor air so that the former can be heated without loss to 21 °C to reach the comfort temperature for humans. In order to properly insulate a building, a material with a low thermal conductivity k must be used. This material often consists of air bubbles within vegetable or mineral fibers, which makes an insulating wool, a thermal insulator.

This thermal insulation must cover the outside of the building's support structure to a thickness of 25 cm in order to reach an optimum U-value (heat transfer) of less than 0.15 W/m2k. Similarly, glass layers are multiplied in double- or triple-glazed window systems to trap thin layers of air between them that function as layers of thermal insulation. However, a triple-glazed window (U-value = 0.6 W/m2k) remains less thermally insulating than a well-insulated wall (U-value = 0.15 W/m2k).

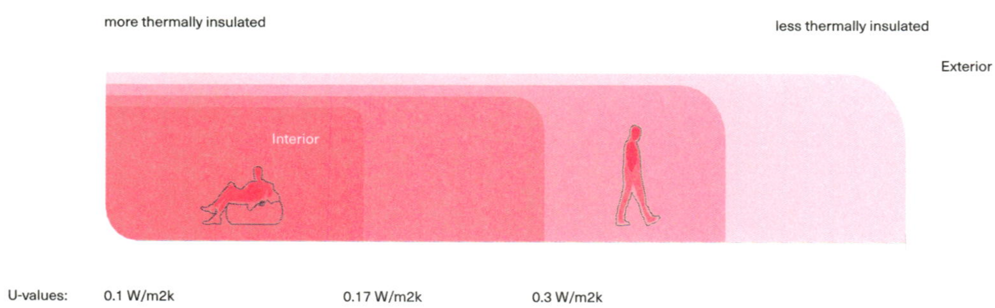

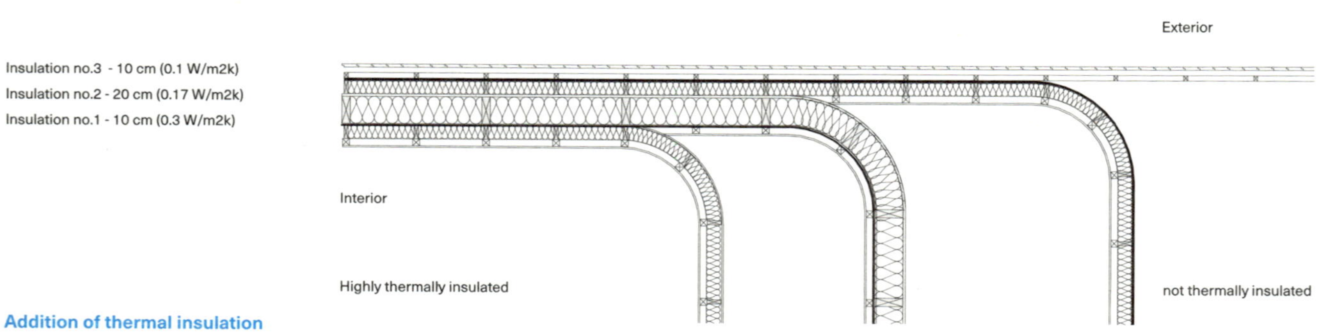

Addition of thermal insulation layers {Fig. 1-72}

The thermal performance required by Swiss SIA standards is 0.3 W/m2K, which corresponds to insulation of about 12 cm. To reach a U-value of 0.15 W/m2K or even 0.1 W/m2K, which corresponds rather to the requirements of a Passivhaus or a Minergie building, it is necessary to add thermal insulation to reach a thickness of 25 cm or 40 cm.

The building envelope thus very literally resembles superimposed layers of clothing to cope with the cold.

Thermal Bridge

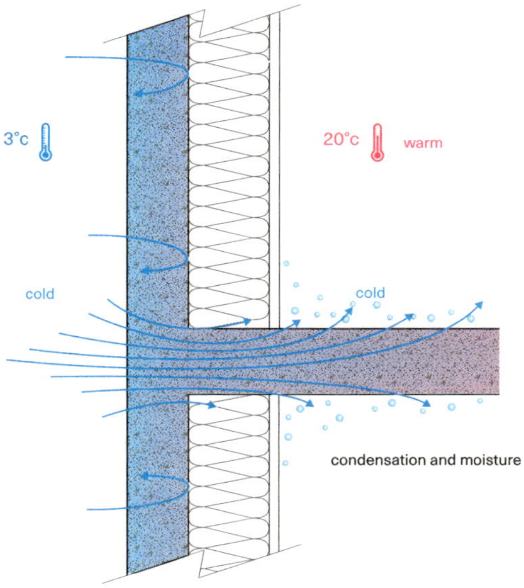

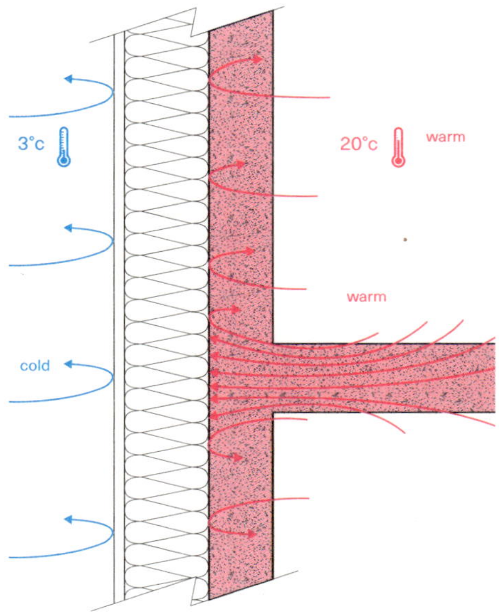

Interior thermal insulation {Fig. 1-73}

The disadvantage of placing thermal insulation inside is the creation of thermal bridges, that is to say a physical and thermal continuity of scarcely thermally insulating materials, such as, for example, between a concrete slab and an outer wall made of concrete or brick, through which the cold will penetrate into the building and cool the inside, leading to energy losses and wasted energy to compensate, as well as problems of condensation and mildew.

Exterior thermal insulation {Fig. 1-74}

By insulating the building from the outside, thermal bridges are avoided as are the risks of condensation. There is no heat loss at the meeting point between the horizontal floor slab and the vertical supporting wall (both remaining protected from the outside cold by being placed within the thermal insulation), which enables saving energy as heating expenditure is kept down.

Double or Triple Glazing as Thermal Insulation

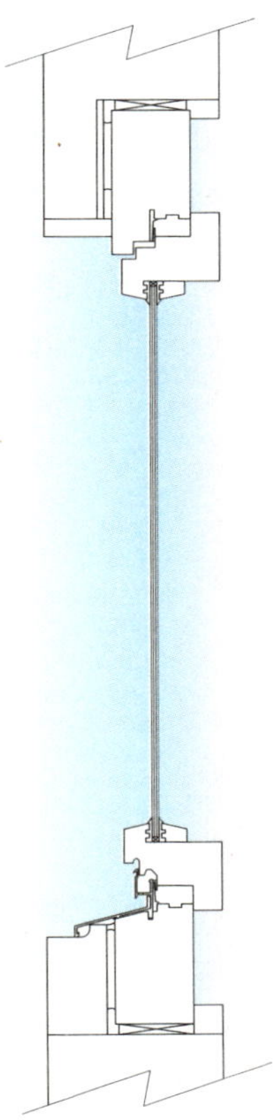

Single glazed window
U-value = 5.2 W/m2k

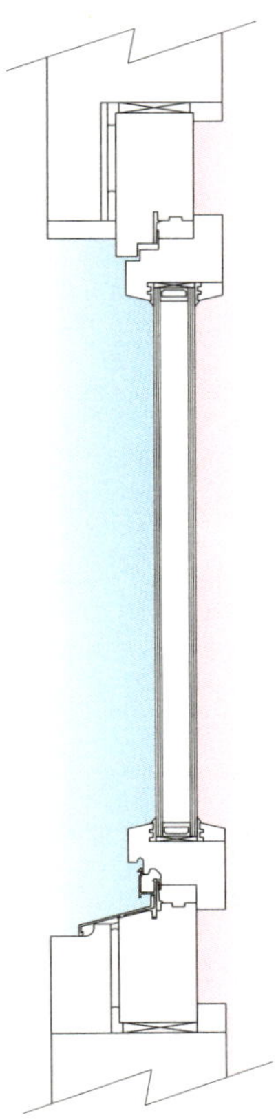

Double glazed window
U-value = 2.9 W/m2k

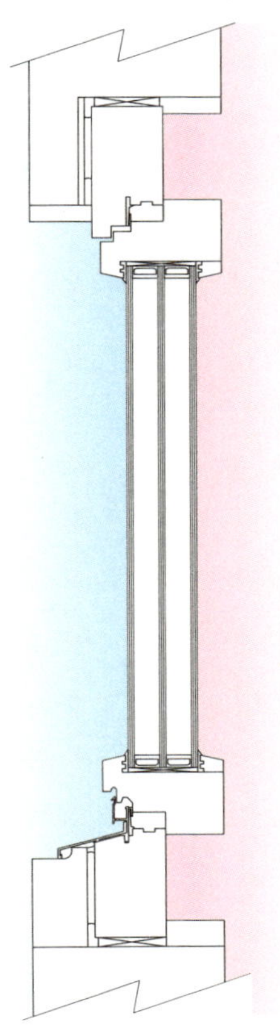

Triple glazed window
U-value = 0.6 W/m2k

U-values for glass {Fig. 1-75}

Multiplying the number of layers of glass in the glazing enables lowering the U-value of a window. This is due to the insulating properties of the layers of air (a relatively poor heat conductor) housed between the glass planes. It is also important to avoid thermal bridges in window frames by integrating a thermal break into the frame.

Thermal Conductivity of Different Materials

k-values of different materials {Fig. 1-76}

Materials with the lowest thermal conductivity (low k value) are the most thermally insulating, preventing heat transfer from one side of the material to the other. This corresponds to less dense, more porous materials containing more air.

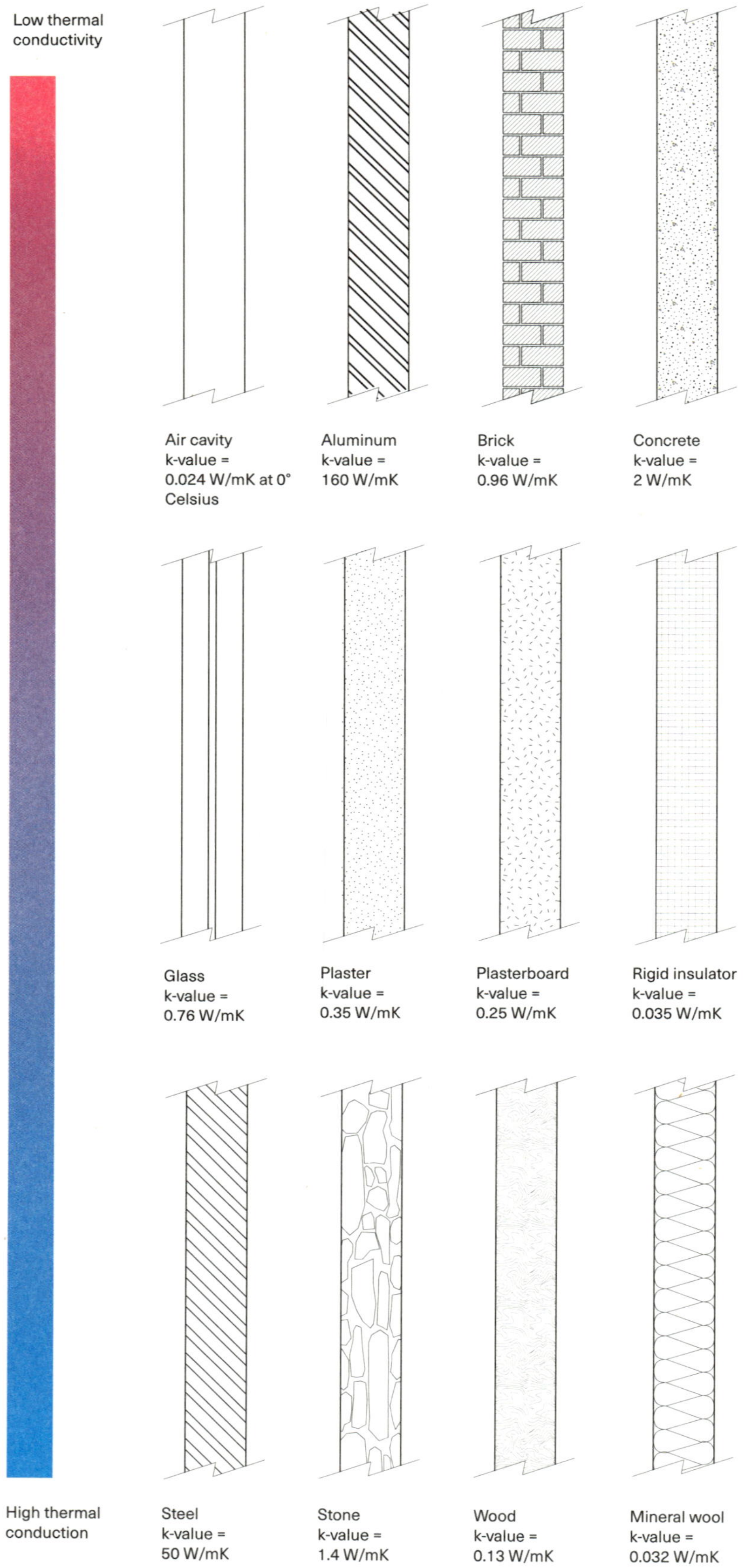

Comparison of the Thickness of Material Needed to Reach a U-value of 0.15 W/m2K

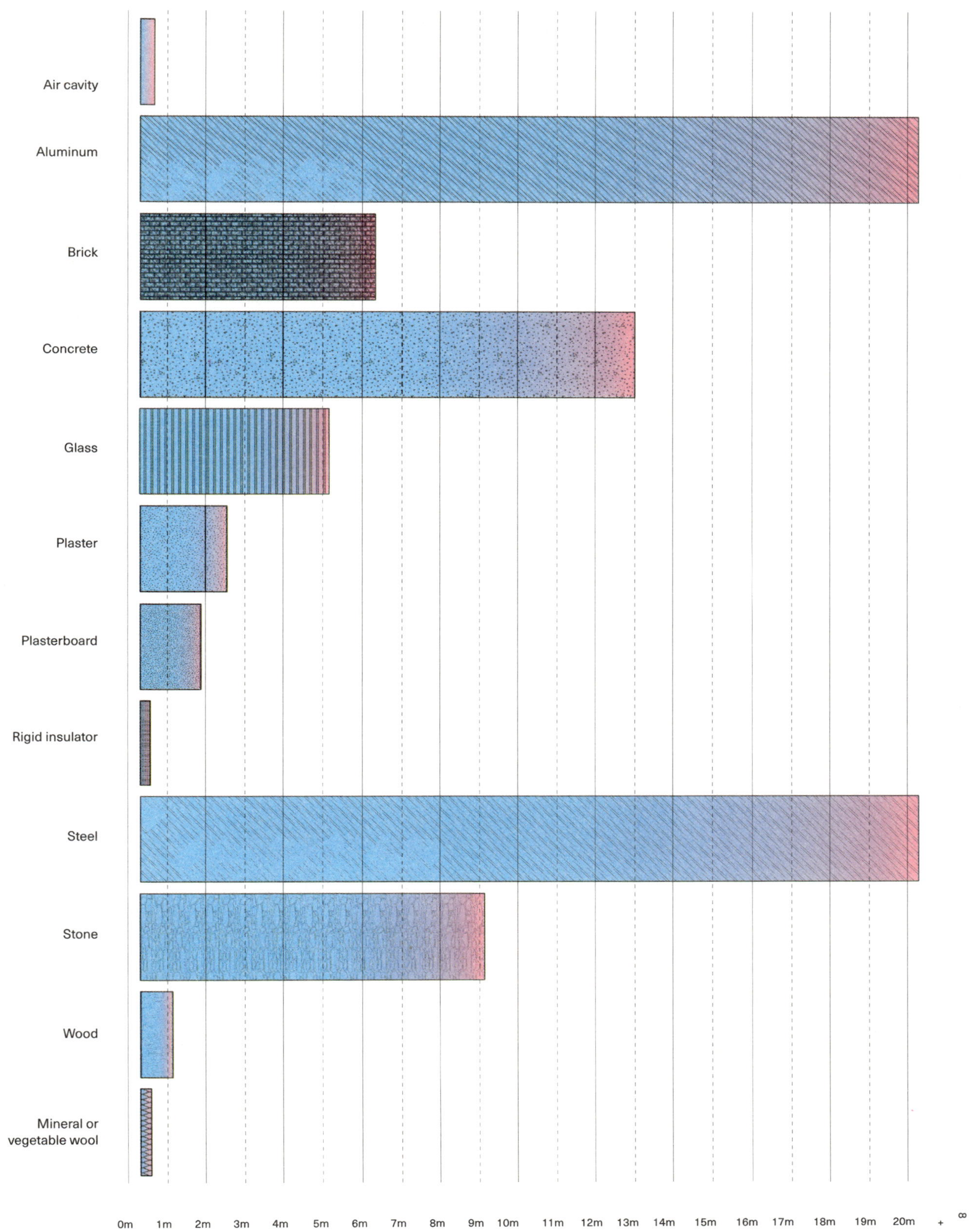

Material thickness required to reach U < 0.15 W/m2K {Fig. 1-77}

To meet energy efficiency standards such as Passivhaus, the heat transfer coefficient (U) of an exterior wall must be less than 0.15 W/m2K.
$U = 1/R_t$ when $R_t = R_i + R_1 + R_2 + R_3$... and R is the thermal resistance of each wall material.

R = D/K when D: material thickness and K: thermal conductivity
U = K/D.

To properly thermally insulate an exterior wall and achieve a U-value < 0.15 W/m2K, a thickness of 25 cm of mineral or vegetable wool must be used, while using only stone, a thickness of 9 meters would be required to achieve the same u-value of 0.15 W/m2K.

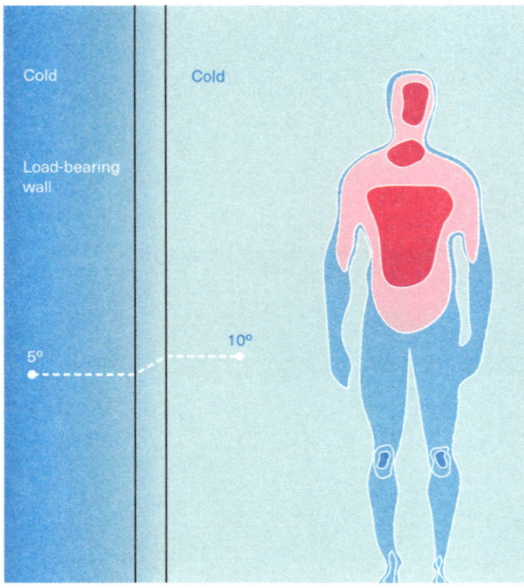

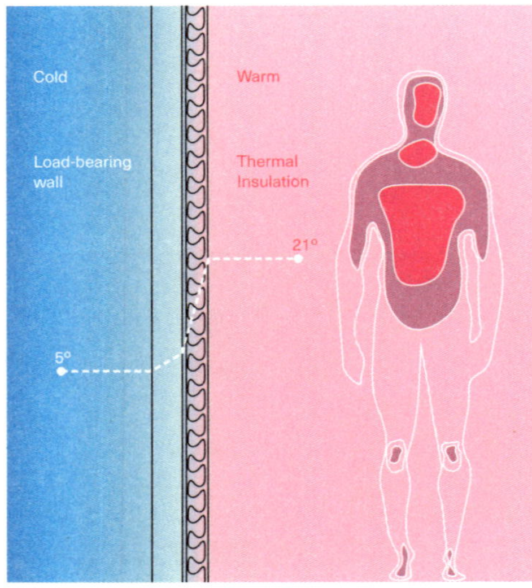

Thermal Insulation principle

Heat transfer without thermal insulation {Fig. 1-78}

Without thermal insulation, the outside cold easily enters the building passing through the walls. The energy used to heat the interior is then exorbitant and wasted to compensate for the constant arrival of cold through the walls, permanently cooling the indoor air. To compensate for this constant cold, it is necessary to increase heating and therefore burn more fossil energy, which leads to more greenhouse gas emissions.

Heat transfer with thermal insulation {Fig. 1-79}

With thermal insulation, the cold remains outside the building, which helps the body and the indoor air to retain their heat, thus reducing energy expenditure since there is no loss of heat by conduction by the exterior walls.

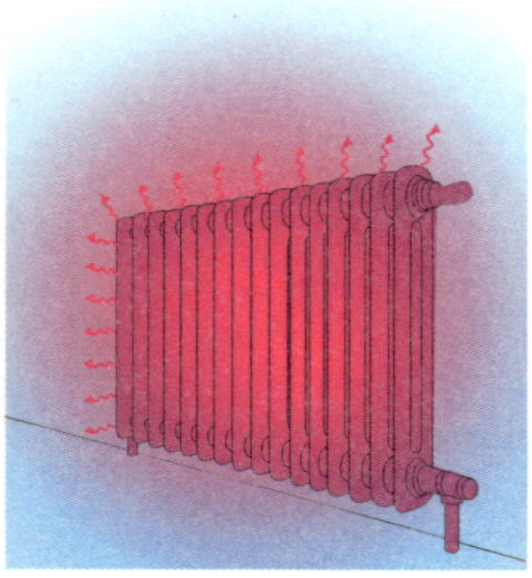

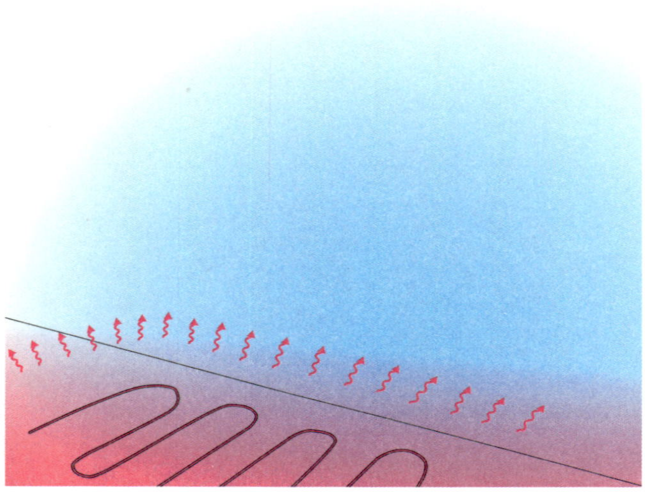

Heating air by conduction

The radiator as heating by conduction {Fig. 1-80}

The radiator heats by convection and radiation. The air flows over the radiator's multiple metal tubes in which hot water flows (usually at a temperature of 60°C). By conduction, the water transfers its heat to the metal of the radiator which (being highly thermally conductive) transfers this heat to the air by convection which in turn heats the room, walls, floor and ceiling and our body in direct contact.

The radiator as a conductive heater {Fig. 1-81}

Coils are placed inside or beneath floor material. Hot water (or cold in summer to cool) heated in the boiler flows through these coils. By thermal conduction, this hot water conveys its heat to the floor material, which will then convey its heat to the air of the room which comes into contact with it and heats it, gradually raising the temperature of the room.

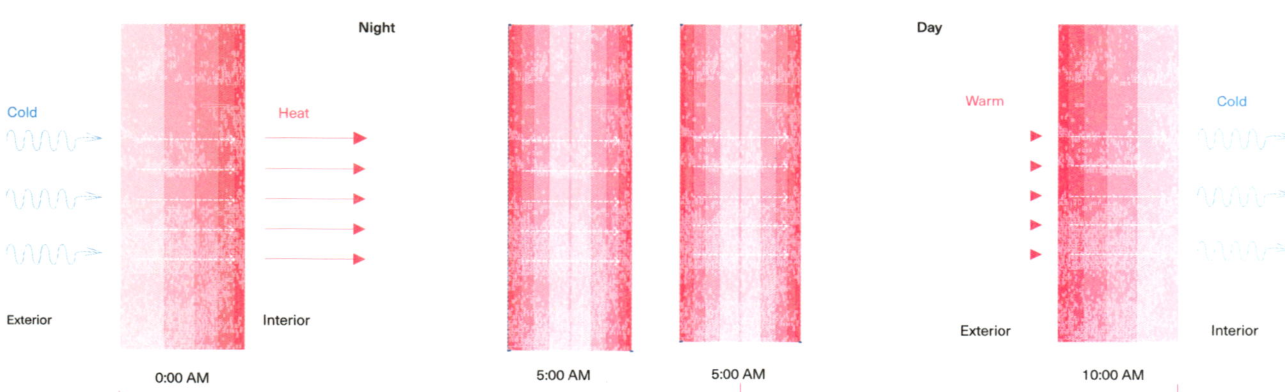

Thermal Inertia

Traditional earth wall 50 cm thickness
Thermal conductivity (W/m.K)
1.21 Density (Kg/m³) 2003 10 h phase shift -
Transfer of energy to within the wall

Inertial phase shift {Fig. 1-82}

Depending on the thermal conductivity of a material, the cold of night will pass more or less quickly through the thickness of the wall to reach the interior of the building within several hours, allowing the cool of the night to return to the interior during the day. This principle is useful in summer or in hot climates by playing on the temperature difference between day and night.

Indeed, the cool of the night will cool down the outer surface of the wall and gradually penetrate and cross the thickness of the wall to finally, during the day, reach the inner surface of the house, thus cooling it during the heat of the day.

V.I Case Study

Thermal Insulation by the Addition of Layers

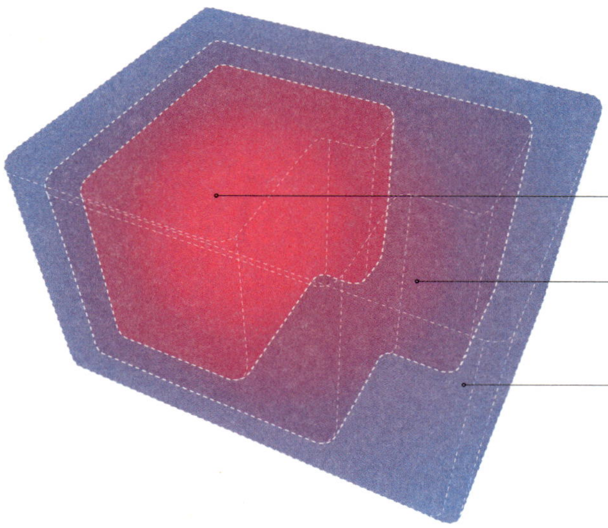

Principle of concentric insulation by the addition of layers {Fig. 1-83}

The layers are volumes made up of thermal insulators, nestled one inside the other, but providing a space between them, and creating a climatic gradation, between a more controlled, stable climate at the heart of the building, and less and less controlled going toward the periphery. The heating source is placed in the core and it is its inevitable losses (even low) that successively heat the following envelopes, from the inside outward.

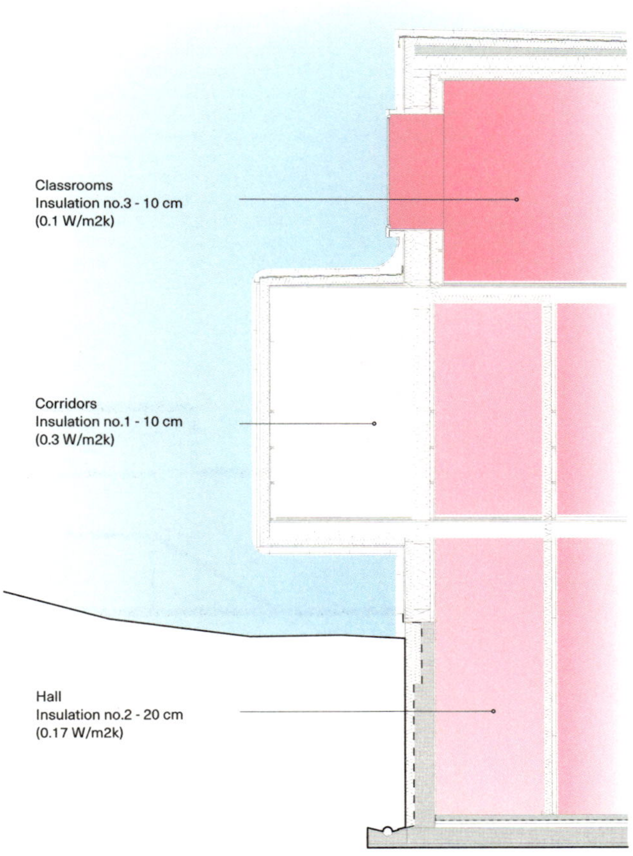

Multi-layer insulation Detail of thermal insulation layers {Fig. 1-84}

The section shows the continuity of the thermal insulation around the envelope, creating layers not only in the plane but also in section. The positioning of the window in the continuity of the insulation is also important to prevent the appearance of thermal bridges in the window frames.

The spaces within one same building have different heating needs and depend on the type of activities and functions they host. This project establishes a gradation per layer of thermal insulation, between the least thermally insulated on the outside and the most insulated on the inside. Thus, the further outwards, we find spaces without presence of people (garages, storage, equipment rooms), spaces where one only passes without staying long enough to get cold (corridors, toilets), and places where people are more physically active - and therefore need less heating because the body produces more heat (due to muscle activity during physical activity)- or more clad or better insulated from the ambient air by clothing. Then we find the spaces where we are a little less dressed but somewhat physically active, then finally those where the people are wearing little in the way of clothing, or even nude, and where thermal comfort must be as high as possible. Gradually increasing the thickness of the thermal insulation (up to 40 cm) around a space will also improve its thermal comfort. One way to maximize the effectiveness of thermal insulation is therefore to create concentric layers of insulation within the building, like in an onion that would provide habitable spaces between two layers. The outermost layers will be the ones that require the least heating because physical activity is more important wherever one remains for a short time, in a staircase or in a corridor, for instance. The central layers will host the activities where the human body is most exposed, that is to say the most temperature-sensitive, or the least physically active. The insulation layers become strategic in the organization of spaces within the building and renew the interior/exterior distinction rather by creating a gradation of interiors.

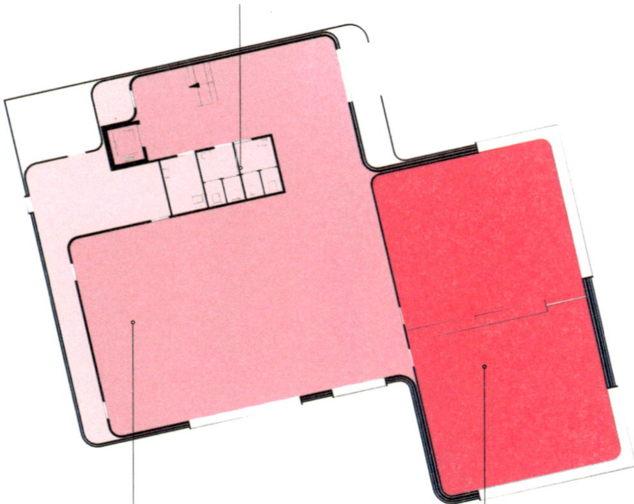

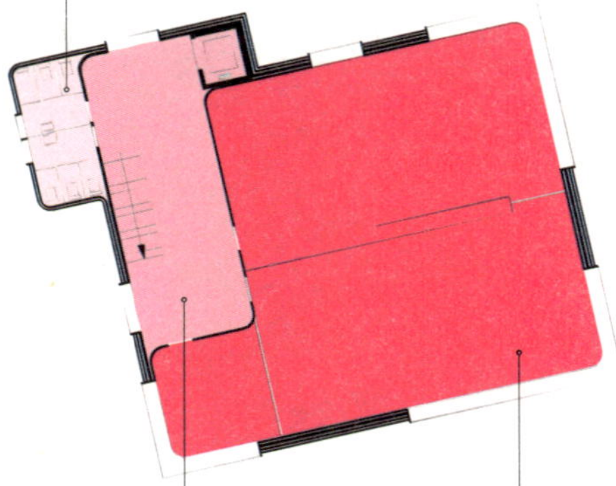

Plans {Fig. 1-85}

The school is arranged like an onion with several thermal layers, according to the temperatures adapted to each activity. The first layer, containing the toilets and technical facilities, is where one spends the least time and is peripheral, making it more prone to external temperature variations.
The second, slightly more central and better insulated layer, accommodates areas of passage and the hall.

Finally, the central layer, protected by 40 cm of insulation by summing the previous thermal insulators, and with a thermal U-coefficient of 0.1W/m2k, hosts the classrooms and provides maximum comfort for the children. The radiators are placed only inside the third insulation layer (no.3), warming the spaces inside insulation layers no.2 and no.1 by way of losses through the thermal insulation.

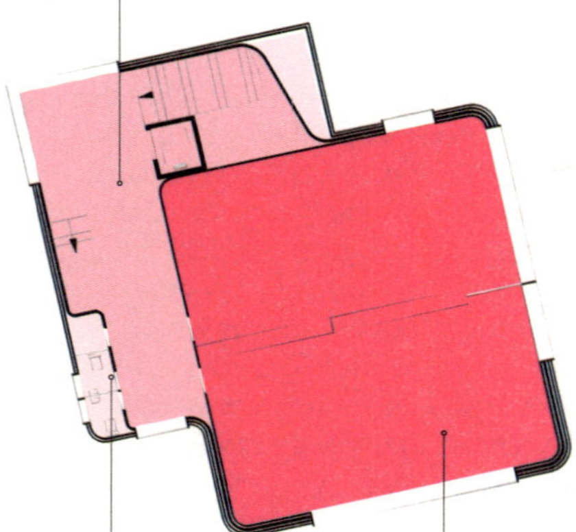

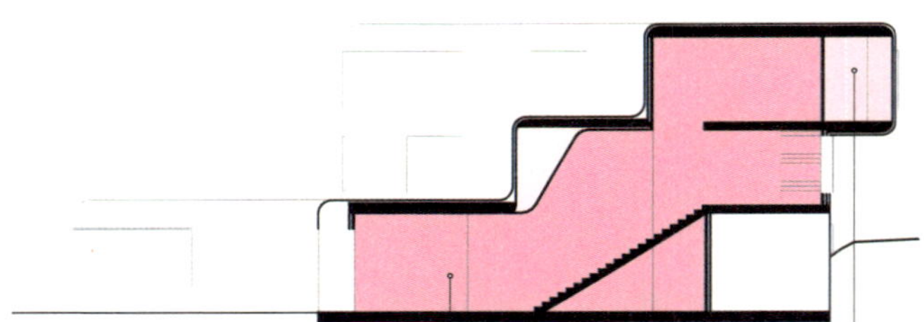

Thermal insulation layers - In-plan layers {Fig. 1-86}

The building envelope literally resembles layers of clothing superimposed to protect against the cold. The addition of layers of insulation in the Minergie-P standard is exactly the same as the accumulation layers of clothes on the body according to the outside temperature but also in terms of activity: a simple t-shirt for running, a light sweater for walking, a thick sweater for being still

Thermal Insulation Layers - Sectional layers {Fig. 1-87}

The programming involves the layers of insulation. The heating, located only at the core of the 3 layers of insulation, will transfer slight amounts of heat to the other layers by conduction, heating the peripheral rooms that require less heat.

The programmatic stratification is closely related to the coefficient of heat transmission.

Inhabiting Thermal Insulation

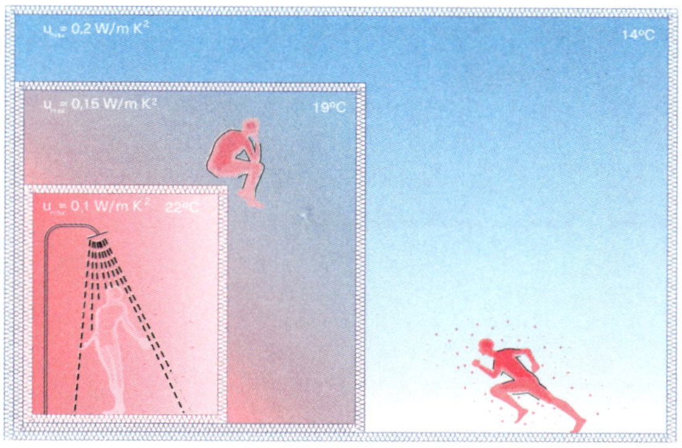

Three thermal bubbles = Diagram of concentric layers {Fig. 1-88}

The different spaces within the same building have different heating needs according to the activities they host.

The building consists of a series of "bubbles" that have both a thermal and a programmatic role. These bubbles are intertwined, enveloping the most thermally sensitive programs in the center (which require more heat in winter) and the least sensitive on the periphery. Each bubble consists of 16 cm-thick thermal insulation fulfilling the Swiss standard. As we approach the center, from one bubble to the next, we add more and more layers of insulation. If in the first we have 16 cm, the second bubble is provided with 32 cm, and the third with 48 cm, that is to say that the most sensitive programs such as changing rooms and showers (which require an air temperature of 22°C) are the most thermally insulated from the outside, allowing increasing the temperature with minimal energy consumption. The three layers of thermal insulation, nested like Russian dolls, respond structurally, visibly and in terms of habitability to the thermal requirements. The principle of thermal decomposition becomes the principle of architectural composition which is reinforced by the staggering of the layers and their distinct openings.

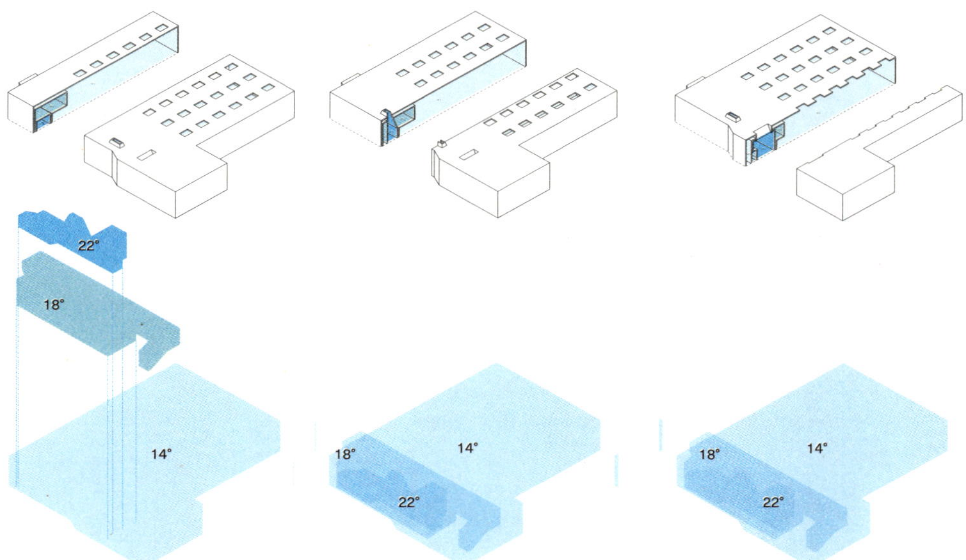

Three thermal bubbles = Interweaving of programs {Fig. 1-89}

The programs requiring the most heat (showers), at 22°C, are located in the heart of the three layers of insulation while those requiring less heat (gymnasium), at 14°C, can remain on the periphery, protected by a single 16 cm layer of thermal insulation fulfilling the Swiss standards.

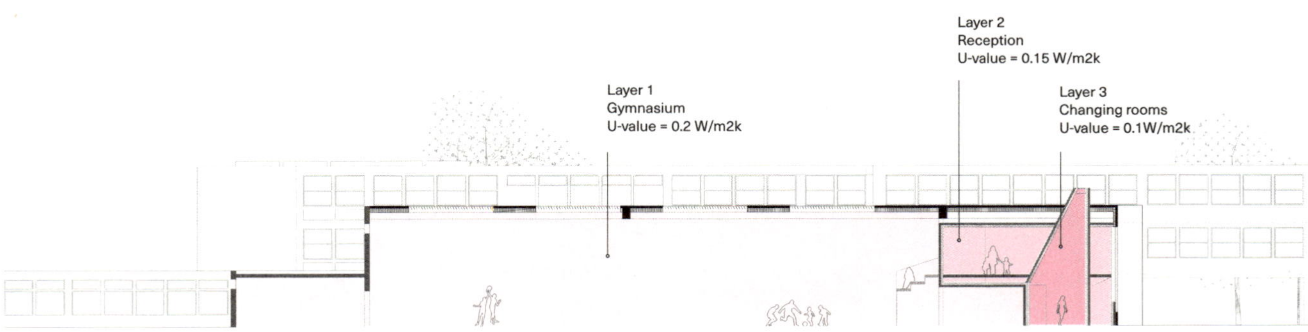

Three thermal bubbles Longitudinal section {Fig. 1-90}

The first layer, the peripheral bubble, has 16 cm insulation and a maximum U-value of 0.23 W/m2K.
It accommodates facilities that require less heating, such as storage, the hall, corridors, and places where the body performs physical activity that heats it. For example, the gym, in winter, can be heated to 14°C.

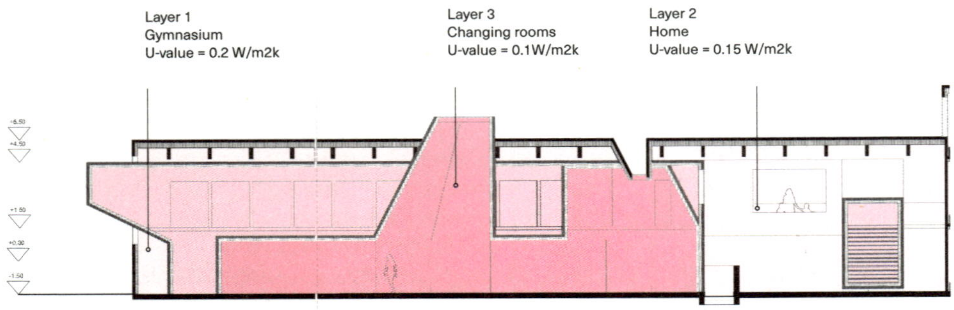

Three thermal bubbles Cross-section {Fig. 1-91}

The second bubble hosts facilities such as the cafeteria, reception and toilets that need a higher temperature than the previous one, between 18 and 19°C. This second layer of thermal insulation has a thickness of 16 cm which is added to the first to achieve 32 cm of insulation, or a U-value of 0.12 W/m2K

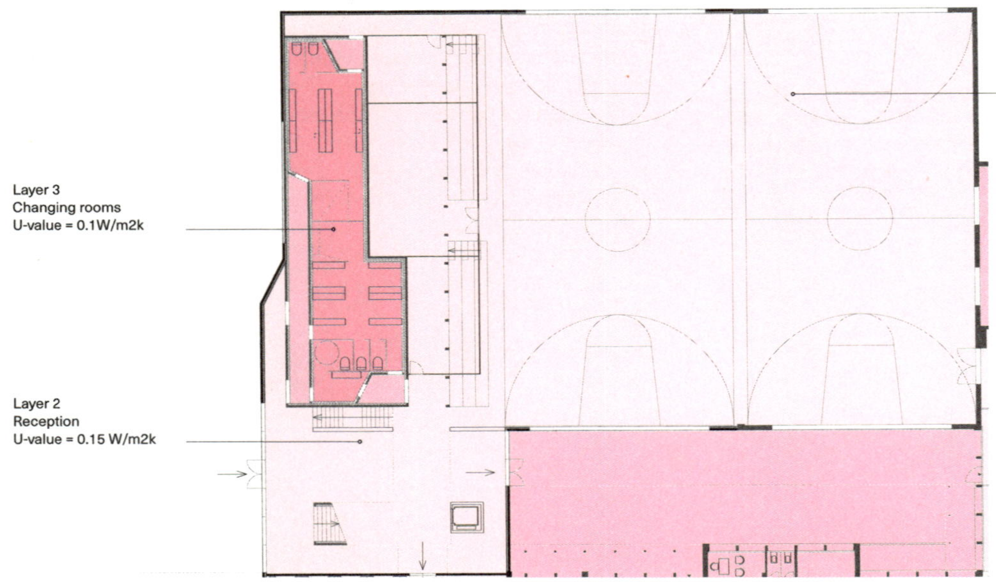

Three thermal bubbles - Plan {Fig. 1-92}

The third bubble, containing the changing rooms and showers, should be even warmer than the other two, at about 22°C. By adding this further 16 cm layer of insulation we reach a total of 48 cm and a U-value of less than 0.1 W/m2K which corresponds to Minergie-P Standard, that is to say the most efficient thermal insulation. This last "bubble", the warmest and most protected, ensures the comfort of naked bodies in exchange for minimal energy expenditure.

A Void as an Unstable Climate Between the Load-Bearing Wall and the Insulation Layer

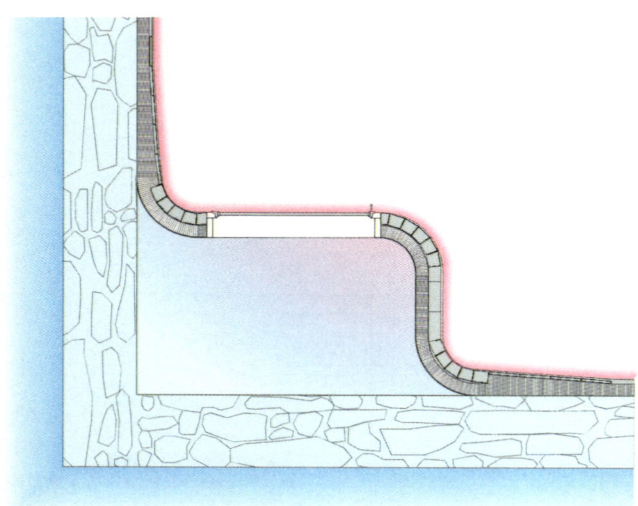

Thermal bubble - Plan detail of insulation and the stone wall {Fig. 1-93}

A new layer of thermal insulation set back from the stone wall of the façade is introduced in order to create an intermediate space with a higher thermal conduction coefficient that is more sensitive to the external climate.

This project to renovate an old dry stone building proposes to leave an undetermined empty space between the new thermal insulation inside and the exterior wall, creating an air pocket with a more fluctuating climate that remains colder than the interior in winter but still remains protected from wind, rain, sun or snow. This type of space can be used as a storage area for food (like the old milk cabinets within thick walls to benefit from the cold of winter as a natural refrigerator) or for all other functions according to thermal and moisture needs. Through this climatically indeterminate space, the architecture maintains a sensual dialogue with the seasons, the outdoor climate, exposure to the sun and the physical quality of the ground. This type of organization recalls older buildings where spaces such as the attic or the cellar played the role of a thermal halfway house between indoors and outdoors.

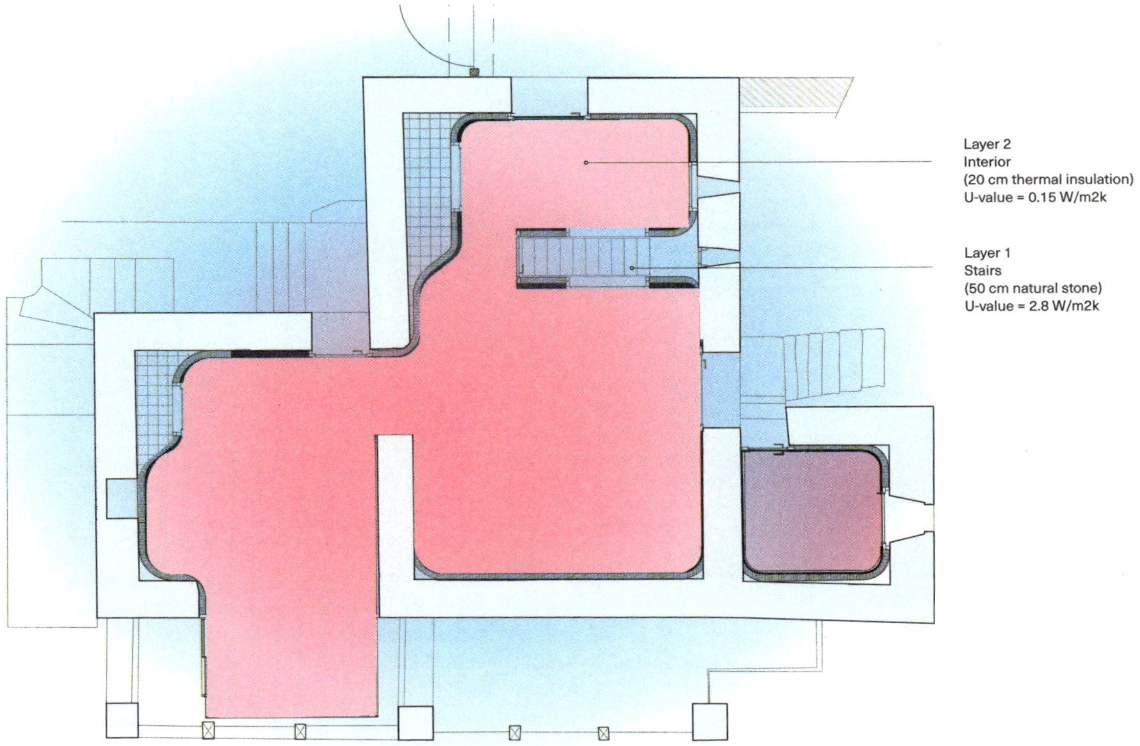

Thermal bubble - Plan 1
Layers of insulation {Fig. 1-94}

A chimney of varying climates ranges from bottom to top like a secret passageway between the new thermal insulation and the existing walls. Less controlled but nevertheless at a moderate temperature compared to the outside, this chimney extends through the house, creating a series of climates, from the coolest and most humid at the bottom, to the warmest and driest at the top.

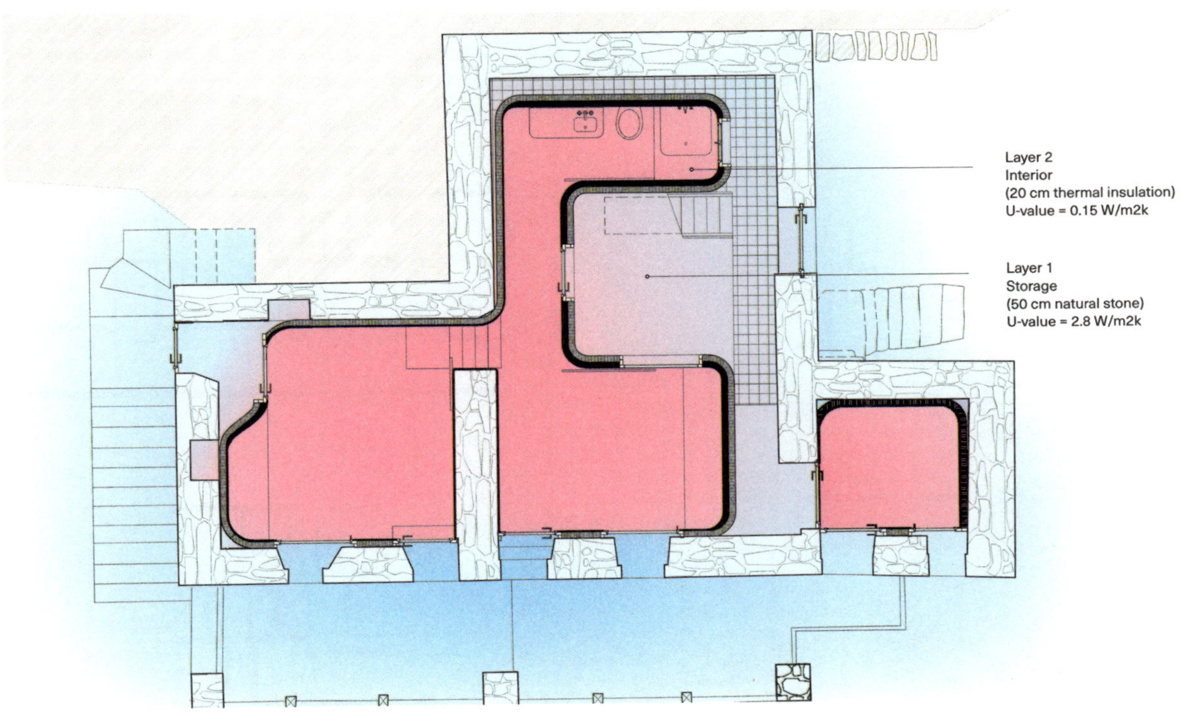

Thermal bubble - Plan 2
Layers of insulation {Fig. 1-95}

It is in these intermediate spaces that garden produce as well as objects and clothing are stored, according to their needs in terms of temperature and humidity. Mint leaves dry at the top, jarred vegetable preserves are stored away from the light, and syrup, broth and apples are stored at the bottom. When crossing the intermediate spaces we become more sensitive to and welcoming of the meteorological and geological variations.

Rooms with Varying Thermal Conductivity

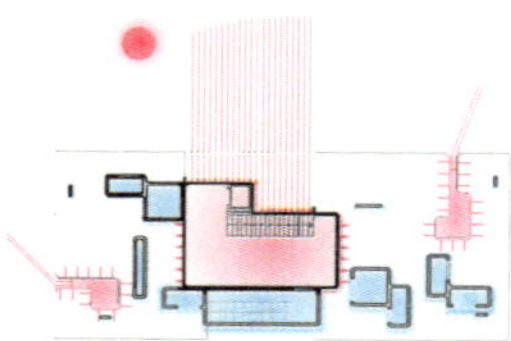

Conductor Condominium - Plan 1
Daytime heat inputs {Fig. 1-96}

The design of this building is based on thermal conductivity. The building's main materials have a very low coefficient of conductivity (wood and insulating hollow fibers, less than 1 W/(m.k)) except for certain parts, which are rendered in concrete or steel. These last two materials, with different coefficients of conductivity, enable recovering the heat or cold from the external environment brought by the rays of the sun or the dark night through the glass façade. (Wich can open in summer).

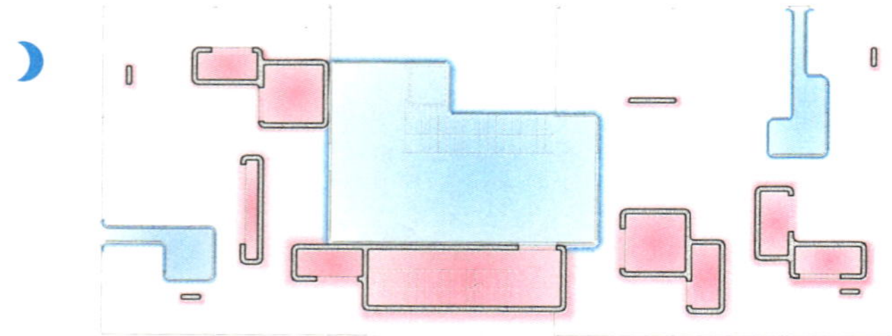

Conductor Condominium - Plan 2
Night time heat losses {Fig. 1-97}

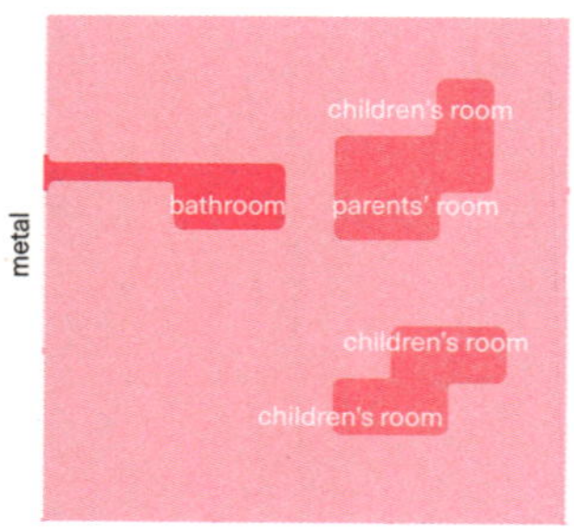

Floor plan of the apartment

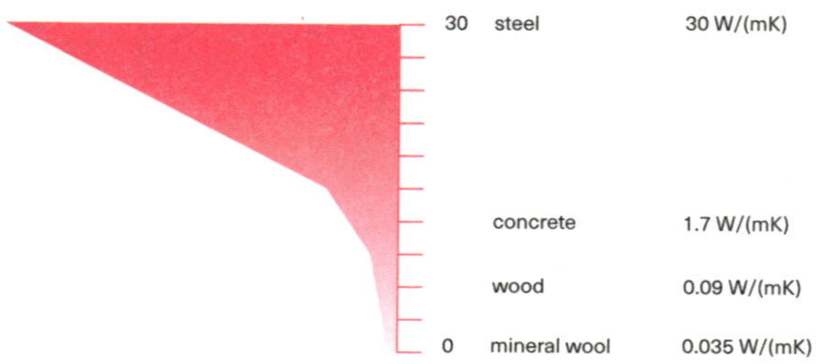

Conduction values of each room {Fig. 1-98}

The gradation of thermal conductivities is related to the materials used. Each room, depending on its geographical position and its purpose, is made of a specific material to heat quickly or to store solar heat and release it at night.

The main purpose of this project is to reintroduce the different climatic qualities in this modern apartment at various times and in different places, like in old houses, by playing with the materials and exposure to the sun. Accordingly, it will be very comfortable to sit in rooms made of concrete in summer because they will have stored the freshness of the night. Conversely, it will be nice on a winter's day to sit close to the metal parts heated by the sun's rays on the façade by the greenhouse effect. The interior collective spaces of the building that accommodate the stairways, the elevator, the laundry room etc., will be designed according to the same thermal principles. Some spaces will be rendered in steel, others in concrete. The programs will be positioned according to the thermal qualities of the materials. The concrete rooms will house the cellars, the laundry room and the bicycle room, for example. The steel rooms, meanwhile, will be reserved for the laundry drying room, to store cereals and buggies, for example. Rather than storing our items and provisions in refrigerators and cupboards, we propose taking advantage of the climatic conditions of collective shared spaces to do so.

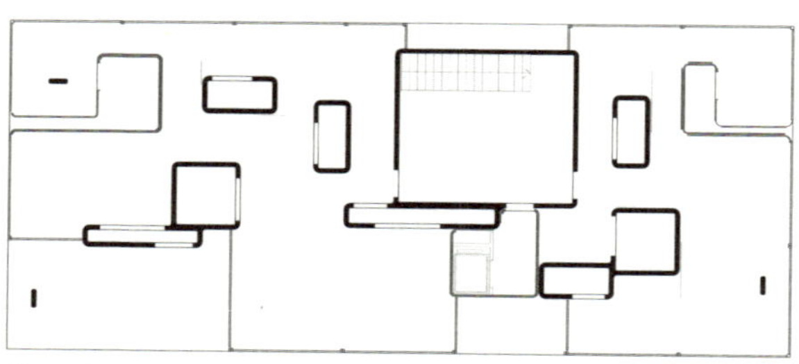

Floor plan

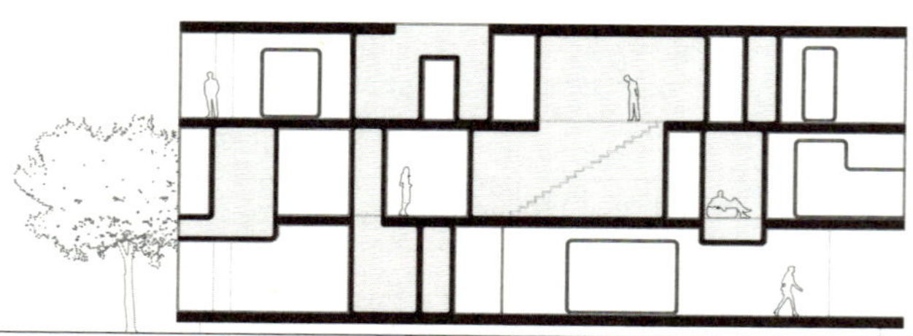

Section

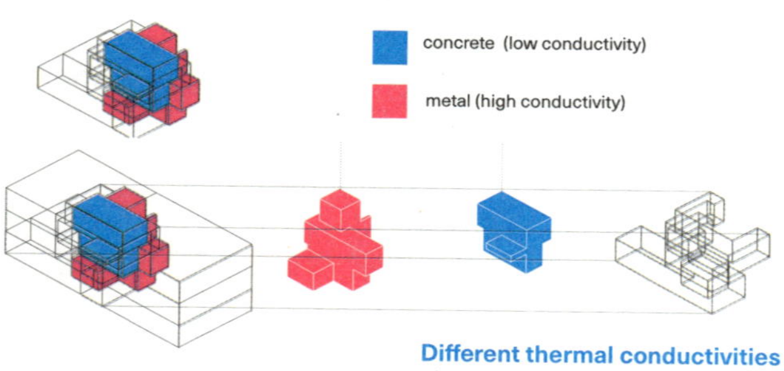

Different thermal conductivities of materials {Fig. 1-99}

The choice of materials is linked to the speed of transmission of solar heat gains which determines use at a specific time of the day.

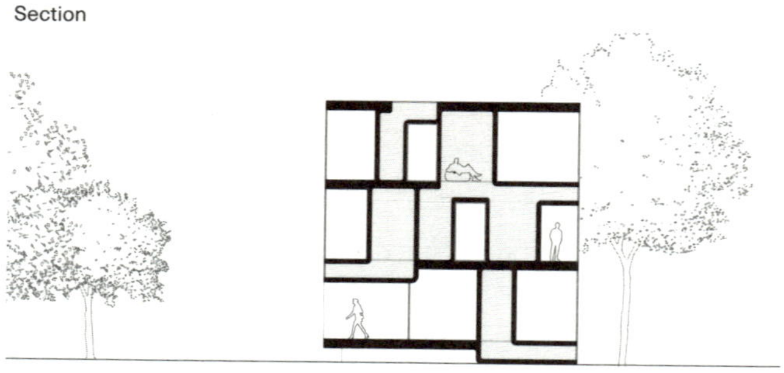

Section

Plan and Sections {Fig. 1-100}

In section, the rooms are connected to the façade to receive, by conduction, the solar heat gains required by the room. If it is rendered in metal, conduction will be immediate, while with concrete there will be a phase shift, according to the principle of thermal inertia.

Inhabiting Double Glazing

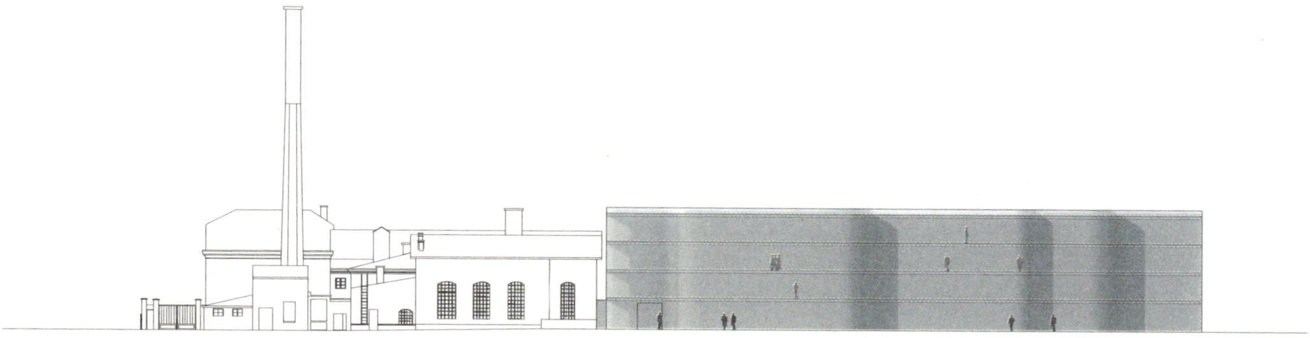

North elevation {Fig. 1-101}

This project for a museum in Poland is based on widening the layer of air that separates the glass in double and triple glazing until this layer reaches several meters and becomes habitable. It uses the same building system as contemporary windows in which several layers of glass enable improving the quality of insulation and lowering the thermal transmission coefficient

U (W/m2.K) (single glazing: U =5, double: U = 1.4, triple: U = 0.8). Our project adds layers to gradually improve the thermal coefficient, one layer after another, in order to provide various temperatures and light intensities. In contrast to the uniformity of the modern climate, we propose a variety of ambiances, lights and temperatures that invite one to move freely.

Depending on the type of activity and the season, we will choose one layer over another, warmer, or further north, with a constant temperature or greater brightness.

Here, architecture finds its own language concerning energy requirements in connection with sustainable development and offers interpretable, extraordinary spaces, freely subject to future behaviors and uses. Inhabiting the space between double glazing.

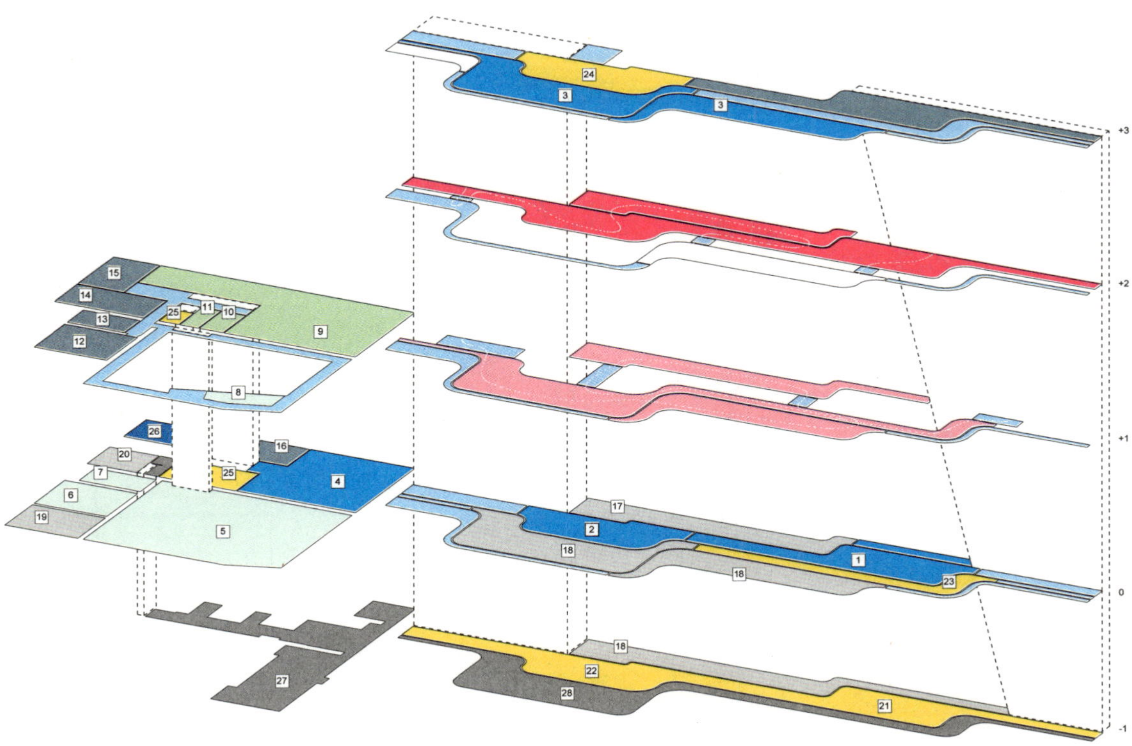

Function follows climate {Fig. 1-102}

The program for the museum is set between layers of glass according to temperature and comfort requirements, the degree of visitors' physical activity or the climatic conditions necessary for the due conservation of works of art.

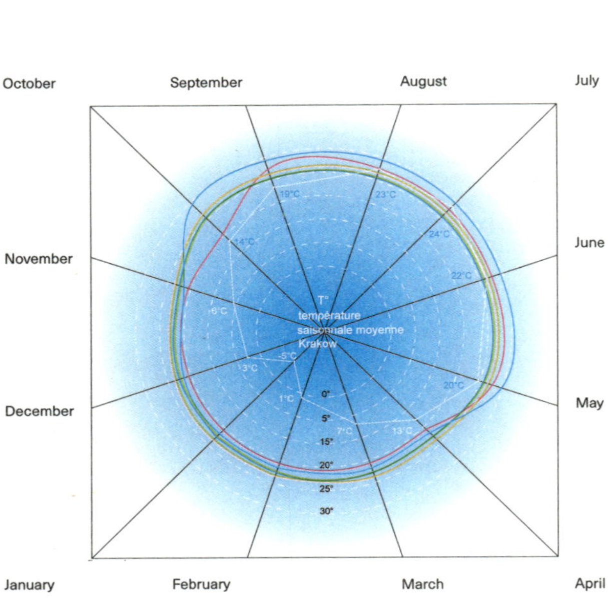

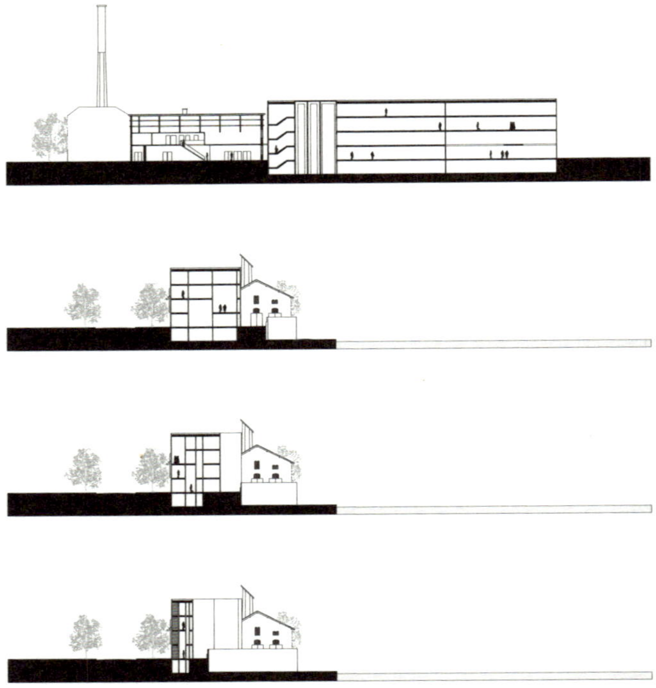

Sections {Fig. 1-103}

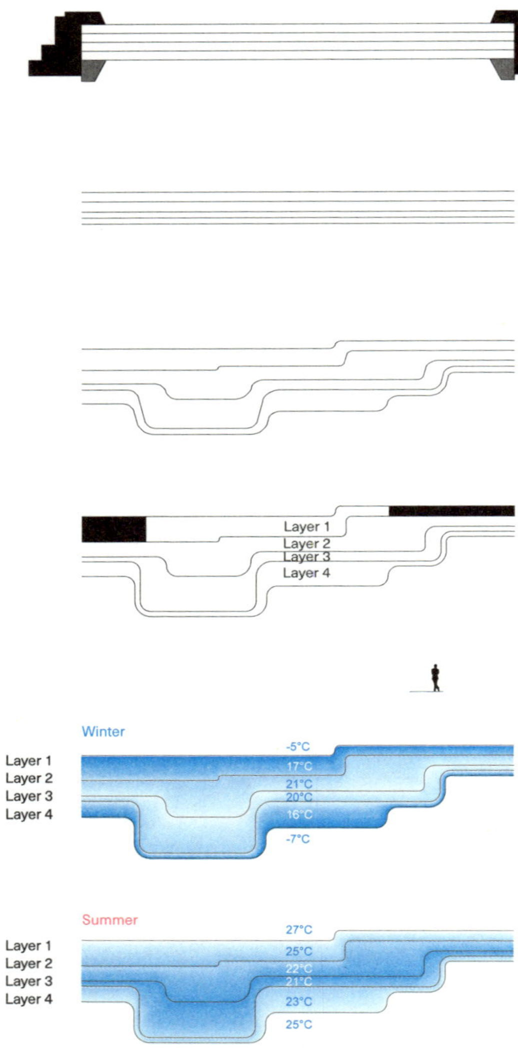

Temperature variations according to season

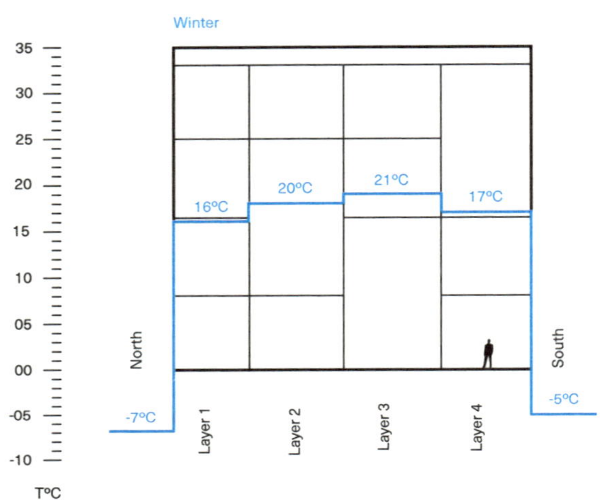

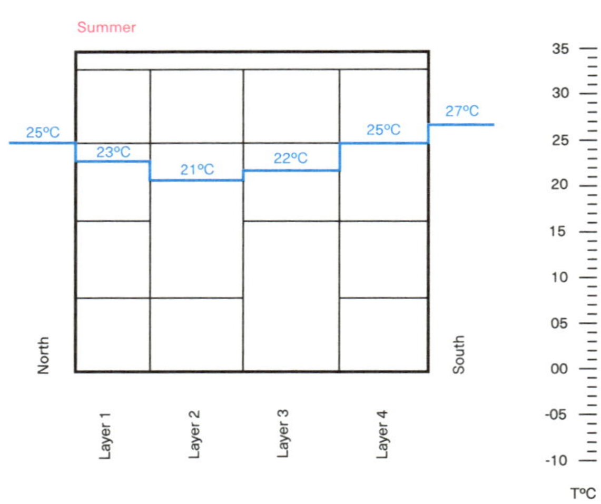

Expansion of spaces between glazing {Fig. 1-104}

The building is literally an expansion of the principle of adding layers of air between glazing, improving the insulation qualities of the windows. Here, the gap between the double glazing is widened to become habitable.

The building is literally a multi-glazed window augmented to the size of the building.

Annual temperature fluctuations {Fig. 1-105}

The temperatures in the different layers of glass will vary throughout the year. However, the temperature will remain constant in the heart of the building, offering uniform, permanent comfort for offices or other permanently occupied spaces.

Different Thermal Conductivities

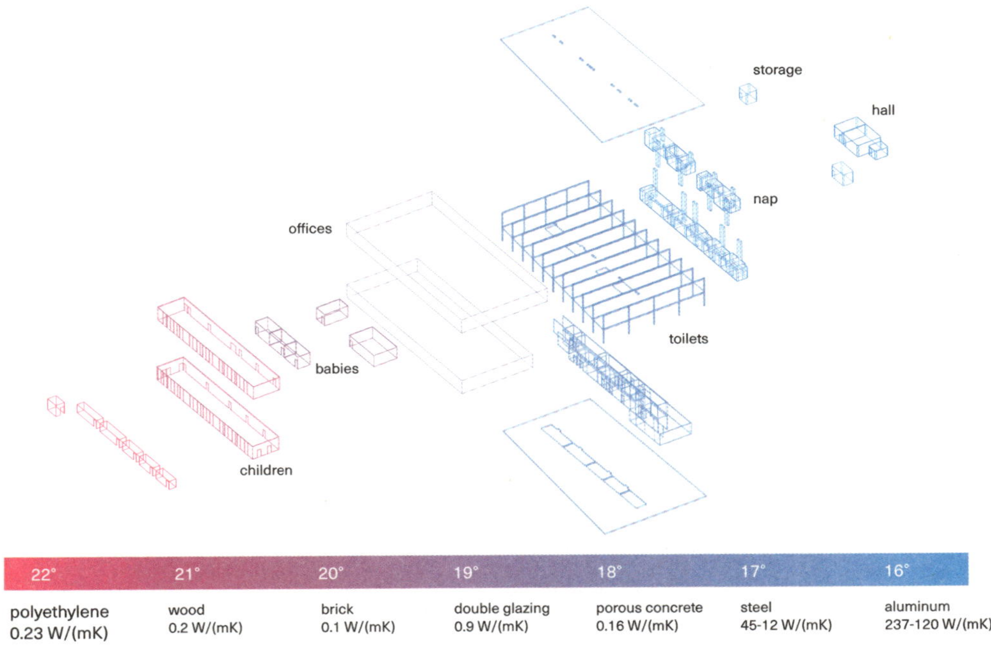

Thermal conductive composition {Fig. 1-106}

Exploded view of the elements according to their coefficient of thermal conductivity. The building consists of rooms each determined by a value of effusivity and of conduction in relation to the desired temperature

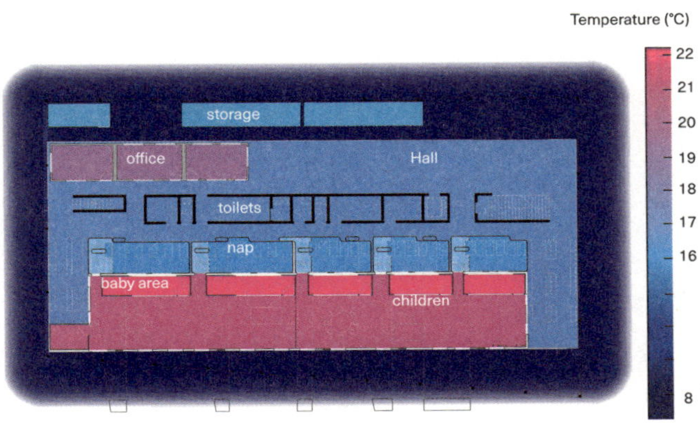

Functions according to thermal conductivity {Fig. 1-107}

The temperature required according to usage (baby spaces must be at 22°C, children's rooms at 21°C, etc.) determines the materials according to their thermal conductivity.

The materials for each of the rooms are chosen according to their properties of being colder or warmer according to their effusivity and heat conduction values (u)

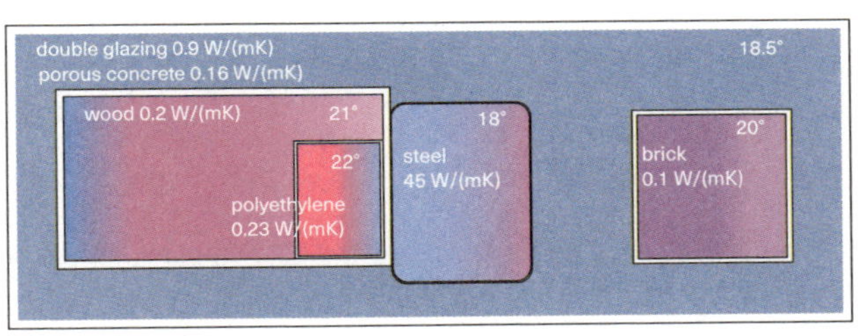

Thermal conductivity program {Fig. 1-108}

Distribution of spaces and materials according to temperature. The least effusive and least conductive materials will be chosen for the rooms requiring the highest temperature, and, following gradual gradation, materials will be chosen which are increasingly effusive, the lower the temperature required.

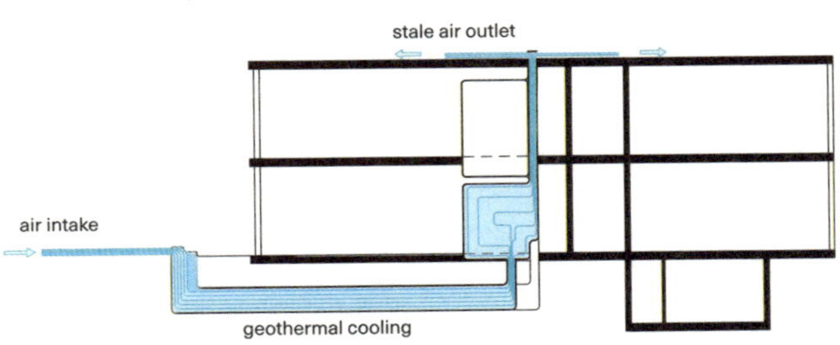

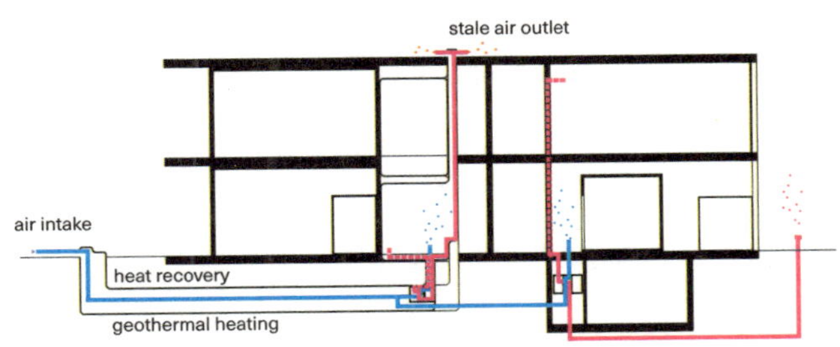

Summer ventilation system {Fig. 1-109}

Natural ventilation for siesta facilities

Winter ventilation system {Fig. 1-110}

Principle of ventilation in winter

This project for a kindergarten implements a series of spaces according to a gradation between the innermost and the outermost, the warmest and the least warm, between warm materials such as wood and other cold materials such as steel, which are chosen according to their physical heat transfer properties. Our project is composed gradually from the outside to the inside, from cold to warm. We first divide the program into different materials according to their thermal requirements for spaces that need to be warmer, made of wood, such as lounges or baby-changing rooms and spaces that can be a little colder, made of brick, such as offices, followed by even less warm ones, made of aluminum, such as nap rooms.

CONVECTION

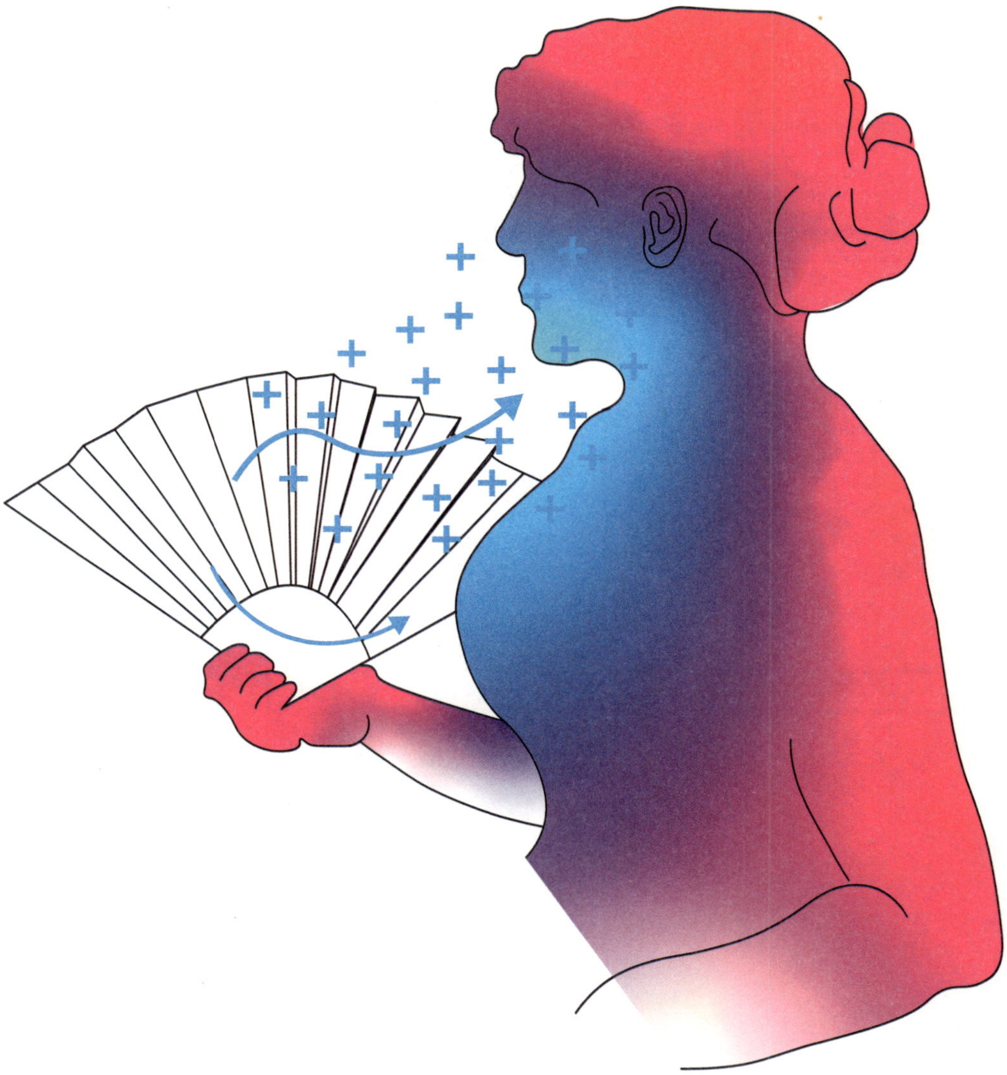

General Properties

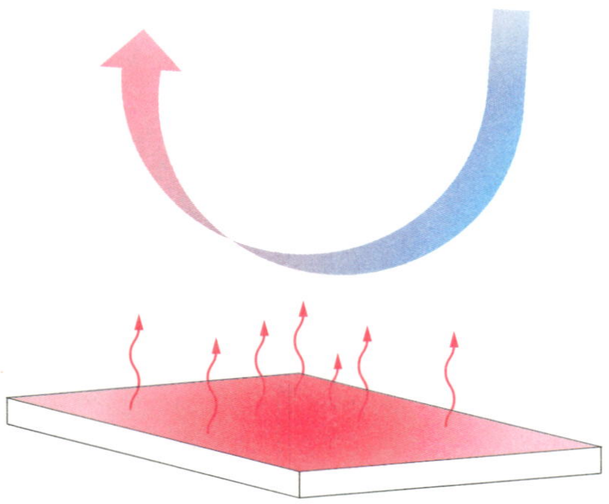

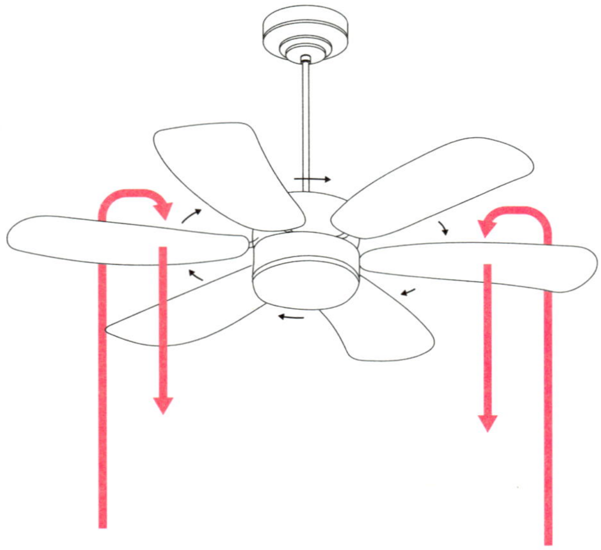

Natural convection {Fig. 1-111}

The heat released by a hot surface causes natural convection, that is to say a displacement of air masses, since the air close to the surface is heated and will rise, creating a convective movement from a simple thermal gradient.

Forced convection {Fig. 1-112}

The ceiling fan creates air movement by pulling the air and then pushing it down. This has a cooling effect on the skin, preventing the accumulation of heat continuously emitted by the human body; the moving air sweeps the heat and renews the air around the skin by bringing in cooler air.

A fan is only useful if the air temperature is below 34°C, lower than the temperature of the skin of 34°C. Above that, a fan is useless, it will have the same effect as a hair dryer

Thermal convection is the transfer of energy within a fluid (water or air) or between a solid and a fluid. A fluid in contact with an energy source will heat up, become less dense and therefore move upwards, repelling denser, cold air, which will fall away from the heat source. The air (or water) current resulting from this heat transfer from one place to another is called convection movement.

There are two types of convection:

Natural convection: fluids are caused to move by temperature variations (thermal gradient) in the fluid itself.
Forced convection: external means (such as a fan or a pump) are used to artificially induce convection movement.

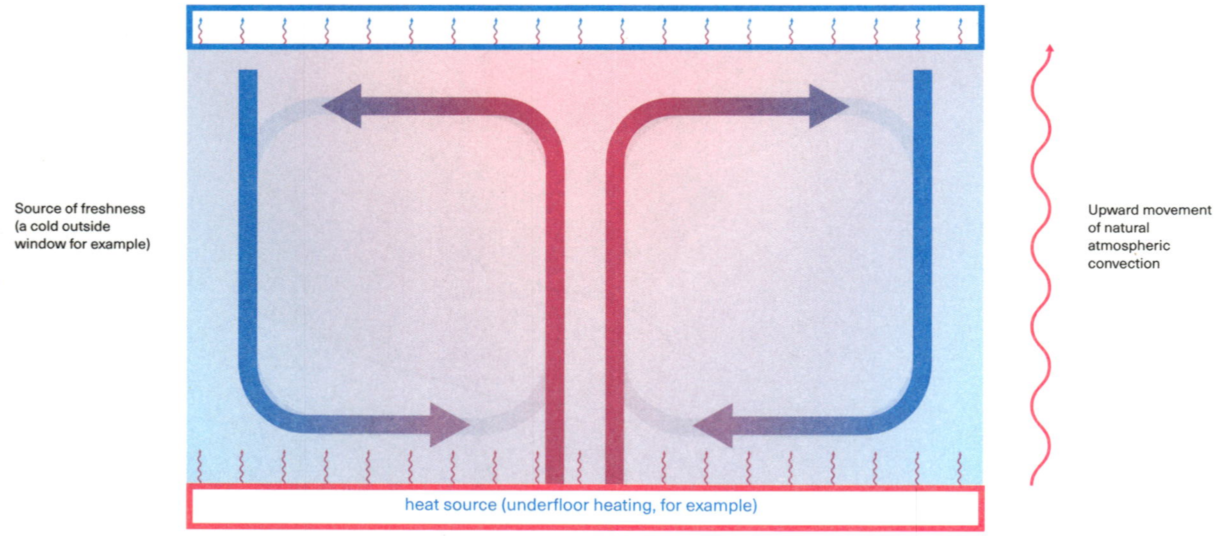

Heat transmission by convection {Fig. 1-113}

Heat is transmitted from a hot object to a cold object via the fluid. The convection current arises from the natural temperature gradient within the fluid.

An air current is then created, the heated air rising and pushing the cooler or cooling air, which then sinks.

In Buildings

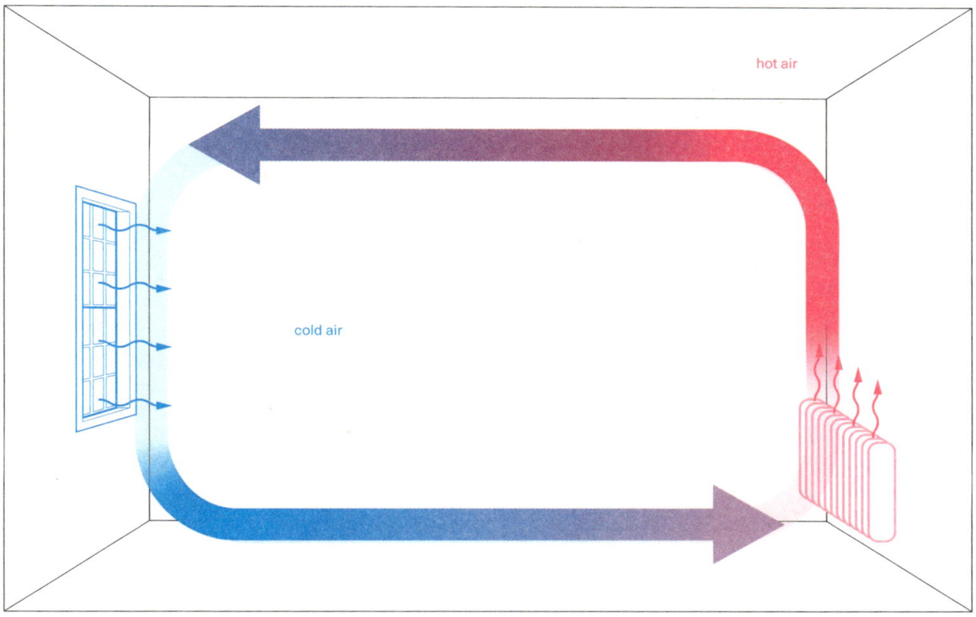

Heat flow by convection in the building {Fig. 1-114}
Natural convection (caused by temperature differences between walls, floors or ceilings) as well as for forced convection by radiators, fans, air conditioners, etc., plays a role in regulating the internal temperature of buildings. Usually, air moves within a room, between the radiator that causes the air to rise and a cooler surface (such as a window or an exterior wall) that will cause it to fall. Thus there is a constant flow of air in the room.

II. Properties of Materials

Convection Flow Equation

Convection takes place within fluids (liquid or gaseous) because they can expand more when heated, unlike solids). Once warm, the heated, less dense particles will float above the cooler, denser particles. According to Archimedes' law: if the buoyant force on an object is greater than its weight, it rises, and if it is lower, it falls. The buoyant force is equivalent to the weight of the fluid displaced by the object.

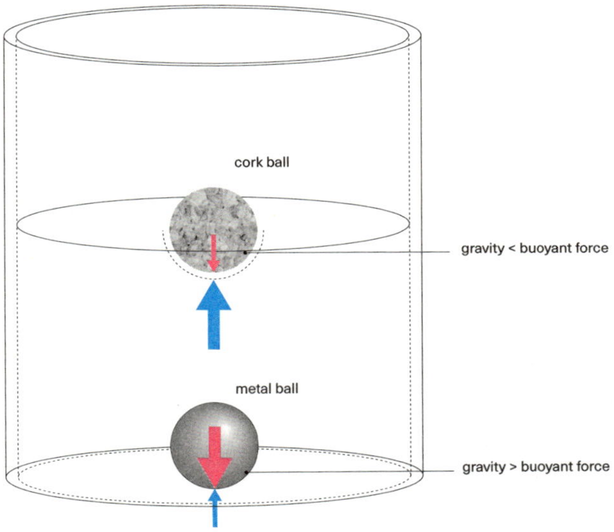

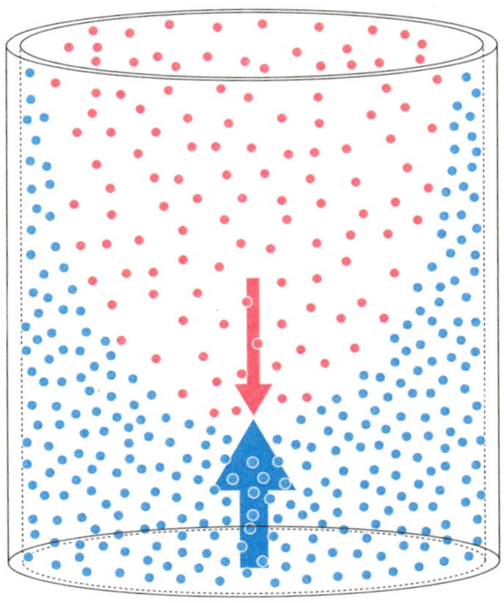

Archimedes' principle {Fig. 1-115}
The principle states that the buoyant force of an object is equal to the weight of the mass of fluid that has been displaced within the container. If the buoyant force is greater, the object floats, otherwise it sinks.

Floatability of hot fluid {Fig. 1-116}
The hotter fluid is less dense and therefore weighs less than the buoyant force of the denser cold fluid below. This leads the hot fluid to rise.

Indoor Convection

In a closed volume of air, as indoors, the temperature variations of wall surfaces (for example the north-facing will be a little colder than the south-facing wall), or induced by the presence of a heat source, will cause a convective air current which makes the hot air rise and the cold air fall. In fact, air molecules are less dense the warmer the air, which makes it lighter and it will thus tend to rise, replaced by the cooler, denser and therefore heavier air which will then fall to the ground pushing to replace it with the warmer air which will then rise.

This indoor atmospheric convection causes large temperature differences between the floor and the ceiling, the air beneath it possibly being 10°C higher than the air at floor level. As we live on the floor, it is energy that is actually lost.

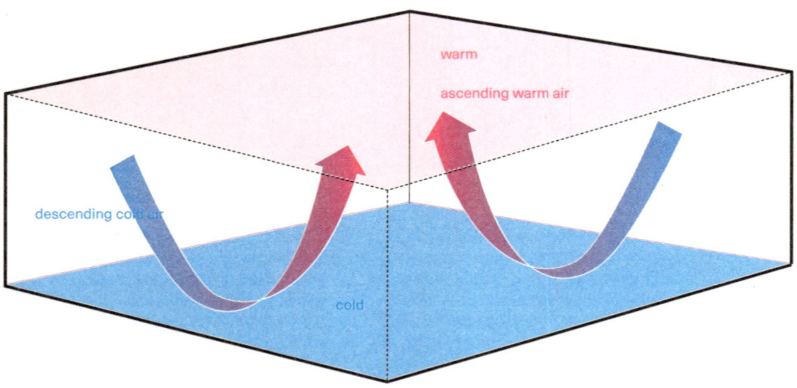

Atmospheric convection {Fig. 1-117}

When a fluid is hot, its molecules are more agitated than those of a cold fluid, and the hot fluid will be less dense and therefore lighter. According to Archimedes' principle of buoyancy - which states that a body of the same volume but of denser composition will tend to sink compared to a body of lower density which will float - hot air will tend to rise while cold air will fall.

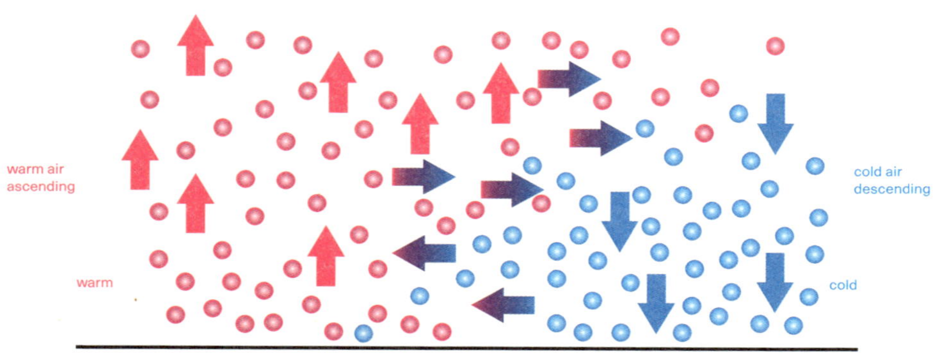

Atmospheric convection {Fig. 1-118}

Variations in the temperature of the walls of the house or the presence of radiators cause a movement of air called atmospheric convection, a kind of domestic wind that moves hot air toward the ceiling and cold air toward the floor.

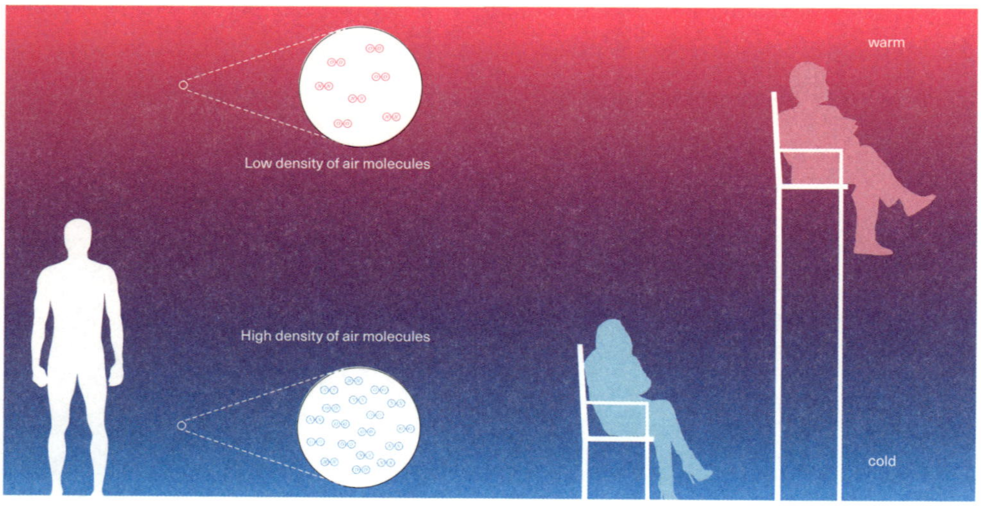

Vertical stratification of air temperature according to the convection principle {Fig. 1-119}

The comfort of the human body depends on the temperature of the surrounding air. According to Archimedes' principle of buoyancy, hot air will rise and accumulate at the top of the room, below the ceiling, while cold air will fall and accumulate near the ground. By lowering a room's ceiling, the human body can enjoy a more comfortable ambient temperature in winter, while by raising the ceiling, hot air will be allowed to escape upwards allowing, lower down, inhabiting cooler and more refreshing air masses.

III. Physiological Scale

Thermoregulation in the Human Body

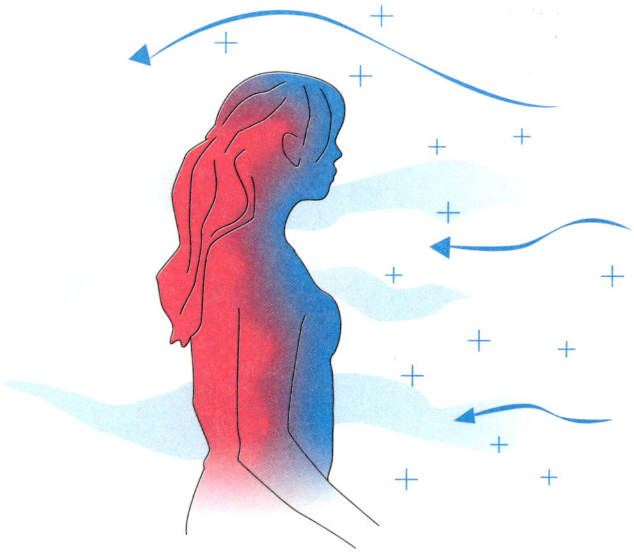

Convection on the human body {Fig. 1-120}

Convection causes a rapid renewal of the air around the body, because when the body is not dressed or is exposed to wind, the layer of hot air in contact with the body produced by the internal temperature of the human body at 37°C is removed, thus cooling the body down.

Convection is a mode of heat exchange between the body and the environment. In winter, when it is cold, the moving air, if it is windy, will cool the skin by convection, which will be uncomfortable. We must therefore protect ourselves from the wind if we are cold, so as to lose heat only by conduction and not, in addition, by convection. Conversely, in summer, when it is hot, heat exchange between the skin and the air must be sped up in order more quickly to dissipate the heat produced by the body, and put oneself in the wind to cool down. Since over time, when the air is still, the body will heat up the air close to its skin by conduction. It thus loses heat when this layer of heated air is carried away by the wind and replaced by cooler, fresh air. The ambient temperature can thus seem much cooler on a windy day because the protective layer of hot air around the body is constantly carried away. This phenomenon is known as wind chill. To keep a thermal balance, our body must lose heat at the same rate as it produces it.

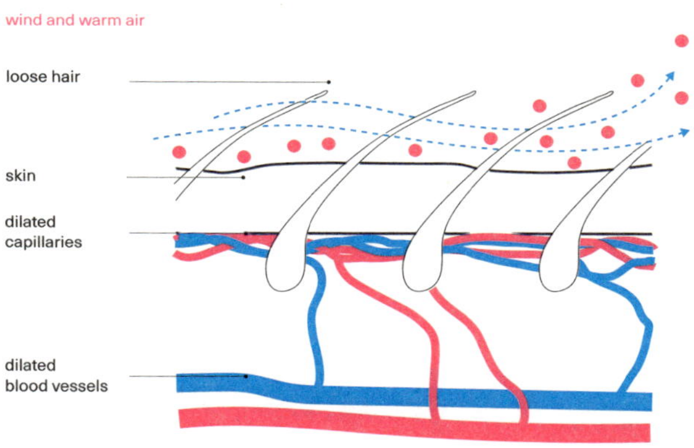

Thermoregulation in a warm environment {Fig. 1-121}

In a warm environment, blood vessels dilate, allowing more blood to flow to the skin and come into contact with the outside air, the excess heat then transferring more widely to the environment. In addition to this heat loss by convection, other modes of thermolysis are involved, such as evaporation, i.e., skin perspiration, which brings water to the skin's surface evaporate, thus lowering the air temperature.

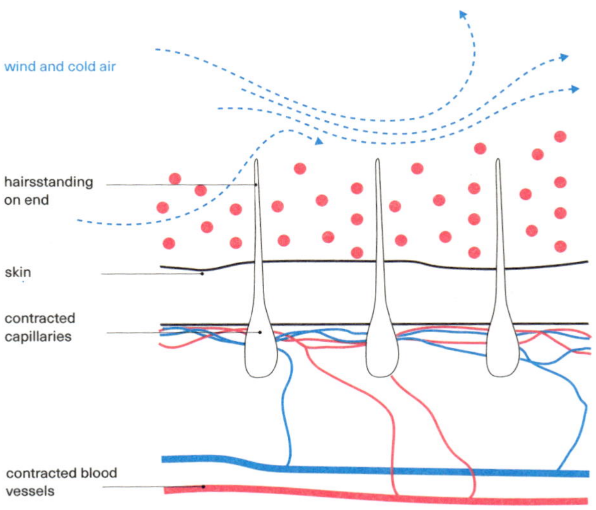

Thermoregulation in a cold environment {Fig. 1-122}

In a cold environment, the vessels will contract to limit blood flow to the skin, keeping more heat within the body and little contact between the cold outside temperature and the blood. Also, the hairs will stand up on end, slowing the wind speed in the immediate vicinity of the skin, thus reducing convective cooling. Thermogenesis also involves shivering, enabling the skin to generate heat through mechanical friction.

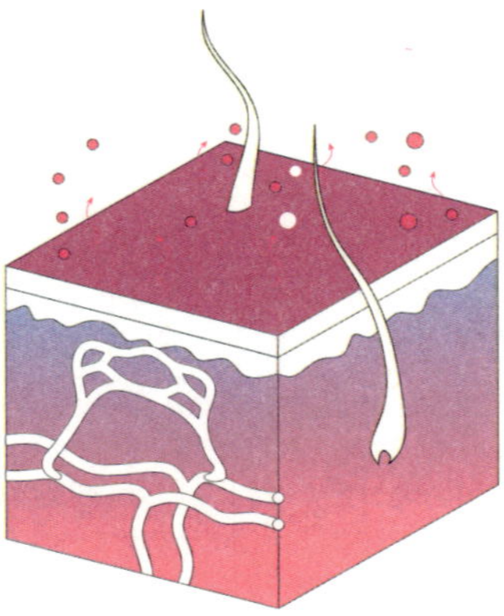

No heat lost through the skin without wind {Fig. 1-123}

The body stays more warm

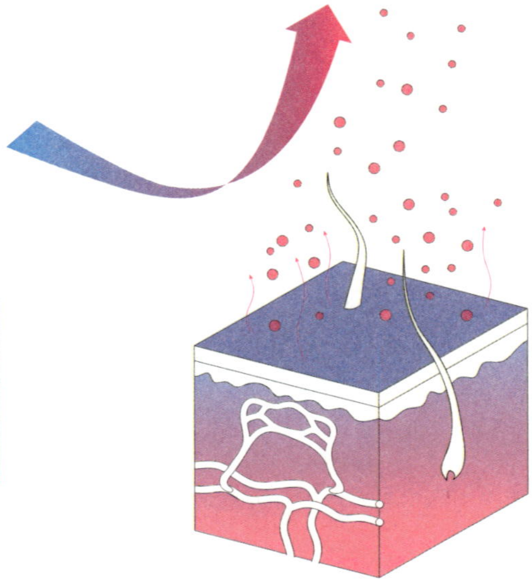

Loss of heat upon contact between the skin and wind {Fig. 1-124}

The body is cooled by the air flow

Dilated blood vessels

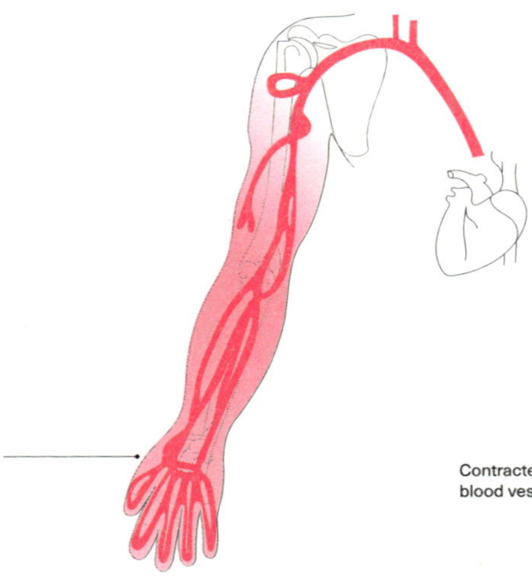

Thermoregulation in a warm environment {Fig. 1-125}

Dilatation of blood vessels to transmit more heat to the environment.

Contracted blood vessels

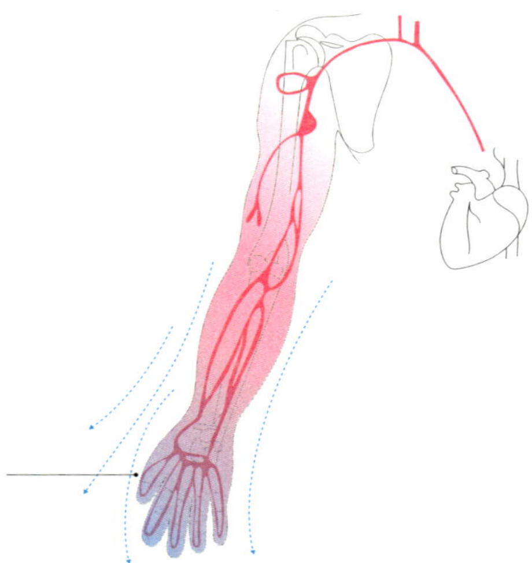

Thermoregulation in a cold environment {Fig. 1-126}

Contraction of blood vessels to transmit less heat to the environment.

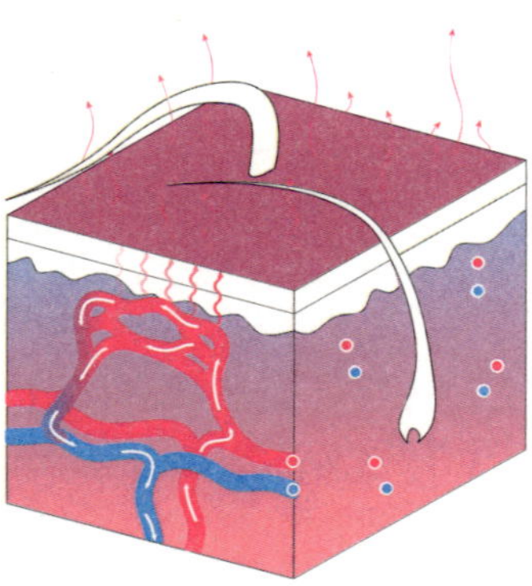

Vasodilation {Fig. 1-127}

Dilation of blood vessels to transmit more heat to the environment.

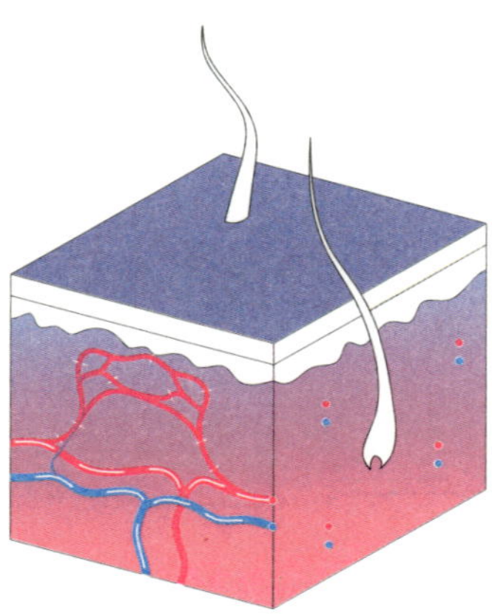

Vasoconstriction {Fig. 1-128}

Contraction of blood vessels to transmit less heat to the environment.

Body Refreshment by Convection

The clothes we wear create a layer of thermal insulation (by lowering the heat conduction value). This insulation consists of a film of air between the fabric and the skin, and a multitude of air bubbles within the fabric, enclosed in the textile fibers. The garment provides protection between the outside air and our skin, also protecting it from the wind that may sweep away the layer of heated air in contact with the skin. Clothing plays the same role as walls and roofs, protecting a small amount of air from the outside atmosphere in order to heat it more easily, and limiting heat exchange between the inside and the outside. The garment here deflects the wind from the skin, preventing the wind from carrying away the heat produced by our body that exits through the skin's surface, preserving this heat beneficially in winter. Conversely, in summer, we will seek currents of air to get rid of the heat produced continuously by our skin, to prevent it from stagnating on the skin and increasing our sensation of heat.

Just as we blow on a hot soup spoon to cool it down, the wind blows on our body and cools it by convection. For this reason, in winter we prefer to wear garments that are closed around the wrists, collar, etc., so as to prevent the passage of air and wind and keep the heat on our skin. However, in summer, we prefer to wear loose, open garments in order to let the wind flow between the fabric and our skin to remove the heat emerging from the skin. The same is applicable with regard to architecture and urban planning. To cool the house or the city, we will circulate air, we will set it in motion we will let the wind and draughts pass to carry the heat away. This cooling effect works only if the air temperature is lower than 34°C (skin temperature). If the temperature of the air is higher, the wind will act as a hair dryer and warm the body. However, in winter, we will protect ourselves from the wind, we will close the windows, we will dress to keep the heat of the air from being carried away.

Summer - Refreshing Effects of Loose Clothing and Sweating

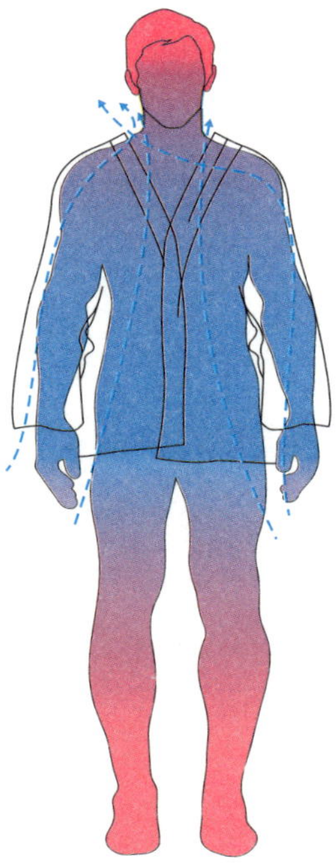

Refreshment with loose clothing {Fig. 1-129}

During summer or in hot climates, the body must be covered by clothes so as not to heat up (or get burned) by the sun's rays. But clothes must create a large space between the skin and the fabric in order to allow the air to circulate, which will remove the heat accumulated near the skin.

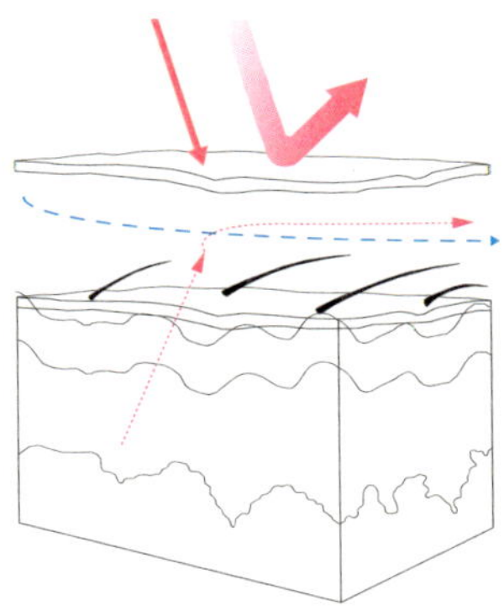

Loss of heat upon contact between the wind and the skin {Fig. 1-130}

Loose-fitting clothing allows the wind to pass between the skin and the fabric to dissipate the heat produced by the body while protecting it from the heat of the sun's rays.

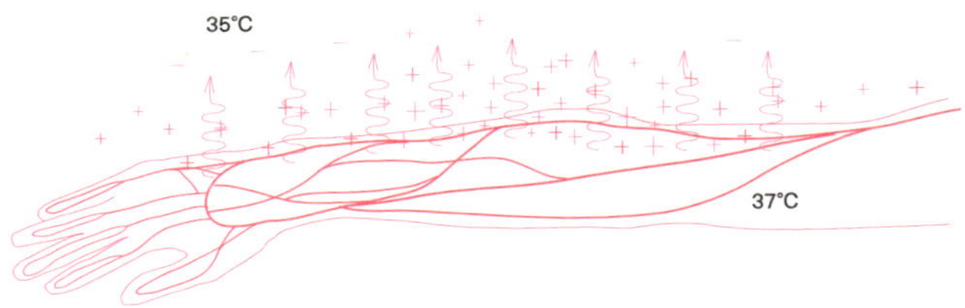

In summer, the skin overheats {Fig. 1-131}

Without the movement of air around the skin, the air touching the skin is continually warmed by the heat of the human body.

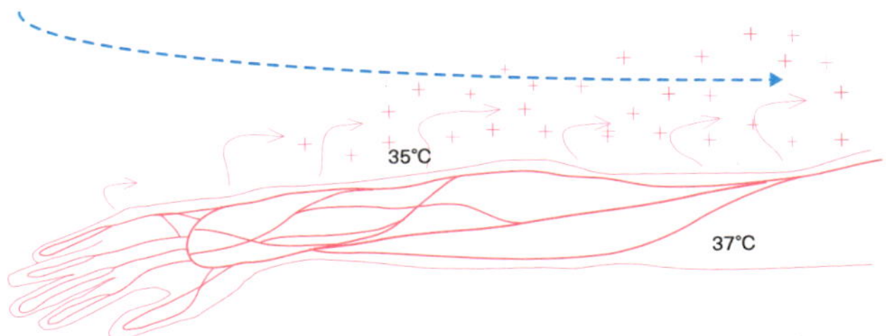

Loss of heat as wind passes over the skin {Fig. 1-132}

The wind helps heat loss by the skin through convection and speeds up the evaporation of sweat, refreshing us by preventing the air from stagnating on the skin which will doubtless increase in temperature if not evacuated.

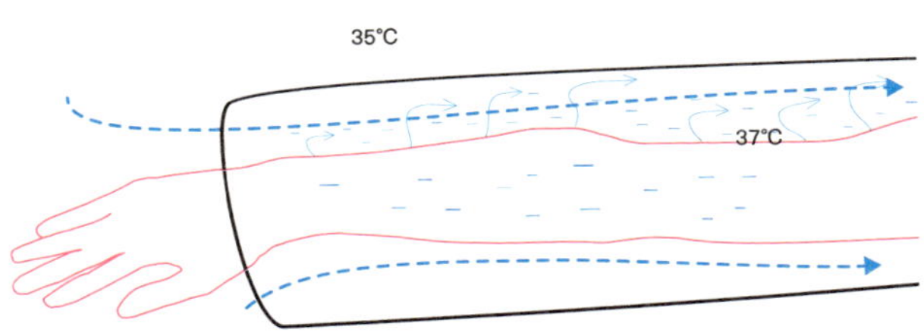

Loss of heat upon contact between the wind and the skin {Fig 1-133}

Loose-fitting clothing allows wind to pass between the skin and the fabric, speeding up the evaporation of sweat and promoting heat loss by convection.

Winter - Warming up in Winter with Tight-fitting Clothes

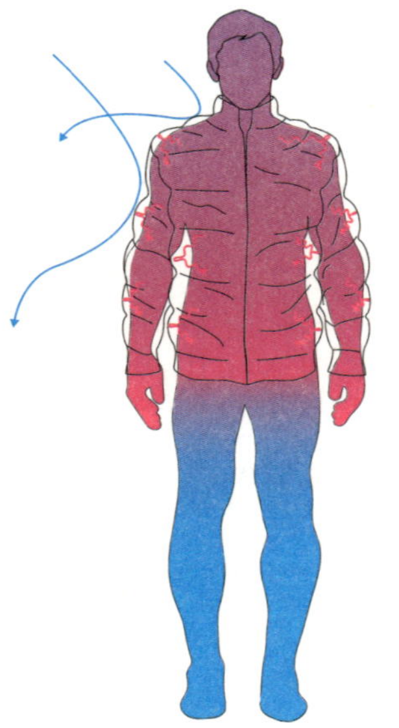

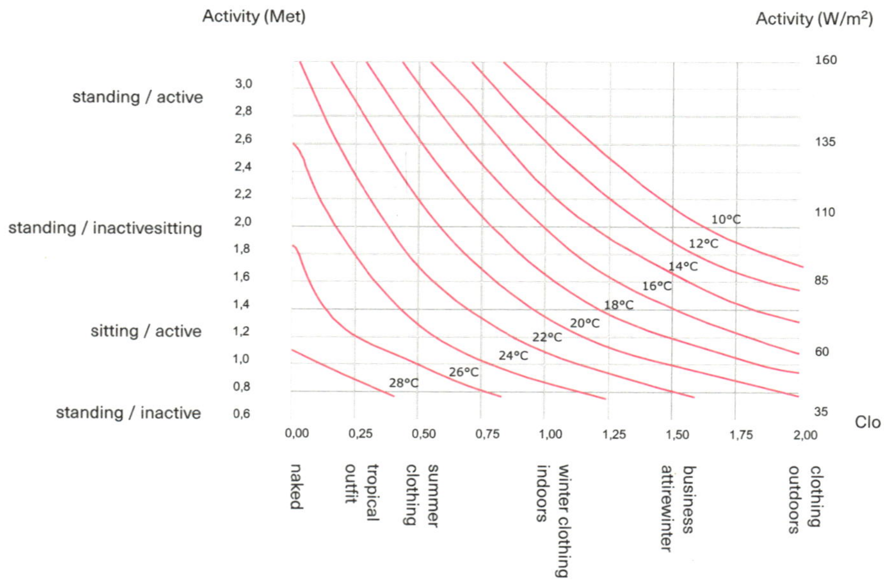

Heat gain by avoiding the breeze {Fig. 1-134}

During the winter or in cold climates, the body should be covered by tight-fitting clothing to avoid forming a space between the skin and the fabric where air could circulate. The absence of air movement over the skin will preserve body heat by preventing heat transfer by convection and the resulting loss of heat. Clothing should be completely closed at the neck and wrists to keep the cold air out.

Ideal air temperature in relation to clothing and physical activity {Fig. 1-135}

The clo is the unit used to measure the thermal insulation of garments. The higher the clothe more heat-insulating the garment is. The more active the body is physically, the more heat it produces, so the lower the air around it may be in order to absorb this excess heat. Thus, the more you are dressed and in motion, the lower the temperature of the air can be in the room.

Conversely, the less active and dressed you are, the higher the air temperature of the room must be to achieve the same degree of comfort. If you want to save energy by heating a room less, you must wear more clothes. The ideal temperature depending on physical activity and clothing is as follows:

Ideal T° = 31°C – P x R, where P is the power in Watts released by the body and R is the thermal insulation of the garment in clo.

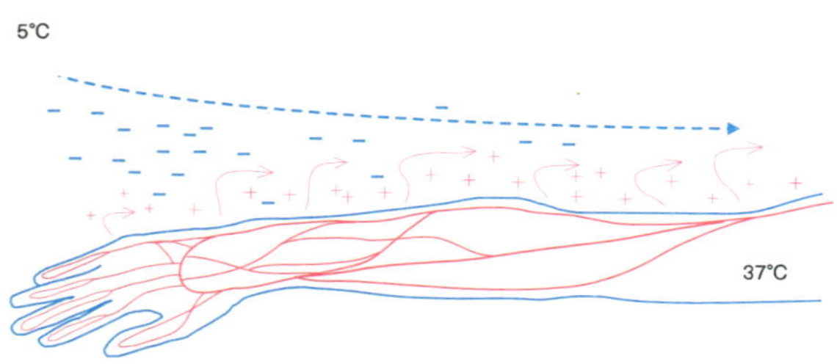

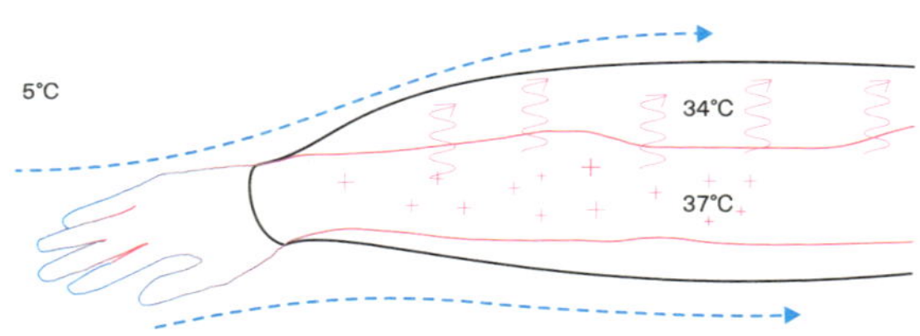

Heat loss through convection {Fig. 1-136}

In winter, the skin loses heat by convection with the cold wind that immediately dissipates heat issuing from the body. To protect ourselves from convective cooling, the skin must be protected from the wind.

Hairs standing on end in winter contribute to this firstly by creating a barrier by putting the wind at a distance to prevent it from directly touching the skin and dissipating the heat.

Clothing to distance ourselves from the wind {Fig. 1-137}

Clothing that fits tight around the neck and wrists keeps the wind from contact with the skin, preventing convective cooling. In addition, tight-fitting and airtight garments enable creating a layer of static air acting as a thermal insulator.

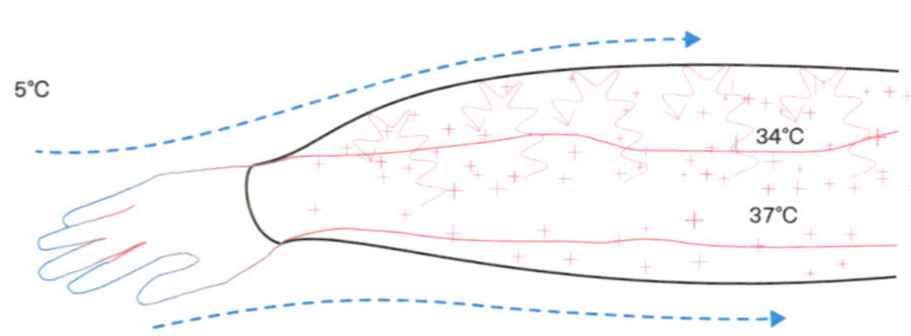

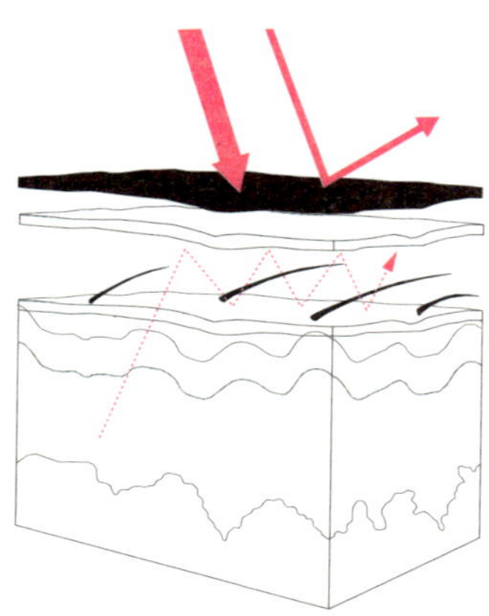

Absence of air renewal {Fig. 1-138}

In the absence of air renewal, the air blocked by the clothes will continuously heat up due to the skin releasing heat, and thus rise in temperature.

Heat production by convection can accumulate beneath clothing {Fig. 1-139}

Close-fitting clothing does not allow air to flow between the fabric and the skin, avoiding heat loss by convection

III. Global Scale

Convection in the Atmosphere

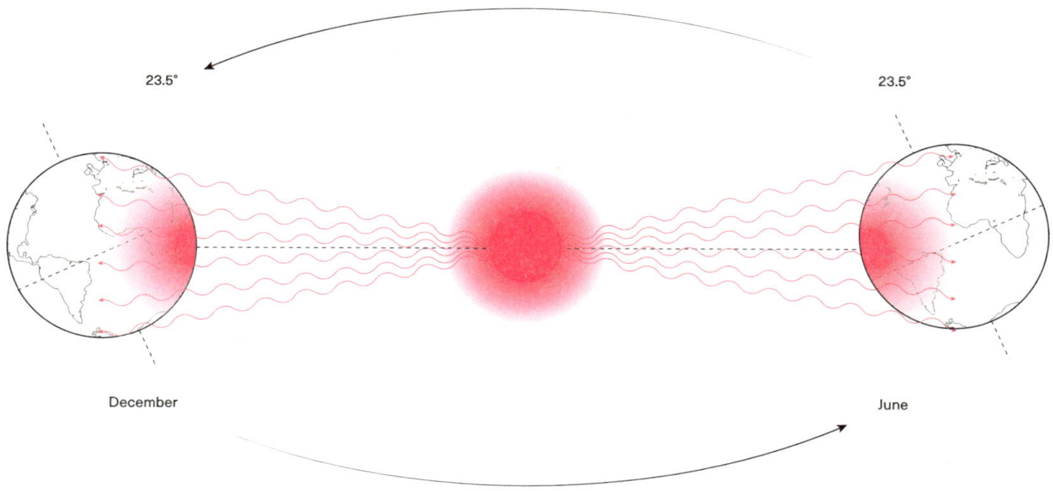

Solar radiation on Earth {Fig 1-140}

The Earth's 23.5° tilt is the root of the seasons and climates as it creates thermal differences. Because of this tilt, the equator receives more solar energy (W/m2) because the sun's rays fall from a higher angle and are more concentrated than at the poles where the Sun's rays reach Earth at a lower angle and are less concentrated, thus diluting the energy. These differences in the angle of arrival of the Sun's rays create irregular heating of the Earth's surface which leads to different temperatures on the planet's surface.

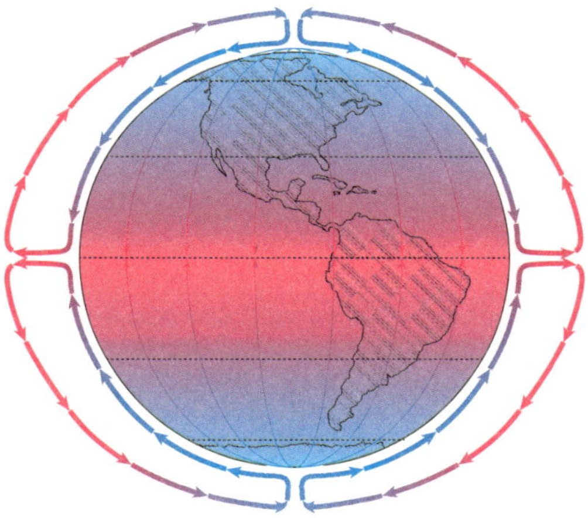

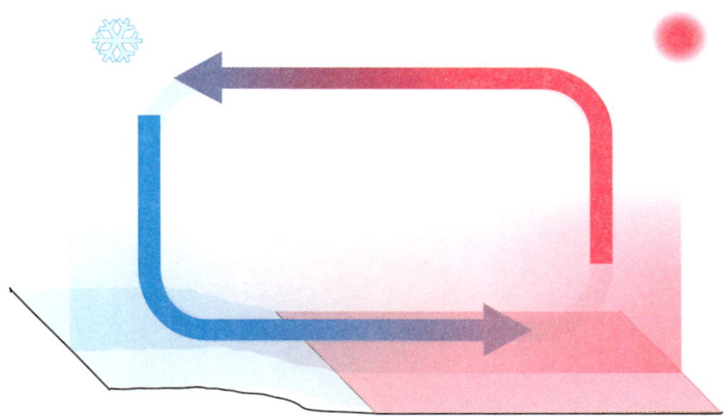

Global convection cells {Fig 1-141}

Convection cells extend on a global scale. The warmer air at the equator creates an upward flow of air that heads toward the poles where it cools and then sinks. The colder air from the poles, pushed downwards by the masses of hot air arriving from the equator, then moves toward the equator, gradually heating up. This phenomenon of atmospheric convection is recurrent and plays a pivotal role in meteorological phenomena.

Atmospheric Convection {Fig 1-142}

The air warms up upon contact with surfaces heated by the sun near the equator, causing it to rise and create an atmospheric convection movement, pushing the masses of cold air that fall and circulate beneath the hot air masses. This phenomenon is the cause of wind.

Convection occurs on a very large scale in the atmosphere, the oceans and the Earth's mantle. Differences in natural temperatures on Earth influence the density and buoyancy of fluids. The heaviest (or densest) fluids fall while the lightest (least dense) rise. These upward and downward movements are at the cause of convection cells. Atmospheric convection occurs when one or more parts of the Earth receive solar radiation, which heats the nearby air. The irregular heating of the atmosphere around the Earth makes certain air masses more or less dense.

The hot air masses rise and are quickly replaced by cold air masses that sink. The Earth receives more solar radiation at the equator (or during the day) than at the poles (or at night), leading to convective movements that cause winds and other weather phenomena. Atmospheric convection raises the mass of hot air at the equator and moves it to the poles where the air cools, becomes dry and sinks, and then returns to the equator by circulating in the lower atmosphere.

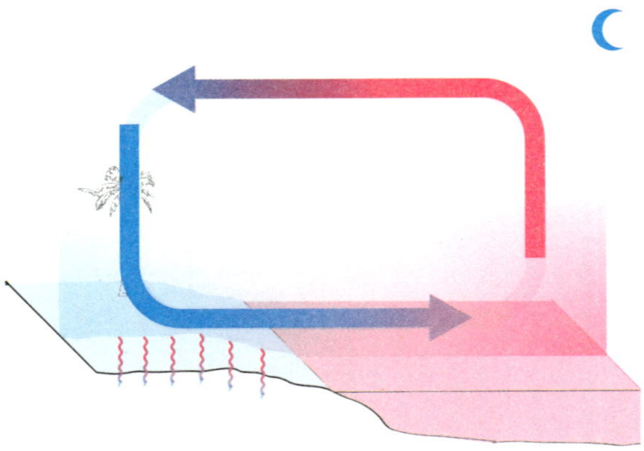

Land breeze {Fig. 1-143}

During the night, the ground, which immediately cools by radiation toward the cold night sky, chills the air in contact, causing it to fall. The ocean, meanwhile, has stored heat during the day and, remaining warmer thanks to its thermal inertia, will heat the air, causing it to rise.

These two movements will create a land breeze, that is to say the formation of a convective movement, a wind blowing at night from land to sea (or lake). {Fig 1-263}

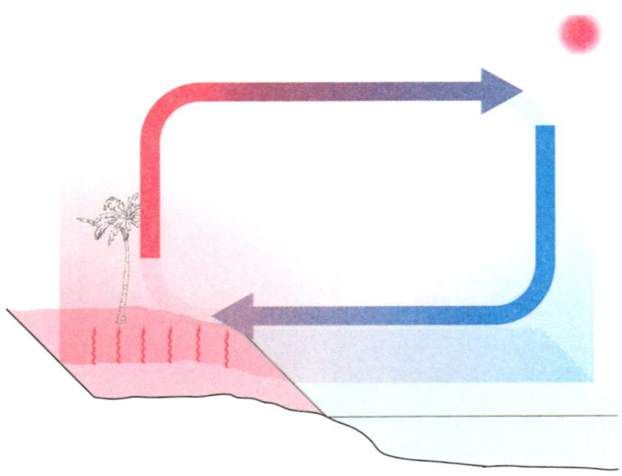

Sea breeze {Fig. 1-144}

During the day, the phenomenon is inverted: the ground is heated by the sun and transmits this heat to the air, which causes it to rise; while the ocean, which has cooled during the night and retains this freshness by thermal inertia during the day, will cool the air, causing it to fall.

This creates a sea breeze, that is to say a wind blowing by day from the sea (or lake) to land.

V. Architectural Scale

Convection in the Building

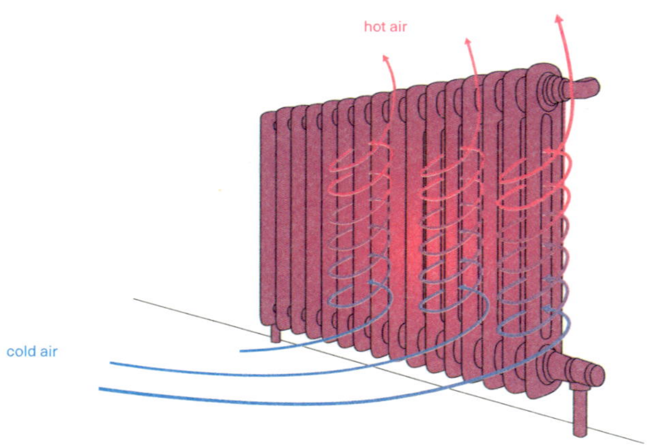

Convective heat from radiators {Fig. 1-145}

As we multiply the folds of a radiator, the cold air heats up on contacting the pipes where the hot water circulates. The hot air will then rise and push the cold air, creating a convective movement.

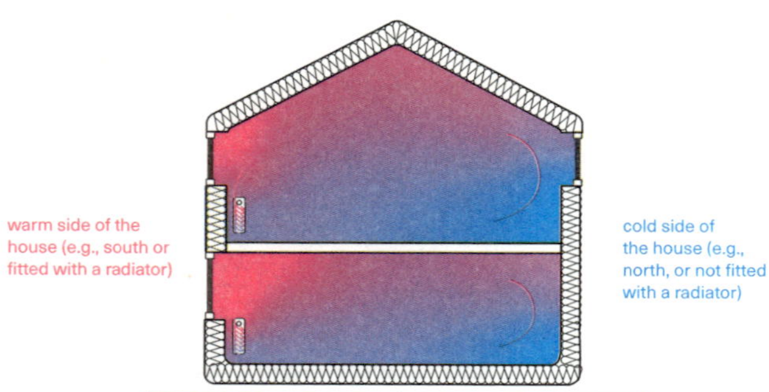

Convective movement in the house {Fig. 1-146}

In each room, a movement of air is created because of the temperature differences between the surfaces of the walls, floors or ceilings, generating various climates with different thermal qualities within the same space.

Temperature differences exist within buildings, whether natural -for example, between the colder north façade, and the warmer south façade- or artificial -for example, between a wall fitted with a radiator and a wall where there is no radiator. These temperature differences will create air movements where the hot air will rise and the cold air will sink. These temperature differences, like a mini geography, between a super-heating equator and an icy north pole, create "winds", air currents, while establishing different climates within the building's volume: warmer and humid at the top, colder and dry at the bottom.

Vertical stratification of heat in the house {Fig. 1-147}

Hot air (less dense, lighter than cold air) will rise naturally according to Archimedes' law, creating a natural vertical stratification of temperatures throughout the house.

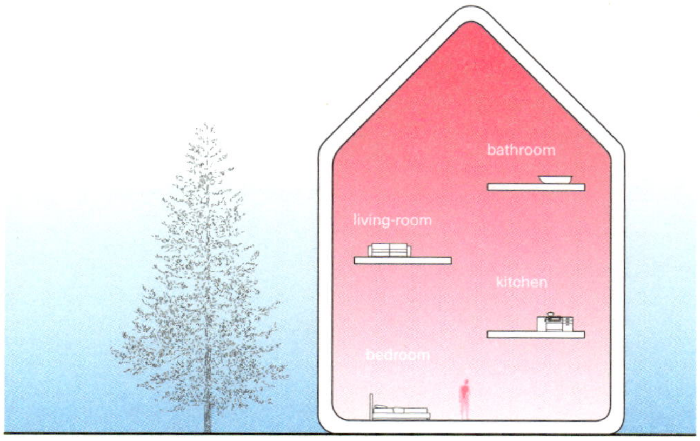

Distribution of functions in relation to heat stratification {Fig. 1-148}

The positioning of the functions follows the vertical stratification of the temperatures, from the coldest at the bottom to the warmest at the top, according to the thermal needs of the inhabitants, their physical activity and their clothing. Thus, the functions where one is most unclothed and least active will be located at the top, and vice versa.

Convective stratification for winter {Fig. 1-149}

Winter rooms must have low ceilings so that the heat stays down, around people.

Convective stratification for summer {Fig. 1-150}

Conversely, summer rooms must have high ceilings so that the excess heat rises, evacuates upward, keeping ground level fresh.

According to Archimedes' law, hot air rises, cold air sinks creating a natural temperature gradation that is established vertically within the volume of the house or in that of each room, leading to higher temperatures at the top and lower ones at the bottom. This creates a thermal landscape where, naturally, it is warmer at the top and colder at the bottom, where the wind rises when the radiator is switched on, or people's warm bodies cause the heat to rise and the cold to fall.

Thus, as Vitruvius already recommended, summer rooms (or hot climates) will have high ceilings so that the heat rises, keeping the bottom of the room cool. Conversely, winter rooms (or cold climates) will be low so that the heat stays at the bottom, at person height so they may benefit from it. These convective principles, applied before the use of fossil fuels, are useful for reducing energy consumption through heating or air-conditioning.

VI. Urban Scale

Convection in Cities

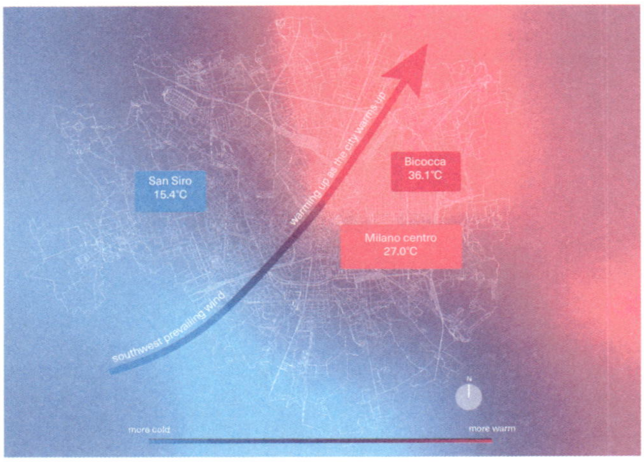

Urban heat island {Fig. 1-151}

Because of dark surfaces such as asphalted roads and pavements, and due to the absence of evaporation by plants or plant cover that, by changing from liquid to gas phase lower the temperature, the city overheats in summer.

To cool it down, among other things, it must be ventilated, the air must circulate, to dilute the heat by convection, as if fanning it or as if blowing on a hot soup spoon to cool it.

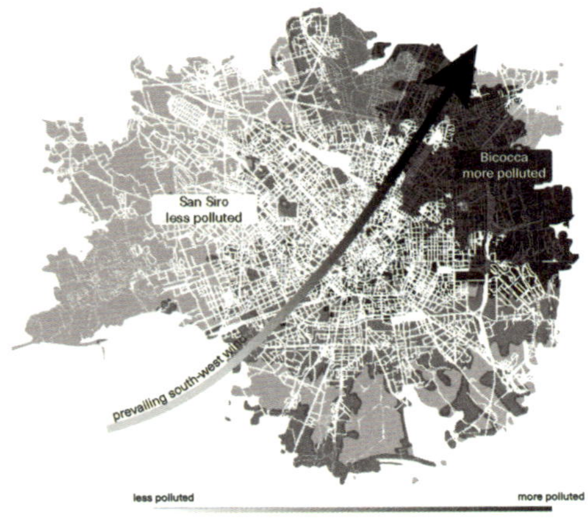

Urban pollution {Fig. 1-152}

Air pollution in contemporary metropolitan areas, especially (carcinogenic) fine particulate matter PM 2.5 and PM 10, largely given off by combustion engines, is accumulating in cul-de-sacs.

It is unequally distributed across the city, with scarce presence where fresh winds arrive (south-west in London, Paris or Milan, for example) and, conversely, accumulating in the north-east, where the wind ends up after having picked up all the pollutants of the city.

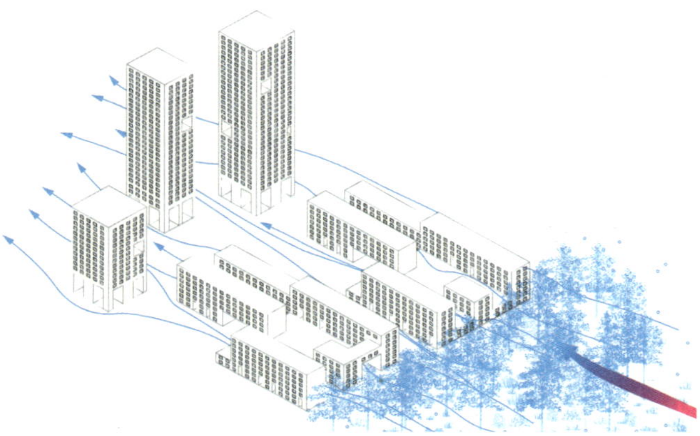

Cooling by convection {Fig. 1-153}

In order to cool urban islets, buildings should preferably be built parallel to the direction of the prevailing wind, allowing it to cross the islet without being halted or interrupted and to carry away the heat accumulated due to the low albedo dark mineral surfaces such as the bitumen of which the streets are made.

Computer modeling of flows {Fig. 1-154}

The new volumes of urban buildings (bottom left) are designed by the winds, allowing them to circulate between buildings and in the streets to remove excessive heat and air pollutants.

In cities, the ground's low albedo (solar ray reflection ratio) will transform solar rays into heat, creating what are called urban heat islands. One of the best ways to fight these heat islands is to prevent the air from heating up too much due to being still. The city must be ventilated, allow natural winds to pass through, to remove heat from the streets and dissipate the air. It is also possible to artificially create a hot spot (low albedo, little evaporation) and a cold spot (high albedo, high evaporation) which, thanks to their temperature difference, in a phenomenon similar to the sea breeze, will create a movement of air and prevent cities from overheating in summer. Creating a temperature difference between two urban poles sets the air in motion, which also enables breaking up and diluting the concentration of air pollutants in the city (and viruses), to avoid the accumulation of fine particles in streets or squares.

Urban Convective Winds

City breeze {Fig. 1-155}

By creating a thermodynamic imbalance thanks to a hot pole (built-up mineral masses) and a cold pole (vegetable park), the convective movement of the air is amplified and enables evacuating the heat from the streets by making it rise into the sky.

Limpidarium d'Aria (Air Limpidarium)

Building winds {Fig. 1-156}

A novel urban planning art may consist of creating city breezes to remove heat and pollution from the city into the sky.

Limpidarium d'Aria (Air Limpidarium)

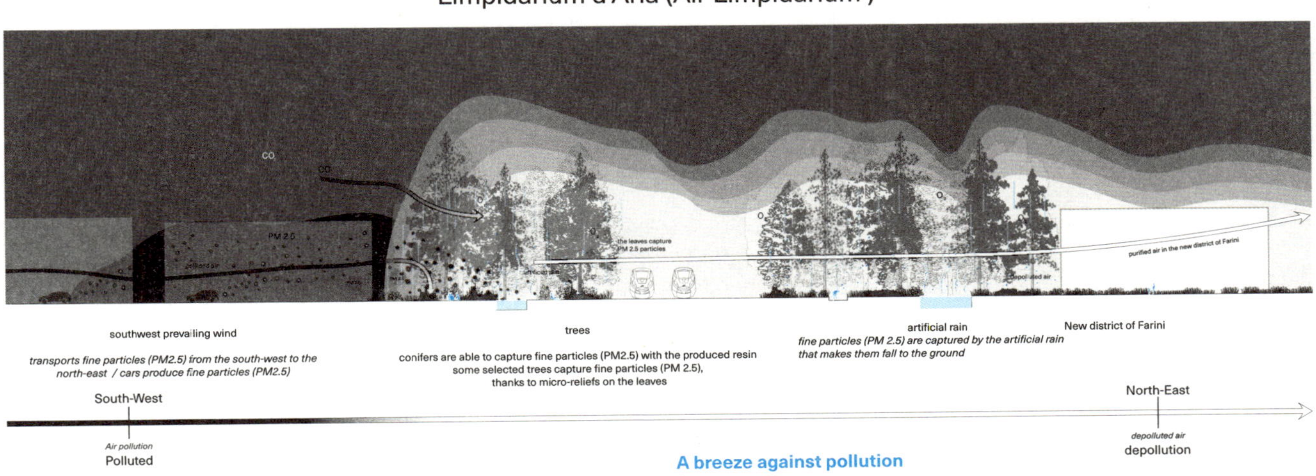

A breeze against pollution {Fig. 1-157}

A filtering zone of bushy coniferous trees as well as artificial rain (which causes the floating polluting fine particles to fall to the ground) make it possible to reduce air pollution, and then allow clean air to blow toward the new neighborhoods.

Urban polarization

Surface

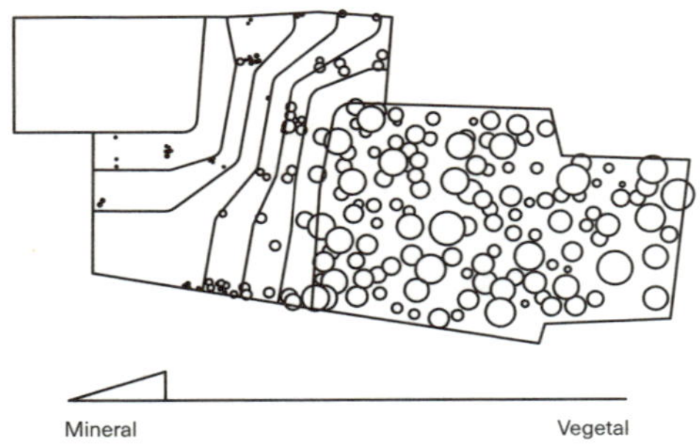

Vegetation

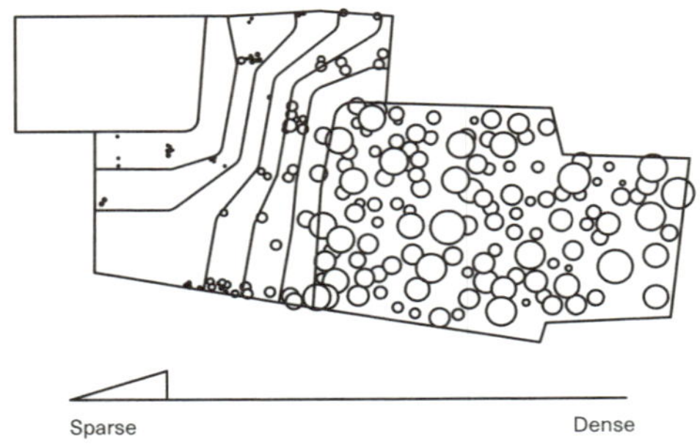

Temperature

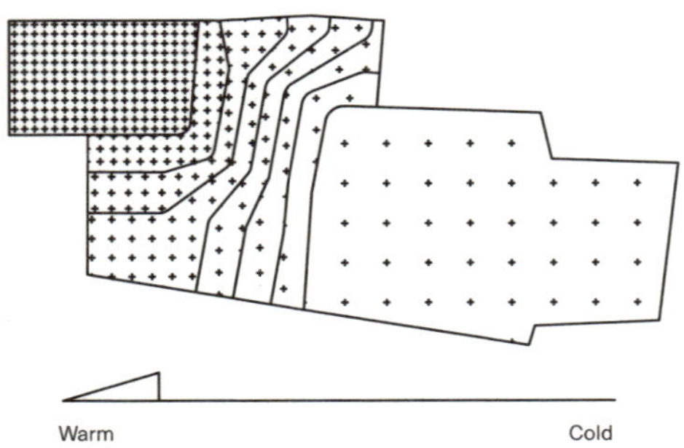

Ground plan

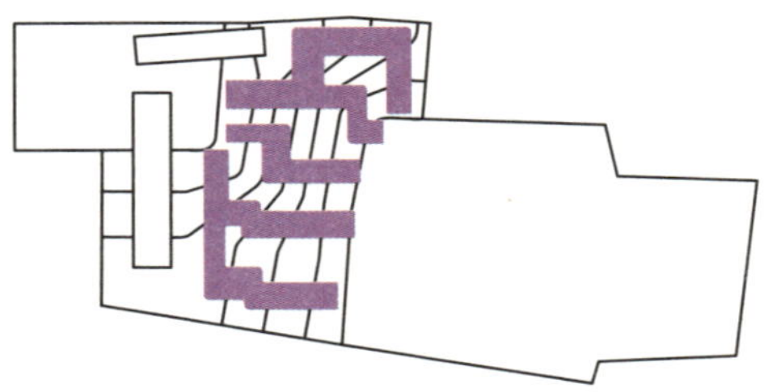

Light

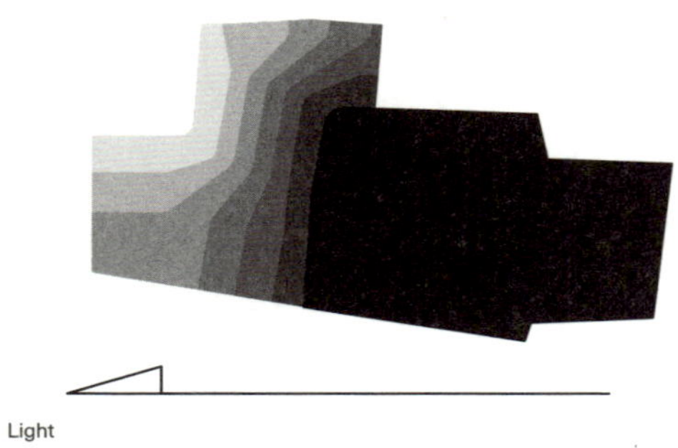

Traffic plan

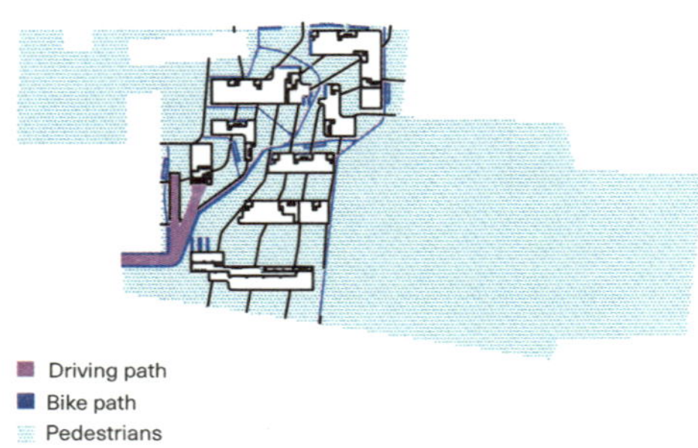

Air renewal system in the accommodations summer situation

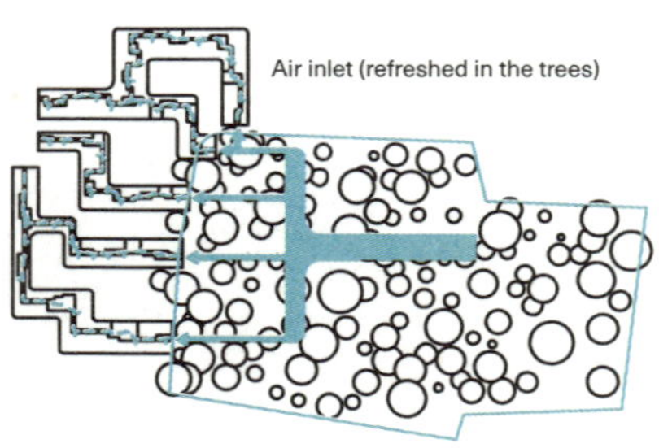

Air renewal system in the accommodations winter situation

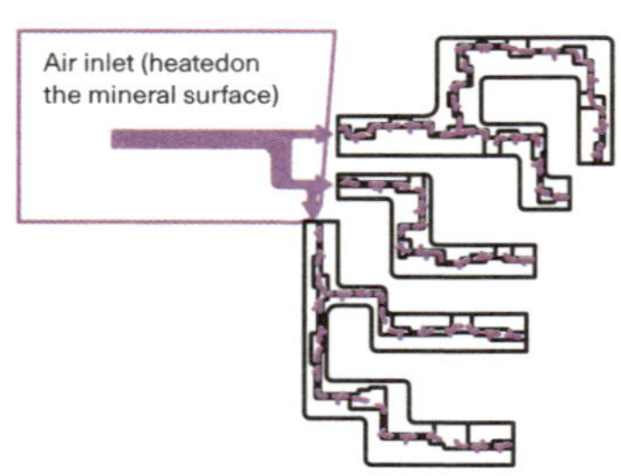

Gradation of urban atmospheres {Fig. 1-158}

The urban agenda of this project for a new residential district is to construct two squares at opposite ends of the site and to build between them. The buildings are implemented between these two public spaces. The corridors of buildings become ventilation ducts augmented to human scale, the ends of which, like nozzles connect to the public squares to draw the ambient air whose thermal quality will have been high at the mineral square to the north-west or lower in the forest to the south-east.

The project is implemented between two thermally differentiated poles, two public spaces built at opposite ends of the site. One is to the north-west, the other to the south-east, one is hot, the other cold, one mineral, the other vegetable, one in the light, the other in the shade, one absorbing the heat, the other reflecting it. From the first space, in the north-west, the warmer, mineral, light and black, we will draw air during winter to supply dwellings with preheated air, in the manner of the Canadian well. From the second place in the south-east, plant covered, filled with trees, shady and cooler, the space of which we virtually extend to the adjacent park, we will draw cooler air in summer, like the Provençal well, to supply all the dwellings in a dual-flow ventilation system. For each of these two poles, we carefully choose the thermal qualities of the materials of which they are made, their albedo, their emissivity, their effusivity, their vegetation, in order to augment the heat of the air in the north-west or to decrease it in the forest in the south-east. These two public spaces, hot and cold, are the sources of air to which the dual-flow ventilation system of residential buildings and the school is connected.

Thermodynamic Urban Planning

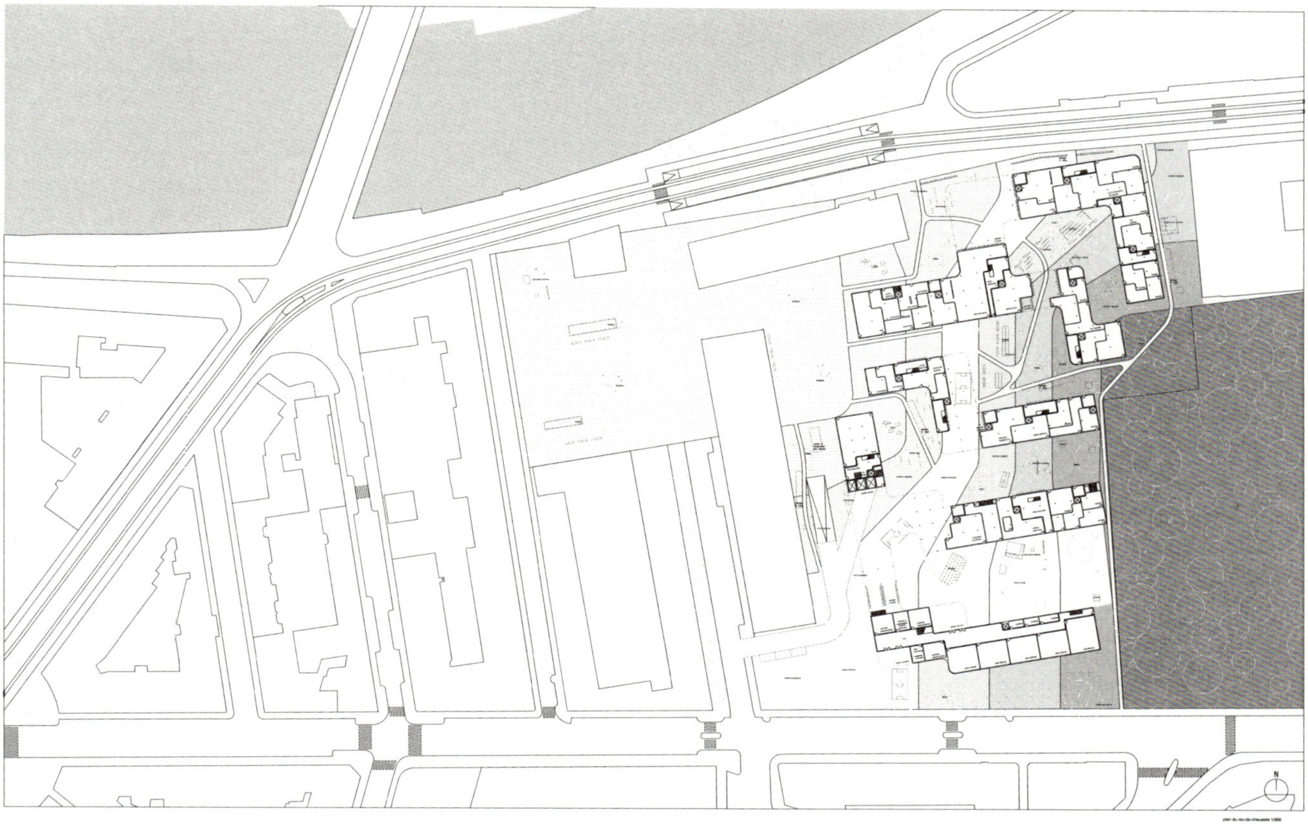

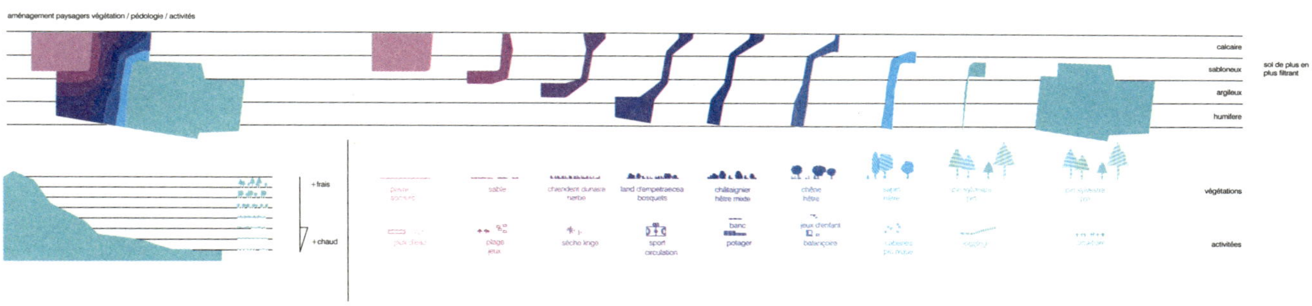

Urban thermal gradation {Fig 1-159}

The district is polarized between two main public squares whose development is based on an initiative to enhance the air, its structuring, where housing is implemented between two poles of air, one hot and one cold. Between these two principal areas, 7 strips of landscape unfold stretching from north to south, which gradually, from west to east, take you from the mineral square in the north-west to the forest in the south-east, from hot to cold, from mineral to vegetation, with alternating strips of different geological and plant qualities, ranging from brownfield land, vegetable gardens, shrubs, fruit trees and forest with which uses are associated that take advantage of the landscape qualities and atmospheres of each of the strips: beach games, gardens, sport on mineral surfaces, games, jogging, walks in the meadows, tree houses.

The Park as a Climatic Gradient

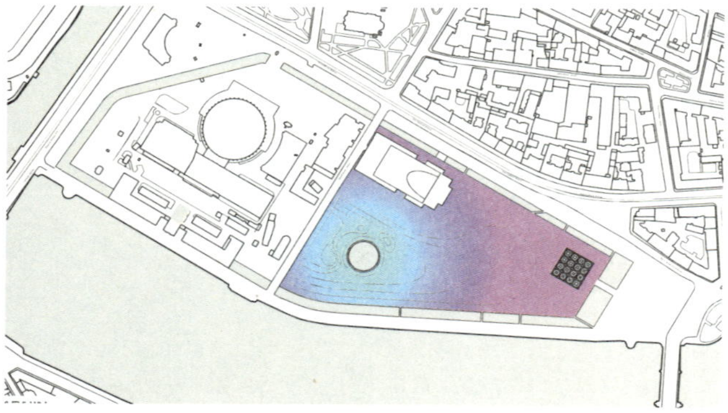

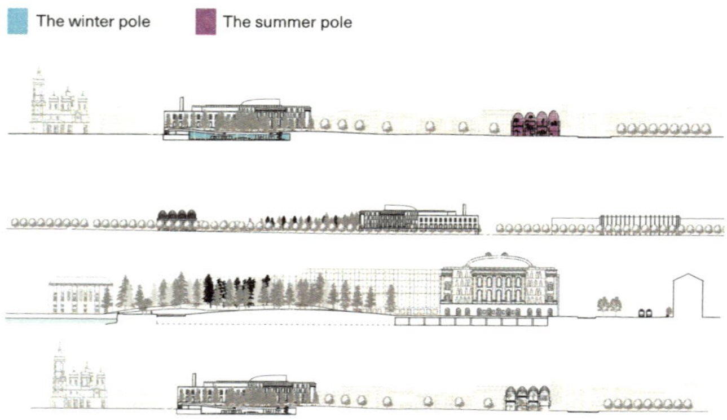

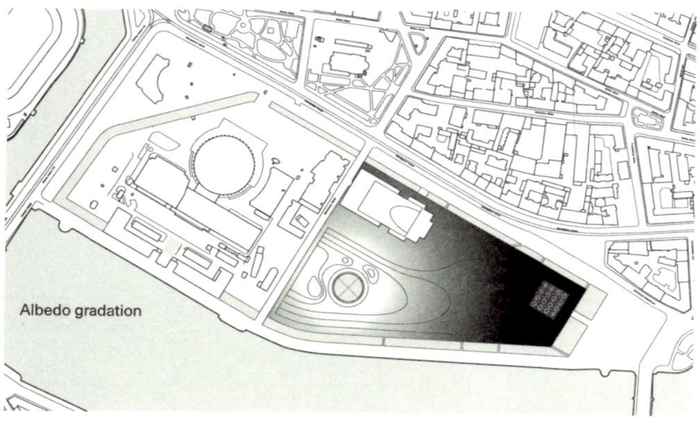

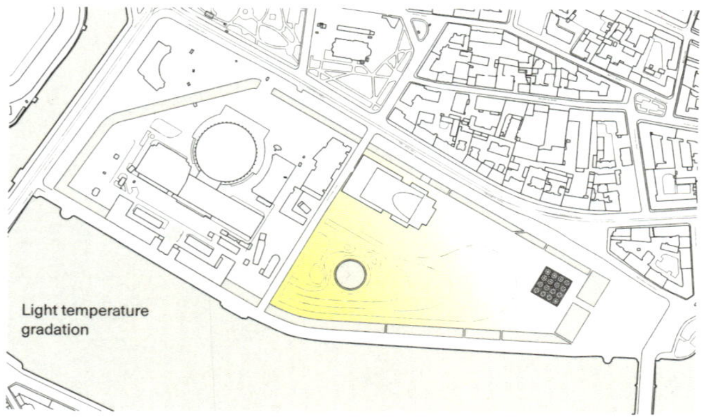

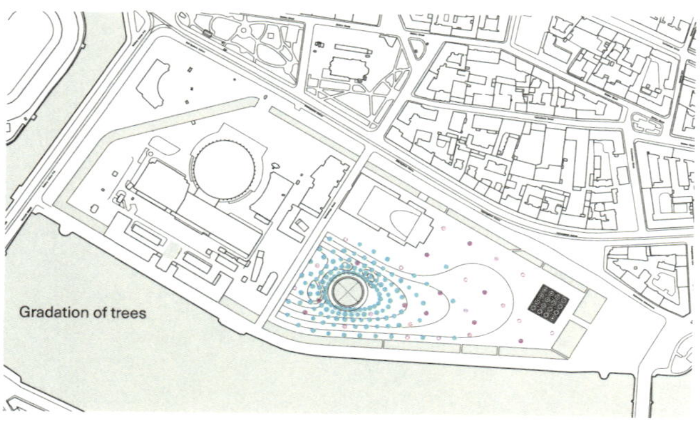

Gradation of landscape atmospheres {Fig 1-160}
The plan for the park establishes a climatic gradation between a cold pole (shady, high albedo, very wooded) suitable for taking shelter in summer, and a warm pole (open, little wooded, low albedo) in which to shelter in winter.

Our project for Buyon Park, designed with the agency Ter, proposes to offer the inhabitants of St. Petersburg twelve months together in a park, in a gradation of atmospheres built according to real physical parameters, passive climatic strategies providing all the comfort, pleasantness and pleasures of a park throughout the year, but with economic and ecological intelligence and thriftiness. To be able to shelter from the rain, snow, cold winds and enjoy the warmth of the few rays of sun in winter. To be able to shelter from the sun, to be exposed to fresh breezes in summer, this is what we propose as a general plan for composition of the park. The park thus stretches between two opposite poles, two opposite temperatures, the winter pole and the summer pole. They constitute the park's two largest public architectural spaces.

The park as a climate gradient

Between these two poles, from west to east, there is a gradation of atmospheres, from coldest to warmest, which allows offering a variety of situations, climates, surfaces, activities, games throughout the year, starting with a cool, shady forest to the west around the winter pole, suitable for summer, to an open, sunny meadow in the east, conducive to winter.

A gradation of physical values to materialize this gradation, we vary actual physical parameters, types of trees and their density, the nature of the soils, furniture.

A gradation of trees

The forest is composed according to a gradient of tree density, high around the winter pole to the west, dispersing and spacing out as you head eastwards. The species of tree follow this gradient. They are essentially coniferous to the west, comprising a winter forest, which are gradually replaced by deciduous trees toward the east, more spaced out, and letting more light pass in winter upon losing their leaves, offering maximum sunshine around the summer pole.

A gradation of light

The light intensity will therefore decrease toward the west, preserving some of the darkness of the night in the winter forest. Similarly, the light spectrum will gradually change from white in the east to orange in the west. Indeed, like that of sunsets, orangey light has the property of not disturbing the diurnal-nocturnal life cycle of animals, protecting them from nocturnal light pollution.

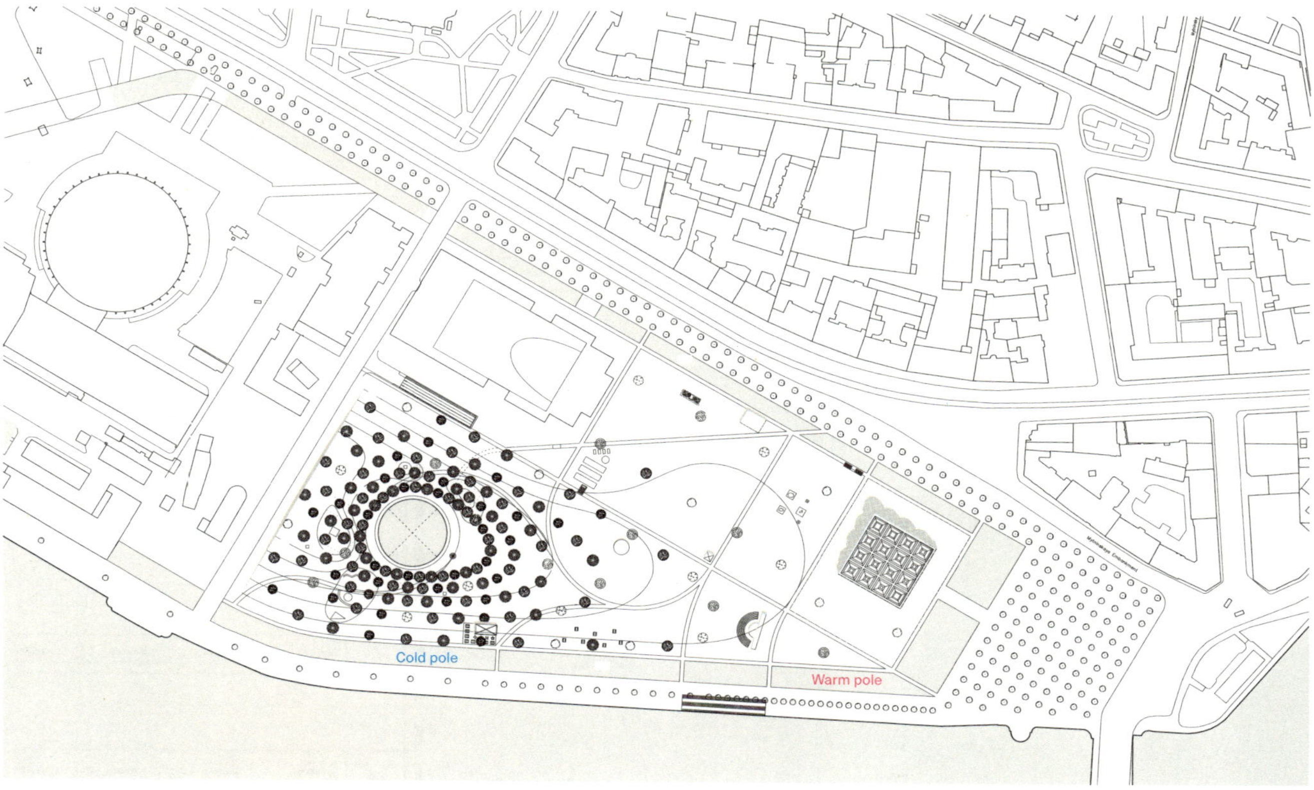

Warm pole

Cold pole

Two thermal poles {Fig. 1-161}

The cold pole is a climatic shelter away from summer heat waves, consisting of an underground cave including a lake and water games, sheltered by a dense forest. The hot pole is a tropical greenhouse, placed beneath the sun in which to take shelter during the cold of winter.

The park extends gradually between these two poles: gradations of tree density, light, and albedo.

VII. Case Study

The Archimedes House

The Archimedes house seeks to establish a relationship between the house and the air, by representing the uses of the house (sleeping, resting, washing, etc.) in terms of air, density, temperature and movement. This house is organized according to the external physiological needs of the inhabitants which vary according to their physical activity and their clothing. This vertical architecture is organized around the precise heat requirements of each room.

Modernity has created uniform, homogeneous spaces, with a constant, uniformly distributed temperature of about 21°C. The aim here, according to the principle of reducing energy expenditure, is to restore the diversity of the relations between the body and space, and promote daily or seasonal movements within the house, migrations between upstairs and downstairs, between warm and cold, summer and winter, clothing and the nudity

The Archimedes House
Convection principle {Fig. 1-162}

The warm floor heats the air and the molecules move away from each other as they are more excited.

Cold flooring or flooring in the shade cools the air, which becomes denser as its molecules come closer together.

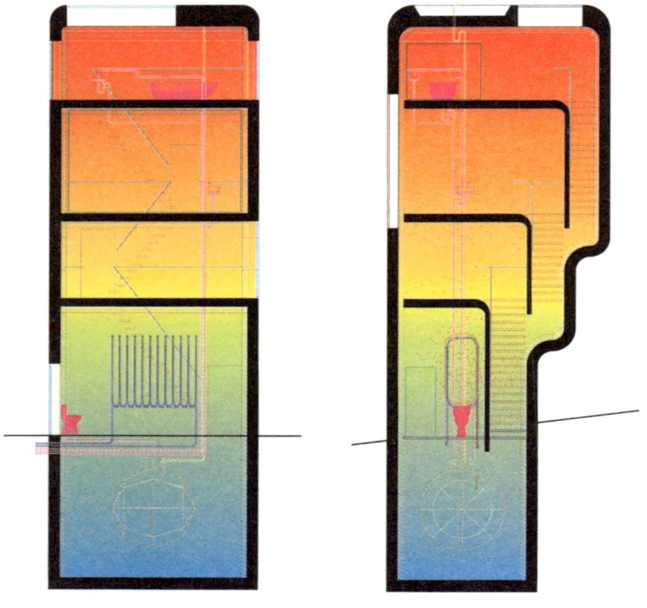

The Archimedes House
Sections BB and AA {Fig. 1-163}

The aim here is to rediscover the diversity of relationships between the body and space, its temperature, and to promote seasonal movements within the house, migrating from bottom to top, from cold to warm. To feel comfortable in a heated room, there must be a balance in convective heat exchange between the body and the air. This balance also depends on clothing, ranging from nudity in the bathroom to the thermal protection of blankets and the light clothing worn in the living room.

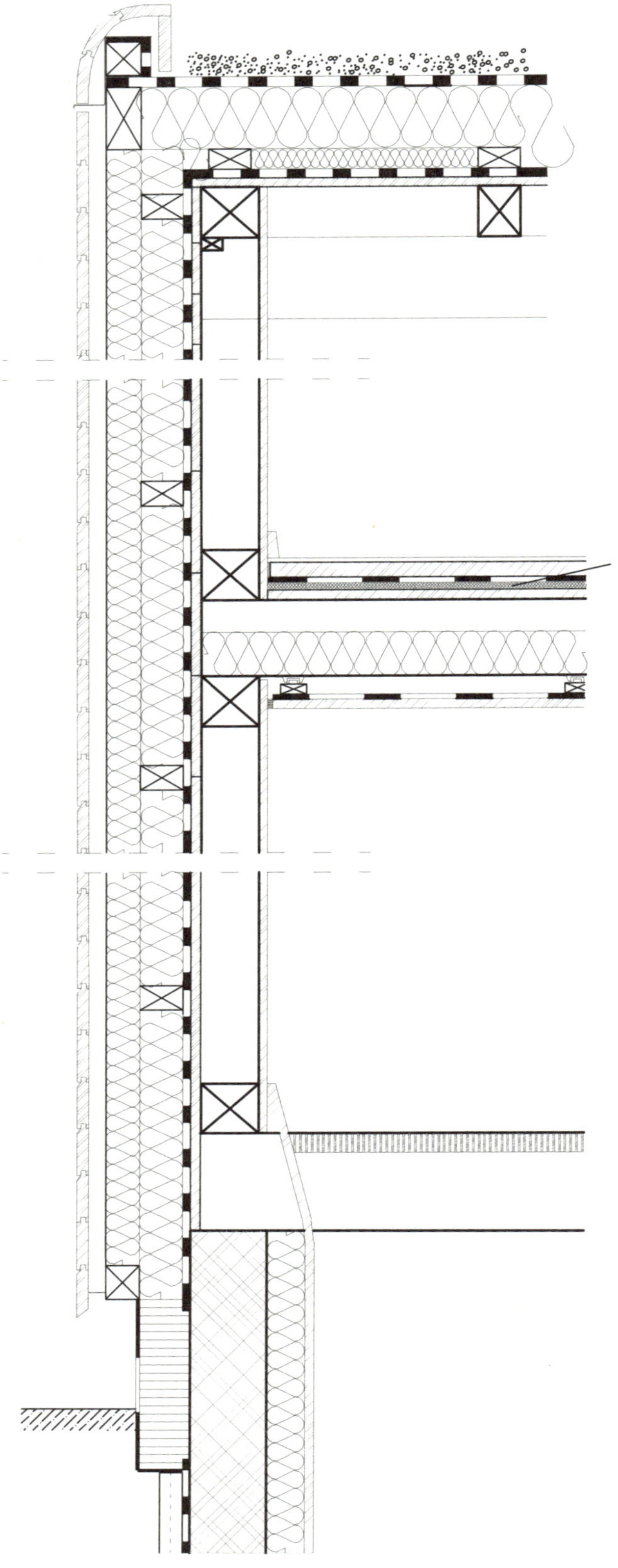

The Archimedes house - Section detail {Fig. 1-164}

The house is organized according to the physiological needs of its inhabitants, according to their physical activity and their degree of nudity. The vertical architecture is thermally structured around the precise needs of each space.

It must ensure the thermal comfort of its inhabitants and project spaces in which the temperature is adapted to the activities of and the clothing worn by the inhabitants.

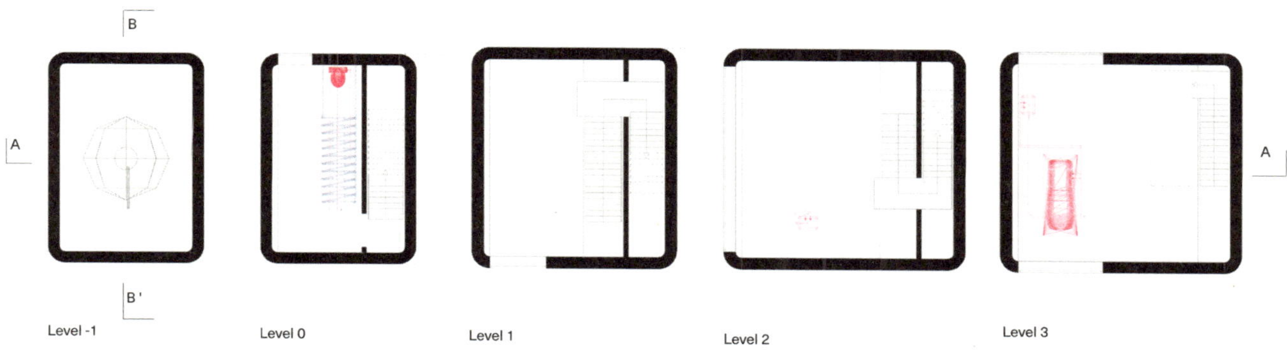

The Archimedes House - Plans
{Fig 1-165}

Of course, the thermal balance is related to clothing: from nudity in the bathrooms, to the thermal protection of blankets, to the light clothes worn in the living room. Today, in view of the need to save energy, we are encouraged for each building, and even each room, to set a precise thermal capacity in order to use only the necessary energy. The Swiss construction standard SIA 3842 gives us the following indications for ambient temperatures:

Type of room	Recommended Temperature (in °C)
Living room	20°C
Bedrooms	16°C - 18°C
Bathrooms	22°C
Kitchens	18°C - 20°C
Corridors, WC	15°C - 18°C
Stairways, Laundries	12°C

The plan and section of the house are designed to follow the path taken by the air throughout the house upward, according to the natural convective vertical movements of the air that depend on its temperature. Since hot air tends to rise, there may be significant differences in temperature between the bottom and the top of the interior volume: for example, 21°C at ground level and 27°C beneath the ceiling. This difference leads to unnecessary energy expenditure when you want to maintain a regular temperature of 21°C and you do not inhabit the area directly beneath the ceiling.

Indeed, according to Archimedes' law, temperature is vertically stratified in the house due to atmospheric convection. So it is possible to design a house where it is: 16°C on the ground floor, 18°C on the second, 20°C on the third and 22°C on the top level. This stratification extends to the uses of the different rooms according to their temperatures, the clothing worn by the inhabitants and their physical activity.

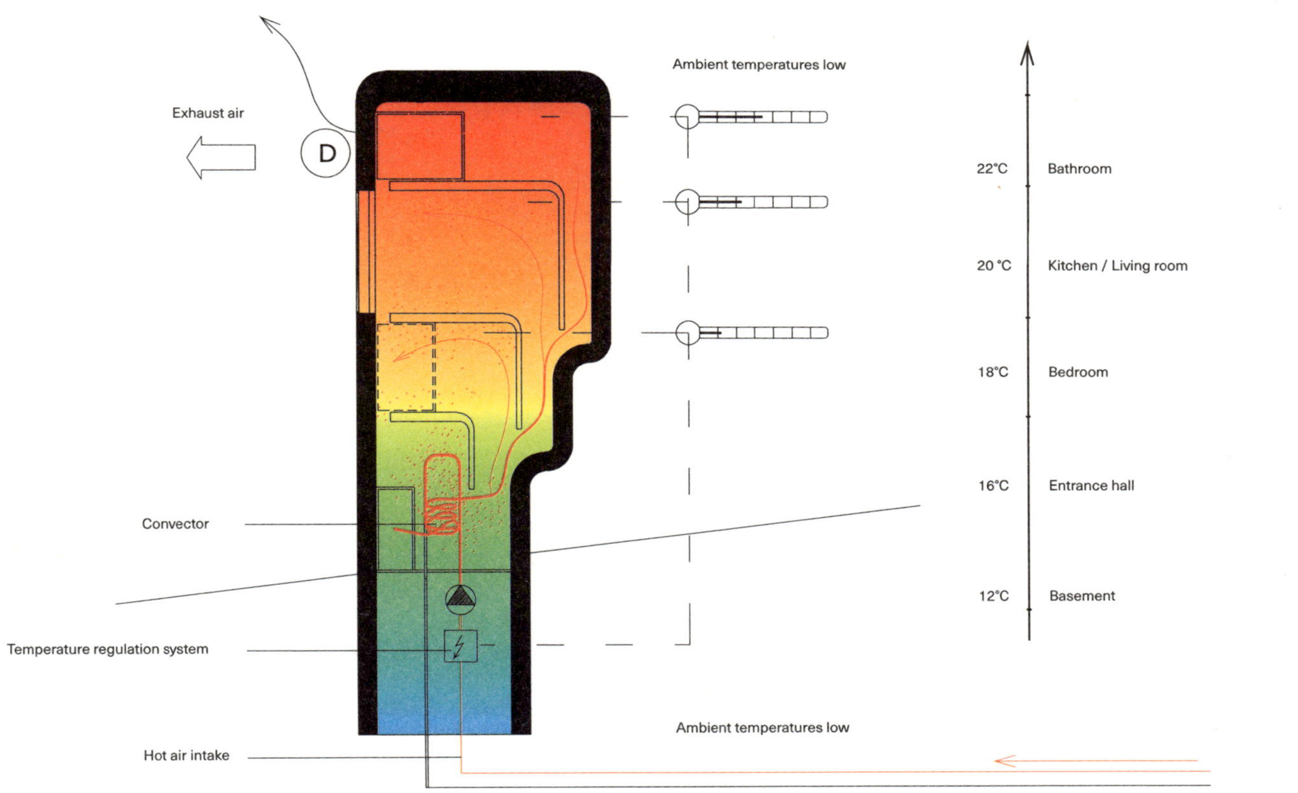

The Archimedes house - Section
{Fig 1-166}

Meteorological Museum

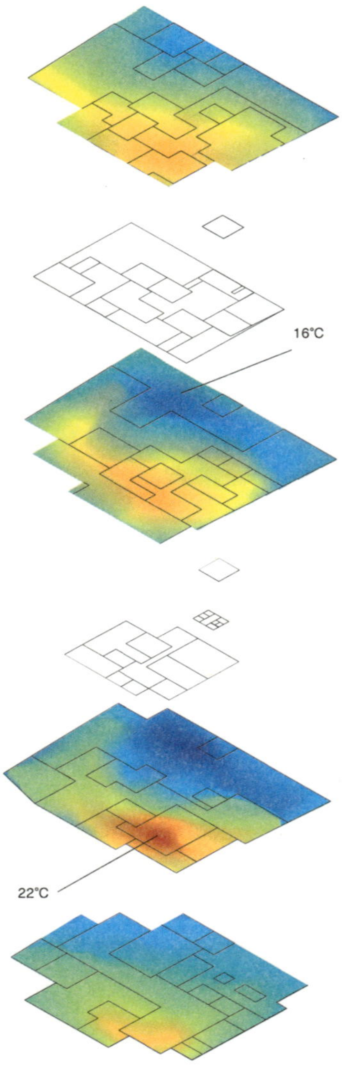

Exploded view

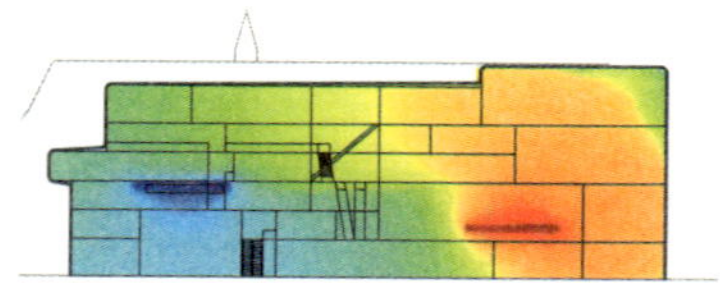

Section A

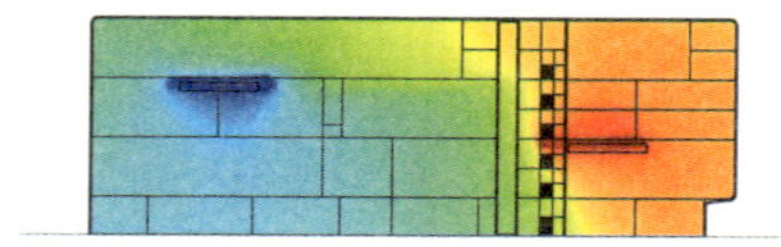

Section B

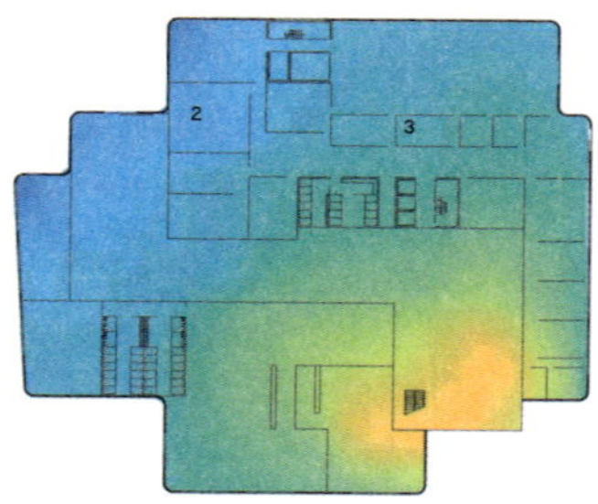

Plan of level +0

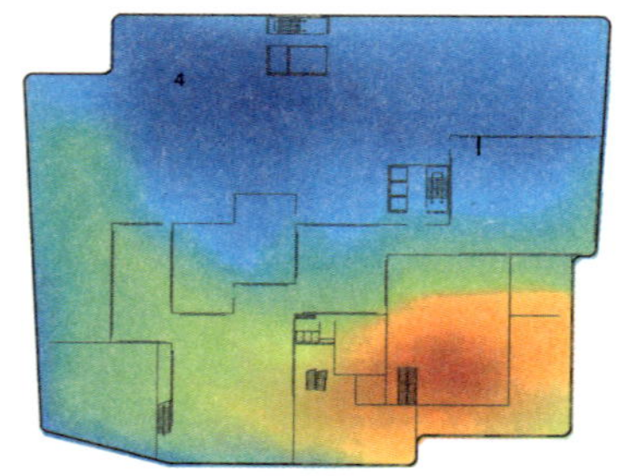

Plan of level +1

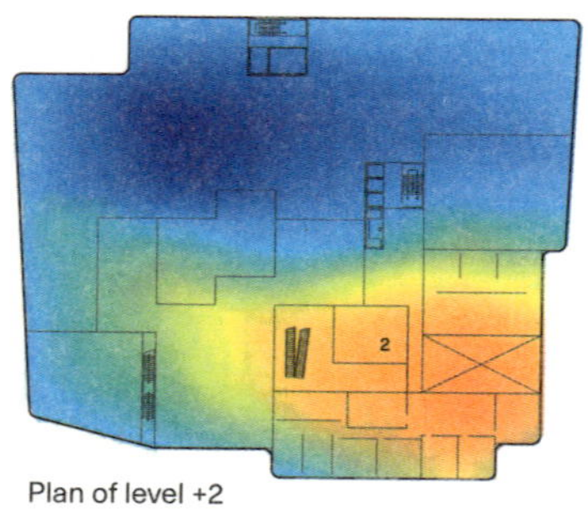

Plan of level +2

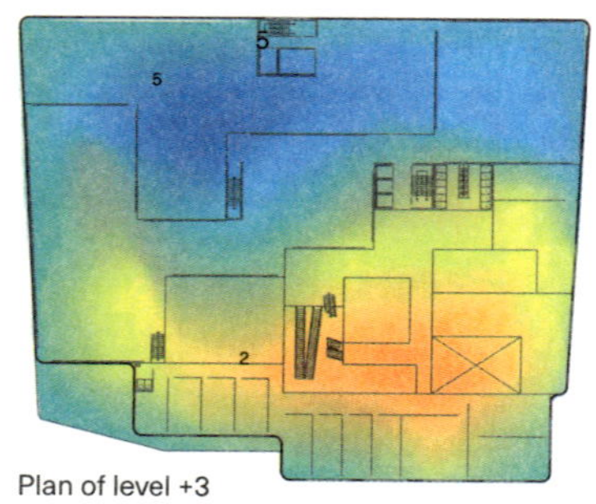

Plan of level +3

Plan of level +4

Plan of level +5

I

II

III

I.Cold areas
1. Home
2. Deposit
3. Equipment room
4. Collection storage
5. Archives

II.Temperate areas
Circulation and intermediate spaces

III. Warm areas
1. Control post
2. Cafeteria
3. Restaurant
4. Children's play area
5. Conference room
6. Offices
7. Photo studio
8. Video and TV studio
9. Multipurpose studio

Convective museum {Fig. 1-167}
Two radiators of different temperatures (22°C and 16°C) placed diagonally in the space distribute the heat irregularly throughout the volume of the museum. The higher you climb, the warmer it is, the more the program requires a constant high temperature. Uses are determined by the different climates.

This project for a contemporary art museum is defined by the introduction of two hot poles (like two large radiators) in the space; one colder at 16°C and one warmer at 22°C. Both temperatures take into consideration the temperature extremes corresponding to the uses of the museum: 16°C for storage and little used spaces, 22°C for the offices. The two poles are implemented to be diagonally opposite in plan and in section; the cold pole is placed higher than the hot pole to create a thermal imbalance. They thus generate the atmosphere of the museum, creating a multitude of climates and flows ranging from 16°C to 22°C, with areas at different temperatures accommodating the uses made of them. The building, as a planetary atmosphere, creates a variety of different climates that then allow different activities, depending on the requirements and the need to minimize energy expenditure by implementing, for each function, at least the recommended temperatures.

Interior Gulf Stream

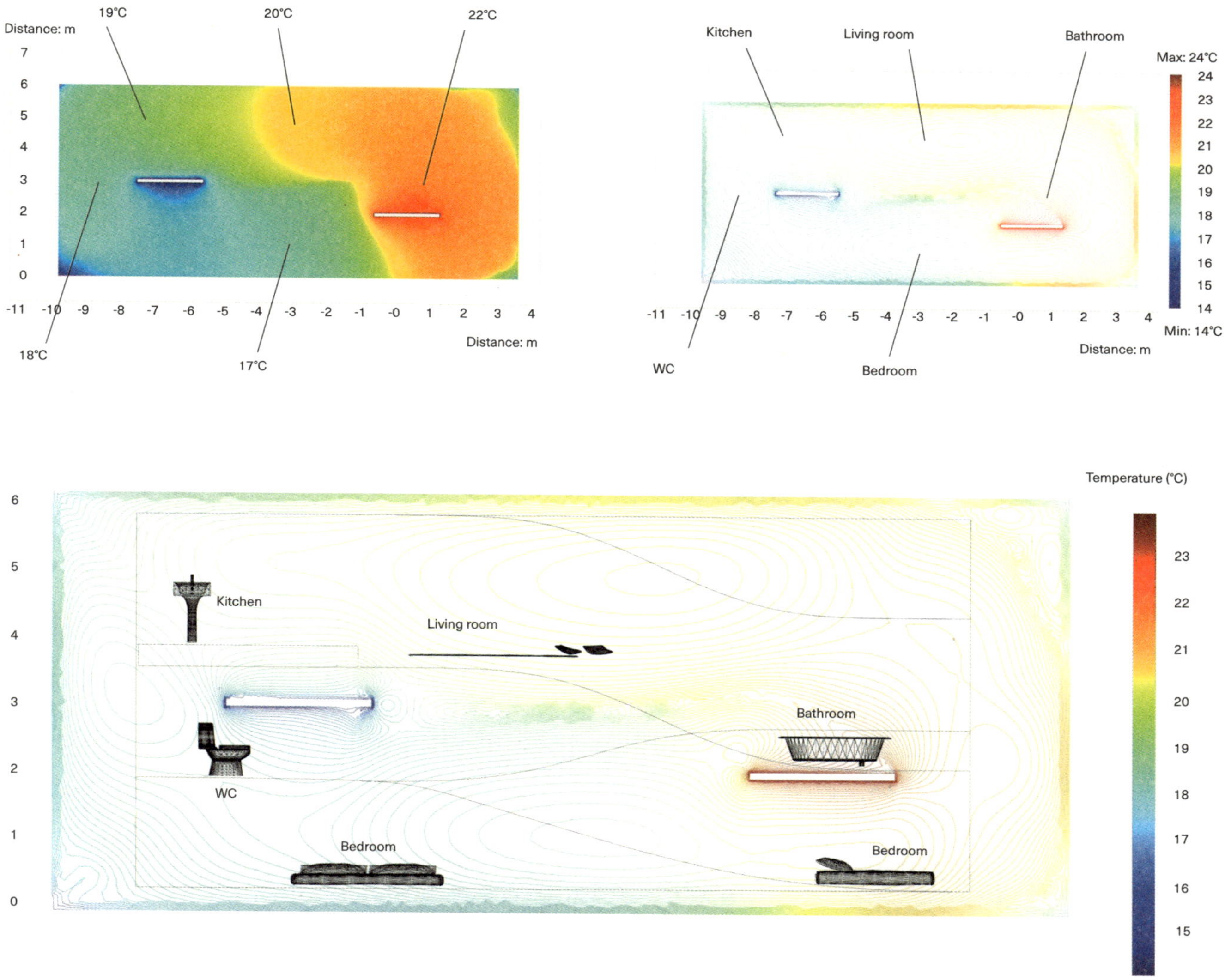

Interior Gulf Stream {Fig. 1-168}
Two heat sources (two radiators) are set at different temperatures. At the bottom right, the radiator diffuses a temperature of 22°C, which is the upper comfort temperature for human habitability when naked, wet or inactive. In the top left, the radiator diffuses a temperature of 18°C, the lower human comfort temperature for short-term activities when dressed or active. The difference in spatial location of the two radiators at two different temperatures creates a thermal landscape in which the various uses are distributed according to their heat requirements.

Instead of heating the entire space to a temperature of 21°C, we propose introducing two heat sources in the house, two differentiated thermal poles that will create a thermodynamic tension within the house: a cold pole at 16°C located in the upper layer of the house and a hot pole at 22°C located in the lower part. Convective movement of air will thus be created throughout the atmosphere of the house, generating a spatial thermal gradient. With the help of CFD (computational fluid dynamics)-type thermal modeling software, it is then possible to map the temperature variations within the space in order to then locate the activities by following the recommended specific minimum temperatures that correspond to them physiologically (16°C for sleeping, 19°C for the kitchen, 20°C for the living room, 22°C for the bathroom). The design process is thus inverted: an internal climate is created first, then the uses are assigned to the zones where the thermal conditions are favorable. The functions follow the climate. The floors and open spaces are designed to follow the path of the air throughout the height of the space, according to the vertical movements of the air and the suggested uses. Then, the location of some uses is immediately established according to their precise needs, from the hottest to the coldest climate. The two poles consist of two horizontal radiators. Like a miniature Gulf Stream, their position creates a movement of air using the natural phenomenon of convection in which the rising hot air will cool as it approaches the cold pole and then descend, and so on, creating a continuous movement of air and heat, an invisible landscape where the qualification of space is also sensual. All flooring and partitions are made of wood, composed of openwork slats allowing air to pass freely through them, without hindering the free deployment of air.

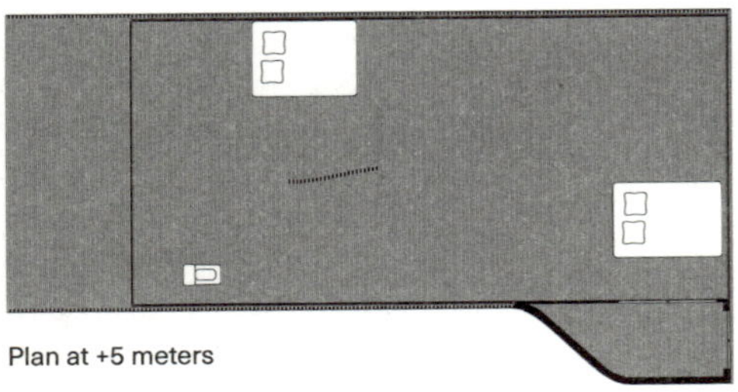

Plan at +5 meters

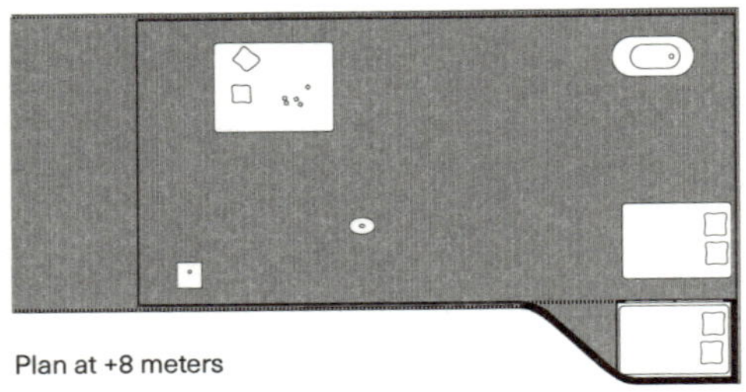

Plan at +8 meters

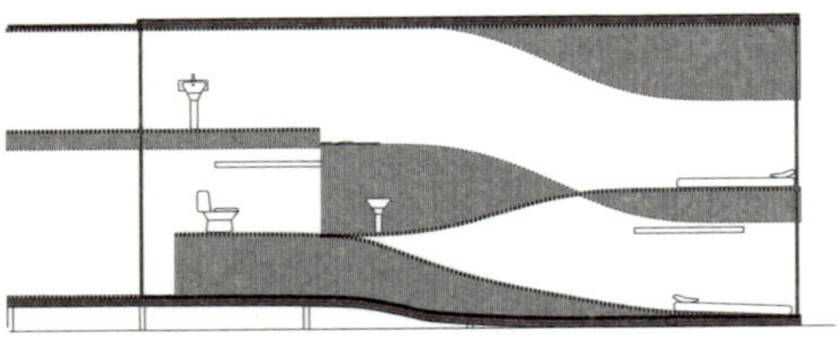

Longitudinal section

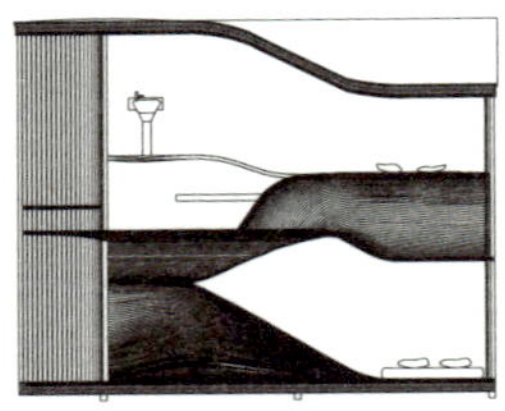

Transversal Section

Plans and Sections

What interests us here is not the creation of a homogeneous space but rather a plastic, dynamic and climatic space where the activation of two poles generates a heat landscape. Here, architecture is literally structured by the movement of air generating a fluid and atmospheric space. This architecture is based on the construction of a meteorology. The shapes of the house are determined by the shape of the computer-modeled thermal movements. The inhabitant can move freely within this invisible landscape between 18°C and 22°C, the temperatures at the two human thermal comfort extremes, by freely choosing the temperatures according to their activity or clothing.

Convective Housing

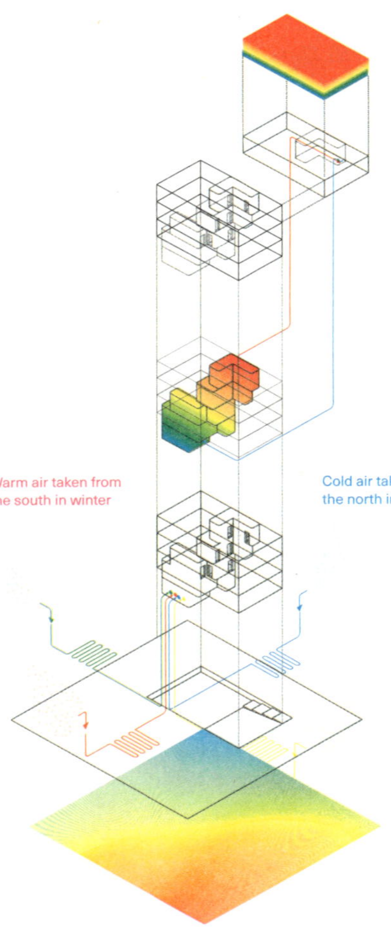

Warm air taken from the south in winter

Cold air taken from the north in summer

Airflow path in summer mode

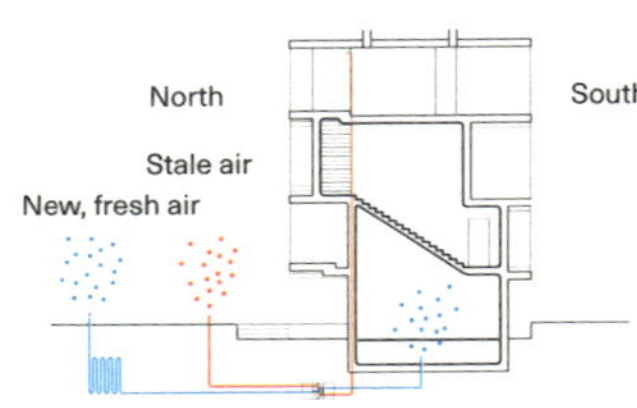

Airflow for summer ventilation between outdoors and shared indoor space

Airflow path in winter mode

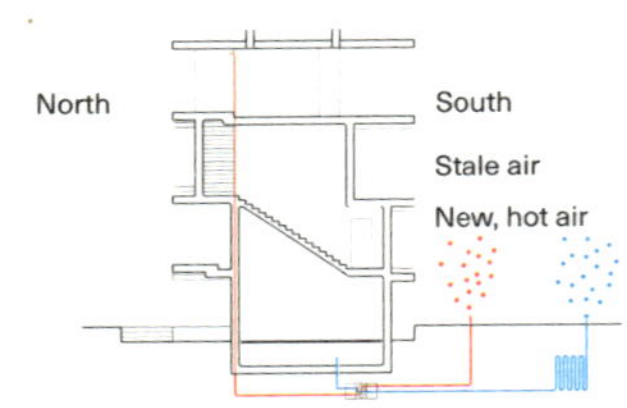

Airflow for winter ventilation between outdoors and shared indoor space

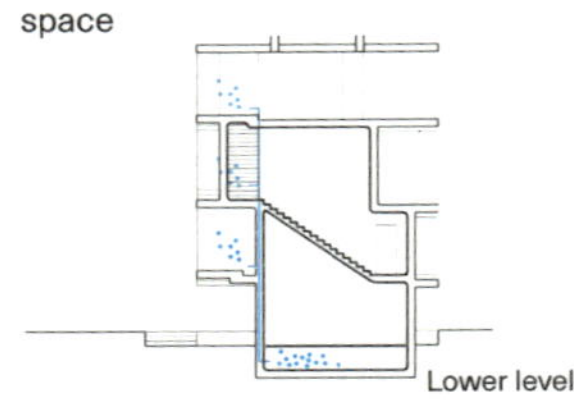

Fresh air at the bottom of the indoor shared space is sent into the apartments

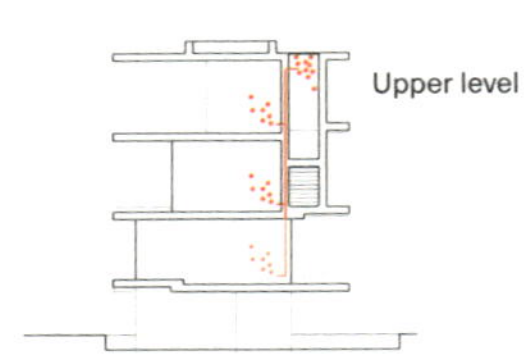

Warm air at the top of the indoor shared space is sent into the apartments

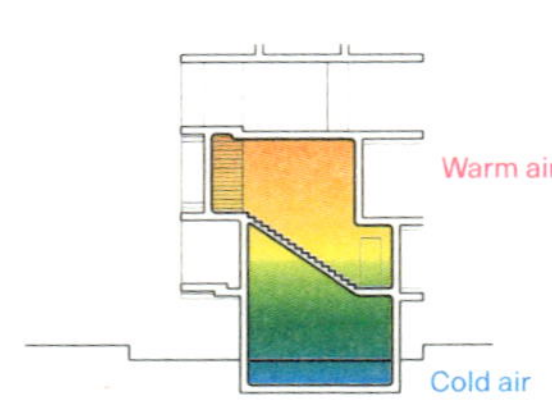

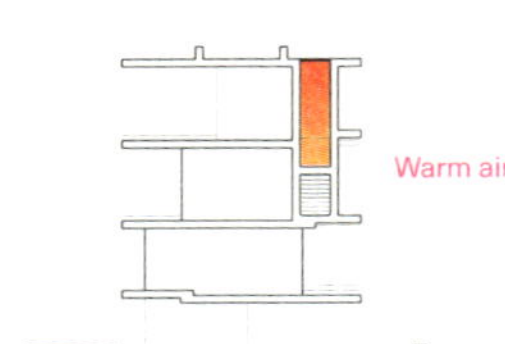

Stairways as a fresh air reservoir {Fig. 1-169}

Here we take advantage of the height of the stairwell to store fresh air at ground level for the summer and warm air at ceiling level for the winter.

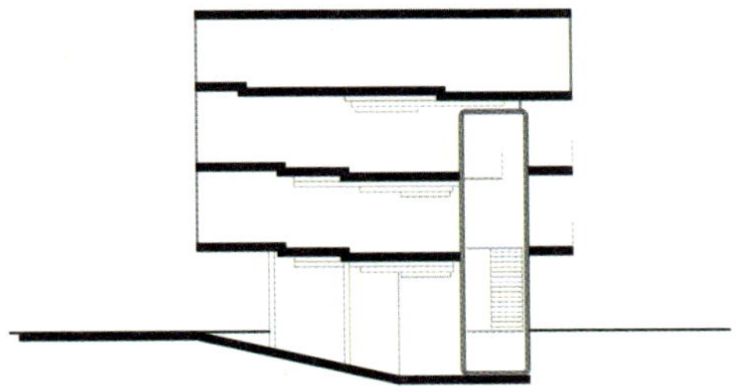

{Fig 1-170} Section

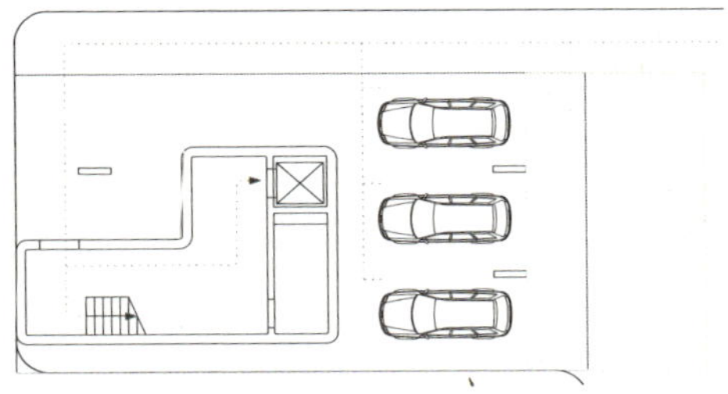

Plan of Level +0

Functions according to thermal areas

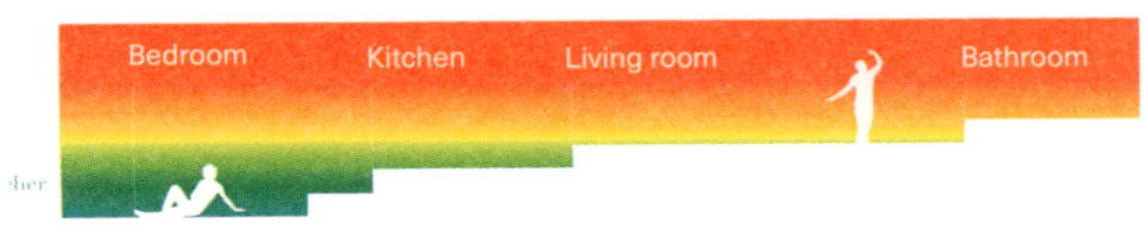

Section

Plan

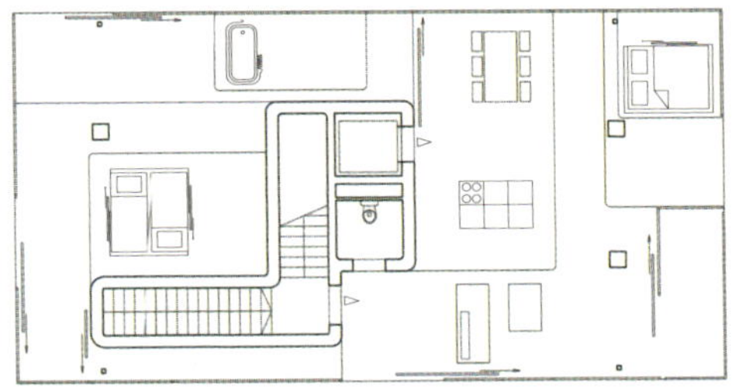

Plan of Level +1

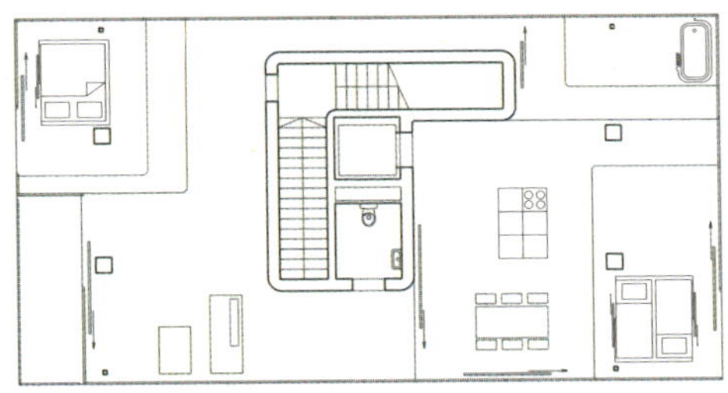

Plan of Level +2

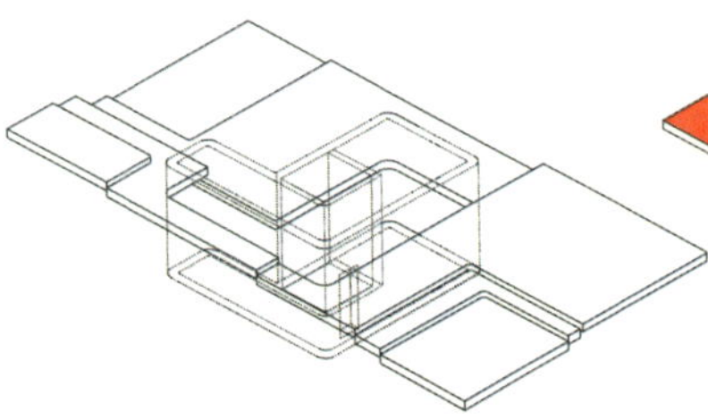

Topography of the apartment

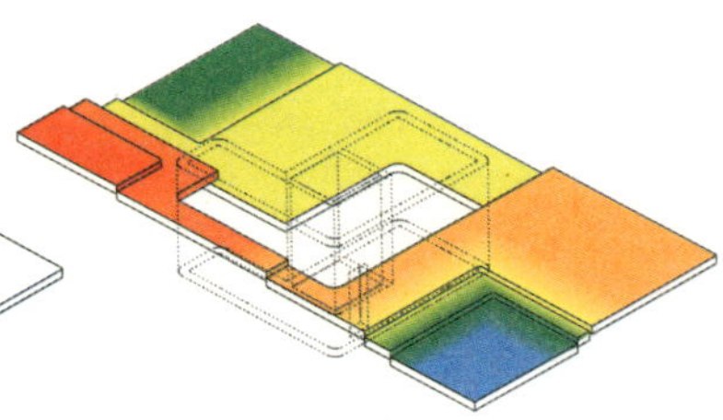

Topography according to thermal areas

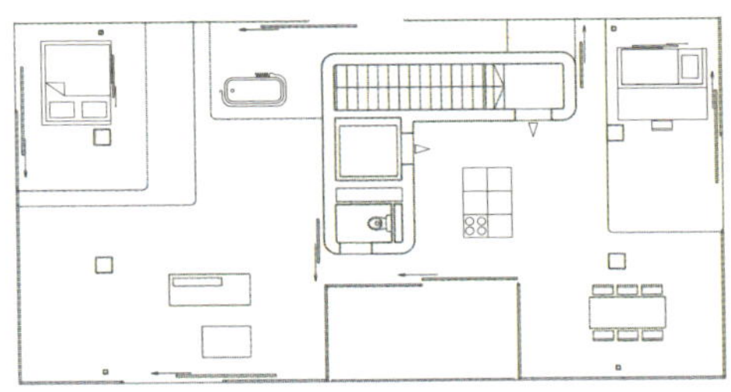

Plan of Level +3 Plans {Fig 1-172}

Vertical heat stratification and the deformation of flooring {Fig. 1-171}

The deformation of the slabs means that the temperature of the different spaces can be controlled: cold at the bottom, warm at the top, with different thermal qualities according to our different physical activities and clothing.

The design of the apartments in this residential building is based on Archimedes' natural law which states that hot air rises and cold air sinks. Often in an apartment, a large difference in temperature of up to 10°C can be measured between the floor and the ceiling. Depending on our physical activity or whether we are wearing more or fewer clothes, the temperature does not necessarily have to be the same in all rooms of the house. If we are in bed, covered with a blanket, the temperature of the room can be lowered to 16°C. In the kitchen, if we are dressed and relatively physically active, we can have a temperature of 18°C. The living room will be closer to 20°C because we are dressed there but sitting still on the sofa. The bathroom is the warmest room in the house because we are naked and wet (water on the body and hair accelerates the heat exchange between the body and the air in the room). Maintaining these rooms at these precise temperatures enables achieving significant energy savings.

Depending on these physical needs, we propose to design the apartment with different depths and heights: the space where we sleep will be lower down while the bathroom will be higher up. The apartment then becomes a thermal landscape with different temperatures, the inhabitant can move freely, as in a natural landscape, in search of suitable thermal conditions, depending on the season and the time of day. By deforming the floor slabs, the spaces have different heights, which allows varying the temperature. The deformation of the slabs is also the reason for the external appearance of the building whose aesthetics follow thermal factors.

Digestible Gulf Stream

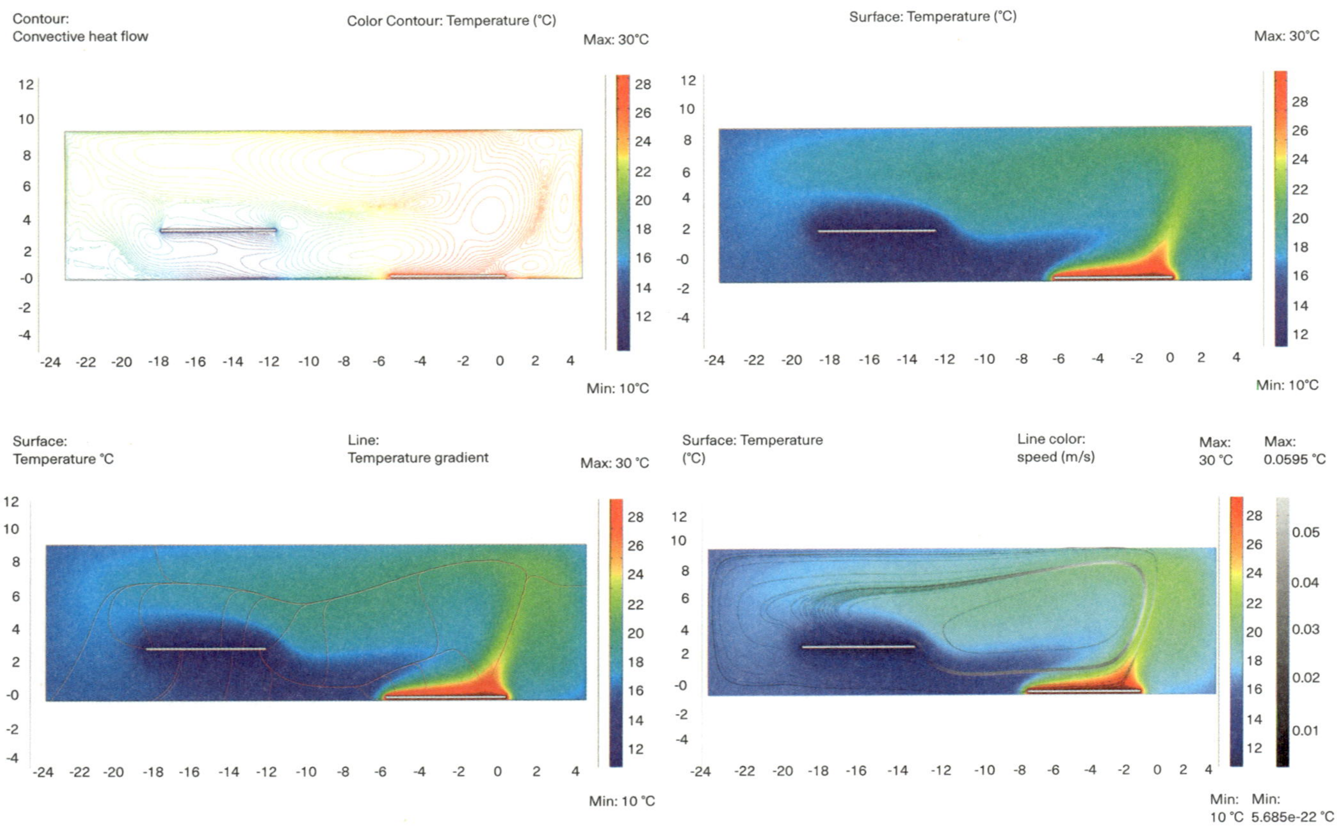

CFD (Computational Fluid Dynamics) models {Fig 1-173}

By placing two radiators at different temperatures, we generate a thermal landscape with flows, shapes and spaces. CFD collaboration: comsol

Architecture should no longer build spaces, rather it should create atmospheres and temperatures. Here, two radiators as horizontal planes made of conductive metal are placed at different heights. The lower one is heated to 28°C and the upper one is cooled to 15°C.

Like a miniature Gulf Stream, their position creates a movement of air by using the convection phenomenon in which the rising warm air meets the cold layer and cools and then sinks, it will warm up again in a continuous heat flow, an invisible landscape. The architecture is structured by air and becomes meteorological.Its inhabitants can move freely between 28°C and 15°C, the extreme temperatures of human thermal comfort, to find the optimal environment according to their clothing, diet, physical activity and social desires. Our architecture spans the from the microscopic to the macroscopic, the gastronomic and the atmospheric, the visible and the invisible. The architecture becomes a "Gulf Stream" that polarizes contrasts at different scales (hot/cold, low/high, dressed/naked, internal/external, resting/active) to conceive the architecture as a convective movement of air that draws the in-between as a habitable thermal gradient.

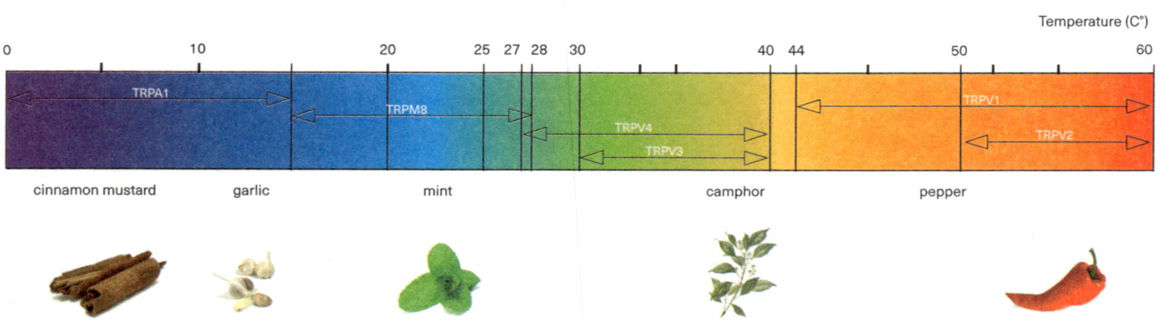

The neuronal perception of heat {Fig. 1-174}

The perception of temperature is as physical as it is chemical, and depends as much on the temperature of the air as on certain molecules which have the capacity, in contact with the skin or the mouth, to relay specific neurological ion channels transferring their information from the skin to the brain and making it possible to perceive different contact temperatures. Thus TRPV1 is the ion channel which informs the brain of a temperature above 44°C but is also relayed by the information given by the capsaicin molecule contained in the chili pepper, in particular.

In contrast, the TRPM8 ion channel carries information concerning cold temperatures of 15°C as well as that of the menthol molecules contained in mint.

TRPA1: Transient receptor potential (TRP) transmembrane ankyrin protein p 1
TRPM8: TRP melastatin 8
TRPV4: TRP vanilloid 4
TRPV3: tTRP vanilloid potential 3
TRPV2: TRP vanilloid 2
TRPV1: TRP vanilloid 1

Domestic Astronomy

Uses according to temperatures related to activity and clothing

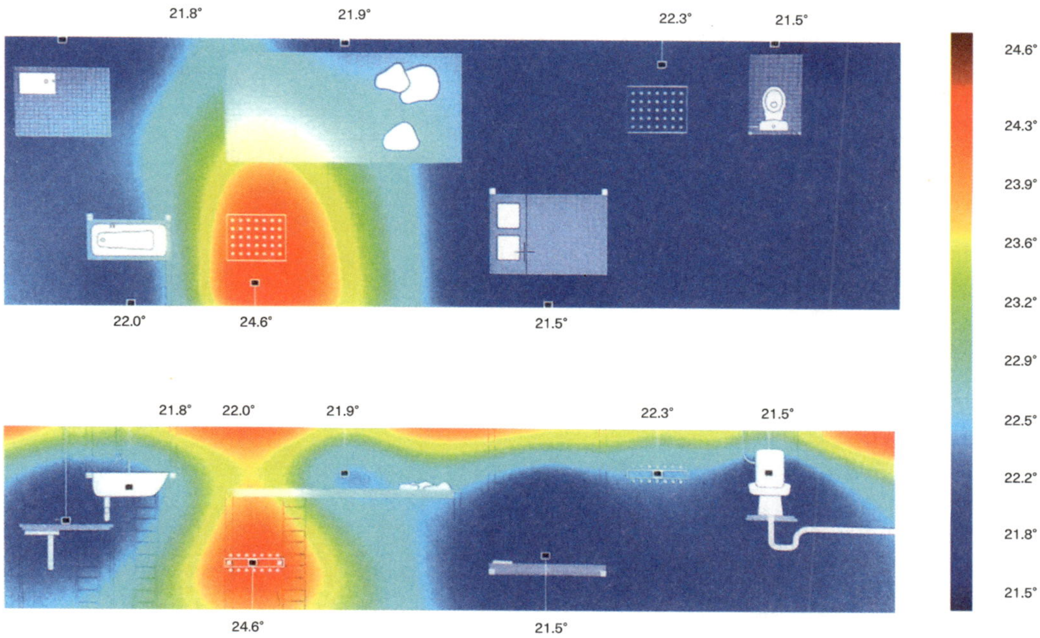

Functions follow heat convection {Fig 1-175}

"Domestic Astronomy" is a prototype for an apartment where you no longer live on the surface but in the atmosphere. Leaving the floor, the uses and the furniture rise and disperse, evaporating into the atmosphere of the apartment, stabilizing according to the temperatures related to the body, its activity and its clothing. According to Archimedes' law, hot air rises and cold air sinks, and this physical reality directly influences the temperature distribution in the apartment. It is possible to measure large temperature differences between the floor and the ceiling. For example, it will be 21°C at the feet and 28°C three meters higher, just below the ceiling. This overheating of the air beneath the ceiling is useless since nobody inhabits that space and it is becoming a problem today in a context of global warming where we must reduce our energy consumption. Indeed, these 8°C above 20°C are a waste of energy. Our goal today is to take these temperature differences into account to transform the way we inhabit space by abolishing the exclusivity of horizontality, to explore verticality, the vertical stratification of temperatures and thermal zones within the dwelling and, for example, raise the bathroom to the warm heights of the space while the bedroom will descend to the cold altitudes of the space.

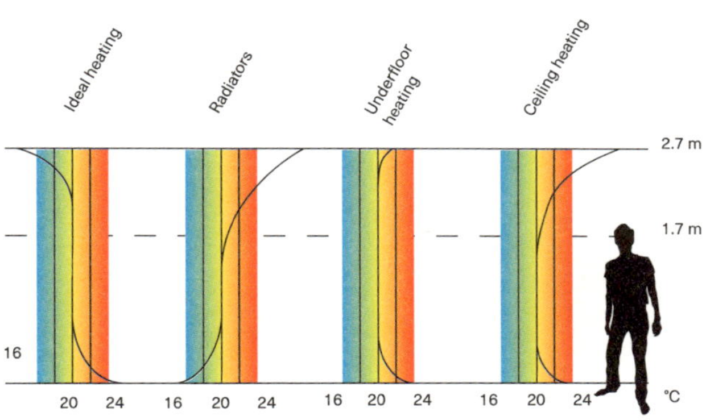

Heat distribution according to the type of heating {Fig. 1-176}

The profile of heat distribution in the space according to type of heating (source: Santé et qualité de l'environnement intérieur dans les bâtiments, Claude-Alain Roulet, PPUR presses polytechniques, EPFL Lausanne, Switzerland, 2004)

Ideal heating {Fig. 1-177}

As our temperature is lower in the legs, we have to compensate for it artificially by a higher temperature lower down, while above the head, the space we do not inhabit, the temperature can be colder.

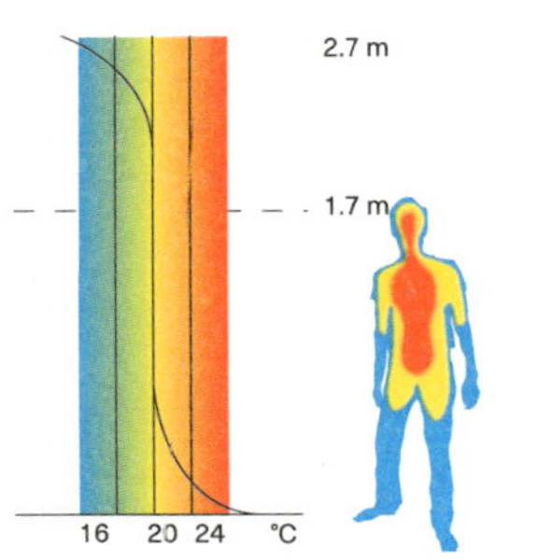

Ideal heating

Thermography of the human body

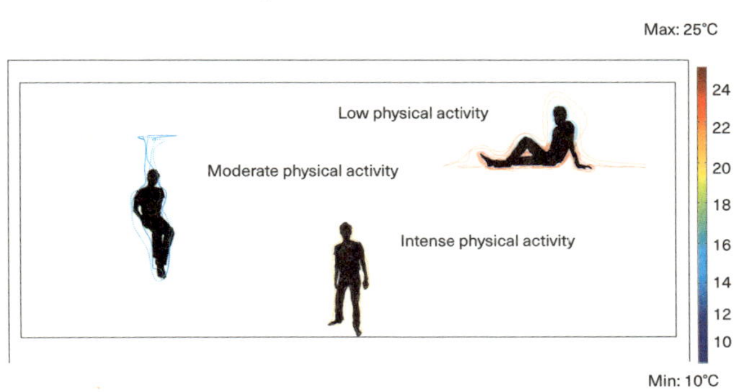

Distribution of heat in space {Fig. 1-178}

According to Archimedes' law, heat rises and accumulates beneath the ceiling. To save energy and avoid economic waste, activities requiring more heat can be shifted to the top of the room, making it possible to heat the lower spaces, that will accommodate activities requiring less heat, to a lower temperature.

The Evaporated Building

Initial situation
The south-facing façade and roof are greatly heated by the sun. The north-facing façade remains cold, especially in the lower part.

Intermediate situation
The warm air in the south is pushed down northward, in the center of the atmosphere. The volume of the empty space must be at least 9m3 so that the hot air can be displaced.

Intermediate situation
Warm air in the middle of the atmosphere

End situation
Heat distribution is more regular in the building. The temperature differences are attenuated with respect to the intermediate atmosphere.

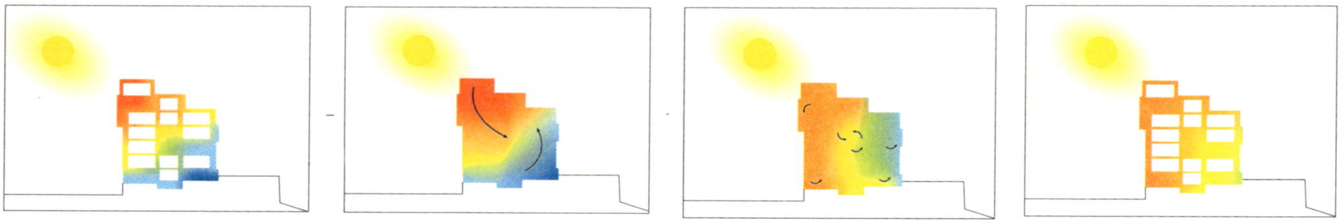

Though the void of interior spaces is usually defined with the help of opaque walls, floors and ceilings, our goal today is to define it no longer by solids, but by the void.

To evaporate structures, to sublimate the walls and the ceiling, to go from solid to gaseous state, to create an architecture of air and light, transparent, open, luminous, fragmented.
A building in its gaseous state, a disintegrated architecture, an evaporated building. Beyond the metaphor, it by basing ourselves on our physical and ecological reality that we must make this change, on the scale of the city and of architecture, but also of the individual, for their well-being and comfort.

Solid partitions are dissipated to allow air and light to pass through the building. The built mass is perforated in two places so that the wind can continue its path.

Balancing convection {Fig. 1-179}
By forcing convective motion, the excess heat moves from the south to the north and descends to better distribute heat to where there is less.

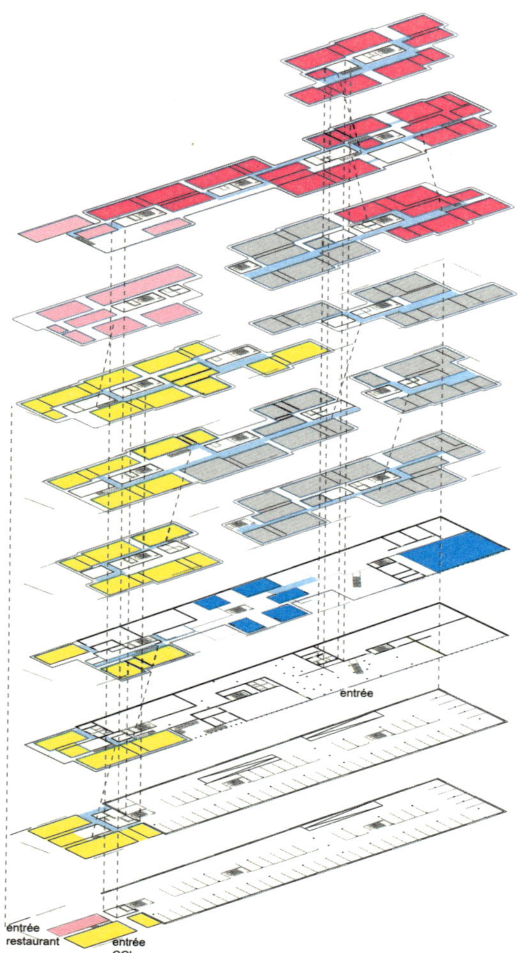

Axonometry of rooms in the Global Atmosphere {Fig. 1-180}
The rooms provide gaps between them that fully communicate with each other, in which the fresh air used to ventilate the premises circulates.

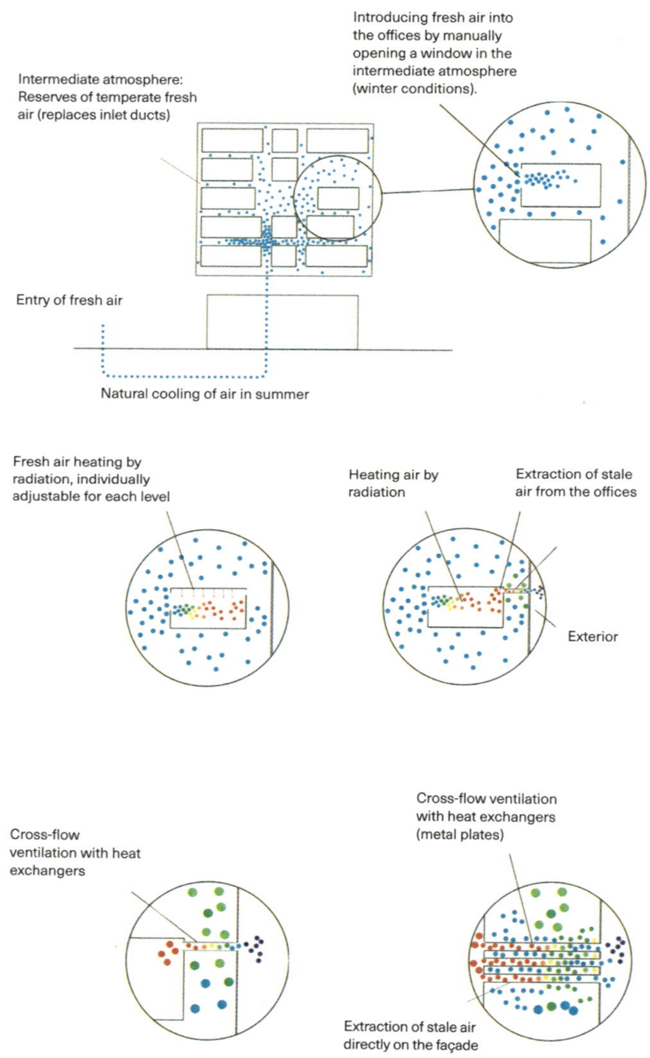

Heat Recovery System

Global Atmosphere and Heat Recovery {Fig. 1-181}
The rooms float in an atmosphere that is a reserve of temperate air that can supply each room by means of an air renewal and heat recovery system.

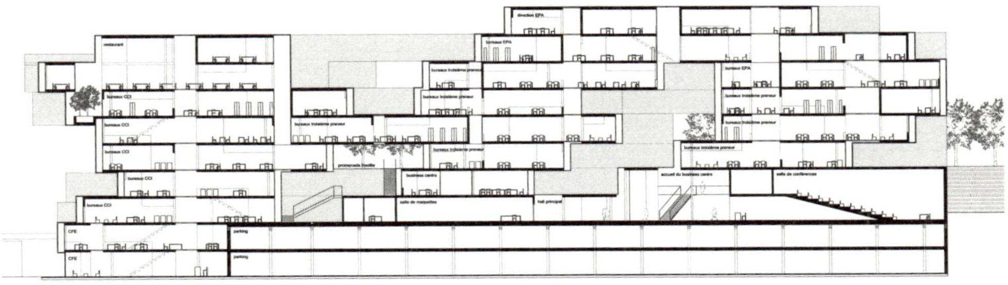

Longitudinal section

Cross-section

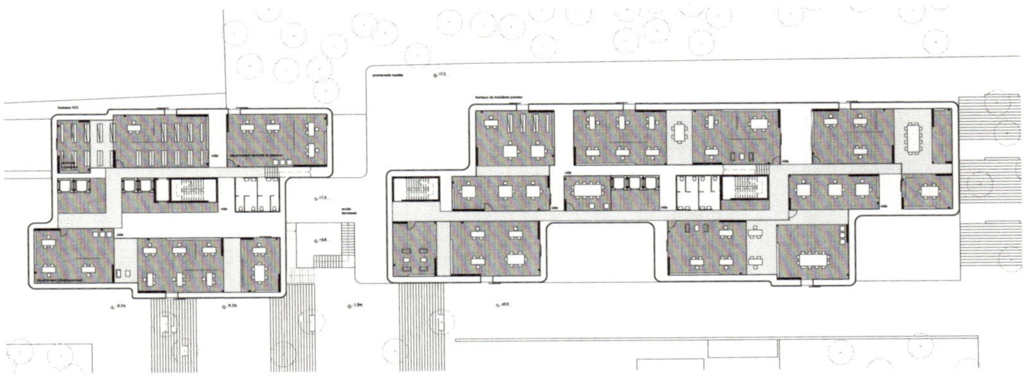

Plan

Sections and Plan {Fig 1-182}

Open Climate

This programmatic pavilion for a School of Fine Arts is a manifest architectural project as an addition/succession of rooms as volumes that are set apart only by variations in light and temperature. There is therefore no visual difference between these volumes, and a priori no specific use.
Only the light and the temperature vary by taking advantage of Archimedes' law, with warmer areas at the top and colder ones at the bottom, brighter areas at the periphery and less bright ones at the center. The pavilion is organized by two parameters that qualify the space: light and temperature, according to two geometric dimensions, one vertical and one horizontal.

In the horizontal dimension, the plane resembles a Russian doll, from the periphery to the center, from the lightest to the darkest, a succession of 4 series of 4 rooms wrapped in each other. Each series has a light transmission coefficient. The first, completely transparent one on the periphery, is wrapped in a glazed wall with a light transmission coefficient of 100%. The second series has a coefficient of 66%, the third 33% (some walls have a coefficient of 66% only because some of the light has already been filtered by the first glazed wall). The last series filters the remaining 33% of light to create a completely black atmosphere.

Vertically, the surfaces develop from bottom to top, according to the principle of the ascent of masses of hot air and vice versa, of the descent of cold air. Thus the rooms located at the bottom are the coldest, and the higher up you go the rooms will be warmer.

The pavilion is a combination of 4 luminosities and 4 temperatures which creates 16 different climates in the form of 16 different climatically but spatially similar ateliers. Depending on the time of day, the season, the artistic practices they host, the inhabitants can appropriate the space and move freely. One particular space, thanks to its reduced, homogeneous light, may correspond to that of 19th century paintings. A warmer space may correspond to a low physical activity (like working on a computer), while a cooler space may correspond to a higher physical activity (like sculpting stone). Dark spaces can accommodate video work. In the end, there is a choice of 16 rooms identical in size but with as many different climates on offer to the inhabitants. Art students are encouraged to choose where to be, whether warm, cold, light or dark with various atmospheres to discover, choose from or experience.

Open Climate

Temperature distribution according to height

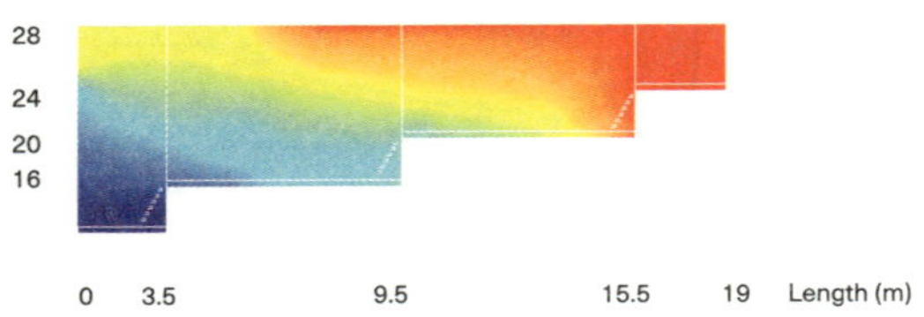

Temperature distribution

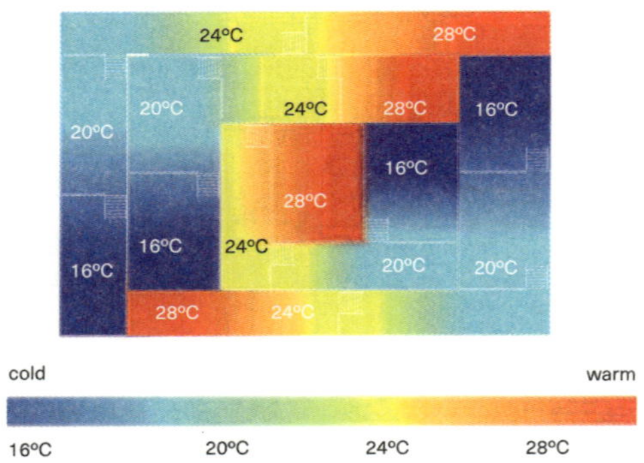

Temperature distribution plan

Thermal stratification {Fig. 1-183}

The higher up, the warmer. Thus, 4 temperatures are suggested to meet different activities and situations.

LT=light transmission coefficient

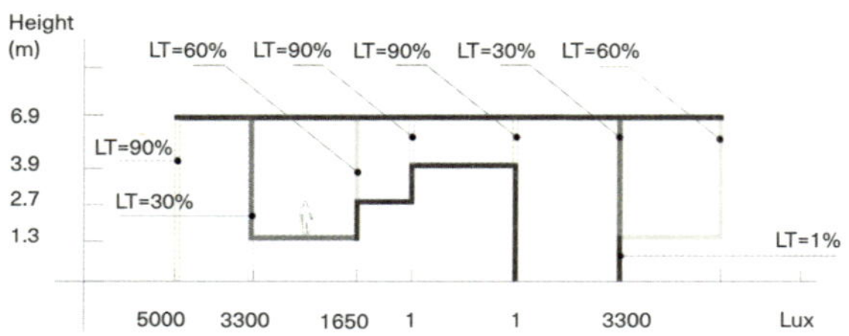

Section of glass distribution by LT

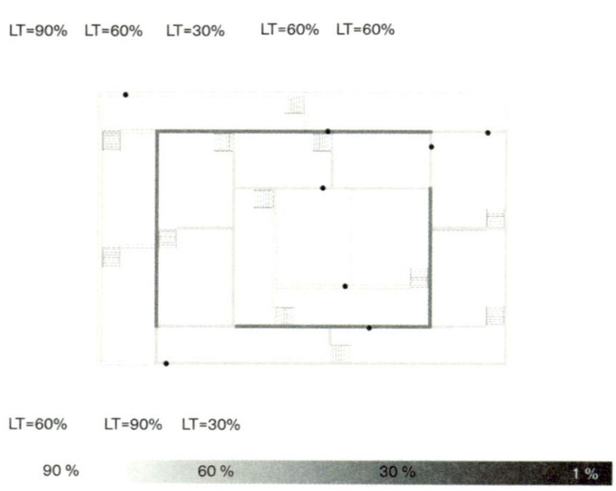

Light transmission coefficient

Plan of glass distribution by LT

Light gradation {Fig. 1-185}

Distribution of the different types of glass according to their light transmission coefficient.

Natural air ventilation

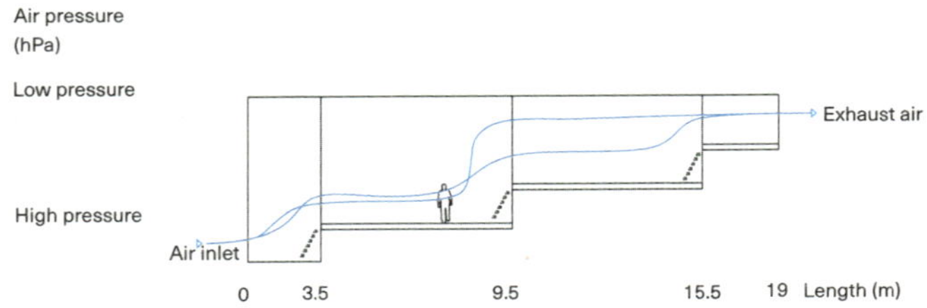

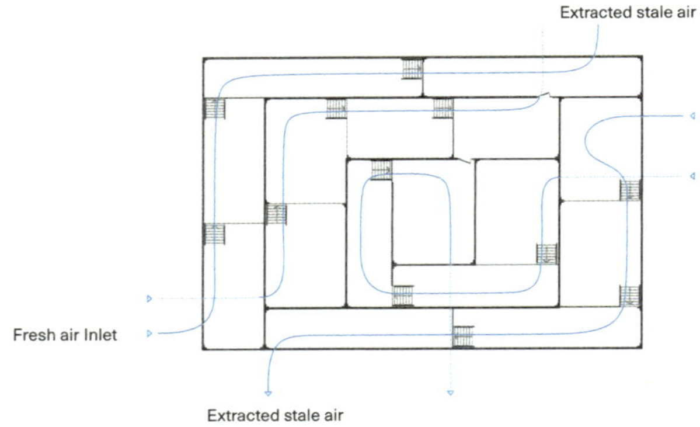

Ventilation flow {Fig. 1-84}

The air movement follows convective motion, bringing the cold, moist air from the bottom (where it is input) to the top (where it is extracted).

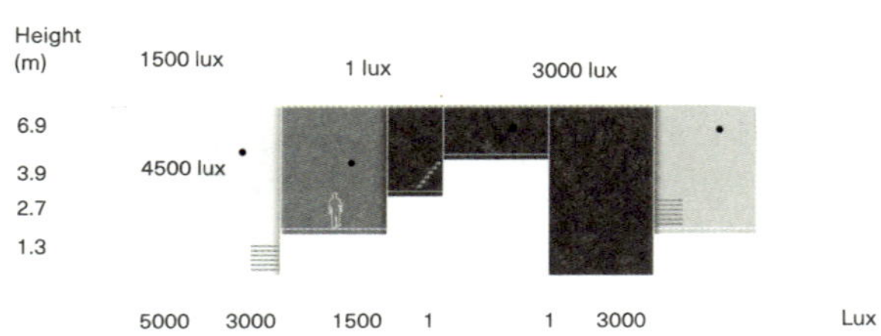

Section of daytime light distribution

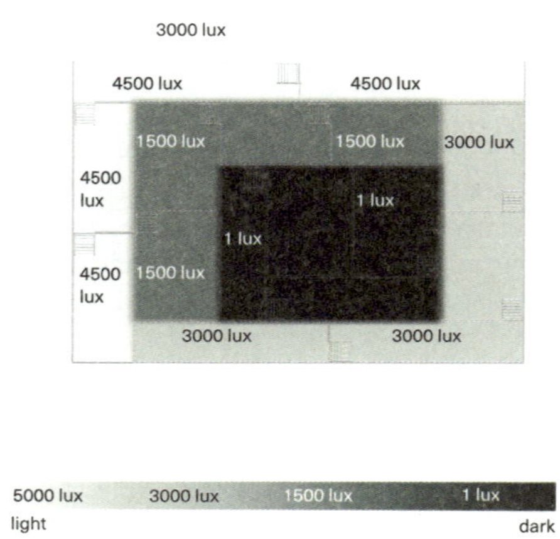

Exterior light intensity 5000 lux

Plan of daytime light distribution

Light intensity variation {Fig. 1-186}

Distribution of light according to the glass transmission coefficient.

EMISSIVITY

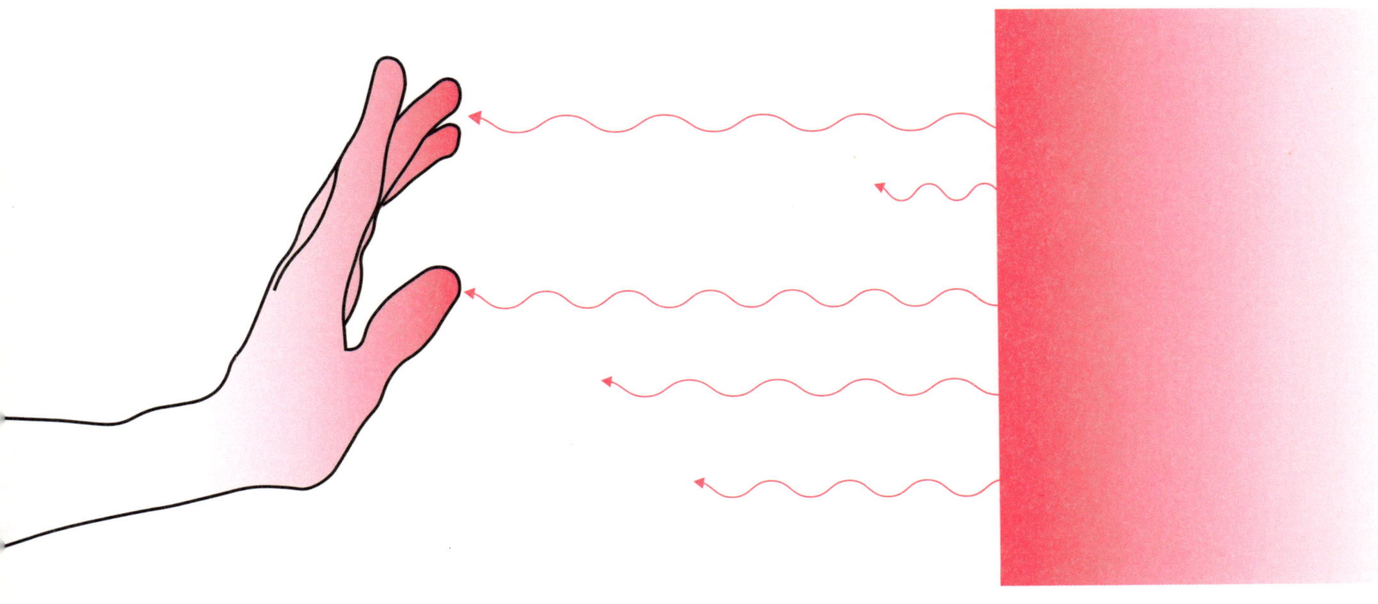

I. Definition

General Properties

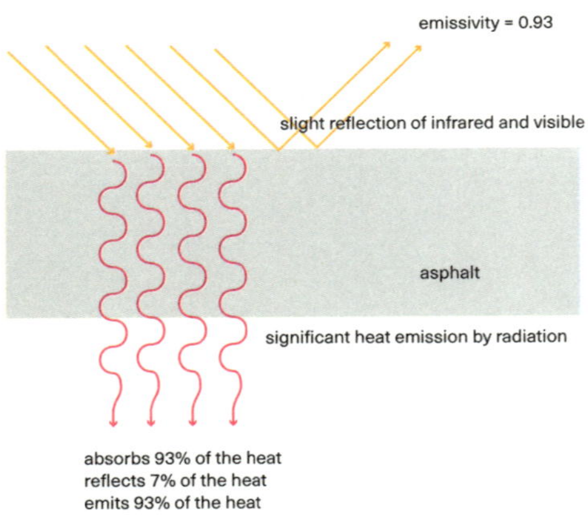

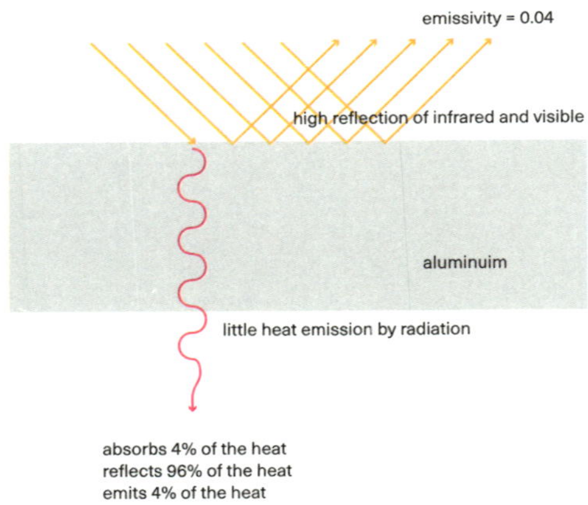

Emissivity and Material {Fig. 1-187}

The emissivity value depends mainly on the type of material. Metal (zinc for example, where e= 0.03 when it is polished and 0.1 when it is oxidized) is usually very poorly emissive, that is to say it will not absorb the heat of incident rays, it will not rise in temperature when exposed to the sun in summer, for example, but only due to the temperature of the air. Conversely, a material like concrete (e= 0.91) or asphalt (e=0.93) will rise in temperature under the sun, gradually reaching a temperature higher than that of the ambient air)

Material	Emissivity min=0, max=1	Material	Emissivity min=0, max=1
Polished copper	0.023-0.052	Concrete	0.92
Aluminum	0.04	Aluminum	0.05
Steel	0.43	Steel	0.59 (inox)
Cement	0.54	Terracotta	0.93
Iron	0.6	Stone	0.96 (calcareous)
Plastic	0.84	Wood	0.9
Epoxy black	0.89	Raw earth	0.91
Wrought iron	0.94	Wool	0.78
Black body	1		

Examples of coefficients of emissivity of different materials Fig. {1-188}

The emissivity value ranges between 0 (lowest) and 1 (highest). The lower the emissivity of a material, the more the material will reflect infrared wavelengths, and the less it will heat under the sun for example. Conversely, the higher its emissivity, the more it absorbs the infrared, and the more it will heat up.

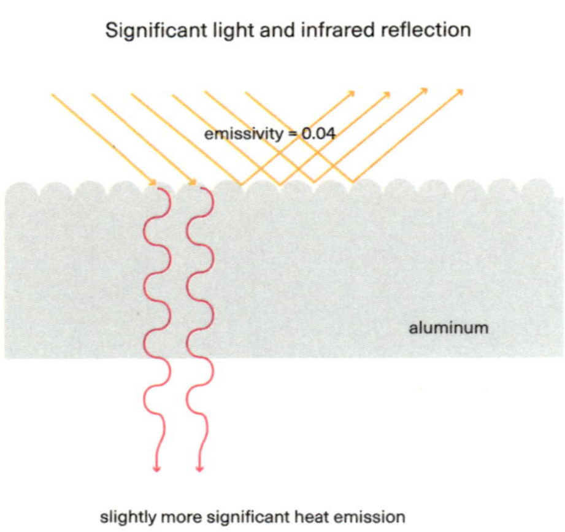

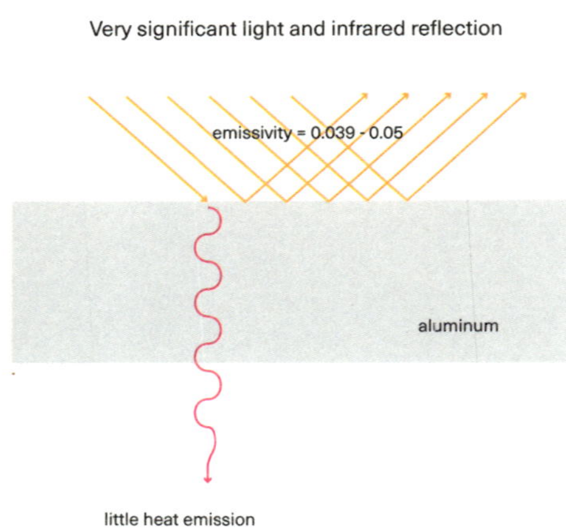

Emissivity and roughness Fig. {1-189}

The rougher the object, the greater its emissivity will be

Emissivity is the ability of a material to absorb and emit thermal radiation in the infrared and in the visible wavelengths. Its value ranges from 0 (low emissivity / neither absorbs nor emits heat by radiation) to 1 (high emissivity / absorbs and emits all heat by radiation). It is measured with respect to a theoretical black body with an e value of 1, which would absorb all electromagnetic radiation coming from an object warmer than itself, just as by radiation it would emit all its heat toward an object before it if it is warmer than the latter. The higher an object's emissivity, the more it will absorb radiation from nearby objects. If the other objects are colder, it will emit all its heat toward these objects until they reach their temperature. Conversely, a low-emissivity material will not be heated by incident visible infrared rays because it will not absorb, but will reflect them. Neither will such a material lose heat by radiation to a cooler object, retaining its heat; it will only exchange its heat with the environment by conduction or convection. Emissivity depends on the material, its color and its texture.

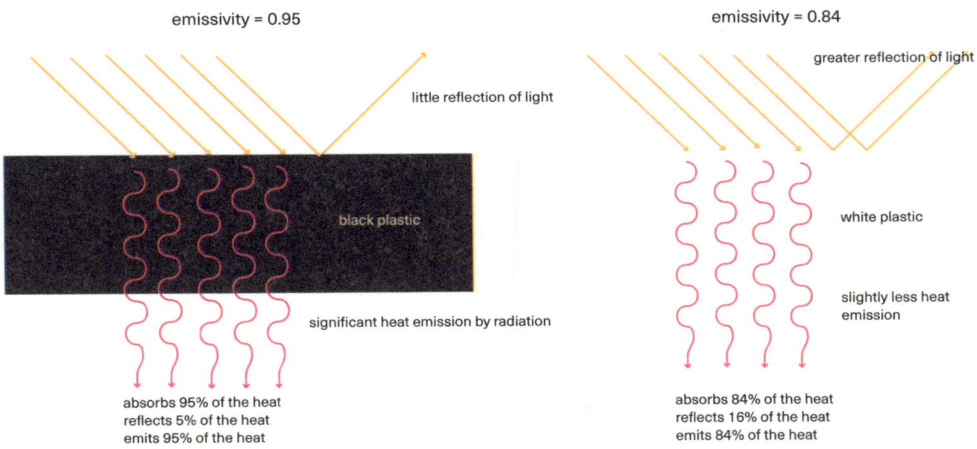

Emissivity and color {Fig. 1-190}

The darker the object, the greater its emissivity because it absorbs the visible part of the incident electromagnetic radiation. This part of the emissivity in the visible field, between 400 and 800 nm, also relates to the albedo, which describes the capacity of a surface to reflect incident solar energy. The lower the albedo, the more it absorbs radiant solar energy, which turns into heat in the exposed surface material and rises in temperature. The higher the albedo of the surface receiving the sun's rays, the more it will reflect them and remain cold under the sun

II. Physiological Scale

Emissivity of the Human Body

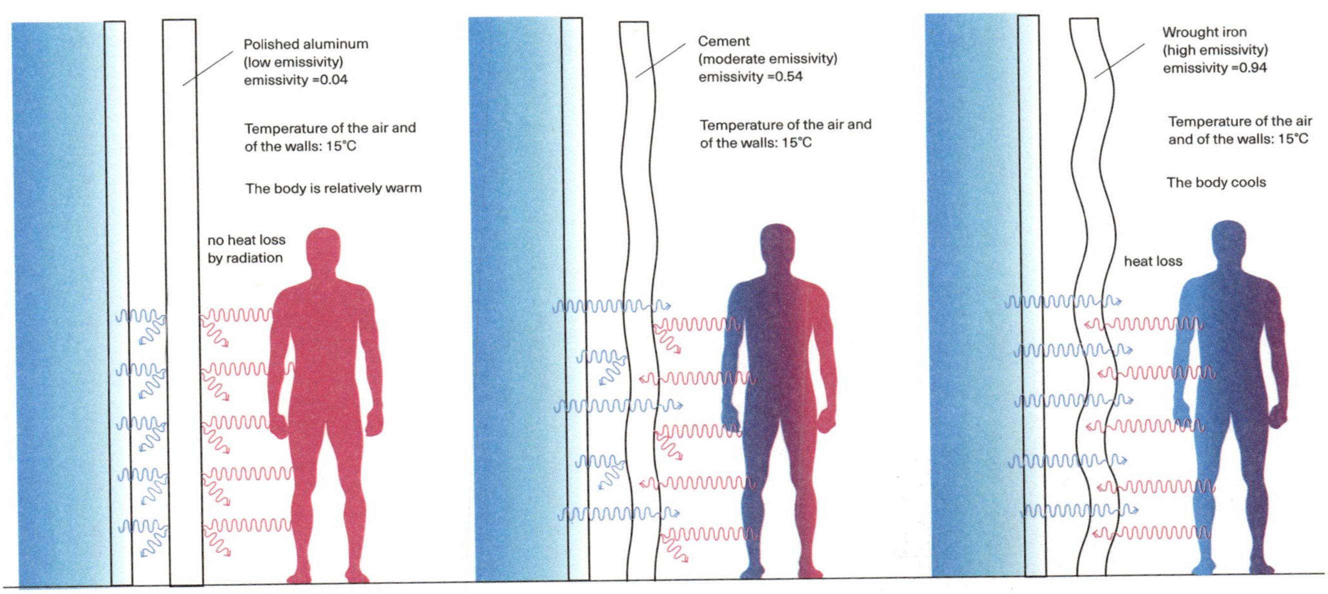

Emissivity and the human body {Fig. 1-191}

Objects emit hot (or "cold") radiation, i.e., absorbing hot radiation from another object). In a room, the walls emit radiation that is cold in winter and hot in summer (if above 34°C corresponding to skin temperature). The human body emits "hot" infrared rays. In winter, placing a low-emissivity material on the wall allows the body to lose less heat by radiation and not "receive" the cold rays coming from the wall. Conversely, a high-emissivity material will be preferred in warm areas (but at a temperature below 34°C) to absorb body heat and refresh it.

Human skin is highly emissive (e=0.98). It absorbs and emits infrared radiation very easily ("hot", because the skin is at 34°C) toward nearby objects. So when we are beside a cold wall made of a highly emissive material, the human body will quickly lose heat and cool down by radiation. Conversely, if the wall is cold but made of a material with low emissivity, the body will not lose heat by radiation in the wall as the latter will reflect the infrared heat without absorbing it. Similarly, if the temperature of the air is above 34°C, a highly emissive material will emit infrared waves, which will heat the skin, doubling by radiation the amount of heat already received by the skin by conduction, that of the air at 34°C. Thus, in an interior in winter, it will be advantageous to have walls, partitions and ceilings whose surfaces are made of a material with low emissivity (as well as under hot air temperatures of more than 34°C) whereas in summer, it will be advantageous to have walls, partitions and ceilings made of materials with high emissivity in order to speed up the loss of heat from the human body by radiation toward these facing surfaces.

Skin Absorption Spectrum

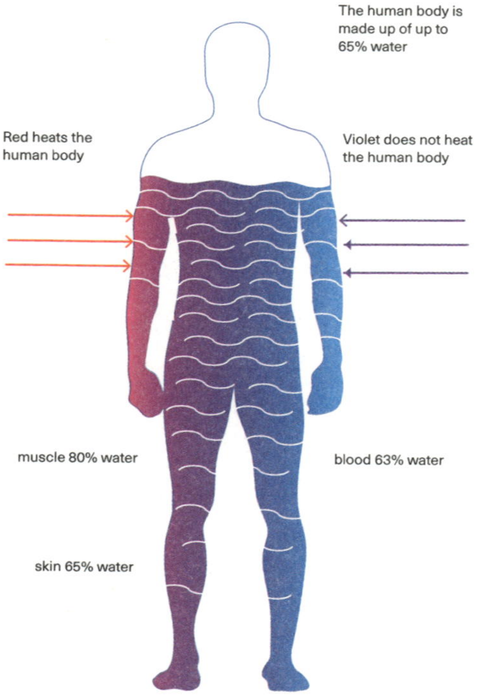

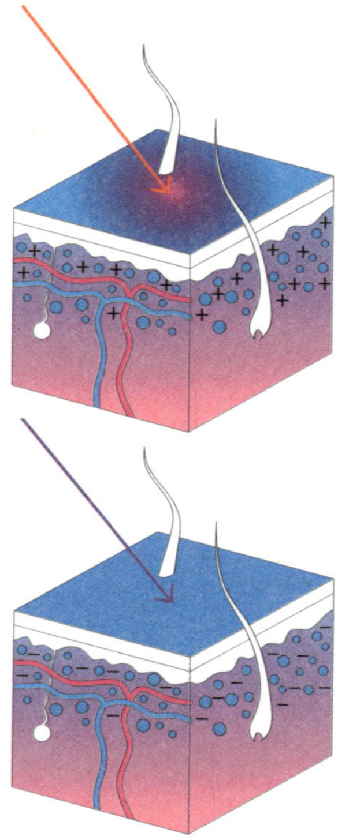

Heat absorption by the skin {Fig. 1-92}

The human body is mainly composed of water and does not heat up when it receives violet light since the thermal absorption of water for this part of the spectrum is not significant.
In front of a violet wall, the body temperature does not vary because no heat will be transmitted by the wall.
The longer the wavelength, toward the infrared, the more the body will heat up, and more so when we pass the visible electromagnetic spectrum to reach the infrared.

Clothing Absorption Spectrum

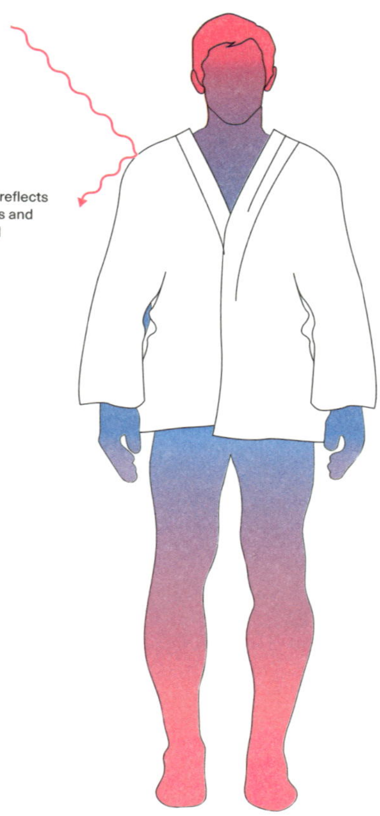

Albedo and emissivity {Fig. 1-193}

The albedo value is relative to visible light energy and emissivity is relative to infrared energy.
The lower the albedo and the higher the emissivity, the more the garment will heat under the sun.
The emissivity of leather is 0.95.
The emissivity of nylon is 0.85.
The emissivity of silk is 0.80.

Albedo and emissivity {Fig. 1-194}

The higher the albedo and the lower the emissivity, the more the garment will remain cool under the sun.
The emissivity of wool is 0.78.
The emissivity of cotton is 0.77.
The emissivity of linen is 0.69.

Energy Gains of Water According to Wavelength

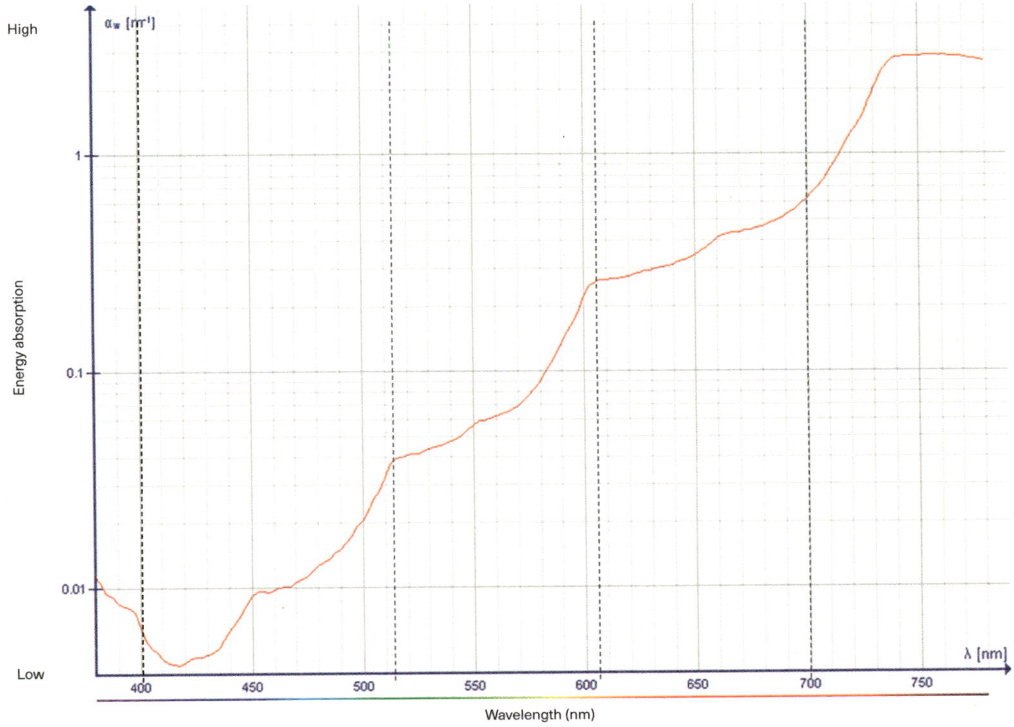

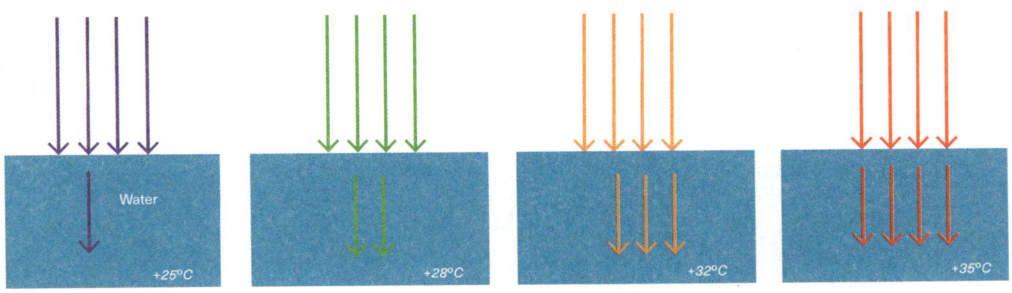

water absorbs less violet light: little heat gained

water absorbs less green light: little heat gained

water absorbs more orange light: more heat gained

water absorbs more red light: more heat gained

Thermal absorption of water {Fig. 1-195}

Water absorbs the energy of red light 100 times more than blue and even less so for violet. Since the skin is composed 70% of water, it will heat more when exposed to long wavelengths (entering the infrared) and less when exposed to short wavelengths (entering the ultraviolet). The more enter the infrared, the more the skin will heat up, which is why we use radiant heaters in the infrared.

These radiant heaters heat human skin regardless of the temperature of the air, since heat exchange with human skin takes place by radiation and not by conduction or convection. This phenomenon can be felt when you are in the mountains in winter in the sun: the air temperature can be negative (below 0°C), your face will heat up by radiation if it is directly exposed to the sun's rays.

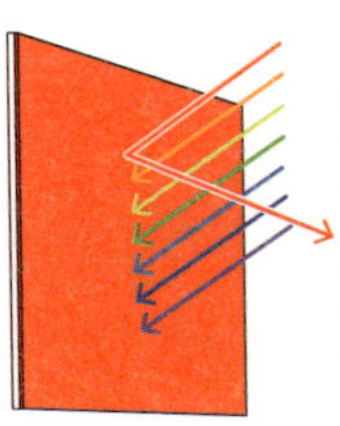

More heat gains in the skin

The red surface reflects the red part of visible light and absorbs all other wavelengths.

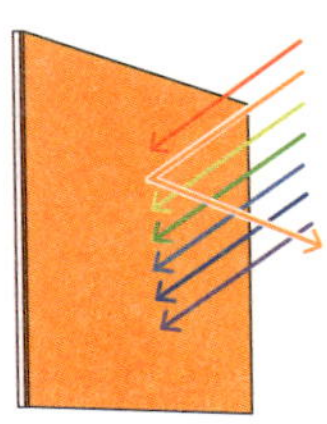

The yellow surface reflects the yellow part of visible light and absorbs all other wavelengths.

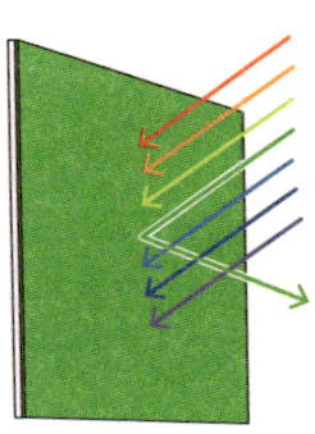

The green surface reflects the green part of visible light and absorbs all other wavelengths.

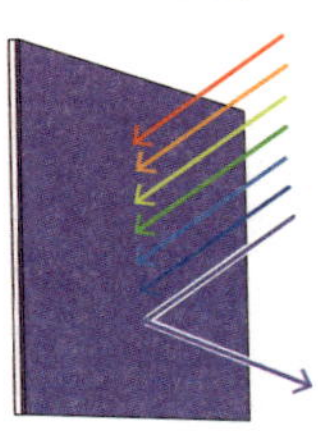

Less heat gains in the skin

The violet surface reflects the violet part of visible light and absorbs all other wavelengths.

Thermal capacity of colors {Fig. 1-196}

A surface of a given color can only reflect that same color, for example red reflects electromagnetic waves at around 700 nm, which explains why we perceive it as red. All other wavelengths are absorbed by the material and are transformed into heat. As 60-70% of our body is made up of water, the light heats our body by heating the water.

The body can heat up quickly if exposed to red and even quicker to infrared light, while it will remain cool if exposed to blue or violet light.

Albedo

roof finishing	albedo
regular weathered roof	0,07
corrugated iron	0,1 - 0,16
colored tiles	0,2
dark tiles	0,35
green roof	0,3
cool paint	0,9
floor finishing and ground	
artificial grass	0,1
black bitumen	0,02 - 0,5
limestone	0,3 - 0,45
white seal coat	0,9
wall finishing	
concrete	0,3
granit	0,35
white paint	0,5 - 0,9
wood panels	0,4
nature	
coniferous trees	0,15
deciduous trees	0,2
grass	0,3
soil	0,3
water bodies	0,1 - 0,6

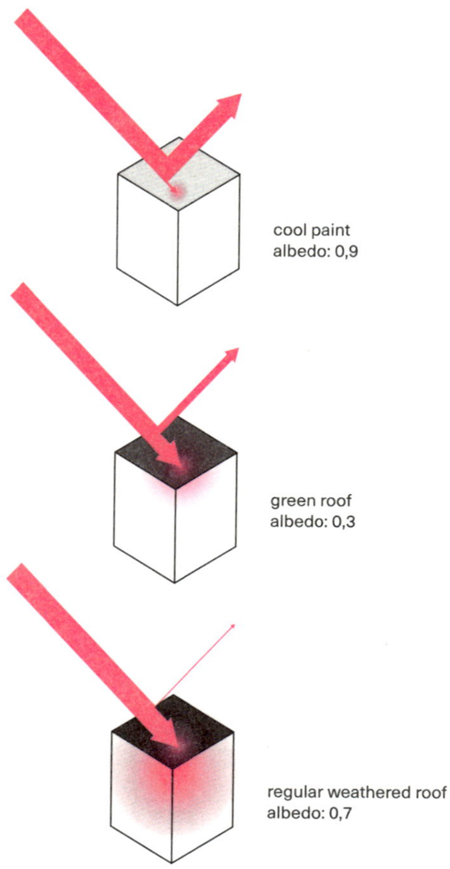

Coefficients of albedo {Fig. 1-197}
The darker the material, the more it absorbs light and turns it into heat, which warms the material and the surrounding space. Conversely, the lighter the material, the more it reflects light and the less it heats, remaining cold.

The visible part of the electromagnetic spectrum contains 50% of the solar energy reaching the Earth. The other 50% is infrared. The sun's rays are reflected or absorbed depending on the color of the materials. The darker the material, the more it absorbs the sun's rays and transforms this energy into heat, thus rising in temperature. Conversely, a lighter material will reflect a large part of the light and will not heat up as much, thus remaining cool. This coefficient of reflection is the albedo. The lower the albedo (that is to say the darker in color) the more it will heat up under the sun, rising in temperature. The higher the albedo (that is to say the lighter in color), the more it reflects the incident sunlight, remaining cool. The consequence is that a black roof or pavement will heat up more in the sun, then send this heat into the ambient air, raising the temperature of the city. Thus, to avoid hot summer overheating in the city, light materials are preferable

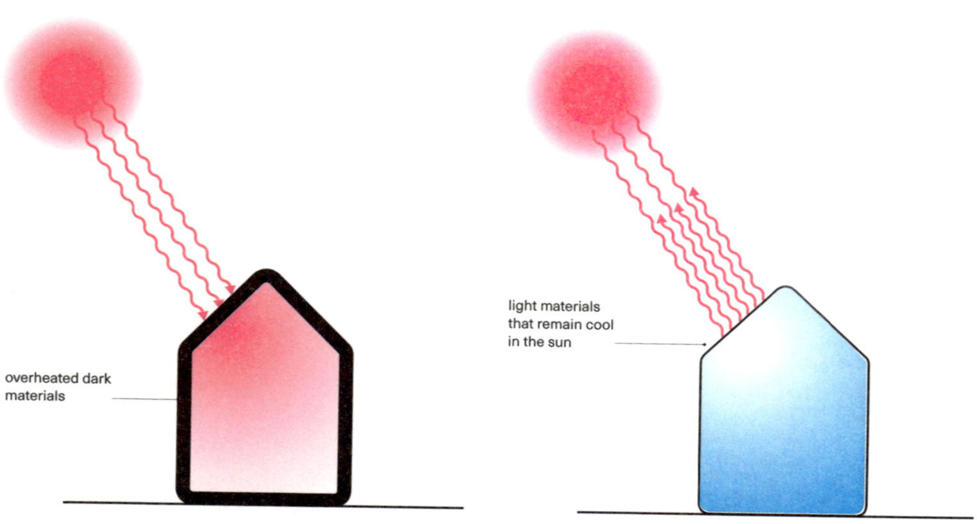

Low albedo and high albedo {Fig. 1-198}
A white or a light-colored roof reflects the sun's light toward the sky, thus remaining cool beneath the sun, also making it possible to keep the living area located under the roof cool. Paint or a black material will tend to absorb and therefore heat it, then sending this heat into the air above as well as to the space below, which will overheat.

III. Architectural Scale

Heat exchanges by radiation

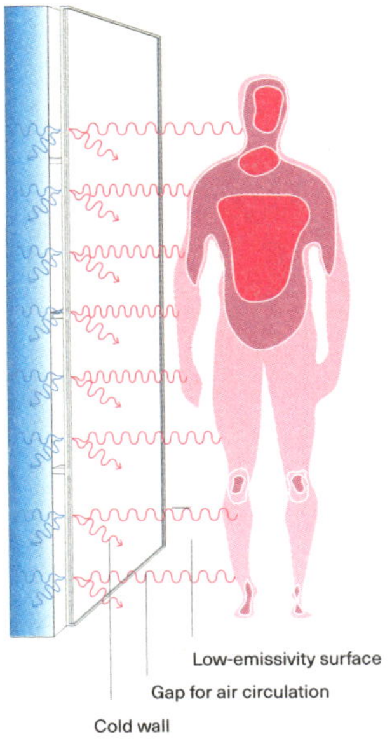

All rays emitted by the body are reflected

Low-emissivity wall {Fig. 1-199}

If the materials of a wall have low emissivity, they will not absorb the rays emitted by the human body, keeping it warm. The same thing occurs for the wall behind, the cold rays are not absorbed.

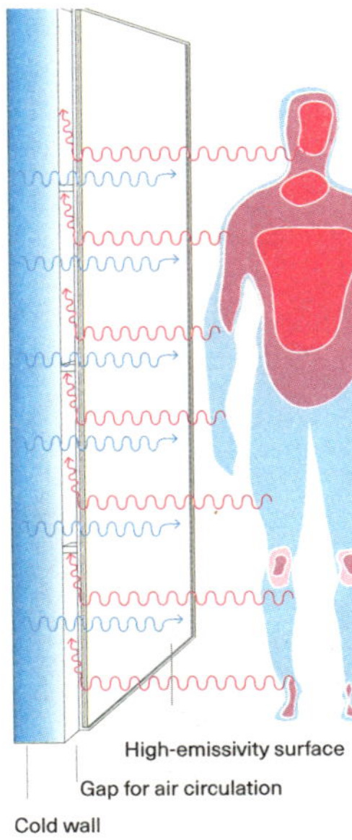

The cold rays from the wall cool the body

High-emissivity wall {Fig. 1-200}

If the materials of the wall have high emissivity, they will absorb the rays emitted by the human body, cooling it. The same thing occurs for the wall behind, the cold rays are absorbed by the walls, enhancing the coldness of the walls.

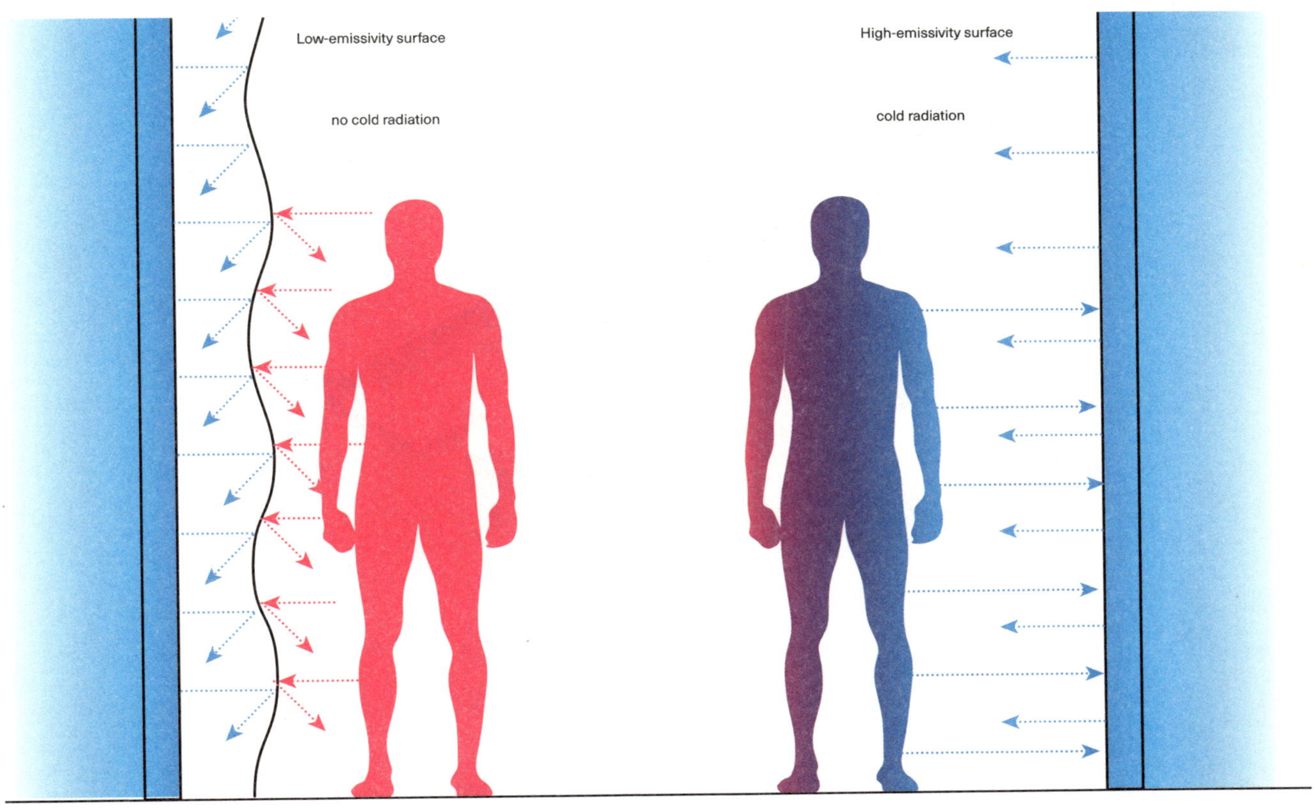

Emissivity as a factor of cooling {Fig. 1-201}

In the vicinity of high-emissivity walls, the human body loses heat and feels the cold. Its thermal energy is absorbed by the walls. By placing a low emissivity curtain, therefore with a low capacity to absorb and to emit infrared, the heat exchange between the human body and the wall is blocked and the human body no longer loses heat by radiation.

The distribution between radiation on one side and conduction/convection on the other, in the heat/cold exchange between a space and the human body is estimated at 50% / 50%. Thus, air temperature accounts for 50% of the perception by the body of the temperature of the space, the other 50% being radiant coming from walls, ceilings and floor. A space is comfortable where the temperature of the air, wall, ceiling and floor surfaces is comfortable. Thus, even if the air temperature of a room is 21°C, if the walls are cold (for example at 15°C) then the human body will lose heat by radiation toward these cold surfaces and will cool down, regardless of the air temperature. For this reason, it is important to thermally insulate the walls to keep their temperature close to 21°C as well, which will prevent the human body from cooling in winter by radiation toward cold walls. Now, if the walls are cold in spite of everything, they must be made of a low-emissivity material.

IV. Urban Scale

Albedo, Emissivity and Urban Heat Islands

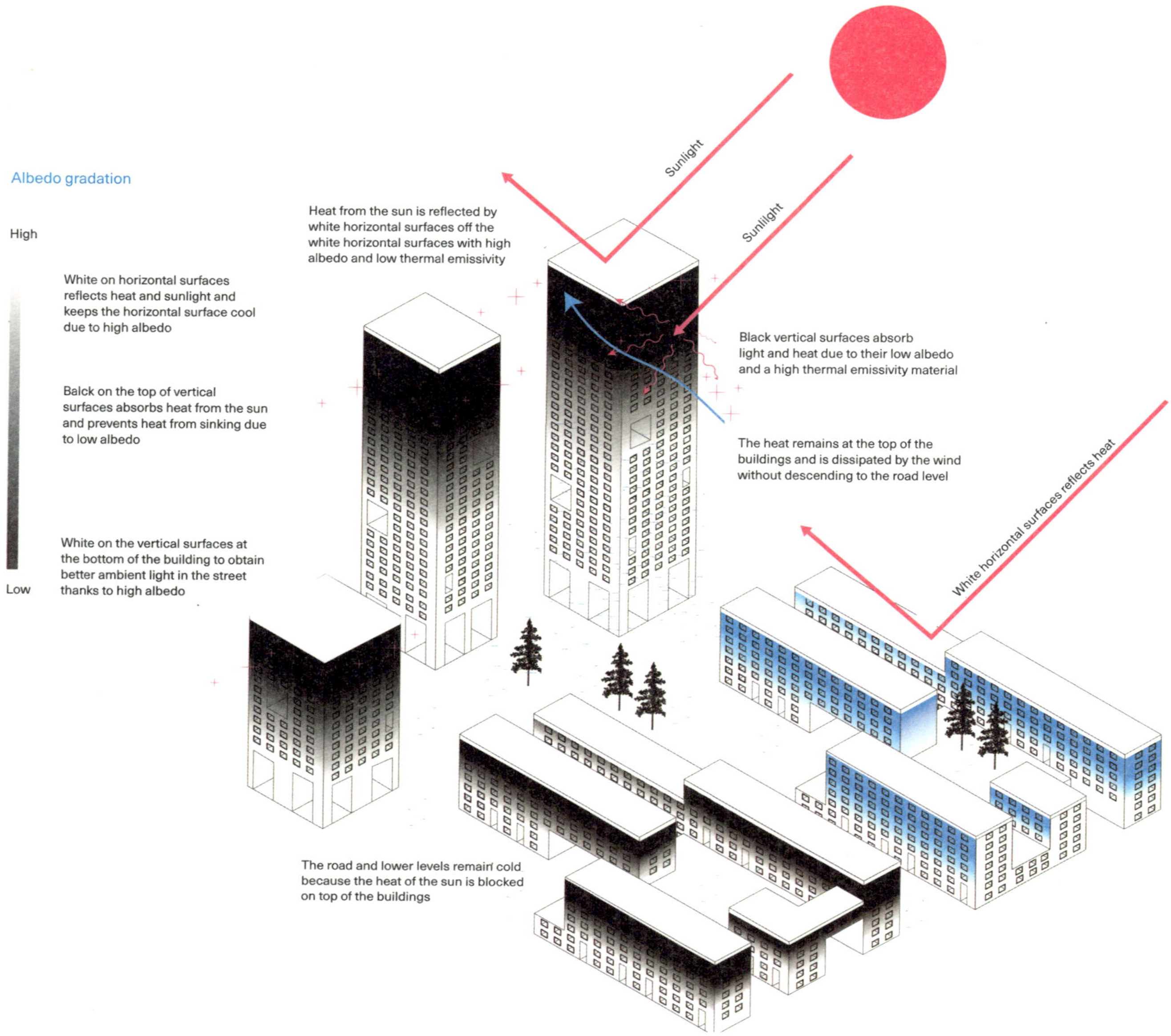

Solar rays and urban heat islands {Fig. 1-202}

To prevent cities from overheating, horizontal surfaces must be white with low emissivity to reflect the rays toward the sky. The upper part of the façades can be dark to absorb the rays before reaching the streets. The heat is thus more easily evacuated by the wind. Vertical surfaces become lighter as we descend to ensure good ambient light.

In cities, the artificial surfaces are often dark, especially horizontal surfaces (the streets are covered with tar and the roofs with bitumen). The albedo of these surfaces is therefore very low, which transforms the solar rays into heat, which accumulates in the materials and contributes to the general increase in the temperature of the cities by giving off their heat to the air by conduction. To combat this phenomenon, the surface albedo must be increased, on the one hand (tending toward light surfaces), and create smooth low-emissivity surfaces on the other, in order to bounce the sun's heat rays toward the sky before they transform into heat and thus avoid summer overheating. One can also consider the albedo of the vertical surfaces of the buildings of the city. Indeed, if the tops of the buildings are dark in color (low albedo), then via a rebound effect, we will prevent the rays from descending further down toward the street and heating it at the level of the passers-by, protecting them (as well as the materials of the pavements and roadways) from the heat of the sun's rays. The vertical crown of the buildings then becomes a trap for the sun's rays; heating high up, but this heat can be evacuated naturally by convection and by the wind. By gradation, the façades of buildings exposed to the sun will be graded from dark at the top to light at the bottom.

Urban Heat Island Effect

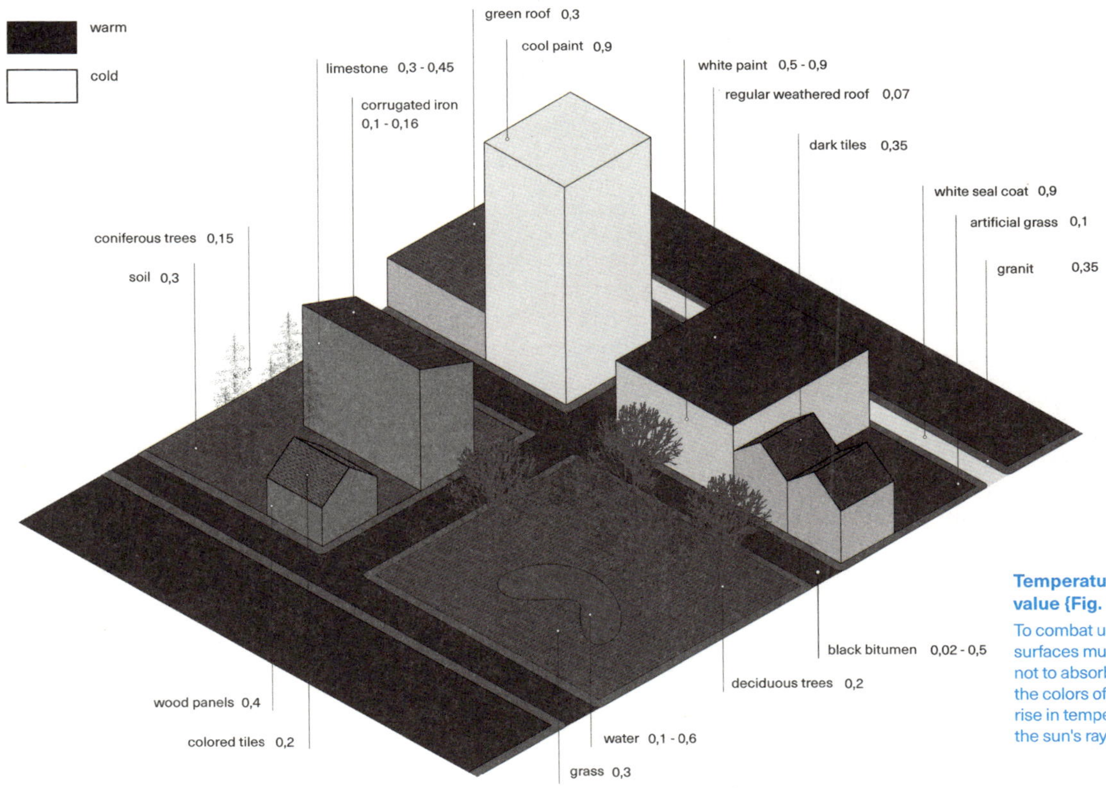

Temperature related to albedo value {Fig. 1-203}

To combat urban heat islands, horizontal surfaces must have a high albedo so as not to absorb solar rays. The lighter the colors of the surfaces, the less they rise in temperature because they reflect the sun's rays and heat.

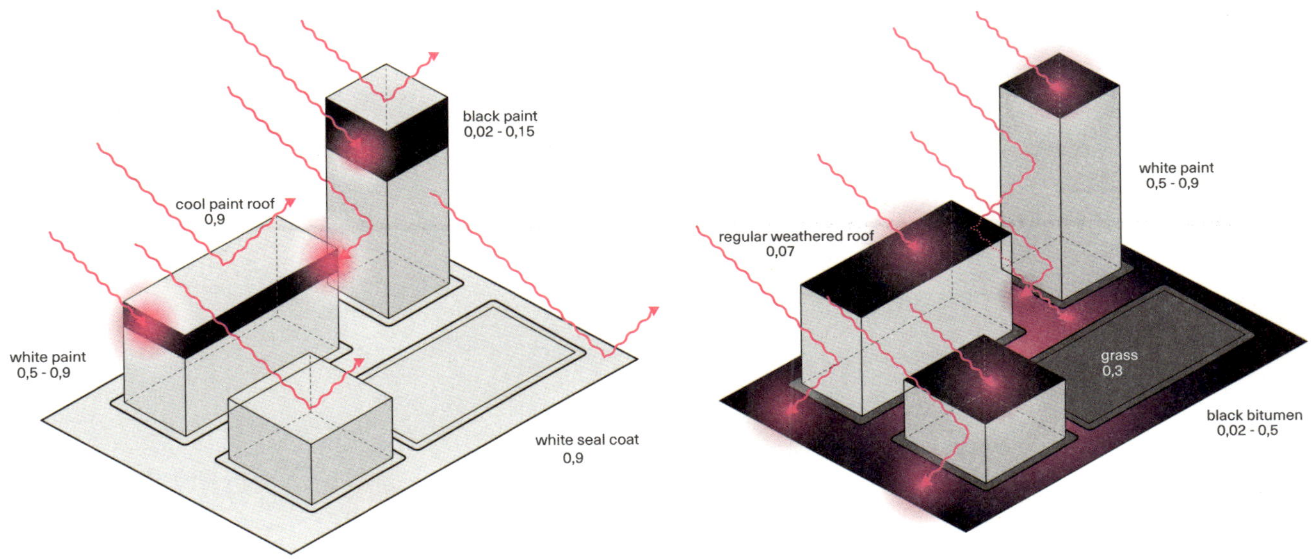

Albedo of vertical and horizontal surfaces {Fig. 1-204}

By introducing dark bands in the upper part of buildings, it is possible to create a heat trap to block solar energy at altitude, preserving the freshness of the streets we walk along.

During the summer or approaching warmer equatorial latitudes, the dark surfaces of the city and more specifically the asphalt roads and black bituminous sealing of roofs absorb the sun's energy and store it in the mass of materials, creating overheating day and night by sending heat to the air and radiating heat to other buildings and passers-by, amplifying heat waves. The temperature of cities no longer drops even at night because of thermal inertia, as the materials overheat under the sun during the day, accumulating this heat in their mass and releasing it at night, which prevents necessary night time cooling.

The albedo value of these surfaces thus plays a crucial role in regulating temperature in the city. Because of the low albedo of the materials usually used in the city (concrete, bituminous mix), their high emissivity and their thermal inertia, they bring about the phenomenon known as "urban heat islands" which see city temperatures rise excessively during the summer compared to the countryside and remain high during the night.

Albedo

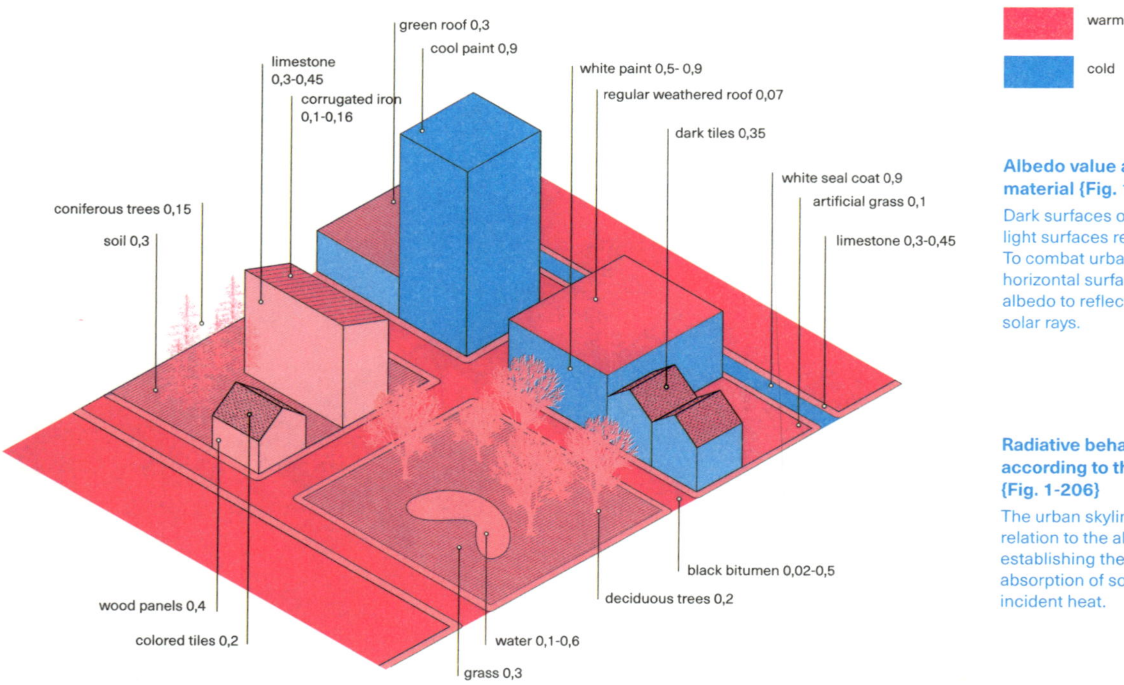

Albedo value according to the material {Fig. 1-205}

Dark surfaces overheat under the sun, light surfaces remain cool. To combat urban heat islands, horizontal surfaces must have a high albedo to reflect the heat energy of solar rays.

Radiative behavior of surfaces according to their albedo {Fig. 1-206}

The urban skyline can be composed in relation to the albedo by studying and establishing the places of reflection or absorption of solar rays and their incident heat.

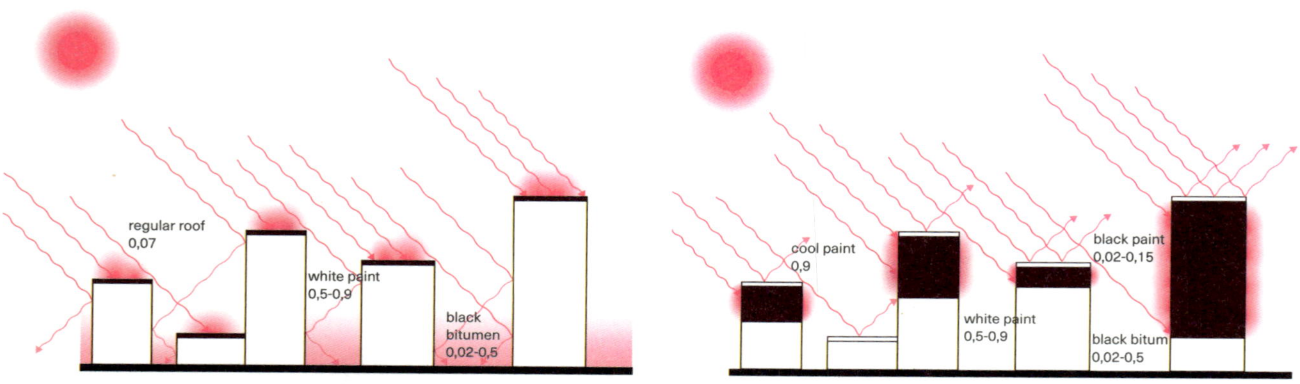

Emissivity of Buildings' Bases

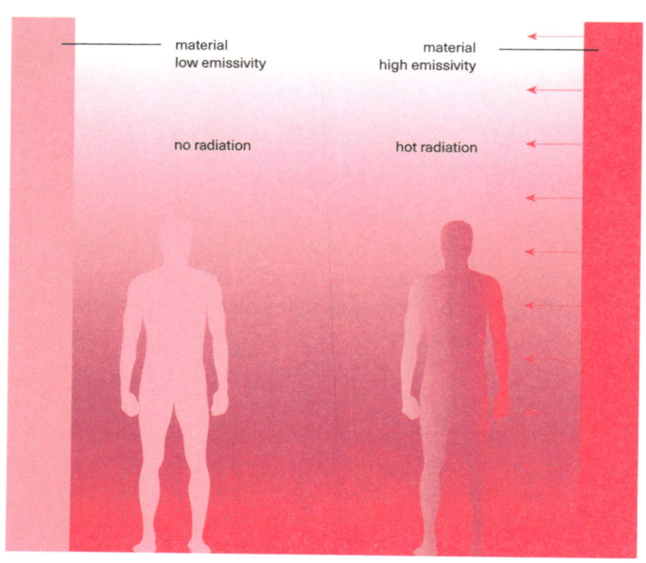

Emissivity as a factor of cooling {Fig. 1-207}

In summer, a wall made of high-emissivity materials will absorb the heat of the air and the heat of the sun's rays, then heat up and re-emit this heat by radiation to the human bodies in the vicinity, which will in turn heat up. Conversely, a wall made of low-emissivity materials will absorb the heat of the air but not the heat of the sun's rays, thus remaining cooler and moreover, will not re-emit this heat by radiation to human bodies in the vicinity, which will remain cooler.

Emissivity and Solar Gains

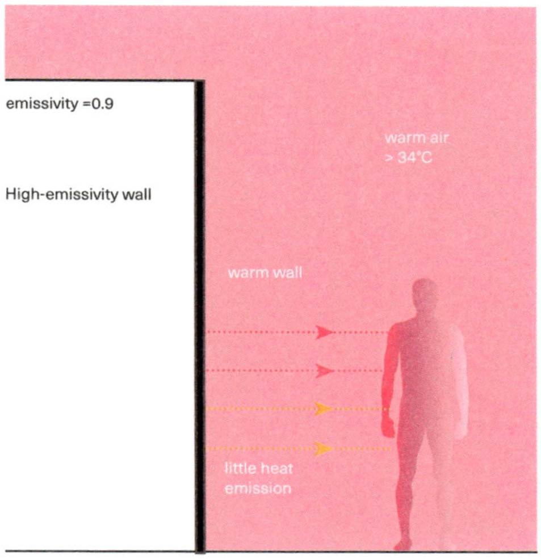

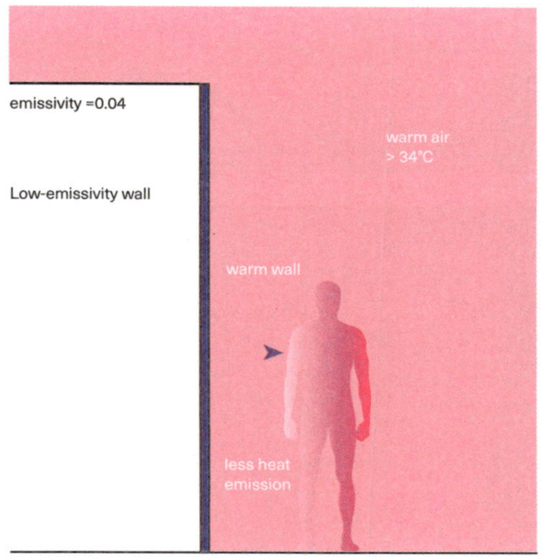

High emissivity as a factor of overheating {Fig. 1-208}
The hot high-emissivity wall (heated by air temperatures above 34°C) emits many infrared rays that heat bodies nearby.

Low emissivity as a factor of thermal neutrality {Fig. 1-209}
The hot low-emissivity wall (heated by air temperatures above 34°C) does not emit infrared rays and does not heat bodies nearby.

The hot solar rays of the summer (>34°C) and the high air temperature heat the walls which will more or less re-emit this heat by radiation depending on whether their emissivity coefficient is higher or lower. The wall is warmest when the outside temperature is high, it is exposed to the direct rays of the sun, it is dark, and its emissivity is high. By creating a building façade whose lower part, at human height, is low emissivity, it will absorb less heat by radiation during the day and also emit less heat by radiation, which will reduce the excess radiant heat at street level and will improve the thermal comfort of passers-by in the vicinity of the wall.

short wavelengths (toward the ultraviolet) long wavelengths (toward the infrared)

Colors and calorific value {Fig. 1-210}
The longer the wavelengths, the more their energy is absorbed by water molecules, increasing the temperature of the water

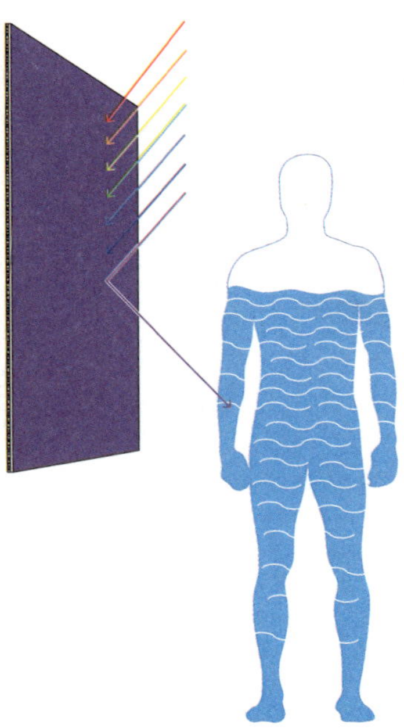

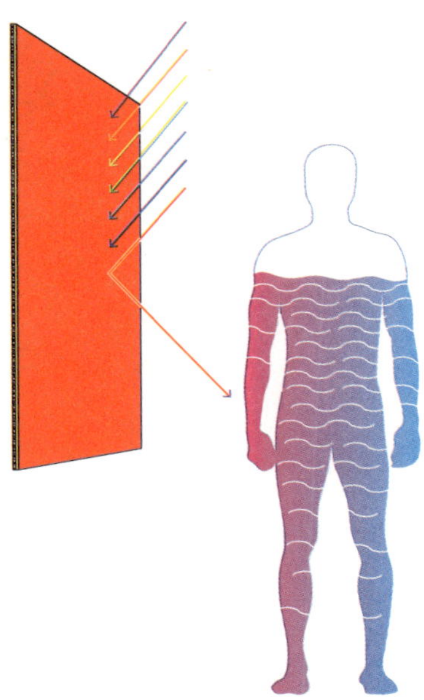

Short wavelengths as cool radiation {Fig. 1-211}
Walls painted violet prevent the body in the vicinity from heating because this color blocks the emission of long wavelengths which alone heat the water of which the human body is made, keeping it free of radiative heat input

Long wavelengths as heating radiation {Fig. 1-212}
Walls painted red heat the body nearby by emitting long-wavelength rays whose energy is absorbed by the 65% of water of the skin.

V. Case Study

Gradation of Emissivity

Simulation of heat transfers

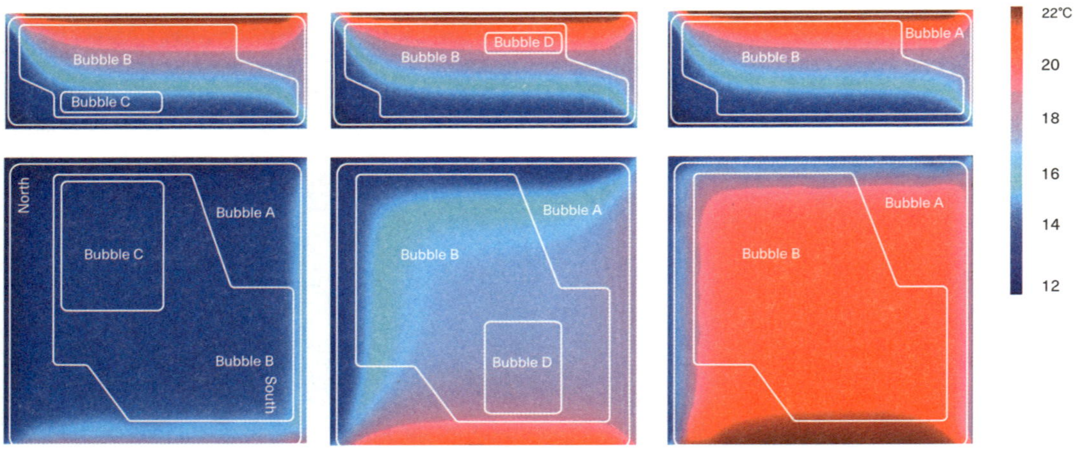

Plan at 1.5 m

Plan at 12 m

Plan at 16 m

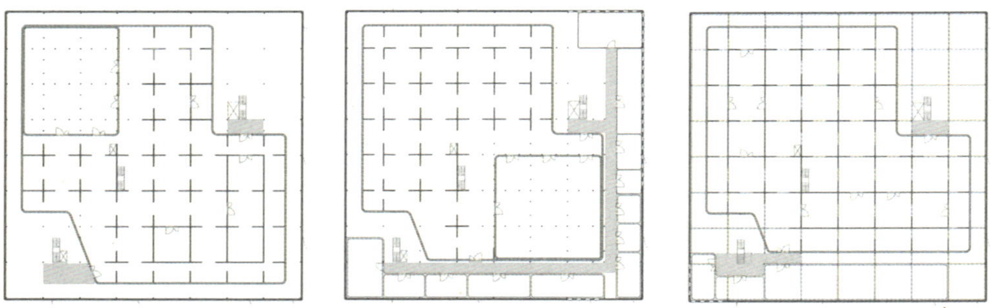

Plan level 0

Plan of level +1

Plan level +2

Plans {Fig. 1-213}

Façade panels

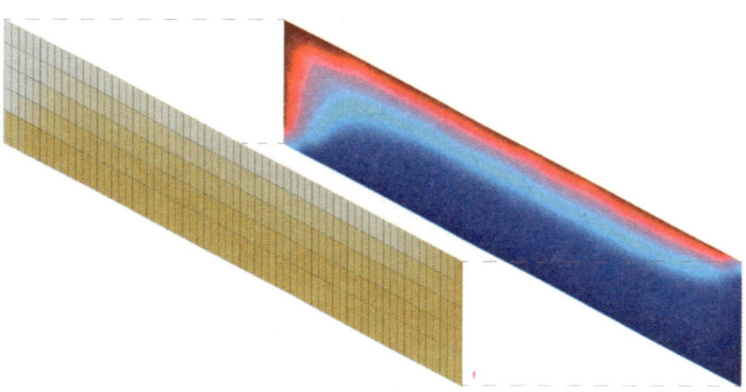

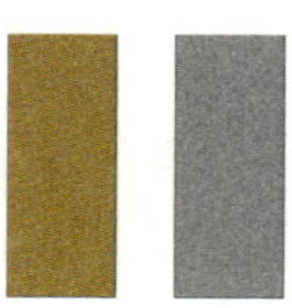

The golden panels absorb infrared which allows heating the coolest areas of the building. Conversely, the silver panels reflect the rays and prevent the building from heating.

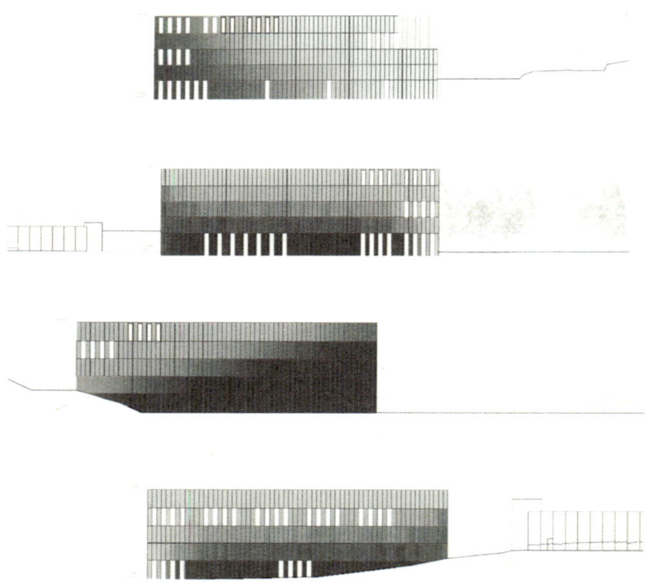

Elevations & Façades {Fig. 1-215}

Emissivity of façade materials {Fig. 1-214}

The materials are chosen according to their emissivity in order to regulate the internal temperature and to heat the parts least exposed to the sun.

The façade is composed of prefabricated elements made of different materials whose surfaces follow a gradation, from less emissive for the highest parts to more emissive, for the lower parts. Like a survival blanket, the golden tiles absorb heat and infrared rays, while the silver ones send them back to the sky. By following a precise gradation of solar radiation during the day and year, according to a computer model, it is possible to place the gold and silver panels appropriately. There will be a greater density of golden panels at the bottom of the façades and to the north of the buildings, while the top of the façades and those facing south are fitted with more silver panels.

Varied emissivity flooring

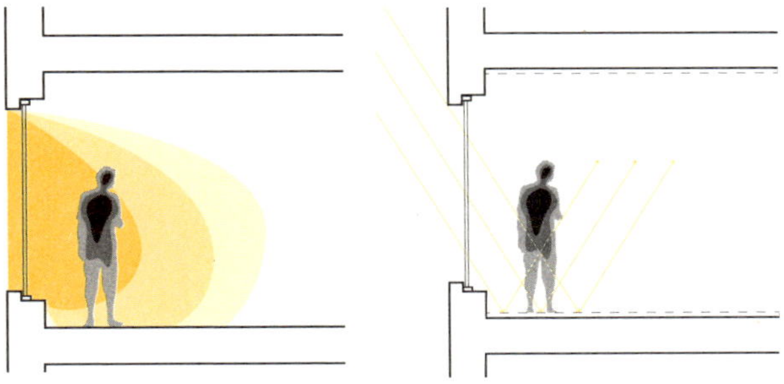

High reflection {Fig. 1-216}

The surfaces of the floor that receive direct sunlight have a high reflection rate (albedo) in order to reflect the light inwards, where it is naturally darker.

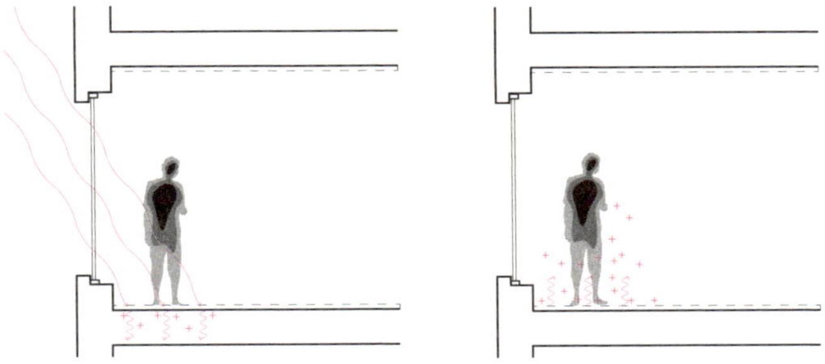

High emissivity, high inertia {Fig. 1-217}

For floors that receive direct sunlight, and therefore infrared light, high-emissivity, high-inertia flooring is chosen in order to absorb and accumulate the heat of the sun during the day and to diffuse it at night.

The materials for the floor are chosen according to their particular exposure to the sun. Where the sun touches the floor directly, a high-emissivity material is chosen to absorb infrared rays and store its heat in the mass of the floor for later diffusion by thermal inertia. Elsewhere, where the sun does not touch the floor, low-emissivity materials will be chosen so as not to emit colder rays. The aim is to take better advantage of the sun's rays while preventing the space from heating by shifting the phase of energy transmission by the inertia of the material.
This project reassesses the art of floor layout, of decorative marquetry of the flooring in relation to natural thermal gains and the heat management of the interior space.

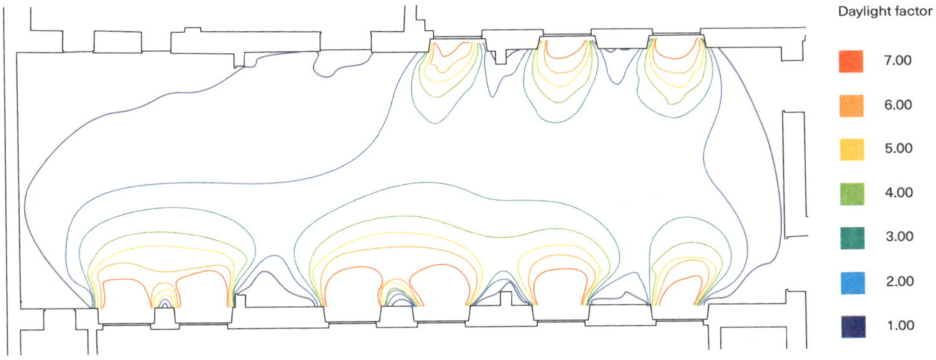

Heat plan - daytime

Façade receiving direct sunlight

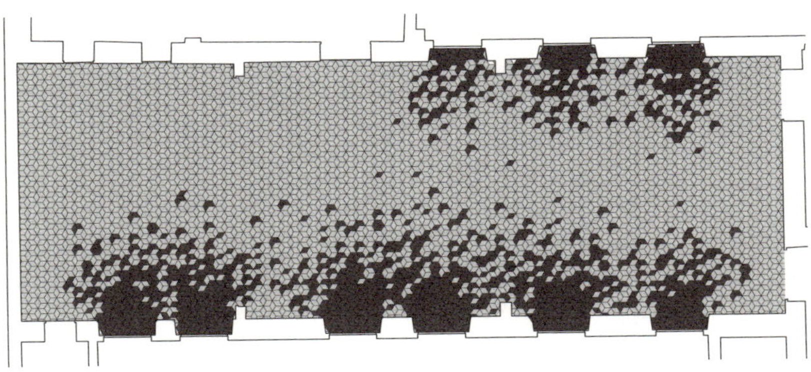

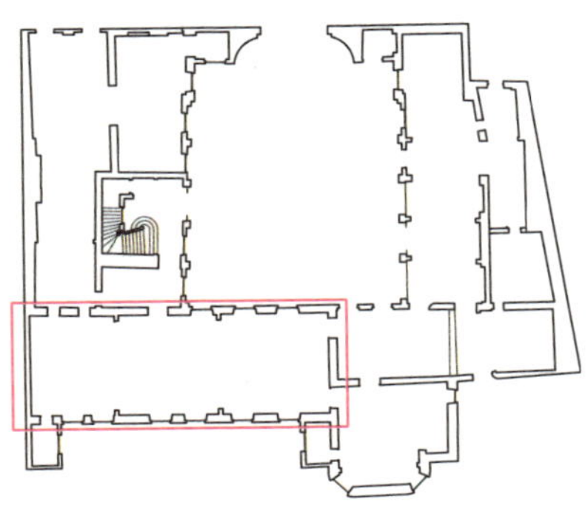

high effusivity and thermal inertia

low effusivity and thermal inertia

Floor plan {Fig. 1-218}

In black, the material absorbs infrared and reflects visible light; In white, the material reflects the entire light spectrum.

Floor emissivity paving {Fig. 1-219}

The areas of the floor directly receiving the sun's rays are covered in a highly emissive, highly thermally inert material.

Conversely, the areas of the floor that never receive sunlight are covered with a material that is low in emissivity and low in thermal inertia.

A Low-Emissivity Ceiling for Winter

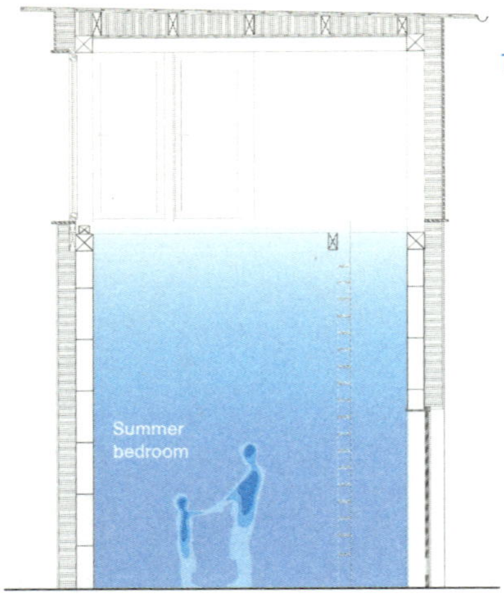

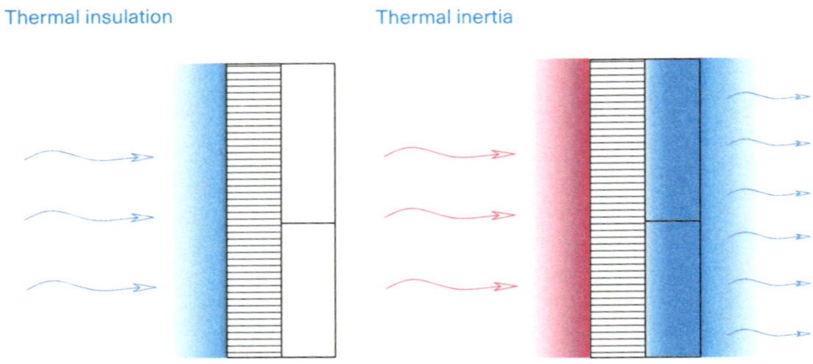

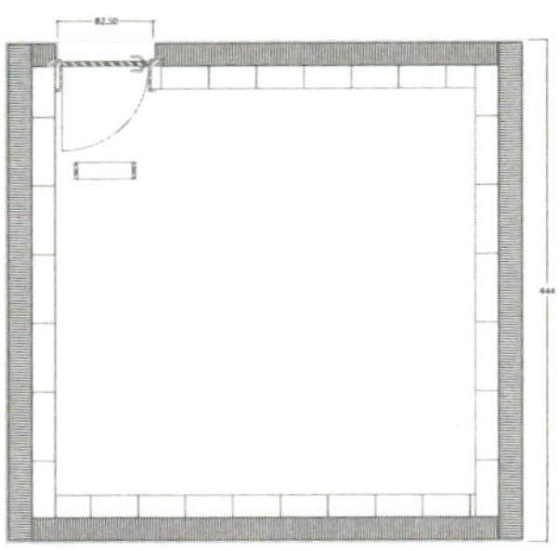

Ground-floor plan {Fig. 1-121}
(summer floor)

Thermal Inertia / Thermal Emissivity. Thermal Effusivity & Thermal Convection Situation in Summer {Fig. 1-220}

The surfaces of the floor that receive direct sunlight have a high reflection rate (albedo) in order to reflect the light inwards, where it is naturally darker. During the day, the stone, protected from the summer daytime heat by thermal insulation, will, thanks to its thermal inertia, radiate the night time coolness accumulated in the stone, inwards, keeping the space cool. In addition, the high thermal emissivity stone of the vertical walls, will generously radiate this coolness to nearby human bodies, thus cooling them by radiation.

As for the heat, thanks to the great height of the space, it rises, keeping ground level space cool. The material of this lower space is stone, chosen for its high thermal effusivity allowing the human body to cool down when in contact with it. The material of this lower space is stone, chosen for its high thermal effusivity allowing the human body to cool down when in contact with it.

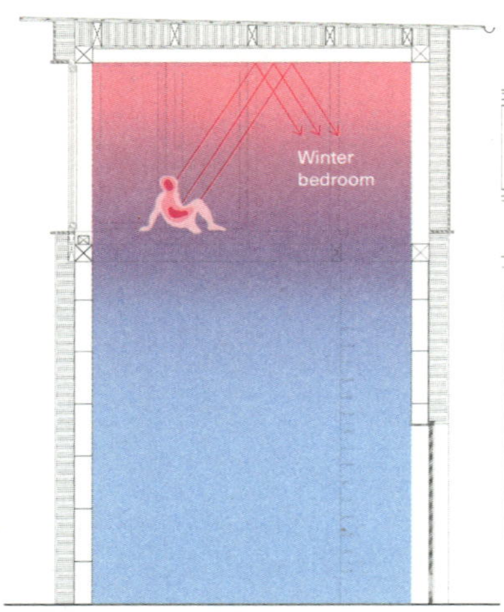

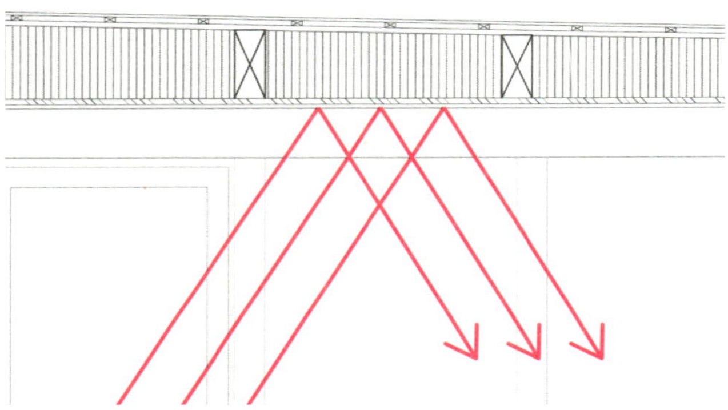

Thermal Emissivity, Thermal Effusivity & Thermal Convection Situation in Winter {Fig. 1-222}

The material of the ceiling is made of copper chosen for its low thermal emissivity allowing the human body not to cool by radiation. The floor material this upper space is wood, chosen for its low thermal effusivity allowing the human body not to cool down when in contact. Because the warm air rises due to thermal convection, the winter living space is at the top, with a low ceiling, allowing living in warmth.

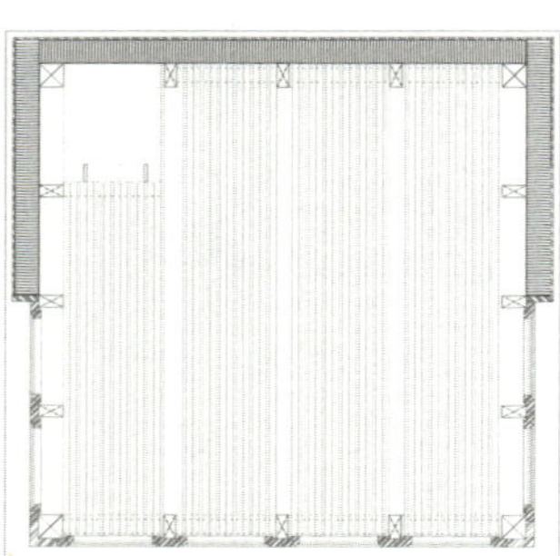

1st-floor plan {Fig. 1-223}
(winter floor)

EFFUSIVITY

I. Definition

General Properties

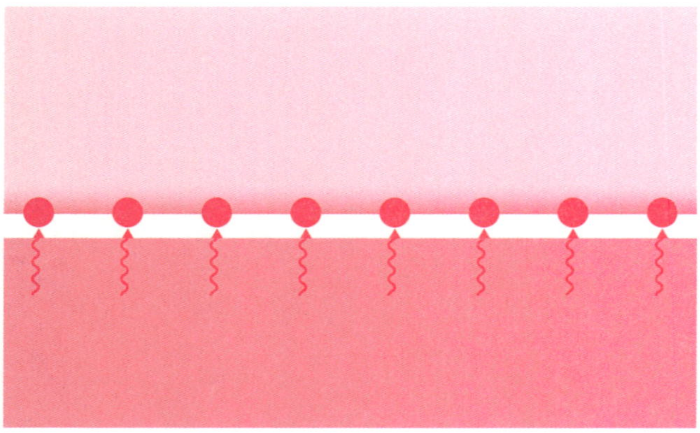

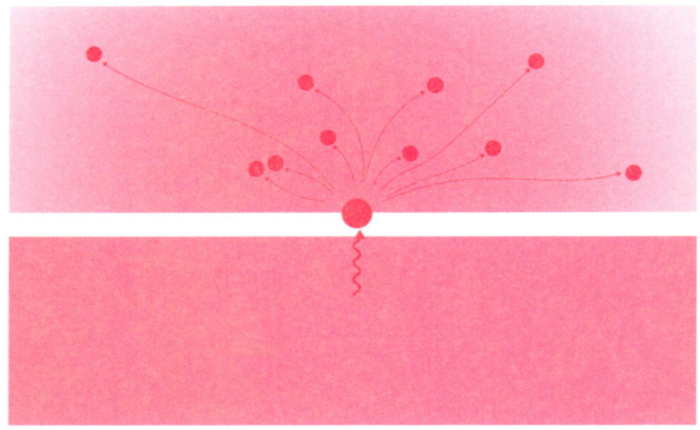

Effusivity {Fig. 1-224}
Effusivity is the flow of heat near the contact surface of two objects at different temperatures. The surface temperature will probably be similar to that of the material with the greatest effusivity.

Diffusivity {Fig. 1-225}
Diffusivity is the heat flux within and through the material. It depends on the material's thermal conductivity, density and thermal capacity.

Thermal effusivity (also known as the heat penetration coefficient), is the rate at which a material can absorb heat. This property determines the contact temperature between two objects that are in contact. Thermal effusivity is the measure of a material's ability to exchange thermal energy with its environment. It is determined by an object's thermal conductivity, density and thermal capacity. Effusivity occurs when two objects come into contact. The greater the effusivity, the faster the heat exchange will take place with the environment. When two objects come into contact, the contact temperature is closer to that of the material of greater effusivity.

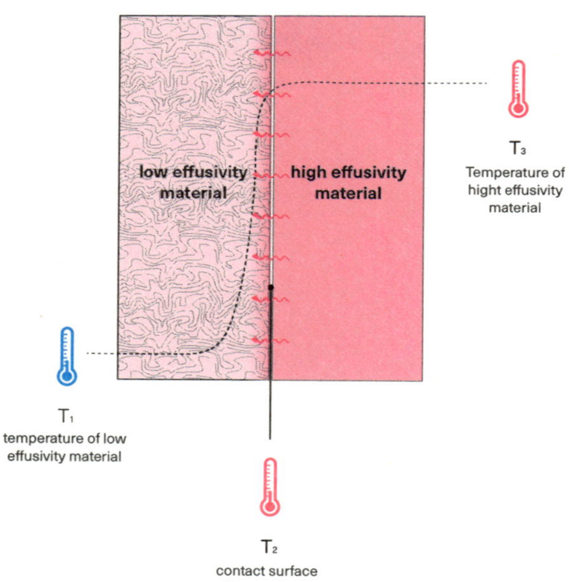

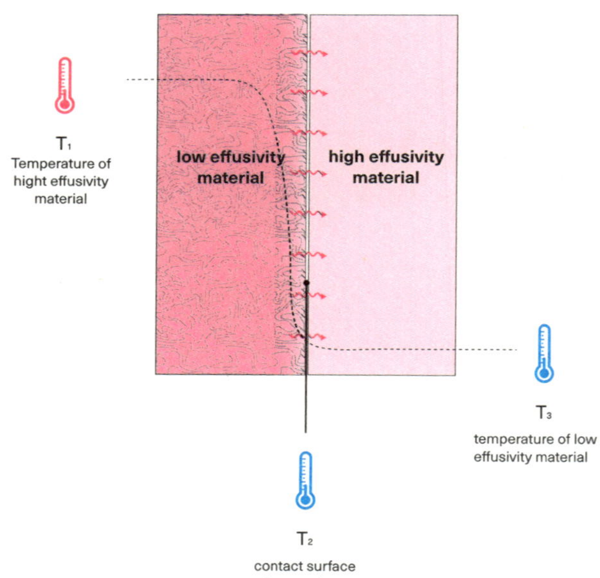

Heat transfer {Fig. 1-226}
If at a high temperature a highly effusive material encounters a low temperature in a low effusivity material, the contact temperature will be closer to that of the former

Heat Transfer {Fig. 1-227}
If at a low temperature a highly effusive material encounters a low temperature in a low effusivity material, the contact temperature will be closer to that of the former

II.Properties of materials

Effusivity

$$E = \sqrt{k \cdot p \cdot c_p}$$

k: thermal conductivity of the material (in W m−1 K−1)
ρ: volumetric weight of the material (in kg m−3)
c: specific heat capacity of the material (in J kg−1 K−1)
E: Thermal effusivity (in J K−1 m−2 s−1/2)

Effusivity is a thermal property that is present in all materials in all forms: solid, liquid, gaseous. Effusivity is the property that determines the contact temperature when two objects of different temperatures touch. It combines thermal conductivity, density and heat capacity in a single value that determines the ability of a material to exchange heat with any substance it encounters. When two materials come into contact, the temperature of the interface will quickly approach the values below and will be closer to that of the material of greater effusivity.

Materials with high effusivity, therefore, have the ability to rapidly cool or heat materials with low effusivity. This is why plastic resin (high temperature, low effusivity) solidifies almost instantaneously when it comes into contact with a metal mildew (low temperature, high effusivity). Objects with low effusivity will take longer to transmit or receive heat.

Examples of coefficients of effusivity

Material	Effusivity (J/m2·K·s0.5)
Copper	35849
Aluminum	23377
Steel	13248
Cement	2500
Stone	1154
Wood	635
Human skin	400
Cork	67
Wool	44

Material

Aluminum has the highest effusivity
Stone has a medium effusivity
Cork has a low effusivity

Aluminum is a sensitive element between the temperature of the body and cement. The surface contact temperature will be closer to the temperature of the material of greater effusivity (therefore aluminum).

T human body = 37 E human body = 400
T aluminum = 20 E aluminum = 23377
T stone = 20 E stone = 595
T wool = 20 E wool = 44

Temperature perceived when touching aluminum
T = (E1*T1+E2*T2)/(E1+E2)
T = (23377*20+400*37)/(23377+400)= 20.28°C

Temperature perceived when touching stone
T=(1154*20+400*37)/(1154+400)=24.38°C

Temperature perceived when touching wool
T=(44*20+400*37)/(44+400)=35.31 °C

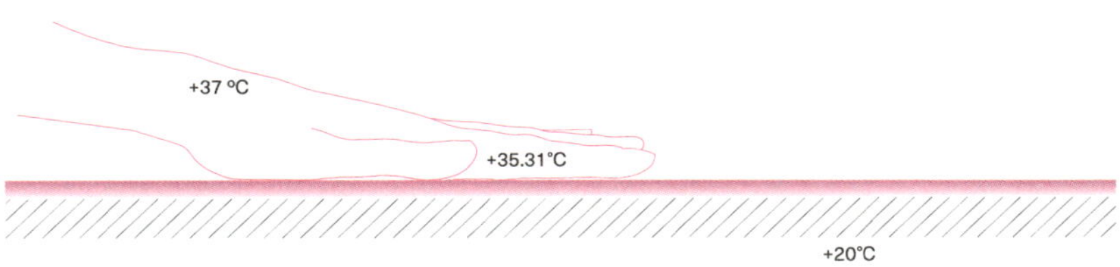

Temperature perceived when touching wool

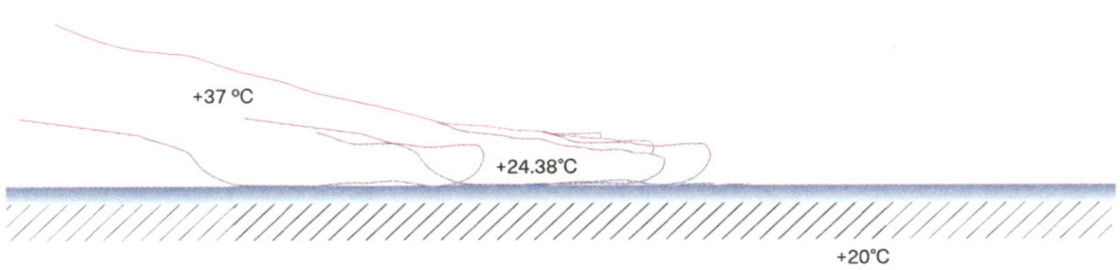

Temperature perceived when touching stone

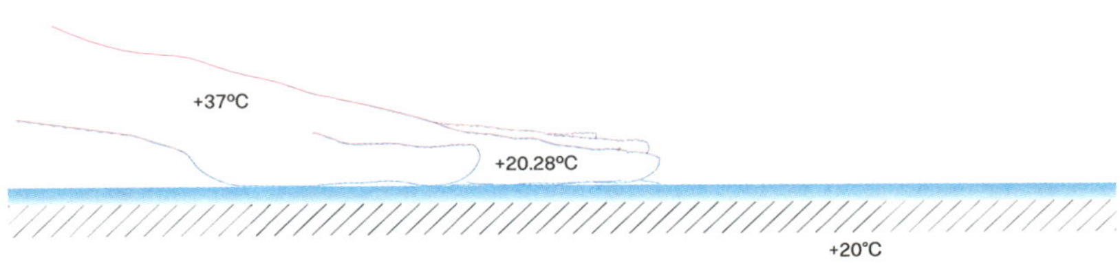

Temperature perceived when touching aluminum

III.Physiological Scale

Temperature and Touch

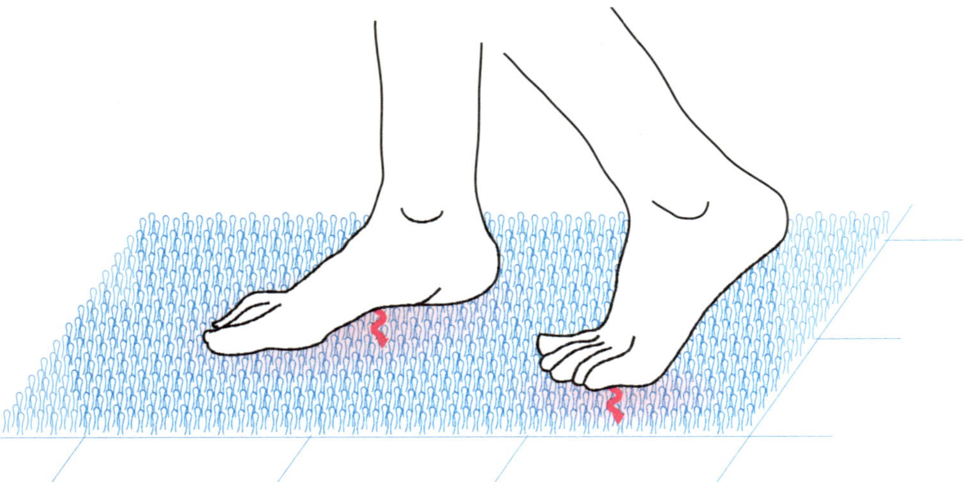

Low effusivity material {Fig. 1-228}

Touching a low effusivity material slows the heat exchange between the two materials and no heat is lost by the skin. The floor feels warm. In the same way, a low effusivity floor (such as wood or a woolen carpet) will very slowly exchange its heat with the air in contact, thus neither heating nor cooling it.

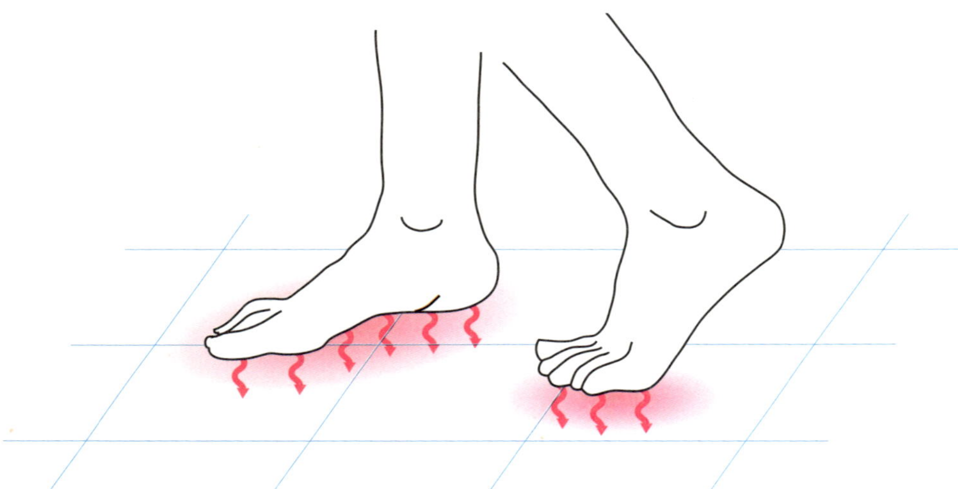

High effusivity material {Fig. 1-229}

Touching a high effusivity material speeds up heat exchange between the skin and the floor resulting in a loss of heat by the skin. The floor feels cold. Similarly, a high effusivity floor (such as marble or metal) will quickly exchange its heat with the air in contact, heating or cooling it, and vice versa.

Our physical sensation of heat is independent of the absolute temperature of the material. The physical process that influences heat perception is the rate of heat transfer known as effusivity. This means that metals (which quickly receive heat from the skin) will feel cold to the touch, while wood (which does not conduct heat as well or as quickly) will feel warmer, even if both materials are at the same temperature. Thermal conductivity (k) is involved in the hot or cold sensation of an object which also depends on the thermal capacity of the specific mass (Cp) and of the density (ρ). These quantities can be combined in a factor called thermal effusivity defined as the square root of the product of the material's thermal conductivity and its volumetric heat capacity.

Even when two materials are at the same temperature, metal feels colder. For example, wool is not a good heat conductor (low effusivity) so it will slowly absorb the heat from our hand, protecting us from conductive cooling.

Metal has a greater effusivity and will quickly absorb the heat from our hand, creating a cold sensation, actually cooling us down. However, our perception of temperature depends on other variables such as absolute temperature and contact time. Thus, at room temperature, wooden objects feel warmer than metal ones when contact is short, but if contact is prolonged, the perceived temperature becomes the same. Nevertheless, the heat exchange between the skin and a low effusivity object will remain low, preventing the human body from cooling by conduction.

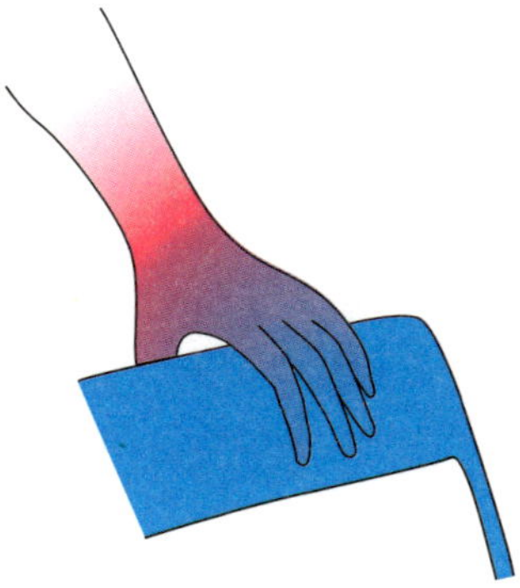

Metal chair

High effusivity material {Fig. 1-230}

Touching a high effusivity material accelerates heat exchange between the skin and the object, resulting in actual heat loss through the skin. The object feels cold.

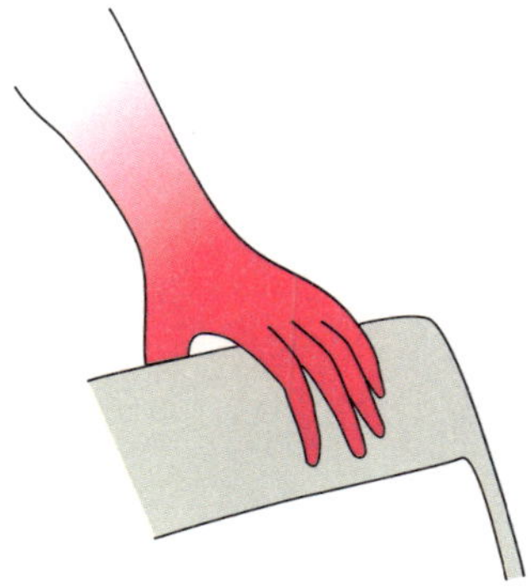

Wooden chair

Low-effusivity material {Fig. 1-231}

Touching a low-effusivity material causes little or no heat exchange between the skin and the object, resulting in thermal neutrality. The object feels warm.

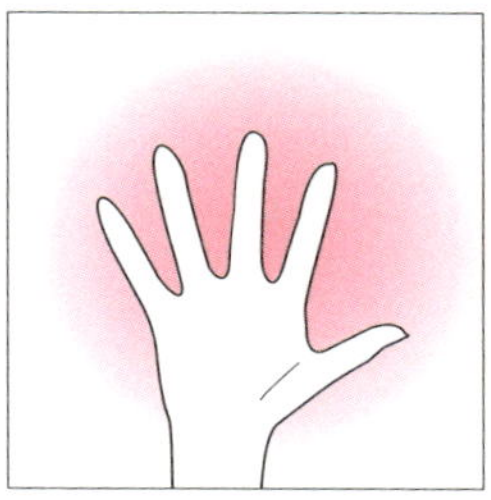

Wool
Perceived temperature: 32.6°C
Temperature of the material: 20°C
Temperature of the hand: 34°C

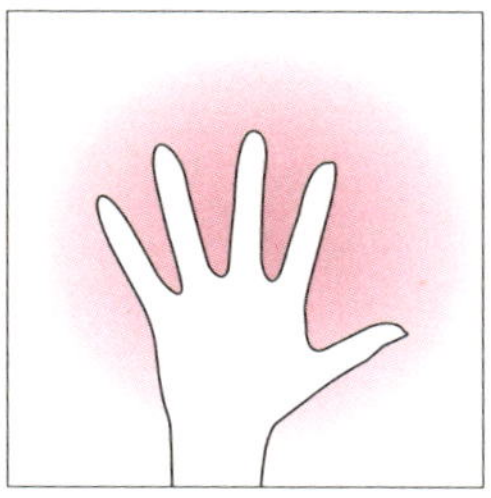

Cork
Perceived temperature: 31.9°C
Temperature of the material: 20°C
Temperature of the hand: 34°C

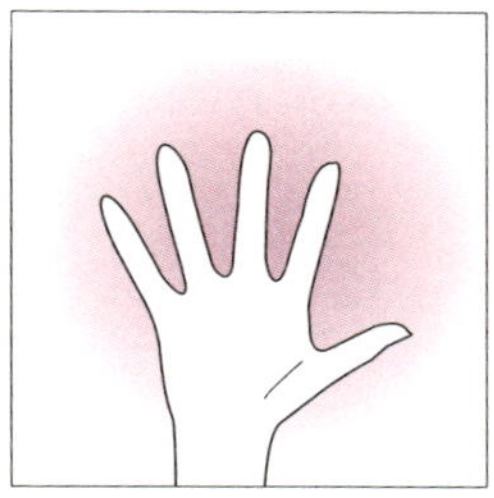

Earth and grass
Perceived temperature: 27°C
Temperature of the material: 20°C
Temperature of the hand: 34°C

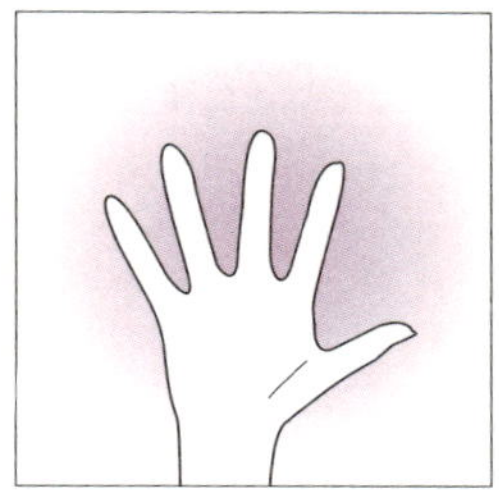

Wood
Perceived temperature: 25.4°C
Temperature of the material: 20°C
Temperature of the hand: 34°C

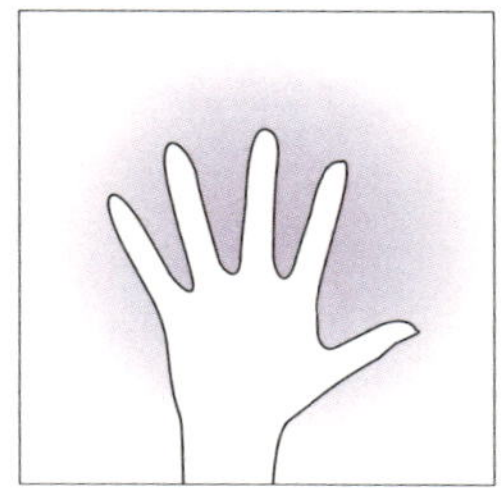

Stone
Perceived temperature: 23.6°C
Temperature of the material: 20°C
Temperature of the hand: 34°C

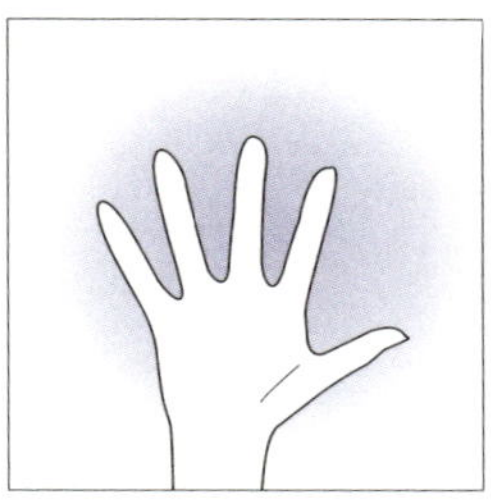

Brick
Perceived temperature: 22°C
Temperature of the material: 20°C
Temperature of the hand: 34°C

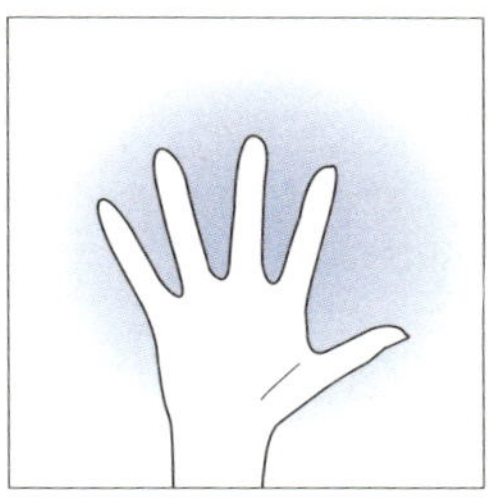

Cement
Perceived temperature: 21.9°C
Temperature of the material: 20°C
Temperature of the hand: 34°C

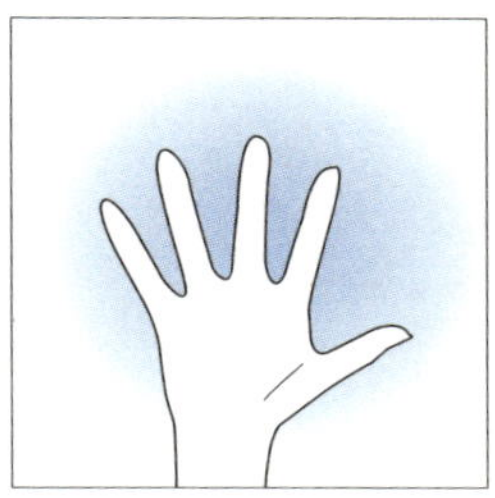

Ceramic
Perceived temperature: 21°C
Temperature of the material: 20°C
Temperature of the hand: 34°C

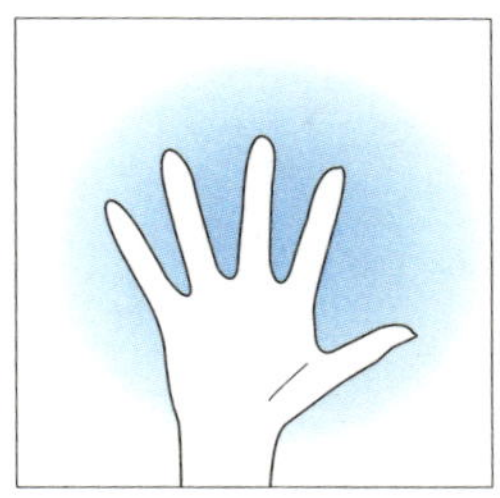

Steel
Perceived temperature: 20.4°C
Temperature of the material: 20°C
Temperature of the hand: 34°C

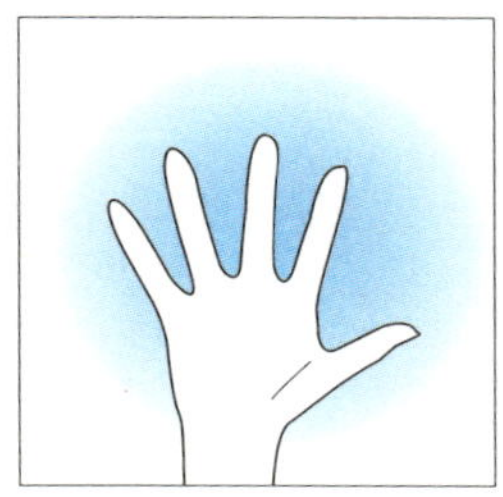

Aluminum
Perceived temperature: 20.2°C
Temperature of the material: 20°C
Temperature of the hand: 34°C

Perception of heat according to the effusivity of the material {Fig. 1-232}

Eh: effusivity of the human body
Th: skin temperature
Em: effusivity of the material
Tm: material temperature
Perceived temperature
Tf = (Eh*Th+Em*Tm)/(Eh+Em)
At the same temperature, a high-effusivity material will feel colder than a low-effusivity material.

This sensation is due to a real heat exchange between the hand and the material. Indeed, by touching a high-effusivity material such as aluminum, at a temperature of 20°C for example, the skin, which is warmer, at 34°C, will actually lose heat to the material and cool down.

Effusivity comes into play in two types of heat exchange in architecture. The first is that of the conduction between the skin (or the soles of shoes) and the ground. We can also mention touching a door handle, or a chair. In winter, low-emissivity materials should be preferred so as not to cool by conduction when coming into contact with the materials. A bus stop in the Nordic countries should therefore be made of a low effusivity materials. Similarly, the floor in bathrooms should have a low effusivity to avoid getting cold when walking on it. Conversely, if the air temperature is below 32°C, and the material is not directly exposed to the sun, it will be preferable to choose a high-emissivity material in summer so that our body can discharge its excess heat. The second type of heat exchange between two materials relating to effusivity is conduction, but with air. Emissive flooring cooled by the subsoil will send its coolness into the air and help refresh the space and consequently the inhabitants immersed in the air. Low-emissivity flooring will remain neutral and will not warm up if the air is hot.

Heat loss or Retention by Effusivity

High effusivity material {Fig. 1-233}

By touching a cold material whose effusivity is high, our body, at 37°C, will quickly lose heat, cooling us by conduction. This can prove pleasant in summer, but not in winter.

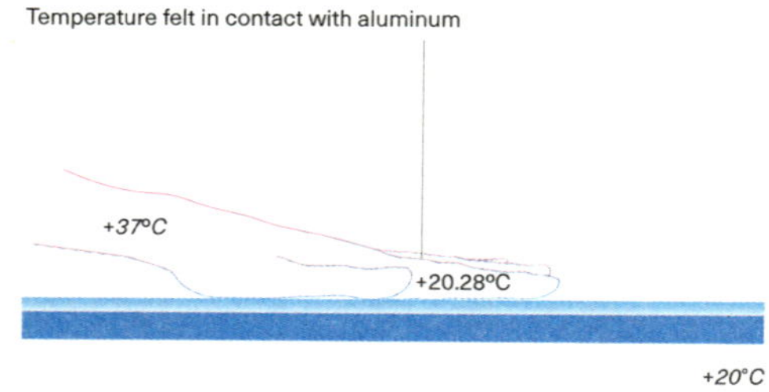

Medium effusivity material {Fig. 1-234}

By touching a medium effusivity material, our body, at 37°C, will slowly lose heat. We will therefore be slightly cooled by conduction.

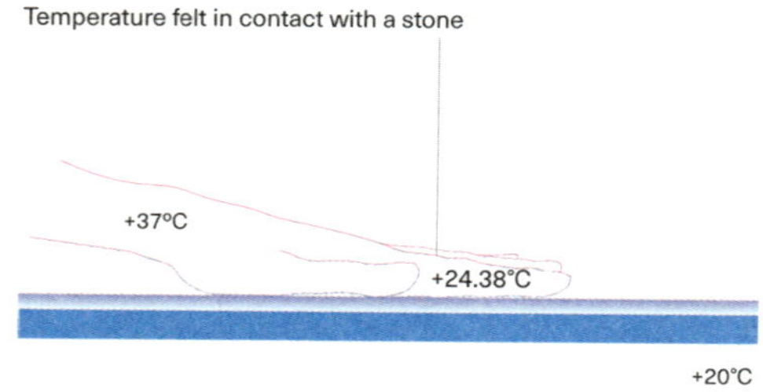

Low-effusivity material {Fig. 1-235}

By touching a low effusivity material, our body, at 37°C, will not lose heat. It can be pleasant in winter, but in summer it can be a source of discomfort.

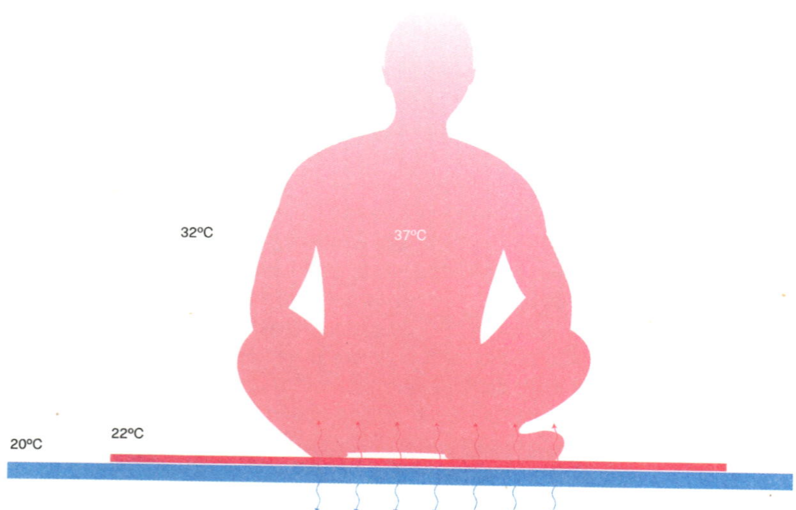

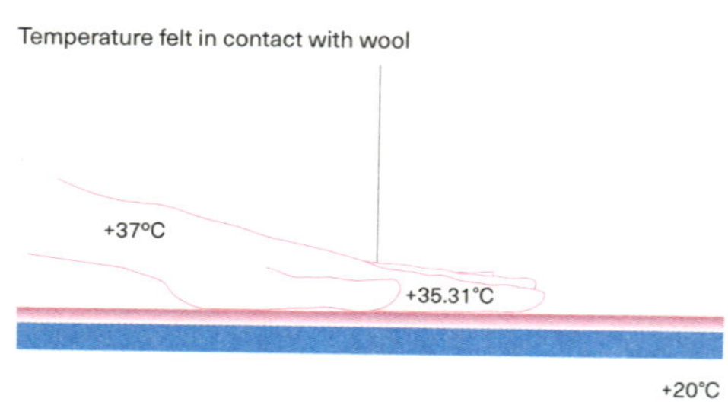

IV. Urban Scale

Effusivity and Urban Heat Islands

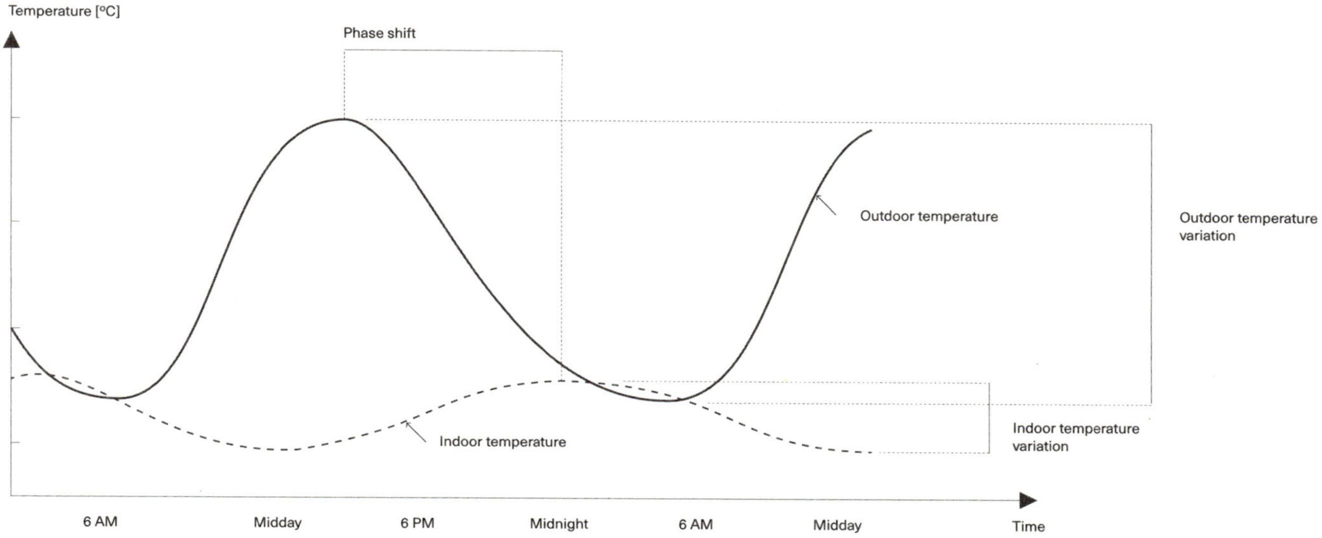

Daytime temperature variations due to thermal inertia {Fig. 1-236}

Thermal inertia shifts the temperature of the day at night and vice versa. The heat wave phenomenon in the city is reinforced by the absorption of heat by the materials of which the city is built (streets, pavement, façades) which will fully absorb heat during the long summer days and remain warm during the night, thus preventing cooling during the short nights.

Thermal effusivity is involved in the principle of thermal inertia. A high-effusivity material will have great inertia because the heat exchange with an object is immediate, whether this object is solid (a foot, a stone) or intangible (air, water). In order to reduce the urban heat island effect, low-effusivity materials should be preferable because their low inertia does not allow them to store heat and release it at night, which allows the city to cool down during summer nights and not create the heat wave phenomenon which involves daytime temperatures above 30°C and night time temperatures not falling below 18°C. High thermal inertia, with a low albedo, lack of wind and little evapotranspiration, is responsible for the urban heat island effect that accentuates heatwaves in the city.

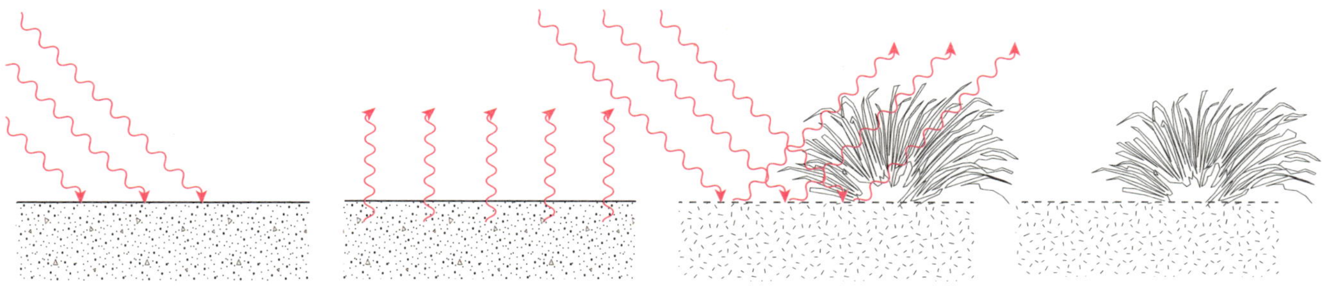

Urban condition: concrete
high effusivity
thermal effusivity:
26 Wh1/2m-1/2.°C-1/2
Daytime: high absorption

Urban condition: concrete
Night time: overheating

Rural condition: soil
(with evaporation by grass)
Lower effusivity
thermal effusivity:
19 Wh1/2m-1/2.°C-1/2
Daytime: moderate absorption
(also due to evaporation)

Rural condition: soil
Night time: cooling

Urban heat island effect {Fig. 1-237}

In the city, dark, high-effusivity materials absorb the heat during the day rather than reflecting it, and continue to emit heat at night.

V. Architectural Scale

Effusivity in Buildings

The physical principle of thermal effusivity, with the principle of diffusivity, is involved in the principle of thermal inertia. This allows a phase shift between the time that the material gains energy during the day - beneath the sun by radiation, or by conduction when the air is hot - and the time, at night, when the energy will be restored - when it is dark and cool. In a Mediterranean climate or in summer, thanks to the temporal phase shift, the higher the effusivity of building materials and their low diffusivity, the more it is possible to cool the interior of a house during the day by the stored night coolness, and conversely, to warm it at night by the stored daytime heat. In a colder climate, in winter, the same principle will be used, harnessing the energy of the sun's rays coming in through the windows to the south during the day and storing this heat in the mass of high inertia materials.

Nevertheless, it will be necessary to pull the curtains at night to keep the heat inside the house, avoiding discharge into the night sky by radiation of the heat stored. In addition, high inertia materials will have to be totally protected from the outside cold by very good thermal insulation placed outside in order to prevent the cold temperatures of the outside air from entering by conduction and cooling the high inertia material. And the windows must be closed so as not to lose heat by mixing with cold air coming from the outside. A high-effusivity, high diffusivity material will very quickly rise in temperature and will restore it immediately, causing the temperature of the room to rise immediately.

Inertia in buildings and energy transfers in floors

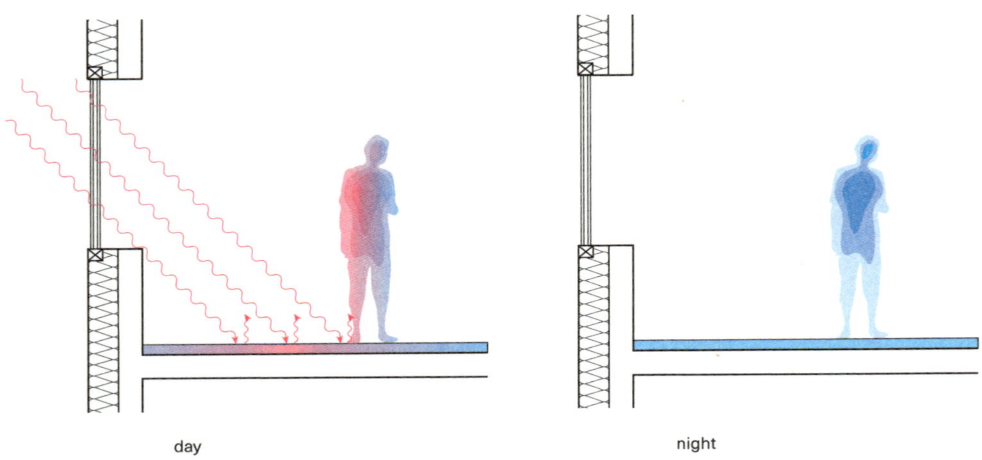

Low inertia material {Fig. 1-238}
A low inertia material will heat up during the day, immediately releasing the heat received. In summer, the material will immediately overheat uncomfortably, and at night will be immediately cold, having stored no heat during the day.

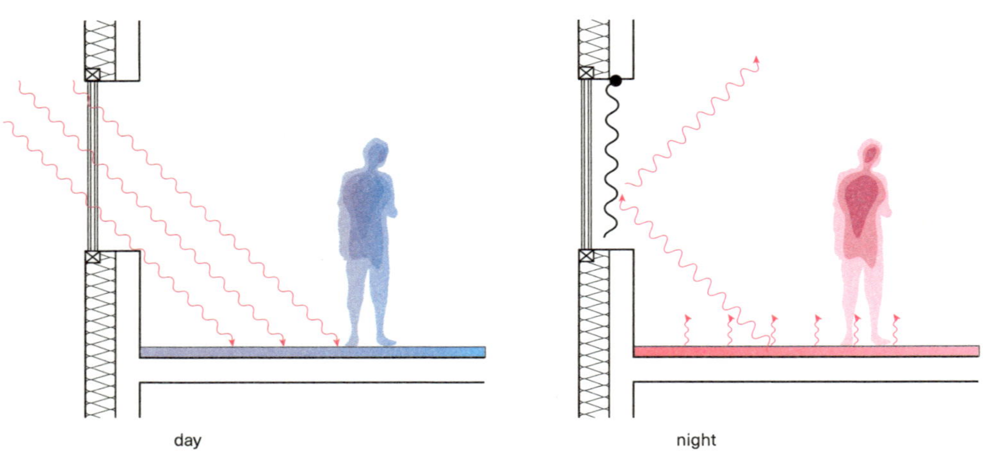

High inertia material {Fig. 1-239}
A high inertia material will slowly absorb the sun's energy during the day and then restore it during the night. It is also important in the evening in winter to draw the curtains to avoid losing heat by radiation to the night sky.

Inertia in Buildings and Heat Transfer in Walls

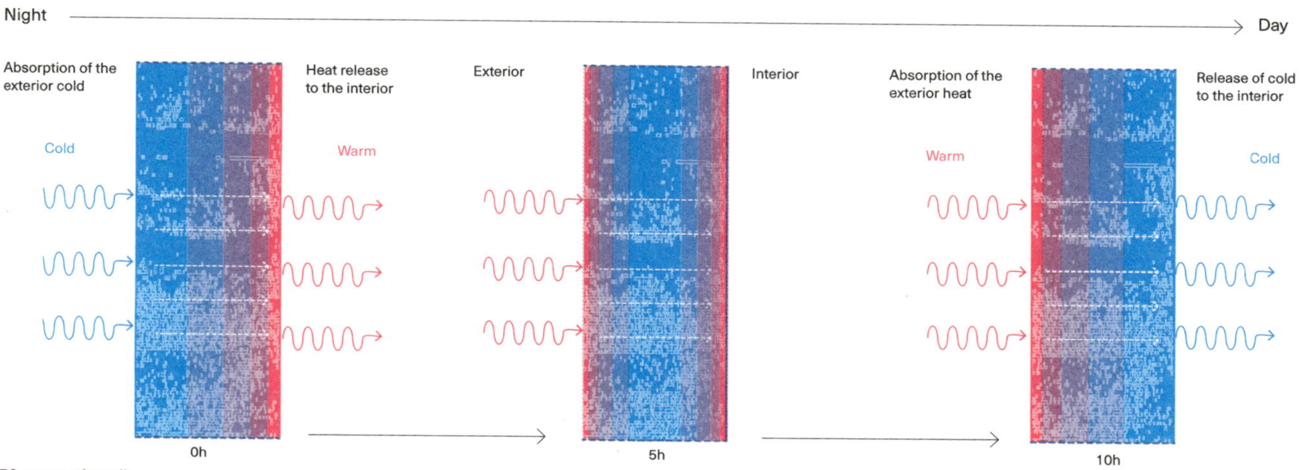

50 cm earth wall
Thermal conductivity (W/m.K) 1.21

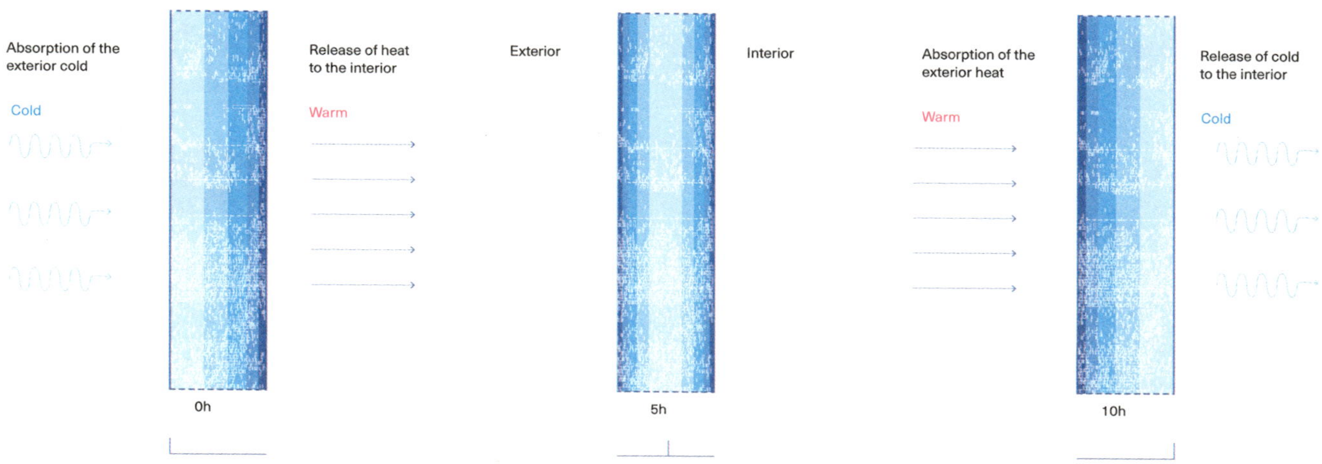

30 cm water wall
Thermal conductivity (W/m.K) 0.58

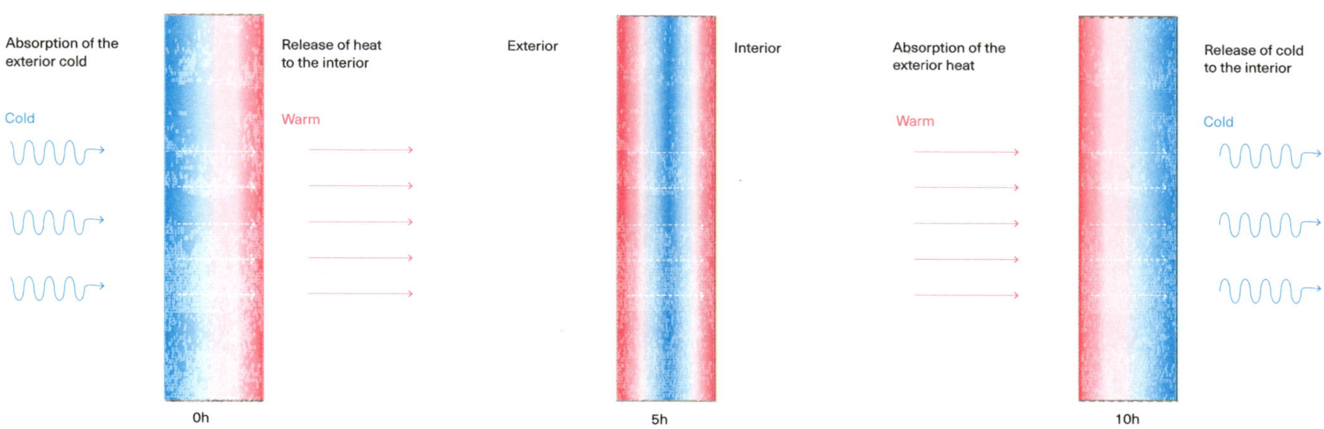

30 cm concrete wall
Thermal conductivity (W/m.K) 0.60

Temporal phase shift by thermal inertia {Fig. 1-240}

In latitudes where the average annual temperature exceeds 20°C (or in summer), the heat transfer through the facade wall can use the phenomenon of thermal inertia to temporally shift the variations of the outside temperature. Thus, by a 10-hour time lag, we can receive inside the house the coldness of the night at noon, and conversely the heat of the day in the evening, making the interior of the house thermally comfortable. This thermal transfer does not work in colder latitudes because the wall poorly insulated thermally lets pass the cold in winter that is present both day and night.

Inertia in Buildings and Thermal Dampening in Walls

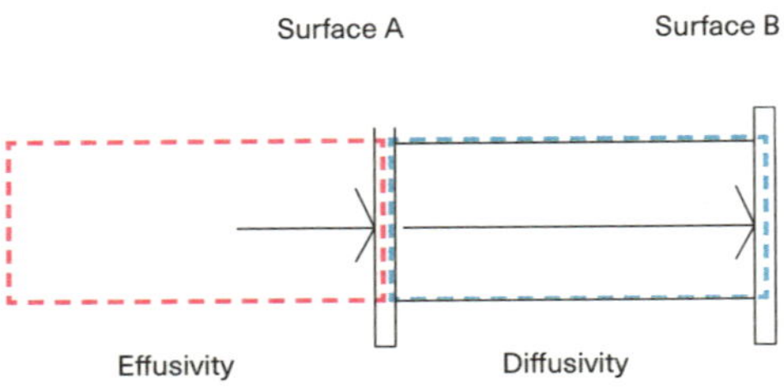

$$E = \sqrt{\lambda \rho c}$$

$$a = \frac{\lambda}{\rho c}$$

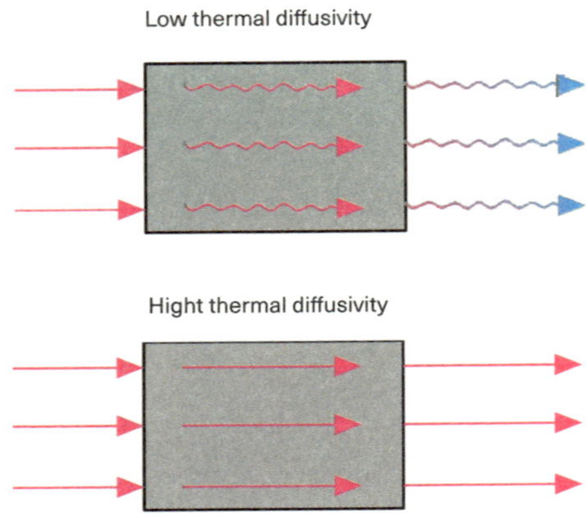

Thermal inertia {Fig. 1-241}
Good thermal inertia requires high thermal effusivity and low thermal diffusivity

$a = \lambda / \rho c$ où

- λ : Thermal conductivity (W/m.K)
- ρ : Density (kg/m)
- C : Specific heat capacity (J/kg.K)

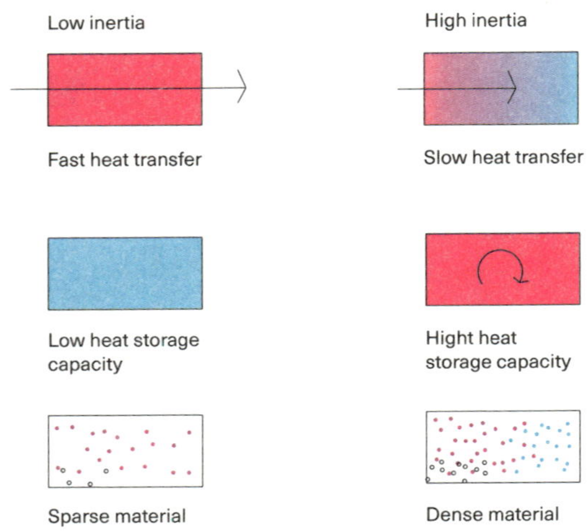

Thermal diffusivity in the material {Fig. 1-242}
The higher the thermal diffusivity of a material, the faster heat will pass through it.

A high thermal inertia material dampens the rise in temperature of the material subjected to conductive or radiant heat since its mass will absorb the heat first before restoring it to the environment. Thus, on a hot day, a high thermal inertia material will dampen the rise in temperature of the room since the excess heat will first pass into the mass of the material. Good (high) thermal inertia, allowing absorbing the summer heat or the winter cold, depends on three factors: 1-the thermal effusivity must be high; 2-the thermal diffusivity must be low; 3-thickness must be significant. Diffusivity (α) itself consists of 3 factors: thermal conductivity (k or λ), the volumetric weight of the material ρ (its density), and its specific heat capacity (C) where α = k/ρC. Thus, for thermal diffusivity α to be low, the thermal conductivity (k) must be low, and conversely, the volumetric weight of the material ρ (its density) and its specific heat capacity (C) must be high. The use of thermal absorption by a high thermal inertia in the building only works, of course, if the material is perfectly thermally insulated from the outside, by being placed inside the thermal insulation so that its temperature is not immediately changed by thermal conduction by air temperature variations. And to preserve the heat acquired by thermal inertia, the windows must be closed.

Examples of effusivity, diffusivity, specific heat, thermal conduction coefficients

Material	Effusivity (J/m2·K·s0.5)	Diffusivity (10-6 m2/s)	Specific heat capacity (J K−1 kg−1)	Thermal conduction (W/m.°k)
Concrete	1257	0.54	880	1.1
Aluminum	23377	98.8	880	237
Steel	13248	13	465	50
Terracotta	1091	0.43	837	0.7
Stone	1154	1.18	890	2.9
Wood	635	0.12	272	0.14
Raw earth	1000	0.2	900	0.75
Wool	44	0.058	660	0.046

Classification of materials according to their thermal inertia

Material	Volumetric weight (kg/m3)	Thermal inertia
Concrete	2500	high
Aluminum	2700	low
Steel	8000	high
Terracotta	1500	high
Stone	2500	high
Wood	450 (fir), 800 (oak)	low
Raw earth	1600	high
Wool	20	low

VI. Case Studies

The Low-Effusivity Rug

Material	Effusivity (J/m2·K·s0.5)
Extruded polystyrene	41
Expanded polystyrene	38
Polyurethane	26

Effusivity of the materials used {Fig. 1-243}

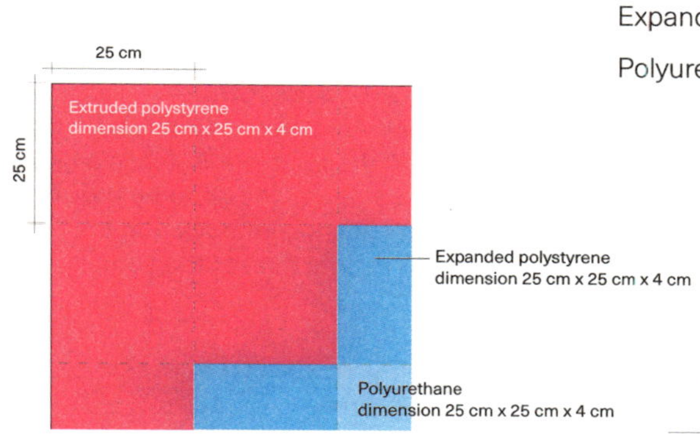

Detailed plan {Fig. 1-244}

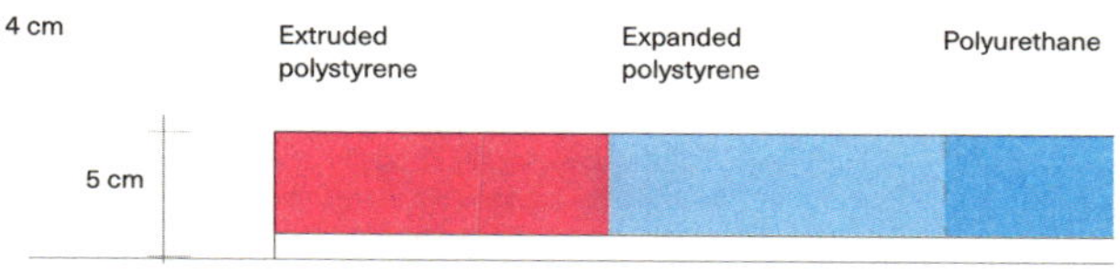

Detailed section {Fig. 1-245}

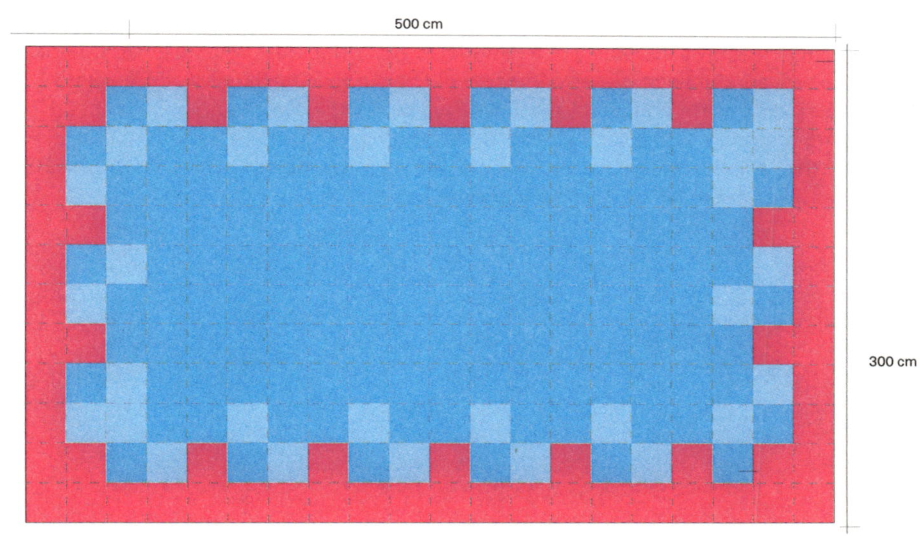

Plan {Fig. 1-246}

The rug functions as a gradient from the outside to the inside where the least effusive material is located.

Elevation {Fig. 1-247}

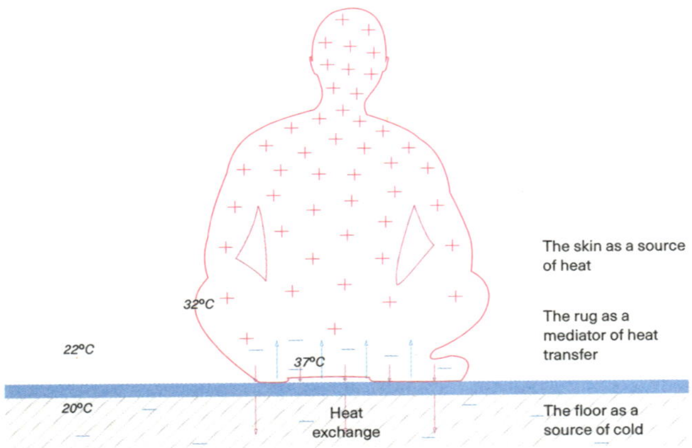

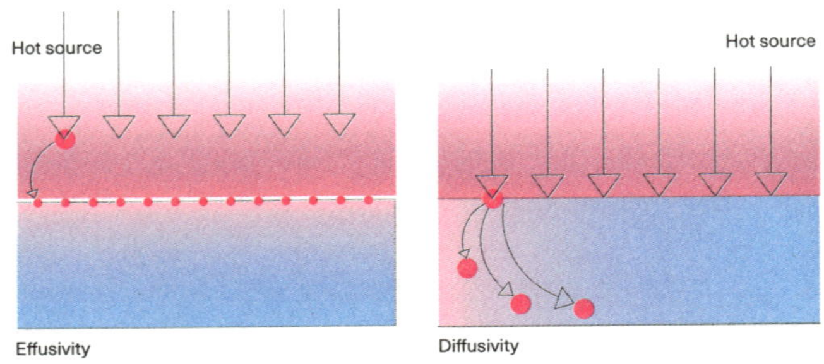

Heat exchanges by conduction {Fig. 1-248}

The skin, with its high effusivity, will exchange heat rapidly in contact with a high effusivity material, until the latter, by conduction, acquires the same temperature as the skin or the human body acquires the temperature of the material.

Effusivity and diffusivity {Fig. 1-249}

The thermal effusivity of a material is its ability to absorb or give thermal energy to another material in contact.
The greater the thermal effusivity, the more heat will be exchanged between this material and the other; the lower the thermal effusivity, the less heat will be exchanged between the two materials.

A material's thermal diffusivity is its ability to transmit heat in itself.
The higher the thermal diffusivity, the faster the transmission will be from one side of the material to the other.

When sitting on the floor, the body is in direct contact with the cold concrete floor. By putting a low-effusivity mediator material between the body and the floor, heat transfer will be slowed down. The woolen rug, for example, will exchange heat with the floor and body very slowly. The body will not feel the same surface temperature of the ground and will thus not lose heat to the ground, keeping the body warm. However, contact by the feet on marble or metal, high effusivity-materials, will cool the body very quickly, leading to a cold sensation.
This contact also takes place between the floor and the air. A low effusivity material, such as wool or wood, will retain its accumulated heat longer if the air in the room cools, while a high effusivity material (such as metal) will change temperature as quickly as the air temperature changes. Thus, wool will prevent cooling in winter, while stone will cool our body and the air in summer (only if the stone is in the shade, protected from the radiant heat of the sun).

Material

Polyurethane or expanded polystyrene have low effusivity and will function as a heat mediator between concrete and the human body. With this additional material, heat will leave the body more slowly than if the body were in direct contact with the concrete floor.

T human body = 37 E human body = 400
T polyurethane = 20° E polyurethane = 261
T concrete = 20° E concrete = 2500
Temperature felt when touching polyurethane
T = (E1*T1+E2*T2)/(E1+E2)
T = (261*20+400*35)/(261+400)= 29.08°
Temperature felt when touching concrete
T=(2500*20+400*35)/(2500+400)=22.06°

Thermal effusivity measures the ability of a material to exchange heat with its environment and depends on its conductivity, density and specific heat capacity. Effusivity comes into play when two objects are in contact. A more effusive material will exchange heat with its environment more quickly. When two objects are in contact, the surface temperature is determined by the temperature of the two objects as well as by their effusivity. It will be closer to the temperature of the material with the higher effusivity.

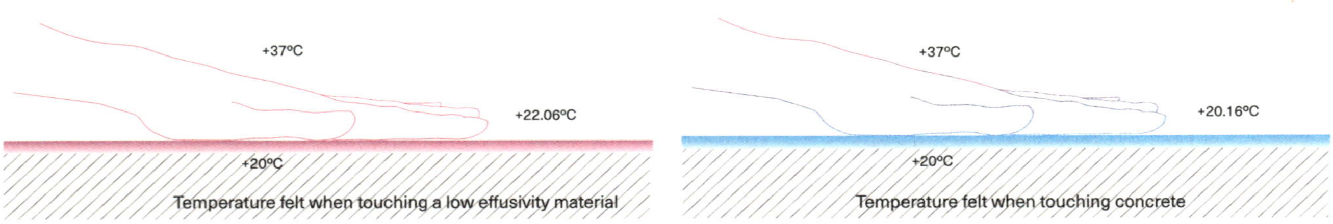

Heat loss {Fig. 1-250}

Depending on the effusivity of the material with which the hand is in contact, the temperature felt will not be the same. The material will feel cold if it is of higher thermal effusivity and hot if it is of low emissivity.

The Rug with Different Thermal Effusivities

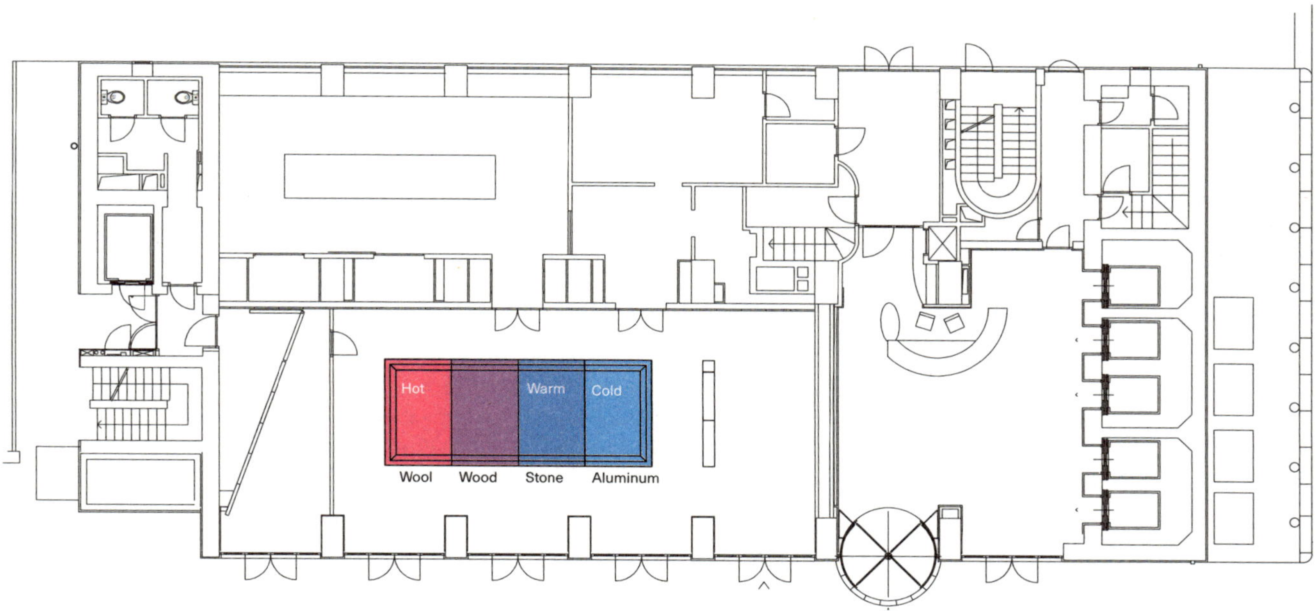

Plan - variable effusivity {Fig. 1-251}

This plan gives the visitor the possibility of trying out different effusivities, by sitting on the material where one is most comfortable.

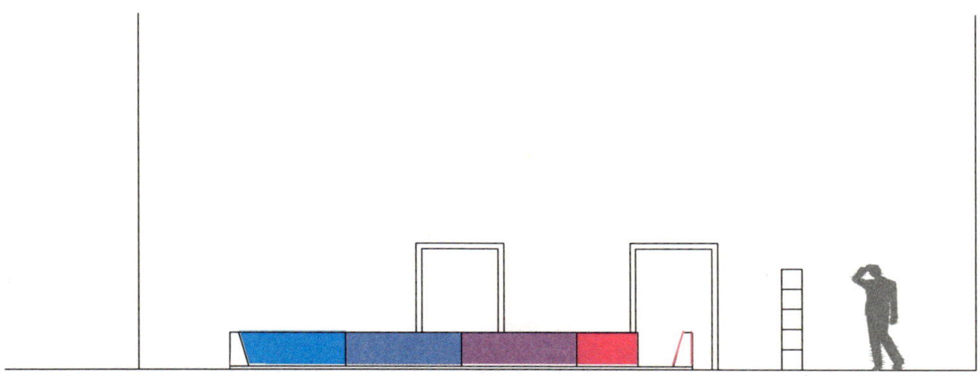

Section - variable effusivity {Fig. 1-252}

The longer the visitor sits on an effusive material, the more they will be cooled by conduction. Low effusivity materials will, however, create a feeling of warmth.

Material	Effusivity ($J/m^2 \cdot K \cdot s^{0.5}$)
Aluminum	23 377
Stone	1 154
Wood	595
Rockwool	44

Coefficient of effusivity {Fig. 1-253}

The variable effusivity rug is the implementation of materials chosen according to their thermal properties. But their choice is no longer made according to analogical or memorial criteria, but according to energy and climate criteria. According to a highly perceivable gradation, we implement four materials according to their intrinsic thermal effusivity properties, i.e., to accelerate or slow down the loss of body heat by conduction when human skin comes into contact with a material. Thus, a progression is established from the coldest to the warmest, between a woolen floor, a wooden floor, a stone floor and an aluminum floor.

The thermal effusivity of wool, that is, its capacity to transmit heat to the environment, is very low, which gives the impression of warmth when touching it. Wood will slightly attenuate this impression. As the thermal effusivity of stone is higher, it will release the cold more quickly to the inhabitants who touch it. At the end of the gradation, we have aluminum, whose extremely high coefficient of thermal effusivity will transmit cold to the inhabitants who touch it.

Gradation of effusivity

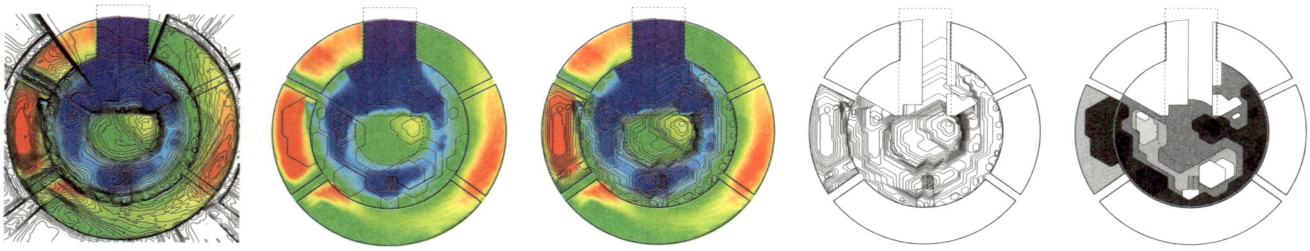

Computer modeling of solar radiation {Fig. 1-254}

From the simulation of sunlight to the drawing of the plan. The sunniest parts are rendered by low effusivity flooring materials that enhance the heat in winter. The parts most in the shade are rendered by high effusivity flooring materials to enhance coolness in summer.

By computer simulation of the sunlight of the Maison de la Radio in Paris (competition, winning project with Nicolas Dorval-Bory), we can map the annual distribution of the sun in the Agora, thus establishing the two poles of a gradient of intensity, from the sunniest to the shadiest, from warm to cold, from matt to gloss, from winter to summer. In the most shaded areas of the Agora (which will be the most pleasant in summer), as the computer model reveals, we increase the feeling of freshness by choosing only high effusivity materials, which will be cold to the touch: polished granite and tin. The areas most exposed to the sun are designed to warm yourself up with a cup of hot chocolate in winter, with a comfortable atmosphere of wool and pine, low effusivity materials that are warmer to the touch and absorb sound. In summer one often goes for a drink at the bar to cool off, in search of shade. In winter the opposite is true, we seek a hot drink in the sun. It is these two fundamental polarities for hot and for cold, shade and sun, hot or cold drinks, that the Agora is mapped and spatialized in an acoustic and thermal progression, starting from the most thermally effusive and reverberant materials to the least thermally effusive and most absorbent.

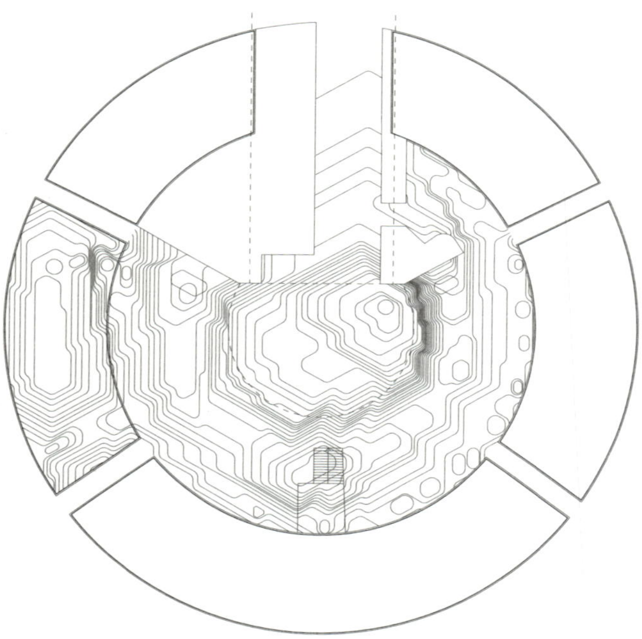

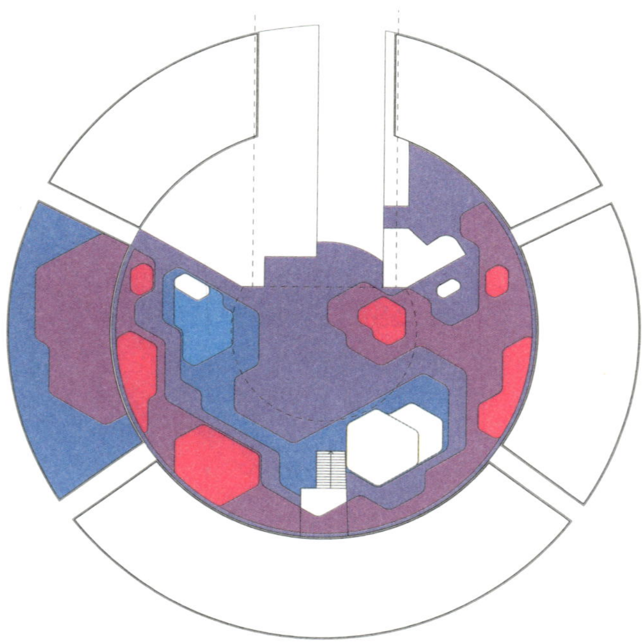

Geometric plan of solar radiation {Fig. 1-255}

Geometric lines are extracted from the computer model of solar irradiation and provide the shape of the gradation for the location of the materials of different effusivities.

Effusivity distribution plan {Fig. 1-256}

The distribution of the materials of different thermal effusivities is performed by following variations in average sunshine

Coefficient of thermal effusivity: distribution by zones

- very high effusivity material: tin (metal)
- high effusivity material: polished granite
- medium effusivity material: rubber
- low effusivity material: pine wood
- very low effusivity material: wood

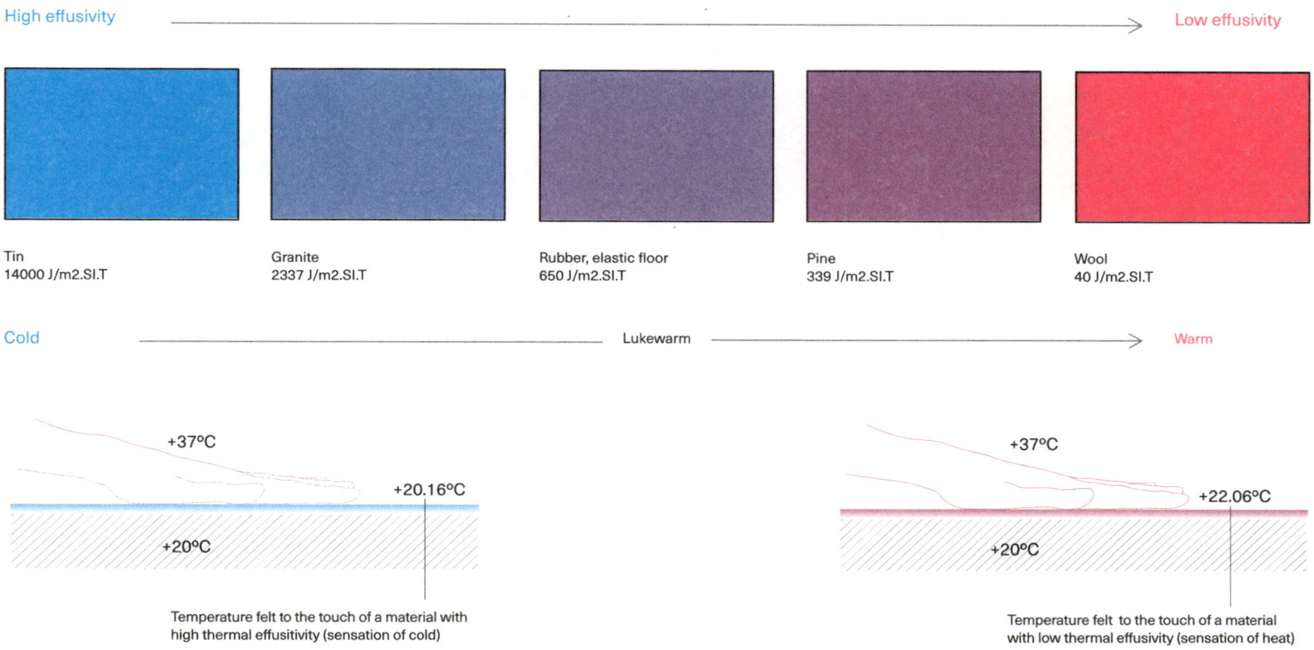

Gradation of effusivity {Fig. 1-257}

The choice of materials depends on their thermal effusivity value, between warm wool in winter and cold tin in summer.

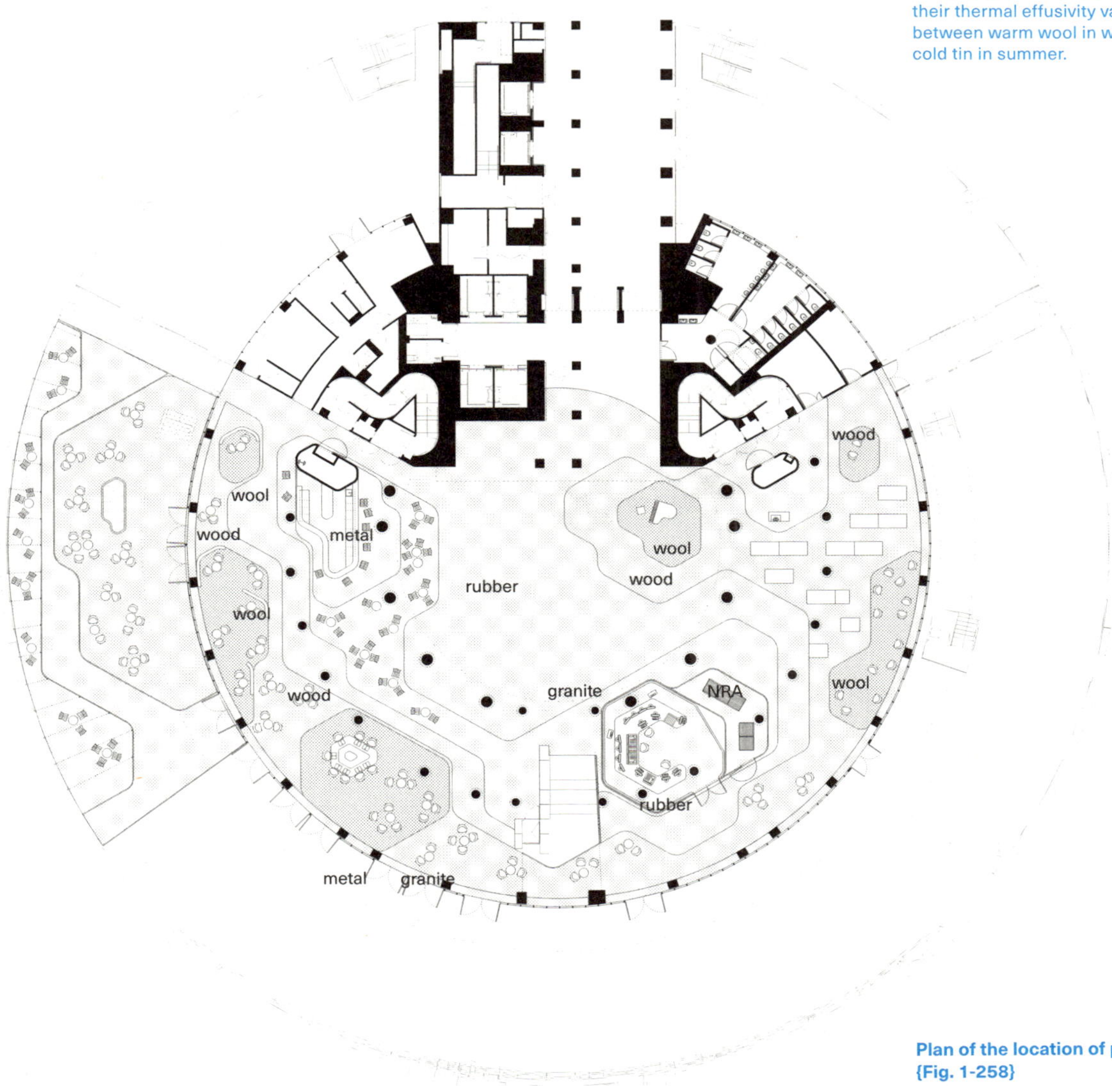

Plan of the location of programs {Fig. 1-258}

The programs are located according to the effusivity values which are located according to average solar light intensities

Gradation of Effusivity

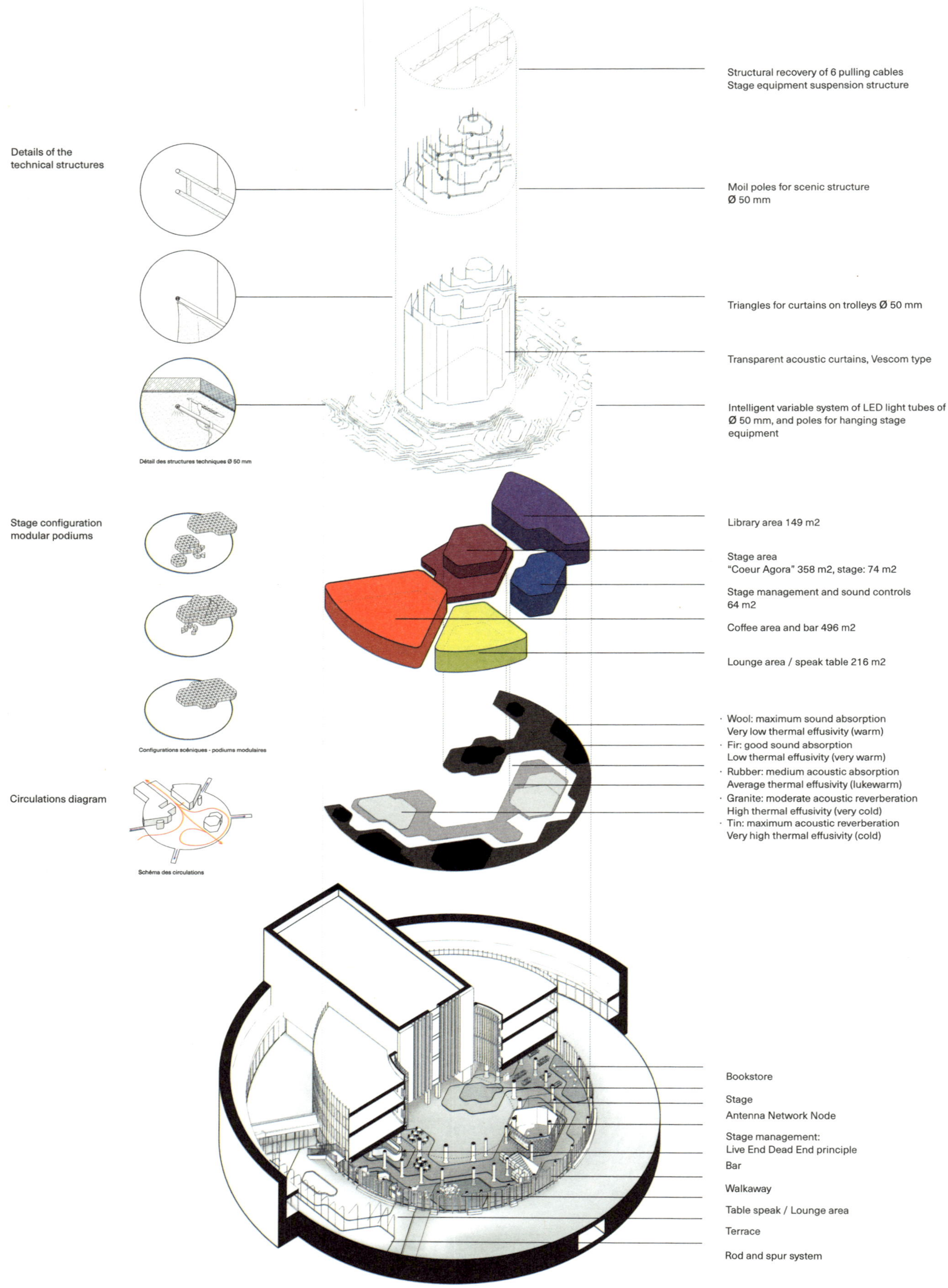

RADIATION

I. Definition

General Properties

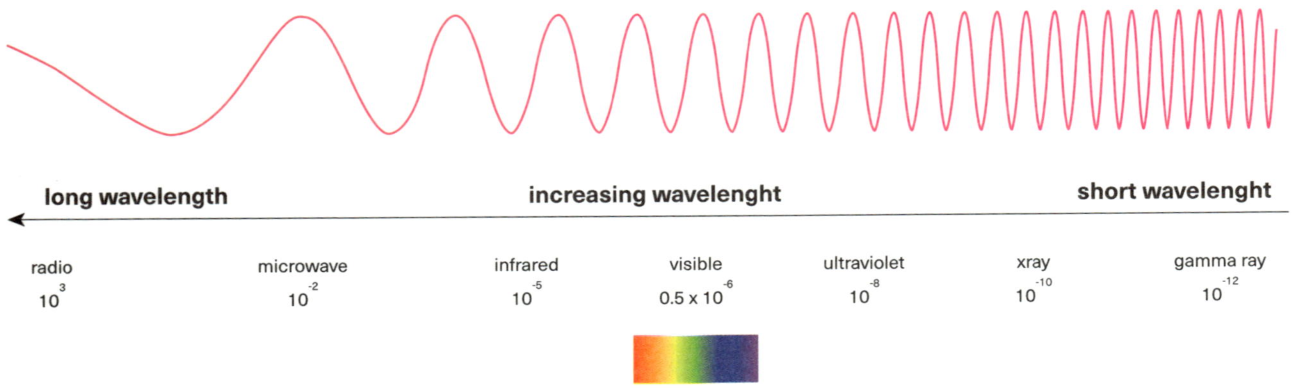

Electromagnetic Spectrum {Fig. 1-259}

The human eye perceives only a tiny part of the electromagnetic spectrum, between 380 and 750 nm. Ultraviolet rays appear below 380 nm, noted by the skin which tans under UV-A and B and synthesizes calcium while germicidal UV-C destroys it. Above the visible wavelengths, infrared are noted by the skin that heats upon receiving them due to its water content.

The emission and transmission of energy in the form of waves and particles takes place in space or within an object and occurs in all directions, outwards from the source. Electromagnetic radiation occurs when an atom absorbs energy, causing one or more electrons to change their position in the atom. When the atom returns to its original position, an electromagnetic wave is emitted. Depending on the type of atom and the amount of energy, this electromagnetic radiation may be in the form of heat, visible light, ultraviolet, or other electromagnetic waves. Electromagnetic rays are the transmission of this energy through time and space, and do not depend on any physical matter to be transmitted.

The electromagnetic spectrum is divided into 7 categories that range from radio waves (very long wavelengths and little energy) to gamma waves (very short wavelengths and a high amount of energy). The spectrum visible to humans is a portion of the electromagnetic spectrum that ranges from 390 to 750 nm. Rays can move infinitely in a vacuum if they do not encounter any obstacle, just as the rays of the sun coming to Earth. Thus, the energy of the sun's rays can be felt on the skin and warm it in the middle of winter up in the mountains, regardless of the temperature of the air: this is what is known as radiant heating.

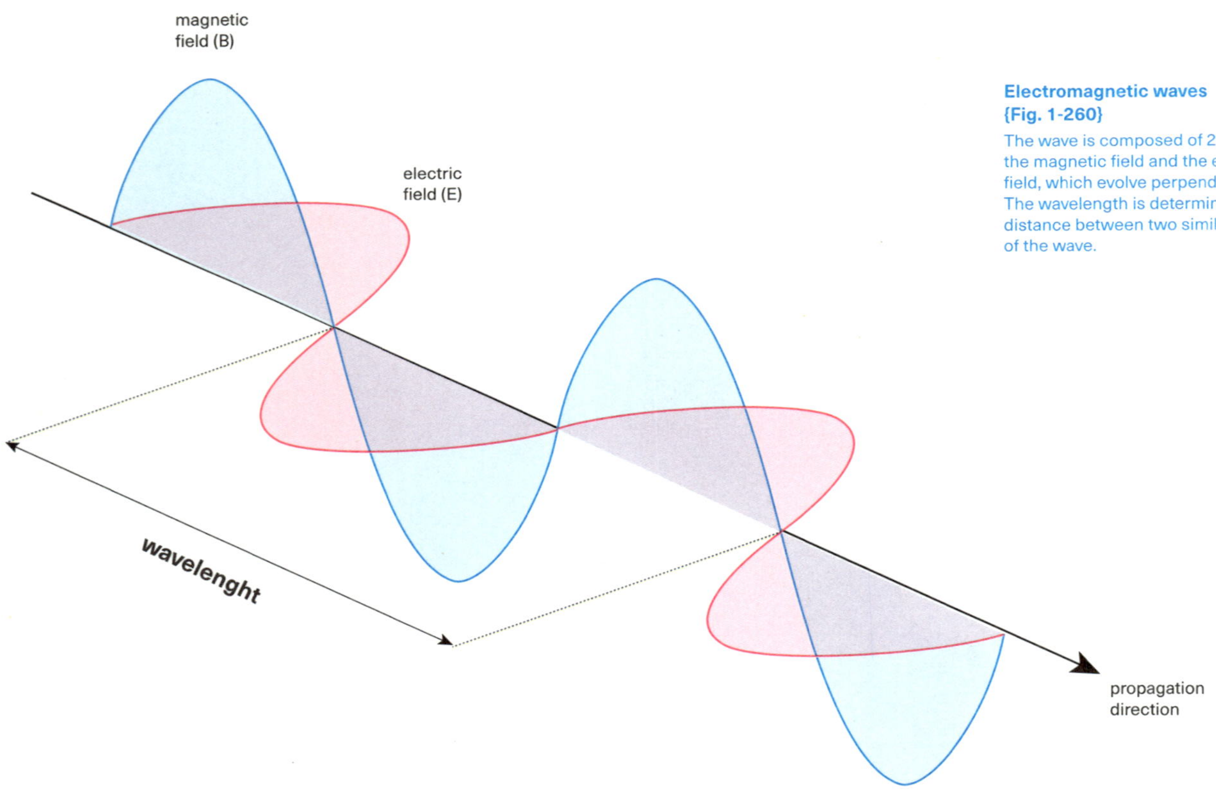

Electromagnetic waves {Fig. 1-260}

The wave is composed of 2 parts, the magnetic field and the electric field, which evolve perpendicularly; The wavelength is determined by the distance between two similar points of the wave.

Thermal Radiation

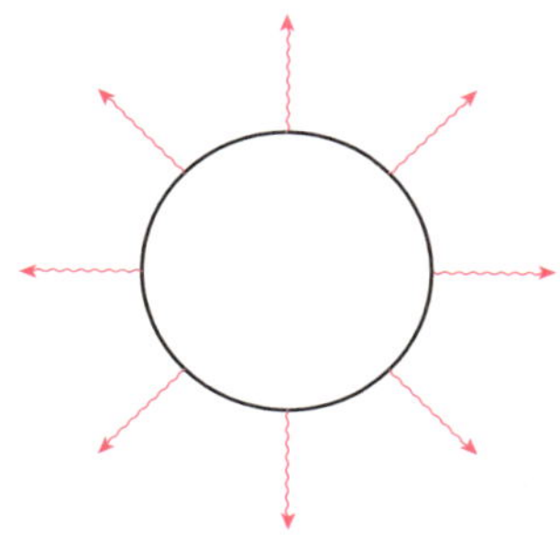

Direction of radiation {Fig. 1-261}
All objects emit radiation in all directions.

Thermal radiation occurs when thermal energy is converted into electromagnetic energy. All materials with a temperature above absolute zero emit thermal radiation. The warmer an object is, the more heat it emits. Thermal radiation behaves like light, to a certain extent. It is absorbed by some surfaces and reflected by others. Radiative heat is transmitted from surface to surface, with little heat absorbed between the surfaces. However, once heated, surfaces can emit heat to their environment by conduction or convection. The human body, at 37°C, also emits electromagnetic waves in the infrared and that is why it cools if in front of a cooler wall, as the latter removes the body's heat by radiation. However, if the energy emitted by the wall or the sun corrsponds to a temperature higher than that of the skin (which is at 34°C), the skin will warm.

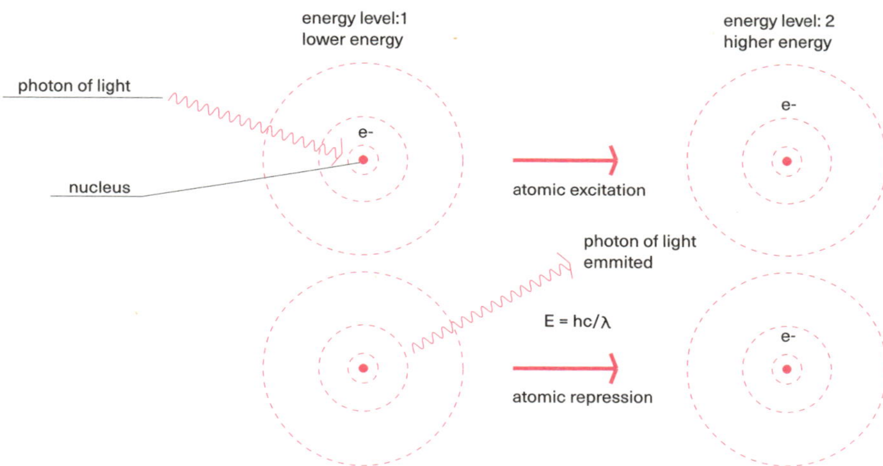

Light, matter, energy {Fig. 1-262}
Under the action of photons, atoms begin to have more energy due to the movement of electrons.

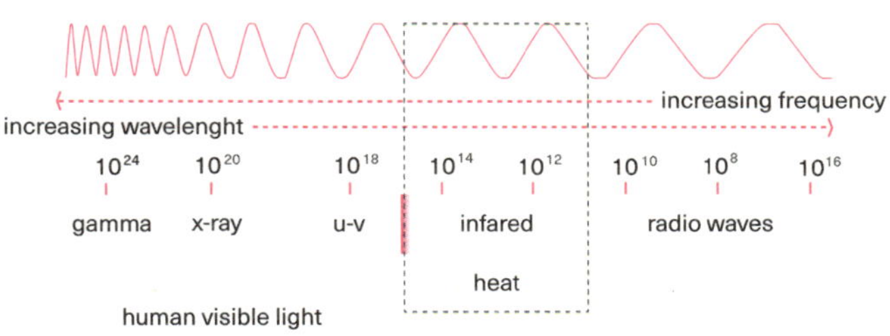

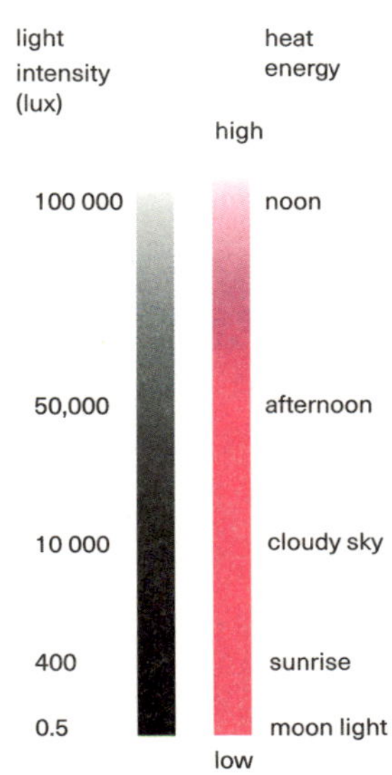

Light intensities of the sun and energy {Fig. 1-263}
The more intense the light, the more energy it contains and heat it carries.

Wavelengths and heat {Fig. 1-264}
The "heating" part of the electromagnetic spectrum is contained in the visible and infrared wavelengths. This is in fact related to the absorption spectrum of water molecules which increasingly react by rising into the long wavelengths, while they will remain stable under short wavelengths, in the ultraviolet.

II. Global Scale

Solar Radiation

The Sun {Fig. 1-265}

The Sun's electromagnetic rays pass through the empty universe without losing any energy. They begin to lose energy as they cross the Earth's atmosphere, diffused by various gases, before being transformed into thermal energy upon touching the dark parts of the Earth's surface.

The light parts of the earth, like snowfields, will rebound a nonetheless weakened part of the rays that will be sent back into the universe, again becoming weaker as they travel through the atmosphere.

The sun's rays are the luminous and thermal electromagnetic radiation emitted by the extremely hot gases present on the Sun's surface. The Earth also emits thermal rays (in the infrared) but far less intense since its temperature is lower. The equilibrium between the heat from the sun and heat emitted by the Earth is the primary phenomenon that determines global temperature. The energy that governs the climate system comes from the Sun. When its energy reaches Earth, it is partially absorbed by different parts of the climate system. The absorbed energy is converted back into heat, which raises the Earth's temperature and renders it habitable.

The absorption of solar rays is not the same throughout the Earth's surface, which gives rise to Earth's different climates, weather and seasonal phenomena. Thus, winds are formed because of the temperature differences caused by the differences in absorption or reflection of the Earth's surface and the angle between the earth and the rays, which leads the amount of energy per m2 to vary. Some of the electromagnetic waves emitted by the Sun are absorbed by the Earth's atmosphere, in particular ultraviolet C rays, which are completely blocked by the ozone layer.

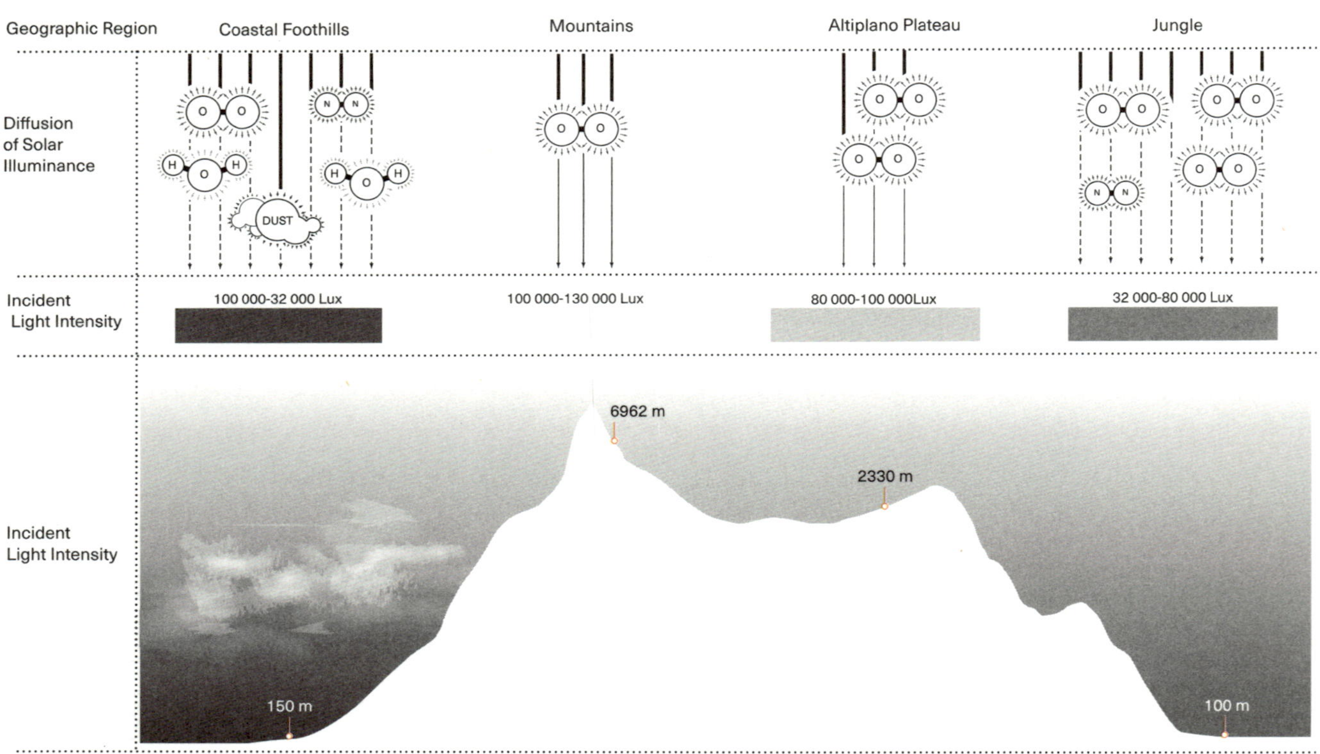

Solar radiation intensity and altitude {Fig. 1-266}

The intensity of the Sun's rays depends on altitude, latitude and time of day. The thicker the layer of atmosphere crossed by the sun's rays, such as at sea level, in the polar latitudes, in the evening or in winter, the more the molecules that make up the air will diffuse certain wavelengths, especially in the blue, and attenuate it, which will make the light yellower and less strong.

However, the thinner the layer of atmosphere crossed by the sun's rays, as at the tops of mountains, in summer, in the tropics or at noon, the more intense and blue the light will be, because it will have lost less energy through diffusion into the Earth's atmosphere.

Light Diffusion

Sunlight Color Temperature Analysis

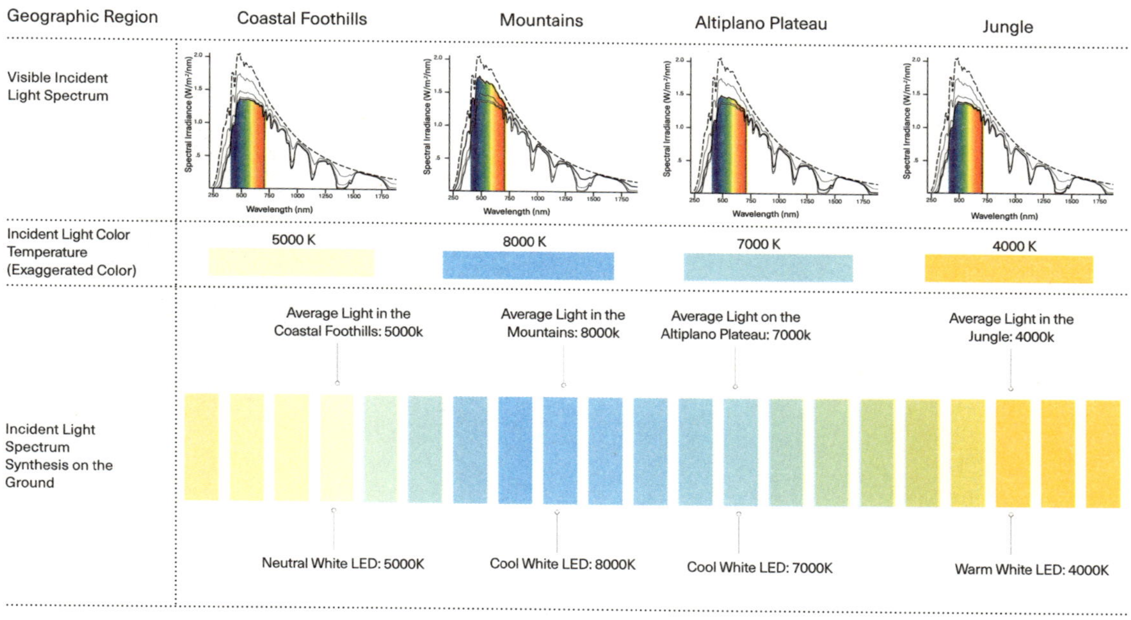

Variations in Solar Light Color Temperature

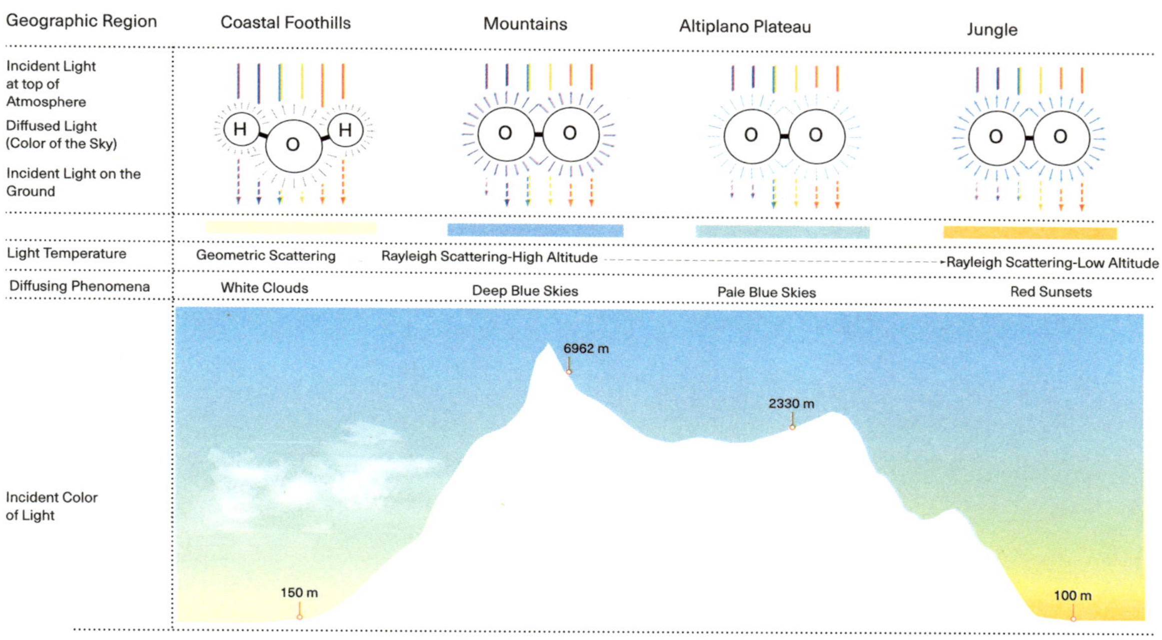

Spectrum of sunlight at different altitudes {Fig. 1-267}

The more we descend in altitude, the more orange the light becomes as it loses its blue part, which is diffused by the gases of the atmosphere. It is also for this reason that in the morning and evening, in areas close to the poles, the light becomes more and more orange while in the mountains and or near the equator the light remains blue. The thicker the atmosphere, the more the sunlight will lose its blue wavelengths. Chronobiology has shown the physiological relationship between changes in the secretion of the hormone melatonin, involved in sleep, and the color of light and its intensity. Indeed, in accordance with the variation in the color of natural light, between blue at noon and red at night, the human body will block the secretion of melatonin under intense blue light, which promotes wakefulness, and on the contrary, under weak yellow, it will secrete melatonin, promoting sleep. Artificial light containing blue emitted in the evening will therefore disturb the natural chronobiological cycle of sleep-wakefulness, inducing problems of insomnia, for example.

Color Temperature

Sunlight at the top of the atmosphere

Low atmospheric thickness, low diffusivity of blue light at the equator or at noon

The sunlight reaching the earth is bluer, cooler

Sunlight at the top of the atmosphere

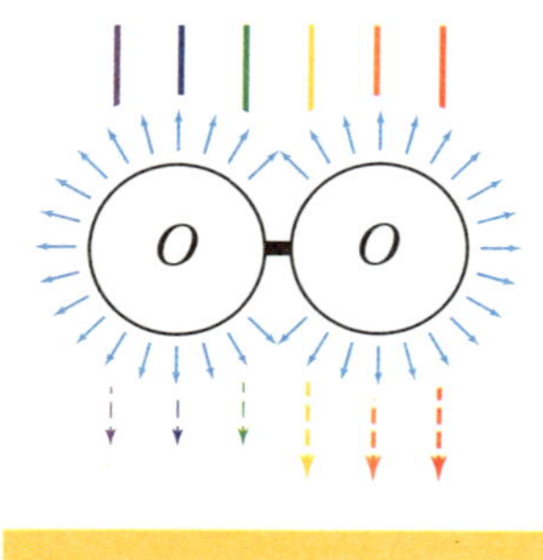

Great thickness of the atmosphere, great diffusivity of blue light in cold latitudes or at sunset

The sunlight reaching the earth is more yellow, colder

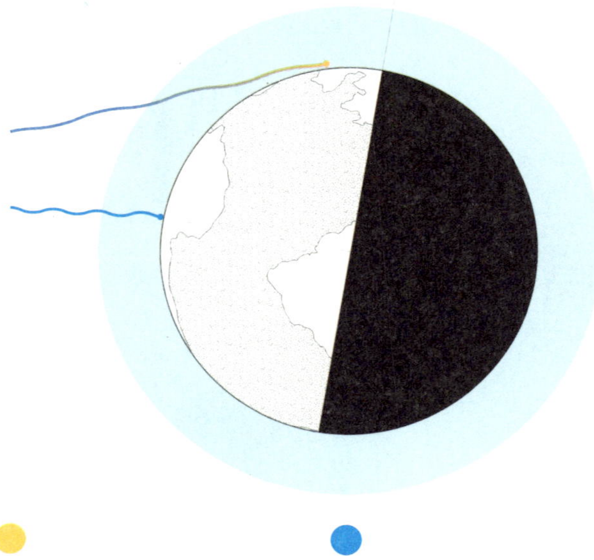

Tangential light rays
Maximum exposure to the atmosphere
Cold latitudes at sunset
The warmest light (yellow)

Direct light rays
Minimal exposure to the atmosphere
The equator at noon
The coldest light (blue)

Solar radiation intensity and altitude {Fig. 1-268}

When crossing the atmosphere, the blue spectrum of sunlight is diffused in all directions by oxygen, nitrogen and water molecules. At high altitude, the atmosphere is less thick than at low altitude and its light will be bluer. Nearer the sea, the distance travelled by the light is greater and the sunlight loses more blue light.

As a result, the light at sea level is more yellow. At sunset, the distance covered by the sunlight is greater, which accentuates the phenomenon of light scattering. As a result, it is stripped of the blue and green spectra, giving a general impression of yellow and red.

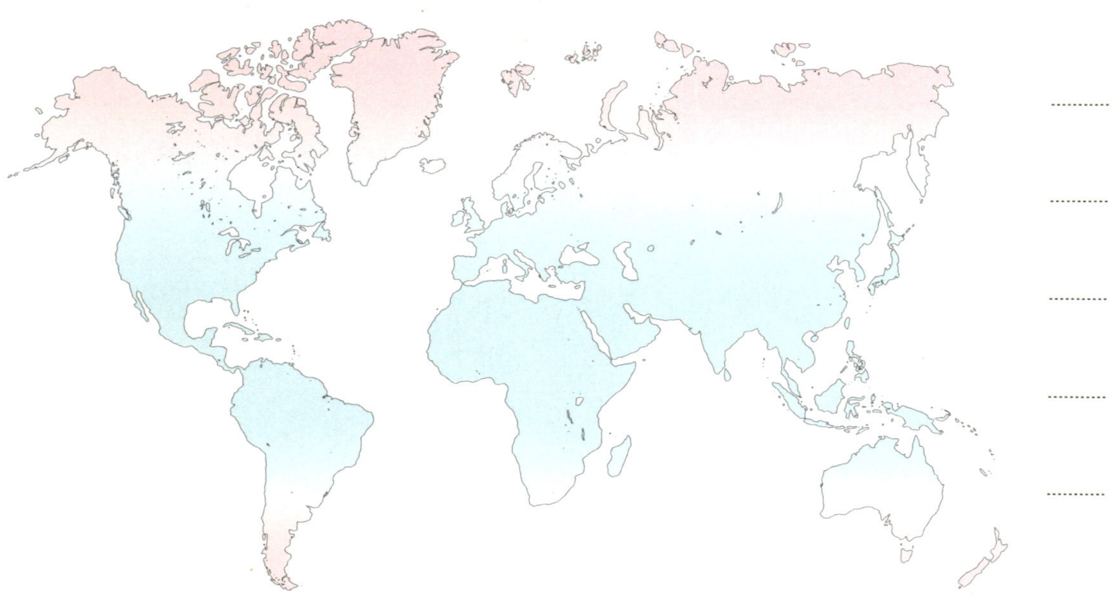

polar light
2500°k

continental light
4000°k

Mediterranean light
5500°k

tropical light
7000°k

continental light
4000°k

polar light
2500°k

Color temperature in ° kelvin

Planetary gradation of color temperature {Fig. 1-269}

The further poleward, north or south, we go, the more orange the light becomes, losing the blue part of its spectrum as it is diffused by the molecules of the atmosphere which then confer the blue color to the sky.

III. Material Properties

Absorption, Reflection, Transmission

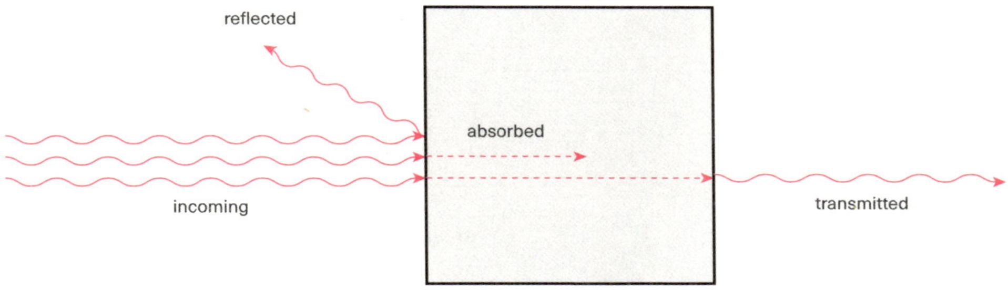

Total energy transmitted {Fig. 1-270}

Not all of the rays reaching an object are absorbed. Depending on the material and other properties, some of the energy may be reflected and transmitted to other objects.

As the temperature of the object decreases, the wavelengths of the emitted rays decrease. Objects at a temperature close to that of the room emit infrared rays, while very hot objects emit in the visible wavelengths. When a ray reaches a surface, energy can be absorbed, transmitted, or reflected (or a combination of the three). The sum of these three phenomena is equal to the total energy of the ray according to the equation $\alpha+\rho+\tau=1$, where α represents the absorption factor, ρ represents the reflection ratio and τ represents the transmission factor.

Color

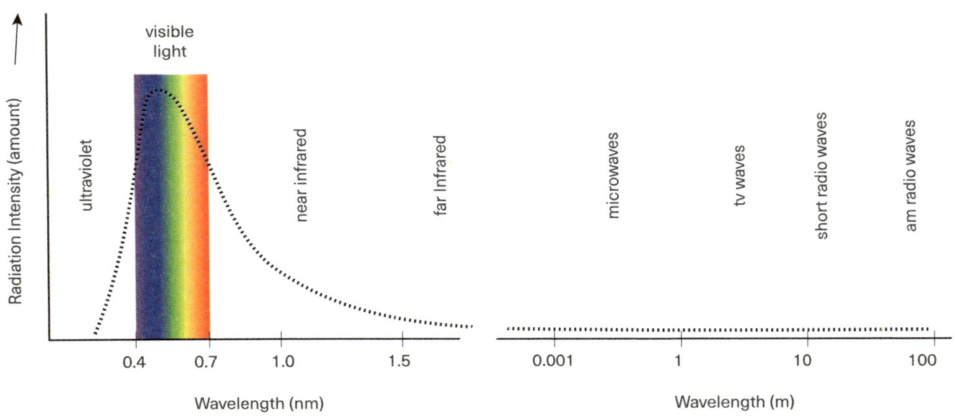

Spectrum of rays emitted by the sun {Fig. 1-271}

Most of the rays emitted by the sun correspond to UV, visible light and infrared. It is estimated that the amount of energy from the sun that ultimately reaches Earth is roughly evenly distributed between visible and infrared, with ultraviolet radiation being much narrower and less intense, as it is largely blocked by the ozone layer at the top of the atmosphere.

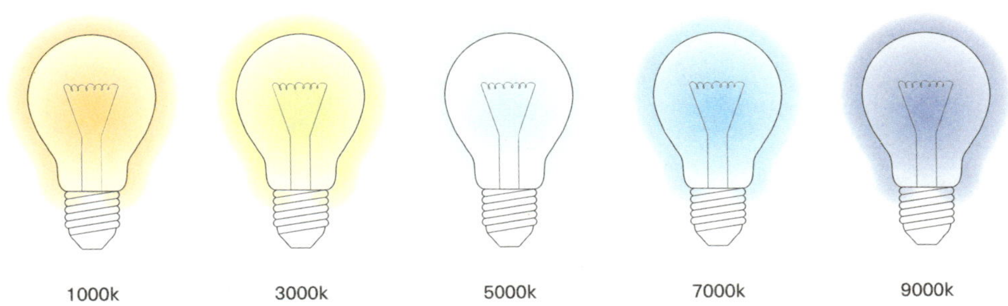

Color Temperature {Fig. 1-272}

"Color temperature" is the name given to the tint of the white light varying between the "hot" (more orange) and the "cold" (more blue), inspired by the variation of natural light. Warm light will tend toward 1000 kelvin, cold light toward 9000 kelvin.

Visible light is characterized by its color. It is both an inherent property of light and an artefact of the human eye. Objects have no colors, rather, they emit light that appears colored to the eye. In other words, color exists only in the mind of the human that perceives it. When an object warms up, it radiates energy in the form of rays of increasingly shorter wavelength, which we perceive as changing colors. For example, the flame of a burner changes from reddish to blue when the power is adjusted. The process whereby energy in the form of heat transforms into light energy is called incandescence.

IV. Architectural Scale

Artificial Lighting

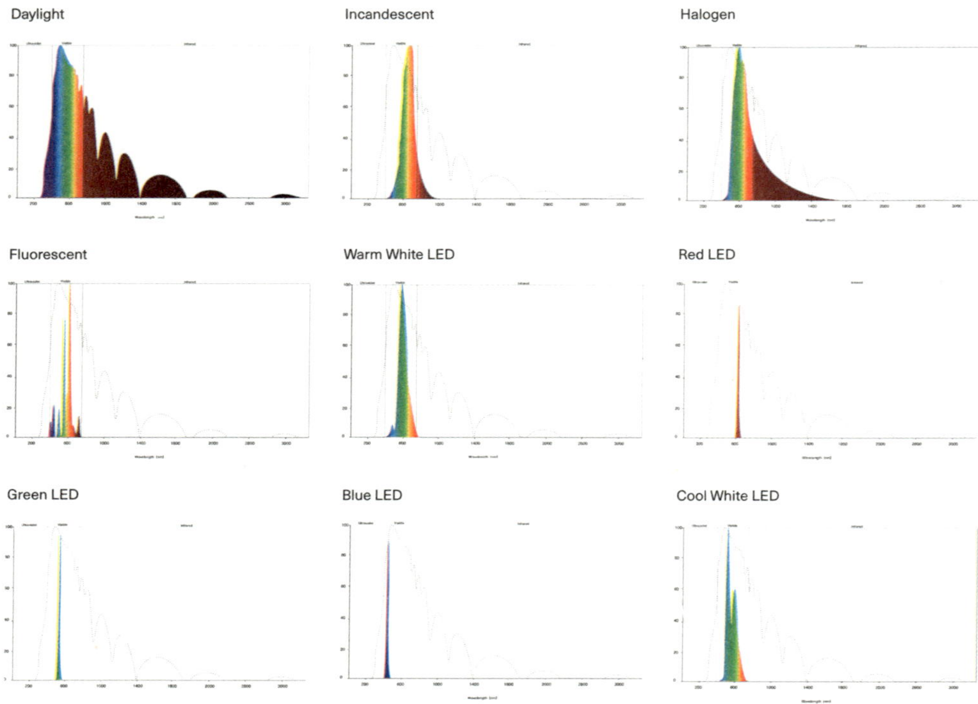

Different light spectra depending on type of lighting {Fig. 1-273}

Incandescent and halogen sources emit a lot of infrared, generating heat rather than light. Fluorescent and LED sources emit wavelengths in the visible spectrum of the human eye only, without emitting in the infrared, which saves energy.

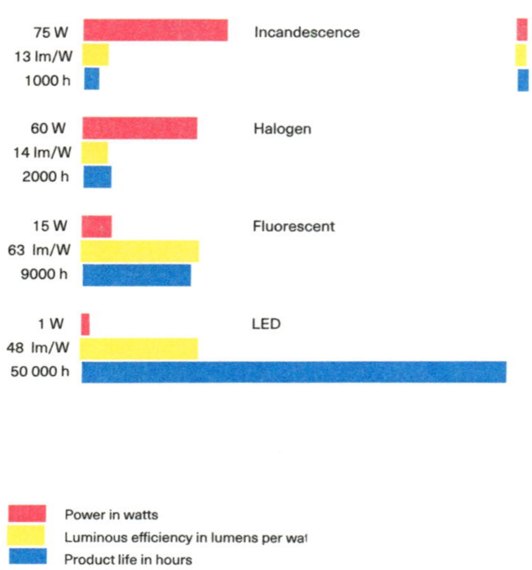

Comparison of the performance of different light sources {Fig. 1-274}

Incandescent and halogen lamps are now being replaced by LEDs because the former consume a lot of energy and produce little light, most of the energy being converted into heat in the infrared.

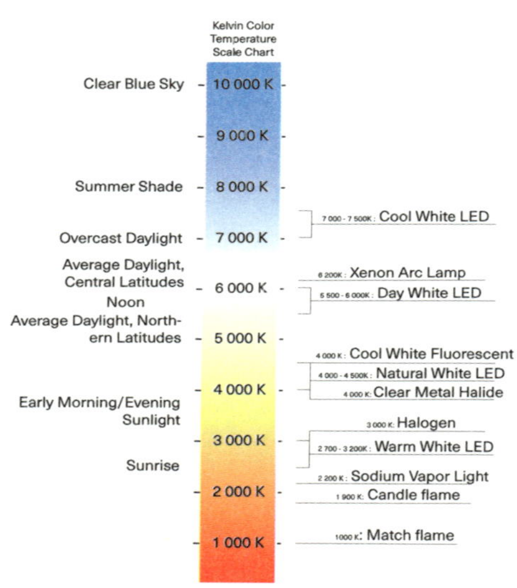

Color temperature {Fig. 1-275}

White light varies between "warm" tones (orange) and cooler tones (blue), depending on the time of day or year. It is possible to reproduce different color temperatures with artificial lighting.

The first electric light bulb invented in 1880 was incandescent, the disadvantage of which is that only 5% of the energy (electrical energy consumed) is transformed into visible light, between 400 and 800 nanometers. The rest of the energy is wasted by emitting infrared, or invisible heat, which represents a disproportionate loss of energy. A halogen bulb has approximately the same efficiency, about 7% energy is converted into light versus 93% into heat, and it emits very long wavelengths. By observing the electromagnetic spectrum of these lights, we note that the two spectra extend well beyond the visible spectrum, emitting mostly in the infrared. In order to save energy, since 1926 we have favored fluorescent light, for two reasons. Firstly, it does not emit infrared. Secondly, fluorescent light, unlike incandescent and halogen lights, does not emit a continuous spectrum but rather peaks in the visible domain, which leads to an energy saving.

While the reduction of the emitted spectrum may be beneficial from an energy point of view, it creates problems from a health point of view because the fluorescent light does not emit in the wavelengths adapted to the human eye, creating visual fatigue. The need to reduce energy consumption in order to combat global warming is leading to a shift away from incandescent and halogen light sources toward LED light sources emitted by diodes of precise wavelengths, which means it is possible not to spend energy outside the light spectrum actually desired. By using three different semiconductors, emitting in the red, the green and the blue, all the color variations are produced.

Light Intensities

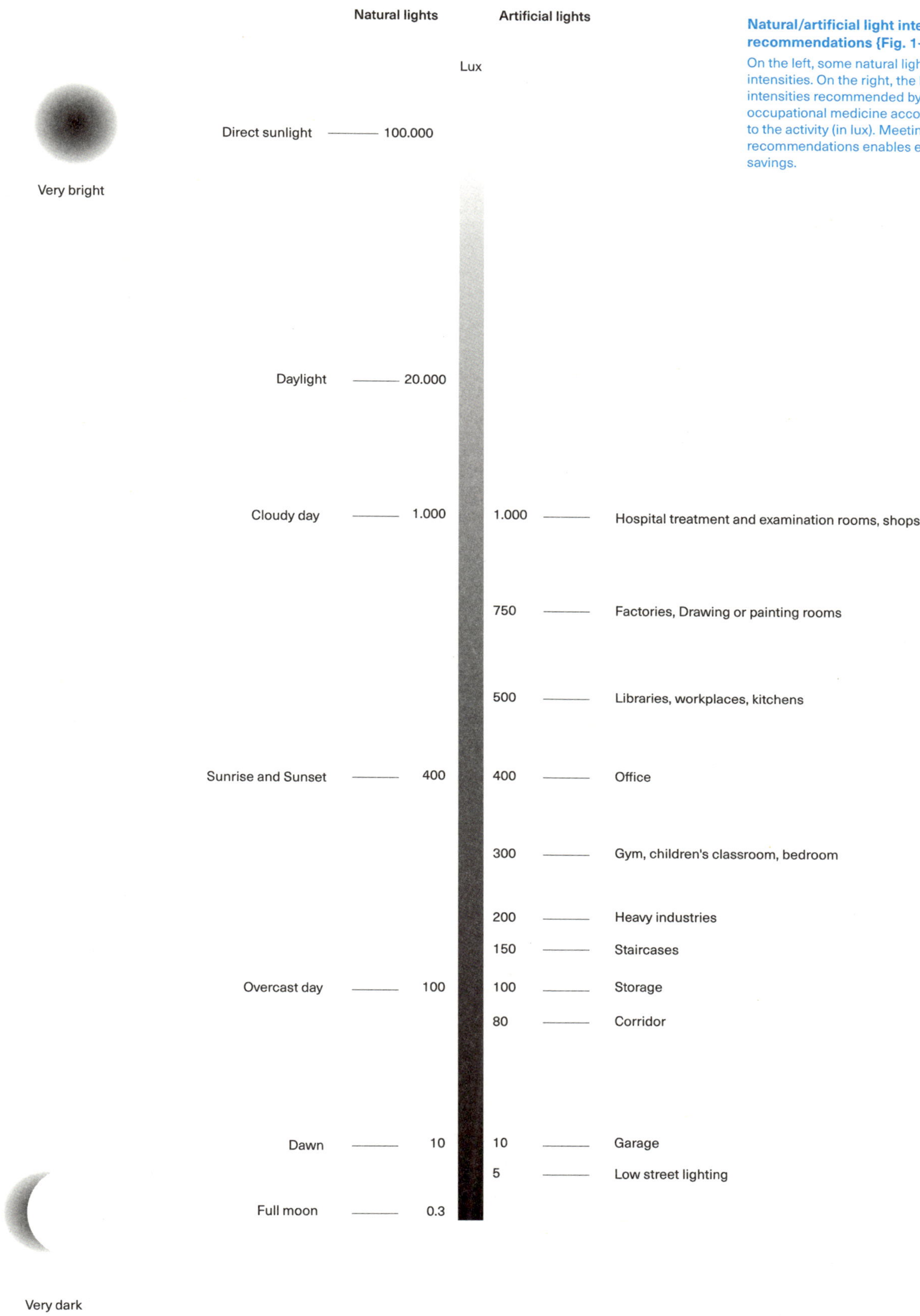

Natural/artificial light intensity recommendations {Fig. 1-276}

On the left, some natural light intensities. On the right, the light intensities recommended by occupational medicine according to the activity (in lux). Meeting these recommendations enables energy savings.

V. Physiological Scale

Thermoregulation in the Human Body

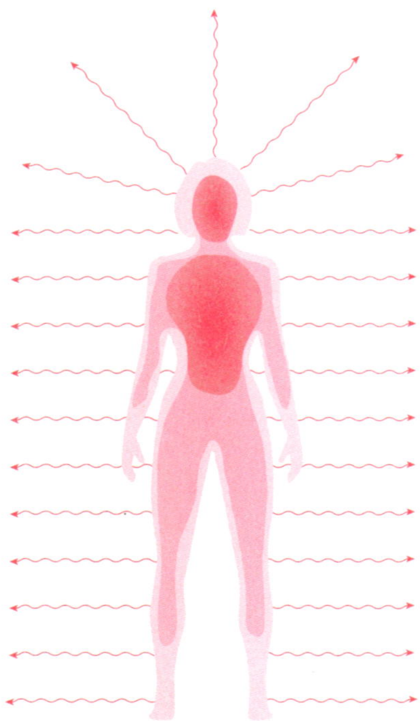

Cooling by radiation {Fig. 1-277}
The body radiates heat in all directions as a means of regulating its internal temperature. Under normal conditions, this is the most effective way to evacuate heat.

Cooling by radiation is a process whereby the body dissipates heat in the form of heat rays. It is a heat transfer without physical contact. The body at rest is the main means by which it evacuates heat. The skin is constantly irradiating in all directions and toward nearby objects such as clothes or furniture whose temperature must be colder. The body can also receive radiation from these objects and heat up if they are warmer than it. At normal temperature, about 60% of our heat losses are in the form of thermal radiation.

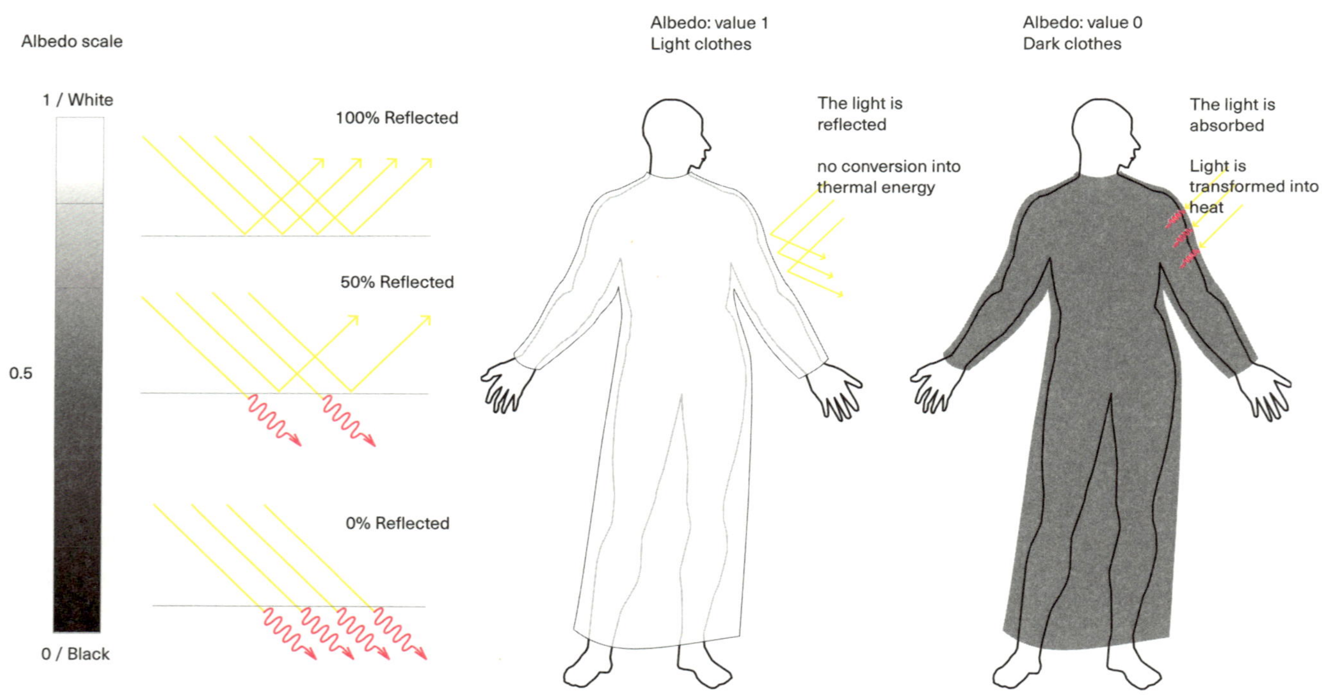

Albedo: reflection and absorption of light {Fig. 1-278}
The albedo is the percentage of sunlight that is reflected, rather than absorbed. The lower the albedo (toward 0), the more an object absorbs the sun's rays and converts them into heat; the color of the object then appears dark and it rises in temperature under the sun. Conversely, the higher the albedo (toward 1), the lighter the object appears, the cooler it will remain under the sun, as its rays are reflected.

Clothing

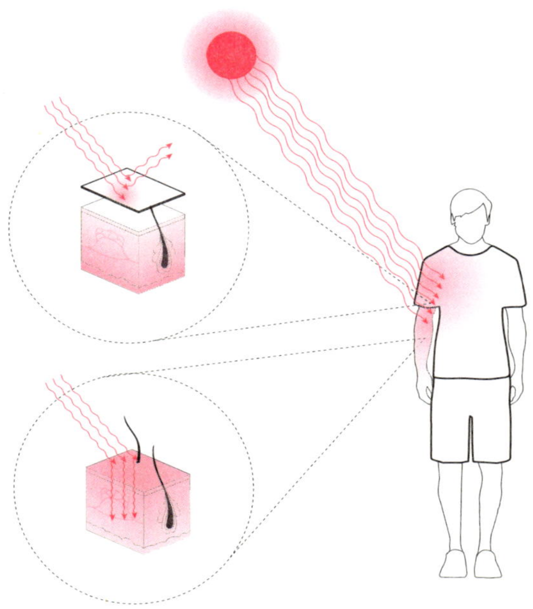

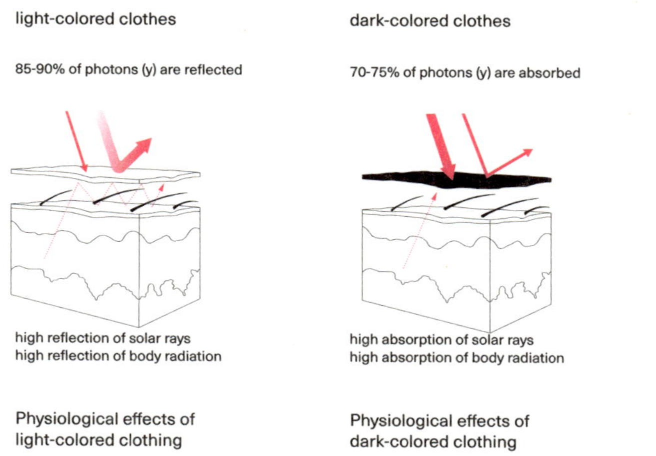

Solar radiation and clothing {Fig. 1-279}

Clothing protects the body from the Sun's rays because it blocks electromagnetic waves. Only part of the heat is transmitted to the skin through the fabric. This prevents overheating by radiation under the sun in hot climates.

Nevertheless, you must wear loose-fitting clothes to allow the sweat and heat emitted by the body to escape with the wind

Albedo and clothing {Fig. 1-280}

Clothing is used in areas that are highly exposed to the sun to protect the skin from its rays. The skin, which is highly emissive, will heat up as much under the visible part of the sunlight, as in the infrared.

A dark garment will heat up under the sun and then transmit this heat to the skin through conduction. Conversely, a light-colored garment will reflect the sun's rays and keep the skin cooler.

The heat radiation flux is approximately proportional to the temperature difference between two exchange surfaces. A layer of clothing between two surfaces will interfere with radiant heat by intercepting the energy flow; the garment will reach a temperature close to the average of the temperatures of the two surfaces, reducing the temperature difference. When the number of intermediate layers increases, the heat flow is reduced. Multiple layers are therefore effective at reducing radiant heat.

The more emissive the material of which the garment is made, the more it will rise in temperature under the sun as well as against the skin by absorbing infrared. Conversely, if it is not very emissive, it will reflect infrared and will not heat under the sun. The color of the garment also influences its temperature in the visible part of the sun's rays, with black absorbing the light and converting it into heat, while white, on the contrary, reflects it and remains cool.

Thermal Radiation in Spaces

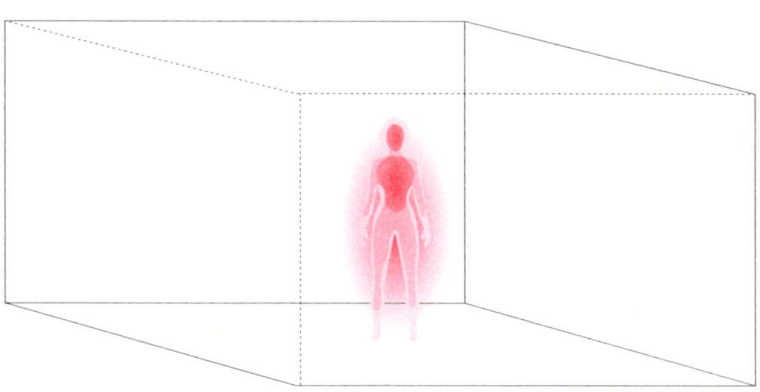

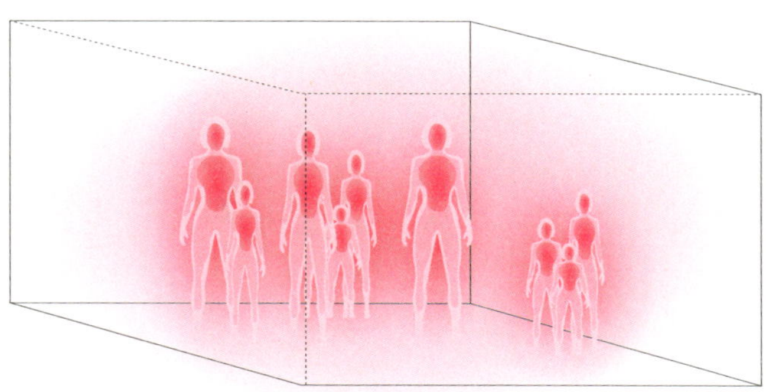

Thermal radiation accumulated with a human body {Fig. 1-281}

In a space with only one person, the heat rays of the human body are absorbed by the walls if they are colder, causing the body to cool.

If the walls are warmer than the temperature of the skin, they will cause the body to heat up.

Accumulated heat radiation with several human bodies {Fig. 1-282}

When many human bodies gather in a poorly ventilated space, the accumulation of thermal radiation from each individual will raise the temperature of the room, which can lead to thermal discomfort.

For this reason, spaces accommodating many people are traditionally larger, and above all higher, to evacuate heat from bodies upwards.

Physiological Impact of Light

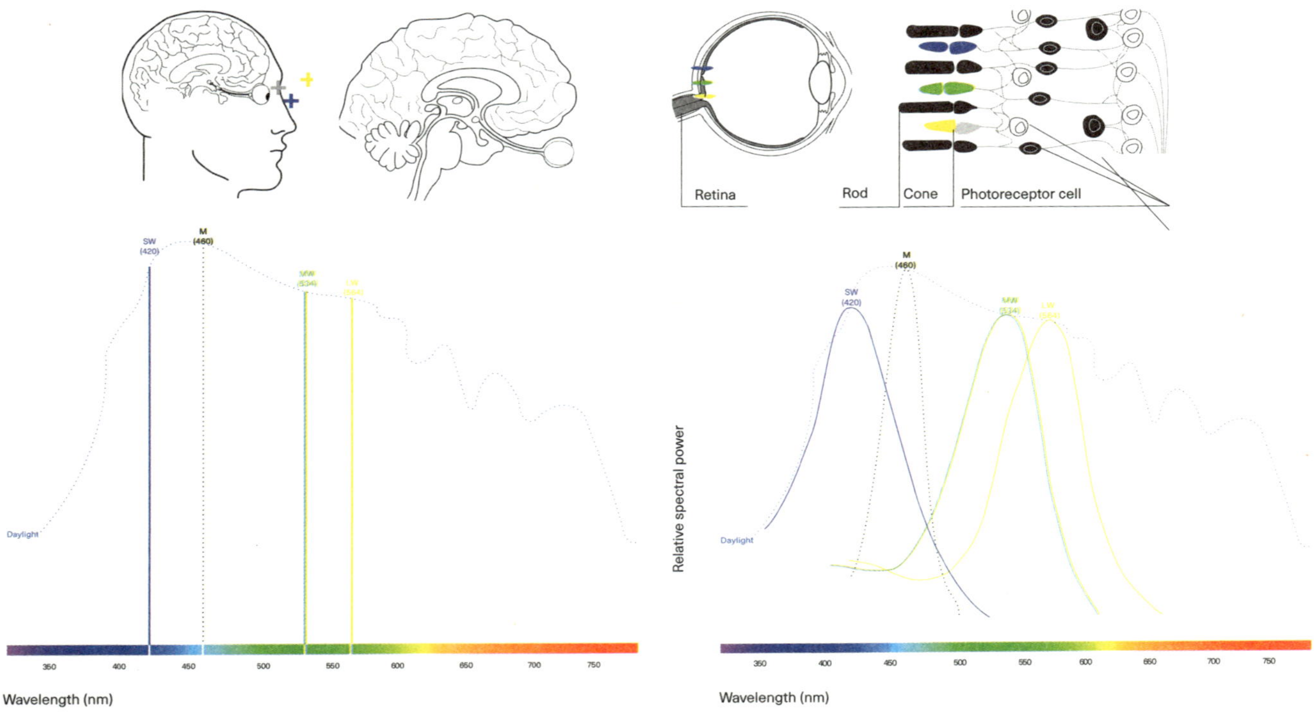

Human vision {Fig. 1-283}

The human eye sees only a small part of the electromagnetic spectrum, between 400 nm (violet) and 800 nm (red). Thanks to the combination of three photoreceptors, the cones, the brain processes the different colors by addition, determining the perceived color.

Human perception of light is phenomenological, as the perception of shapes, colors and light intensity, and endocrine, as the activation of certain hormones according to the intensity and electromagnetic spectrum of light received by the retina. Light intensities are perceived by photosensitive rods, which only distinguish variations in luminosity. Colors are perceived through the 3 photoreceptor cones, each of which is sensitive to a specific electromagnetic spectrum (420 nm, 534 nm, and 564 nm around the blue, the green and the yellow). Covering a part of the visible electromagnetic spectrum of natural light, these three cones give the perception of color to the world by superimposing their information in the brain. A fourth wavelength is perceived by the human eye, around 460 nm, which is the one that blocks the secretion of the hormone melatonin, which is responsible for sleep. Blue waves keep us awake while yellow waves let natural sleep come to us. Indeed, above 460 nm, electromagnetic waves no longer affect melatonin, thus allowing the human natural circadian cycle to pass, in 24 hours, from an increase in the secretion of melatonin (under yellow light in the evening) to the disappearance of these secretions (under blue light in the day). Rays of less than 420 nm should be avoided as they cause macular degeneration in sensitive subjects.

Circadian Cycle

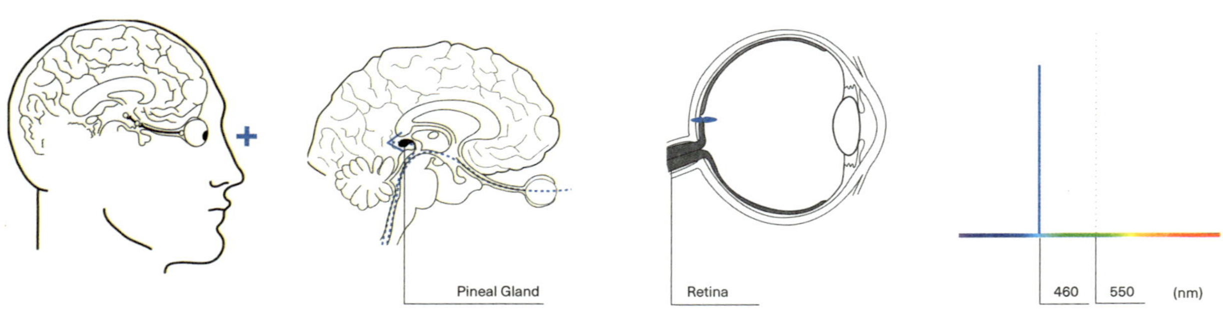

Melatonin cycle {Fig. 1-284}

When light at a wavelength of 460 nm is perceived by the photoreceptors, the lymph node cells send information to the pineal gland, which blocks melatonin secretion. When humans are no longer exposed to this light, as during the evening and night, exposed to wavelengths greater than 550 nm or to certain artificial lights without blue light, the gland secretes melatonin. Though the cones are no longer sensitive to colors at night, the rods, on the other hand, perceive night light with a peak around 498 nm.

Light spectrum allowing melatonin secretion

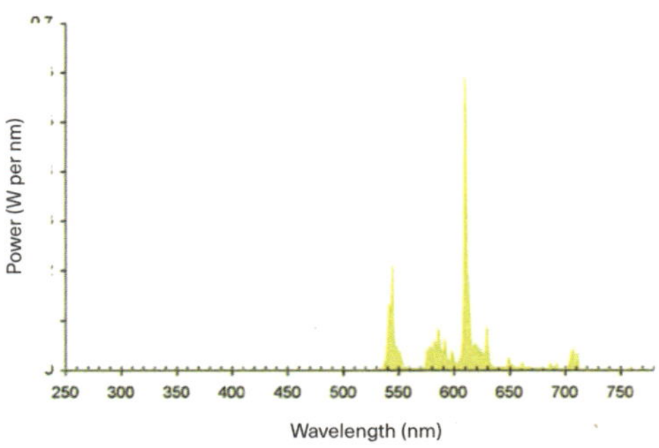

Light spectrum blocking melatonin secretion

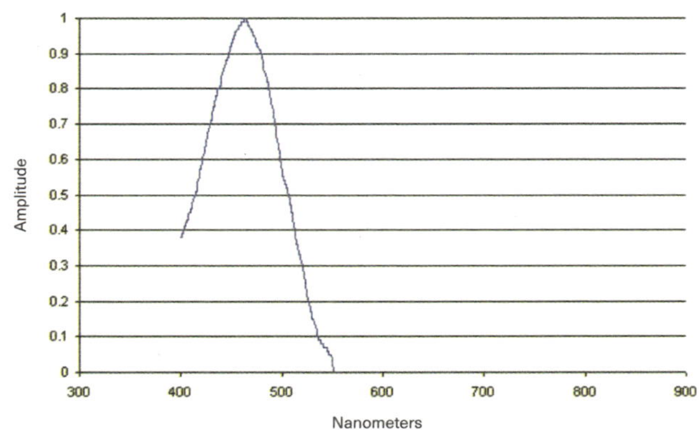

Wavelengths and Melatonin {Fig. 1-285}

On the left, the electromagnetic spectrum greater than 550 nm (orange, yellow, red) which does not influence the natural rhythm of melatonin, thus allowing hormone secretion at night or in the evening. On the right, the electromagnetic spectrum (below 550 nm, in the blue range) of pineal gland response, which induces the suppression of melatonin hormone secretion and stimulates arousal.

Blue light blocks the secretion of melatonin, the hormone that induces sleepiness. During the day, natural light contains blue and melatonin secretion is blocked by this wavelength. However, at the end of the day, when the sunlight loses its blue and becomes yellow, when the emitted spectrum is above 550 nm, when night falls, the pineal gland secretes this sleep-regulating hormone.

On the other hand, research by medical scientific institutions into chronobiology has highlighted the health risk of artificial lighting due to the fact that the electromagnetic spectra emitted do not correspond to human physiology. For example, people who work under a reduced spectrum of light, especially fluorescent light, may suffer from headaches and insomnia - which does not occur when working under natural light. Indeed, the spectrum emitted by fluorescent light is discontinuous with respect to the natural spectrum, broad and regular, covering the entire visible spectrum. The few emission peaks of fluorescent lights are not all perceived directly by the human eye, forcing it to interpret the wrong wavelengths. This recomposition of unperceived wavelengths tires the eye and causes headaches.

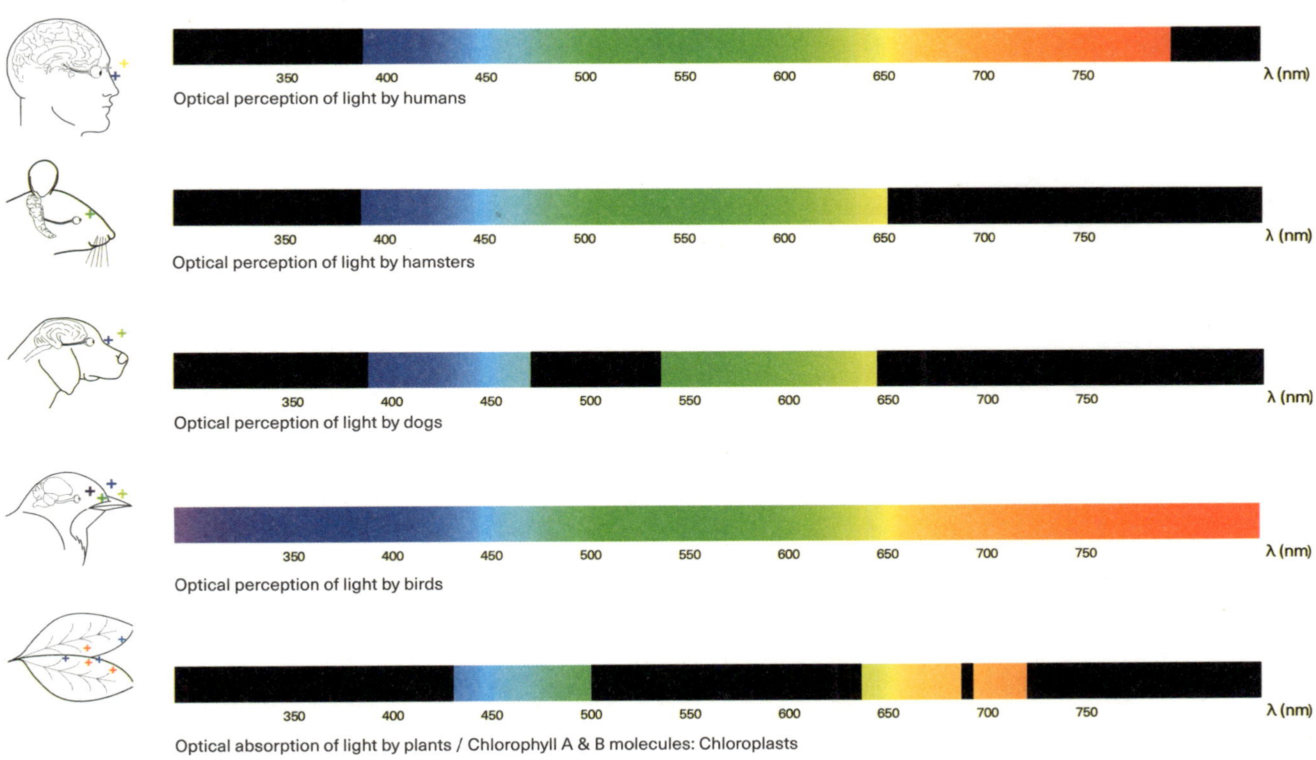

Perception spectrum of different animal species {Fig. 1-286}

Mice, which have only one cone in the retina of the eye, see the world in shades of green. Dogs, which have two, see the world in two colors, while birds, which have a fourth cone in the ultraviolet, see colors that we do not know.

Plants absorb blue and red through chlorophyll to make energy, and reflect green, which has no purpose in photosynthesis.

VI.Urban Scale

Light Pollution

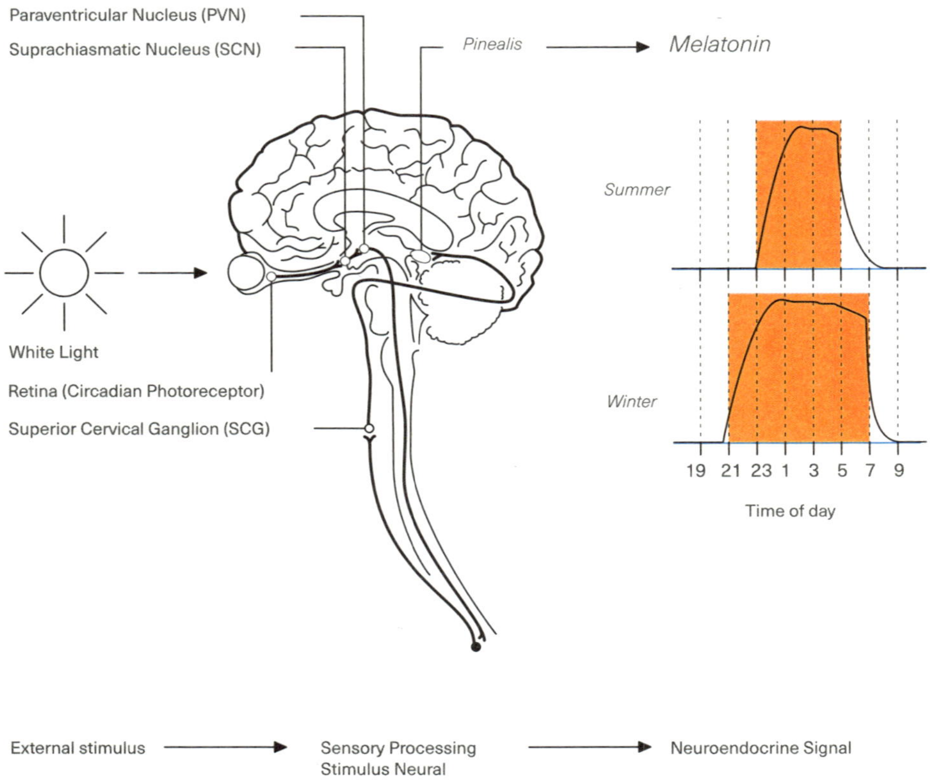

Black or orange night {Fig. 1-287}

The absence of artificial light outdoors at night or light in wavelengths above 460 nm (in yellow-orange or red) allows animals (including humans) to follow their biological rhythms naturally, without disturbance. Conversely, a city lit by white or blue light will disrupt their internal biological rhythm, desynchronizing it from the natural astronomical rhythm of alternating day and night.

Since the 19th century, urban landscapes at night have been polluted by artificial light. Artificial light is considered a serious form of pollution in modern life because of its unnatural nature and its physiological effects on people, as well as on animals, insects, flora and fauna, which depend on natural variations in light intensity and color for navigation, sleep, food and survival. Human circadian rhythms control body coordination, cardiovascular efficiency and muscle strength, the digestive system, libido and, most importantly, wakefulness and sleep habits. The pineal gland of the central nervous system (CNS) controls the secretion of melatonin; melatonin is the hormone that causes drowsiness when it is secreted. It is inhibited by light on the retina and permitted by darkness. Light stimuli are directly detected by a small number of retinal ganglion cells containing melanopsin and are transmitted to the CNS by the retino-hypothalamic tract.

In addition, the human response to light is wavelength-dependent: short wavelength light has been shown to be more effective than longer wavelengths at suppressing nocturnal melatonin and retarding the rhythm of melatonin. Thus, artificial blue light (460-480 nm) is most effective at preventing melatonin secretion and thus desynchronizing human circadian rhythms from the Earth's rotation that produces day and night alternations. Artificial blue light emitted at night is, therefore, the most "polluting". Strategically, it will be advantageous in the city to implement orange light night lighting (640-660nm) which does not block the secretion of melatonin naturally at night, thus not disturbing human and animal biological circadian cycles, the natural alternation of wakefulness and sleep. Or to switch off lights.

Gradation of Light Pollution

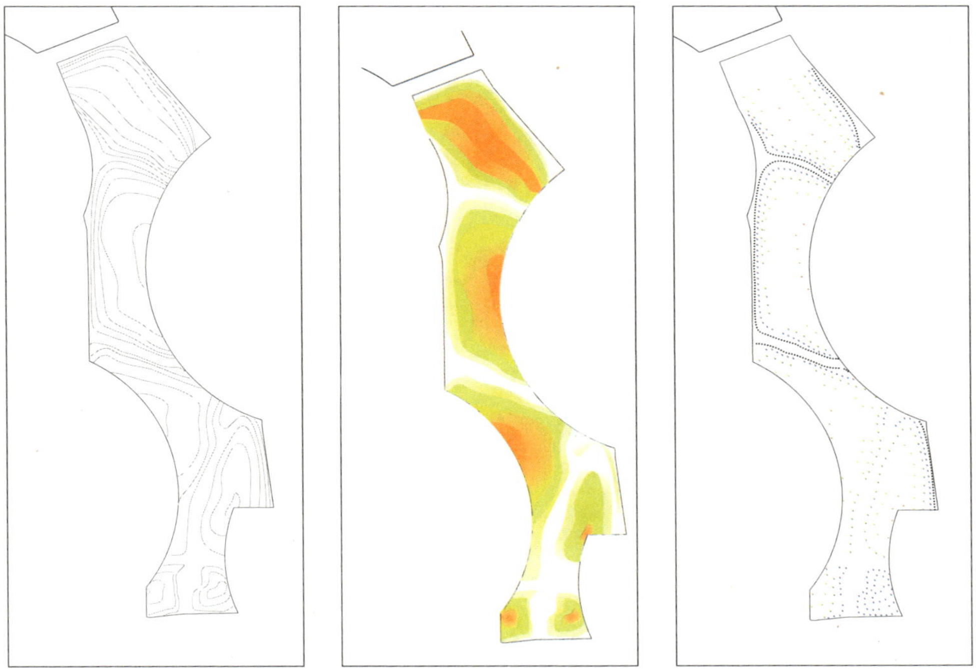

Light intensity gradation lines

Light temperature gradation

Implementation of lighting masts

Pollution-free areas away from roads

Polluted areas close to roads

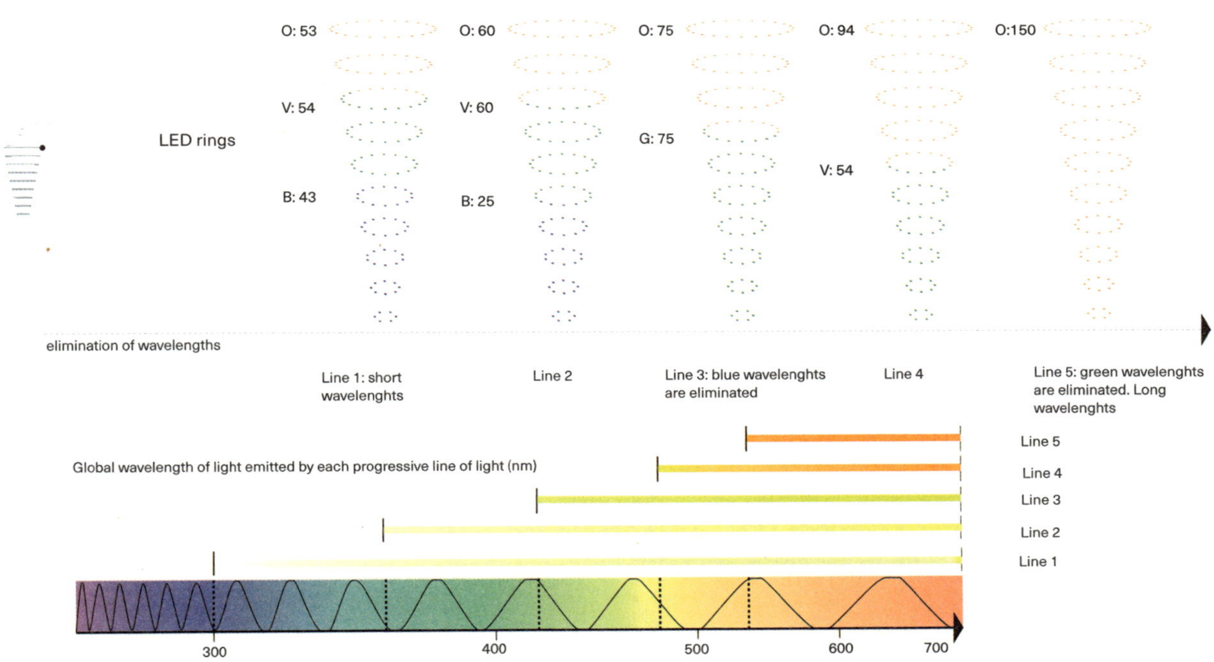

Short wavelengths of light (blue) have the greatest effect on melatonin levels (blocking its secretion), while long wavelengths (orange) have the least effect and are therefore considered to be the least "polluting". For Taichung Central Park in Taiwan, shorter wavelengths of light emitted by lamps are gradually eliminated the further one moves from dwellings and roadways, and the more one enters the heart of the park in order to preserve the natural circadian biological rhythms of animals. The color temperature of the LEDs of the street lighting thus varies between the coldest (white) on the periphery of the park and the warmest (red) in the heart of the park.

One Kilometer of Shade

Plot plan

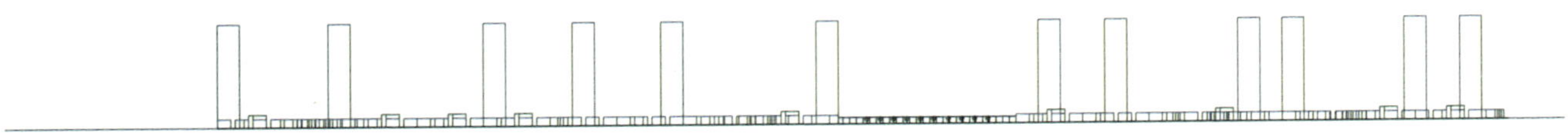

South elevation

1: The kilometer of shade
(canopy 1450m x 60m x 7m in height,
rendered with 31,434 100 cm x 160 cm photovoltaic panels)
2: Towers (height: 120 m)
3: Low urban fabric, 10 m high, with narrow streets.
4: Arvand Rood river
5: Tree square

One kilometer of shade {Fig. 1-288}

Between the new built-up district and the river, a canopy of photovoltaic panels 1,450 meters long by 60 meters wide, provides a public space to accommodate all kinds of leisure activity in this temperate climate through shading.

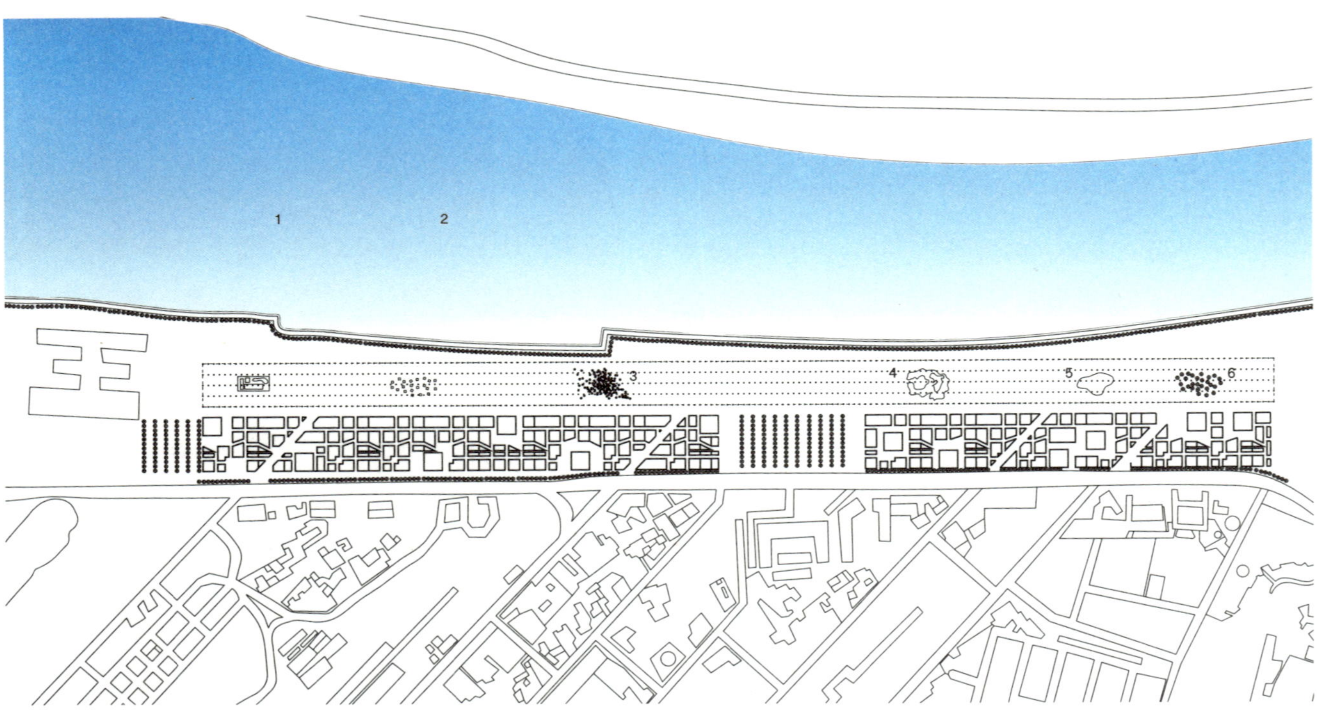

1: THE UNDERGROUND COLD SQUARE The conduction cooling square.
2: THE WIND SQUARE The convection cooling square (tbc)
3: THE MIST SQUARE The evaporation cooling square
4. THE CAVE SQUARE The inertia cooling square
5. LAKE SQUARE The cooling square
6. THE FOREST SQUARE The evapotranspiration cooling square

Shade hosting different types of refreshment {Fig. 1-289}

Beneath the canopy of a kilometer of shade, different islands of freshness take their place offering many meeting places doubly protected from the heat.

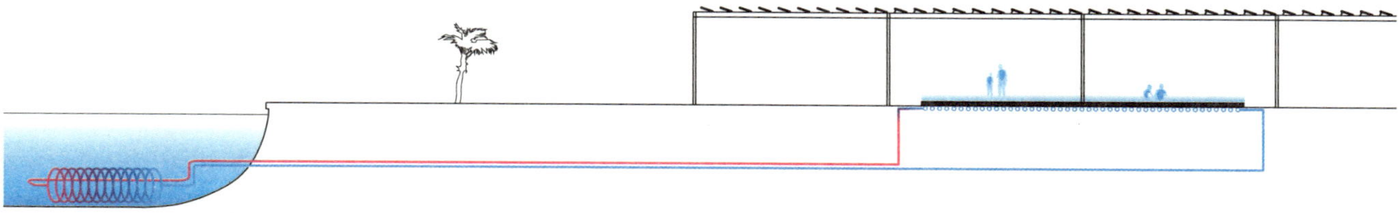

Cooling by geothermal energy {Fig. 1-290}

Beneath the shade of the kilometer of shade, a place is defined by ground in which cooler water circulates, taken from the river, cooling strollers by radiation and conduction.

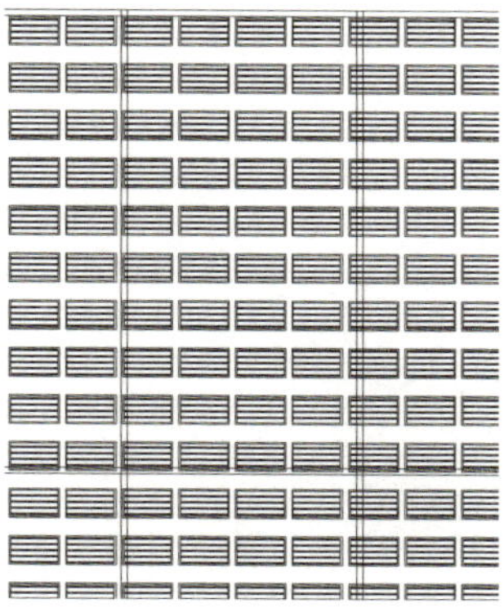

Photovoltaic power production {Fig. 1-291}

The 87,000 m2 of shade, 7 meters high, are covered with photovoltaic panels to produce electricity for the new district while creating shade to offer a cool public space, conducive to strolls, leisure and meetings.

Plant shade {Fig. 1-292}

Beneath the shade, a palm grove is proposed, for example, to double the shade effect constituting an island of freshness in the promenade beneath the kilometer of shade.

The project is the urban redevelopment of the former oil port of Basra in Iraq into a new mixed-use district (shops, housing, offices) overlooking the river. The urban development principle, between the river and the new buildings, is to constitute a huge canopy of shade 1 kilometer in length as a public promenade. This kilometer of shade is a roof raised to 7 meters in height, 1,450 meters long by 60 wide, made of photovoltaic panels, providing a refreshed climate as it is sheltered from the intense solar radiation of the region. Beneath this canopy, islands of coolness are established, as well as many other spaces cooled by convection, conduction, evaporation, thermal inertia, etc., which are added to the radiative cooling generated by the shade.

VII. Case Study

Open Climate

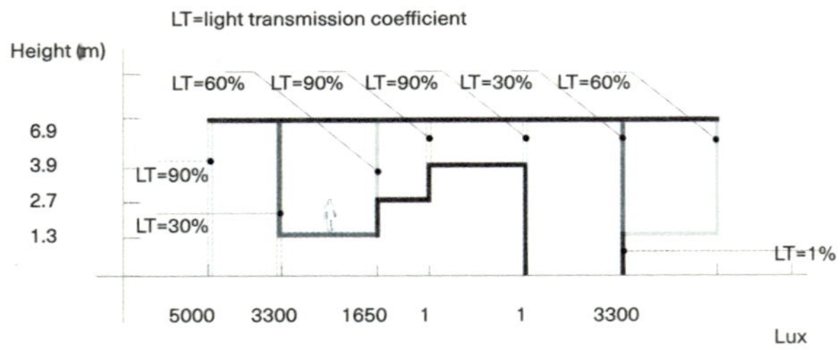

Section of glass distribution by LT

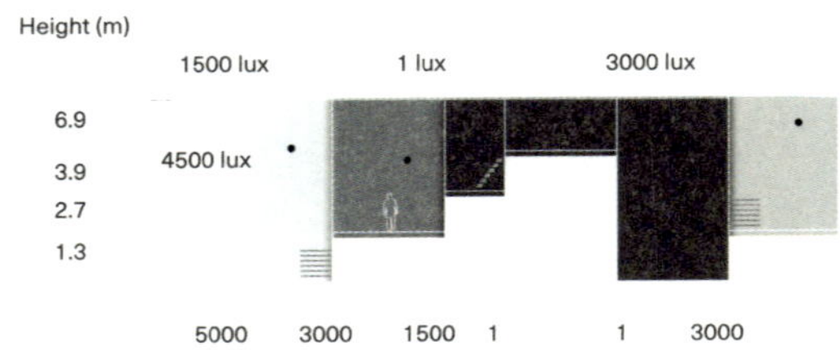

Section of daytime light distribution

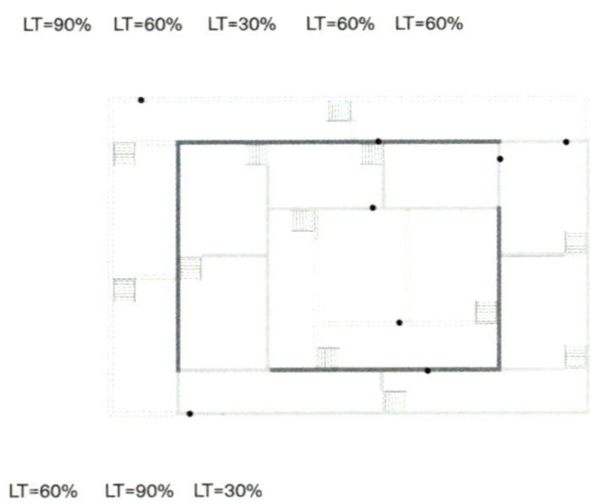

Plan of glass distribution by LT

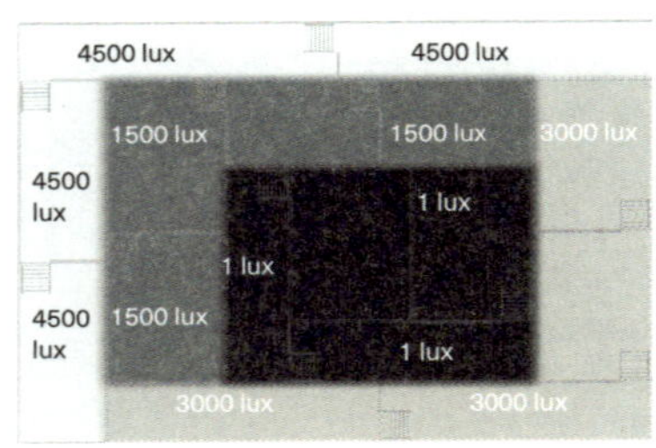

Plan of daytime light distribution

Exterior light intensity 5000 lux

Light Gradation {Fig. 1-293}
Distribution of the different types of glass according to their light transmission coefficient

Variation of light intensity {Fig. 1-294}
Typical distribution of light according to the glass transmission coefficient

This pavilion for a School of Fine Arts in France consists of the addition/succession of rooms as volumes that are differentiated only by variations in light and temperature. There is, therefore, no visual difference between these surfaces and, a priori, no specific use. Only the light and the temperature vary by taking advantage of Archimedes' law, with warmer areas at the top and colder ones at the bottom, relying on the progressive loss of brightness as one goes toward the interior. The pavilion is organized by two parameters that qualify the space: light and temperature, according to two dimensions, one horizontal and one vertical, in plan and section. In the horizontal dimension, the plan resembles a Russian doll, from the periphery to the center, from the brightest to the darkest, a succession of 4 series of 4 rooms wrapped in each other. Each series has a light transmission coefficient. The first, completely transparent on the periphery, is wrapped in a glazed wall with a light transmission coefficient of 100%. The second series has a coefficient of 66%, the third 33% (some walls have a coefficient of 66% only because some of the light has already been filtered by the first wall).

The last series filters the remaining 33% of light to create a completely black atmosphere. Vertically, the surfaces develop from bottom to top, according to the principle of the ascent of masses of hot air and vice versa, of the descent of cold air. Thus the rooms located at the bottom are the coldest, and the higher up you go the rooms will be warmer.

The pavilion is a combination of 4 luminosities and 4 temperatures thus creating 16 different climates. Depending on the time of day, people can appropriate the space and move freely. Thus, one space may be given over, thanks to its reduced, uniform light, to 19th century paintings. Another, in the warmer climate, may be used for to working at a table. A cooler space will correspond to the workshop of a sculptor, whose work is more physical. The completely black space can host video. In the end, it is a choice of 16 rooms identical in size but with as many different climates. Art students are encouraged to choose where to be, whether warm, cold, light or dark with several atmospheres to discover, capture or succumb to.

Black Day

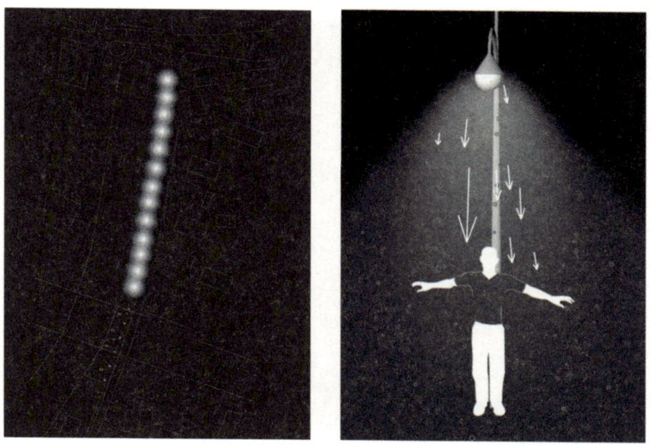

Nocturnal street lighting {Fig. 1-295}

At night, street lighting is a miniature reproduction of small daytime suns that illuminate the street, as in daylight, distorting the natural alternation of day and night, leading to a sleepless night.

Diurnal street lighting {Fig. 1-296}

During the day, the inverted public lighting of the Black day project proposes mini, cold, black night skies, bringing about a black day.

The introduction of street lighting in cities during the 19th century was the reason for one of the major social and political revolutions of urban planning and the shape of the city. The goal was demiurgic: to create daytime at night. Public lighting is behind new urban typologies (such as boulevards, for example) but also of novel behaviors, such as noctambulism, strolls at sunset and dance balls. The goal today is similar, but more contemporary, more ambiguous: to create night time during the day. Our project for the development of a street in Danzig, Poland, is to invent a negative urban street lamp, allowing physically to create night time during the day. It is a perverse response to the eternal day created by modernity, the Internet, and contemporary globalization. After the "White night" we seek to invent the "Black day". After noctambulism we would like to propose diurnism. To create night time during daytime in the streets, we work on electromagnetic wave emissions, as if we were dealing with a traditional lamp, but inverted. If the traditional lamp, like a mini sun, emits visible light radiation (artificial electromagnetic radiation), our lamp emits invisible, cold electromagnetic radiation, like that emitted by the night sky.

Our lighting consists of a diffuser similar to a miniature night sky, which is cooled by a flow of glycol water at a temperature close to 0°C. Its surface is black, absorbing the entire light spectrum. As at night, our body, at 37°C, will lose energy in the form of infrared toward this cold sky. It will thus cool according to the thermal equilibrium phenomenon.
A warm body always loses heat to a cold body. Our outdoor lamp produces night time, that is to say, a cold, black light corresponding to the electromagnetic spectrum of the night sky. Five beds are installed in the street as street furniture. They are made of black, flexible metal sheets. At the head, two loudspeakers play the "Nocturnes" by the Polish musician Frédéric Chopin, but also other composers such as Debussy, Glinka, Satie and John Field, the inventor of this form of music conceived at the beginning of the 19th century for the night.

Expanded house

Light Humidity Temperature

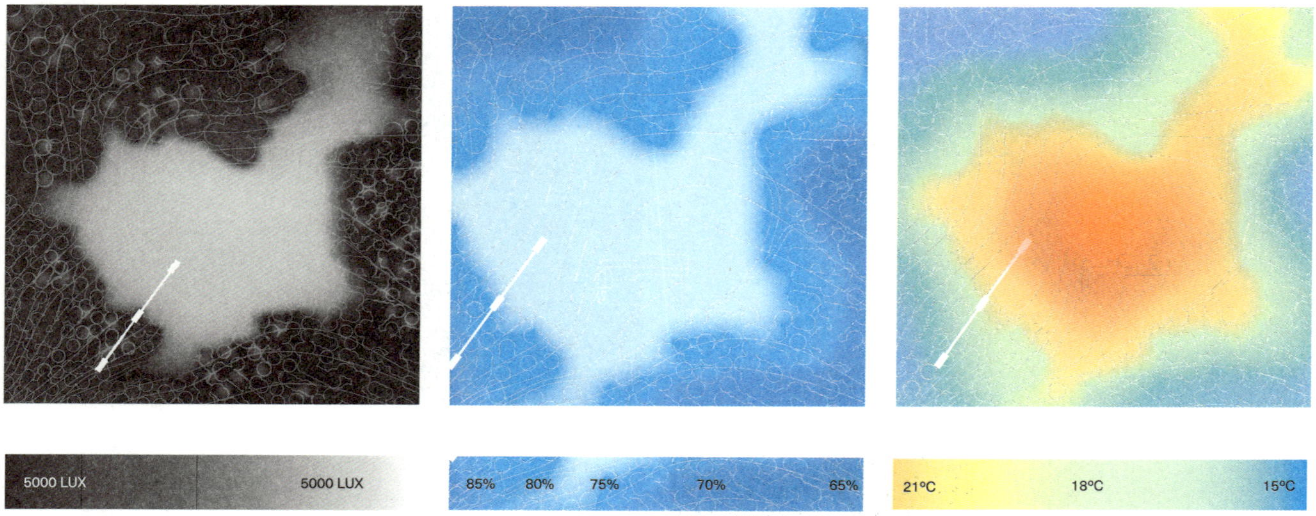

Microclimatic qualities of the sites {Fig. 1-297}
Naturally, the site offers a variety of climates, each of which is adapted to a time of year or a time of day.

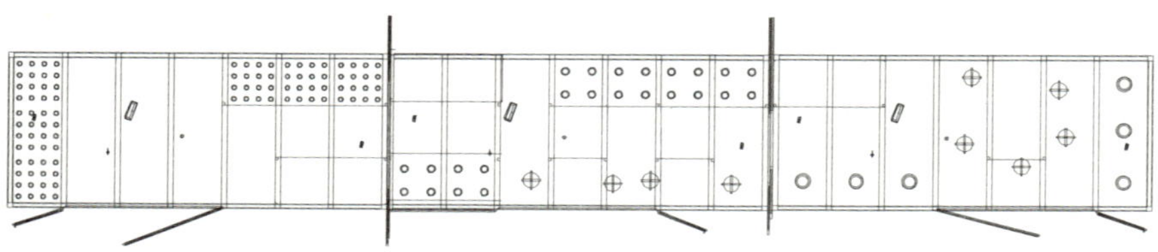

Compact plan

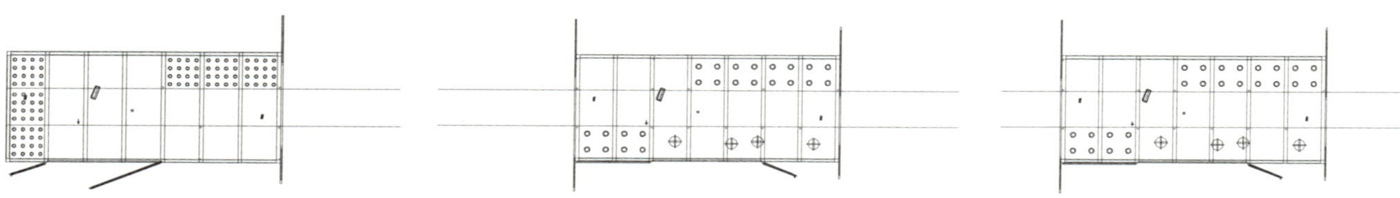

Expanded plan

Expansion of the house {Fig. 1-298}
The house spans over different microclimates, providing a variety of natural environments of different qualities.

Here, the idea is to expand the house in a landscape, to disperse the rooms throughout the site, choosing particular areas according to their climate. The house no longer resembles a closed, compact object, it is exploded, and its rooms scattered throughout the site, multiplying the views, climates, situations and atmospheres. This project for an artists' residence in the north of England occupies the entire site. It scatters uses in the different spaces, climates and qualities of light, humidity and temperature based on the times of the day or the seasons. Three sites are selected: the prairie, the border between fields and the forest, and the forest itself. Each site is determined by specific climatic qualities: light, temperature and humidity related to the trees. Depending on the time of day or season, the interior conditions of the rooms vary. The uses of each room will be closely linked to these different qualities: the warmth of the forest at night or of the fields on winter days, the coolness of the border in spring, the warmth of the meadow in winter beneath the sun or, conversely, the coolness of the forest on a summer's day. Its inhabitants are then free to move around according to the season, according to the time of day, to enjoy the natural qualities of the climates offered by the local geography.

Climatheque

Winter

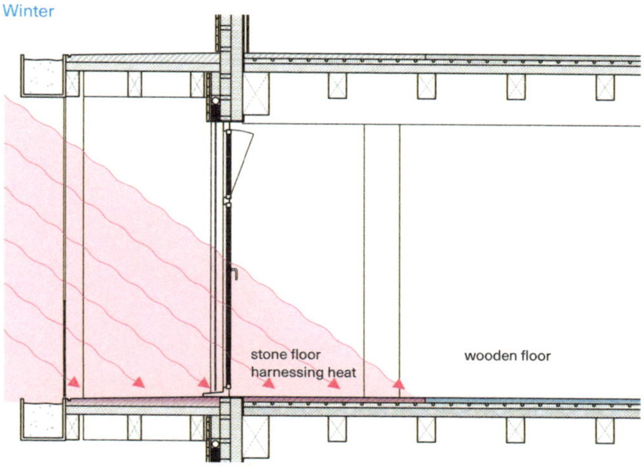

Summer

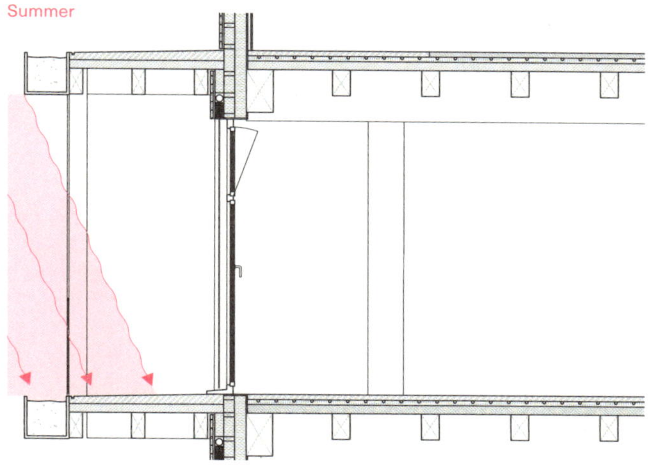

Sun screens {Fig. 1-299}

The balconies are sized to block the intense summer light and let it pass in winter. The direct rays of the sun provide both light and heat, in the visible and infrared spectra.
In summer, we will seek to get rid of the high radiant power of light and infrared (971 W/m2), to avoid both glare and heating, favoring diffuse natural light rather than direct sunlight for lighting. We do this by providing balconies all around the building acting as a sun screen, blinds, and plants at the ends of the balconies.

Number of hours of sunshine in June

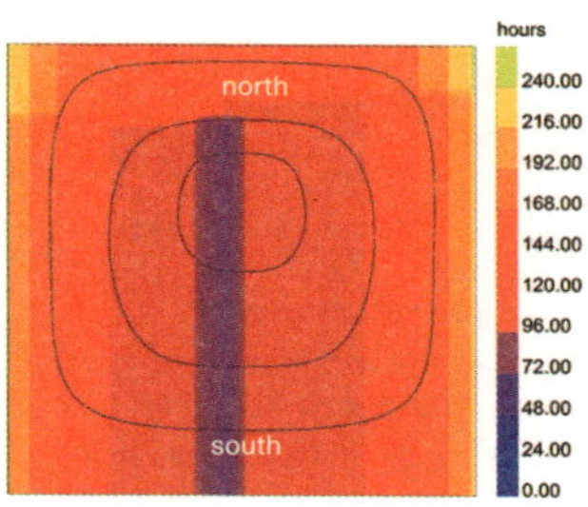

Direct sun hours

Number of hours of sunshine in March

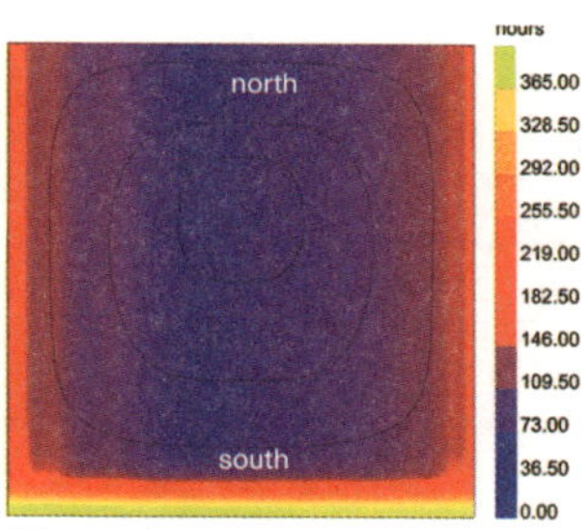

Direct sun hours

Number of hours of sunshine in December

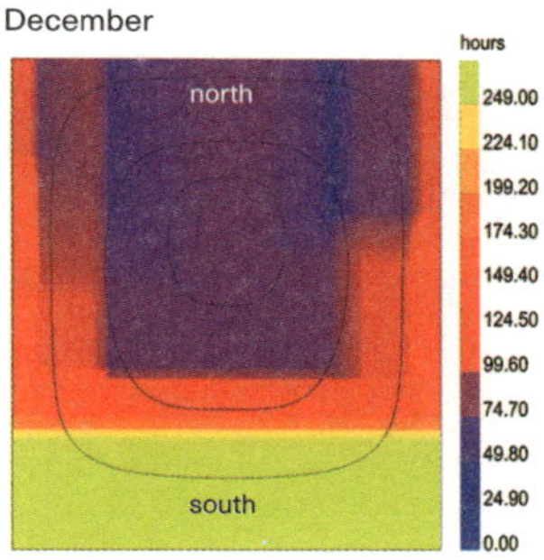

Direct sun hours

- strong natural light
- moderate natural light
- weak natural light
- no natural light

Overall luminance

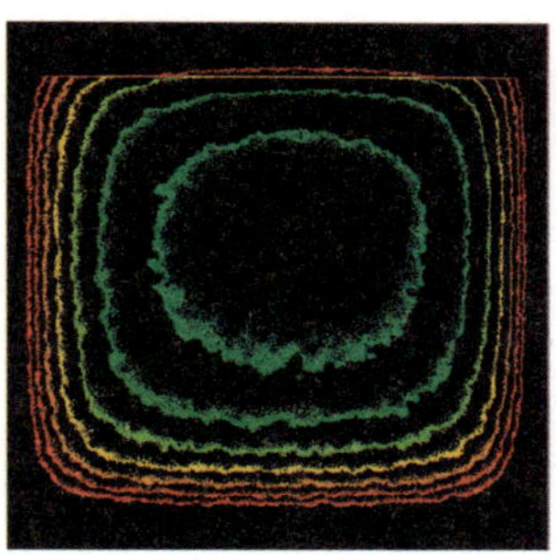

- strong natural light
- moderate natural light
- weak natural light
- no natural light

north

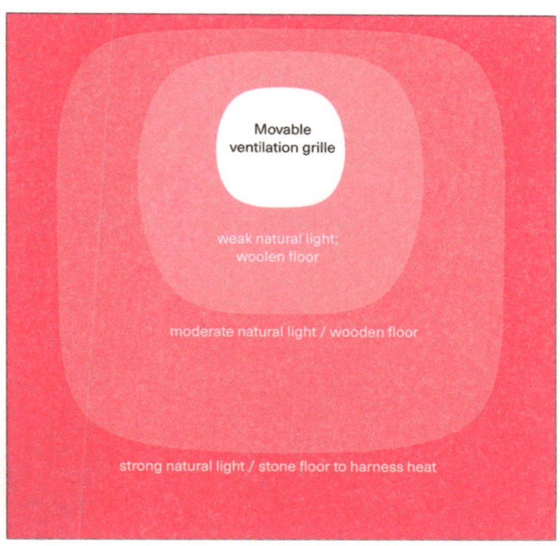

south

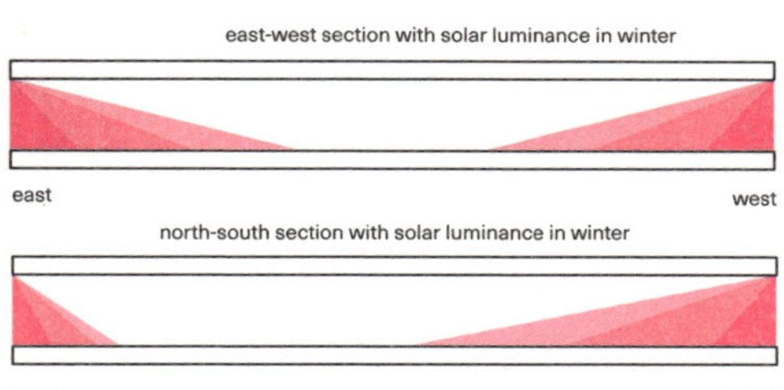

A floor plan modulated according to the sun {Fig. 1-300}

To capture the sun's heat in winter, we first analyze the winter sunshine and the average for December. Where there is most sun (yellow on the graphic above), we choose a stone floor with a thickness of 6 cm in order to fully harness the solar energy, store the radiant heat of the sun's rays in the thickness of the material to gradually release it, again at the end of the day once the sun of the short winter's days has already set, into the air by convection and to users by conduction. The parts of the floors where the sun's rays reach less, (in red and violet on the diagram below), which will harness less radiant heat from the sun, we render in wood and natural wool, in order to reduce the effusivity and, therefore, coldness in winter, in the manner of the old Persian rugs that were placed on the stone slabs of the castles in the Renaissance to improve comfort and suffer less cold in the feet.

Multiple Lights

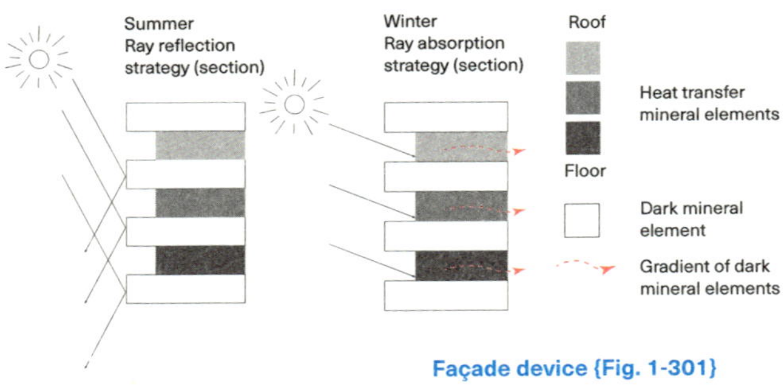

Façade device {Fig. 1-301}
The light stones reflect the rays in summer when the sun is highest, while the dark stones, set back, absorb the lower incident rays in winter to slowly heat the building.

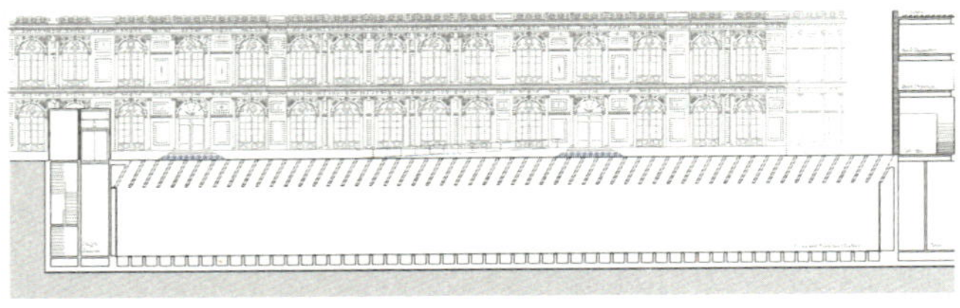

Longitudinal section

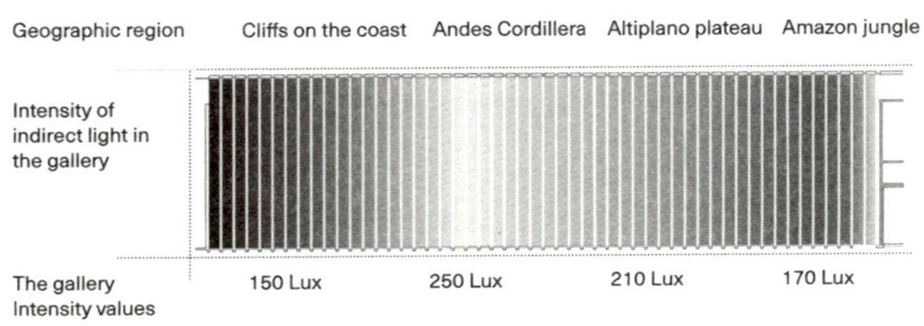

Light temperature variations across Peru

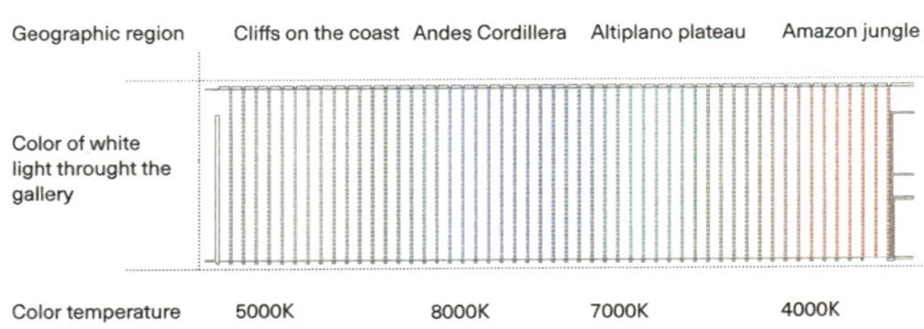

Light intensity variations

The gallery, a journey through the multiple lights of Peru {Fig. 1-303}
Light intensity and temperature vary to recreate the different luminous atmospheres of Peru, from the coast to the jungle, passing through the mountains and highlands of the Andes.

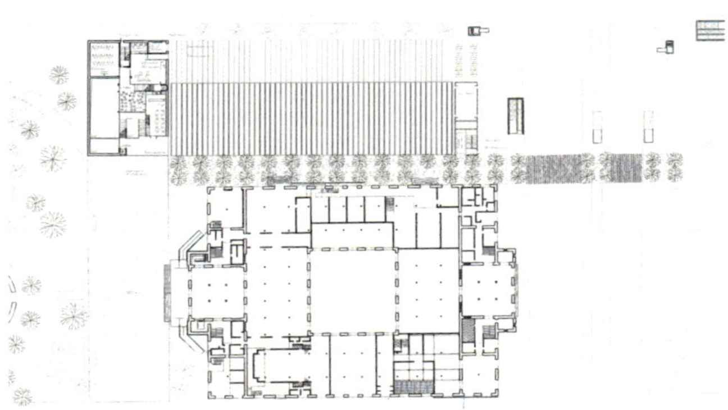

Ground floor plan {Fig. 1-302}

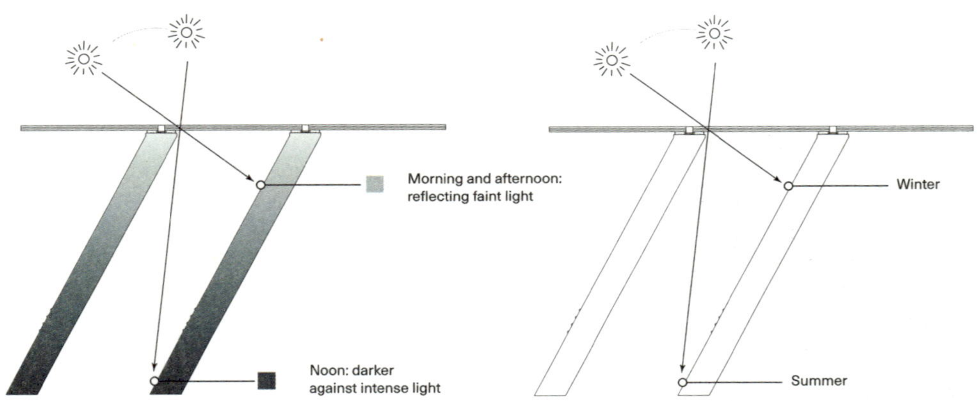

Sun screens {Fig. 1-304}
To obtain indirect light throughout the year, sun screens combine a geometric effect and a light reflection effect by a gradation, from the brightest at the top to reflect the rays of the less luminous winter sun, to the darkest at the bottom to absorb the excessive brightness of the summer sun rays

The thickness of the façade of this museum project in Lima, Peru, is made of mineral panels, ranging between 45 and 60 cm deep to create a thermal phase shift and to obtain more moderate, more constant temperatures in the walls and interiors. These 60 cm elements on the façade are made of white stone in order to reflect the sunlight and heat in summer, preventing the museum walls from heating. They alternate with 45 cm thick recessed elements made of dark stone in order to absorb the sun's rays in winter and to take advantage of the heat input. These stones, which are dark at the base of the building, become lighter as you go up. This makes it possible in winter to absorb more heat at the base of the building and to avoid heating in the upper part. The architecture thus responds to the vertical stratification of the temperature in the building.

A museum requires specific light intensities ranging from 150 to 250 lux. The light should be white, with a temperature of between 4000°K and 8000°K. Our project proposes working with these subtle variations in the intensity and temperature of light to recreate the different atmospheres found naturally in Peru, from the neutral light of the coast, through the cold light of the Andes, to the warm light of the Amazon. In the linearity of the exhibition space we propose organizing a promenade through the different lights of Peru, from west to east, starting from the Pacific coast with a temperature of 5000°K and a low intensity of 150 lux to reach the Amazon where the light is warmest at 4000°K and of medium intensity at 170 lux, passing through the Andes where the light is colder, 8000°K, and the high light intensity is at 250 lux.

EVAPORATION

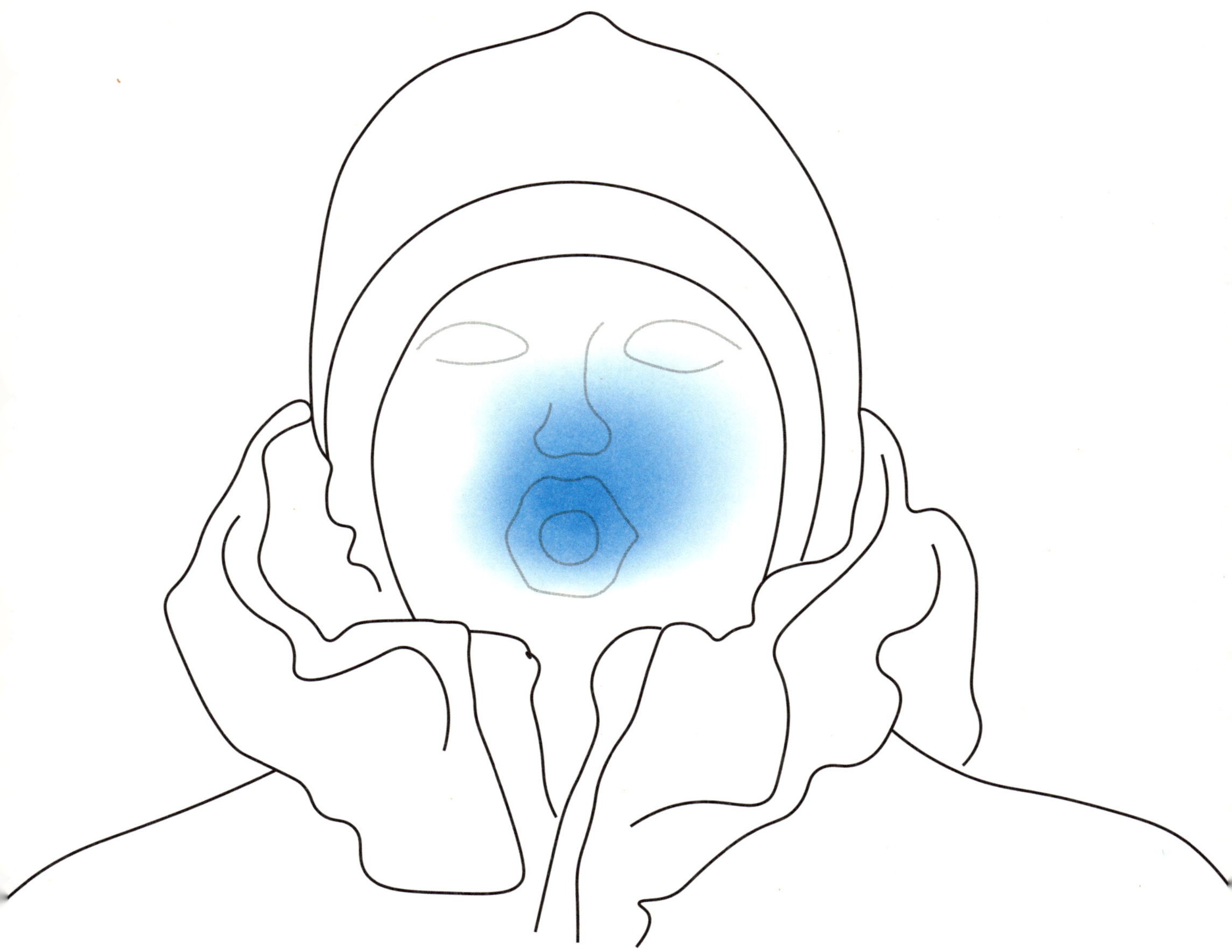

I. Definition

General Properties

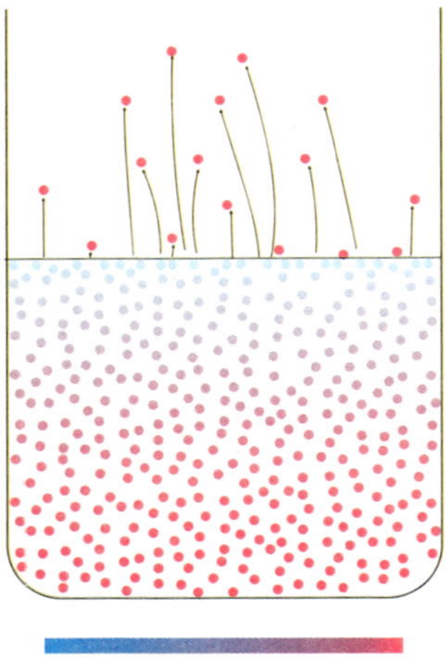

Water evaporation {Fig. 1-305}
Water gradually evaporates into the air at any temperature above 0°C.

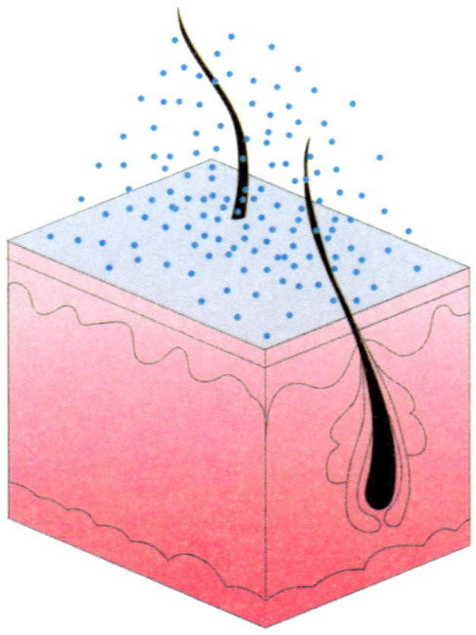

Perspiration {Fig. 1-306}
When the human body is too hot, the skin will perspire liquid water (sweat) to create an evaporation phenomenon that seeks to cool the body.

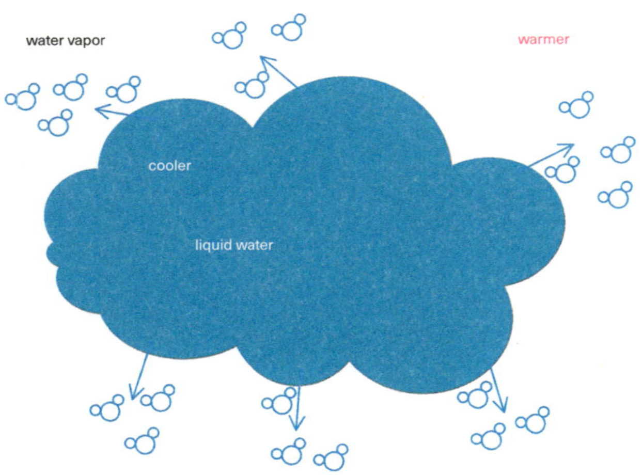

The evaporation process, from liquid to gas {Fig. 1-307}
Depending on its temperature, air may contain more or less water vapor. If the air cools, clouds form. If the air warms up, then the liquid water can evaporate again, becoming invisible.

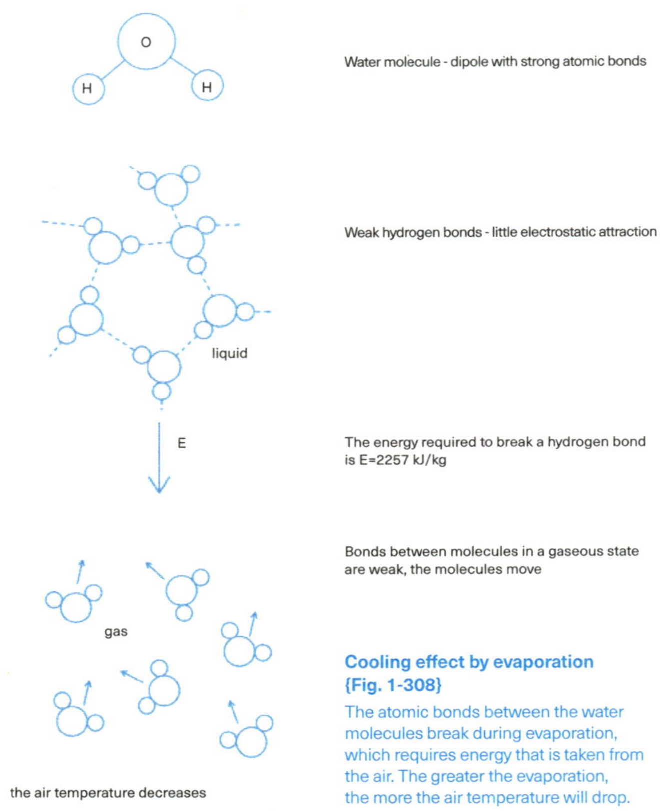

Cooling effect by evaporation {Fig. 1-308}
The atomic bonds between the water molecules break during evaporation, which requires energy that is taken from the air. The greater the evaporation, the more the air temperature will drop.

Breathing by humans, plants and animals, but also certain everyday activities, such as washing or cooking, cause the emission of water particles into the atmosphere.
This water will evaporate, turning into a gas, accumulating in the air in the form of water vapor. The physical process through which water passes from a liquid state to a gaseous state consumes energy, which is supplied by the air, which causes a drop in temperature, and therefore cooling by evaporation. This is why fountains and other water features are useful for cooling the air in summer or in hot countries. If the air temperature is above 34°C (skin temperature), the human body can no longer lose heat by conduction/convection, and can only cool itself by sweating.

However, if the air is already saturated with water vapor, the sweat will notevaporate, leading to death from overheating. However, at a given temperature, the air can contain only a certain amount of water vapor. If the water vapor content exceeds a certain threshold, known as air saturation point, the water vapor will condense and return to its liquid state. This is the main reason why we ventilate interior architectural spaces, because the water on the one hand causes the appearance of mildew and fungi, which can be harmful when breathed in, and on the other hand ruins the building materials.

I.Physiological Scale

Respiration / Transpiration

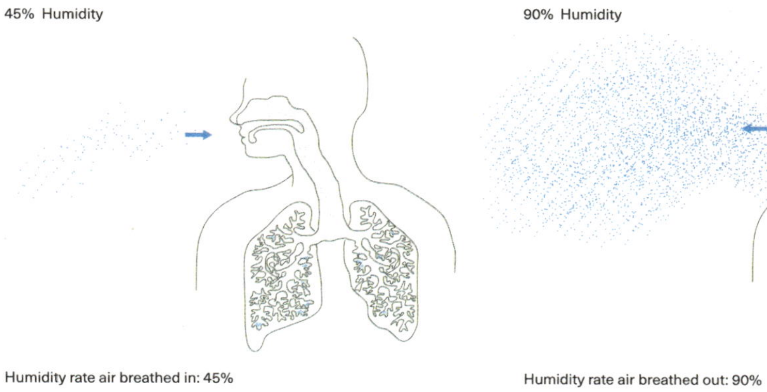

Inspiration / Expiration {Fig. 1-309}

With each breath, our lungs emit molecules of water in the form of vapor into the air, which helps to slowly increase the moisture level of the air to the point of saturation when this water condenses, if there is no mechanical or natural renewal of the air.

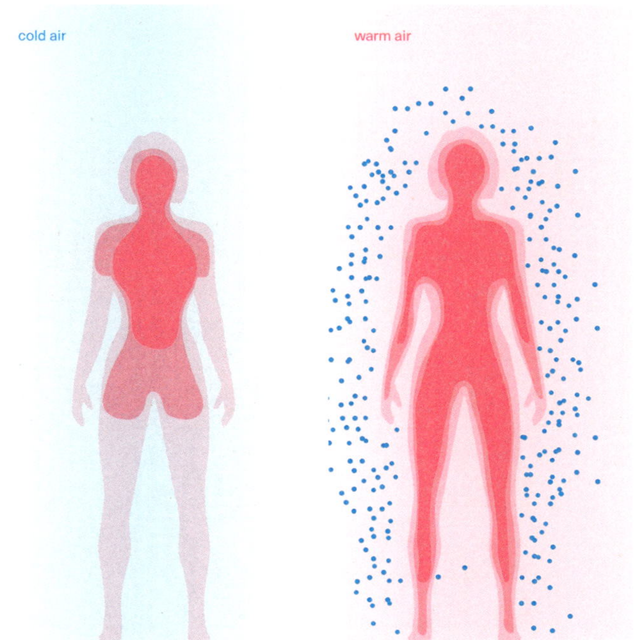

Cooling of the body by transpiration {Fig. 1-310}

The warmer the skin, the more it will sweat to cool by evaporation, gradually charging the air with water vapor, while lowering the temperature of the skin by changing from liquid to gaseous phase.

Plant transpiration {Fig. 1-311}

To raise the sap that brings the mineral salts to a plant, the leaves transpire, ejecting water and cooling the air, which then leads to the formation of clouds.

Since the human body has to maintain its temperature at a constant 37°C, it must regularly evacuate heat into the surrounding air which must be at a temperature between 20°C and 28°C. If the temperature is higher, the body can no longer naturally lose enough heat and must activate other thermoregulatory mechanisms, known as thermolysis, such as vasodilation and perspiration. If the body is hot, it will sweat more to promote the evaporation of water from the skin, thus cooling it. But for the sweat to evaporate properly, the surrounding air must be dry, because if it is moist, it will not be able to receive any more water in gaseous form and the water from the perspiration will remain liquid on the skin, producing no cooling effect.

Plants also transpire, to increase the concentration of mineral salts, which refreshes the surrounding air by evapotranspiration according to the same process as human perspiration, removing energy from the air in the form of heat to pass from liquid to gaseous form. Just as trees transpire, grassy soils evaporate water.

III. Architectural Scale

Vapor, Humidity, Condensation, Ventilation

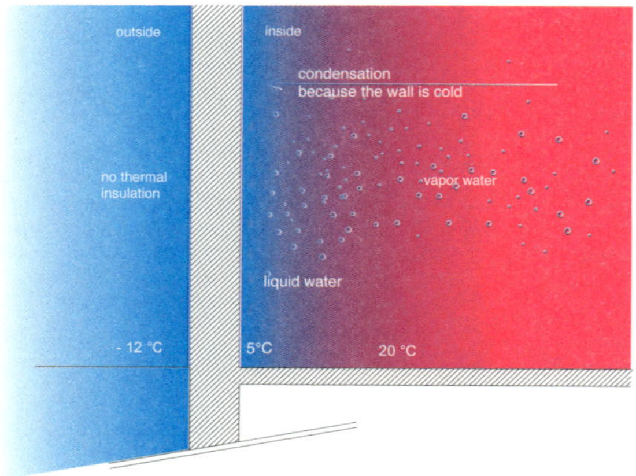

Principle of condensation on walls {Fig. 1-312}

Poorly insulated walls will conduct the outside cold inwards by heat conduction. Since the layer of cold air in the immediate vicinity of the wall on the inside cannot contain as much water vapor as the rest of the interior air in the room which is warmer, the water vapor will condense, damaging the walls.

Principle of thermal insulation to prevent condensation {-Fig. 1-313}

Properly isolating a wall from the outside will keep the wall structure warm, which will not lead to the phenomenon of condensation on the warm side of the wall. In addition, to ensure that air moisture does not penetrate the wall, it is recommended to install a waterproof vapor barrier inside.

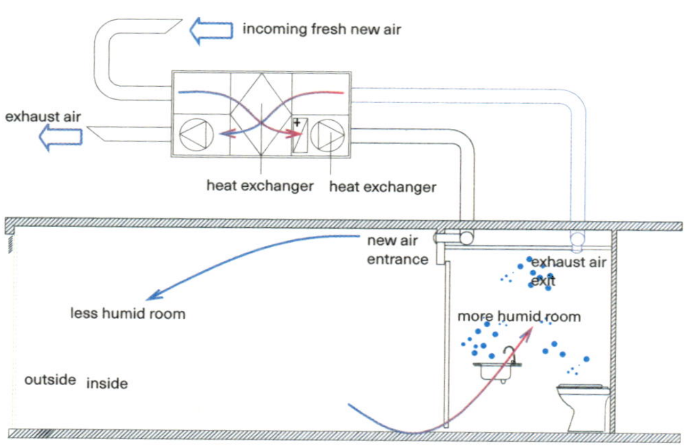

diagram of a dual flow ventilation system with heat recovery

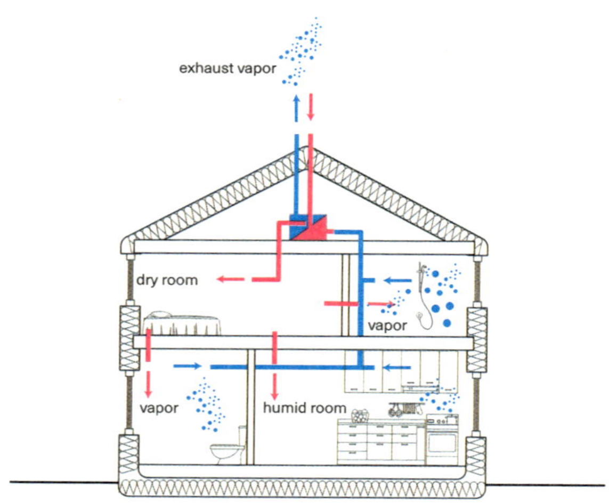

Cross-flow ventilation {Fig. 1-314}

In order to minimize energy losses by heating the fresh air entering the house, a current is created throughout the building rather than in each room individually.

The stale air extracted from damp rooms then crosses the dry incoming fresh air in the heat exchanger and supplies it with some of its heat, reducing artificial heating energy inputs.

Cross-flow ventilation {Fig. 1-315}

Ventilation is necessary to prevent water vapor from accumulating in the air indoors. The principle of ventilating premises is to inject fresh air from the outside into the driest rooms, where less water vapor is emitted (such as bedrooms for example), and to extract air from the wettest rooms such as the toilets, bathroom and kitchen.

Thus, composing the different rooms according to their moisture content, following the path of the ventilation air current, enables reducing the quantity of incoming fresh air and therefore the quantity of energy necessary to heat it in winter, while limiting the number of ventilation ducts, the spaces themselves acting successively as ventilation ducts.

Humans produce water vapor by breathing out and sweating, but also during certain domestic activities such as cooking or taking a shower. If this water vapor is not evacuated, the air can become saturated with water vapor, reaching saturation point, also called dew point, that is to say the threshold at which the air can no longer receive any water vapor, and causing the water to condense. This can create mildew on the walls, and in particular on exterior walls that are poorly thermally insulated, because they are colder. In fact, the lower the temperature of the air, the lower the dew point, the less water the air can contain in the form of vapor, and condensation is quicker.

Condensation therefore takes place primarily against cold and poorly insulated walls in winter. The temperature in the vicinity of the walls is lower than in the rest of the room, which causes the water vapor to condense precisely there, as the rest of the air in the room, which is warmer, can receive more water vapor. To combat water condensation, it is necessary to properly insulate the outer walls so that their surface temperature is not colder than the rest of the air in the room.

The House as a River of Air between Dry and Moist

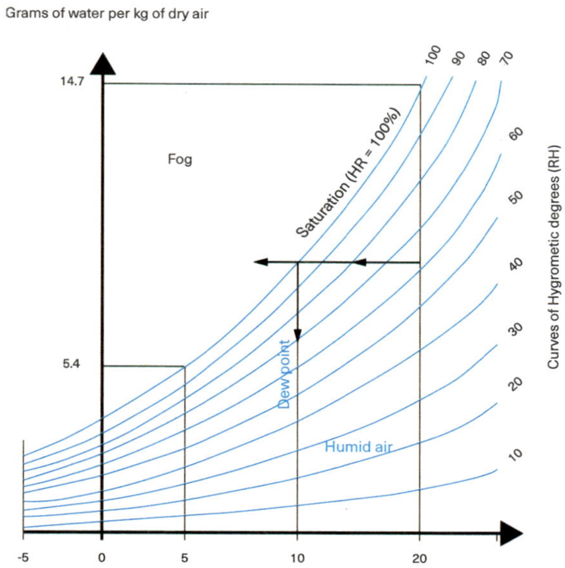

Saturation: from vapor to liquid {Fig. 1-316}

Depending on its temperature, air may contain more or less water vapor. If the water vapor content exceeds a certain threshold, it condenses and regains its liquid form, which is the reason for the appearance of clouds in the sky, of mist just above the ground, of fogging on windows or when you exhale through your mouth in winter.

Thus, after taking a shower, for example, it is necessary to open the windows in order to reduce the concentration of water vapor in the air by introducing fresh air from the outside, which is drier, diluting it.

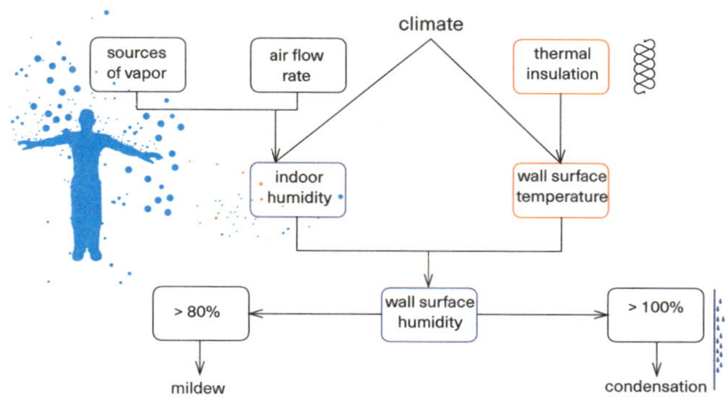

Diagram of the conditions for the onset of mildew {Fig. 1-317}

The appearance of mildew on the walls, or paint flaking off due to moisture, is directly related to the amount of water vapor in the air as well as to the temperature of the walls. If the walls are cold (poorly insulated), the water vapor condenses on them, damaging the building and damaging the health of the inhabitants.

To prevent this, it is necessary either to increase the air renewal rate or to thermally insulate the walls in contact with the outside to keep them at a temperature equal to that of the air in the room. (Source: Santé et qualité de l'environnement intérieur dans les bâtiments, Claude-Alain Roulet, PPUR presses polytechniques, EPFL Lausanne, Switzerland, 2004)

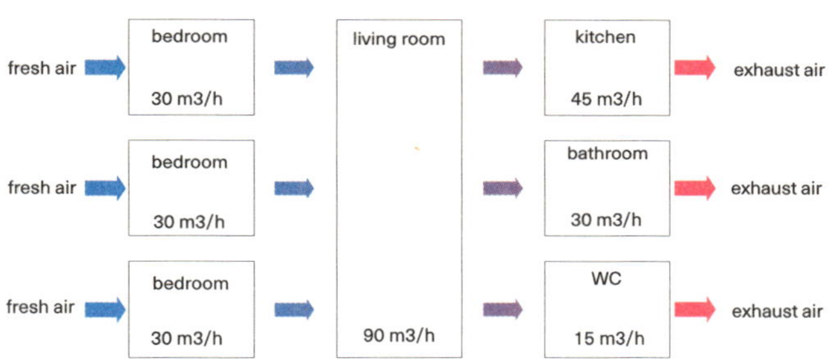

Path of ventilation for a 4- or 5-room dwelling {Fig. 1-318}

In order to minimize energy expenditure for heating the air entering the house, a continuous flow of air is created in the house, from the driest to the wettest rooms. The airflow created by ventilation always follows the same path, from the driest to the wettest rooms.

In order to determine which rooms are the wettest, the amount of water vapor emitted during the various activities must be studied. (Source: Minergie, Switzerland)

Thus:

A person asleep produces 40 grammes of water vapor per hour

A person awake produces 150 grammes of water per hour

Using the kitchen produces up to 500 grammes of water vapor in 20 minutes

Using the bathroom produces up to 800 grammes of water vapor in 20 minutes

Then:

The bedroom is placed at the start of the ventilation flow, at the fresh air inlet

Followed by the living quarters, where due to more intense breathing than when sleeping, we emit more water vapor.

The water vapor released by cooking pots and pans quickly increases the relative humidity of the air

The shower produces a lot of water vapor, which is why it is located at the end of the air flow, toward the stale air outlet

The primary reason for ventilating buildings is to evacuate water vapor from the air in order to prevent it from increasing, which, when it exceeds the dew point, would lead to condensation on the walls and the appearance of mildew and fungi, which are harmful to the health of the inhabitants. The second reason for ventilation is to dilute and remove air pollutants such as volatile organic compounds, fine particulate matter (PM 2.5, PM 10), viruses and bacteria, or other harmful spores.

The last reason is to bring in oxygen and evacuate the CO2 given off when humans breathe. Following energy efficiency logic, it can be concluded that the house can actually be understood as a river of air, as the sketch of an interior wind between the incoming drier, fresh air and the exit of the polluted air, which is wetter as it is charged with the moisture released by human activities in the house.

IV. Urban Scale

Vapor, Humidity, Condensation, Ventilation

Transpiration of trees and cooling {Fig. 1-319}

Trees cool a given space in two ways. On the one hand, the leaves of the canopy create shade, stopping the Sun's rays from reaching the ground, preventing it from heating up. On the other, transpiration by the leaves gives off tiny droplets of water that evaporate naturally in the form of gas, removing the heat from the air, thus helping the temperature to drop.

Fountain water evaporation and air cooling {Fig. 1-320}

Water in liquid form continuously evaporates in the form of gas, which lowers the air temperature due to the energy required for this phase shift from liquid to gas. The smaller and more numerous the droplets, the greater the surface area of liquid water in direct contact with the air, the greater the evaporation, the more the air temperature will fall.

Therefore, a flat, calm expanse of water, such as a pond, will cool the surrounding air less as it provides a lower surface area of water for evaporation than a jet of water that sends thousands upon thousands of tiny water droplets into the air which each rapidly evaporate

Fountain as a cooling device {Fig. 1-321}

Fountains, water jets, waterfalls, and to a lesser extent lakes, which we find in parks, cities or courtyards vaporize liquid water which, by evaporating, lowers the local temperature of the air, bringing freshness.

Evaporation and Convection {Fig. 1-322}

In the city, evaporation by trees and plant soils in parks and promenades produces a localized drop in air temperature, causing a thermodynamic imbalance with the surrounding, warmer built-up sites.

These temperature differences lead to the emergence of a convective breeze between the two opposite thermal poles. This breeze will set in motion the accumulation of hot air in the mineral parts of the city, thus cooling it.

Fountains and trees emit fine droplets of water which, when naturally evaporating at ambient temperature, pass from liquid to gaseous state using the energy in the air, thus lowering the temperature of the surrounding air. Thus, the leaves on the trees produce microscopic water droplets that evaporate in contact with the air causing the nearby air temperature to fall. Similarly, grassy soils refresh the nearby air. Fountains and water features operate on the same principle, refreshing the surrounding air thanks to the evaporation of water.

An Adiabatic Quarter

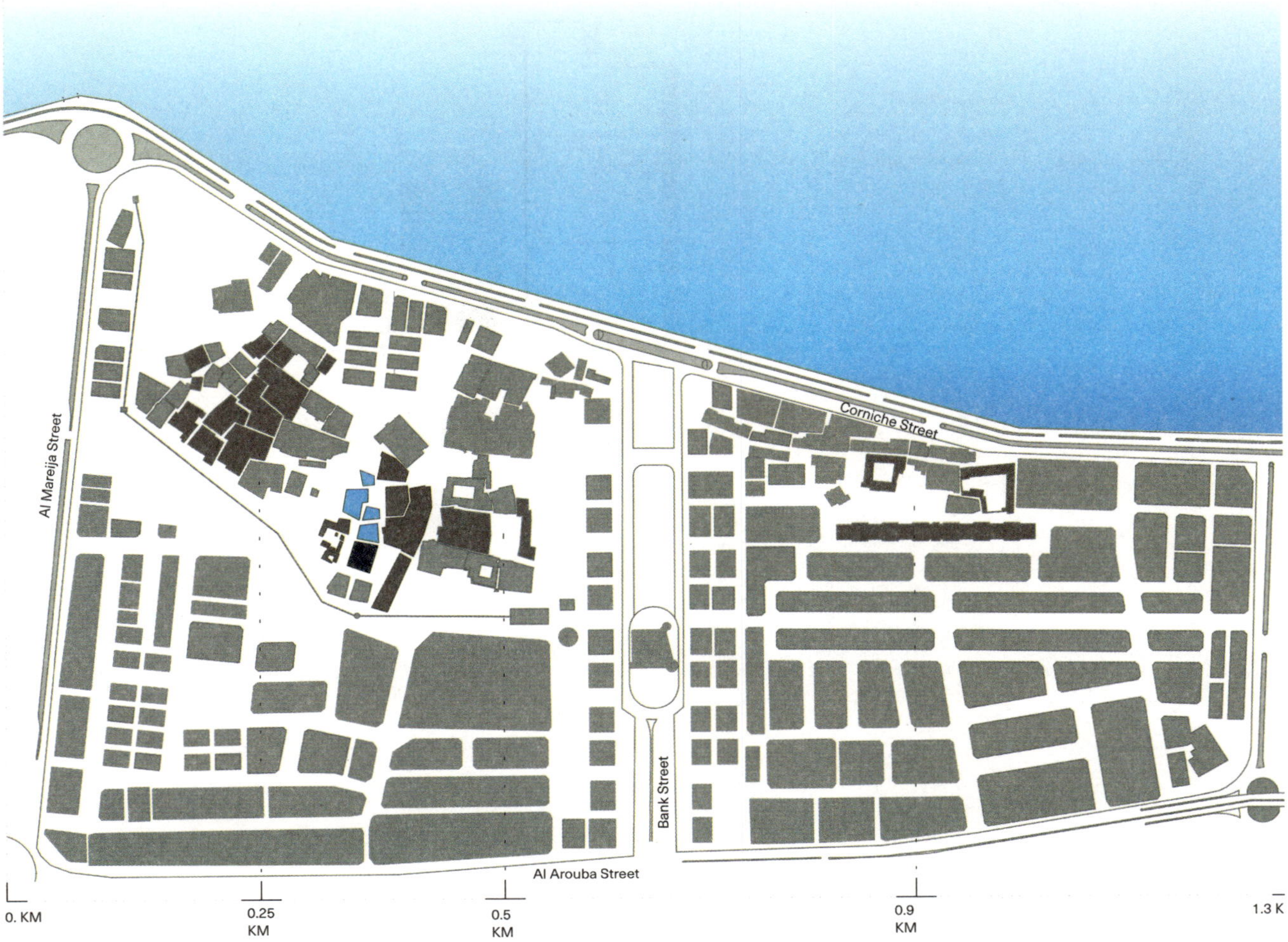

An adiabatic quarter {Fig. 1-323}
The new buildings of the Sharjah contemporary art museum are all based on foundations comprising interior lakes with fountains and water jets that cool the new incoming air by geothermal energy in the basement.

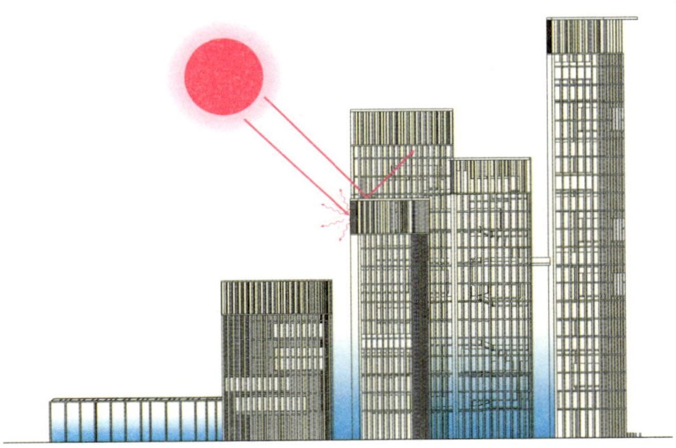

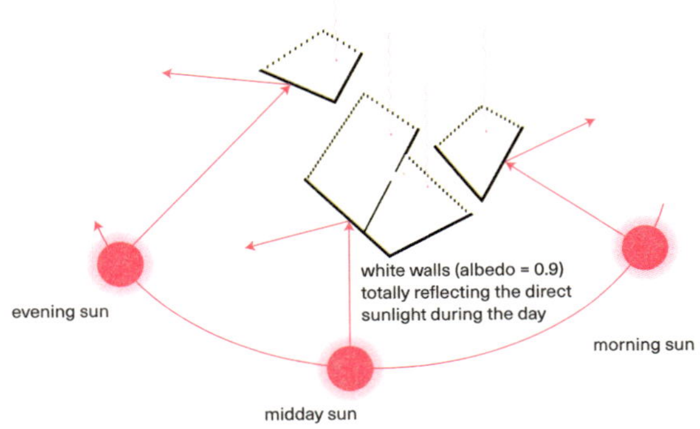

Principle of opening façades in relation to the path of the sun {Fig. 1-324}
The façades of the buildings of the museum are completely closed and opaque to the south so as not to let in the direct hot rays of the sun. In the north, the façades are open to allow the free circulation of the air initially cooled underground by geothermal energy and by the fountains

This project of a new museum of contemporary art in Sharjah proposes a cooling system free of electrified mechanical air-conditioning based solely on the cooling of the air by geothermal energy and the adiabatic principle: by the evaporation of the sea water which constitutes small inner lakes, on the ground floor, whose water, which is vaporized in fountains and water jets, lowers the temperature of the incoming air, already cooled by geothermal energy, required to ventilate the spaces of the museum and its visitors. Precise measurements of only indirect light also make it possible to avoid any radiant overheating.

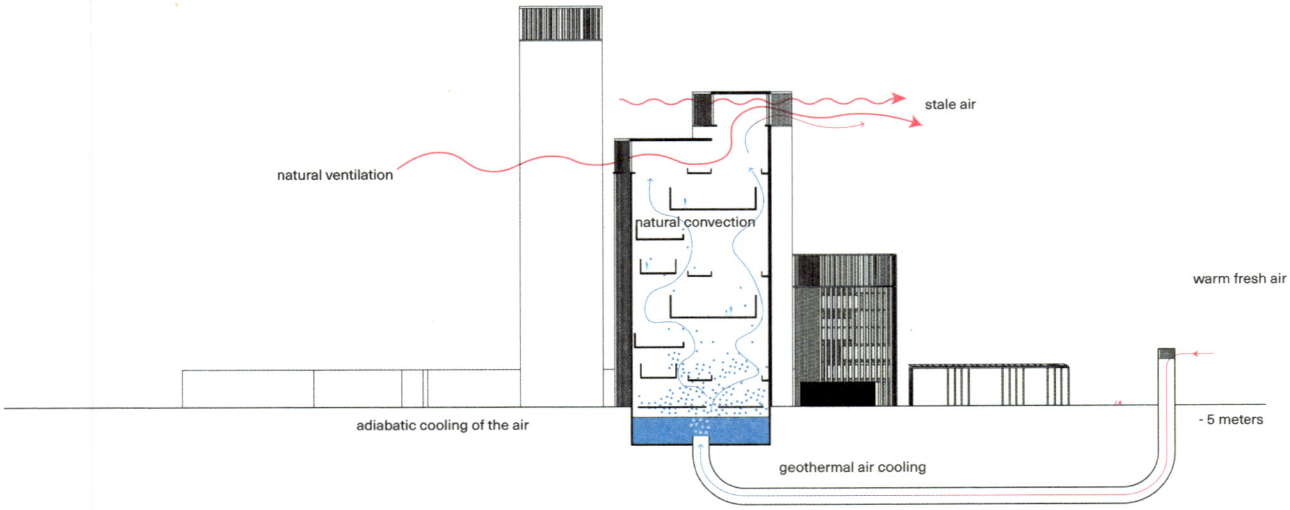

Adiabatic cooling {Fig. 1-325}

The air entering the building is cooled by geothermal energy, by circulating the air underground at a depth of at least 5 meters, in order to reach a constant ground temperature throughout the year of 27.5°C. Introducing fresh outdoor air at 27.5°C instead of 45°C saves a lot of energy and therefore reduces greenhouse gas emissions.

The cool air is introduced at the bottom of the building, further cooled by adiabatic cooling by passing through water fountains. The hot air rises by convection, and is evacuated beneath the roof by the passage of the wind, which creates suction, allowing the air to be naturally sucked in through the geothermal well.

Thus, the floor slabs of the buildings create vacuums and are themselves micro-perforated to allow air to rise throughout the building and to cool spaces and people by convection.

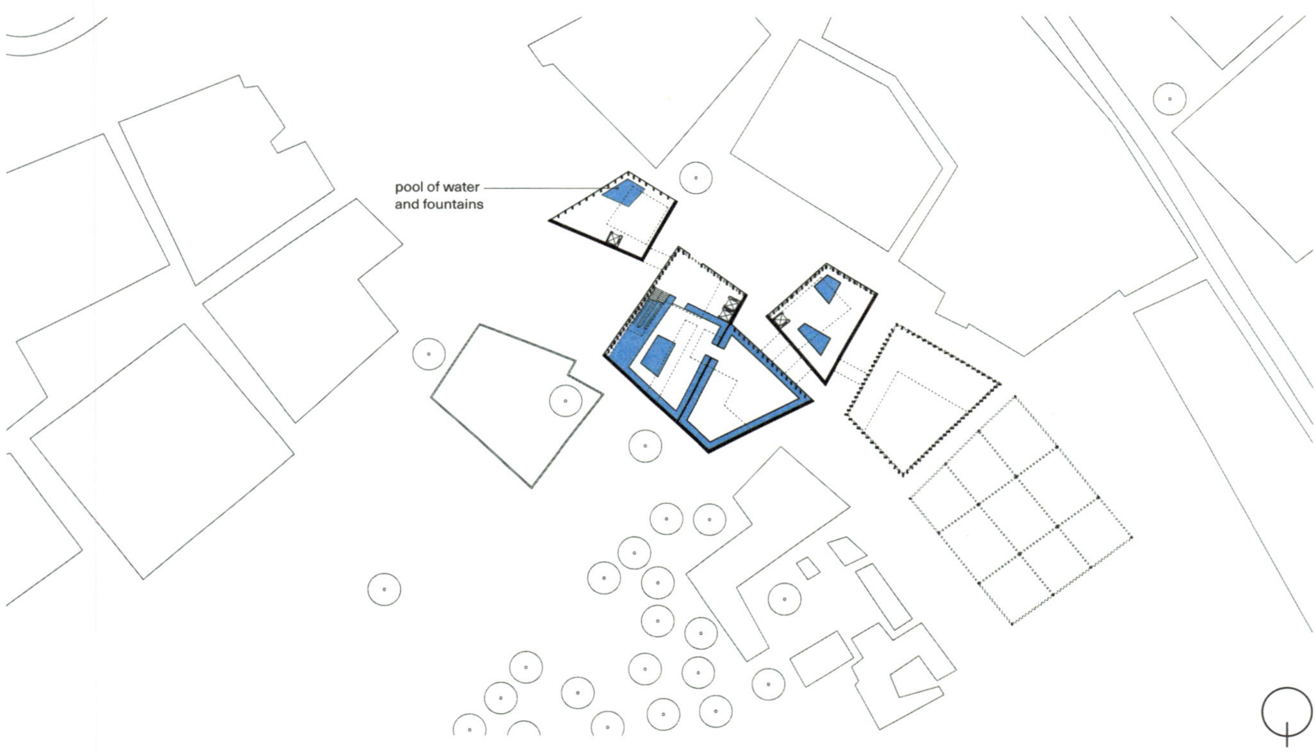

Layout of adiabatic buildings {Fig. 1-326}

Small lakes of evaporating water form the base of each building, from which the thus cooled fresh air is drawn, with the energy required to evaporate the water being taken from the air, lowering its temperature.

V. Case Study

Evaporated Building

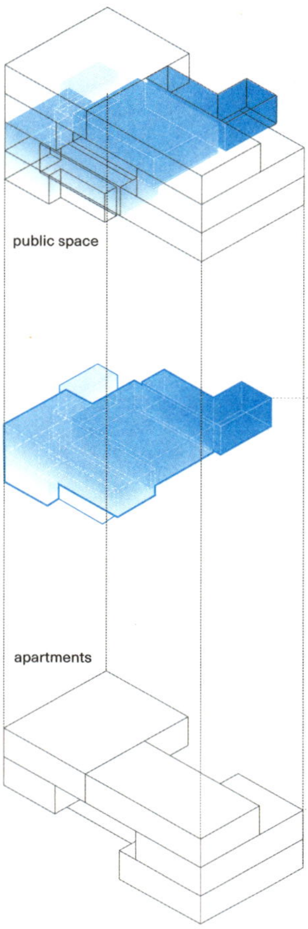

Floorplan

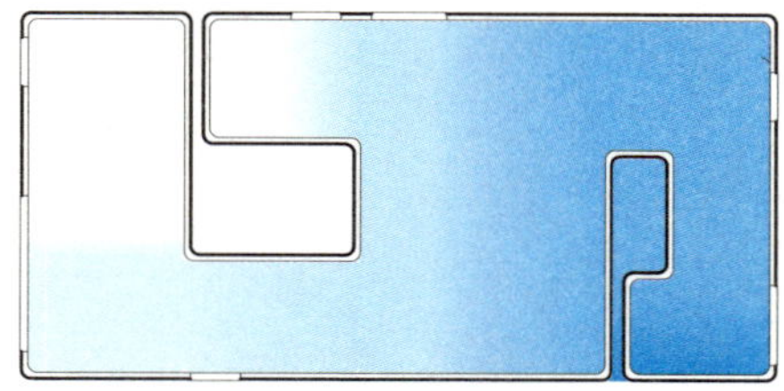

horizontal stratification of humidity

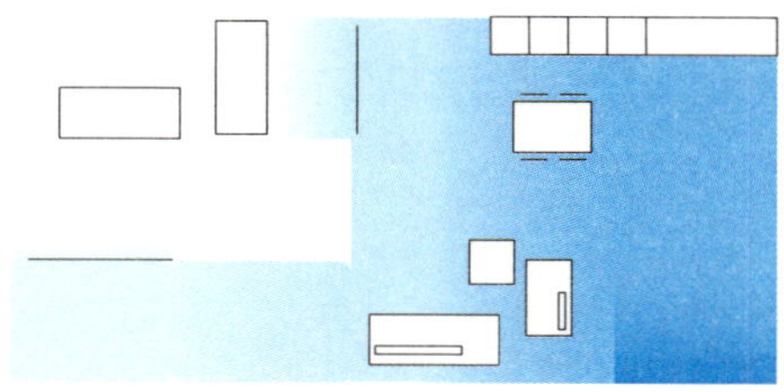

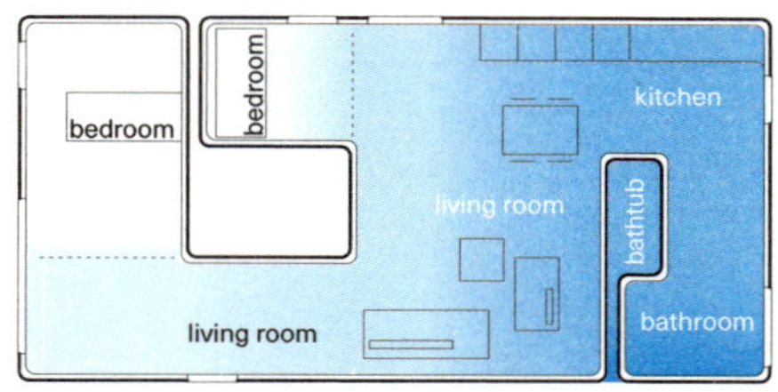

uses according to humidity

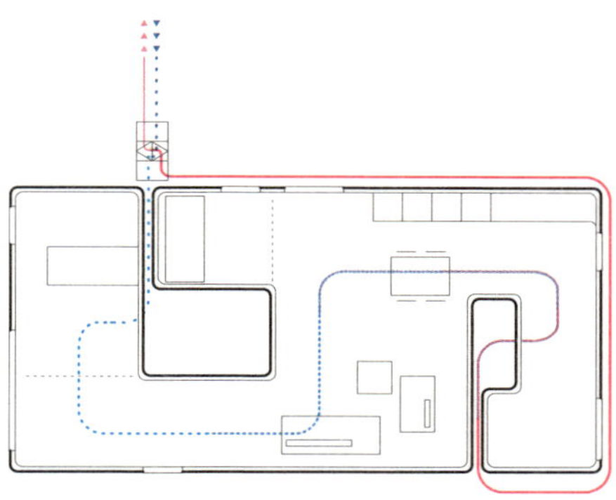

dual-flow ventilation with heat exchanger

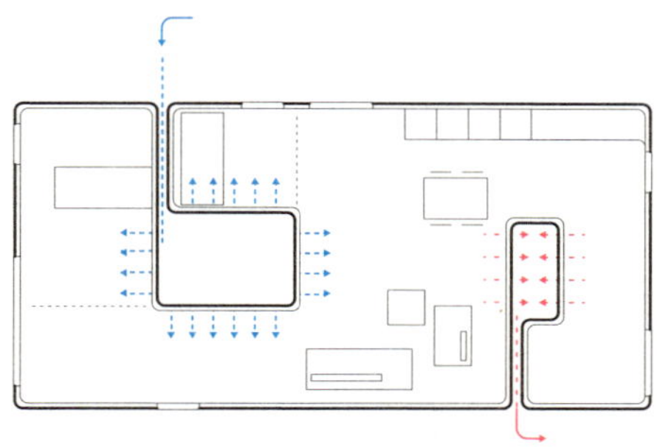

fresh/stale air

Functions follow the water vapor ratio {Fig. 1-327}

The plan extends between the intake of dry, fresh air and the outlet of humid, stale air, distributing uses according to a map of the humidity and according to the quantity of water vapor produced. Thus, the bedrooms are the driest parts, followed by the kitchen, and then the bathroom.

The design of this building is based on the path of vapor in the house. The new dual-flow ventilation system (one of the most effective solutions for reducing energy consumption) can only work if there is a single flow in the building. Air renewal begins in the driest rooms and ends in the moistest; there is thus a slow flow of air throughout the house, driving all the moisture in the same direction. This path is determined by the amount of water vapor produced by the various activities carried out in the dwelling. One person produces 40 grammes of water vapor per hour while sleeping, 150g/h when awake, 1500g/h when cooking, and 2400g/h when washing. The layout of the apartment is designed according to the interior moisture of the rooms, generating a more sensitive ratio in the rooms. This solution reintroduces a variety of climatic zones to the modern apartment. Instead of responding solely technically, we wish to transform the problem, poetically and sensitively. The apartment then becomes a natural landscape with latitudinal variations in relative humidity. An inner geography, between a miniature desert and an inner tropical jungle, between 30 and 90% relative humidity in which the inhabitants can move freely.

The public part of this building extends from the cellar to the roof with a continuous movement of air from the lowest to the highest spaces. Since moist air is lighter than wet air, the upper level of the public space will be more humid than the lower one. The common spaces will be located on this vertical vapor geography. The spaces for drying clothes will be in the upper part of the building, while the lower levels will be dedicated to the storage of potatoes, apples or wine. Just as in an old house, with its warm attic and humid cellar, used specifically to store such and such a food or such and such an object according to the conditions most suitable for their preservation. The public spaces will not only contain the functional distribution, such as the stairways and elevators, but also diverse shared spaces where social life can take place. The different humidities of these public spaces will be used as a kind of collective refrigerator or shared pantry, scaled up to the size of the building.

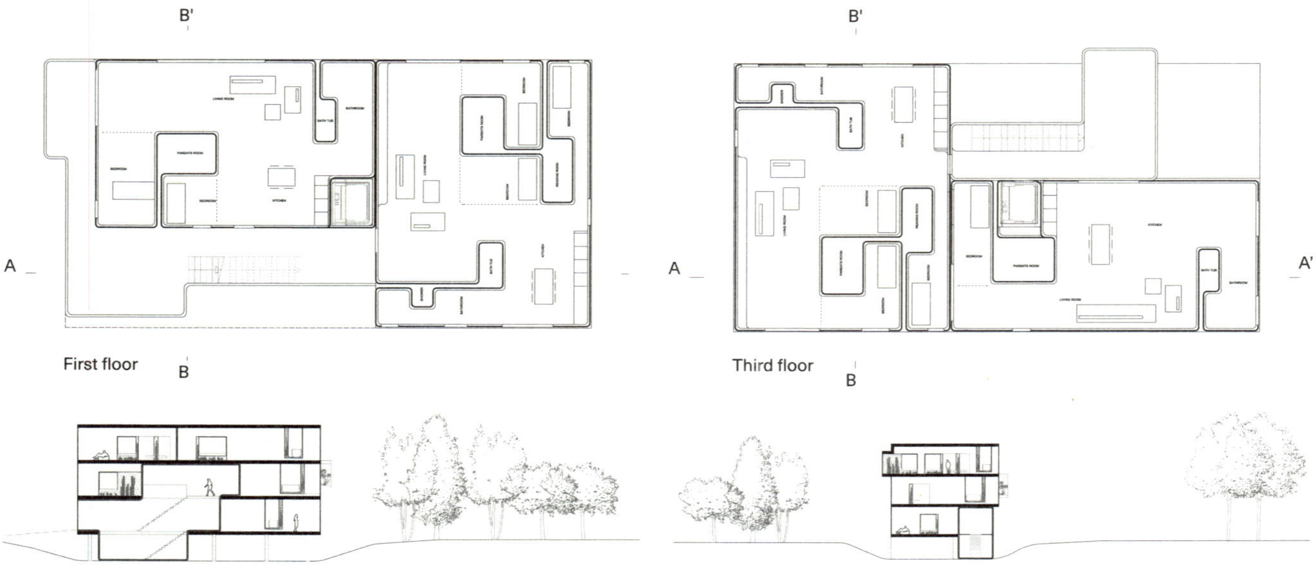

Plans and Sections {Fig. 1-328}

A Gradient of Climates

Plan

Section

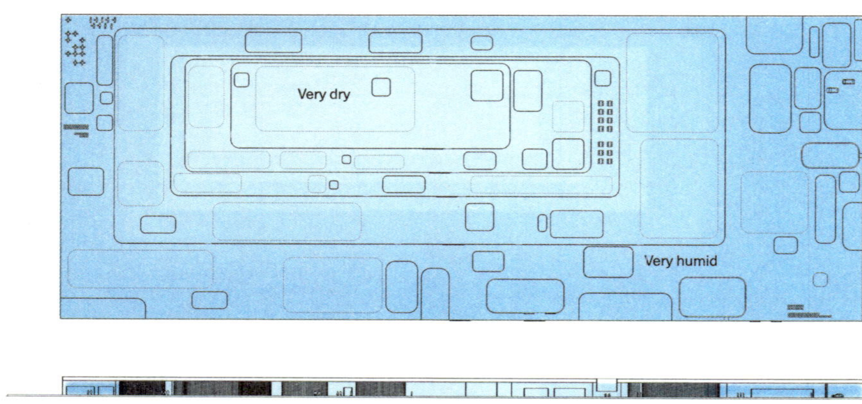

The layout as a gradient of the humidity of the air {Fig. 1-329}
The layout goes gradually from the wettest, in the periphery, relative to the natural environment, to the driest spaces in the center of the building. The plan stratifies itself in this progressive loss of moisture in the air, offering the various criteria essential for the proper conservation of works of art according to each material.

Relative humidity, temperature variation, light intensity and its spectrum are all elements highlighted by this architectural project for a museum. It is not the use of spaces that will determine the form of the building, but contrarily it will be the form that will bring about multiple practical interpretations. Here, in reality, architecture is nothing more than a gradient of climatological values such as the degree of humidity, the quantity of UV rays, or the intensity of the light. The aim is to satisfy the museum's needs for the preservation of its works over time by protecting them from natural physical and chemical conditions. Storage conditions depend on the nature of mineral or organic exposure. For example, metal must be stored at very low humidity between 15 and 30% in order to prevent it from oxidizing. Organic materials, on the other hand, require higher degrees of humidity, of up to 60%, to avoid drying out. But humidity must not exceed 75% to prevent the appearance of mildew. Similarly, each material can withstand a different amount of light. For this reason, paper must be kept at a maximum intensity of 20 lux while wood and metal can withstand much greater luminosity. The museum is thus organized as a synthesized natural climate, progressing rigorously and rhythmically from the outside to the inside, from the moistest to the driest, from the lightest to the darkest, from the strongest ultraviolet to the weakest.

The layout of the museum is organized according to a series of concentric glass envelopes, which are gradually crossed to go from the corrosive natural environment to one that is more attenuated and chemically neutral. From the outside to the center, five climatic layers with different humidities succeed each another, while the humidity gradually decreases.
At the same time, the intensity of natural light decreases as the visitor goes deeper into the museum, from 5,000 lux to 10 lux in the heart of the museum. The various rooms are located in one or the other of its climates according to the chemical conservation needs of each. The character of the architecture materializes in this journey through the invisible. The visitor is invited to participate in a cultural but also a sensory experience. The works are no longer captured by the eye alone but also by the physiology of the space and of the body.
By descending into darkness, the perception of the increasingly dry air. A new way of visiting a museum has been invented, based on the very physical phenomenon of conservation, where chemical de-generation gradually disappears, and in which the visitor begins to feel an architectural form of permanence.

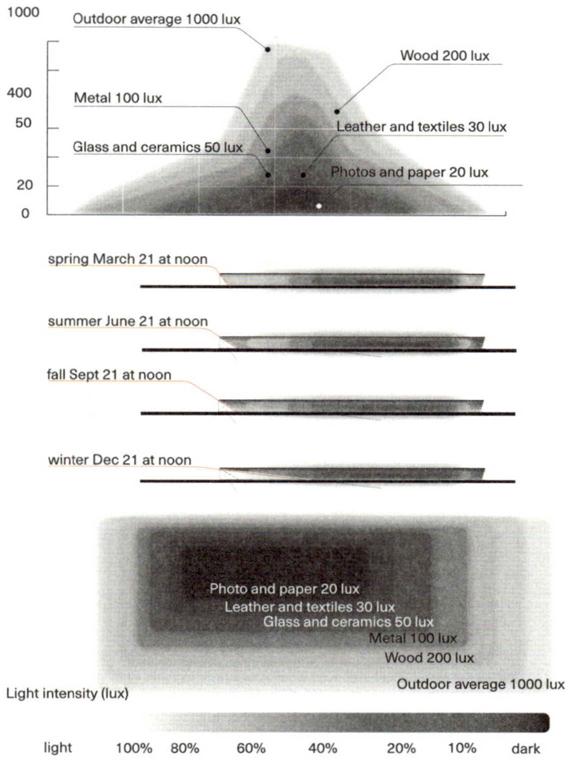

Light intensities and preservation of the works {Fig. 1-230}
Recommended light intensity levels to preserve artworks according to their material.

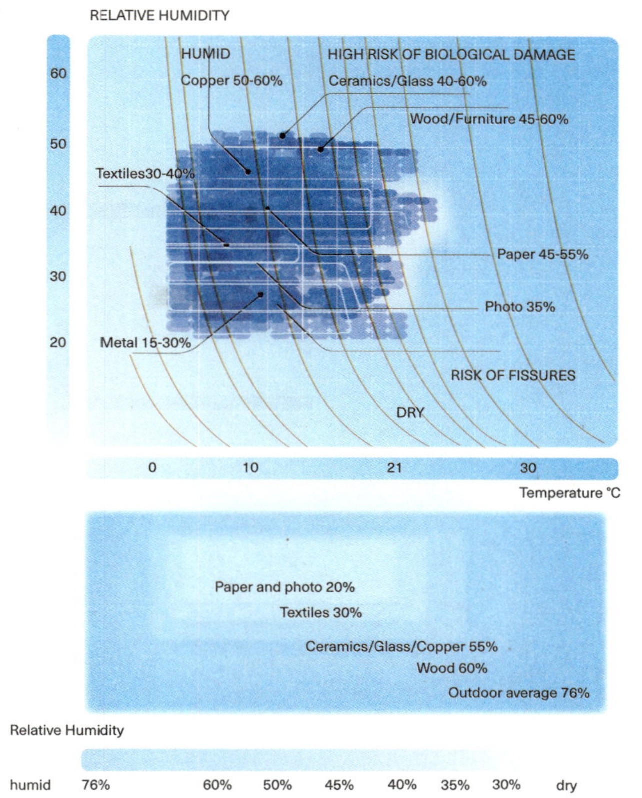

Humidity rates and preservation of the works {Fig. 1-231}
Recommended humidity levels to preserve artworks according to their material.

Mollier Houses

An individual in an enclosed space produces water vapor, not constantly, but according to the activity that is carried out in the room. The presence of water vapor in the air is naturally due to human breathing and the use of hot water, leading to the risk of condensation and damage to buildings. Whereas today the only solution to combat excessive humidity is mechanical ventilation, in this project we propose redefining space based on water vapor, in order to create a complex and profound relationship between the inhabitants, their body and the space with its physical and chemical characteristics. Therefore, our architecture is conceived, and the spaces designed, according to variations in relative humidity, from the driest to the most humid, from 20 to 100% relative humidity. By being capable of containing water vapor, architecture finds its quality in the real and physical immersion of the inhabitants' bodies in the humid, variable space. Our project establishes a stratification of different degrees of humidity within the space. A person asleep produces 40g of water vapor per hour (room), while someone who is awake produces up to 150g/h (living room). Using the bathroom produces 800g in 20 minutes and the kitchen 1500g/h. Like Russian dolls, the living areas are arranged according to the ventilation path within the house, going from the driest to the most humid, from the coolest to the most used, from the bedroom to the bathroom.

But our project will not define spaces according to specific activities. It creates spaces that are more or less dry, more or less humid, which can be freely appropriated according to the weather and the seasons. The layout of the house is a spatial representation of the Mollier diagram, creating new programmatic reciprocations in which a space can accommodate different uses that do not habitually share the same space. The driest room, between 0 and 30% relative humidity (RH) can be a drying room or a sauna. The next one, between 30 and 60% RH, can accommodate a bedroom, an office or a living room. The third, at 90% RH, can be used as a bathroom, living room or kitchen. Finally, the last one, between 90 and 100% RH can be used as a living room or a swimming pool. But here, none of the rooms are specifically defined by a use. They remain freely appropriable, depending on the desired degree of humidity.

Mollier Houses

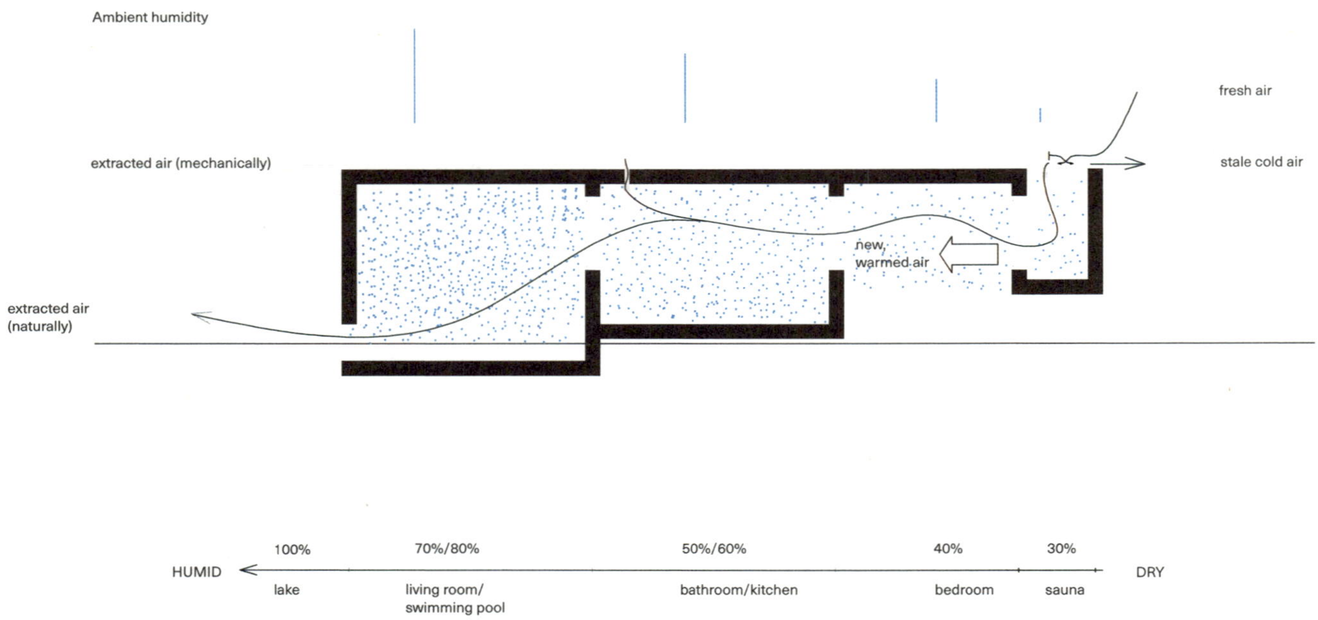

Relative Humidity and Section {Fig. 1-332}

The house is organized according to a flow of air between the dry and the humid, accompanying the different uses that generate increasing humidity.

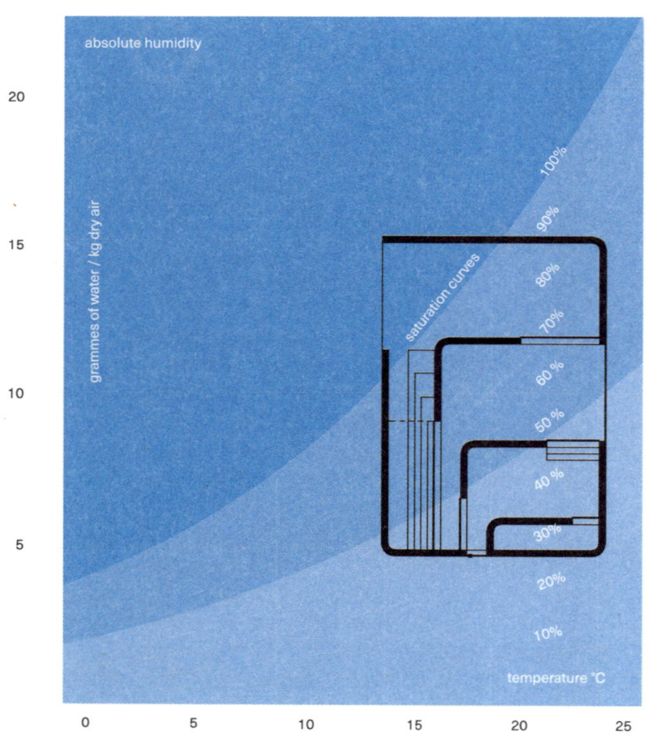

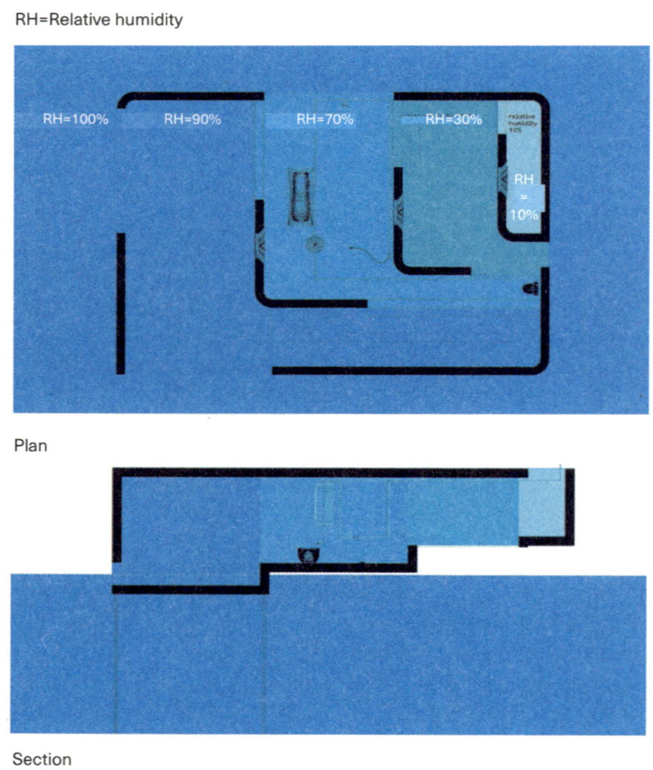

Mollier graph and layout {Fig. 1-233}

The house is no longer organized according to a (public-private) social convention but according to the degree of humidity in the air

The layout and section dictated by Relative Humidity {Fig. 1-234}

The house goes from driest at the top, to most humid, to eventually plunge into the lake.

Water Atmospheres

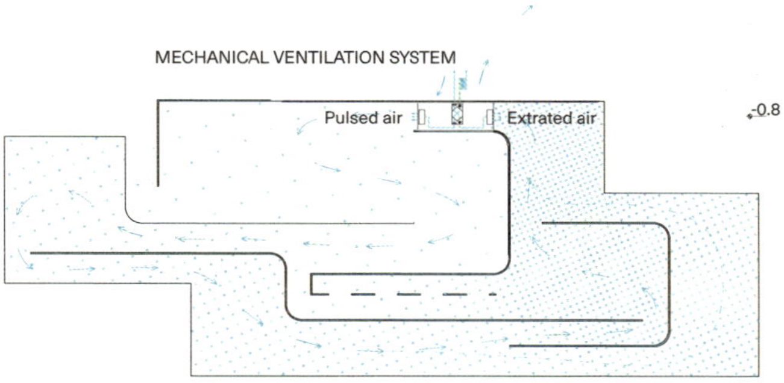

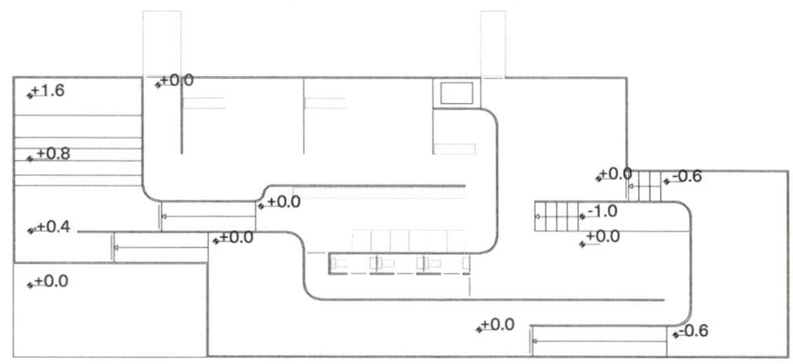

Linear gradation of relative humidity {Fig. 1-335}

Humidity control arising from a context of sustainable development is transfigured here to invent a new typology where architecture becomes a linear horizontal variation of humidity, from the driest in the north to the moistest in the south.

Controlled dual-flow ventilation with heat recovery (Minergie, Switzerland) creates a continuous flow of air from the driest to the moistest areas, where the air is extracted.

The function follows the humidity level {Fig. 1-336}

The building, designed as a habitable ventilation duct, is organized around the linear displacement of air. The programs are positioned by following the gradual increase in the humidity level of the air, placing the ones that emit the least water vapor, at the inlet, until the one that emits the most water vapor, at the outlet.

Relocated Barn

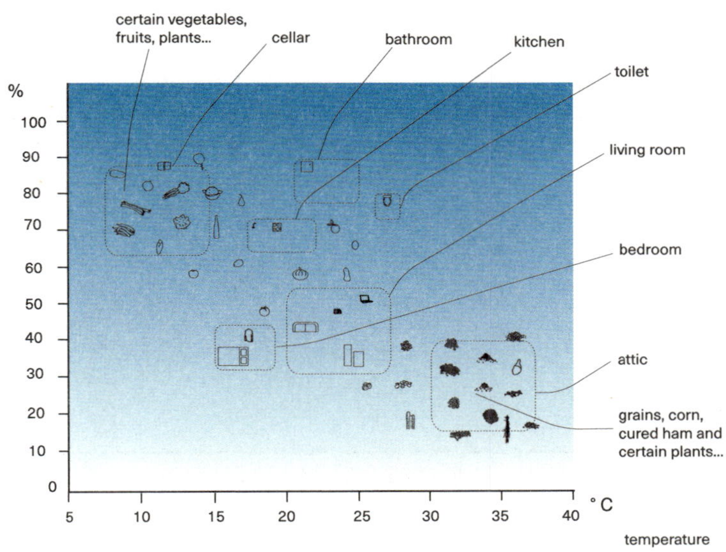

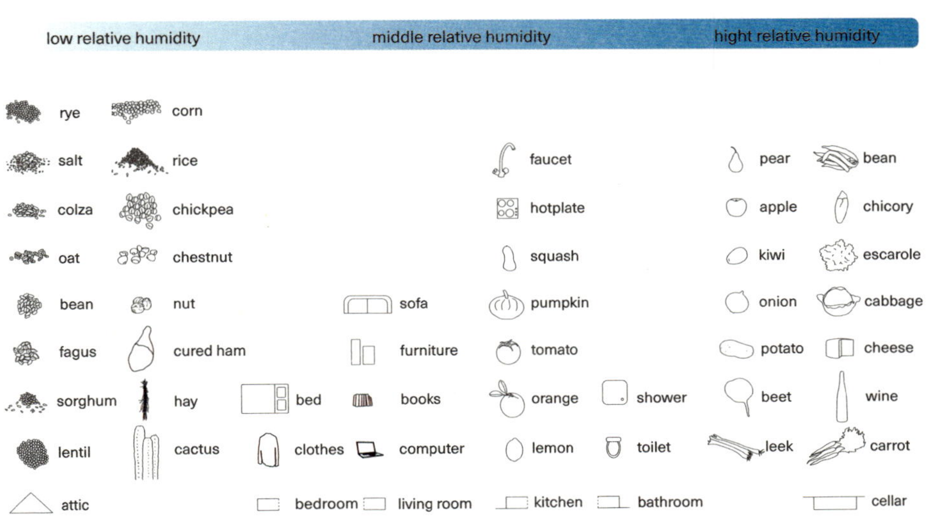

Categorization of functions according to the humidity of the air {Fig. 1-337}

Controlled ventilation with heat recovery, recommended to save energy, inevitably leads to a reconsideration of the layout of rooms and programs in the home according to acceptable air humidity levels. The house can become an internal geography, embodying its uses in the variations of the quantity of water vapor.

Relocated Barn

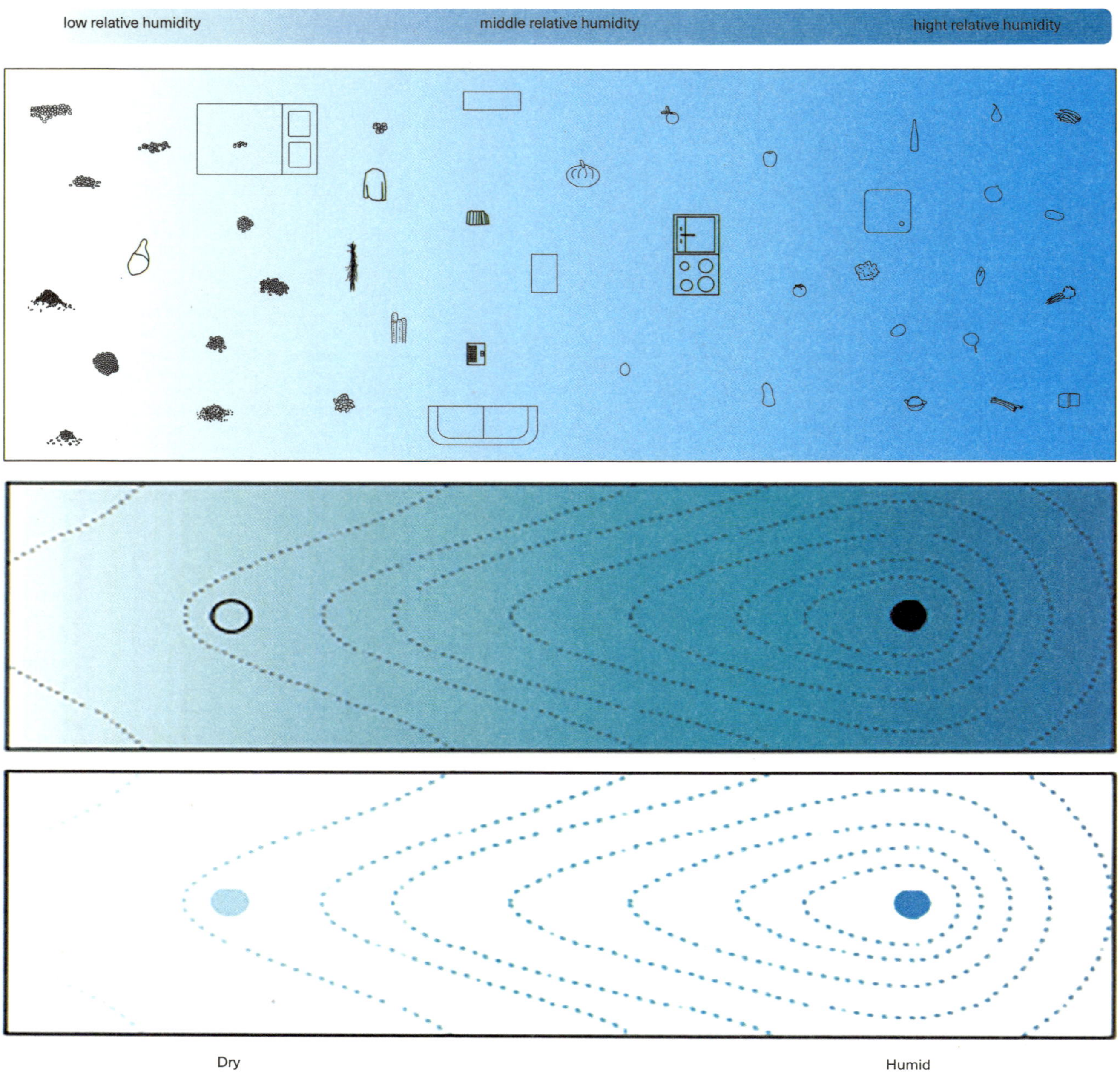

Organization of the house according to the humidity of the air {Fig. 1-338}
The layout is organized between the fresh (drier) air inlet and the air outlet (which has been humidified by absorbing the water produced by human activities: shower, kitchen, etc.). Programs, objects and food naturally find their place in this hydric geography.

The plan is arranged between two devices on either side of the room, at medium height. The first device is an ultrasonic atomizer producing water vapor as a fine cloud, a sort of cellar vapor. On the other side of the room, the second device is a dehumidifier that dries the surrounding air, providing a kind of attic atmosphere. Forming an ecological loop cycle, the water thus passes continuously from one side of the room to the other, from gaseous to liquid phase, in the form of vapor between the atomizer and the dehumidifier, and in the form of a liquid between the dehumidifier and the atomizer. Our target would then perhaps be to re-situate contemporary housing within a poetry of space, to retain today's space from the cellar to the attic, from the atomizer to the dehumidifier. Our proposal seeks the synthesis between conservation and consumption, places for storage and places for living, between modern rationality and the poetic imaginary, by reorganizing the space of the house according to a climate map, by coming out of the refrigerator and pantries, the two polarities of the cellar and the attic, to redeploy them spatially in the living space of the house. In this new free layout based on the variation of relative humidity rates, the different foodstuffs are arranged according to the storage needs at the same time as the items in common use today, shower, cooker, computers, books, clothes, also taking place in this landscape according to the water vapor they generate or that they withstand. As for the inhabitants, they encounter a more sensual and tactile and perhaps more poetic way of living, between more humid, mistier areas where they take a bath or a shower or cook, and drier, transparent areas where they rest, chat or work.

Climatheque

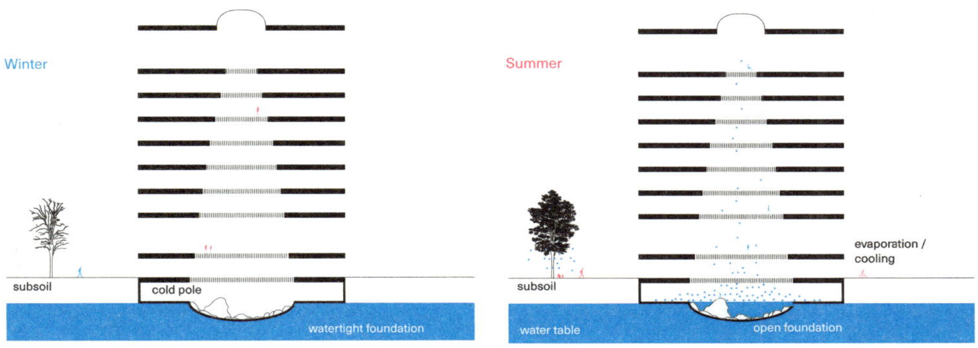

The Frigidarium: an underground summer lounge to build fresh air during hot weather {Fig. 1-339}

In the basement, we propose to implement a volume of water and cool air, the Frigidarium, cooled by the earth's thermal inertia, by the coolness of the groundwater which we propose to input into the building in summer to cool the incoming ventilation air which is then diffused throughout the office spaces. Accessible and public, the Frigidarium, which also houses heating or ventilation equipment facilities, becomes a climate refuge, a public summer lounge where you can escape naturally from the scorching heat, just as historically provided by the shade of the tree or the opaque inert mass of Italian churches.

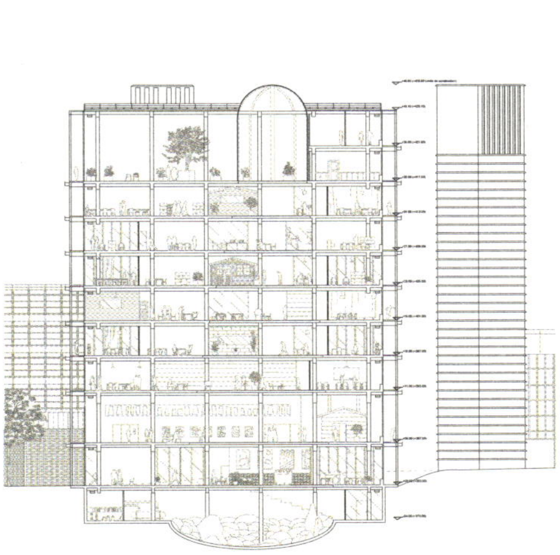

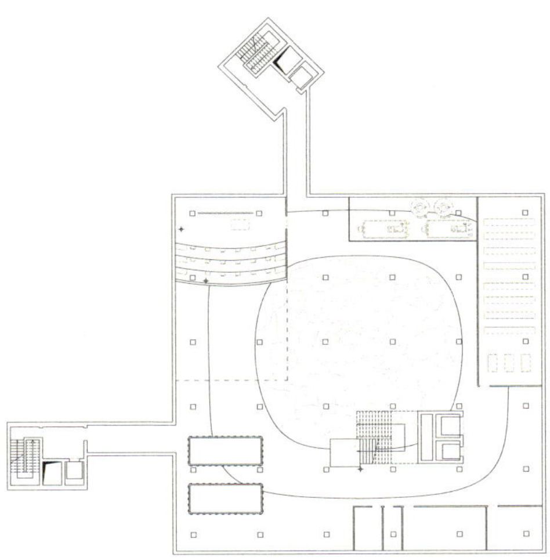

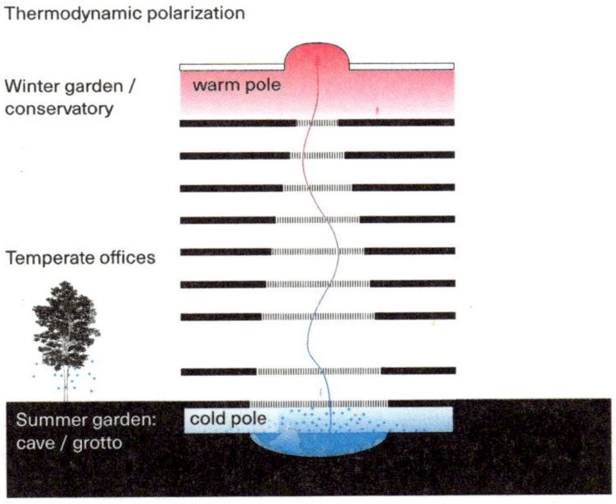

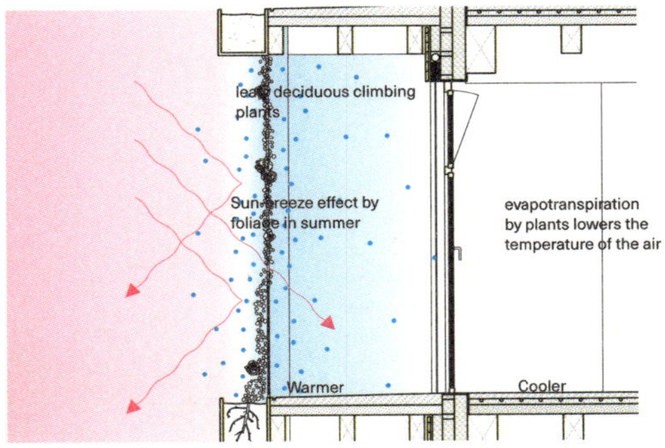

The Cold Pole: underground cooling by thermal inertia and evaporation {Fig. 1-340}

At a depth of 3 meters, thanks to thermal inertia, the temperature of the subsoil remains at a constant 11.1° C throughout the year in Lausanne; this is also the temperature of the water of the water table present on the site at this altitude. We propose taking advantage in summer of the coolness of the naturally chilled basements on the site and of the groundwater to cool the air entering the new humanities building of the University of Lausanne while creating an underground public space, a kind of habitable cave called the Frigidarium, whose thermal quality can be appreciated as a climatic refuge or a cool room, a summer lounge, most appreciated on a hot summer's day. In summer, the warm air entering the building from the wind tower where it was harnessed to the north, will take advantage of this geothermal cooling in the basement, before returning to the atrium of the building and flooding all the spaces of the building with freshness.

Roses to shade and refresh summer afternoons {Fig. 1-341}

At the ends of the balconies surrounding the building, on each floor, uninterruptedly, flower boxes measuring 50 cm thick by 50 cm deep filled with plant soil are arranged, allowing planting several types of flowers chosen for certain health and climatic properties, such as the climbing rose, to provide the building with shade in summer with its deciduous flowers and foliage and to cool the air near the façade by the fine evapotranspiration of the foliage which, by allowing fine droplets of water to form during photosynthesis, lowers the temperature of the air by natural evaporation, taking the energy during the shift from liquid to gas phase.

PRESSURE

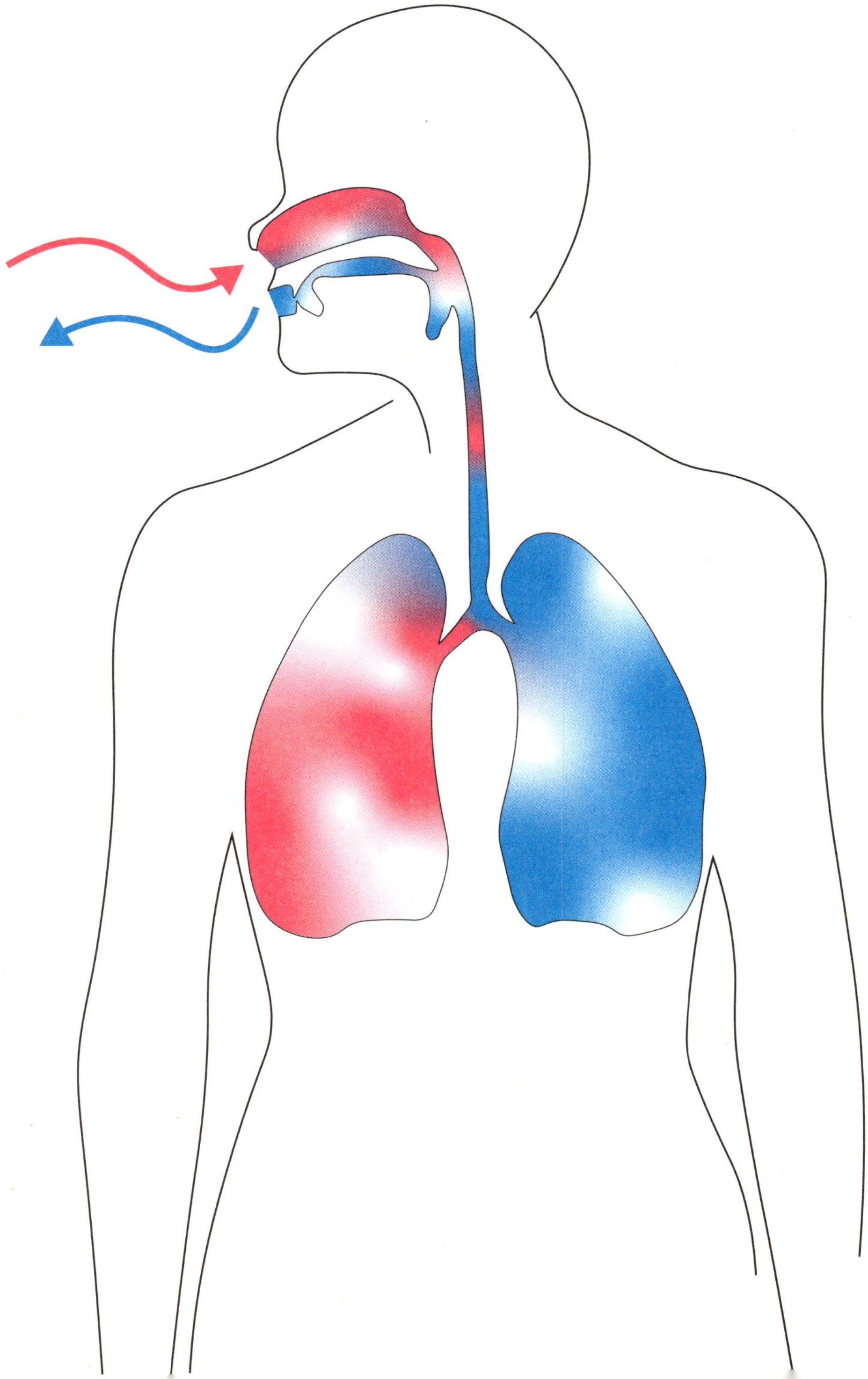

I.Definition

General Properties

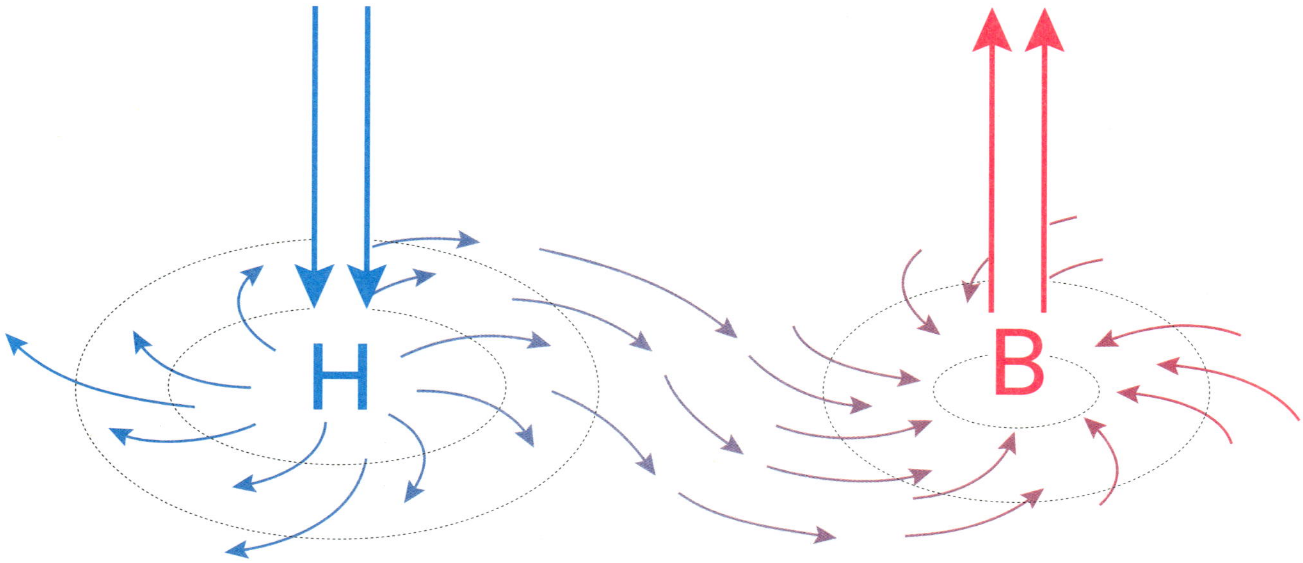

High pressure system {Fig. 1-342}
There is an area of higher pressure in the center. The air moves away from the center.

Low pressure system {Fig. 1-343}
There is an area of lower pressure in the center, the air is sucked inwards.

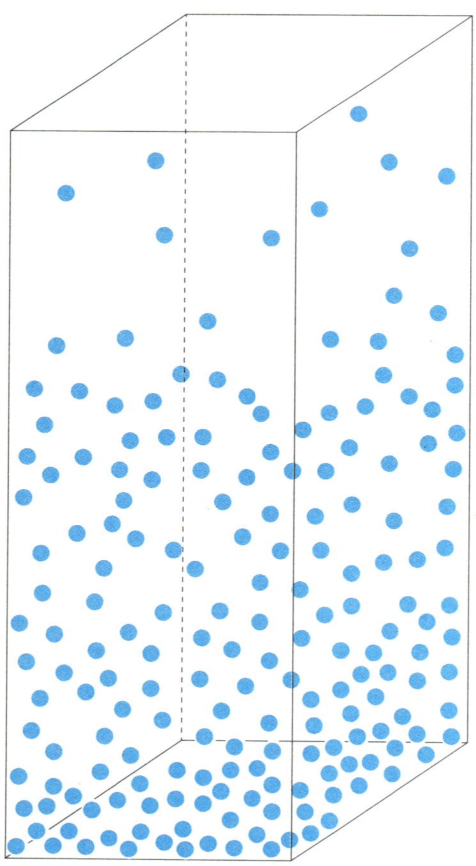

Air pressure in a volume {Fig. 1-344}
Pressure tends to be greater in the lower than in the upper part. This is due to gravity and to the accumulated compression of the air layers.

Air pressure is the pressure in a container due to the compression of atmospheric gases. Air tends to move from high pressure areas to low pressure areas until these pressures become equal. Low pressure means that the atmospheric pressure in the region is lower than the area around it, the opposite is applicable to high pressure.

Thus, the designation of high or low pressure is only relative, as is the designation of peaks and valleys in mountainous areas. Pressure differences are also due to temperature differences between two regions. As hot air is less dense than cold air, an air current will be created between the two regions by a thermal pulling effect from the cold to the hot.

II.Global Scale

Winds

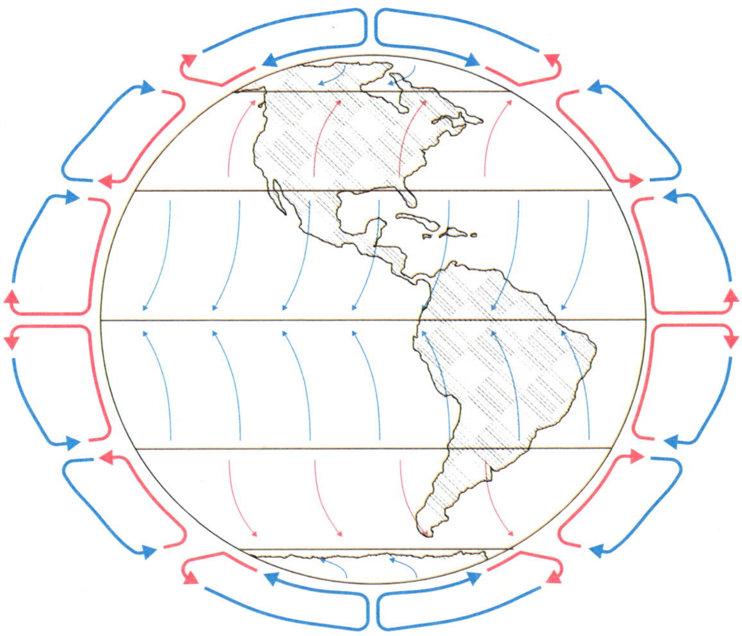

Pressure system and convective movements on Earth {Fig. 1-345}
The movements of air on the Earth's surface are due to temperature differences that create convective movements; wind.

In the atmosphere, as soon as there is a pressure difference between two points, the air will move toward the lower pressure point. Because of this phenomenon - which is mainly due to temperature variations - the pressures on the ground and at different points in the atmosphere are different.

Winds are thus created when the air is pushed toward an area of higher pressure than that in which it is located. The greater the difference in pressure, the stronger the wind.

Weather

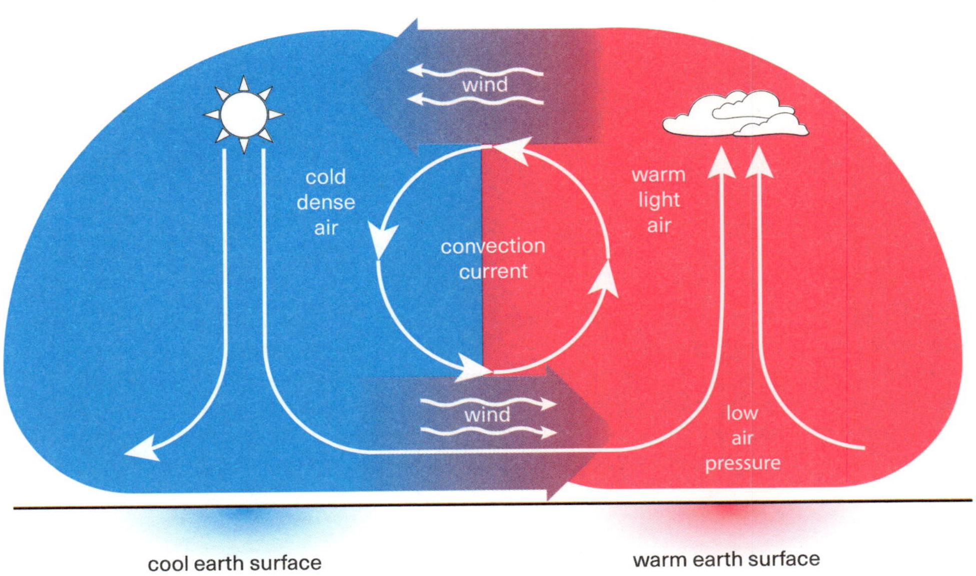

Weather according to pressure systems {Fig. 1-346}
The movements of air on Earth are due to differences in surface temperatures which create convective movements: wind.

III. Physiological Scale

Oxygen Consumption

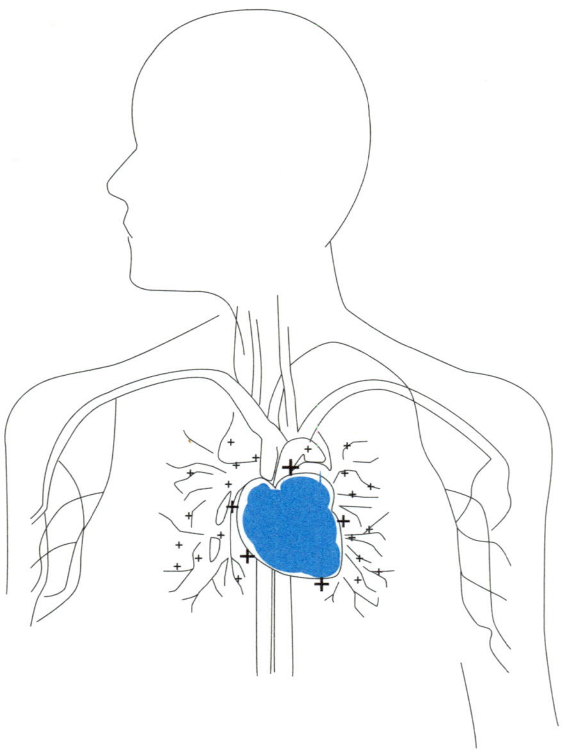

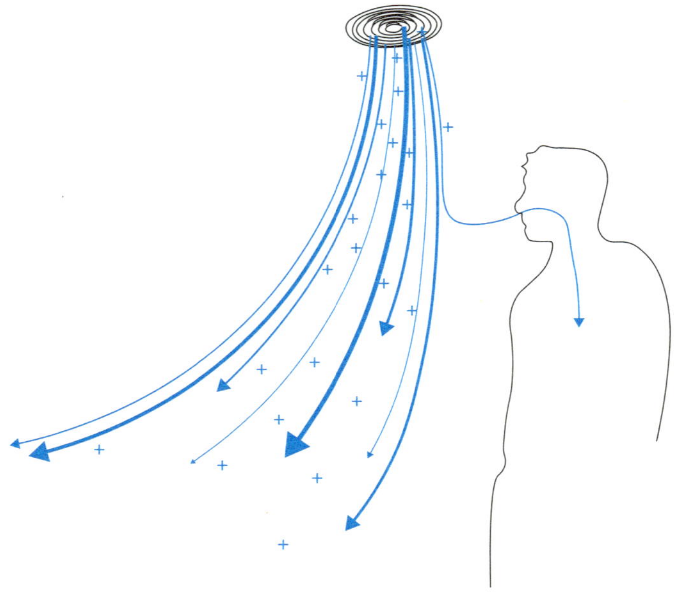

Heartbeat {Fig. 1-347}

Increasing the heart rate leads to an increase in oxygen consumption. The more physically active the body, the faster the heart beats and the more oxygen it consumes.

Ventilation {Fig. 1-348}

The constant consumption of oxygen (O_2) by the human body requires a constant supply of fresh air that will be inhaled and the extraction of the stale air containing the CO_2 expired during breathing.

Rest

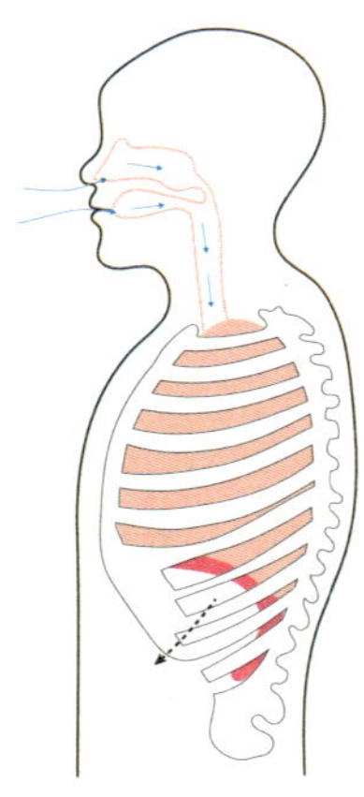

Inhaling

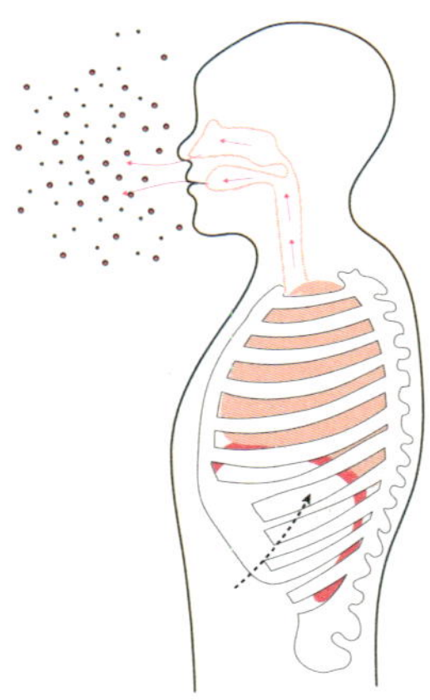

Exhaling

Exercise

Oxygen Consumption {Fig. 1-349}

On average, every minute, humans absorb 6 liters of fresh air at rest. This is the amount of air that must be renewed by ventilation. This amount of air can reach 200 liters during intense physical activity of the human body. Thus, the air of a room that hosts sports activities must be renewed much more than that of a room without any physical activity being conducted by its inhabitants.

During hard physical activity, the muscles of the human body require 500 times more oxygen than at rest. The greater our physical activity, the faster we breathe, the faster our heart beats, the more oxygen we consume and the more CO_2 and water vapor we release. Since indoor spaces are hermetically sealed and the indoor air does not mix with the outdoor air, we need to control the renewal of air so that the stale air (CO_2 and water vapor) is replaced by fresh air (oxygen). Thus, a space that hosts sports activities will have to renew its air more often. It is the role of ventilation (natural or controlled) which is used first to evacuate excess water vapor, then the CO_2 and indoor pollutants, and to bring in fresh, drier air, thus renewing the oxygen content of the indoor spaces.

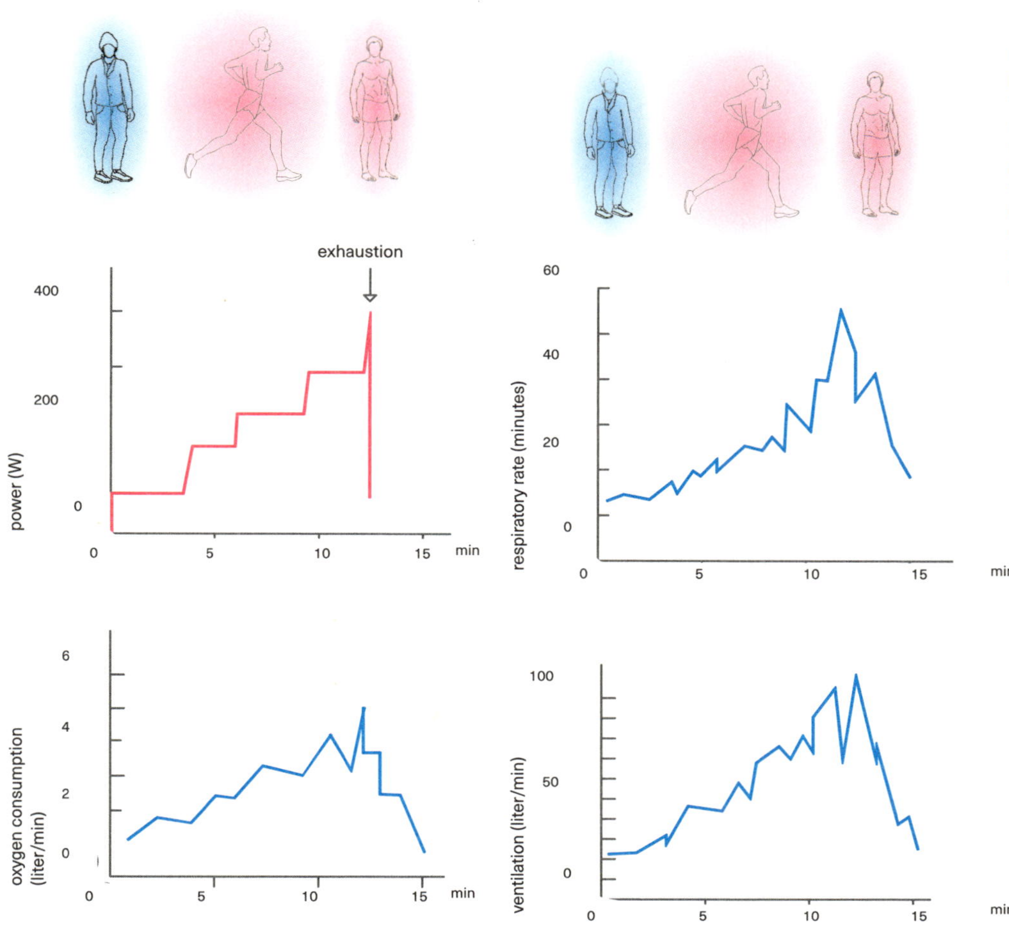

Physiological variations during physical exercise {Fig. 1-350}

The greater the physical effort, the more the breathing rhythm increases, the more the oxygen demand increases, the more moisture and CO_2 we release. For this reason, rooms with intense physical activity need more air renewal than rooms with low physical activity. (Source: atlas de poche de physiologie, Flammarion, 2001)

1/3 glucose	250	Coronary flow Qcor (ml/min)	600	1/7 glucose
1/3 free fatty acid	0.12	Arteriovenous difference in O2b (l/l blood)	0.15	1/5 fatty acid
1/3 lactate	30	Consumption of O2 (ml/min)	90	2/3 lactate

Flow, Oxygen and mineral consumption of the cardiac muscle

Workshop, with physical work	60 m3/h/person	Public shower (e.g., swimming pool)	220 m3/h/person
Office, without physical work	25 m3/h/person	Conference room	30 m3/h/person
Store (sales)	30 m3/h/person	workshop (light physical work)	45 m3/h/person
Office	15 m3/h/person	Nursery	15 m3/h/person
Living room	15 m3/h/person	High school class	18 m3/h/person
Bathroom	30 m3/h/person	Office	18 m3/h/person
Bedroom	15 m3/h/person	Restaurant	30 m3/h/person
Kitchen	45 m3/h/person	Gym	25 m3/h/person

Flow of new air according to the activity {Fig. 1-351}

Depending on the use made of a room, the air renewal rate will be adapted to the number of people present and their physical activity. The more intense the physical activity, the more people there are, the greater the need for ventilation. In order to save energy, this air renewal must be kept to a minimum so as not to heat (in winter) or cool (in summer) excess air coming from the outside, which will not be breathed by the inhabitants. Every liter of air blown in whose temperature is changed by heating or cooling, every liter consumes energy.

IV. Architectural Scale

Air Renewal

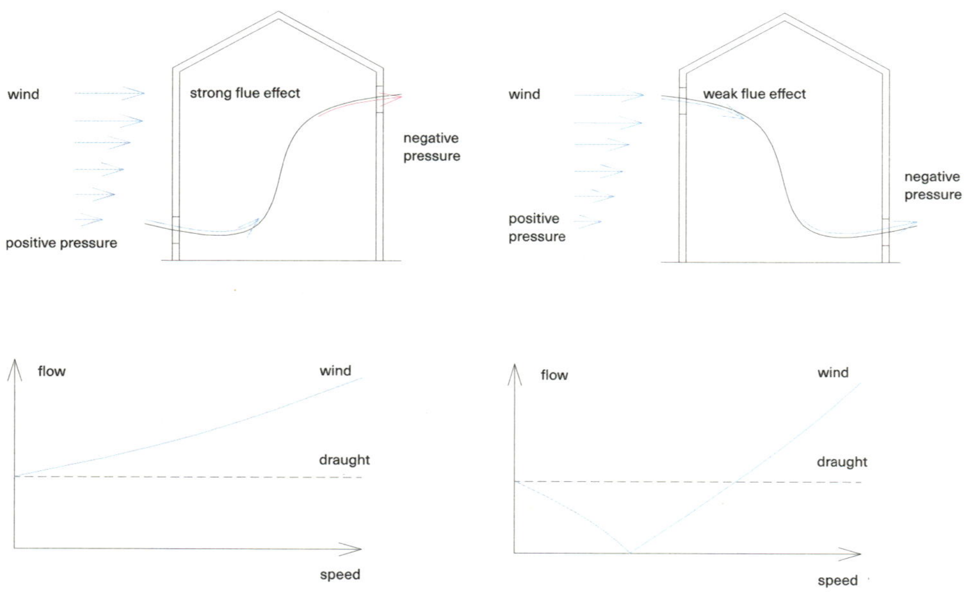

Effects of wind and flue effect {Fig. 1-352}

The façade exposed to the prevailing wind is under positive pressure, while the opposite, downwind façade is under negative pressure. The air will naturally pass through the house. This natural ventilation is reinforced if the fresh air is introduced at the bottom and leaves at the top because the displacement of the air between the high pressure and the low pressure is accelerated by the natural temperature difference in the same space, which is stratified into a lower temperature at the bottom and a higher temperature at the top.

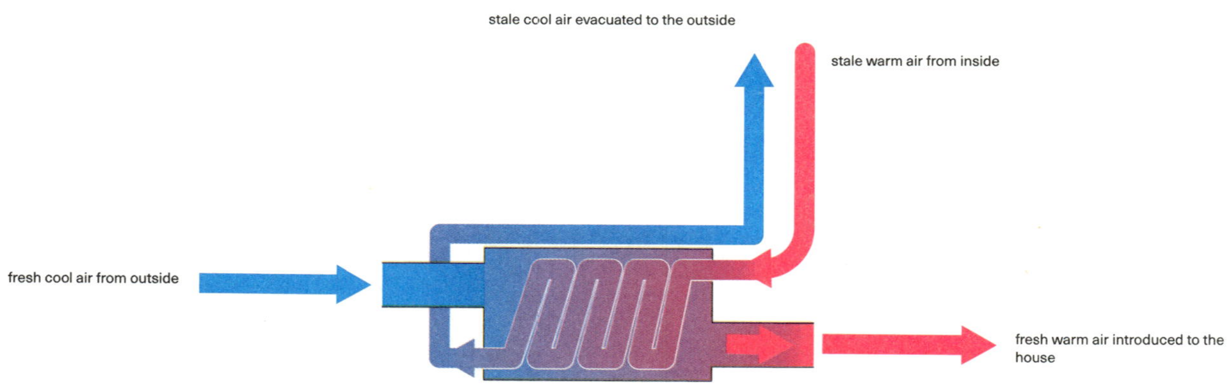

Diagram of a heat exchanger {Fig. 1-353}

The outside air is heated before entering the house by passing through a heat exchanger, where it crosses warm stale air before being discharged to the outside. This avoids losing heat. The heat exchange between the incoming and the outgoing air takes place by means of numerous aluminum plates which are highly thermally conductive, making it possible not to bring the incoming air into chemical contact with the outgoing air.

Air pressure in architecture is the pressure necessary to ventilate the space, the current created between the inlet of new air, under positive pressure, and the outlet of stale air, under negative pressure. Because we must renew the indoor air. Because the inhabitants inhale the indoor air, consume it and dirty it, taking its oxygen, releasing carbon, water vapor and bad smells. In the past, poorly rendered joins in door and window frames etc., were sufficient to renew the air in houses, but in an uncontrolled and sometimes uncomfortable way because drafts appeared to the point where the inhabitants had to protect themselves from them by using curtains, screens or interior doors. Moreover, the lack of control over the amount of air entering the house led to unnecessary energy consumption because more air was entering the house than was required. Today, to avoid these energy losses, the house must be hermetically sealed so that the hot air does not escape through cracks in the envelope and the cold air does not penetrate.
The interior becomes like a bubble whose air inlet and outlet are precisely hermetic and controlled in order economically to maintain the inside temperature at 21 °C.

The amount of air we inject into a room depends on the physical activity conducted and the number of people in it. Where people are physically active and numerous, it will be necessary to renew the air more often than in spaces where people are stationary and less numerous.
The interiors of the buildings, the shape of the spaces and their continuity can then be designed as rivers of air with their tributaries, their lakes, their rapids, their torrents with as many variations in speed of the air that flows, fashions and shapes like a geological agent, architecture.

V. Case Studies

The Building as a Form of Air

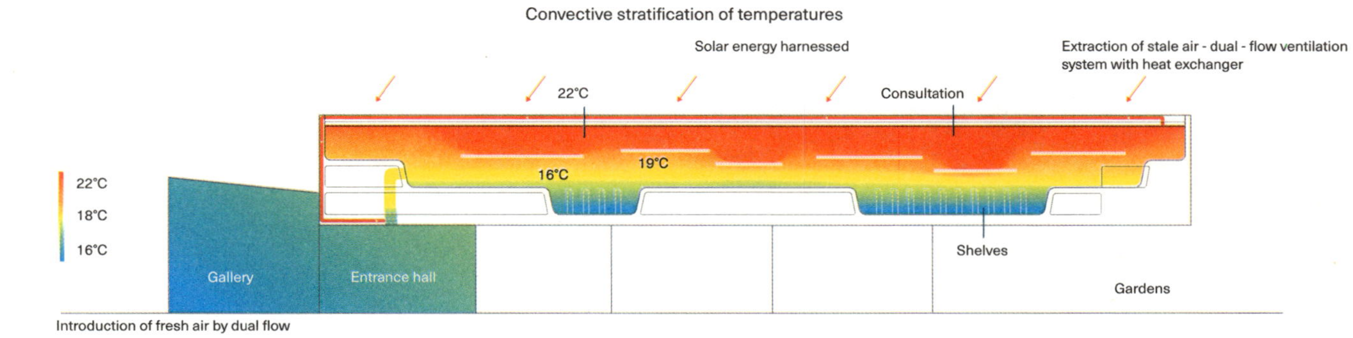

Cross-section of thermal stratification {Fig. 1-354}

As heat rises naturally, the public is placed at the top of the space, while the deposits, which one does not need to heat, are placed at the bottom.

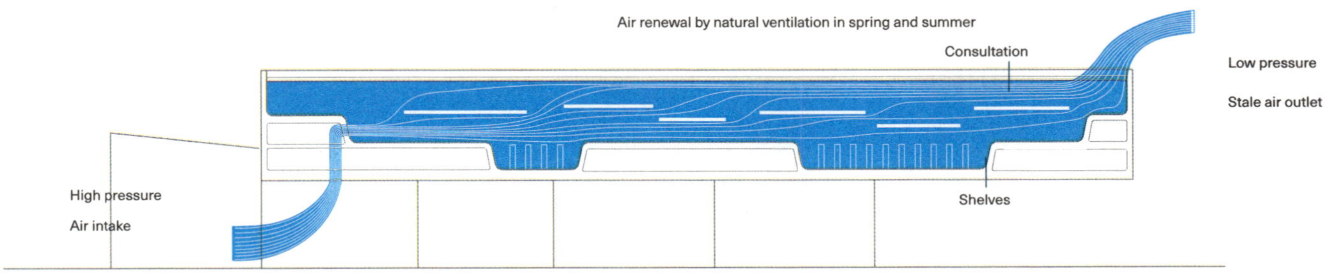

Cross-section of internal ventilation {Fig. 1-355}

The internal shape of the building is designed according to the movement of the air, narrowing to accelerate the air flow where the public is more numerous, widening to slow it down where the public is less numerous, according to the amount of air needed for the ventilation of the premises according to the program, the activities and the number of people occupying the premises.

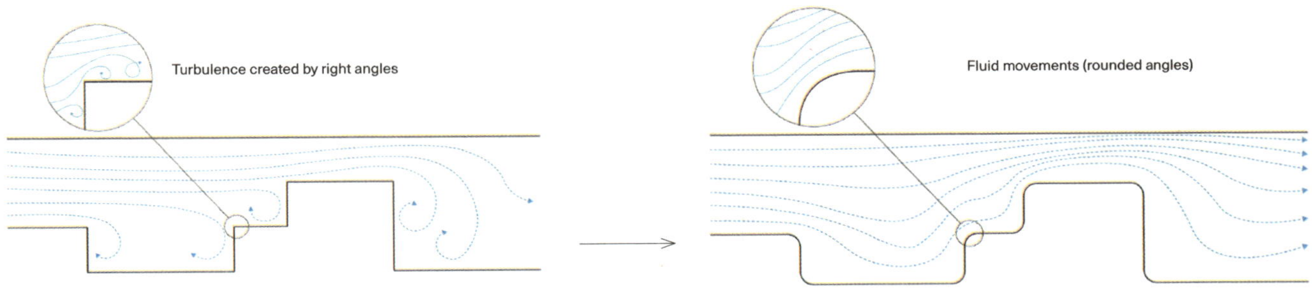

Detail of the treatment of angles {Fig. 1-356}

As in ventilation ducts, the corners of the spaces are rounded in order to fluidize the movement of the air and avoid turbulence and noise accordingly.

The Building as a Form of Air

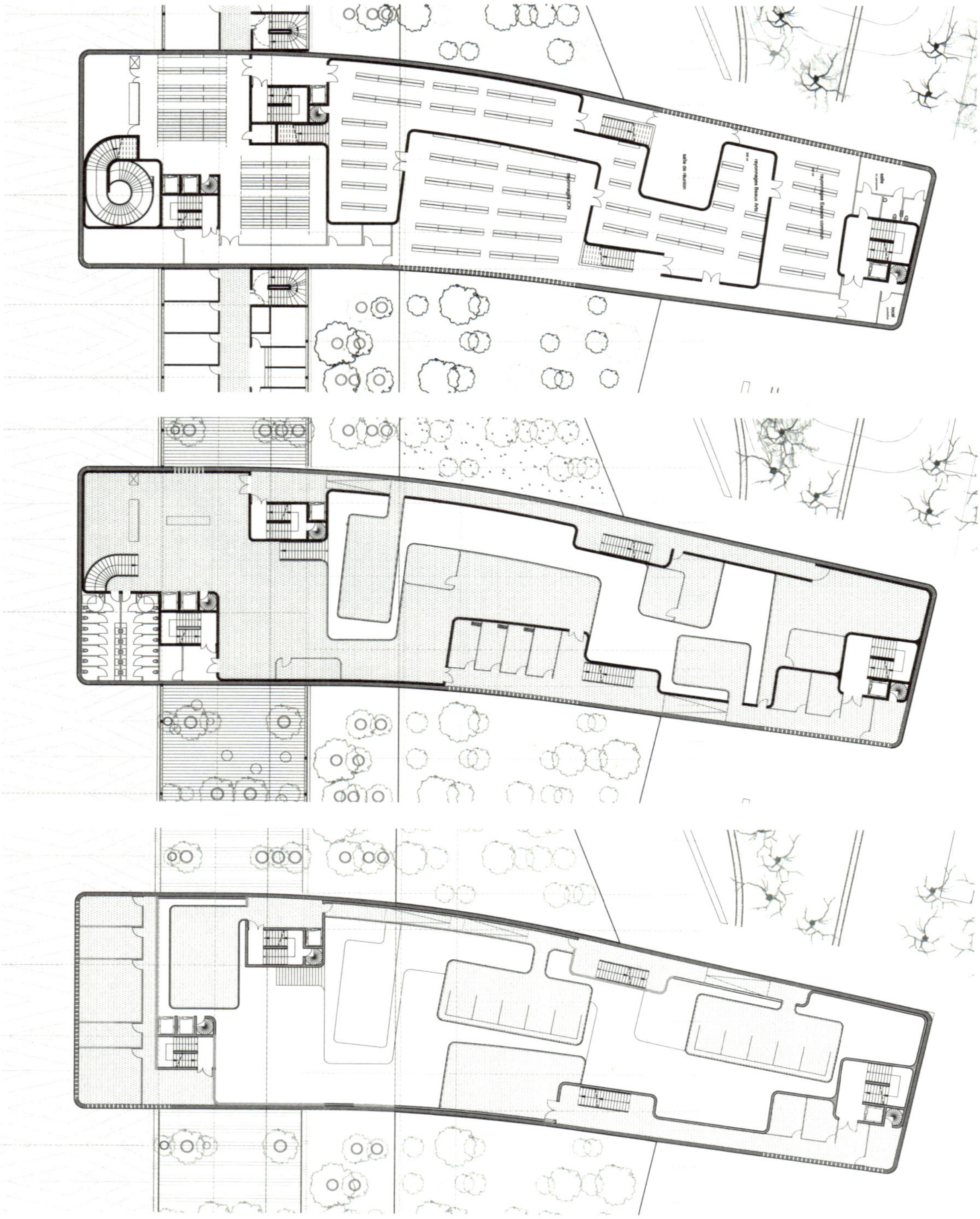

Plans {Fig. 1-357}
Like a river of air, the building and its spaces are deformed to sculpt the flow of ventilation air, speeding it up depending on the program in order to make energy savings

To define its form and spatial organization, this project for a Library very literally follows the minimum recommendations for the ventilation of premises in terms of the quantity of air per hour in order to reduce energy consumption in buildings. In a volume fully thermally insulated with 35 cm-thick insulating wool, meeting today's most stringent thermal requirements, the library takes shape in layout and in section along the air ventilation path and on the natural ascent of hot air according Archimedes' law.

The goal is to minimize heating and the amount of renewed air (as a high rate maximizes heating costs) by taking advantage of the heat naturally present at the top to locate the public, while the books are at the bottom in the cold, and, by deforming the architectural form to speed up the air where the breathing public is located, and slow it down where the books, that do not breathe, are located. The shape of the building literally follows the shape of the ventilation airflow; the architecture is presented as an expanded ventilation duct to become habitable.

Inhabiting a Ventilation Duct

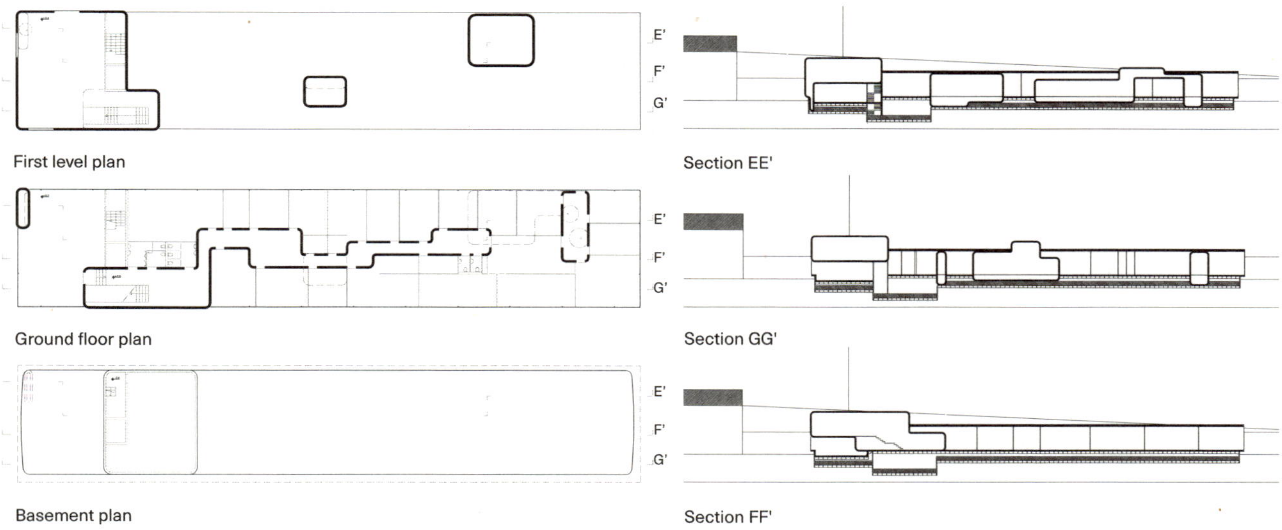

Plans and sections {Fig. 1-358}
The central corridor distributes both users and fresh air by becoming the main ventilation duct of the building, widened to become practicable.
The stale air is returned to the façade.

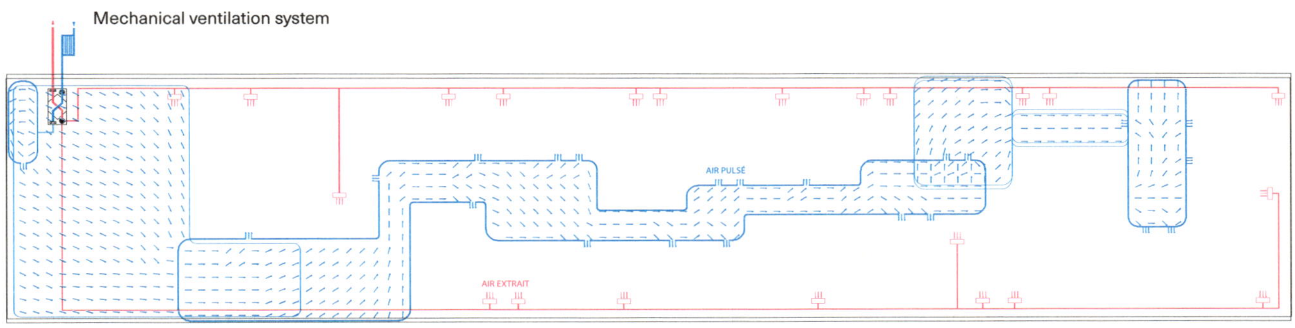

Ventilation system plan {Fig. 1-359}
The principle here is to eliminate ventilation ducts by turning the building itself into a ventilation duct. The corridor distributing the spaces is thus a ventilation duct expanded to become practicable

This floating office building on the Seine recreates a terroir indoors, in view of the need for building techniques, air management, aeration with heat recovery, water management, rainwater recovery, solar panel domestic water heating, purification. They will be natural terroirs but indoors characterized by the soil and subsoil, a climate, an exposure, which will convey its fragrances and its taste to the air and to the water. Various kinds of limestone extracted from the soils crossed by the Seine from its source as far as the sea, will lend their taste to the air, while glass, polyethylene and stainless steel will remain chemically neutral in this architectural, olfactory and gustatory composition. The composition of the building here is based on the requirements of air management (gentle, automatic ventilation with heat recovery / Minergie, Switzerland) in order to reduce the energy consumed in the building by a factor of 4.
Our project themes this supply of forced air in the building, transforming a technical problem of an internal duct and the routing of air in the rooms into an architectural, spatial and poetic principle. The size of the forced air ventilation duct is augmented to the living space that slips, deforms, compresses and expands throughout the entire length of the building. Rather than being made of galvanized steel sheet, this duct is made of solid load-bearing limestone, from west to east encompassing the length of the geological course of the air, from the sea to the Paris basin, from Le Havre to Paris, reproduced in miniature. The architectural quality we provide is that of an atmosphere, an odor, a perfume, an air, an air of Paris, coming from the sea, from the west which, in the heart of the building, will fill with the fine scent of limestone. A few oaks are planted along this route, their presence adding a touch of vegetation to this fragrant air.

Inhabiting a Ventilation Duct

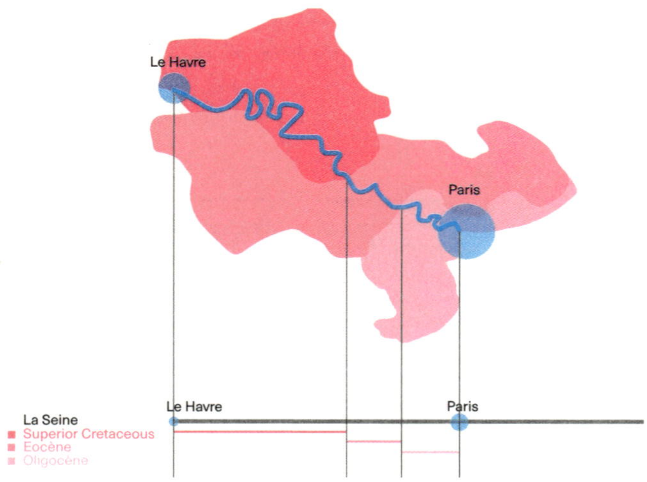

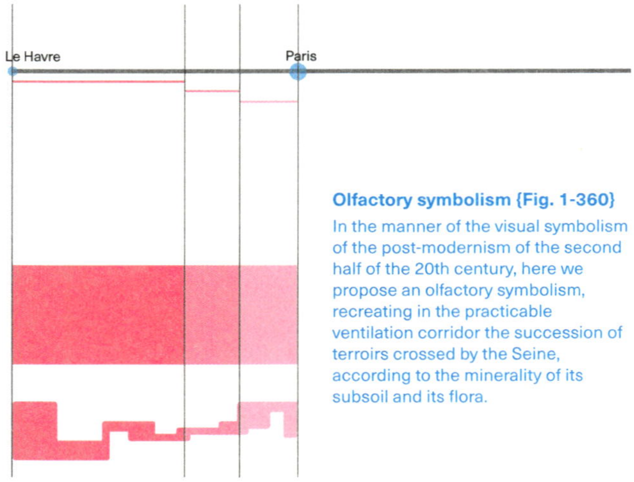

Olfactory symbolism {Fig. 1-360}
In the manner of the visual symbolism of the post-modernism of the second half of the 20th century, here we propose an olfactory symbolism, recreating in the practicable ventilation corridor the succession of terroirs crossed by the Seine, according to the minerality of its subsoil and its flora.

A Ventilation Duct Augmented to the Scale of the Building

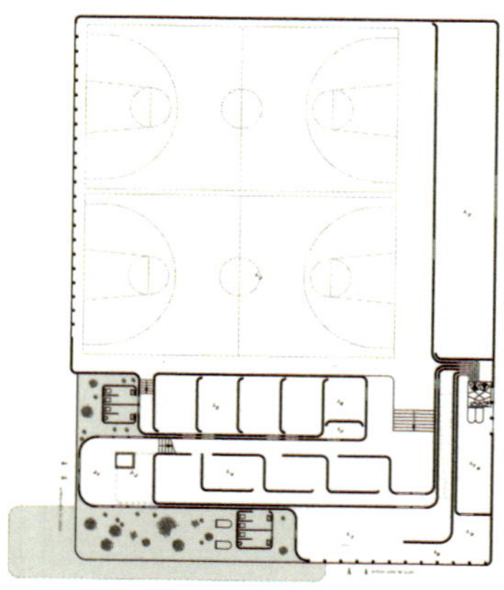

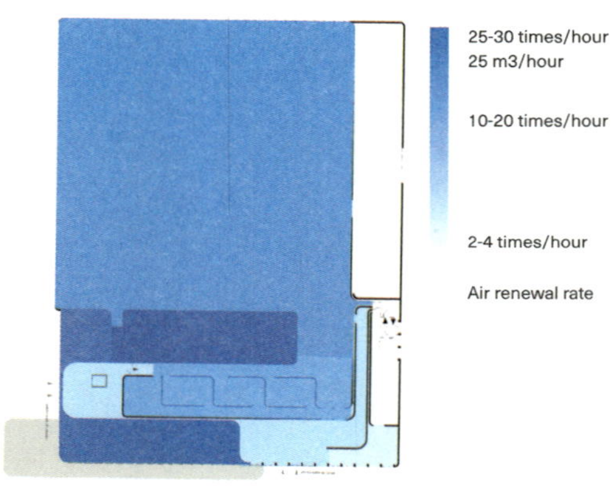

Plan {Fig. 1-361}

Air renewal rate {Fig. 1-362}
The spaces follow on one after another in a continuum of ventilation air movement, between the lowest (Entrance hall) and the highest (for the showers) air renewal rates.

What is the deep link between the space and sports activity? It is undoubtedly the question of the supply of oxygen required for the athlete's muscles, the consumption of which is multiplied by 500 during significant muscle work compared to muscle rest. At the same time, the organism must evacuate metabolic waste products, such as H+, CO2 and lactates. During physical activity, the ventilatory flow rate rises from about 7.5 liters per minute at rest to 120 liters per minute. This increase is possible thanks to an increase in the respiratory rate which then requires a quicker exchange between fresh air and stale air. There is, therefore, a precise relationship between the form of the space and the amount of air to be supplied and renewed, again posing the question of the ventilation of the building as the main efficient strength of the architectural form. Our project consists of, is ordered, is indeed formed precisely from the management of air in the building, by means of its path, its speed, its volume, its renewal rate.

The architecture thus becomes a mere ventilation system and the spaces literally take the form of ducts in which air flows at low pressure. The building becomes a continuous movement of air, with its tributaries that increase the amount of air brought to the space. The architecture is gradually composed, linearly, between spaces with a low rate of air renewal to spaces with a high hourly air renewal rate. So, first the spaces where one spends the least amount of time, where the body is little active in terms of burning muscle, where the least amount of water vapor is emitted, then gradually, everything increases. Our project, therefore, is gradually composed like a stream of air which, joined by others, gradually turns into a tributary, then into a river, then into a sea. The architecture we propose is thus a habitable flow of air, a wind that is designed and augments according to the average hourly rates of fresh air renewal depending on the type of room.

Air River and Tributaries

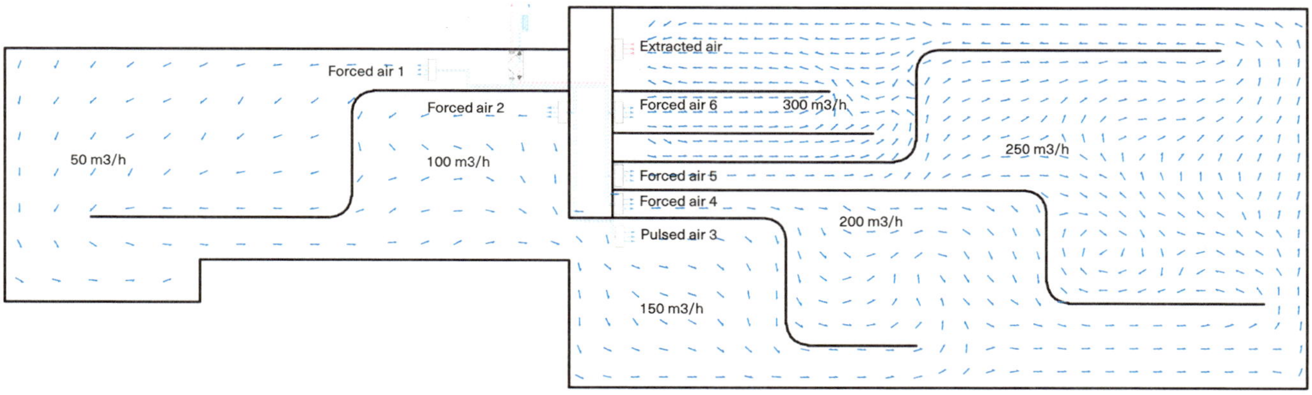

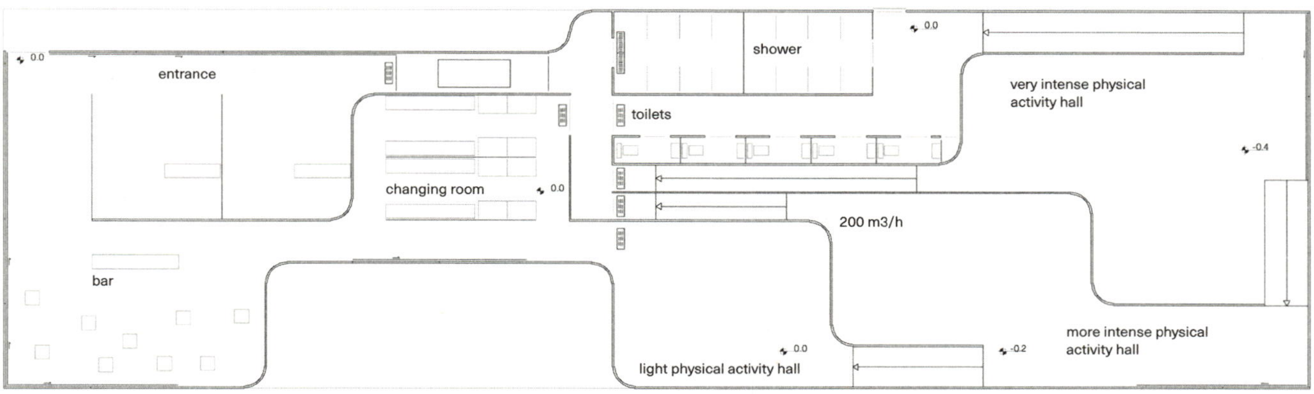

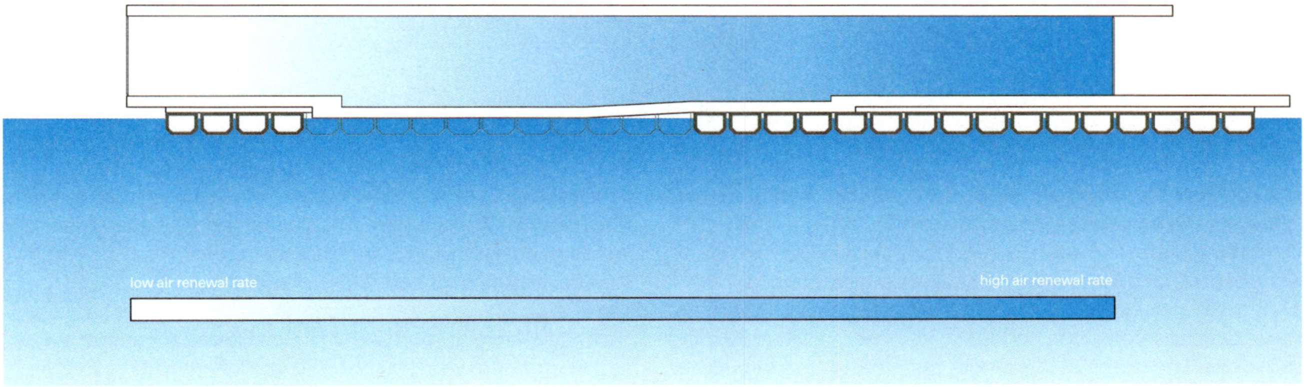

Mechanical ventilation system with heat recovery {Fig. 1-363}
Forced fresh air accumulates from space to space like tributaries join to form a river, increasing the amount of air in relation to the physical activity and the rate of water vapor release in each of the spaces.

Plan of the program following air flow {Fig. 1-364}
The program follows the amount of forced fresh air, from the more gentle activities with less fresh air flow at the beginning of the airflow, to the most physically demanding activities or those generating more humidity, as it progresses toward the removal of stale air at the end.

Section diagram of the principle of ventilation {Fig. 1-365}
The more intense the physical activity, the more you sweat or when you take a shower, the greater the amount of incoming fresh air required in order to renew the stale air more quickly.

The architecture is formed like a wind, a movement of air whose flow gradually increases according to the increase in the respiratory frequency of its occupants according to their activities and the amount of water vapor released.
In the end, the air is recovered only in the showers to pass through a dual-flow heat exchanger that extracts the heat from the exiting stale air to transfer it to the fresh, incoming air.

Thus, the air impulses accumulate to reach the following rates:

Entrance date	4 times / hour
Light sports halls	15 times / hour
Intense sports halls	25 times / hour
Sports changing rooms	20 times / hour
Showers	30 times / hour

Interior Wind

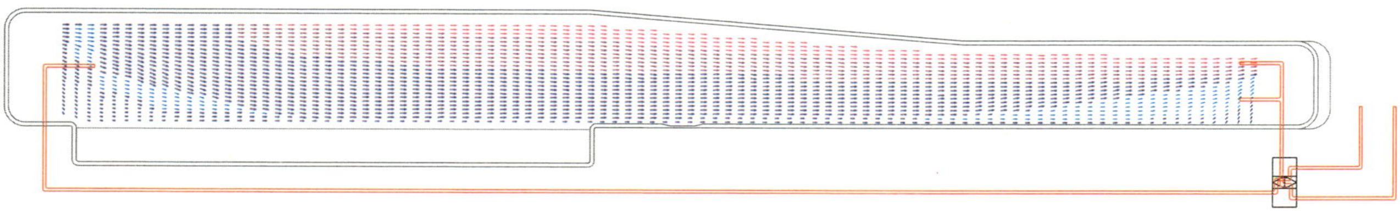

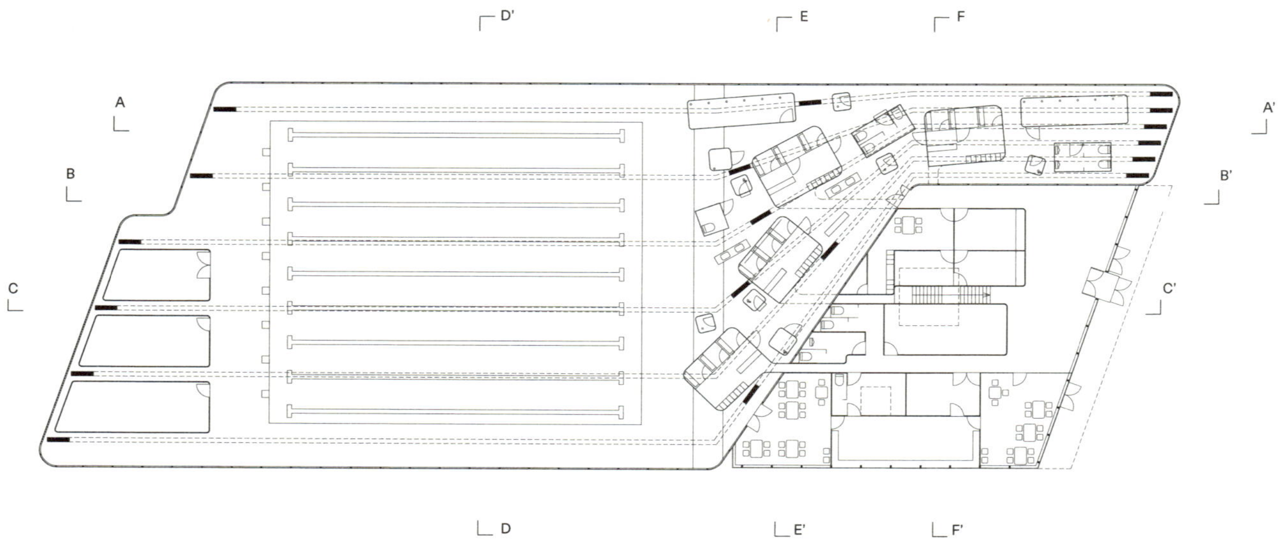

Ventilation section {Fig. 1-366}
Mechanical ventilation with heat exchanger in winter, following the same direction as the prevailing winds.

Plan {Fig. 1-367}
The shape of the building blows the wind onto the walls of the changing rooms, to cool them in summer and warm them in winter using the warm air of the swimming pool.

Our project for a new swimming pool considers environmental issues as the end and means to its architecture.
Our architectural forms and solution thus follow the climate. They are designed and arranged, taking meteorological phenomena into account, air and wind movements, water vapor displacements, the conduction of materials, temperature variations. The shape of this indoor pool follows the direction of the prevailing summer winds by taking advantage of their natural strength to ventilate the pool during the summer.
Our architecture slips into the path of the wind, gently interfering with its flow, bringing it in throughout the length of the building, taking advantage of this natural source of ventilation in summer. Our building is constructed around the wind, it follows its forms, it is designed along its path, it accompanies it for a few dozen meters, harnesses then releases it, contracts then expands it and allows it to follow its course. The shape of the building and its supporting structure are defined by following the guiding principles of wind direction, from south-west to north-east, accompanying it, deflecting it slightly, without interrupting it, conducting it along the length of the building. The structure supporting the roof is made of wood. It is made of long longitudinal beams that channel the wind from south-west to north-east by gradually lowering it toward the changing rooms in order to evacuate the stale air in summer and, in winter, to input heat. In winter, a dual-flow mechanical ventilation system with a heat exchanger controls the path of the wind in the building.

The fresh air is thus guided between the beams, then it is sent down toward the changing rooms due to the shape of the roof, which enables driving to the floor the heat of the hot air above the pool that has tended to rise and accumulate beneath the ceiling. So we use the heat of the pool's ambient air to heat the changing rooms. But to avoid introducing moisture to places where you get changed, we "chemically" separate them from the pool space, connecting their air supply to the entrance hall. The heating of these changing areas, therefore, is by conduction from outside the cubicle. The hot air coming from the pool "kisses" the highly heat-conducting aluminum walls and therefore, by conduction, sends the heat inside the cubicles. Conversely, ventilation is rendered from the entrance hall to avoid inputting the water vapor coming from the pool. The heat contained in the stale air is removed by a dual-flow heat exchanger before exiting the building. The glazing is triple and the roof is perfectly insulated.

Interior Wind

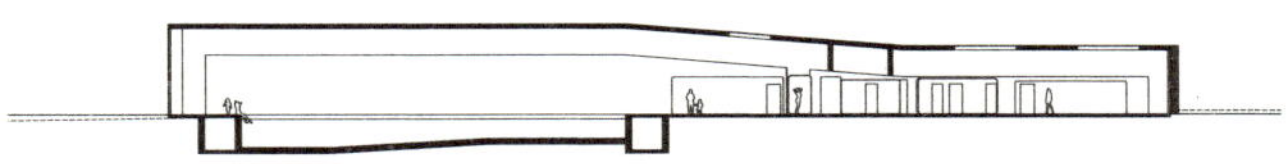

Section AA'

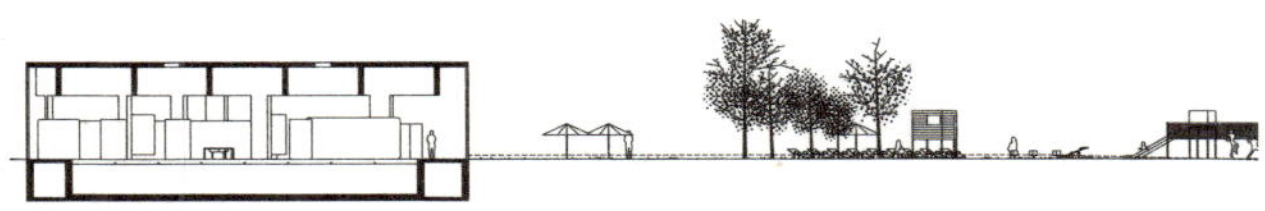

Section DD'

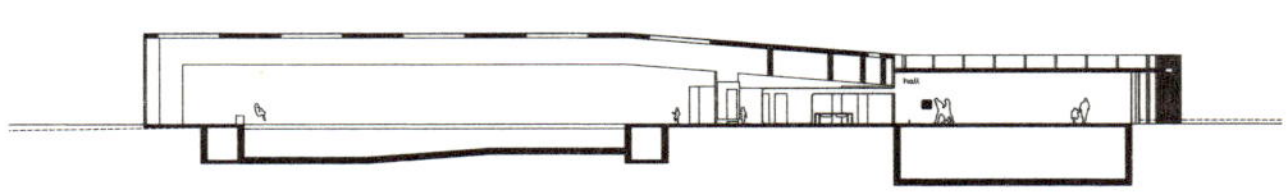

Section BB'

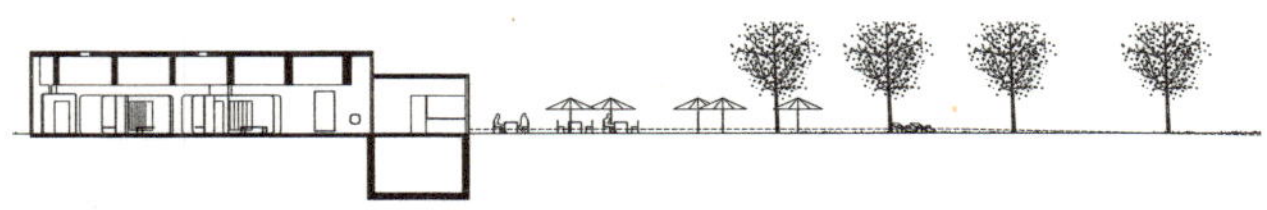

Section EE'

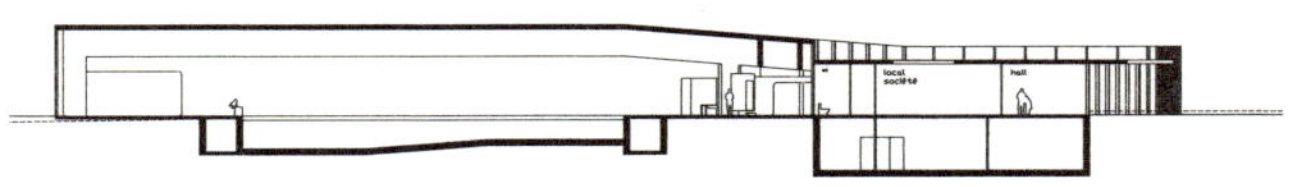

Section CC'

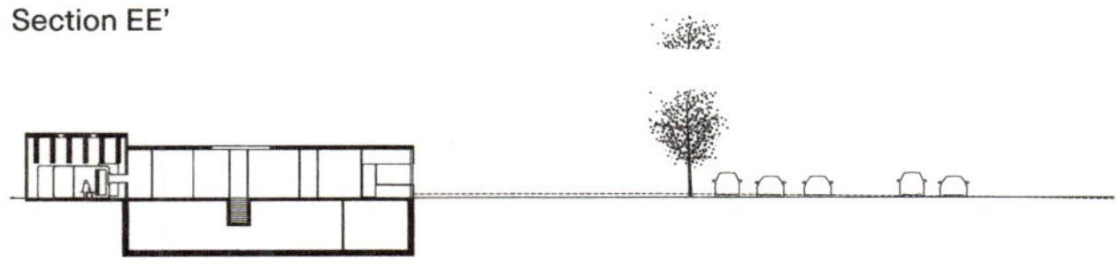

Section FF'

Sections {Fig. 1-283}

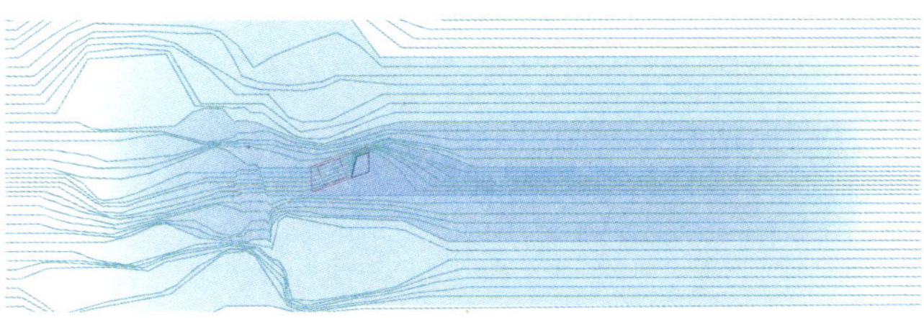

Predominant wind {Fig. 1-368}

The building is set in the path and the direction of the summer wind, to naturally cool and ventilate the pool and its spaces in summer.

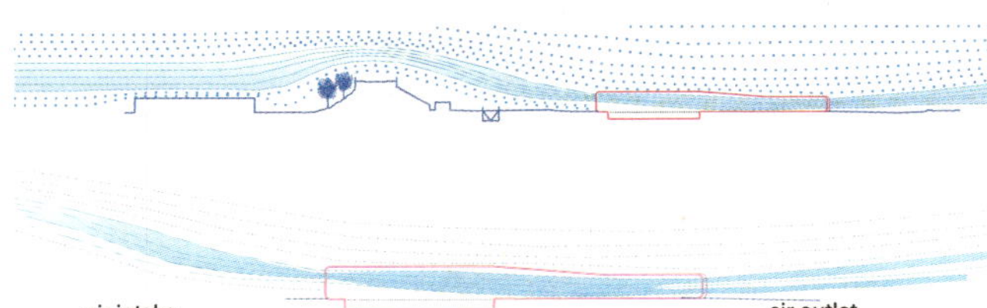

Section of the building in the wind {Fig. 1-369}

The building receives the prevailing wind to naturally ventilate all interior spaces in summer. In winter, mechanical ventilation replicates the direction of the summer wind reproducing it artificially.

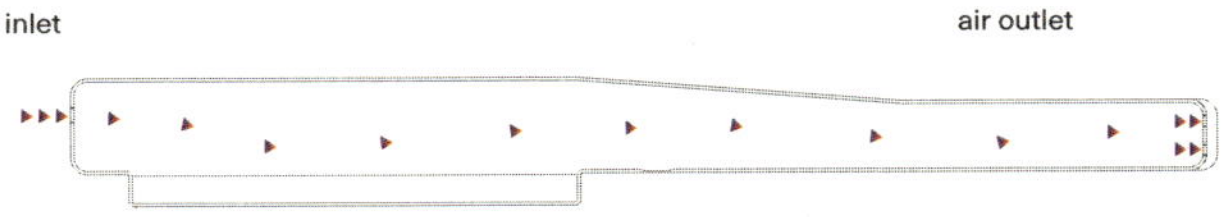

Summer strategy / natural ventilation {Fig. 1-370}

The building opens to the wind. The supporting structure thus follows the lines of the wind direction, allowing itself to be entirely traversed by the cooling, cleansing air current.

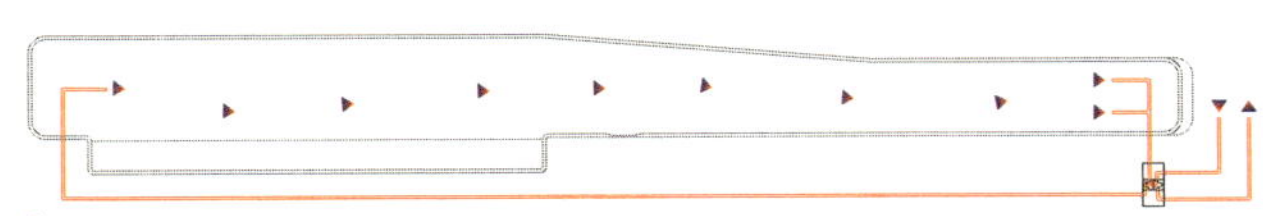

Winter Strategy / dual-flow controlled ventilation {Fig. 1-371}

In winter, ventilation is controlled and heat is recycled through a dual-flow exchanger. It replicates the direction of the summer wind to dissipate moisture and water treatment pollutants.

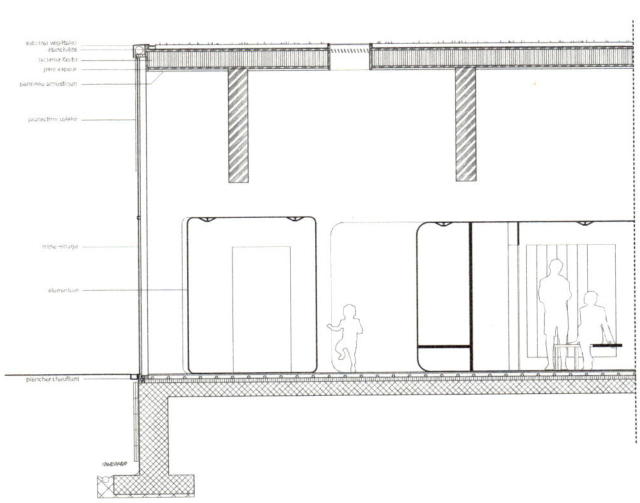

Section of the cubicles {Fig. 1-372}

Changing room cubicles are removed from the humid ambient air to benefit from drier air coming from independent ventilation. However, thanks to the use of a high thermal conductivity material, they benefit from the heat of the pool which communicates by conduction from the outside to the inside of the cubicle.

Interior Wind

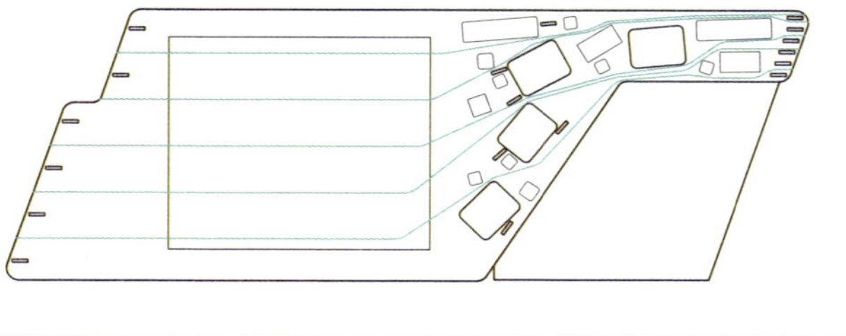

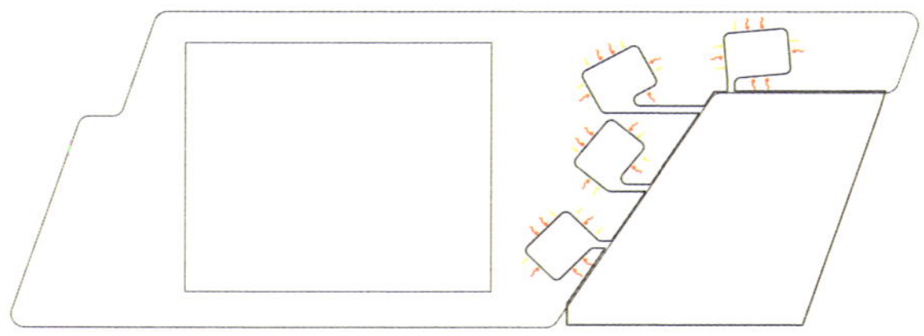

Movement of wind in the building {Fig. 1-373}

The long load-bearing beams of the roof are arranged in the direction of the wind so as not to hinder the movement of the ventilation air and guide it toward the outlet.

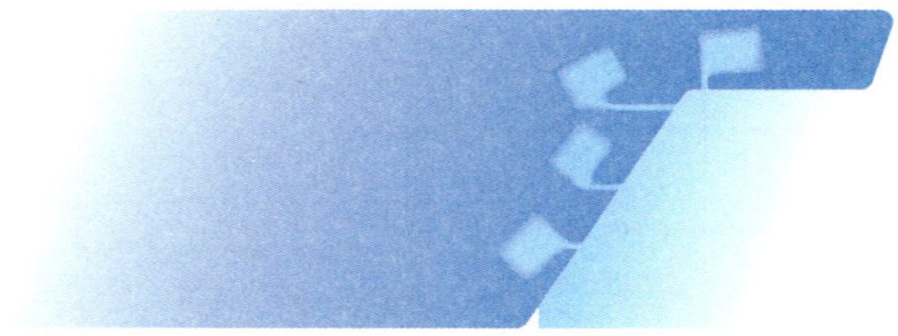

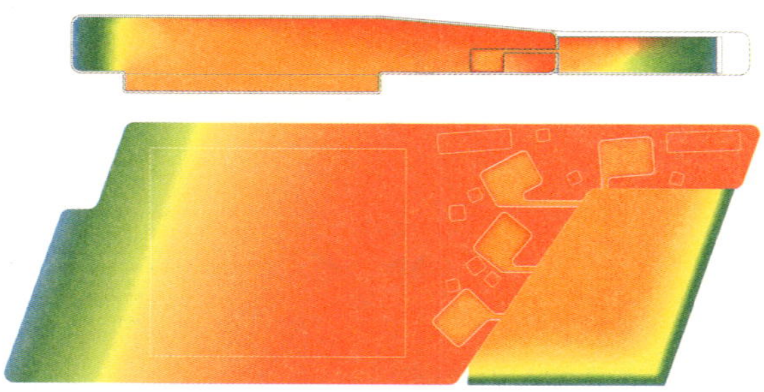

Winter Strategy / dual-flow controlled ventilation {Fig. 1-375}

Changing room cubicles are sealed against the pool water gases and moisture. However, they take advantage of the heat for warmth using the warm air of the pool.

Heat distribution {Fig. 1-374}

The changing room cubicles benefit from the heat of the pool air (which must be at 28°C) which is transmitted by thermal conduction through the aluminum walls.

Façade {Fig. 1-376}

Windtrap

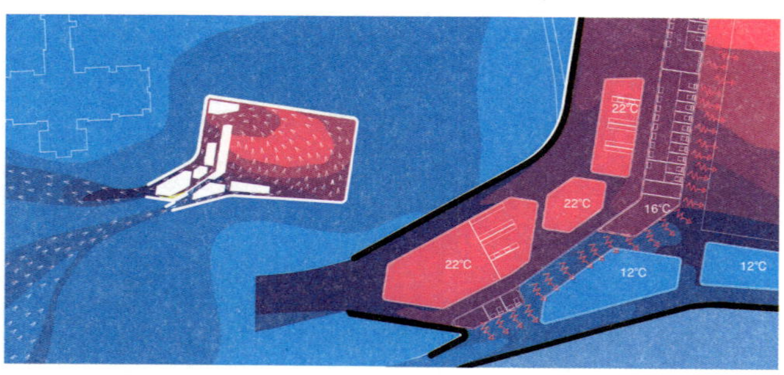

Windtrap site plan and detail drawing {Fig. 1-377}

The corridors and toilets operate as a boundary through which heat exchange takes place between incoming cold air and outgoing hot air.

This project for a sports hall is a spatial expansion of the dual-flow ventilation system with heat recovery - a system that is usually the size of a large machine - to turn the entire building into one. All of the rooms of the building are involved in the exchange of heat, according to the needs related to their intended use. The hottest rooms, such as showers and changing rooms, whose temperature is kept at 22°C, are placed toward the stale air outlet, while the coolest spaces such as storage rooms, at a temperature of 12°C, and then the gym, at 14°C, are at the fresh air inlet end.

The toilets and corridors form a sealed aluminum chemical boundary between the incoming fresh air and the outgoing hot air, though it is thermally conductive. The exiting hot air thus preheats the incoming cold air by means of the aluminum partitioning wall and finally the whole building serves as a stream of air where the activities and uses are placed according to their specific required temperature and returns on itself to recover the heat before it leaves.

Climatheque

Winter / Hot summer

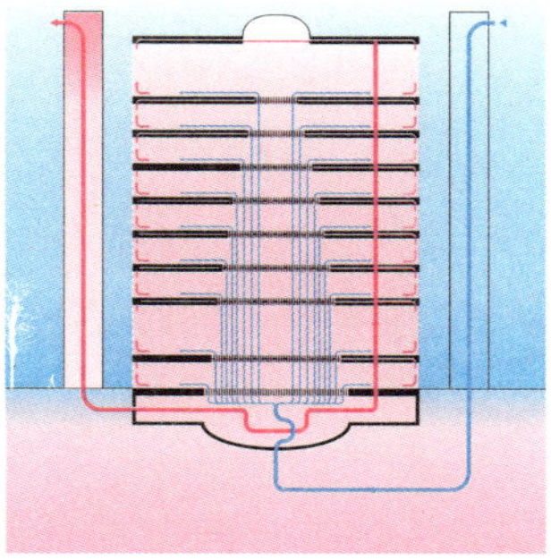

A building as a habitable ventilation duct {Fig. 1-378}

In winter, the cool, fresh air is harnessed in the wind tower in the north, it descends to the basement where it is slightly heated by geothermal energy. The fresh air then passes into the dual-flow heat exchanger to be preheated by the exiting stale air. The fresh air is then distributed throughout the building by passing through the slats of the central well in the heart of the building.

The fresh air is recovered in the edges of the flagstones being sucked mechanically for recovery, at ceiling level, of the stale air. The stale air is conducted via the peripheral duct surrounding the buildings on each floor to descend to the basement, pass through the heat exchanger to lose its heat to the cold incoming air and finally be sucked in by the solar updraft tower to the south.

Spring / Summer

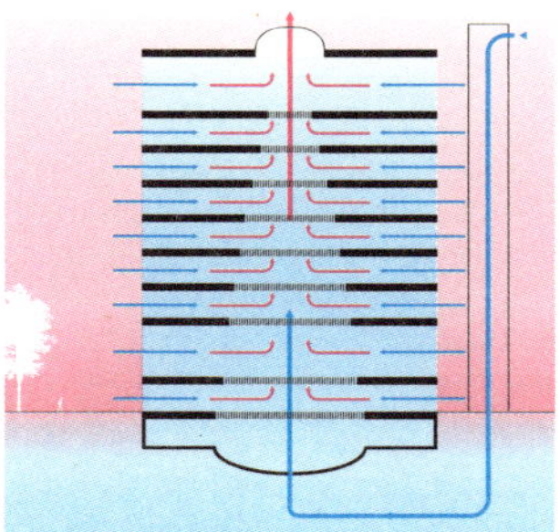

The heart of the building as a thermal chimney {Fig. 1-379}

In spring and summer, the air is renewed by manually opening the windows while the central well acts as a thermal chimney (hot air rising according to Archimedes' law), thus drawing in air from the windows up to the dome on the roof from which the hot air will be expelled naturally.

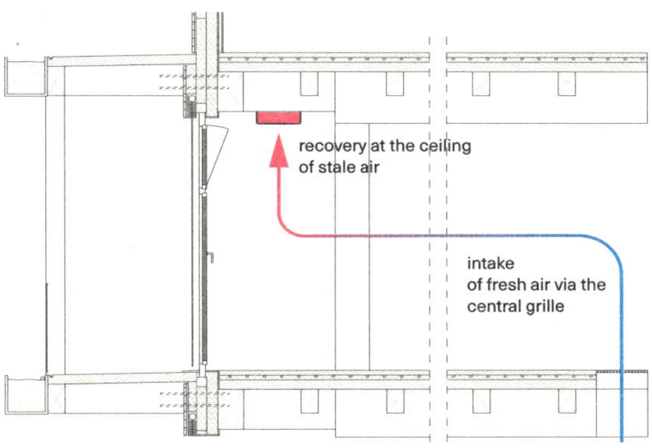

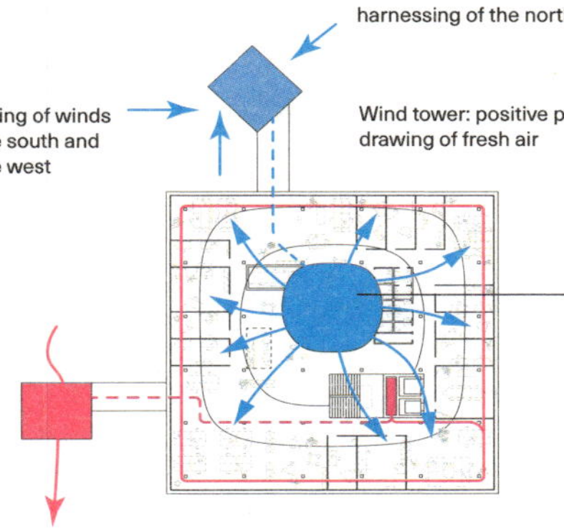

Controlled ventilation system {Fig. 1-380}

The fresh air is not distributed through a ventilation duct, but by the architecture itself. The stale air is returned via the periphery by a ventilation duct in the ceiling of each floor.

Fresh air distribution plan {Fig. 1-381}

Summer night time ventilation: by thermal inertia, the cool of the night is stored in the stone floor and provides freshness during the day.

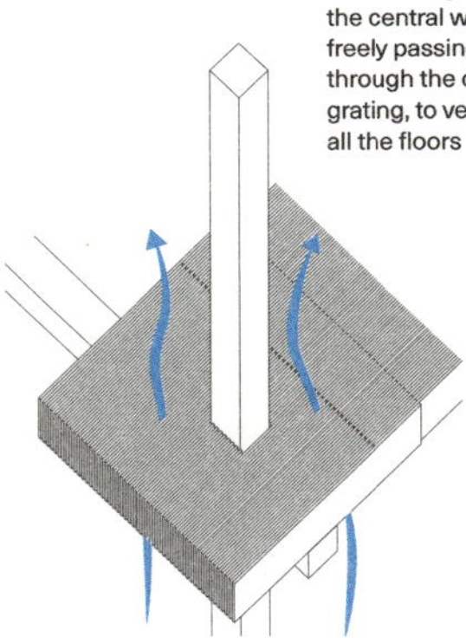

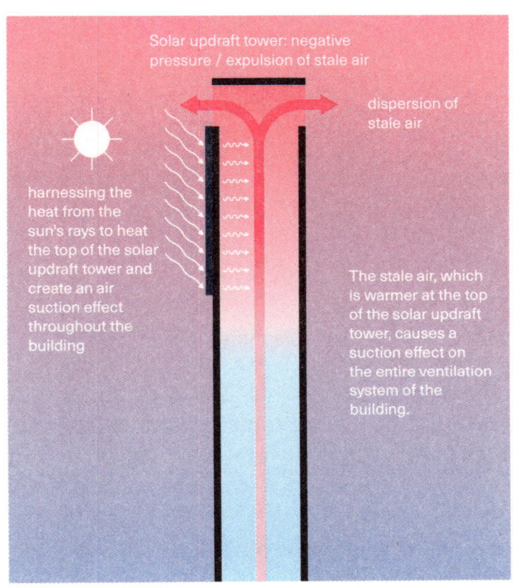

System for raising fresh air through floors {Fig. 1-382}

Concrete slatted floors allow fresh air to rise in the heart of the building. In addition to being flame retardant, the use of concrete is also suitable here due to its role in the inertia of the building, compensating for the very low inertia of wood and, on hot days, allowing further cooling of the incoming air.

Wind tower system for extracting stale air {Fig. 1-383}

Harnessing the heat from the sun's rays to heat the top of the solar updraft tower and create a negative pressure effect sucking the stale air and bringing in the fresh air and setting the entire ventilation system in motion.

Indoor Seasonal Shelters

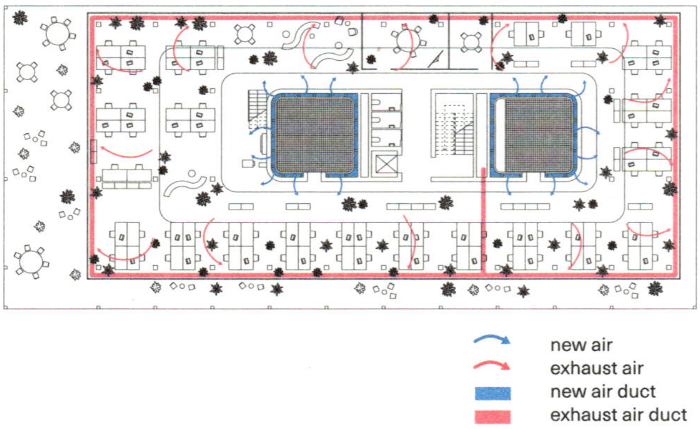

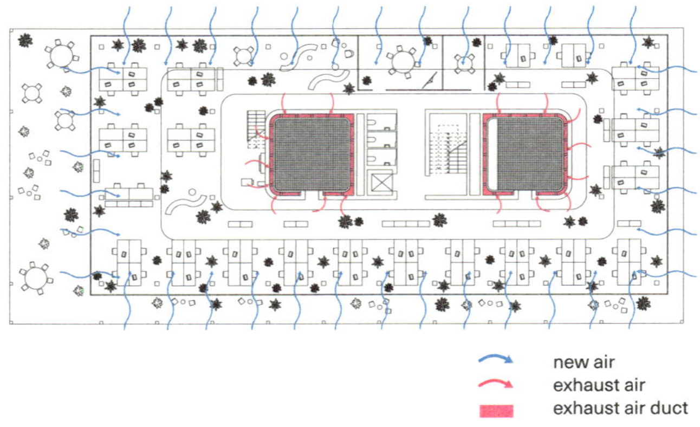

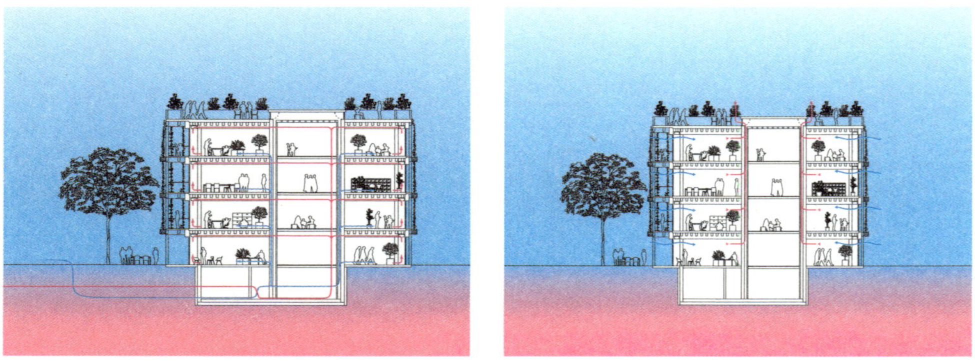

Principle of controlled ventilation in spring / summer {Fig. 1-384}

In spring and summer, the air is renewed by manually opening the windows while the two central wells act as a thermal chimney (the hot air rising according to Archimedes' law), then drawing the air from the windows to the roof domes from which the hot air will be expelled naturally.

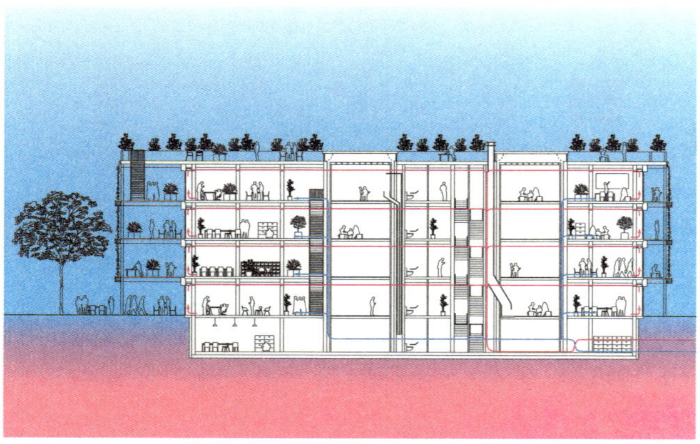

Principle of controlled ventilation in winter / hot summer {Fig. 1-385}

In winter, the cold, fresh air is harnessed in the park, it descends to the basement where it is slightly heated by geothermal energy. The fresh air then passes into the dual-flow heat exchanger to be preheated by the exiting stale air. The fresh air is then distributed throughout the building by passing through the slats of the two central wells in the heart of the building.

The fresh air is recovered in the edges of the flagstones being sucked mechanically for recovery, at ceiling level, of the stale air. The stale air is conducted via the peripheral duct surrounding the buildings on each floor to descend to the basement, pass through the heat exchanger to lose its heat to the cold incoming air and finally be sucked in by the solar updraft tower to the south.

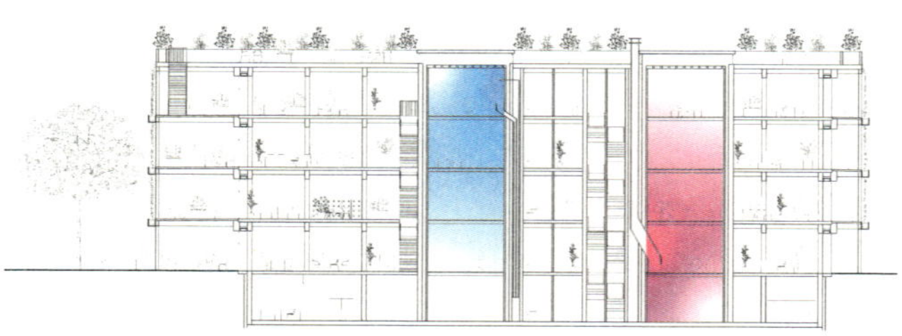

Climate Chimneys {Fig. 1-386}

Two shared collective spaces are envisaged in the heart of the building, one for summer, the other for winter. The first, rendered in stone, is coupled at the top with an evaporating fountain to cool the air in summer. The second, made of wood, is coupled with a fireplace to warm the air in winter.

This office building project is based on the creation in the heart of the building of two climatic chimneys in 50 cm thick Auvergne stone in order to constitute two thermal masses like two reservoirs accumulating heat in winter and coolness in summer by thermal inertia. The first provides a summer climatic shelter, in the manner of traditional Auvergne 'burons', or mountain cottages, which were seasonal shelters for shepherds in summer, with inner walls lined with white polished granite, which are cooled at night to store the low temperature of the air which will be released during the day.

At the top, it features a fountain / water jet, whose freshness created by evaporation falls into the chimney to cool all floors. The second chimney acts as a winter climatic shelter, like a winter 'buron', lined with wood, in which a wood fire is proposed at the bottom, whose heat will rise and warm all floors. The principle is to re-establish the social and collective value of space based on thermal properties. Ventilation, mechanical in winter and natural in summer, uses these two chimneys to provide fresh air in winter, and to expel stale air in summer that will escape naturally by convection at the roof, like a thermal chimney.

CHAPTER II.

SENTIMENTAL METEOROLOGY

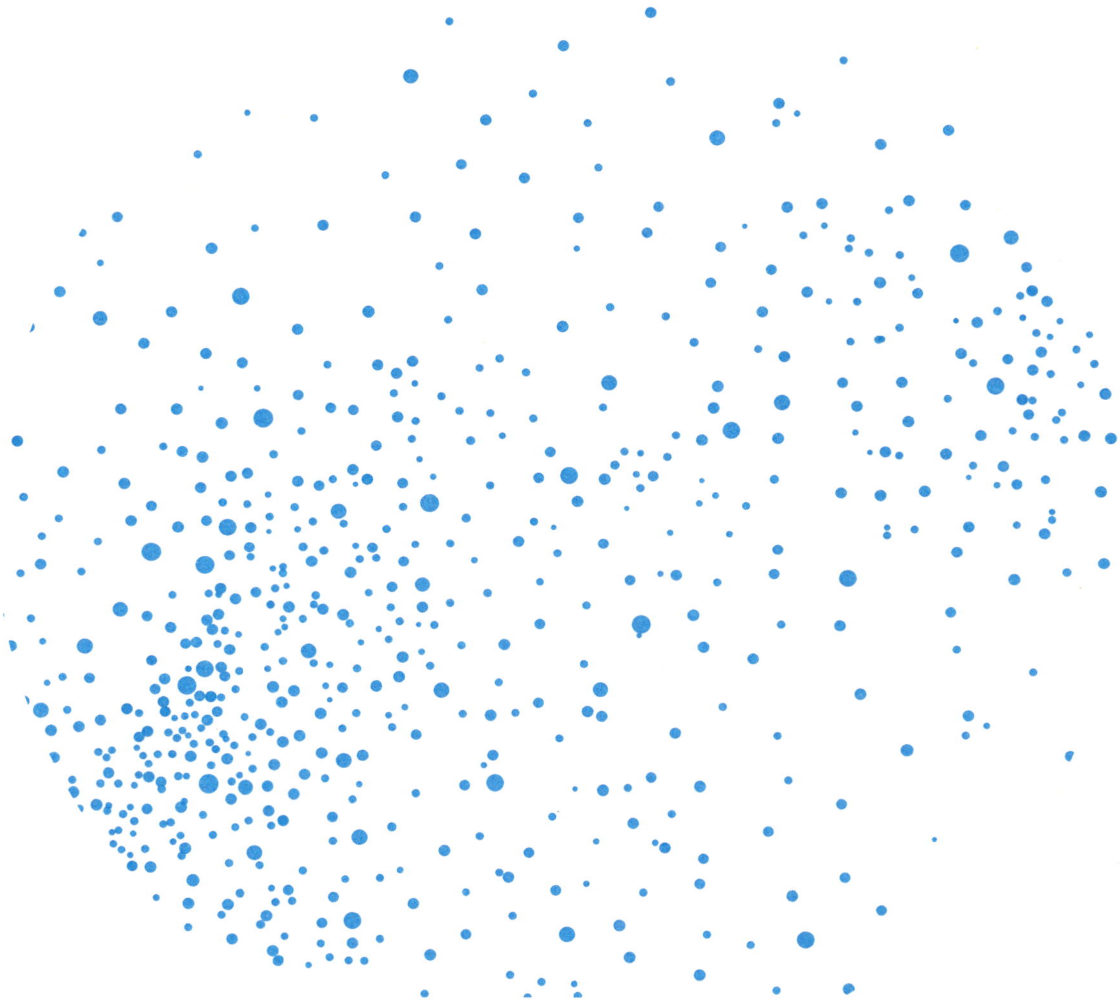

Texts by Philippe Rahm
English translation by Shantel Blakely

"We are warm."
{Fig. 2-387}

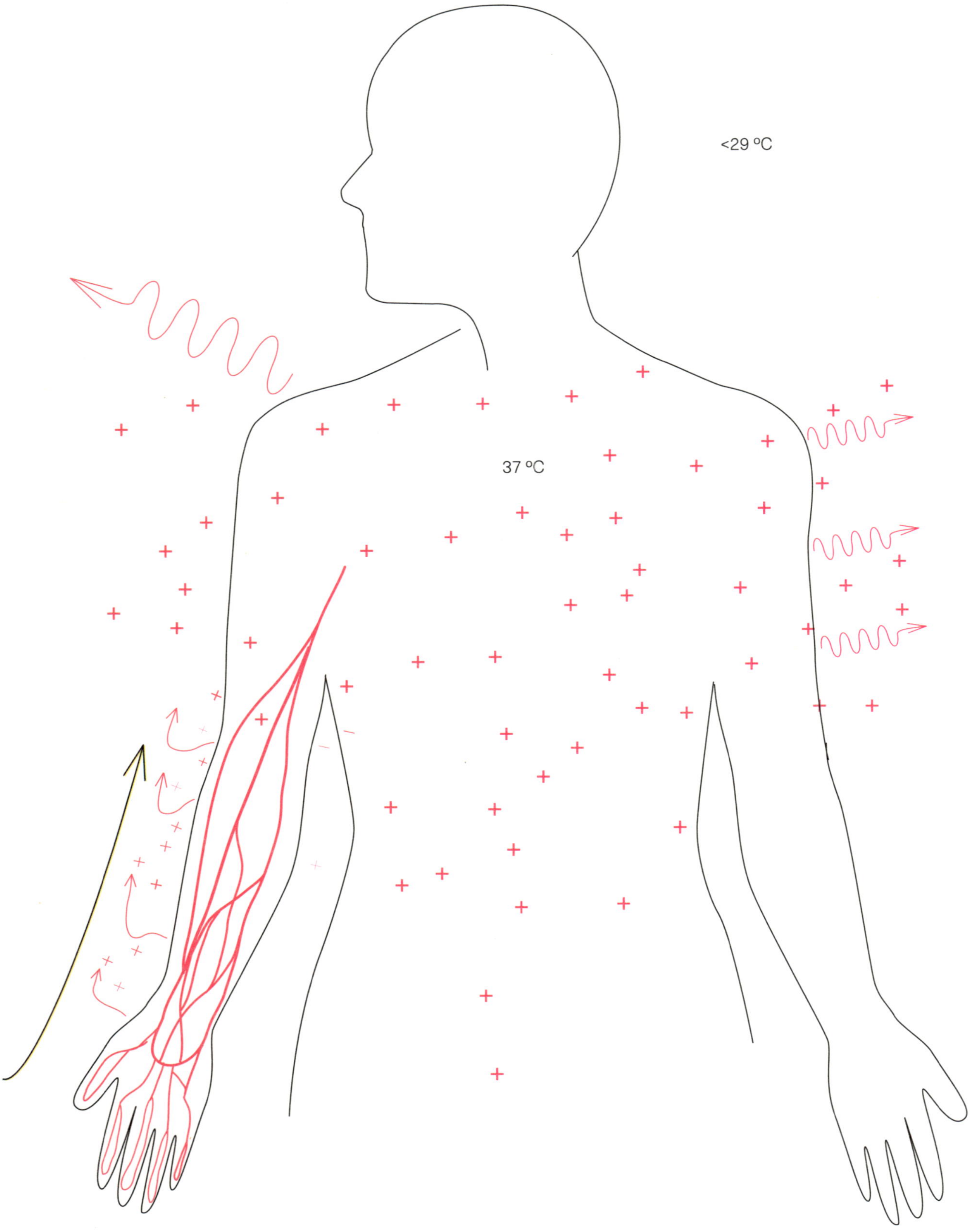

II.I. Radiation

It is warm. It is summer. We are in Switzerland, above the lake. It is 1:15 p.m., the afternoon has barely begun. There is not a sound in the town, not a breath of wind. I stand, leaning back slightly, my back against the car, feet pointed south on the sidewalk of a street at right angles to the slope. The sun emits an intense light, multiplied by the sparkling lake water and the town's mineral surfaces which have a higher light albedo. The fragments of quartz that bloom in bitumen and the glass of apartment houses in a certain orientation reflect the sun's rays in crystalline wavelets that reach my eyes from this refracted multitude of sources. They blind me. I close my eyelids and tilt my head in an unsuccessful attempt to escape them. My eyes hurt, not so much because of the light that I see but because of the immense invisible spectrum of ultraviolet rays that reverberate and strike my retina. One sees ultraviolet rays, though they produce no image in the brain. They are present in the form of a pain that is more or less strong as a function of their intensity, their wavelength. Could one paint a canvas not with visible colors but with the ultraviolets that produce certain pains as the eyes cross this tableau? A blue discomfort here, a yellow one there, green irritations here and there draw a certain figure. For protection one can wear sunglasses but I have none. Apparently there are some whose lenses remain transparent, blocking ultraviolet only, suppressing the pain of the sunlight without reducing its intensity.

There is no one on the streets. The neighborhood's inhabitants rest in the darkness of their homes, at a distance from the windows, under the shadows of their roofs. After lunch, they drift to the north and slumber against the walls of rooms across which they will breathe rapturously, by inverse radiation, the mountain's coolness and humidity. Their sleepiness is a form of thermolysis that affords them the limited endogenous production of heat by reducing muscle activity. But they digest, which augments the heat of the sun inopportunely. Burning in the stomach, the foods raise the body's temperature. There are thus two sources of heat in the early afternoon, the after-lunch, like two suns. One is exterior, astronomical, of an intensity of 70,000 lux, attenuated somewhat by the lowered awnings. The other is interior, digestive, produced by consumption of proteins during the Krebs cycle to maintain the constant body temperature of 37 °C. For my part, I haven't eaten. I suffer only the heat of the first sun, in front of me, near my skin, the distance being the shortest during the summer and in early afternoon. The horizontality of the ground and the rays of the sun form almost a right angle. The sun's rays traverse the layers of the atmosphere almost vertically and therefore encounter few water droplets, fog, and clouds that might otherwise have mitigated their intensity and relieved me. The sun's combustion propagates a radiant heat made of white and visible, infrared, and ultraviolet waves (and other invisible ionizing or non-ionizing that do not pass through the ozone layer). The electromagnetic waves are propagated in space without losing heat, crossing the terrestrial atmosphere until they finally strike solid surfaces, objects encountered by accident: the mica in a piece of granite, the foam of the water on the lake's surface, the metal of an automobile chassis, or the skin of my jaw. Upon making contact with these objects, the light waves are partially transformed into heat and rays by conduction, the general quantity of heat in the object. Some wavelengths nevertheless rebound back towards the sky: from my jaw, those found in red; from the stone, green; from the water's white foam, the whole spectrum; and from my uncle's car, red (about 750 nm). And then there is this rain of infrared that one does not see, that inundates the heat of the earth and thus the intensity now elevated is going to diminish progressively until the sun sets. But there is no question of that at this moment; at 1:20 the afternoon has just begun.

I am fifteen. My friend Eric is with me on vacation and we have accompanied my architect uncle to a construction site. Holding his keys, not knowing what to do with ourselves, we wait by the car. In this silent neighborhood high up in the city there is no activity—no commerce, no one passing. The homes here are all newly built, with modern apartments whose large expanses of windows are open to the warm southern light, the sky, the mountains in the distance, and the lake's immense, glittering landscape. Eric closes the car door, which clunks heavily, without resonance, its muted sound immediately absorbed by the environment. A car that passes slowly behind us produces a rubbery sound. Does bitumen lose all hardness under the sun? Will it become cotton-soft in the heat?

We hear birdsong from the trees of nearby gardens and would love to take advantage of their gracious shadow, a free public cooling to protect us from the direct rays of the sun, but in the street, unfortunately there is not a single tree. We are warm. Is there anything to drink? That is another possible mode of thermolysis: cooling from inside, excess body heat evaporated through sweating and evaporation. We have nothing to drink here in the sun, standing by the car that belongs to my uncle, who does not return.

At some point, one of us arrives at the conclusion that there is no point in trying to be less hot because we have no means to cool off. In the end our freedom, our liberty to act is the power to choose to be even warmer. A search of the trunk yields a sort of coverlet. We get into the car. The interior air is boiling due to the greenhouse effect that bends the sun's rays to the steel and glass interior, preventing them from escaping the car after having let them in by the glazing and transforming them into heat against the fabric seats and plastic surfaces. The doors are closed, the last open windows put up, the ignition turned on, the heat turned up to maximum. We put on our sweaters, which we had left on the back seat, and bundle ourselves under the coverlet.

II.II. Inertia

We are seated, Pierre and I, on one of six beds in our room. There is plenty of light, the morning being well underway. The single window faces north and not a single ray of direct sunlight ever passes through it. The sun is for our neighbors across the street, who live above the cheese shop. Inevitably, having been forbidden to go out since the previous morning, we have watched them a lot. From time to time, according to a precise angle in a partially opened window in their home, a ray of sun reflects on the pane of glass and comes to us, degraded in intensity by many thousands of lux. It glances on the hard and soft surfaces of our room and is deformed on the greenish linoleum floor and on the woolen covers of a bed, whose color it changes from a dark gray to a lighter brown. The wool heats up. Apart from brown, the light rays of other colors, more specifically red and orange, are transformed into heat. The tiny fibers of the wool stand up, illuminated from behind, liberating thousands of millions of dust particles that flutter across the cylinder of light between the bed and the window. The electric light fixtures are on. The spot of light formed by the sun, cooler and bluer, contrasts with the warm white of the artificial incandescent light that tends to orange. Suddenly, the spot disappears; across the way, someone has closed the window.

On the bed, I read aloud The Secrets of the Aura, by Lobsang Rampa, while Pierre slowly turns his hand before his eyes, looking for his aura. He really sees something, he tells me: a faint haze of some millimeters surrounds each of his fingers. I concentrate in my turn to see this haze grow larger and change colors. David, for his part, had ventured into the hallway a few minutes before in spite of the interdiction. We are being punished, all three, excluded from activities and confined for three days in our room. It is the second time in the course of July, 1978. We are at summer camp at Haute-Savoie, in the mountains. David and I are eleven years old but Pierre is ten, born the same year as us but in October. We tease Pierre by telling him he will not be able to come to the cinema with us when, in two years, we will be able to see films forbidden to children under thirteen. Two days before, the director of the camp had moved us from the room we had occupied on the south side of the building to a room on the north. We must stay there the entire day, in the shade, in the inertia of the damp, cold walls. There is no large balcony linking all the rooms to one another on the north side, unlike on the south. During the night, we had gone out onto the balcony to enter adjacent rooms and frighten our fellow campers. But the monitors caught us. The punishment imposed by the director was not to put us in the darker and damper northern part of the building; that was a collateral effect. He had put us here to deprive us of the balcony. But it is the absence of direct light that I resent, this cold darkness as if the sun did not really rise in the morning—a bit like us in the end, deprived of the right to go out, constrained to stay inside all day, out of the sun.

The door to the room opens soundlessly and David enters. He closes the door delicately and comes towards us making large gestures with his arms, eyes wide open, opening his mouth broadly but without making a sound. Breathless, in a low voice, he tells us he has just seen Corinne in her room and she is naked. Our monitor would now be in the shower. I would never have dared enter the monitors' rooms, which were at the end of the hallway to the east, on the same level as ours. David takes us by the hands to incite us to get up and return with him.

Corinne is beautiful. I don't know her age; in my eyes she could be 18 or 28. She is the sole female monitor at camp, the two others being men. One of the two is a hippie with long blond curly hair whose jeans and attitude I really like. On a hike in the mountains a few days earlier, he had asked us to collect tiny mushrooms that grew on cow dung. He planned to prepare a tea for the evening, „like Tibetan lamas," he said. He had read Lobsang Rampa too (unlike a second monitor who, after leafing through our book The Secrets of the Aura, had accused us of only making a show of reading it). Pierre asked if we could taste his tea, but he refused: it would be possible if we were in Tibet, but not in France. That afternoon we were a dozen children in the field, looking for cow dung from which we could collect mushrooms to bring back to our monitor, who would delicately pack them up in a handkerchief. At the end of the afternoon, as it was late, we returned by way of a wide paved road that descended toward camp. David, Pierre, and I walked next to Corinne. We asked questions about her life, her boyfriends. David asked if she had made love yet. When she was younger, she replied, she told her parents that she was going to sleep at a friend's house but in reality she slept with her boyfriend. Corinne had white skin and long, slightly wavy red hair. I looked at her partially open mouth and blushed, and this redness spread from my face into my legs and seemed to melt me. She loosened her hair with her hand, exposing her throat where I wanted to place my lips and render her skin liquid. She wore a generously sized white T-shirt whose neckline descended well below her shoulders. The fabric deformed on her breasts which I never saw but nevertheless imagined. I breathed quickly and with more effort, dispersing the oxygen in my body more rapidly. I lacked air. Blood flowed to the back of my cranium, inflating my brain, pressing it against my skull, pulsing in my ears with a more and more rapid rhythm. On impulse David asked if she would allow us to kiss her on her mouth, which none of us had ever done. She smiled without giving a firm response. He asked if we could meet her in a hiding place, one night, when everyone slept, behind the small wood to the southeast of the camp. She smiled but did not say no. That had been last week. It had not yet happened.

We finally agree to follow David to Corinne's room. We open the door noiselessly. No one is in the hallway, which is plunged in darkness. Our campmates have all been out for hours, accompanied by the two other monitors. Corinne has stayed in her room because it is her day off. The hallway is even darker than our room. Nothing is visible except the orange lights of switches in the darkness. Their luminosity intensifies at the rate and to the degree that our eyes habituate to the blackness, until they are reflected in the plastic floor. We do not turn on the light. We wait a few seconds until our pupils dilate and then, progressively, the lines of transition between vertical and horizontal surfaces appear and draw the form of the hallway. We are in socks. Without making a noise, we close the door of our room behind us before setting off toward the east. We move twenty or so meters, passing three doors similar to ours, to reach the end of the hall. David stops in front of a door, presses its handle and opens it softly to enter Corinne's room. We advance a few steps and then the room appears on our right, its blazing light instantly dispersing the darkness and cold around us. Rays of sunlight bounce from the windows, rebound on the floor, the white walls, the ceiling, and blind us for a moment. Before us now are three large windows that open on an immense blue summer sky without a single cloud. Turning to our right, we discover Corinne stretched out on her bed, in a deep sleep. On seeing us, she pulls up a folded white sheet to completely cover her nude body. Her head tilts back slightly, propped on her ear, over which her long hair tumbles, nearly transparent in the light. She looks at us and smiles without saying a word. We approach her, surround her, sit on the edge of the bed. Her skin is very light pink, her throat nearly white. She speaks to us softly of sleep, of sleeping in, of summer. David then catches the top of the drape and slowly pulls it down.

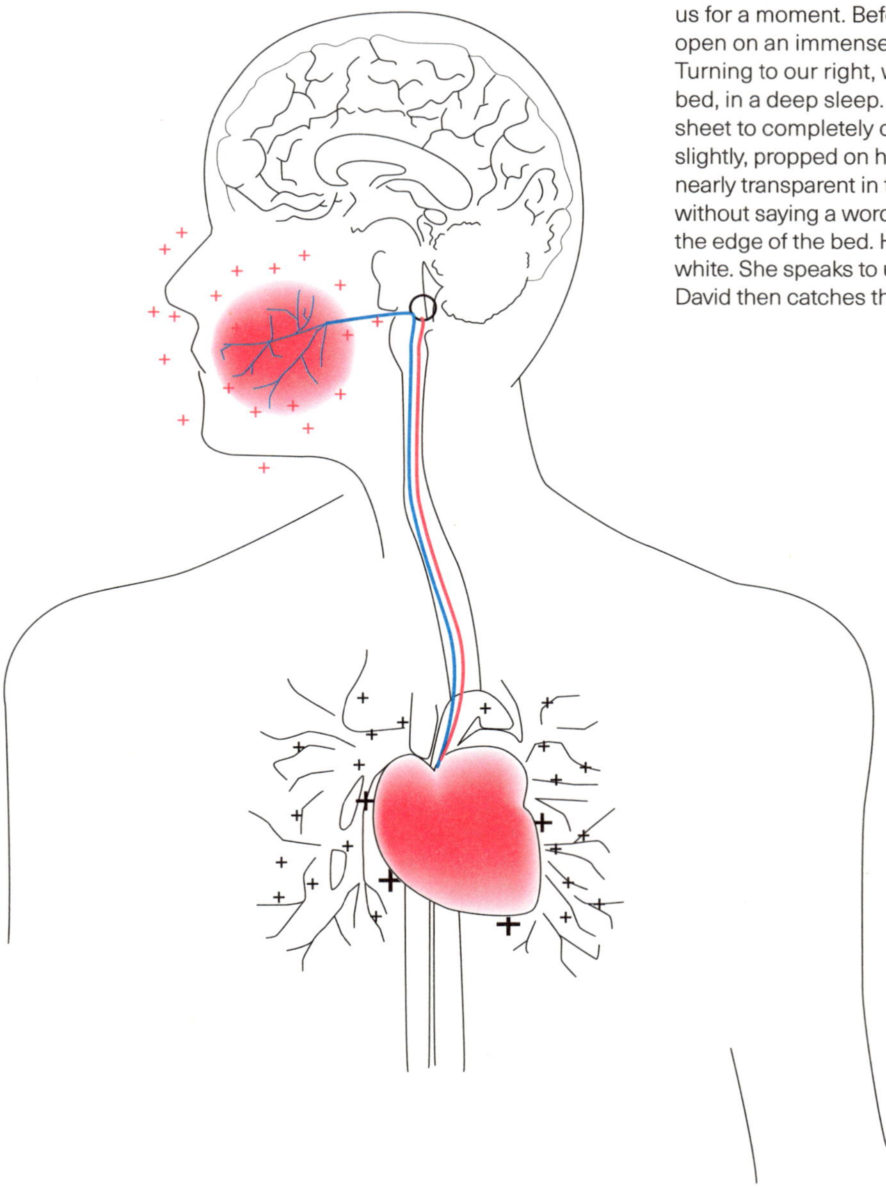

"I looked at her partially open mouth and blushed..." {Fig. 2-388}

II.III. Altitude

There are correspondences between history and geography. Voyages in time may be enacted by movement in space. It is spring. I cross the Rhône Valley at 120 km/h. The sky is blue, without clouds. The highway passes level with the sun in a field of apricot trees, perfectly horizontal in the direction of the valley. Around me, covering the plain, millions of white flowers. Petals of a pale, almost crystalline green, scintillate under the influence of the wind and my displacement in the bright sunlight. In the distance, mountains rise high above this silver horizon. Veiled by microscopic droplets of water in suspension in the air, they appear blue, in tones more or less somber or bright. To either side of the car, they frame the landscape and darken it below.

The thermometer in the car reports an exterior temperature of 20°C. It said 14°C an hour ago, as I left Lausanne. I can definitely turn off the heat now—the first time since last November. But the air from outside progressively permeates the car's interior and the two temperatures, once different, now synchronize. The smell is no longer the same inside and outside: no apricot flower reaches me, but something—fabric or warm plastic—gives off a light odor. Little by little, I cease to smell anything in particular.

It is noon. I am on my way to Italy. The music of the radio is barely distinct above the sound of the motor, which is louder than usual, so I make sure all the windows are closed. A few days earlier, spring had arrived. Since then, the climate had warmed progressively until reaching an agreeable temperature in harmony with that of my body, today at noon. As summer approaches, the temperature surges in the morning and continues to rise in the afternoon. The exterior air is now neither too cold, as it was in the morning, nor too warm, as it will be in a few weeks at the same hour. It is perfectly tepid. It can invade my body by my mouth, my nose, the pores of skin that barely breathe. I take off my sweater and feel fine in just a T-shirt. Soft spring light, barely 20,000 lux, arrives on my right from an angle still sharply to the south, and rests lightly on the gray shoulders of the road that absorb most of it. The muscles of my eyelids relax, my pupils repose halfway, neither contracted nor dilated. With few cars on the road my attention on driving relaxes and evaporates, gliding.

There are correspondences between history and geography. Voyages in time may be enacted by movement in space. It is spring. I cross the Rhône Valley at 120 km/h. The sky is blue, without clouds. The highway passes level with the sun in a field of apricot trees, perfectly horizontal in the direction of the valley. Around me, covering the plain, millions of white flowers. Petals of a pale, almost crystalline green, scintillate under the influence of the wind and my displacement in the bright sunlight. In the distance, mountains rise high above this silver horizon. Veiled by microscopic droplets of water in suspension in the air, they appear blue, in tones more or less somber or bright. To either side of the car, they frame the landscape and darken it below.

The thermometer in the car reports an exterior temperature of 20°C. It said 14 °C an hour ago, as I left Lausanne. I can definitely turn off the heat now—the first time since last November. But the air from outside progressively permeates the car's interior and the two temperatures, once different, now synchronize. The smell is no longer the same inside and outside: no apricot flower reaches me, but something —fabric or warm plastic—gives off a light odor. Little by little, I cease to smell anything in particular.

It is noon. I am on my way to Italy. The music of the radio is barely distinct above the sound of the motor, which is louder than usual, so I make sure all the windows are closed. A few days earlier, spring had arrived. Since then, the climate had warmed progressively until reaching an agreeable temperature in harmony with that of my body, today at noon. As summer approaches, the temperature surges in the morning and continues to rise in the afternoon. The exterior air is now neither too cold, as it was in the morning, nor too warm, as it will be in a few weeks at the same hour. It is perfectly tepid. It can invade my body by my mouth, my nose, the pores of skin that barely breathe. I take off my sweater and feel fine in just a T-shirt. Soft spring light, barely 20,000 lux, arrives on my right from an angle still sharply to the south, and rests lightly on the gray shoulders of the road that absorb most of it. The muscles of my eyelids relax, my pupils repose halfway, neither contracted nor dilated. With few cars on the road my attention on driving relaxes and evaporates, gliding on the music. I reach the elevation of the village of Saxon. During architecture school I had a classmate who came from this village, which amused me because Saxon was also the name of a heavy metal band whose song "Dallas 1 p.m." I liked a lot at age 14. The summer of our first year, he invited me to his home to review our course materials. We did not stay on the plain; we climbed along the valley's northern slope, to Sapinhaut. His family had a second home just a few hundred meters from the first but at a higher altitude. He explained to me that the villages in the valley were often double, with one below (Sapinbas) and one above (Sapinhaut). Residents would possess a home in each village and could split the year between two altitudes, two climates, two states of ambiance slightly separated in time. By moving their bodies in transhumance, they could freely choose to step up toward the alps if they were too warm or to descend if they were cold to the plain, with animals, fruits, vegetables and legumes. Sapinhaut was not quite alpine but the difference of a hundred meters in altitude relative to Saxon meant a corresponding lowering of temperature by a few degrees. We spent a week up above, in air effervescent with bees, butterflies, and insects, eating exquisite salads of dark greens that we gathered in the garden, sleeping in a wooden cabin. We worked inside under the rays of a sunlight softened and lightly greened by the trees' thousands of leaves, escaping the plain's punishing high-summer heatwave.

The highway becomes less precise at the rate and measure at which one rises in the Rhône Valley towards the river's source. It is interrupted, begins again, is interrupted again, and leaves me finally on a little road that winds and wrinkles until the mountains enclose us. The valley is narrow on the river, rising in altitude, but I have stopped following it and slant toward the south to climb the mountains. The car barely goes up; I downshift and press the pedal harder. The motor's sound intensifies in fits and starts. I accelerate and brake. The ascent zig-zags with a succession of hairpin turns to soften the incline. The sun is poised on one side and the other of the car until it disappears definitively behind the mountains.

After the tenth turn, my side window gives onto crystalline blue, green, or pink variations of granite. I rise now in the shadow of the mountains, above the luminous plain. As I gain in altitude I can see it more completely below, in flower, extending farther and farther along the valley. But at my level, the vegetation grows darker. There are no more flowers here, not even buds, just desiccated stems still in winter, barely green. The leaves give way to needles, the bark of trees acquires the oily, brown-orange luster of resin, the green turns to black, the shadows are deep and cool. The higher I rise, the more the external temperature falls. I turn toward the Simplon Pass to get to Italy, the other side of the Alps. The road is wide and well-drawn. It follows a little valley now, directly to the south, perpendicular to that of the Rhône. Trees are scarce. Below the road are meadows of yellowish, barely thawed out, close-cropped plants.

The exterior temperature has fallen to the level of that morning in the plain, 14°C. It continues to descend as I climb on. It drops to 12. . . to 11. . . to 10°C. I do not feel this falling temperature; I read it on the console. Inside the car, the temperature has remained the same as below, 20°C.
In short, I have crossed the mountains taking with me a pocket of spring air in the cabin of the car while my surroundings have frozen and retreated in time towards winter. The few rays of sun that come through the sealed windows continue to heat my clothing, the fabric seats, my skin, the asphalt of the road in front of the car. The countryside becomes mineral, fragmented. The outcroppings of rock have an adventurous geometry: crystalline polyhedrals and rhomboids, hard rocks among which the road was constructed with increasing effort. Brusquely, I enter the profound darkness of a first tunnel.
The intensity of the motor's sound is amplified and multiplied, resounding against the rock. The great countryside disappears and the car is engulfed by the mountain. During the long minutes of the approach to the saddle, the tunnels follow one another almost without interruption. The rock, hollowed out and coated with concrete, is damp and glistens greasily in front of me, reflecting the car's headlights. At some points, the tunnel crosses the course of underground rivers. Their water runs along the walls and pools on the ground, darkening the asphalt. The temperature drops again and, this time, there is neither infrared radiation nor the greenhouse effect to maintain a comfortable temperature inside the car. Again I turn on the heat, still set to 20°C. My pupils dilate in the dark.

A white spot in front of me, the end of the tunnel enlarges as I advance toward this light, whose contrast with the subterranean darkness progressively renders the shapes around me imperceptible. Then the exit from the tunnel, the horizontality of the saddle, the sky, everything going white. The sound of the car dissipates and the space around it is suddenly expanded, bleached. In the surrounding landscape, snow covers everything. My pupils contract uncomfortably as the winter sunlight, reverberating on the snow, springs back from all sides with renewed intensity. The sky is immense, cloudless blue. As the car slides soundlessly over the cut road in the snow, I open the window. Exterior air enters the car violently and beats my eardrums. I breathe the cold, dry air that dispels in an instant the last spring warmth of the plain, which had remained in the car. I am in winter once again and badly dressed for it.

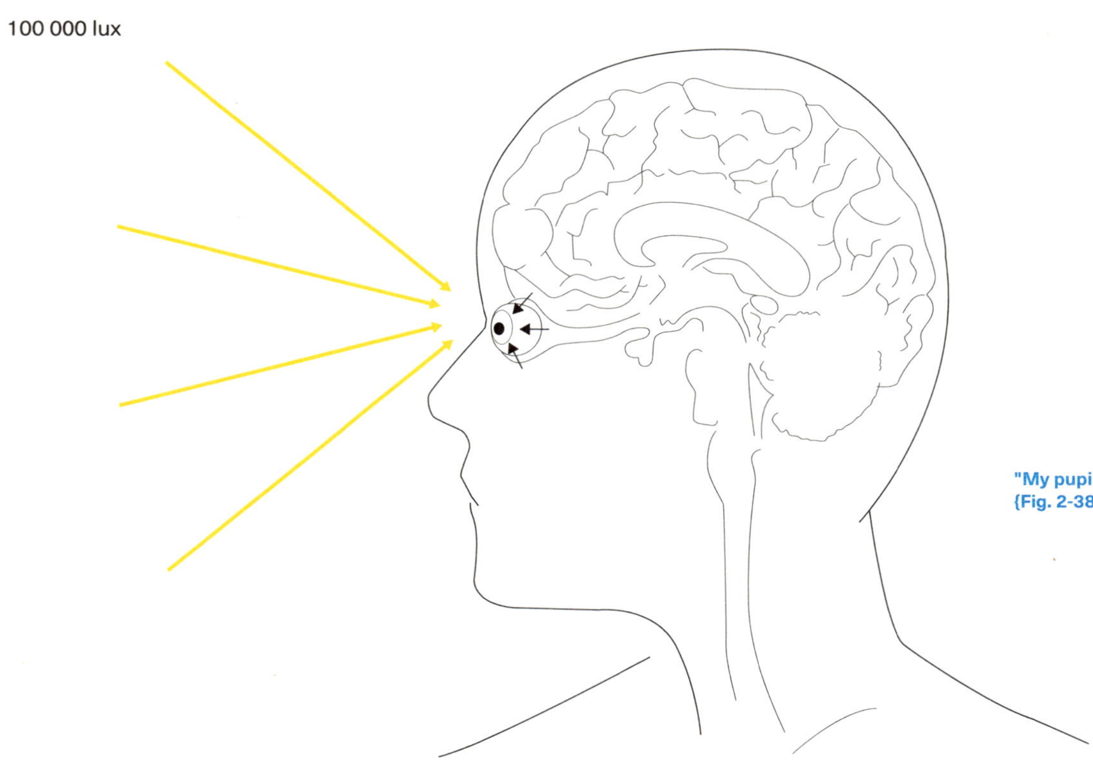

"My pupils contract..."
{Fig. 2-389}

II.IV. Diffusion

It is soon 4:30. I refold the newspaper and produce the change to pay for my coffee and sparkling water. The light is soft and colored. The sun has lowered more than 90%, touching the top of the walls along the road, which have absorbed its wavelengths between 350 and 650 nm and 700–800nm, to reflect only those between 650–700nm—orange. Not a single ray of direct sunlight reaches the sidewalk; the straight road no longer aligns with its axis this afternoon. I'm in shirt sleeves. The temperature is warm but comfortable with a light wind. Walking toward the intersection, I regain the sun—the full ensemble of its wavelengths and 100% of its heat—at the same time as the noise of automobiles. There is not a cloud in the sky. Water vapor has dispersed weakly into the air. I wait for the green light before crossing the street and returning to the small street's orange quietude.

My son is at the crèche today. He is four-and-a-half and with me in Rome for one month. He does not speak Italian and I ask myself if he will learn it during these three weeks among children who speak no French. My son had saved my life when his mother and I first separated, just over a year before. I was in despair and in a deep solitude. For many months I awoke every morning at 4:00 a.m. and recalled this rupture with a fierce anxiety. During the day I sought from his mother any signal, any promise of reunion, in the minutest intonations of her voice. My son, whom I saw every other weekend and every Wednesday, was my only happiness. The days we spent together, which I planned entirely for him, are for me even today among the most beautiful of my life. I no longer know what led me to make this comparison, but at some point I realized that I had to consider this as a period of mourning, not for a relative but for my wife herself, for her existence, her presence. Without seeking to understand, without having to accept the separation, I had to act as though she was no longer there. I ceased to see her, to speak to her, to read her text messages in which I had searched unceasingly for signs. It worked. My unhappiness dissolved, my son remained. Today, I see her and speak of her entirely without resentment, with the love one feels for an estranged member of the family.

The crèche is in a parallel road, slightly lower. I stop at the door among the other parents and embrace my son, whom I take by the hand. We head down the road toward the west, staying on the northern sidewalk in the diffuse, orange light, in the comfortable, shadowed warmth of summer. I hold his right hand in order to place myself between him and the few cars that pass along the road. Slightly farther along, we stop in front of a small frozen yogurt shop. My son chooses one scattered with candies and chocolate sprinkles.

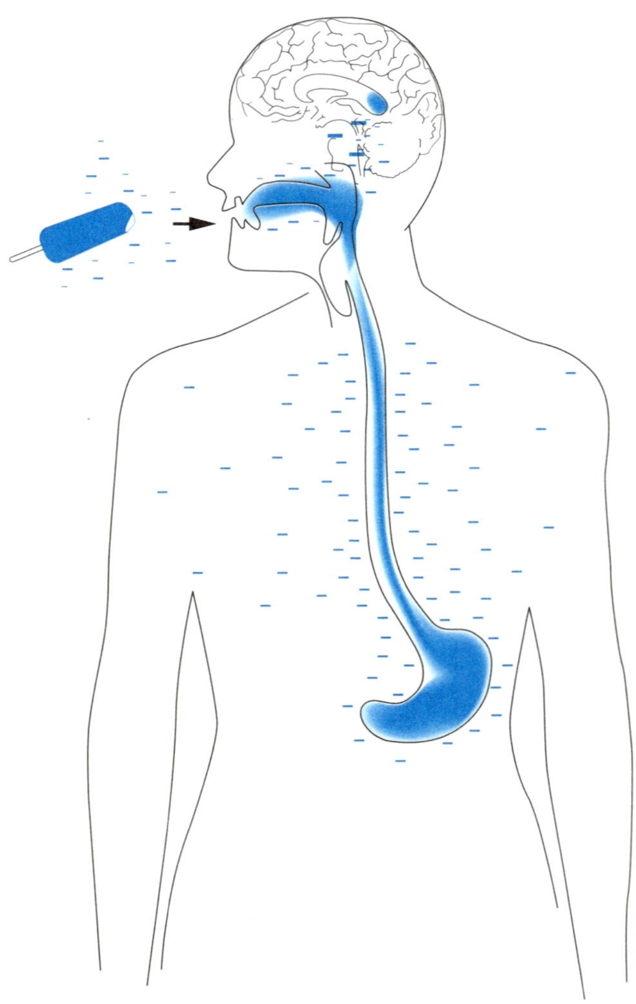

" [...] of a small frozen yogurt shop." {Fig. 2-390}

II.V. Afterglow

In the hotel lobby, we ascend a few steps whose thick red carpet absorbs our footsteps. I say "red" but I exaggerate because the carpet color is visibly closer to a dark brown in the weak electric light, at best 100 lux, that emanates from the single illuminated lamp perched on the reception desk to our left. The light is divided in two: there is the incandescence that suffuses the red paper of the lamp shade, giving a warm and reddish radiance to the hotel's front hall; there is also the white light that rebounds from the shade and is reflected in the wood of the counter, tinting the higher part of the space a light yellow. There is no one in the hall or behind the counter. All is silent. It is around 1:30 a.m. and I ask myself where the bar we are seeking could possibly be.

We are there to meet the professor of architecture whose fourth-year atelier I am in, at his invitation. I find myself with Élisabeth, one of the assistants, a brilliant Zürich architect of about 30. The students and assistants had celebrated the end of the year together at a restaurant in the town center. Our professor had not shared the meal with us. In the course of the meal, I had moved to sit next to Élisabeth, a very pretty woman—small, with short blond hair—on whom the crow's feet that bracketed her blue eyes on either side made her, in my eyes, even more desirable. As dinner ended, nearly midnight, some of us wanted to go home while others proposed to go for a drink.At this hour there was not much to do in the town. The assistants proposed to descend to the lake's edge to meet our professor in a jazz bar. Élisabeth had not understood exactly where the bar was, but she thought it must be the hotel where the professor was staying. The assistants were not staying there because of the price. We decided to go there. Some went by car but I proposed to Élisabeth to descend on foot. She dressed well. I knew she lived with a wealthy Zürich businessman. She accepted. We arrived at the lake's edge a half-hour later. There was no one, few lighted windows, all the restaurants closed. A single car passed.

There is no bar in the hotel. Élisabeth approaches me and murmurs in my ear that perhaps what we are looking for is a drink in a room. A large wooden staircase in front of us leads to the upper floors. We register it without turning on the light and step up side by side, grimacing and putting our hands in front of our mouths to stifle our laughter. We have the feeling of having illicitly entered someone's home—no one must see us—and this amuses us. We put one foot in front of the other very softly to avoid any creaking of the stair treads. The light, which comes only from the reception desk, weakens little by little, losing 10 lux with each new step. Halfway up, the stair turns on itself and darkness is nearly complete. Now on the first level, we distinguish the hallway, perpendicular to the stairs, that would take us to rooms on either side. We hear not a single sound except our own breathing. In my vision, the geometric forms of the hall and doors are blurring with a sort of dark halo on the first level, composed of small, more or less black dots that vibrate, densify, regroup, drift, and break as a function of the movements of my body, my head, my eyeballs and blinking eyelids. I no longer distinguish vertical from horizontal, near from far. We separate to go listen at the doors of rooms—Élisabeth to the right, I to the left. In order not to hurt my hands on some unseen object, I raise my arms and slowly embrace the air in front of me with supple, generous movements of extension. The images in my field of vision diffract and contract in a multitude of opaque, aberrant forms that surge from time to time from the first floor and against which I dread banging my nose.

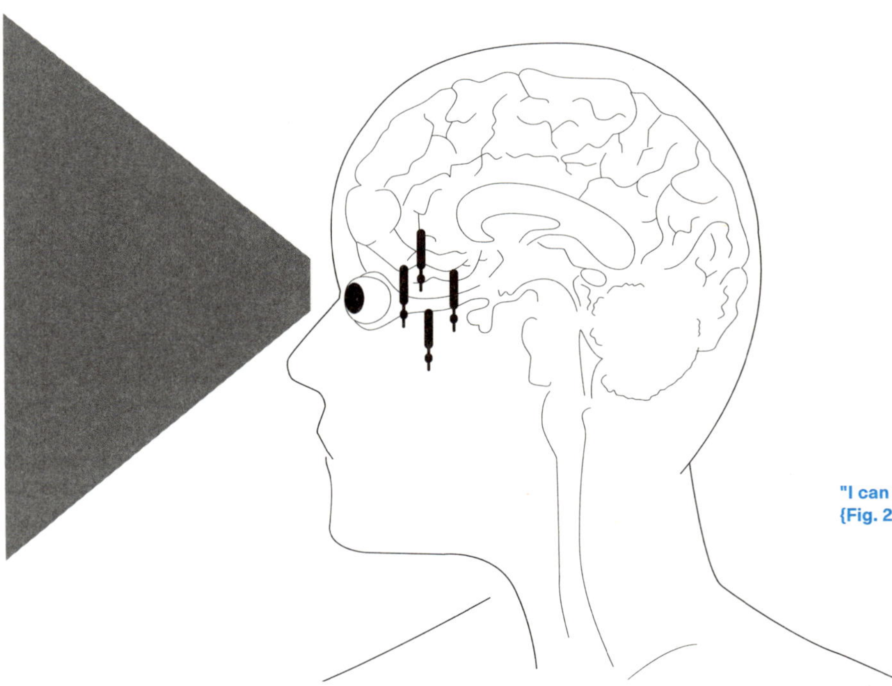

"I can now not see a single shape."
{Fig. 2-291}

There is no bar in the hotel. Élisabeth approaches me and murmurs in my ear that perhaps what we are looking for is a drink in a room. A large wooden staircase in front of us leads to the upper floors. We register it without turning on the light and step up side by side, grimacing and putting our hands in front of our mouths to stifle our laughter. We have the feeling of having illicitly entered someone's home—no one must see us—and this amuses us. We put one foot in front of the other very softly to avoid any creaking of the stair treads. The light, which comes only from the reception desk, weakens little by little, losing 10 lux with each new step. Halfway up, the stair turns on itself and darkness is nearly complete. Now on the first level, we distinguish the hallway, perpendicular to the stairs, that would take us to rooms on either side. We hear not a single sound except our own breathing. In my vision, the geometric forms of the hall and doors are blurring with a sort of dark halo on the first level, composed of small, more or less black dots that vibrate, densify, regroup, drift, and break as a function of the movements of my body, my head, my eyeballs and blinking eyelids. I no longer distinguish vertical from horizontal, near from far. We separate to go listen at the doors of rooms—Élisabeth to the right, I to the left. In order not to hurt my hands on some unseen object, I raise my arms and slowly embrace the air in front of me with supple, generous movements of extension. The images in my field of vision diffract and contract in a multitude of opaque, aberrant forms that surge from time to time from the first floor and against which I dread banging my nose. The black shadows they draw are figures on the ground of other black shadows, one of which seems to suddenly approach me so that I am frightened and actually move my whole body backwards. Then they disappear.

Apparently everyone on this floor is sleeping. There is not the slightest sound that would lead one to think of a party. At the end of the hallway I turn around and distinguish the vague radiation of very low yellow light from the stair that is immediately troubled and darkened into a darker spot. I understand with immediate self-reproach that it is Élisabeth's backlit silhouette. We meet in the middle of the hallway to ascend yet another level without speaking, making no sound, leaving the hotel in silence. Then we hear steps. The lights come on at the lower floor, rising swiftly through the stair, rebounding on the hallway wall in front of us and allowing to appear, against the balustrade, a large couch. We hold our breath. The beat of the steps does not stop. We sit cautiously on the couch, Élisabeth on my right. The light goes out. I can now not see a single shape except a blue line that blocks the lower portion of my vision. This blue bar has a tendency to drift to the left of my field of vision, only to recover its place at the right before drifting again. Little by little it fades in the weak, retreating glow.

We speak in whispers. Her words are not carried by her voice, only by the modulation of breathing. They are articulated at the last moment, just before leaving her mouth, and indistinguishable from her perfume. They form a warm and scented wind that caresses my throat, slides in the sinuous shapes of my right ear, and lodges itself in me. I tremble lightly under this wind; I imagine it transformed into a kiss, as if the words were the advance guard of her lips then her tongue, that will soon touch my skin. We debate during the long minutes without my knowing where her body is, at what height exactly is her mouth, to which part of the darkness she turns her gaze.

We suddenly hear steps again. The light goes on at the first floor, then in the stair. Someone is coming up now and discovers us on the couch. It is the night watchman, who demands to know what we are doing here and asks us to leave. I reply to him, but what troubles me much more is to discover, as the light illuminates the hallway, the position of Élisabeth's body on the couch. I had held myself well to the right, feet on the ground, sunk against the back of the couch and the left armrest, but Élisabeth had raised her arms and stretched them out along the right side of the couch, propping herself on her left elbow. Her body leans toward mine, her head inclined, her mouth less than five centimeters away from me.

II.VI. Sunstroke

I have an appointment with the Pro at 2:00 for my first golf lesson. The whole family is there because our house, formerly surrounded by agricultural fields, had found itself ringed by a course of 18 holes after the property was sold. I have been told I have a natural predisposition to the sport: I am flexible, I make good body movements. A particularly good whack of the club on the ball reveals immediately whether the stroke is good, and I have already succeeded in knocking one or two balls that follow a pretty trajectory. With my father and brother, I play in the night when the members have left. We haven't paid the exorbitant "green fee" that would permit us onto the grounds. We play too poorly to be admitted anyway.

Golf has simplified the landscape and made it accessible to us. It has also transformed the hamlet where our home is located. Some of the farms, with pigs, chickens, and the old car restorer, have been transformed into club buildings, restaurants, dressing rooms, and "Pro-shop." The construction site had been an entertaining destination during my summer vacation. I would drag myself there in late afternoon and gather discarded planks of bright, aromatic fir. The caps from beer bottles abounded among the bits of plaster and I sought the various brands as though they were bits of gold. The days were long except when friends came to visit for a few days and we pulled out the tent, made of old, bleached-out fabric marked with traces of moisture, to plant in the field. We slept in the glow of a broken glass lantern around which gathered mosquitoes, which we would slap one by one, until we recovered the silence of the night and the cry of an owl. In the morning, more than the light it was the unbearable heat and smell that would awaken us. The air trapped in the tent would gradually heat up, releasing the odors of earth and damp grass on which we had put this simple cover for sleeping. The first rays of sun, striking the cloth, would transform it into a heat that was transferred to the air in the tent until it too was reheated. Perhaps carbon gas heats more quickly. There was not a lot of oxygen in the tent; the slow breathing of three or four children would distribute much of the oxygen among our bloodstreams in the course of the night. In the afternoon we might go to the beach of the Teine, where we would gather with our feet, crossing the churning water of the lake, the large mussels that littered the dark sand due to erosion of the yellow rocks of the Jura. We would depart from time to time to stroll among the cool valleys of summits, scattered with large stones and with small trees whose forms were torqued by the altitude, the weak soil over the rock, and the chalky soil's inability to hold water. On special occasions, once per year, we would go to eat the Valangin butter cake that we loved above all else.

But we have no visitors at the moment. My parents have offered me a first course in golf. I came ahead in the early afternoon to practice, a training strip along the edge of the woods. It is fairly warm, due more to the direct solar radiation than to the air temperature, which stays comfortable. We are at a slightly elevated altitude, the sky blue and cloudless.

It is August, mid-day. We place ourselves under the sun, the strip in front of us marked with yellow balls. The Pro wears a cap; I have none. He corrects the position of my arms and feet and places my hands on the grip in a way that suppresses any natural sensation of holding the club. When I hit the ball, nothing about it is correct. The sun is strong; its powerful electromagnetic radiation traverses the universe until it strikes my head and is transformed into heat. I must stay the course, execute the movements following the Pro's directives. But my vision blurs, the sun is too much. I lose consciousness for a second and fall to the ground, and will rest on the bench for a few minutes, in the shade, before going home.

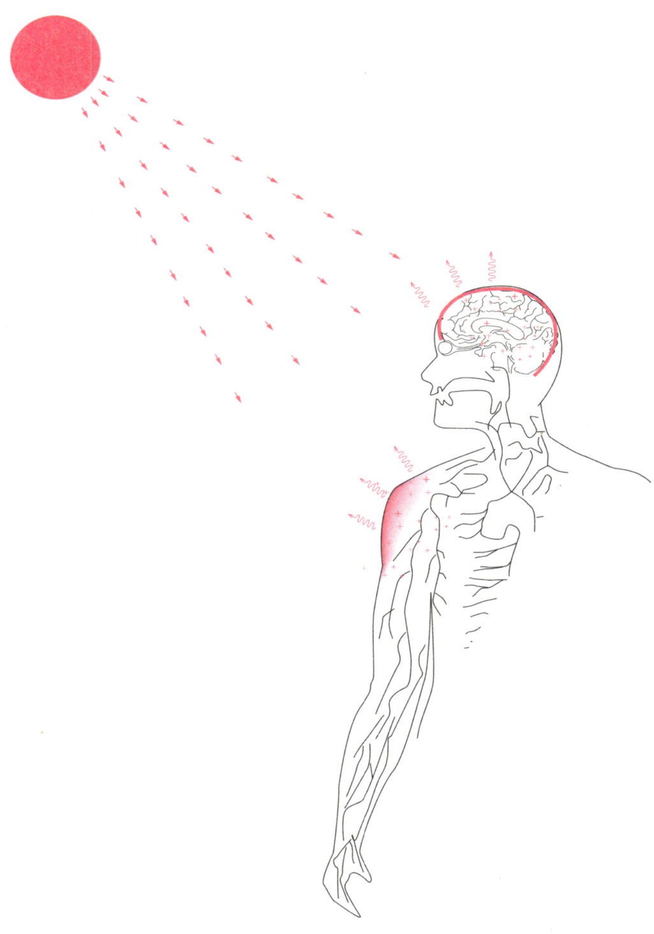

"the sun is too much."
{Fig. 2-292}

II.VII. Oblique

I am awakened by the sun. I panic and leap out of bed in a single bound. I've missed my plane, which boards at 7:30. A taxi was supposed to collect me at 5:00. I don't understand why no one woke me, why my alarm did not go off. Outside, morning is well underway. My window opens on to the sea; we are about an hour north of Copenhagen. It is a very beautiful sunny day. I pick up my phone, which reads 4:30. Can the time on my phone be wrong? That does not seem possible. Did I oversleep and awaken at 4:30 in the afternoon?

I knew the same temporal disorientation during my military service in Switzerland. During a general exercise of several days that simulated a wartime attack and involved all the schools of recruits, I was to quickly sustain a wound (I was given a plastic "wound" to wrap around my belly). The medics had to transport me from the battlefield to the hospital to care for me, under a time constraint—my wound would only leave me a few hours to live. At the recruiting school, where we rose early and went to bed late, one slept during every period of free time, during trips by truck, awaiting orders in the mountains. On the stretcher, in the middle of a field, I fell asleep oblivious to everything, until I awoke, still wearing my wound, nestled in bed in a row of other beds, in a military hospital. There was no window in the sleeping-hall; the light came from electric ceiling fixtures. I was in an underground anti-atomic hospital. I got up. Without a watch, I had no sense of what time it was, nor time of day. The military doctors and nurses spoke in Swiss German. I dressed and walked around the hospital in search of the exit, which I found only by accident after numerous blunders. A garage exit, it followed a concrete ramp that rose toward a light that was almost blinding to me until it resolved into the many shades of green in the trees on the mountain in front of me. At the top of the ramp, several soldiers were smoking. It was daytime but I was quite incapable of knowing if morning or afternoon. What troubled me more was to not know whether it was the same day or whether I had slept until the next day or the day after that.

I realize in time that it is really 4:30 in the morning, that the summer night is extremely short in high northern latitudes. I pack and head out to meet the taxi that awaits me.

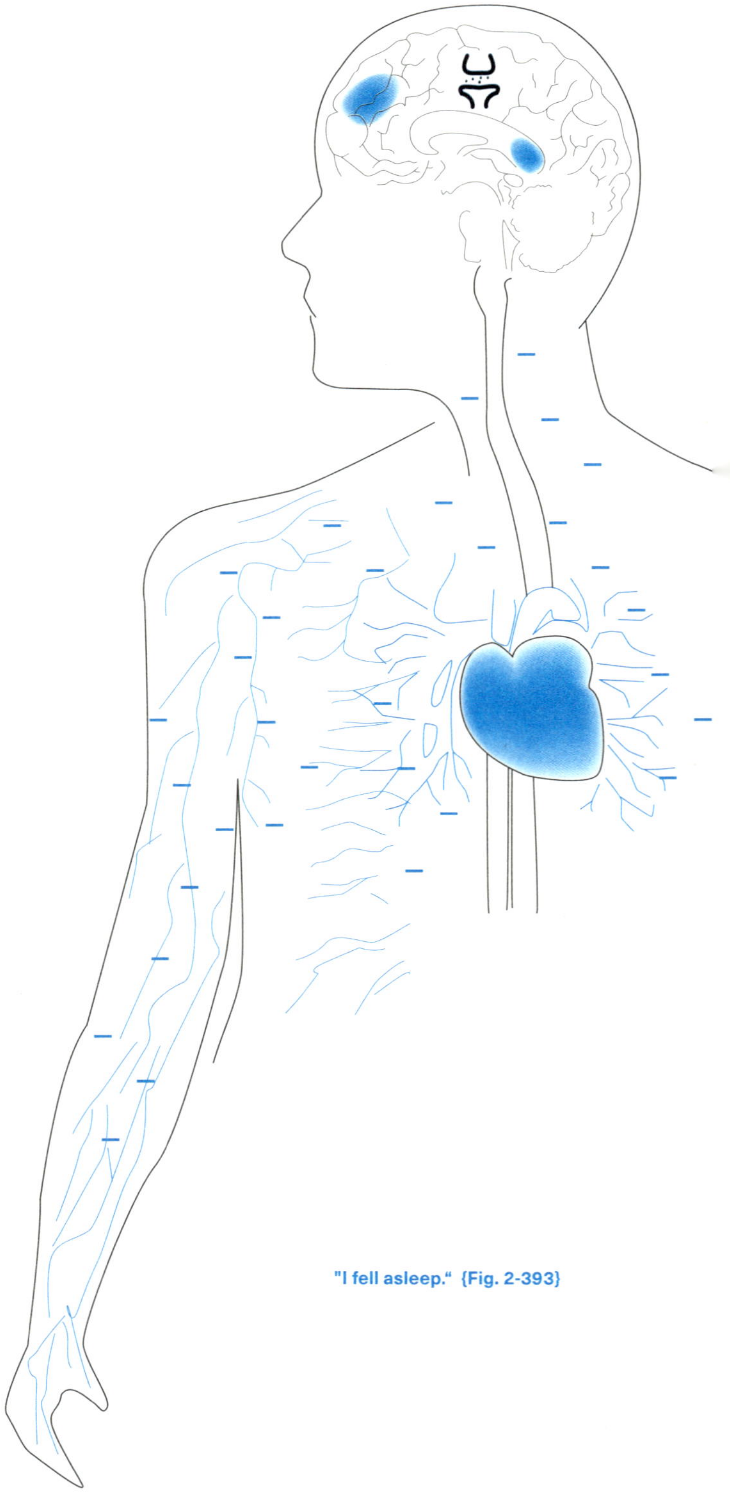

"I fell asleep." {Fig. 2-393}

II.VIII. Obfuscation

The house leans over the garden, its principal facade oriented to the south-east. In the morning, sunlight penetrates the rooms and slides among the wood cross-pieces of ancient windows. It is refracted already, having crossed the leaves of a large linden plant with their multiple hues of green, some meters in front of the house, on the terrace of light gray gravel. On the raw wood in wide-boarded bands, in the room, the light creates a moving composition of spots of all sizes, in more or less intense, more or less compatible colors, whose undulating chopping movements draw more or less distinctly the dark outline of each leaf. The shadow of a cross appears and dissolves according to the movement of wind in the tree. The window is open; the heat of summer enters the room and dispels the surplus of carbon gas produced during the night. A bubbling fountain is heard. Its water falls irregularly, from a height of about 35 cm, on the mirrored surface of a basin, making concentric waves that continually extend from the center and flatten to the point of disappearance. The sound of a bell, heard through the leaves for a moment, covers the fountain's bubbling sound.

I am outside now, on the terrace. We are in Switzerland, above the Lake of Neuchâtel. The day is nearly done, it is 5:00. The setting sun moves in the sky until, as it disappears behind the vertical line formed by the principal facade, this line suddenly throws a shadow that darkens the terrace's gravel surface except for a very small triangle at the southwest corner. Seated in the sunlight, on a bench against the house, I am reading the paper. The paper's white background, striated with characteristic black lines, darkens all of a sudden, attenuating the contrast between the paper and letters and making reading more difficult, little by little; but by the time I notice it, the sun might have slipped past the southwest corner ten minutes ago. I am more conscious of being cold as reading becomes more difficult, and finally there is no more sun. I close my newspaper, keeping the page with the fingers of my right hand. I stand up and walk along the house toward the west. Behind the facade's southwest corner, an annex has been added to the house very recently to the south side, the same height but slightly shallower. The result is a light reinforcement of the facade, on the order of four meters, that lengthens the terrace. I cross the shadow line and find myself in full daylight, tilting my head to avoid the direct rays of the low late-afternoon sun. In my household we call this particular part of the terrace, which is always sunny at the end of the day, "the little Provence." My mother is seated there on the flat, green-and-white-striped cushion of a white-painted metal chair. In front of her, on a white table, are a wide, light-green plate, a teapot, and several cups on saucers. I sit in the sun, unfold the paper on the gravel of the terrace, and pour myself a cup of a clear, steaming, pale orange liquid that I determine to be an infusion of linden.

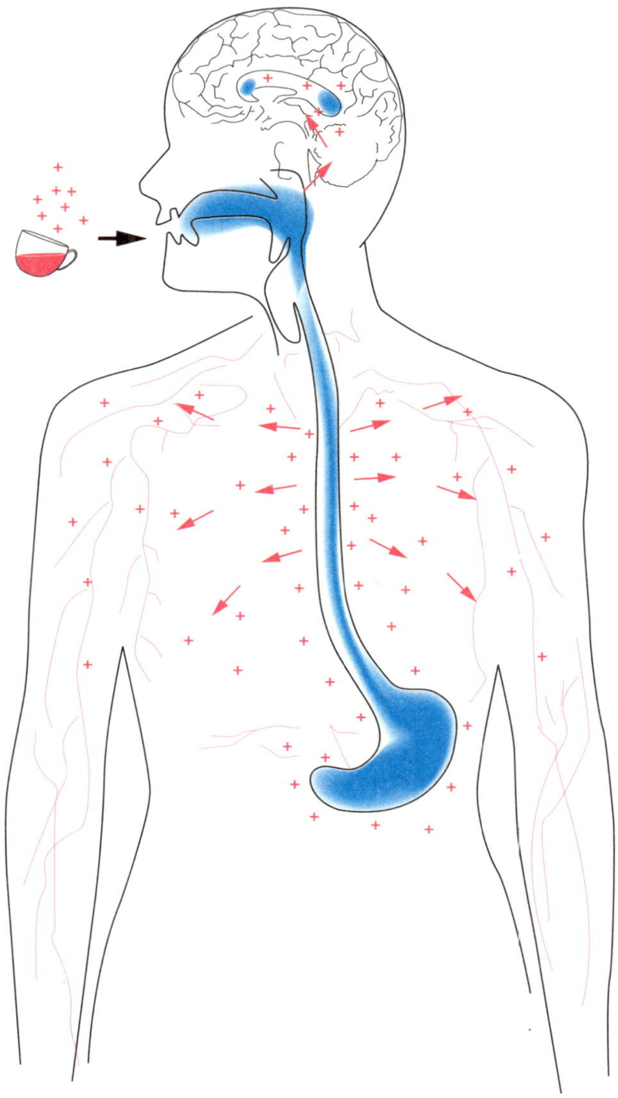

"I determine to be an infusion of linden." {Fig. 2-394}

II.IX. Condensation

We must each, separately, spend three entire days under an umbrella, in the south garden in front of the camp.
As the watch, during an excursion in the mountains, David had rung the bell of the village church. Pierre and I had broken a traffic signal by throwing rocks at it; and, together, we broke into the hut of a ski lift that was closed for the summer.
We were caught by the director after a telephone call from an inhabitant of the village. I sit on the ground, pull up a stalk of grass and suck from its stalk a liquid that is very lightly sweet. Then I place the stalk in front of an ant, creating an obstacle that forces it to make a half turn after trying several times, without success, to climb over it. It mounts another blade of grass, slips, and falls back to the humid black earth.
As the sun gradually moves, the shadow of the umbrella turns and induces me to shift a few meters to the west in the course of the day. The parasol projects a more and more oval spot of shadow on the ground and I revolve in it completely, without letting it move off of me by a millimeter. The spot's movement and deformation, however slow, compels me to constantly readjust my position in the course of the day, retracting myself at noon and detaching from it only at the end of the afternoon, when I leave it completely. The green of the lawn is dark green under the umbrella and darker still in the shadow of my body.
As the rays of the sun are stopped by the fabric of the umbrella, the earth underneath is shaded, its temperature is lowered, and it seems immediately to give off a light odor of dampness.

In itself, the temperature of the ground is colder than the air temperature in summer because of the earth's inertia, which keeps the coolness of night throughout the day.
When exposed to the sun, the earth reheats at its surface and to a certain depth. But when the earth passes into shadow, the immense cool of its interior rises again to the surface.
The ground, colder now by many degrees Celsius, radiates negatively, thermally speaking. It draws the 37°C of my body and produces a slight sensation of cold by thermodynamic equilibrium. At the same time, because this surface is colder than the air temperature, a slight phenomenon of condensation ensues, drawing out the fine moist odor that I can barely smell, just now. I am not too hot. In the mountains, the air temperature never rises too high, staying lower than on the plain even in full summer. At high altitude, the air dilates and loses heat as the atmospheric pressure decreases. It is above all because of the infrared radiation of the sun that one can be too hot in the mountains. The invisible waves depart from the sun, traverse the universe, and penetrate the earth's atmosphere when there are few clouds, before they finally, in contact with the skin, take the form of heat. Under the umbrella, almost no waves from the sun reach me. The only heat I feel is that of the air, very agreeable. The day passes until the shadow of the umbrella is covered completely by the giant shadow of the mountain, irrevocably abandoning all the bodies to a cooling air. The director gives us permission to leave the vanished shadows of the umbrellas and reenter the building.
At night, in our room, after this first day under the parasol, we plan to adopt a nocturnal rhythm of life. Instead of being awake and bored during the day under the umbrella, we will reverse the sleep-wake cycle in relation to day and night. We calculate that in terms of surface area, the space under the parasol is not very different from that of a bed. When night comes, we will awaken from under our umbrella to begin to live at night.
But on that first day, around 11:00 p.m., we fall asleep.

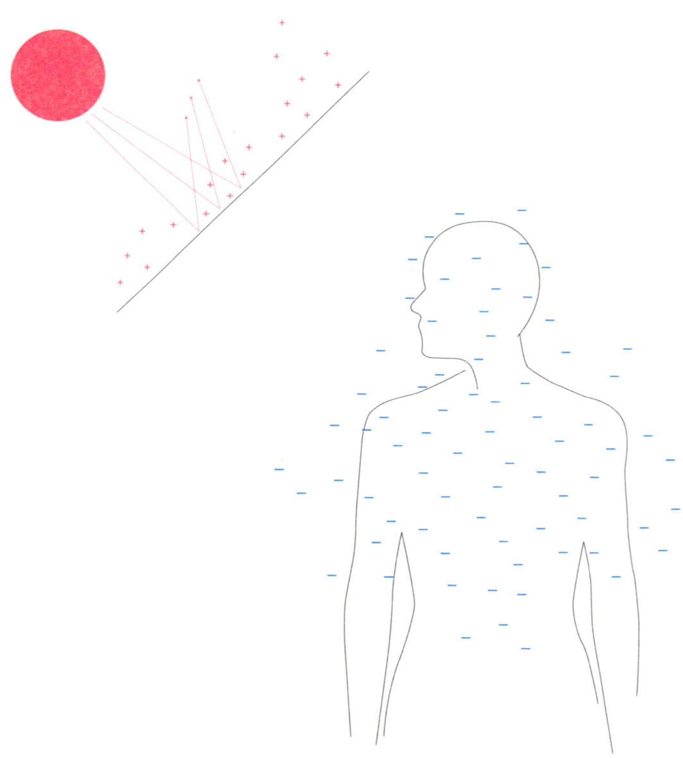

"Under the umbrella, almost no waves from the sun reach me." {Fig. 1-395}

II.X. Cooling

I sit on the embankment of a cemetery. It's 10:00 at night. I'm slightly cold but it is summer and I persuade myself that the cold is bearable, and settle backwards. My back crushes the slightly cooled grass on the embankment and I stabilize in this position. As I lay my head down, the grass creases against my ears, tickles my neck, and imprints its coolness by conduction. The visible forms of the town disappear beneath my field of vision, until nothing of it remains except the luminous halo of electric lighting that forms an orange zone in the lower part of my view. Above, the orange is dark to the point of spookiness in which, one by one, stars begin to shine. The completely cloudless night sky forms an immense cupola of absolute, glacial blackness. Its temperature of -273°C draws, by negative radiation, the heat the earth had received during the day and that of my body, at 37°C. We cool progressively. I can't sleep. The wisest thing would be, nonetheless, to sleep, for in the morning I will board a bus and depart this small East Coast town.

Before sitting on the embankment, walking the deserted roads of the town center I had felt this light coolness already and sought an apartment entryway, a stair, any protected place to shelter and pass the night. But all the building entrances were closed. I was also told it would be dangerous to spend the night, though I thought the young woman who warned me had seemed dangerous too, perhaps rightly since she told me so. As I passed a French bakery with a terrace, closing for the day, the young woman had appeared, accompanied by two boys. Each had a bottle in his hand. She approached me but I did not understand what she said to me. She offered me a drink. Her friends left and she stayed with me until night fell. She loved Baudelaire; she liked to dress in black and wore a steel cross around her neck. Her perfume, which I breathed at each breath until nauseated, was extremely strong and sweet. She wanted to see me again and asked me to meet her later in the evening, at 11:00. I begin to feel the cold again. I would like to put something on my back but I have nothing with me, having lost my backpack some days before. It had disappeared, along with all my possessions, on the bus from Chicago to Niagara Falls, USA, while I took another bus to Niagara Falls, Canada. Since then, I had crossed the country in a T-shirt, carrying only a plastic bag in which I had collected a newspaper, a chocolate bar, and a banana.

The embankment gives on a small unfrequented road without streetlights. On my side is the embankment, a barrier, and the cemetery. On the other side is a field. I consider sleeping in the cemetery, where I would have had complete shelter, but I prefer to stay on the edge. A car passes in the distance. To my left, some trees hide me from the main road. I feel secure and close my eyes. I sleep.

The cold awakens me, a profound cold. I open my eyes. The bright part in the bottom of my field of vision has disappeared. Except for the minuscule points of light scattered in the sky, nothing is left but an infinite night that pierces me with cold. Could I warm myself with the fire of the stars? Could I orient some uncovered parts of my body toward those faraway sources of heat and let their infrared radiation warm me? I look at my watch, turning my wrist to different positions until the face with its weak luminosity is visible. 1:30 a.m. My muscles contract to produce a superficial warmth that is instantly sucked away by the dark night. My body reacts physiologically to keep my core temperature at 37°C. I grow stiff, my blood vessels contract.

Teeth chattering, I straighten my back. Pushing against the embankment with my right hand, I stand up and jump up and down to warm myself. I pull on my clothes, trying to cover the exposed parts of my body. I fumble in my plastic bag, take out everything I have, and lay it in front of me. I eat the chocolate bar but keep its packaging to put under my shirt. I tear the bag, lay it on the embankment, and sit on it. I take the newspaper's pages and wrap them one by one around my arms, my legs, my torso, and my back, passing them under my clothes and closing as best I can any opening where air could slide under my clothes with the paper, which I press into my shoes and under the belt of my pants as well. I use the bandana, which I unfold to obtain the largest possible surface area, to cover my neck. The only thing left to do to avoid the cold is sleep a little more—to disappear—for the unconsciousness of sleep is the only warmer place where I can find refuge at the moment. I close my eyes, relax against the embankment, curl into myself, and fall back to sleep.

The cold wakes me up again. 3:30a.m. I have nothing more with which to cover up. All of me and everything around me is chilled and damp. My clothes, the newspaper, the bandana have lost their capacity of thermal isolation now that they are wet and the same temperature as the air. I get up, fully refrigerated. Pulling off these now useless covers, I walk toward the main road which I take in the opposite direction from the town. There is not a single sound, not a single light. I walk five minutes and this physical muscular activity counteracts the mounting sensation of cold. The road rises gradually and a bridge appears in front of me. Looking down, I see a highway carved into the landscape, previously invisible to me, with no specific exit to link it to the town where I have found myself. Approaching, I distinguish farther along, to the left of the bridge, the neon light of a gas station. I hop over the barrier, drop down the highway embankment, and turn my feet toward the building. The interior is lit, with people inside. I push the door open. A mass of heat escapes towards the outside, covering me and reheating me instantly as my eyelids flutter in the bright fluorescent light. A few moments later, I drink a warm coffee that the station attendant lets me take without asking for payment.

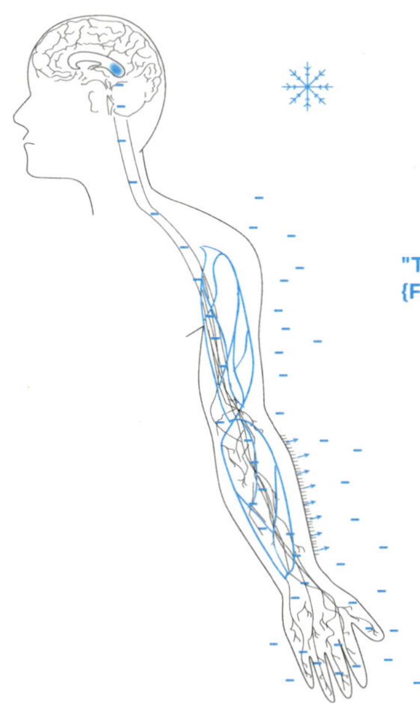

"The cold wakes me up." {Fig. 2-396}

II.XI. Greenhouse Effect

It is warm, nighttime. I am inside and the park of the villa is plunged in darkness. There isn't a single public light. I know there are some electric lights on a little farther away, those of other residents. The others' pavilions are strewn in the park but I do not see them; the dense foliage of trees and bushes screens them completely. The windows of a grand hotel glitter above this terrestrial blackness. The power of electric light bulbs, 70 watts each, produces light a weak 90 lux light that does not counterbalance at all the darkness that reigns down here, but I press the outdoor switch to turn on a small orange light in the garden. It is spring in Rome but for me, just arrived from Switzerland, it's the equivalent of summer. I am in shirtsleeves and the doors and windows of my residence are wide open all day. I inhabit the inside and outside of the house indifferently, crossing the threshold without transition, noting all the same a slight acoustic difference between the confined and silent interior and the more open, reverberant exterior. I take by the hand the beautiful Taiwanese girl to lead her toward the greenhouse that adjoins my residence. She follows, smiling when I look at her.

We had met a few hours earlier when she was sitting with a friend on the terrace of a café on the Campo de' Fiori. I was amusing myself in the city with an acquaintance who was in Rome for a few days. I was walking around with an acquaintance who came to spend a few days in Rome. We liked each other and drank cool white wine. As night approached, I suggested to the two young women to have a drink at my place. They accepted. I claimed to live in a palace, which was true. The four of us had come by taxi. The minuscule entrance was just a bare opening in a corner of the immense door inset in the stone foundations of the gigantic facade that dominated the city. The opening was so small that one had to bend one's back to enter. One found oneself then in a large deserted hall plunged in darkness and silence, surrounding the statue of an ancient king. We then registered the wide stairs leading to the gardens. The trichromatic cones no longer came to our aid with any information in this weak light; we navigated by the weak illumination still transmitted, in black and white, by the photosensitive rods in our eyes. My two invitees seemed both impressed and amused by the solemnity of the place. They held hands and spoke in low voices, pretending to see ghosts. "There are ghosts," said the beautiful Taiwanese. A little bit higher up, in the extended, yet blacker stair that followed the grand staircase, I passed behind her back and took hold of her waist. She let herself be guided by my movements in the near darkness. My hands slid lightly from her waist to her hips, released and repositioned. My nose and mouth were drawn behind her long hair, inebriated by the vapors of her perfume. We left the palace for the upper garden, which we crossed to reach my pavilion.

I push open the glass door of the greenhouse. Entering first, leading the young woman behind me, I am greeted by a warm odor of dried flowers. The interior air, rarely renewed, is heated by the greenhouse effect during the day and befouled for lack of ventilation. It is the same warm air that has remained there for days, perhaps weeks, while no one has come here. But this warmth is agreeable. Dry, it does not make our bodies sweat uncomfortably, as far as one can tell, because the water exuded by the pores of the skin evaporates instantly, in gaseous form, without the least liquid trace. We step over pieces of white plaster molds and fragments of roman statues produced by the elderly inhabitants. We slip between two long tables on which a multitude of plants and pots of different sizes have been placed. I pull by the hand the young Taiwanese woman, who laughs now, cautiously passing a leg over a bush. I take her by the waist, her face towards mine, lift her, sit her on one of the tables that I imagine, judging by the noise, to be covered with white dust, dead petals, small plaster pieces, and terra cotta. The table is high, not made for sitting down at but for working on one's feet. The feet of the young woman no longer touch the ground. They balance in the void. I catch one to slide off the shoe which I let fall on the ground. Her foot is naked. I pull it towards me, incline it, kiss it. She puts her hands behind her and lets her head fall back. A short while later, in the silent darkness of the park, walking back to the villa, I ask her if she really believes in ghosts. "Yes," she replies, "but I love ghosts."

"I am greeted by a warm odor of dried flowers." {Fig. 2-397}

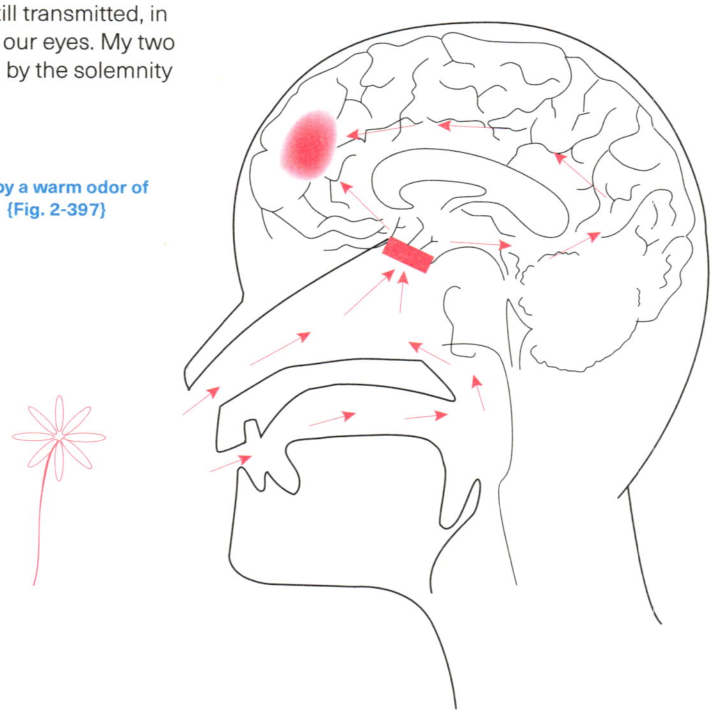

II.XII. Conduction

The temperature in the room is glacial. A cloud of steam forms in front of my mouth at each exhalation. I must undress before retiring but, apprehensive about the cold that will surely seem even more vivid to my naked body, I lack the courage to make the first move. Staying at my friend's house, I have been ill for several days, feverish. Snug in bed, under the covers, I continue to be cold. We have turned the heater on the wall to "maximum," but it does nothing.

It is impossible to elevate the air temperature in the house above 16°C. Aside from the astronomic, seasonal reason, there is the inclination of the axis of rotation of the earth, which determines that for part of the year, in winter, the Northern Hemisphere will be farthest from the sun. Because of the lengthening of distance between the Sun and the Earth, the latter receives weaker rays. The freezing polar currents descend into more temperate latitudes and cool the ambient air below zero. At the same time, this inclination of the north-south earth axis is responsible for the reduced length of the day. The sun rises later and sets earlier, which shortens the warming period and, by the same token, diminishes the quantity of energy the earth receives during the day. The winter nights become so long that they deplete the earth of whatever feeble warmth it may have acquired in daytime. There is also an intermediate reason, proper to the mode of construction of the building and relative to its poor thermal isolation coefficient. In effect, the walls of the house are exposed concrete (which performs this task poorly) without supplementary isolation. The house is effectively plunged in a bath of freezing air that crosses the small dimension of the walls, bit by bit, to radiate negatively toward the interior and cool everything. The little radiator on the wall, whose electric resistance draws in the air to blow it out reheated, is powerless in the face of the enormous mass of cold that encircles our living space.

I had arrived in Japan a week before, at the height of winter. I was badly dressed but nonetheless stayed outdoors for the first few days, venturing into the cold in my cotton undershirt under a fine wool pullover. My body temperature was lowered constantly during the first day and, despite an accelerated combustion of protein in my stomach to compensate for this drop in ambient temperature, I was unable to keep my body from cooling to the threshold where bacteria and viruses could enter without resistance. I nevertheless tried to rebound in the course of the two following days. I adopted the multiple dissociated, circumscribed modes of heating and that one finds in Japan: drinking warm tea, plunging my body in a hot bath, placing in my pockets, my shoes, against my back, small tissue paper packets containing a material that warmed up when rubbed. I had bought myself a coat but at no great expense and without great results; it was wool, but the absence of a lining and the imprecise mode of closing by three buttons on the front could not really generate an air pocket impermeable to the air and themically isolated in which my body could be protected from exterior cold. I had also bought a black woolen hat—efficacious but insufficient to compensate for the inefficiency of the rest of my gear.

For two days, I intermittently entered shops to heat up and I took taxis and subways, perpetually seeking the warmest spatial pockets where I could hope to lay down my arms for a few moments in this physiological struggle of my body against the cold. I toured the Kyoto area, visited temples in the snow, crossed dewy forests, walked on icy paths in small shoes. I was accompanied by a friend and two young Japanese women he had met in Tokyo and invited to spend a few days at his home for the New Year. He confessed that he hoped to have an adventure with one of them. I flirted with the other one without asking myself if I really liked her.

After a day spent in bed following these two days of tourism, the fever has fallen slightly but I continue to feel the cold inside me; it is impossible to get rid of and I never invested in a better coat. My new Japanese friend invites me to spend a few days at her home in Tokyo. Seeing a means of escaping from this refrigerated mansion, I accept with pleasure. We take the train to Tokyo in the morning. After an afternoon in the city, we arrive at her parents' small, ordinary Tokyo house. Her mother welcomes us kindly in the main space, whose dimensions she encloses around us by pulling two sliding panels. It is cold here too. The wood-and-paper walls seem ridiculously thin and totally inefficacious against the rigors of winter. I notice that there are no radiators except for a single, small, non-functioning space heater against the wall.

A short time later we sit on the floor, passing our legs under a low table whose feet are surrounded by an isolating quilt. I have the pleasure of discovering that it is warm underneath: an infrared lamp, protected by a metal grille, is attached to the table's underside and radiates agreeably on our legs. My friend brings out a second infrared lamp, on feet, which she switches on and aims at our faces. Her mother serves us food. We are now in a complex climatic condition: my legs are warm, almost too warm. The left side of my face receives the infrared rays of the other lamp and is heated, while the right side of my face and body, except for my back, remains cold because the air itself in this space is not heated at all.

The only mode of heat in the room is the pair of lamps which, by direct infrared radiation, heat only exposed skin and clothes, without elevating the temperature of the room air. For this reason, steam rises from everything that was warm: my cup of tea, my bowl of soup, my friend's mouth, reddened by infrared; her mother's mouth, and mine, from which steam passes in front of my eyes at every breath, blurring my vision. The meal appears to me clearly to be a form of complementary heating (which it is, in fact, since it provides nutrients that are burned in the stomach and transformed into energy for maintaining our homeostatic metabolism at 37°C)—one that is distinctly tastier than an electric radiator and releases a wider variety of aromas than a wood fire.

After dinner, I am effectively warmer. I hold in my hands a cup of steaming tea that I lift regularly to my face to drink a mouthful or to feel the radiant heat once again on my jaw. We stay a moment more in this room, then my friend takes me to her room, which she will lend me for the night. It is colder still, but she shows me how to turn on the heated mattress of her bed and how to regulate the temperature. Here too, the experience will be one of sleeping practically outside, in full winter, but in a bed heated from inside, and happily protected from the wind and snow by the roof and the wall cavities. It is the opposite of modern Western heating, which raises the air temperature of a room to a comfortable level at which one breathes warm air. Here, the air breathed stays icy and it is only against the body, in the body, at specific, circumscribed places, that heat is applied.

I am on my feet, hesitant to undress. It occurs to me that I must still go to the toilet, which is at the bottom of the wooden stair. I leave the room without closing the door, perhaps imagining that the heat of my cup of tea downstairs could rise to my room. At the bottom of the stairs, I open a door on the left and enter the bathroom. The window is open; I close it immediately, shivering. The seat of the WC is already down. Poised to undo my trousers and lower them, I dread the moment when my rear end is in the air, where the cold will strike my newly naked skin. But I dread even more the moment when I will place my naked rear end on the toilet seat, which I can only imagine as glacial. I must hurry and do all of this quickly to not give the cold time to rise through my nerves from my skin to my brain.
I unbutton my pants, lower them just to my knees, and catch my underwear and lower it onto my pants. Folding my legs, I sit on the seat. It is warm, very warm, heated by electrical resistance integrated in the plastic, and this heat is communicated throughout my entire body, by conduction, to finally reheat me entirely, fully.

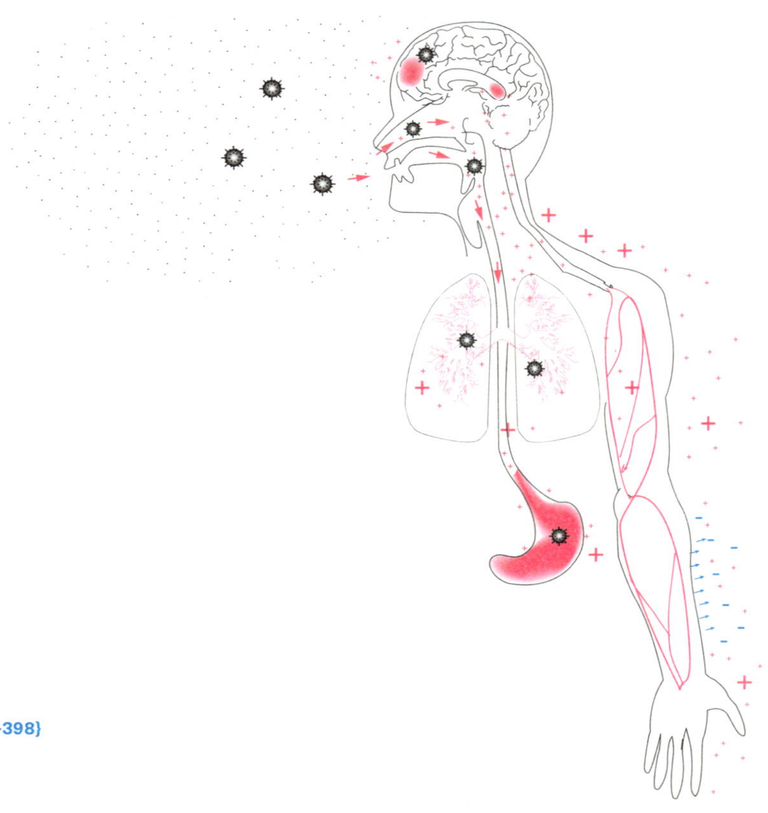

"I have been ill." {Fig. 2-398}

II.XIII. Rotation

It is a beautiful winter morning. The sun has come out over the distant apartment buildings, and it emits to my skin long rays of white light that a striped pane of glass, partly covered by vertical trails of dust and old water spots, has preliminarily filtered and refracted. I am on the train headed to the airport. We've just exited a tunnel and emerged in a northern suburb of Paris which I can't see clearly because of the speed and the first blinding rays.

I am now in the airplane. It takes off with the sunrise and turns to the west, toward the United States. We fly in the blue sky without really changing latitude. Below, one distinguishes clearly the sea and waves. The Sun, immobile, stays low in the sky behind us. Time passes but its position does not change. The morning is prolonged relative to what my watch shows, as if we are treading water. I ask myself whether the 800 km/h reported on the screen in front of me represents the speed of the plane or that of the earth, which turned as we stayed immobile, flying at zero km/hr and waiting patiently for the earth to roll the Atlantic Ocean beneath our feet until New York. Or perhaps there are masses of air which, bound to the earth, we keep over Paris, which would explain the necessity of flying 800 km/h not to move, but to struggle against the pressure of the moving atmosphere. Perhaps the earth does not turn fast enough toward the east and we must nonetheless advance a little bit toward the west. Our plane then had, in absolute terms, a speed well below the stated 800 km/h, profiting from the movement of the earth in the opposite direction. That would be to say we would land in the New York night, having retreated in time. The truth is difficult to comprehend: the plane has the task of effecting a new curved trajectory toward the north in order to always be two hours or less from an airport to be able to land in case of technical difficulties. We therefore fly in a direction opposite to the earth's rotation. But, with respect to the Sun, we are still not moving but fixed at 8:00 a.m. The 8:00 a.m. hour slot extends for two, three, six hours until New York, where we land in a beautiful winter morning.

I must change planes to go to Portland, on the West Coast. The customs formalities are complicated because of the visa and I miss my direct connection. I am offered another flight with a stop in the middle of the American continent. The plane takes off at sunrise and heads west. We fly in a cloudless blue sky. Below, one distinguishes lands, fields, towns. The sun stays low in the sky behind us, immobile. Time passes without its rising. The morning is yet more prolonged. We still fly against the rotation of the earth. Locked into 9:00 a.m, we stay in the time slot that extends to two, three hours, then land in a beautiful winter morning. I have less than one hour to connect in the airport, then I board the plane to Portland. It takes off during the sunrise and turns toward the American West. We fly in a cloudless blue sky. Below, one distinguishes clearly the earth, mountains, fields. The sun stays low in the sky behind us, immobile, without changing position, though the hours slip by on my watch. The morning has extended into an entire day. When I finally arrive in Portland, the day has not advanced a single hour. I ask myself whether the air I breathed this morning in Paris had also drifted as we did toward the west, and if it was not finally the same Parisian air that I breathed on disembarking from the plane. Or whether, because this air is certainly heavier, a little less "viscous" than our plane, it might have drifted more slowly and presently flooding New York during an afternoon that I no longer know.

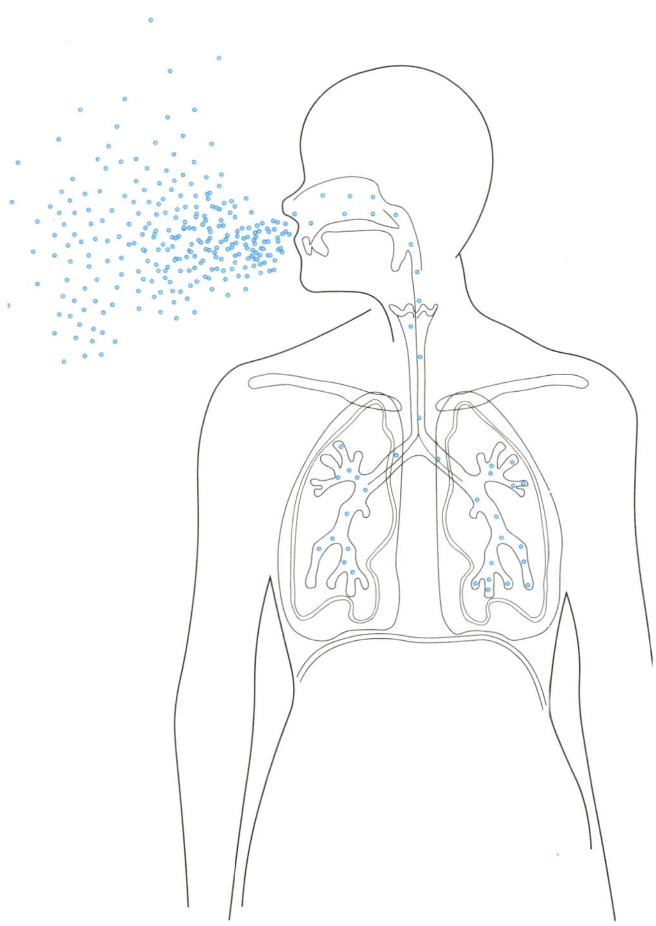

"The air I breathed this morning in Paris." {Fig. 2-399}

II.XIV. Pressure

The concert is boring. I fail to recover the enthusiasm I had for the album. I'm seated at the back, on the mezzanine, in a room of the cinema. The audience must stay seated—surprising at a rock concert. The sound is loud; it warps the mass of stagnant air in the room, which takes aberrant forms in the space; it twists, gathers, distends my ear drums, makes my skeleton and the building structure vibrate. But this air pressure is pretty pointless and I feel as if under a giant vent whose wind sends leaves and dust flying and bends branches; I think only of finding shelter behind a wall or under a rock.

The concert comes to an end. After the last song, the group calls the musicians from the first part back onto the stage, each holding in hand a small percussion instrument. They are numerous now, forming almost a crowd in front of us.
It is impressive. The singer begins the piece alone, with guitar. Then suddenly, the pianist pressing his hands on the lower end, the piano raises a giant wave of air that floods the room and tumbles the spectators who rise one after the other to applaud. I am submerged in this bass that pushes me to stand up in my turn, thighs shaking, and climbs my shoulder to rap on my neck, raising the temperature of my body by a few degrees within seconds. I mistakenly interpret this bass as a freezing wind.

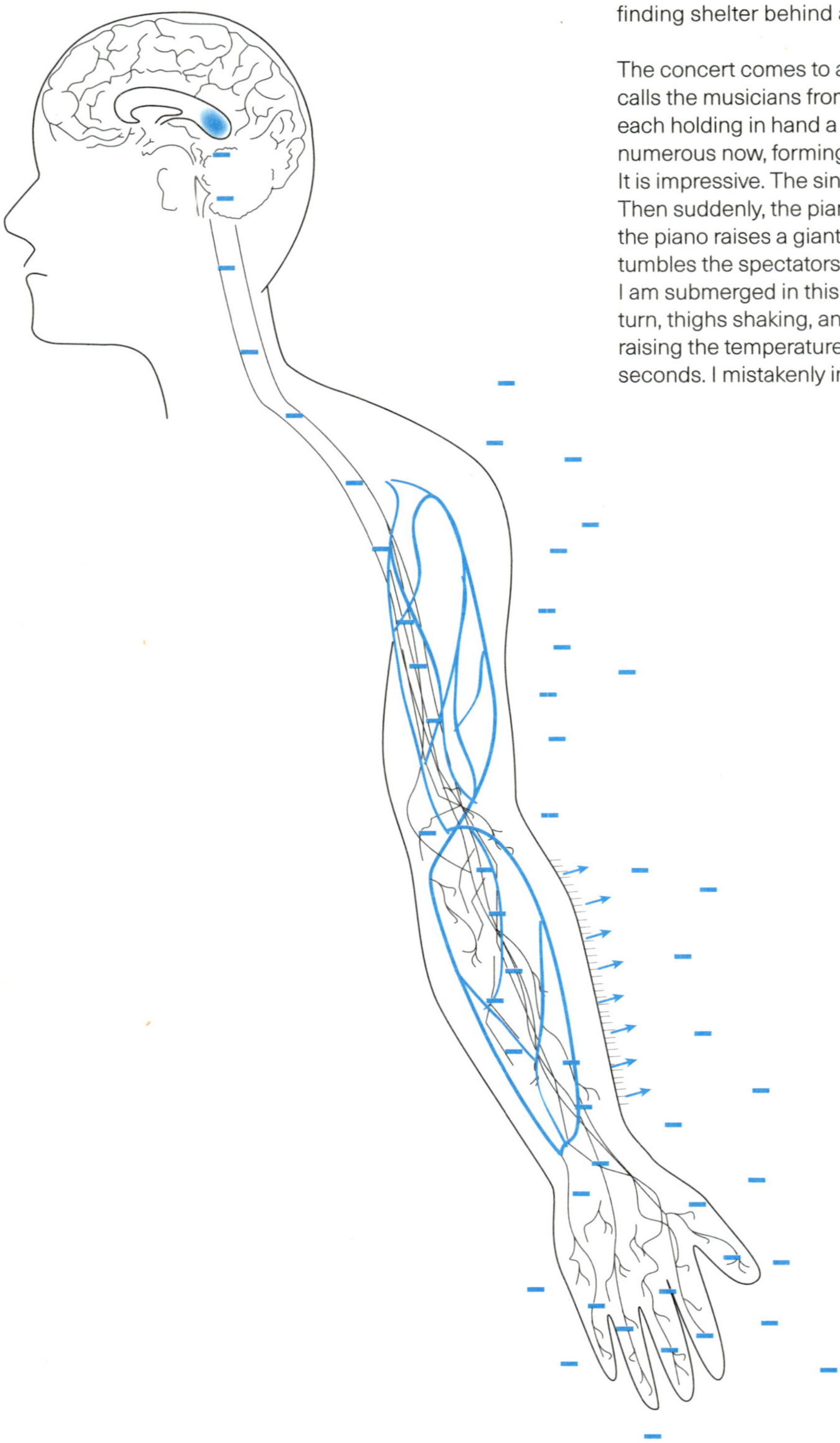

"Thighs shaking." {Fig. 2-400}

II.XV. Anaerobic

I try to stay on the surface. I see in the distance the shore, and when the current lifts me high enough on a wave, I distinguish the tiny heads of other bathers. I know how to swim a little, but I am too far away from the beach, the current is too strong, and I am too small and weak to be able to do what must be done. I am eight. My mother, who came to bathe at the same time, is nearsighted and left her glasses on the beach, so she did not see the wind that caused me to drift. I held with my arms the back of a little orange plastic canoe that served as a raft and made me lighter than water, though still slightly heavier than air (or I would have flown). Beating my arms, I ventured easily and happily into the sea. A short time before, I had been close to shore—until the wind pushed me away. The wind had lifted the nose of the canoe and overturned it violently all of a sudden, plunging me into the water. I raised my head above the water and beat my arms and legs to remain at the surface. The orange canoe bounded farther away, blown by the wind, and is already a small orange form in the distance.

I grow tired of beating my arms and legs. The salt burns my eyes and the water veils my retinas, keeping me from seeing the shore clearly. The heads of bathers are little more than tiny black specks in the blue. I try to keep my face above water but the waves and the wind submerge me in fits and starts. I call my mother, I cry out and swallow more water than I spit out. I lack strength; I breathe too fast and run out of breath. When I open my mouth, I breathe a mixture of air and water. The water does not support me; its molecules are too weakly linked to one another to carry my body whose own molecules, much more linked, stretch even more the space between the molecules of water in which it sinks. My feet cannot push off of the water, which, being more solid than the air (whose molecules are not at all linked to one another) seems to support me better, but weakly, only a few seconds more than the air and for far less time than sand or ice, not nearly long enough for me to stand and cease to beat my arms and legs. If the water is gaseous in the air above me, liquid around me, could it be solid below me? There would then be ice in the depths on which, should I let myself sink on the water, I could put my feet, rest, recover my breath. I sink now. The horizon rises above my field of vision, the blue of the sky darkens, and the shore disappears as I close my eyes to avoid the salt. The sounds are absorbed and condense in my ears, which are deafened as the water replaces the air which escapes in the form of bubbles. These bubbles rise to the surface, taking with them the sound of churned water and abandoning me to silence. I sink for a few seconds until I suddenly feel the water density around my torso, surround me and lift me up to the surface. I reopen my eyes to the sky and open my mouth to breathe air, turned on my back by the lifeguard who swims me back to the shore. In the small beach cabana, my mother will seem shocked by what she neither saw nor heard. Years later, I could not recall the least emotion of fear or panic that I would have suffered in those moments. I retained neither anxiety nor trauma, nor fear of water nor of the sea.

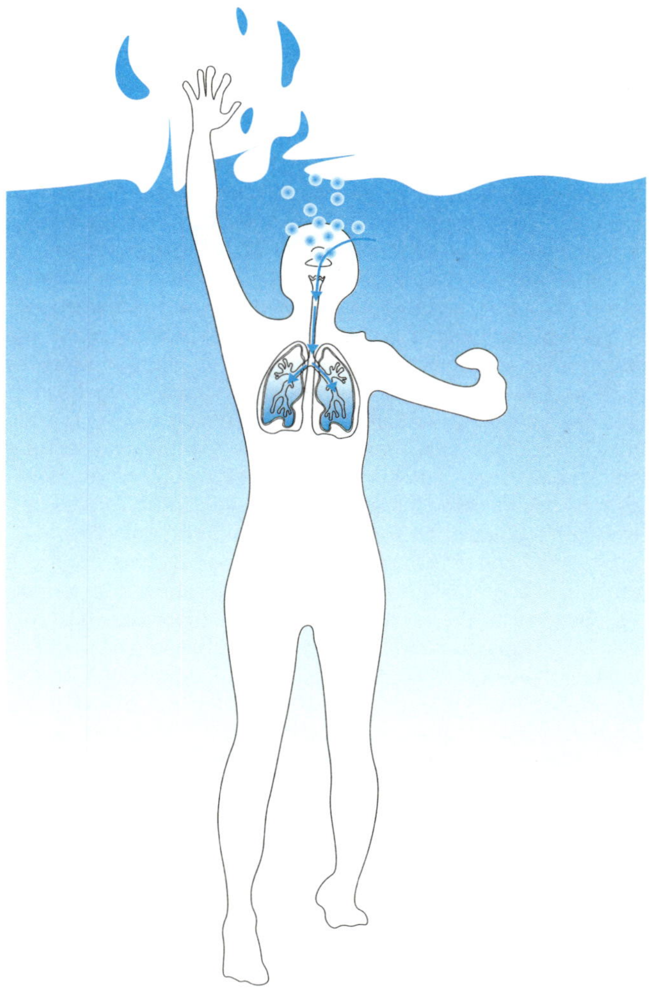

"When I open my mouth, I breathe a mixture of air and water." {Fig. 2-401}

II.XVI. Evaporation

It is noon in Copenhagen, on the waterfront, just at the beginning of spring. The white, intense light does not yet influence the temperature of the air. It is cold. This is also due to the wind that blows from the north, along the basin that traces a northwest-southeast line, and continually draws to my face the heat that rises from my body's nerve center.
The wind also accentuates the coldness of the air, preventing it from stagnating and thus from warming up by conduction in contact with the quays and facades that are exposed to the sun. By the same token, the wind disperses the very slight elevation of temperature that occurs on my skin, produced as the rays of sun touch it, preventing me from profiting from their warmth. But as soon as we turn and leave the north-south road for a southwest-northeast one, the wind no longer reaches us. The sensation of warmth increases immediately and at once the situation becomes more comfortable. For this reason, the smokers who have come from nearby restaurants linger there despite having come there only to light a cigarette. There is a world on the quays, an agreeable place where fishing boats are moored and the old renovated buildings each harbor a restaurant at ground level. Across the canal are neither tourists nor restaurants. Oriented to the northeast, those facades receive no sun after mid-morning; one walks in shadow and cold for the rest of the day. The tourists understand this and do not go there. They stroll elsewhere, to the sun, on the northeast side of the canal and, a bit later in the afternoon, on Frederiksberggade, a popular old commercial street in Copenhagen where shops and restaurants line both sides of the roadway. Oriented east-west, the lower part of the road corresponds precisely to the axis of the sun's rays between 5:00 and 6:00 in the afternoon, which render it formidably luminous, warm, and welcoming to people who are getting out of work or school.

By the canal, at noon, the light is blinding. The spring sun, low in the sky, practically never disappears from one's field of vision, even with head tilted down toward the ground, which also reverberates. It is therefore necessary to look at a black surface such as the coat my friend is wearing or, through the glass, into the darkness of shop interiors, to rest the pupils for a moment. But my eye is caught by a boat covered by a satiny plastic tarpaulin for protection against the rain, which reflects the light intensely. The tarp looks white to me because it reflects all the wavelengths of visible light; they rebound on an oblique angle that enables them to follow us to the interior of the restaurant we enter now. The room is very bright. Sunlight, which enters the room directly through the roof's large glazed opening, is reflected on the ceiling and on the light wood floor, which tints it faintly red, still shining at many hundred lux.

We sit lightly on high seats, near a fireplace that gives off a strong, distinct perfume of wood. "Pear wood," my friend tells me. Perhaps the proprietor changed the scent of wood every day, as he changed the plat du jour? Friday there would be fish on the menu and seaweeds burning in the fireplace. In reality there is fish every day: herring, salmon, a multitude of species finely differentiated from one another by more or less subtle gradations of color, odor, temperature, flavor, texture, linked to the chemical and thermal variations of the waters and seas of the North, and to modes of conservation by cold, smoke, or vinegar.

To accompany the fish, my friend orders beer and schnapps. The beer, he explains, is really the water of our meal, water whose conservation was only possible by being made into an alcoholic drink—a weak alcohol at that—to avoid its chemical and bacterial alteration. The flavors of our meal appeared then as consequences of treatments against rotting of food and drink by oxygen, heat, light, humidity, or bacteria. Other flavors then form in our mouths and noses according to the different sorts of aquavit we are served, in each case the products of evaporation of particular plants, vegetables, and lichens collected in different parts of the cold lands of the North. Inebriation progressively dissolves my brain, which I feel evaporating now in the strange perfumes of vegetables and lichens, of flesh salted by seawater. Flooded with white light that crosses my retina, my thinking dissolves among the products I ingest, in the light of the north, and takes on a marine taste. Different forms of inebriation must exist, specific condensations of chemical qualities of different lands: a hot and dry Mexican inebriation, a French inebriation that takes root in the chalky soils along the Gironde . . . The richness of aromas and flavors intensifies in the course of the meal. By the time we finally leave the restaurant on the quay, I am no longer paying attention to the cold.

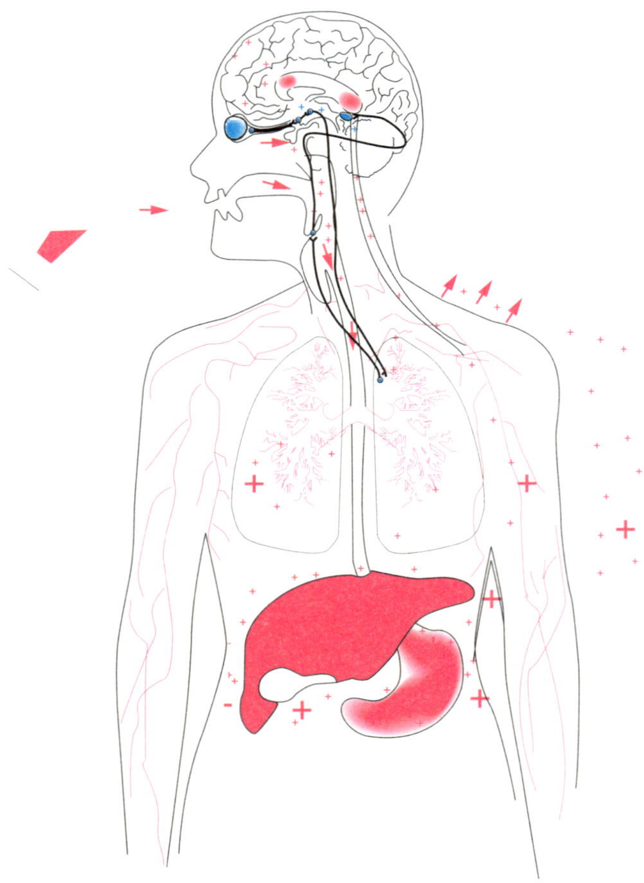

"Inebriation progressively dissolves my brain." {Fig. 2-402}

II.XVII. Emanation

We've been driving for hours. The night fell as we arrived in the Jura. There are few people on the road. We cross forests blacker than the sky, whose slightly lighter tint appears to me above the small colored diodes on the speedometer, between the two black masses of fir trees on one and the other side of the windshield. The interior of the car is a bubble of somber, tranquil, warm air. The car glides in the cold air of a landscape of imprecise forms that permit our attention to be drawn by the stereophonic music of the Pink Floyd cassette. The Christmas lights of the villages we pass surge up from time to time, marvelously. They are the prelude to those on our own fir tree and the gifts arranged at its foot. There is no one on the roads of the villages and little light in the windows, except the suspended electric garlands, which seem attenuated by the orange lighting of streetlights as we pass a Town Hall, school, or church. Then we plunge once again into a blackness even blacker after the blinding brightness of Christmas, blacker still when we look toward the ground, which absorbs without reflection the sun's last electromagnetic rays.

No one sleeps. My father drives, my mother does not say anything. My brother sits quietly beside me in the back.
I recognize the road, the turns, the hills and descents all the more precisely as we approach our home. Now comes the last hill, a road my father takes at great speed because there is never anyone on it and because he knows its course by heart and we are impatient to arrive. My brother and I lean toward the left window, vying to be the first to see our home and the lightless windows of the living room, kitchen, and our bedroom above the hedge, between the leafless silhouettes of lindens and fruit trees. We make a final hairpin turn, then only a shallow slope of some dozens of meters stands between us and our home. It is very cold outside. Once the headlights are off, no more than the few dozen lux of the celestial vault illuminates us. We get out of the car and run to the entrance. My mother puts the key in the door and raises the latch, pushing on the lever with her thumb and pulling the door until we hear the metallic noise we are waiting for, that resounds against the stone walls. It is completely black in the house; we advance a few meters by touch to reach the electric switch whose location we could find with our eyes closed.
The odor of home suffuses me, this odor of the ground floor that diminishes as one rises to the upper levels. It is a damp, cold, fleeting smell. This smell of earthly effluvia rising between the floorboards, of frozen dusts that rise with the air current of the open door, of particles that come from old boxes fraying with age, of chairs with seats of straw, has progressively filled the space of the hall, in weak quantities but over several months, until it has troubled the air's limpidity. As I advance towards the switch the perfume has vanished already, attenuated by my almost immediate adjustment to it and covered by the odorless outdoor air. The smell will not really come out this year, only in the next year. Today it comes out sometimes, just as fleeting, just as minimal, at rare and sudden occasions, brief and unexpected, as in moments of intense joy.

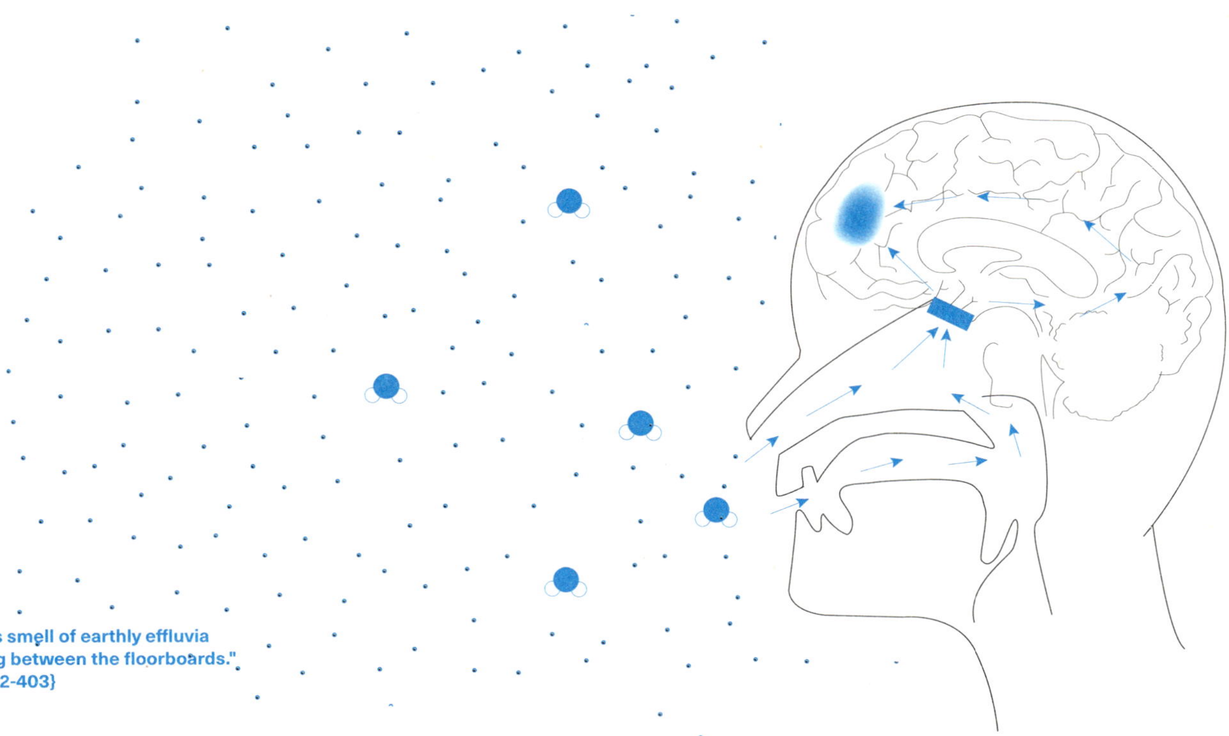

"This smell of earthly effluvia rising between the floorboards."
{Fig. 2-403}

II.XVIII. Isolation

That is the month that I wear hiking shoes. They are climbing shoes, lined inside with wool or a similar synthetic material. They are equipped with thick crampons of thermally isolating rubber that maintain a space that allows a reduction of the surface area of contact between my feet and the extremely cold concrete of our office. We are on the first floor in a large, beautiful volume. The ancient office below us is dilapidated, certain parts of the walls have fallen in, the glass of the windows is broken and the air rushes in and cools all the more for the fact that it receives not one ray of sun, only the continually cooling inverse rays of the earth. Our office is in the town center, but on the bad side of the mountain, the northern slope, the one that only rarely receives sunlight. Our building falls into ruins, the stone walls are covered with climbing plants and some trees have begun to grow there, taking advantage of a flat place created by a partially detached stone. In reality it is the entire quarter that has been abandoned for dozens of years, waiting for a project of urban ism that neglected to note the faults of its geographic positioning. In the meantime, a few of us inhabit more or less legally the last offices between the wastelands: a fashion designer, a vintage clothing dealer, a bookbinder, an artisan frame maker and a leather worker, some artists, and down below, in the constriction of an ancient river, a café where we play pinball. But it is cold in the café too, and my associate and I are in the habit of regularly spending half an hour in a large department store to fully reheat ourselves, to the point that in reality, on some days we do not manage to stay in the office for more than an hour. The cold being unbearable, we need these interruptions to raise our general level of body heat. There is an oil-burning stove in our space; one refills it by pouring in oil, which runs over our fingers onto the ground because of the weight of the bottle and the spout that has deformed through force of use. We have an electric heater but we rarely turn it on because we think it costs too much. At any rate, nothing is isolated here. The windows with their simple glazing close badly; the walls and floors have no thermal insulation, the space is large and the ambition to heat its air is fairly vain. What is needed is to reheat just a small space around our body, as in the interior of a sleeping bag, a bubble of air in which we can sit and from which only one's head and right hand would protrude. We work without removing our anoraks, whose materials we chose precisely: Gore-Tex, polar fleece. We keep our hoods on and wear a glove on our right hands. My associate and I are happy for this large space for which we pay no rent.

Upon returning from one of the warm places in town, we understand that, air being a bad conductor, our oil-burning stove must heat not air but water. It will communicate its heat quickly and without loss directly to our skin. So we draw a sort of sitting-bath whose upper surface would be isolated by a sort of quilt from which our heads and hands could protrude so that we could work. We will move the office a few months later.

"They are equipped with thick crampons of thermally isolating rubber." {Fig. 2-404}

II.XIX. Acclimated

It was cold despite the morning, having been autumn in Switzerland for several weeks. The trees are losing their chlorophyll-emptied leaves. We began by wearing a vest, then a pullover under the vest, and the rain began some days after that. Until the following spring the pullover would not be taken off, only layered with undershirts and changing material from cotton to polar fleece, from nylon to Gore-Tex. I had returned to Switzerland in September, reinstalling myself at my studio in Lausanne after six months spent in Rome during which every day was beautiful and warm and I had fallen in love with Irene. Rome had begun as a more precocious, brilliant, luminous spring than that of Switzerland, which I had fled from, already drowning in the atmospheric humidity that blocked the sun's rays under clouds and let winter linger in the ground until well after March 21st.

On this October afternoon, I board a plane at Geneva in the rain, in the cold. The plane rises in the sky and crosses the clouds in the turbulence, bouncing in an immense blue sky traversed by blinding light from the sun, the sun at the same altitude and to the southwest of the plane on which my porthole opens. I am for the first time surprised by the passage overhead of clouds which, abandoning the climates local to their gravity, carry toward the sun and sky a pocket of spring air at 21 °C. I land two hours later in the darkness of a new night. I take the train and then a taxi directly to Trastevere, where I will meet Irene. She waits for me at the intersection of the road and the pedestrian road that leads directly to the Trastevere church. The taxi stops. She is held there, surrounded by the world, more beautiful than ever in the orange light of Roman nights. She wears a black cotton T-shirt that clings to her shoulders and chest, a short skirt also of black cotton, no vest, nor pullover, nor fleece, nor Gore-Tex. I get out of the taxi in a beautiful summer night at a warm, comfortable temperature and I notice that the air in the taxi, colder and similar to that of my airplane, is acclimatized at some degrees below that which envelops me now. I take Irène by the hand and the brief contact of her skin with a few of my fingers electrifies the nerves of my hand, my arm, my body. She raises her face, holds it toward the night, closes her eyes. I kiss her. We walk away from the cars hand in hand. It is good. Everyone is in a T-shirt or light shirt and walks exactly as one month ago, happily, between the open shops, the terraces of restaurants full of the world and the moving merchants of bags of flour, pirated CDs, and lighted electrified toys. Nothing has changed here. It is still summer and I kiss Irene exactly as one month before, as if the shift of five degrees of latitude, between Switzerland and Italy, had been in reality a fold in time, an immediate return to the torpor of summer; as if the first month of humid, cold autumn hand been no more than a bad dream of some hours in a Roman sleep from which one awoke in full summer, in love for life.

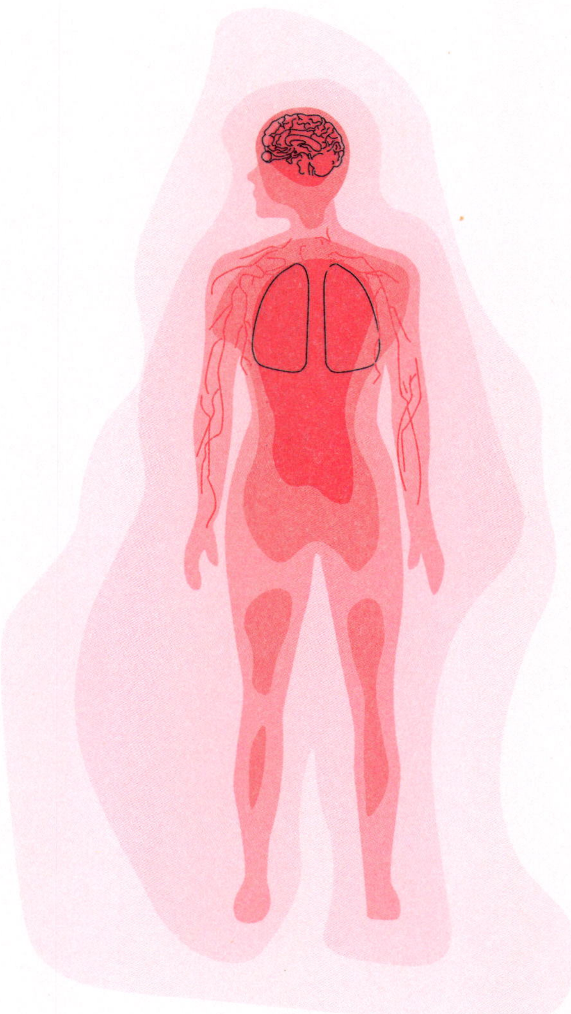

"a warm, comfortable temperature..." {Fig. 2-405}

CHAPTER III.

BUILT ATMOSPHERES

III.I. Climatorium

Taichung, Taiwan, 2011-2020

East-north view of the Climatorium with the yard created by widening the gap between the property line and the waterproofing membrane {Fig. 3-406}

The main entrance to the Climatorium {Fig. 3-407}

At the back left and in the middle front, the courtyards located between the physical and the waterproofed barriers {Fig. 3-408}

East view of the Climatorium with the space created between the physical and the waterproofed barriers {Fig. 3-409}

Circulations are created in the building by pruning the gap between the waterproofing and thermal insulation. {Fig. 3-410}

Inside the thermal insulation, the living spaces are covered with a textile, a material with high emissivity allowing to radiate the coldness of these walls cooled by the air conditioning towards the bodies of the inhabitants {Fig. 3-411}

The inner surface of the waterproofing is covered with aluminum, a low-emissivity material that does not overheat by radiation the human bodies passing through these intermediate spaces {Fig. 3-412}

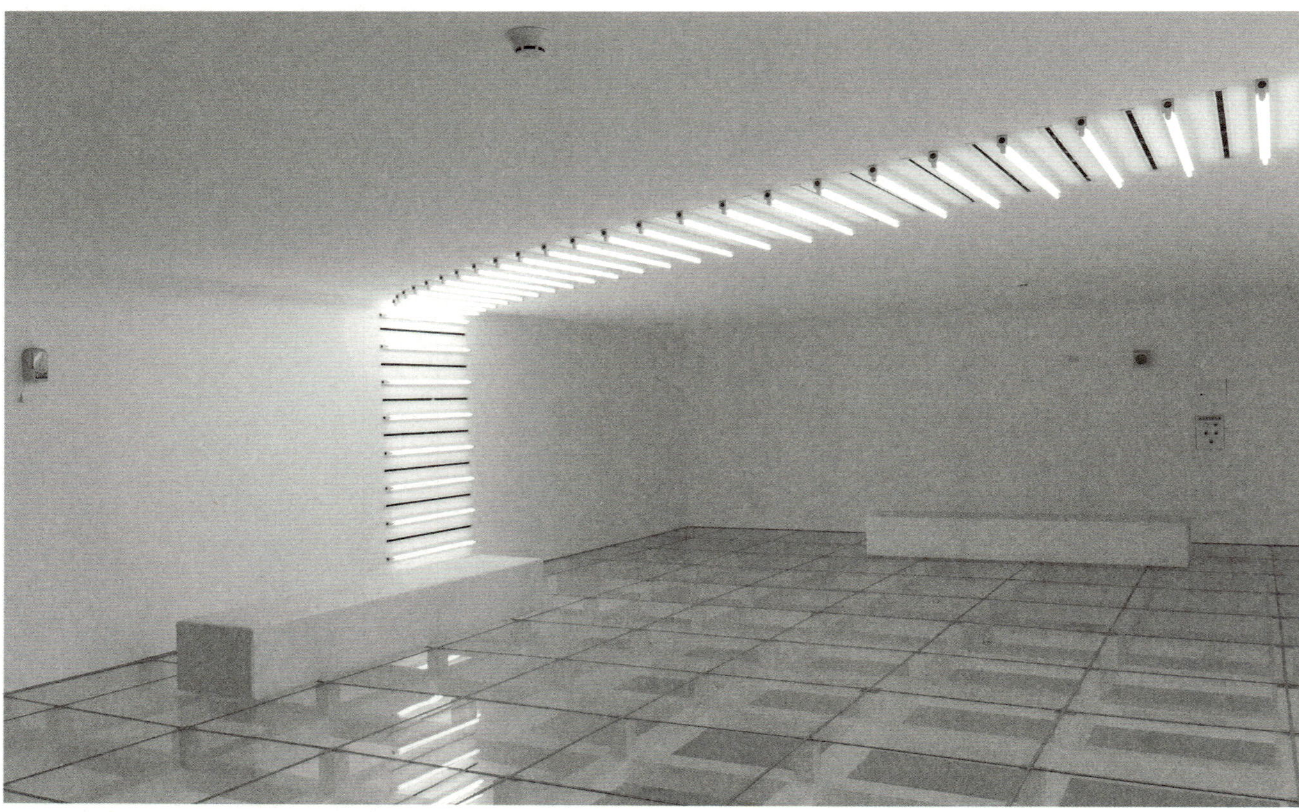

The Dryium of the Climatorium. The drying reproduces every day of the year the climatic characteristics (temperature, air humidity, light) and their variations during the day, of the same day that is repeated endlessly, that of November 21, which is usually the driest day of the year and therefore the most comfortable in a tropical climate because the human body can sweat better to cool itself. {Fig. 3-413}

The Dryium of the Climatorium. The course of the sun on November 21, with its variations of position in the sky, between the zenith of noon and the inverted zenith of midnight is reproduced in its entire course; the night variations being visible through the transparent glass floor. {Fig. 3-414}

The Clearium reproduces Taichung's climate prior to the Industrial Revolution, before 1812, by blowing in air from the top of the room – at a temperature that is 1.2 °C lower than the outside air and cleaned of pollutants (PM2.5, PM10, NOX, O3).

Because this cold air is heavier than the warm air, this "uchronic" air (as though global warming and industrial air pollution had never happened) will remain at the bottom of the room, even though there is no roof to keep it there. {Fig. 3-415}

The Coolium reproduces the climatic characteristics (temperature and luminosity) and their variations during the year of the mountain village Alishan, always cooler than Taichung because in altitude. {Fig. 3-416}

Climatorium

Climatic Layers

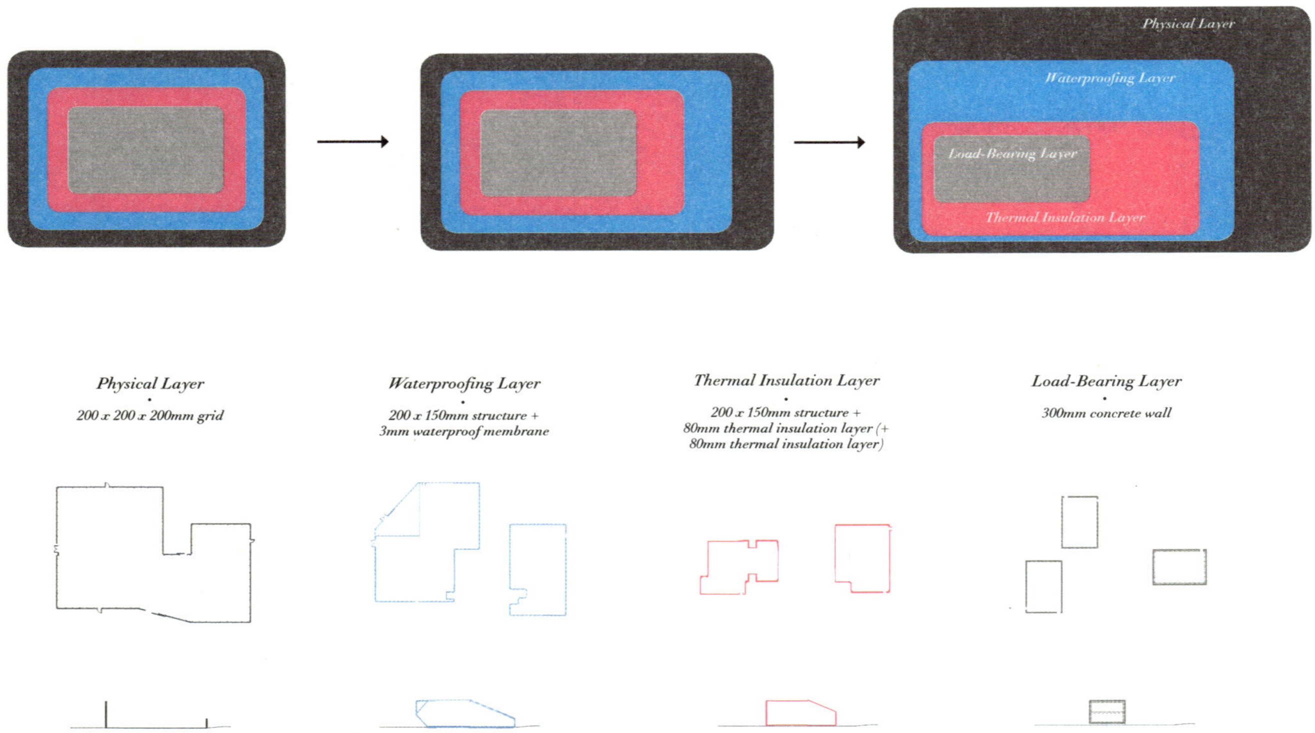

Climatorium

Albedo

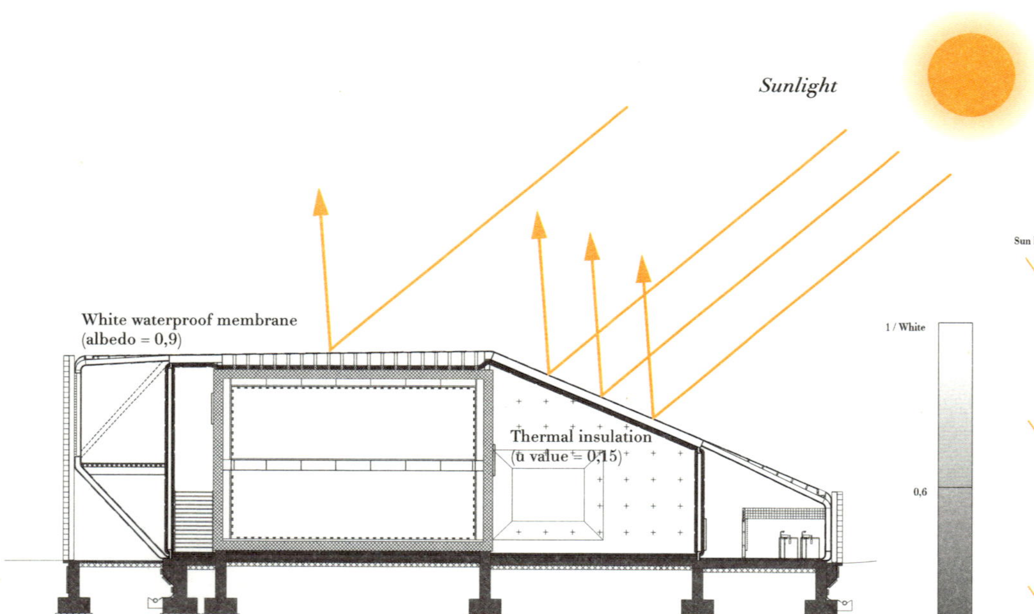

Section of the Climatorium {Fig. 3-417}

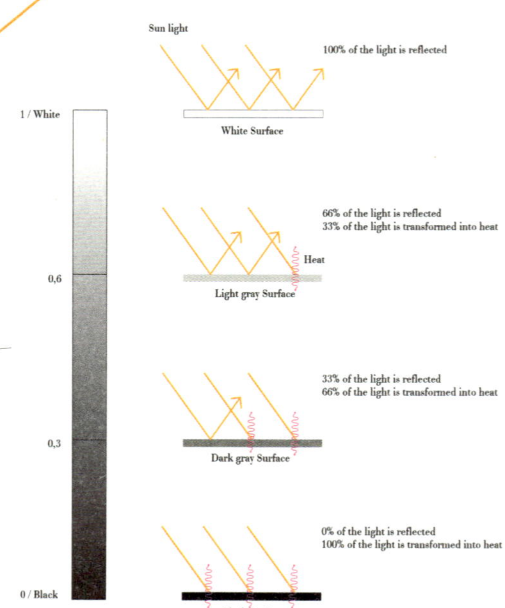

Albedo of different colored surfaces

Climatorium

Plan - Ground Floor Plan

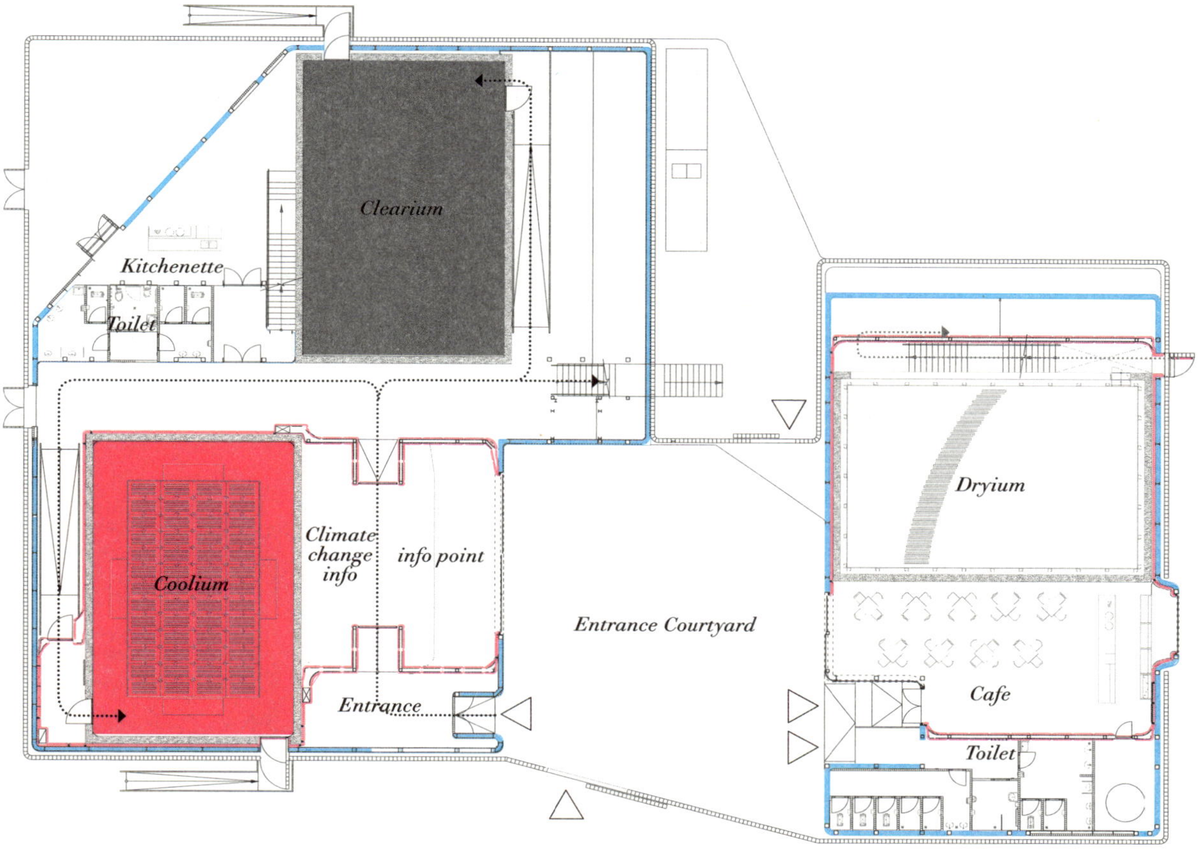

Climatorium

First Floor Plan

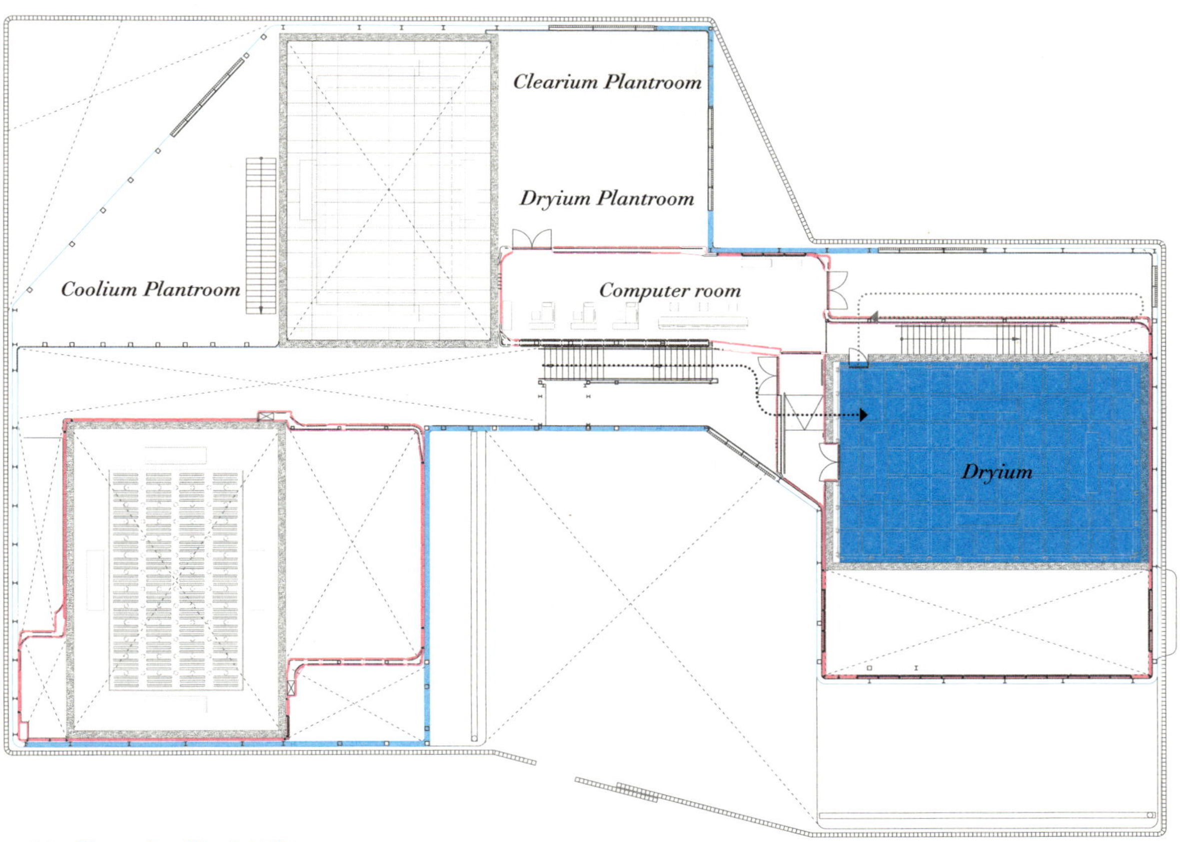

Plans of the Climatorium {Fig. 3-418}

Climatorium

Coolium

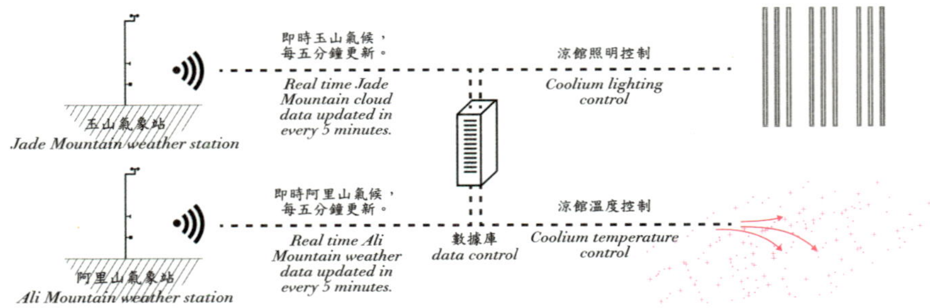

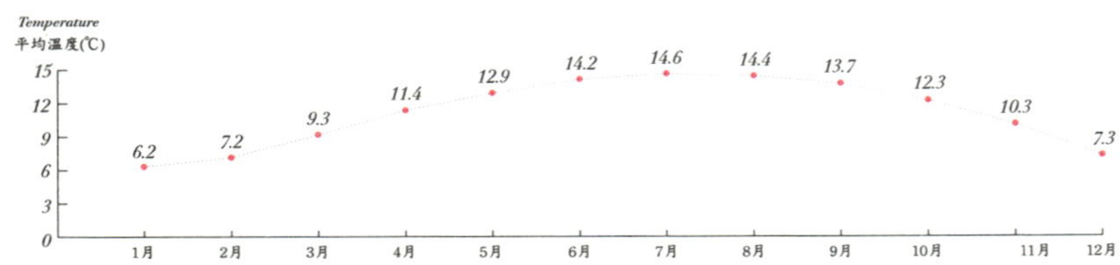

交通部中央氣象局阿里山氣象觀測站
1981-2010年月平均值資料

Ali Mountain weather station, 1981 - 2010 average data

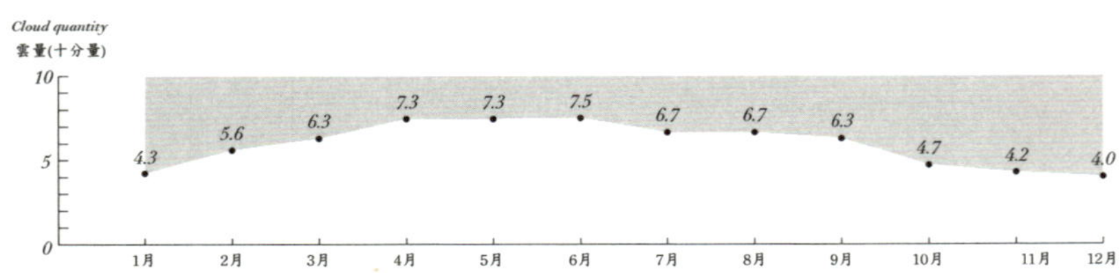

交通部中央氣象局玉山氣象觀測站
1981-2010年月平均值資料

Jade Mountain weather station, 1981 - 2010 average data

Climatorium

Coolium

Information:
Heat and Cold related to Altitude
Sensual Experience:
Real time Alishan temperature (average 6.2-14.6°C) and Jade mountain cloud intensity.

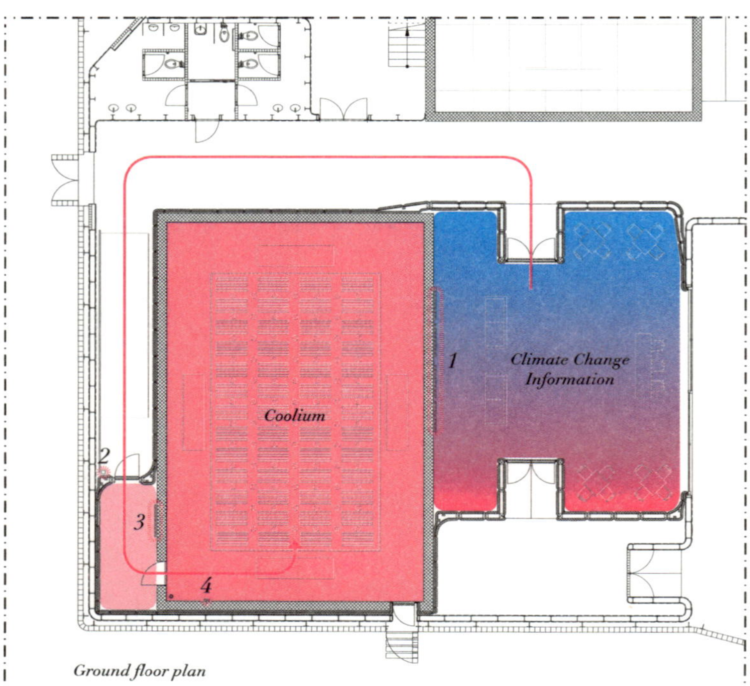

Climatorium

Coolium

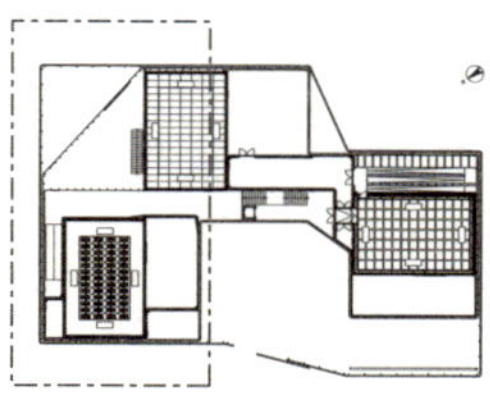

1:400

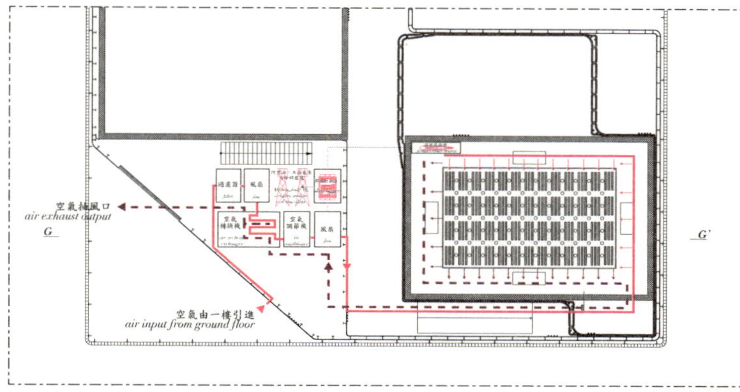

1:200

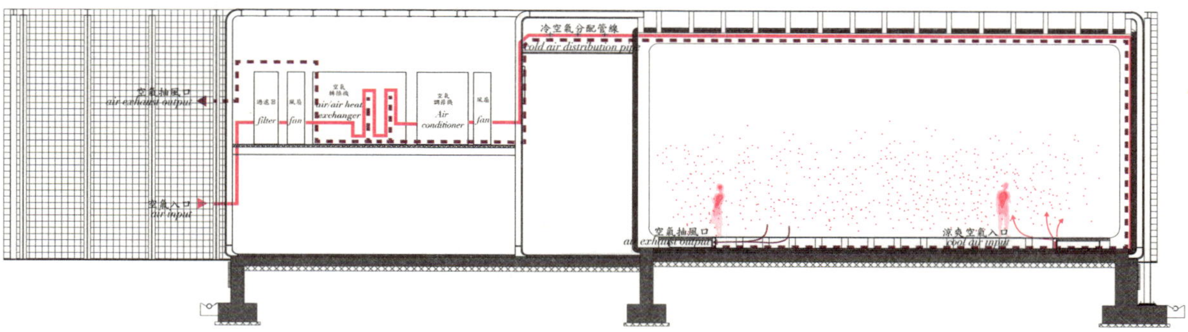

Climatorium

Dryium

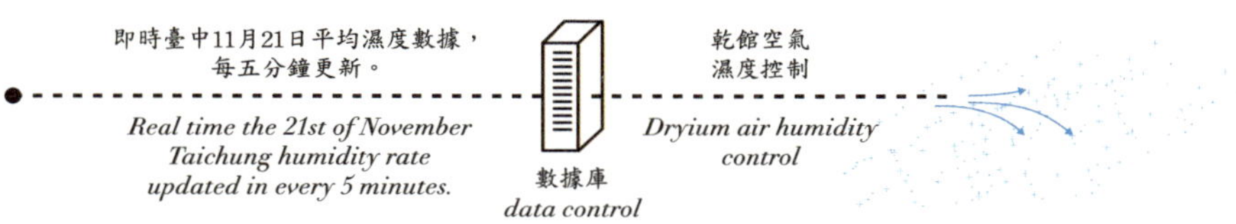

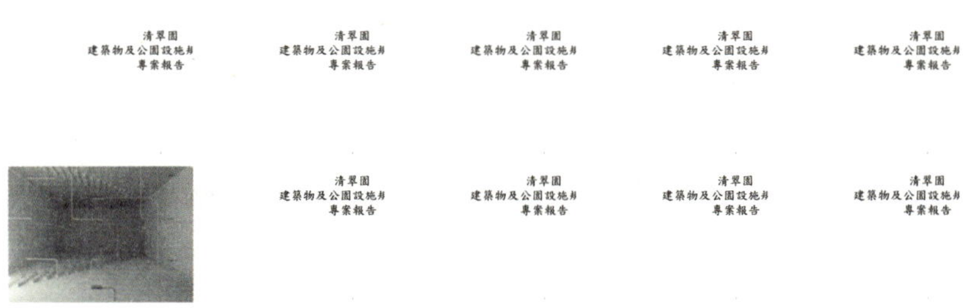

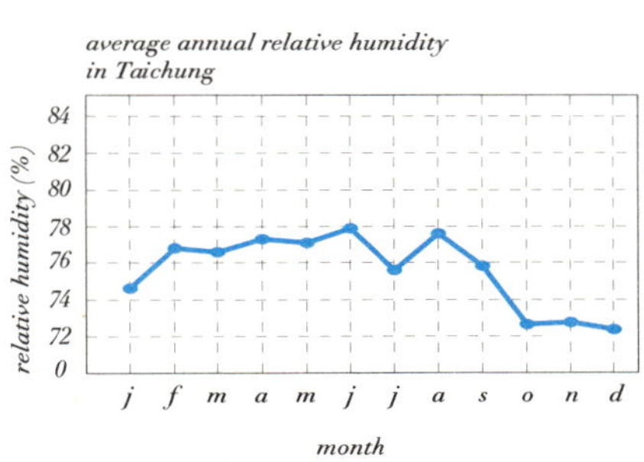

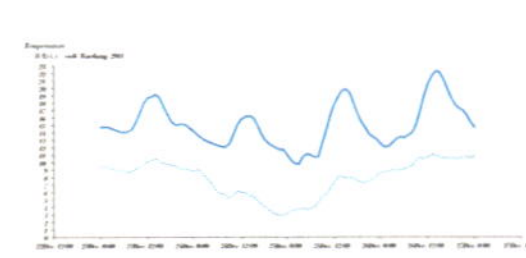

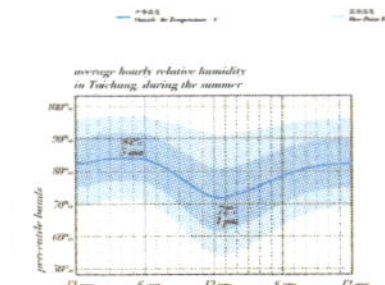

Climatorium

Dryium

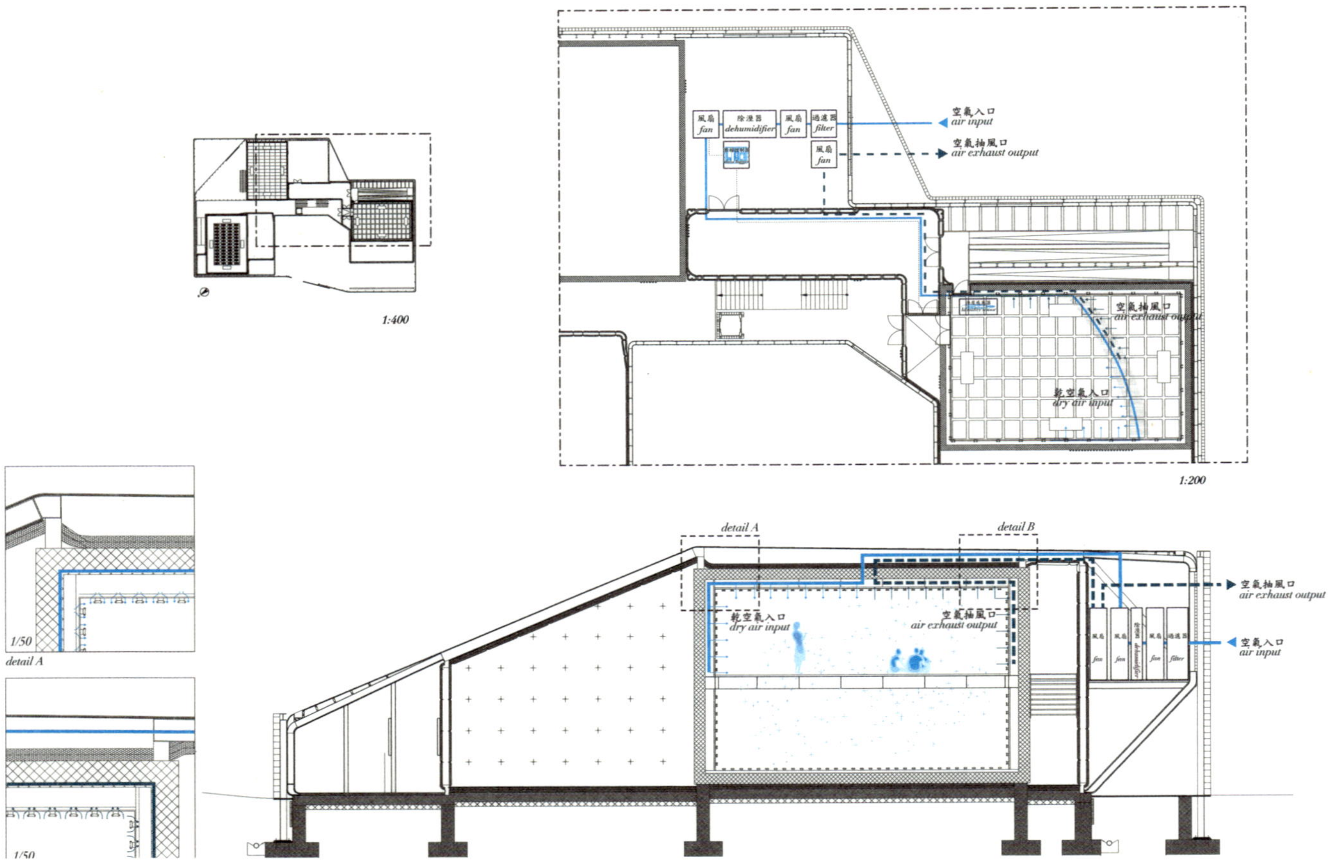

Climatorium

Dryium

Information:
Humidity of the air related to seasons
Sensual Experience:
Real time humidity of Taichung's driest day 21st November all year round.

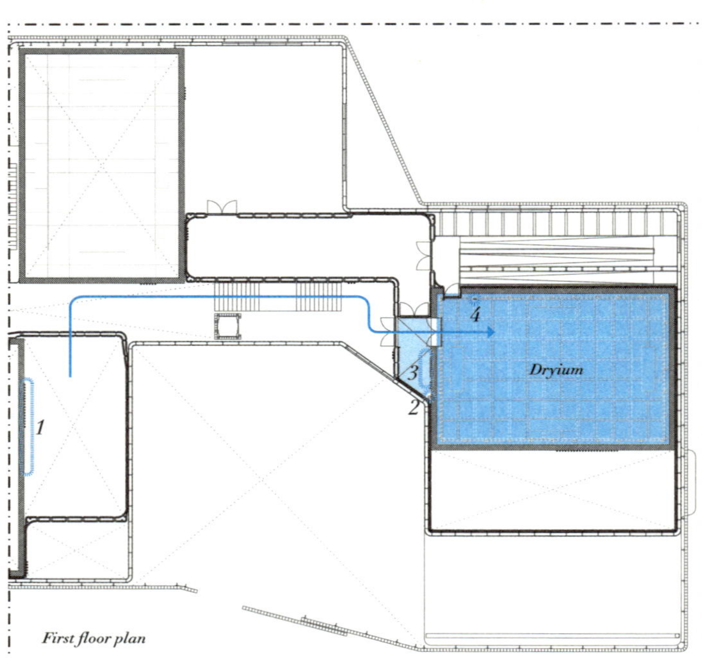

Dryium {Fig. 3-419}

Climatorium

Clearium

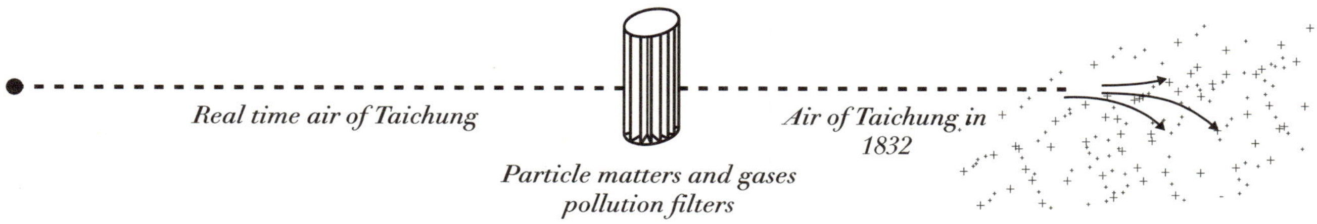

Climatorium

Clearium

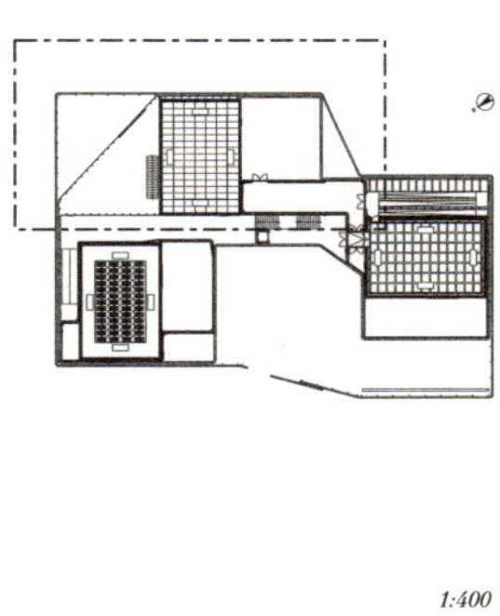

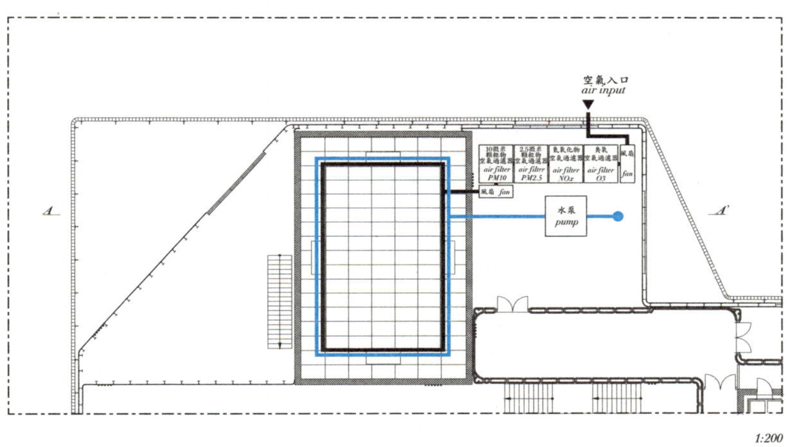

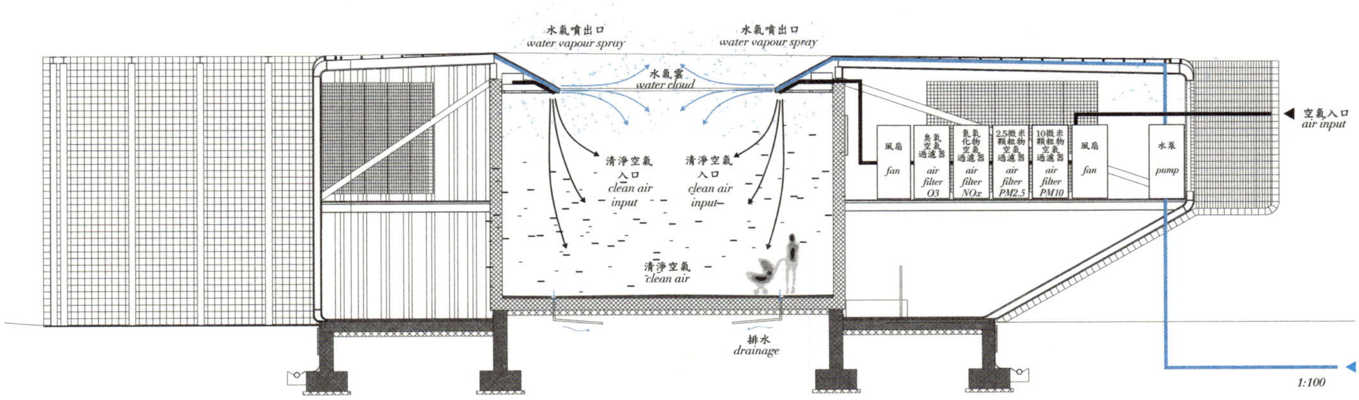

Clearium {Fig. 3-420}

Climatorium

Dryium

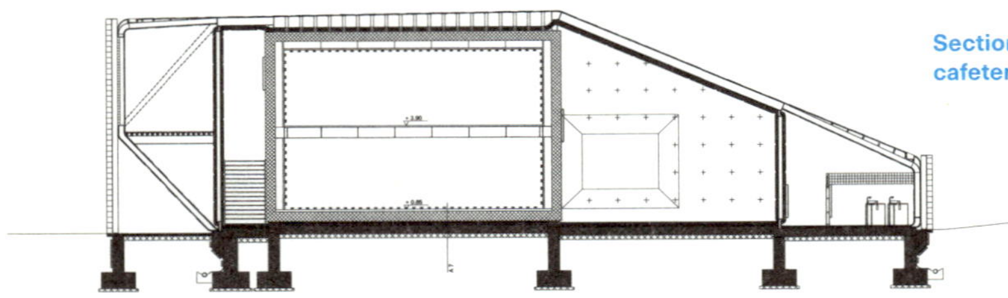

Section in the Dryium and cafeteria {Fig. 3-421}

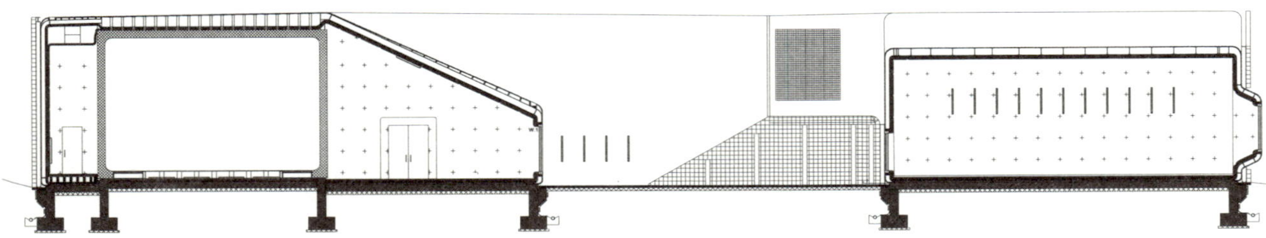

Sections

Section in the Coolium, the data room, the courtyard and the cafeteria {Fig. 3-422}

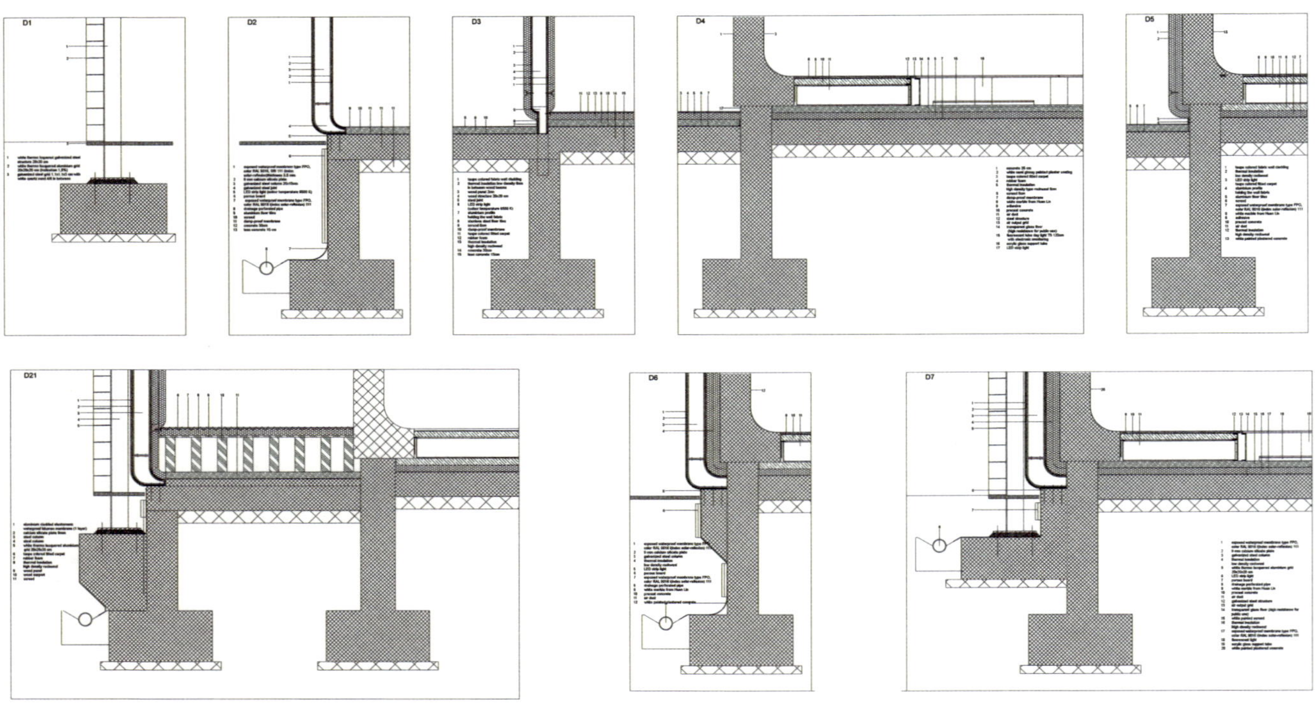

Construction details

The Climatorium's 3,500 m2 host Taichung Central Park Information Center. It is located in the central-eastern part of the park. The program consists of a cafeteria, an information center and a museum on raising awareness of global warming, comprising 3 climatic rooms themed around heat, humidity and pollution. Its architecture focuses on architecture's prime climate-related purpose. It is made up of 4 concentric layers creating a climatic gradient, one within the other, from the most peripheral to the most central, from the outermost to the innermost, from the least to the most comfortable. Each layer fills a particular role in climatic filtering. The first is a white aluminum grille whose purpose is to create a physical barrier and protect from the sun, while remaining visually transparent. It is rendered by a 20 cm x 20 cm mesh, creating an initial degree of privacy, and can be opened to visitors during the day and closed at night. This layer hosts the covered areas as well as the open areas that can be used by the cafeteria as terraces. A first interior perimeter is thus delimited, even if the rain, the wind and the heat can still penetrate this layer, constituting a climatic gradient from the outside toward the inside. The second layer consists of a hermetic polymer membrane that blocks out the rain, the warm wind and the humidity of the ground. This white membrane is used differently for the walls and for the roof, with a high albedo to reflect the

sun's rays to prevent them from being converted into heat. This second hermetic layer is either placed against the first, or is detached. It stretches and shrinks to create spaces where the outdoor conditions, like rain, can be felt by the visitor. Inside this layer is a third layer that is thermally insulating and consists of a 20 cm layer of wool covered in fabric. This layer is detached at times from the previous one to open spaces that are protected from the sun and rain, but are not insulated and receive natural ventilation. These slightly more comfortable spaces are intended to accommodate the corridors, access points, toilets, storage facilities, and other rooms where one does not remain long enough to experience discomfort. The centermost layer, surrounded by the insulating layer, accommodates the most comfortable spaces, the only air-conditioned ones, at a temperature of about 25°C and are completely insulated from the outside heat. These spaces are home to the data areas, museum, cafeteria and offices. Finally, in the heart of the building, within the three layers and inside the last, structural layer, rendered in concrete, are the three so-called Climatoriums (Coolium, Dryium, Clearium).
Each of them generates a particular climate: the first being cooler (Coolium), the second drier (Dryium), and the third less polluted (Clearium). The climates generated are three ideal conditions of heat, humidity and pollution for Taichung, and enable these chambers to be experienced by visitors using the senses. They become climate oases where you can take shelter in the event of excessive heat, humidity or pollution in the city. These are constructed atmospheres, built climates, in which the visitor is immersed both physically and sensorially.

Coolium

The Coolium is a spatial reconstruction of a cool climate.
It reproduces, in real time, the weather of the Jade Mountains (Yushan), located in the center of Taiwan Island. The climatic conditions of two zones, the foot of the mountain and the summit, are reproduced. The meteorological data at the foot of the mountain are recorded at the village of Alishan, whose temperatures vary between 6 °C in winter and 15 °C in summer. The village is located at an altitude of 2,190 m above sea level and has a subtropical mountain climate. Given its geographical position, the village has cold winters as well as low temperatures in summer and fall. To recreate these conditions, the interior temperature of the Climatorium is controlled by a temperature sensor located in the Coolium's equipment room. Data from Alishan's weather stations are transmitted to the network in real time and compared to the indoor temperature. The cold air then circulates throughout the space to lower the indoor temperature and, in real time, reconstruct the temperature of the village of Alishan. The top of the mountain, at an altitude of 3,000 m, is often bathed in the reflections of the sun on the snow. As a reminder of this effect, a series of fluorescent white lights imitate daylight and recreate the effect of the sunlight reflected on the snow at the peak of Taiwan's tallest mountain, Jade Mountain, whose summit shines like jade. Again in real time, light sensors on the mountain transmit the intensity of the sunlight, as well as the cloud cover, controlling the Coolium's fluorescent tubes (T5 electronic ballast) so that the light is similar to the light at the summit. A false floor made of transparent glass covers the network of fluorescent lights, allowing visitors to walk in the light, as if on snow. The Earth is represented by white marble, which surrounds the rectangular acrylic glass panels.
The walls and ceiling are matt white to amplify the reflection of light from the ground, creating a white mountain veil.

Dryium

Taichung has a warm, humid subtropical climate. As a result, moisture persists throughout the year and excess water blocks the body's perspiration process, trapping heat in the human body on hot summer days. It is only in October, November and December, when the humidity is relatively low at around 70%, that the body feels a temperature lower than it actually is. According to statistics, November 21 is considered the most pleasant day in Taichung. It is relatively dry and cool, an ideal climate for strolling outdoors. Is the Dryium a spatial construction of this least humid day? It is a space in which the conditions of November 21 are recreated and repeated throughout the year. To do so, we first recorded the changes in relative humidity and temperature in Taichung on November 21, from November 20 at midnight to November 21 at midnight in 2018. This information is then transmitted by computer and repeated every day of the year, at the same times. The sensors of the Dryium communicate with air-conditioning units to accurately reconstruct the air humidity and temperature of November 21 at all times. Changes in temperature and humidity are measured in real time and adjusted. The interior of the Dryium is a white, oversized space, interrupted only by a glass slab that divides the space in two. Imagine the transparent glass floor as a skyline, above which is the Taichung sky. Located on the walls and ceiling, the many rows of fluorescent lights simulate the path of the sun in Taichung on November 21. The fluorescent lights switch on depending on the different times of day and the position of the sun. For example, at noon, only the 5 lights located in the middle of the ceiling are lit; at sunset, the 5 lights on the western wall located closest to the glass floor come on; at midnight, the 5 lights located in the middle of the floor are lit.

Clearium

The Clearium is an unpolluted built environment. Since the Industrial Revolution, emissions of fine particles by vehicles and industrial processes have become far too much for the natural the water cycle and the wind to filter. Similarly, the accumulation of particles in the atmosphere is far too great for humans to filter by breathing. Short- and long-term exposure to these harmful substances in the air can damage the health and can cause asthma, nausea, decreased cognitive ability, impaired reproductive organs and even cancer. But it is possible to return to Taichung's pre-industrial environment (before 1832) by blowing in clean air and occasionally creating a cloud that covers the open sky of the Clearium. In the equipment room of the Clearium, two air filters collect and filter the air, then blow clean air into the space. One of these filters is intended to filter NO2, O3 and SO2, while the other filters fine particles suspended in the air. The temperature in the Clearium is 2 degrees Celsius lower than outside, as in 1832, when global warming had not yet increased the global temperature by 2 degrees Celsius. The roof of the Clearium is an open void, through which visitors can contemplate the Taichung sky; by regularly releasing water vapor into this void, the cloud layer is intensified, recreating pre-1832 environmental conditions when global warming had not limited the planet's evapotranspiration and there were more clouds in the sky. In the Clearium, visitors can relive Taichung's environment before global warming and air pollution.

III.II Meteorological Urbanism

Taichung, Taiwan, 2012-2020

Meteorological Urbanism

Central Park, Taichung, Taiwan / Philippe Rahm architectes, mosbach paysagistes, Ricky Liu & Associates, 2011-2020

The park's site plan is designed to optimize visitor comfort by naturally reinforcing the cooler, drier or less polluted micro-climates of certain regions. These eleven climatic islands have a higher density of specific trees and receive special soil treatment to strengthen the natural micro-climatic qualities.

The 4 coolest areas of the park, which are most exposed to winds from the north, are called Coolia and accommodate recreational programs. The species of trees for the Coolia are chosen according to their ability to create dense, refreshing shade, with broad or abundant foliage, such as *Cinnamomum camphora.* The materials used for the ground are white and highly inert, thus keeping the Coolia cool even under the sun. The 3 areas of the park that are furthest from the water points and protected from the humid winds blowing from the south-west are called Dryia. They are home to trees such as *Diplopterygium chinense* or the ficus, that draw moisture from the air by capillarity action thanks to their epiphytes or their aerial roots that grow on the branches. The ground materials are porous to quickly drain rainwater and keep it dry. The Dryias host sports activities.

The Clearias are the least polluted places as they are furthest from roadways and noise. The trees that compose them have decontaminating properties. They play host to children's games.

The park is crossed from north to south by three different paths that connect these micro-climates and offer different experiences to strollers. The cool path, connecting the 4 Coolias, provides a quiet, refreshing, slightly undulating walk. The dry path, connecting the Dryias, is the sportiest and most challenging as it passes over the summits of the various hills. Finally, the clean path connects the four Clearias and offers a completely flat walk suitable for babies and the elderly.

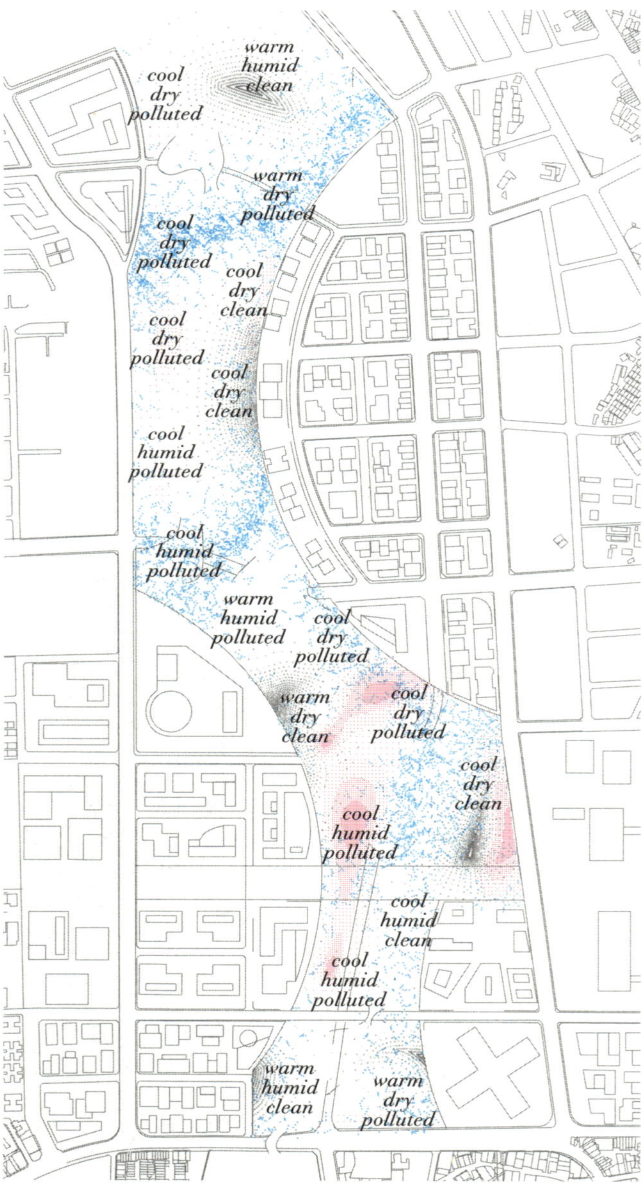

Park microclimates {Fig. 3-423}

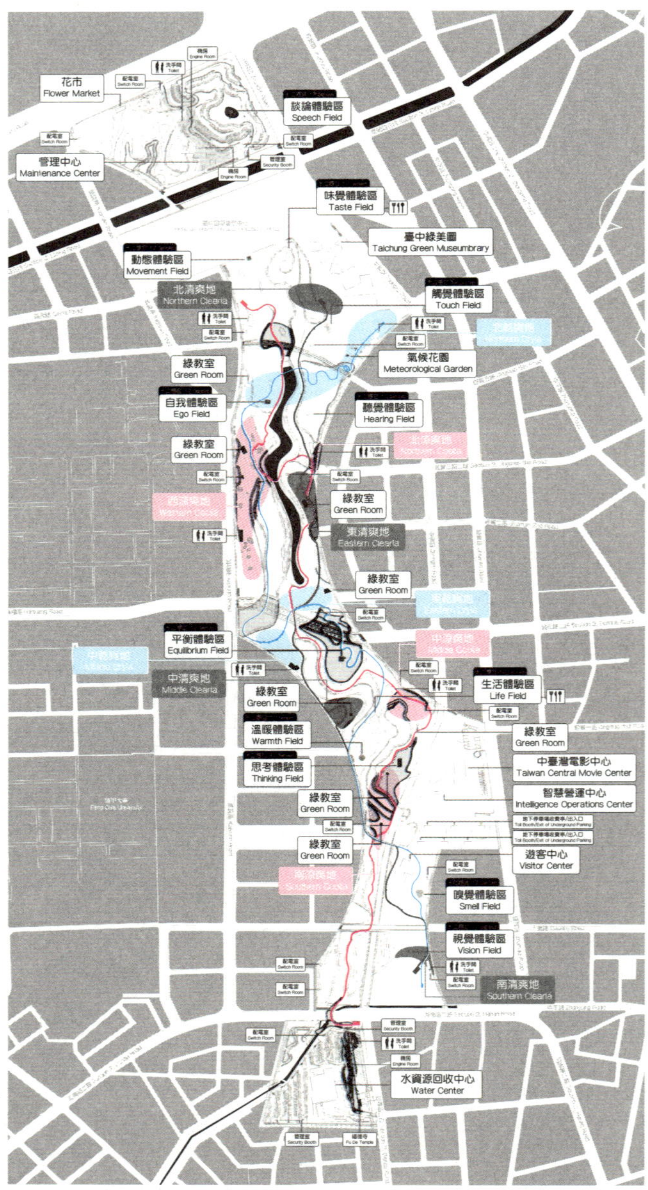

Park program {Fig. 3-424}

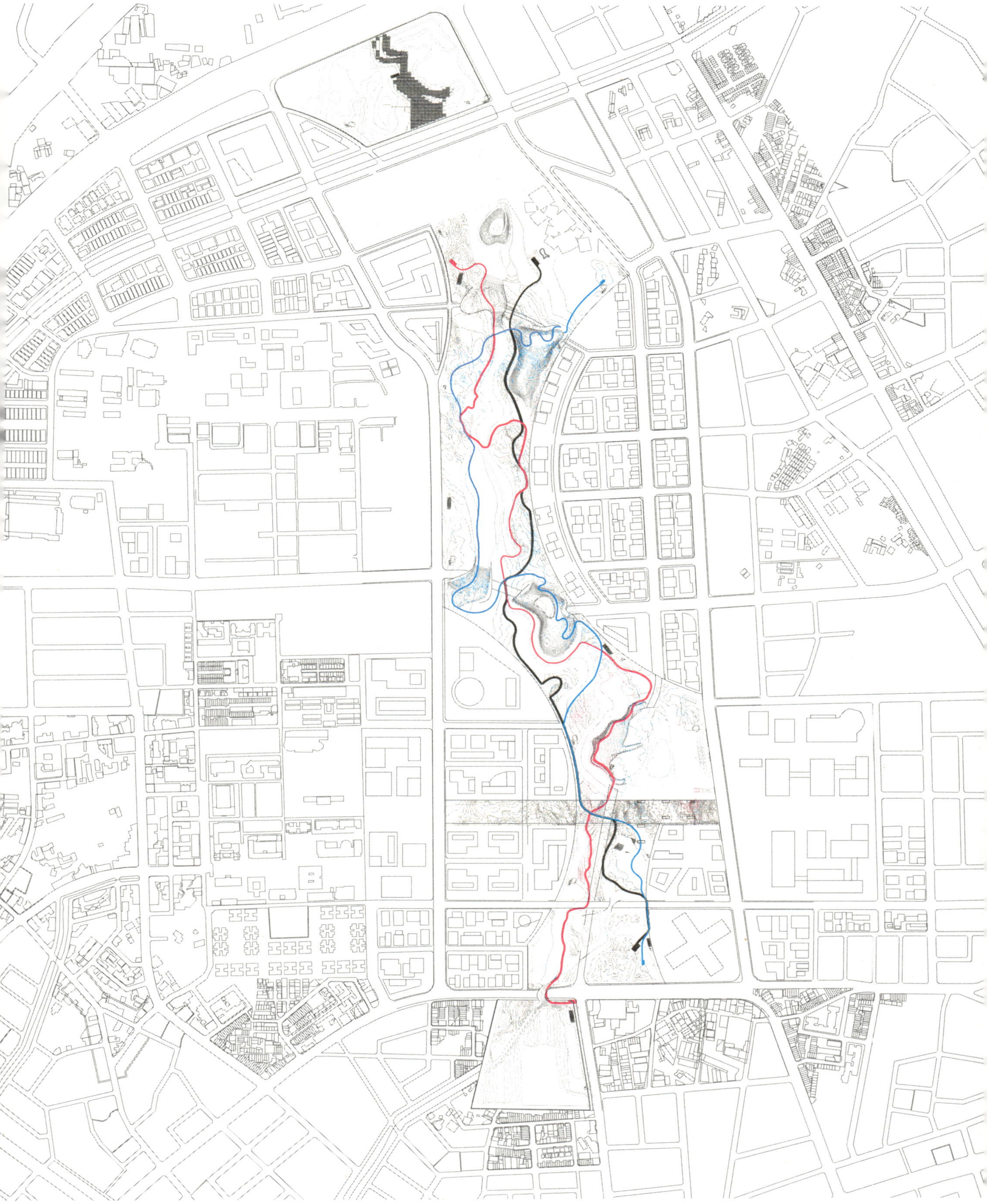

Site Plan with the 3 climatic paths connecting the coolest (in magenta), least humid (in cyan), and least polluted (in black) regions. {Fig. 3-425}

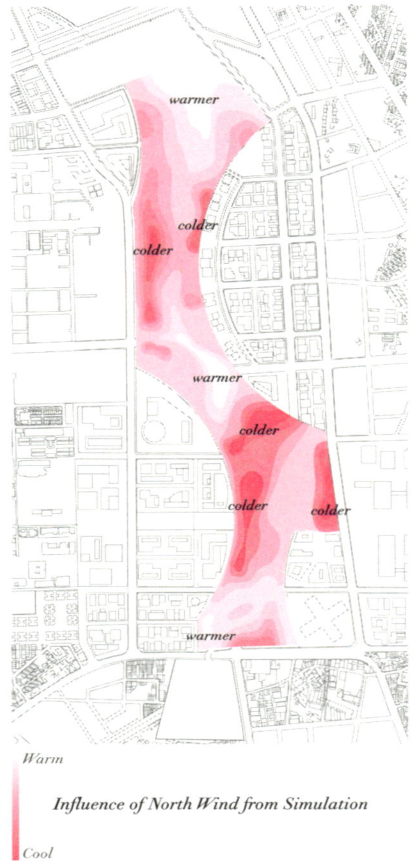

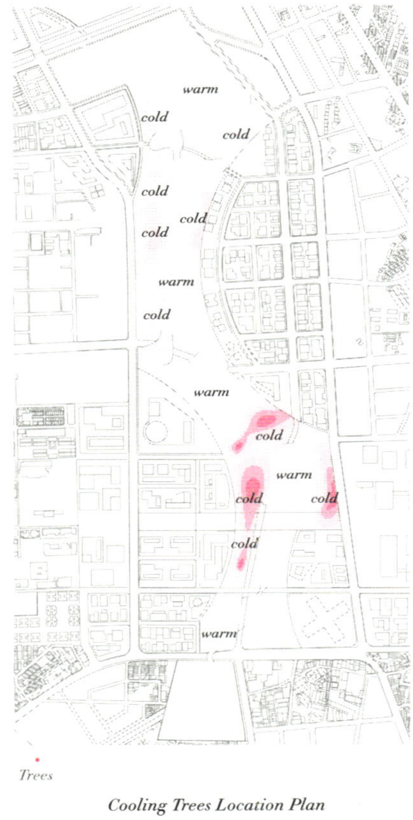

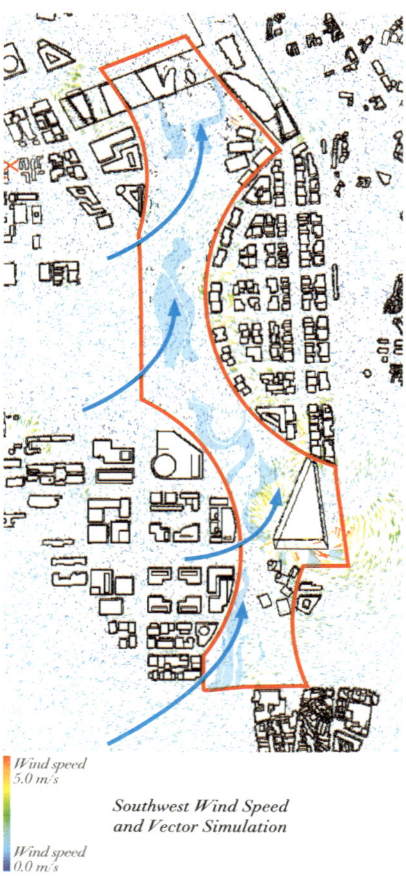

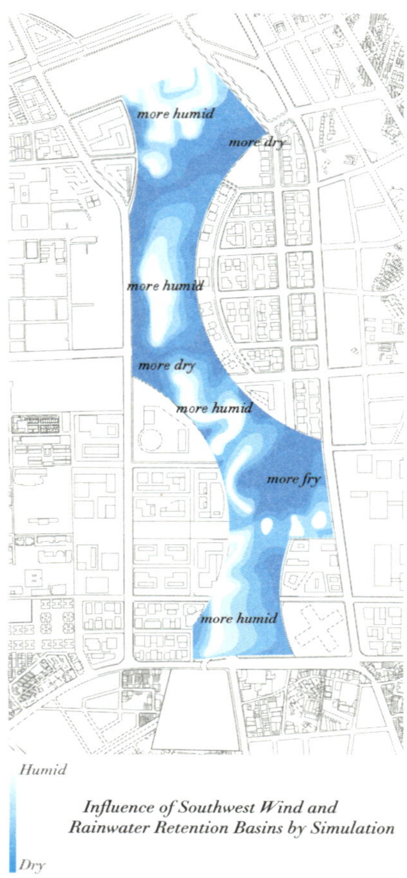

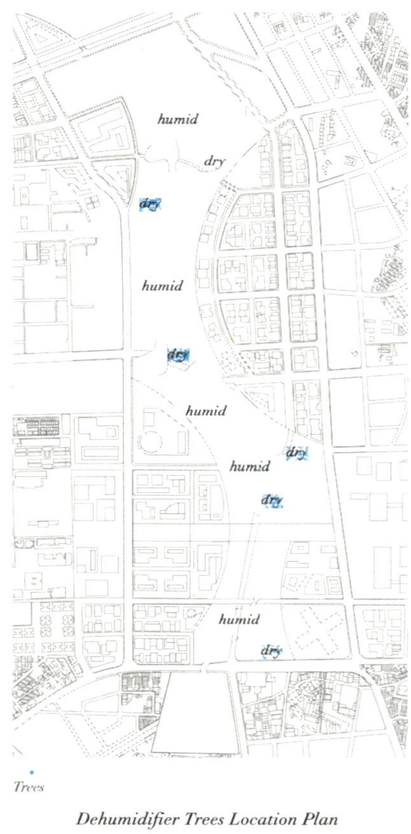

CFD model of wind on the site: the form of the trees location follows the climate map, reinforcing the cooling effect of the wind and welcoming the leisure areas. {Fig. 3-429}

CFD Model of humidity on the site: the form follows the gradations of humidity of the air and in the driest parts are placed the sports program {Fig. 3-427}

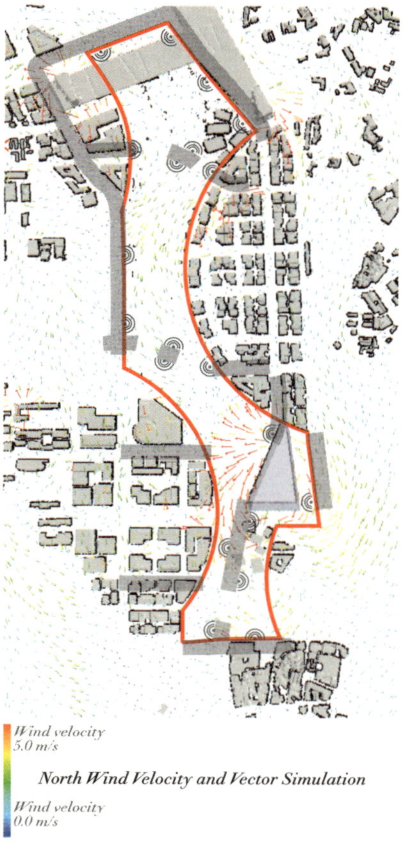

North Wind Velocity and Vector Simulation

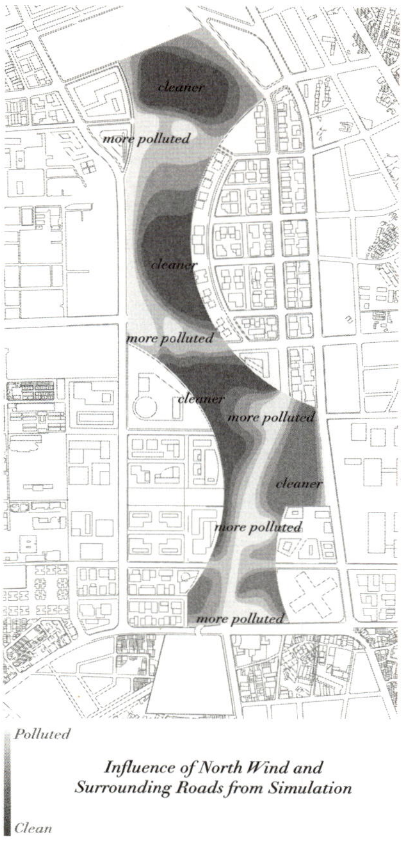

Influence of North Wind and Surrounding Roads from Simulation

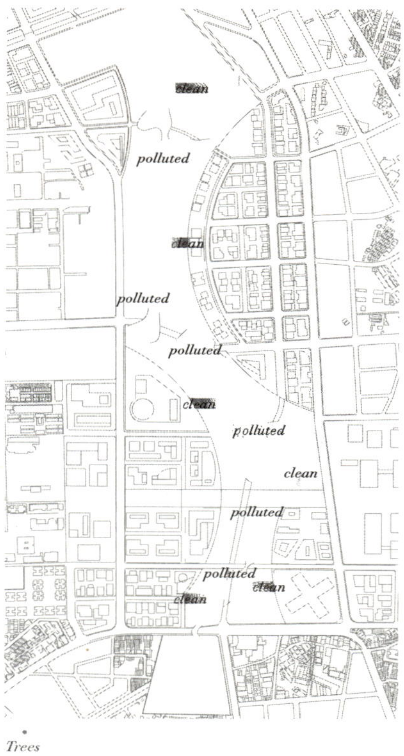

Depolluting Trees Location Plan

Diagram of pollution on the site:: the location of the trees follows the air pollution gradients, and the children's games are placed in the least polluted areas. {Fig. 3-428}

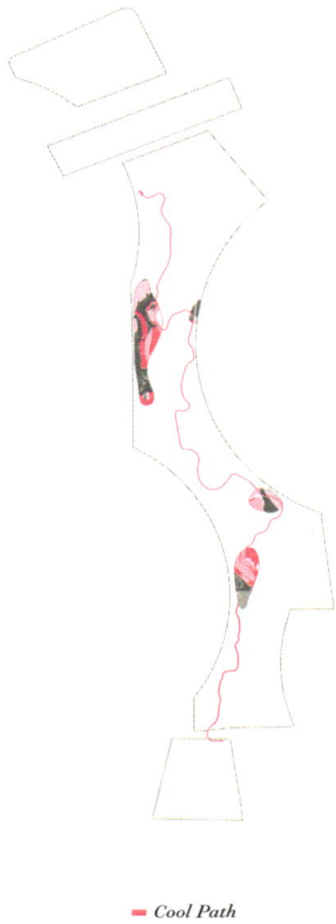

Map of the 3 climatic paths connecting the coolias, dryias and clearias {Fig. 3-426}

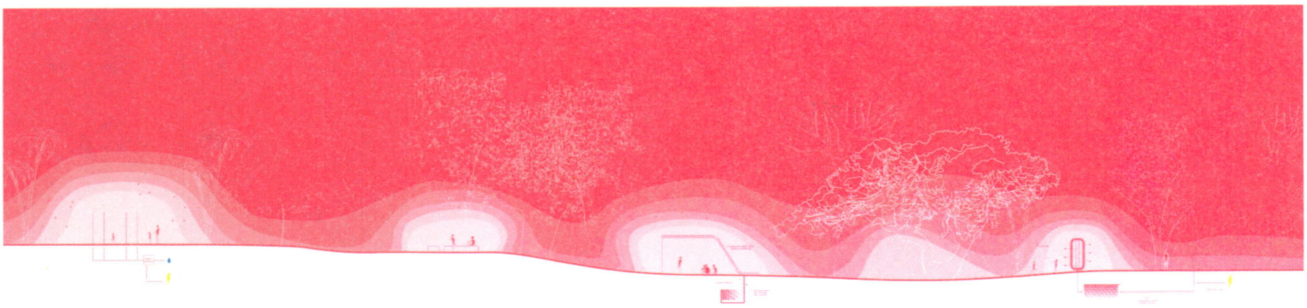

Dryia

Clearia

Heat, humidity and pollution sections {Fig. 3-430}

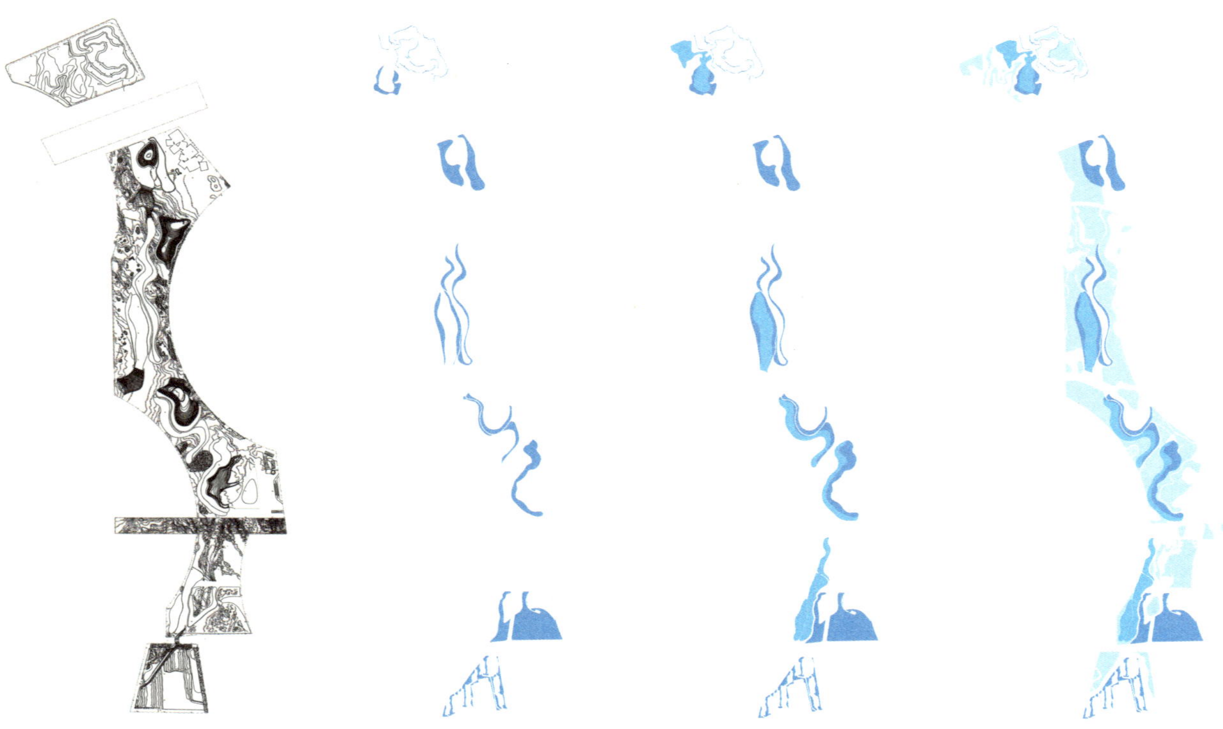

Topography creating detention basins

General flooded area
17% of the park
Amount of rain: Most of the time

Average flooded area
25% of the park
Amount of rain: Heavy rain

Special flooded area
80% of the park
Amount of rain: Once in a century

Rainwater management {Fig. 3-431}

Rainwater management and constitution of the topography as a water retention basin {Fig. 3-432}

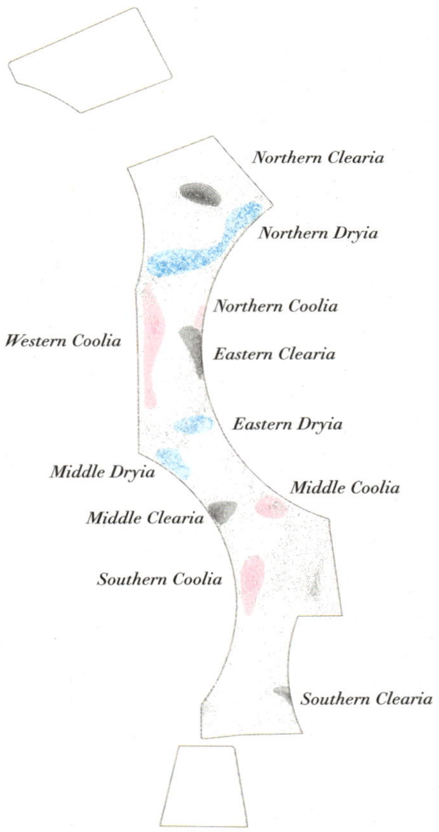

Climatic plan of soils {Fig. 3-43}

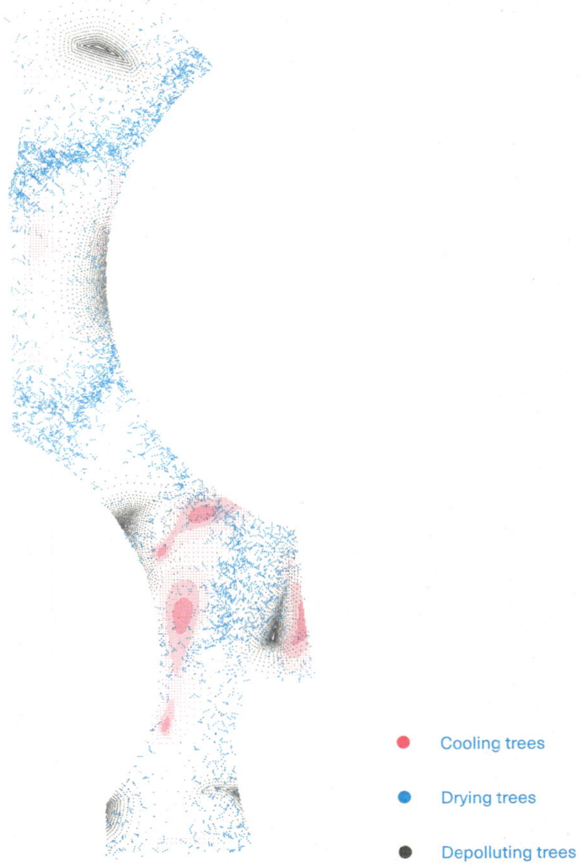

Climatic trees {Fig. 3-434}

III.III. Light Gradient

Taichung, Taiwan, 2015-2020

Light Gradient

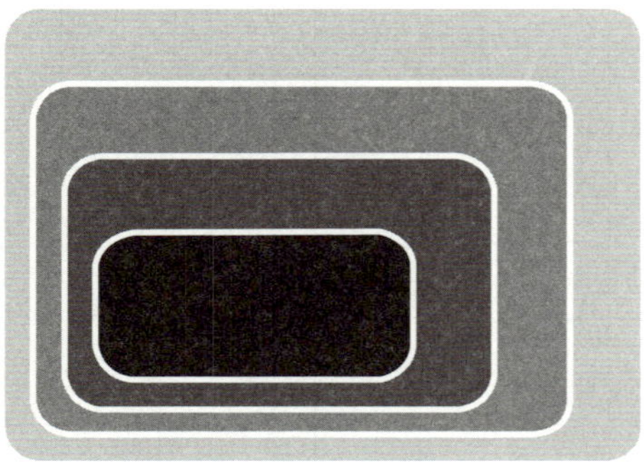

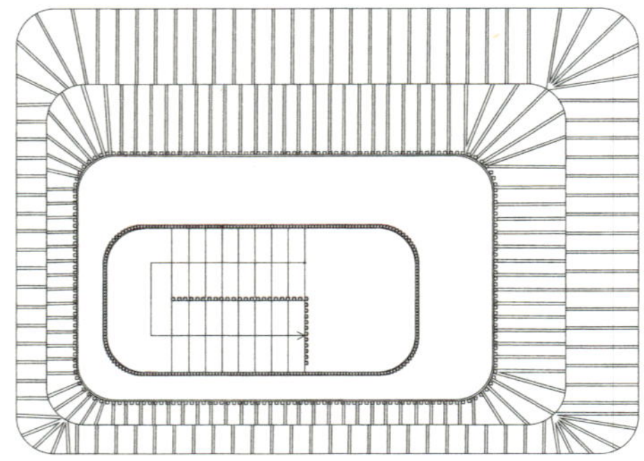

Plans {Fig. 3-435}

The plan consists of 4 tiers of wooden posts whose spacing decreases as you enter the building gradually diminishing in brightness and the connection with the outside.

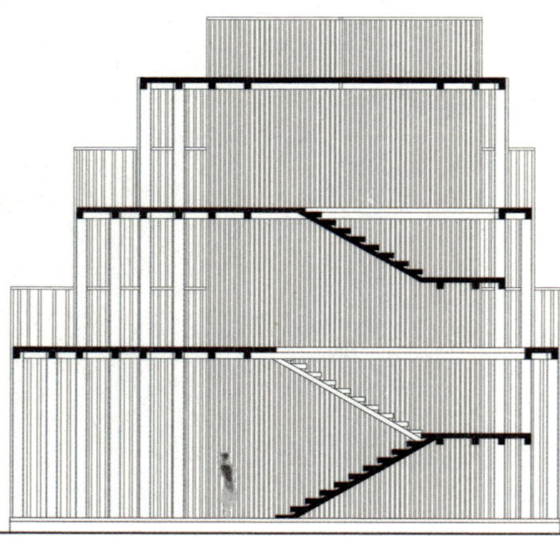

Section {Fig. 3-436}

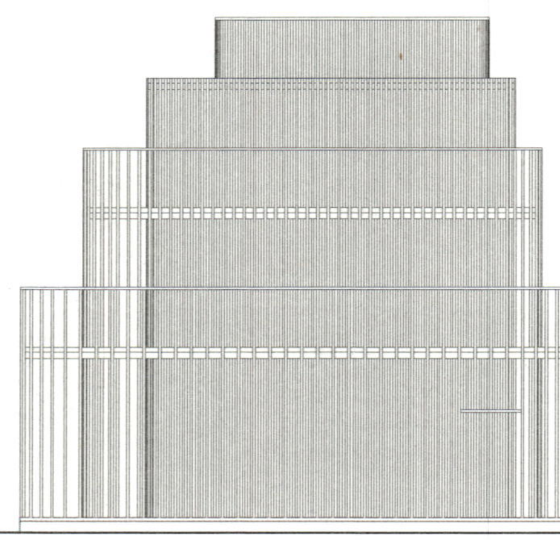

Elevation {Fig. 3-437}

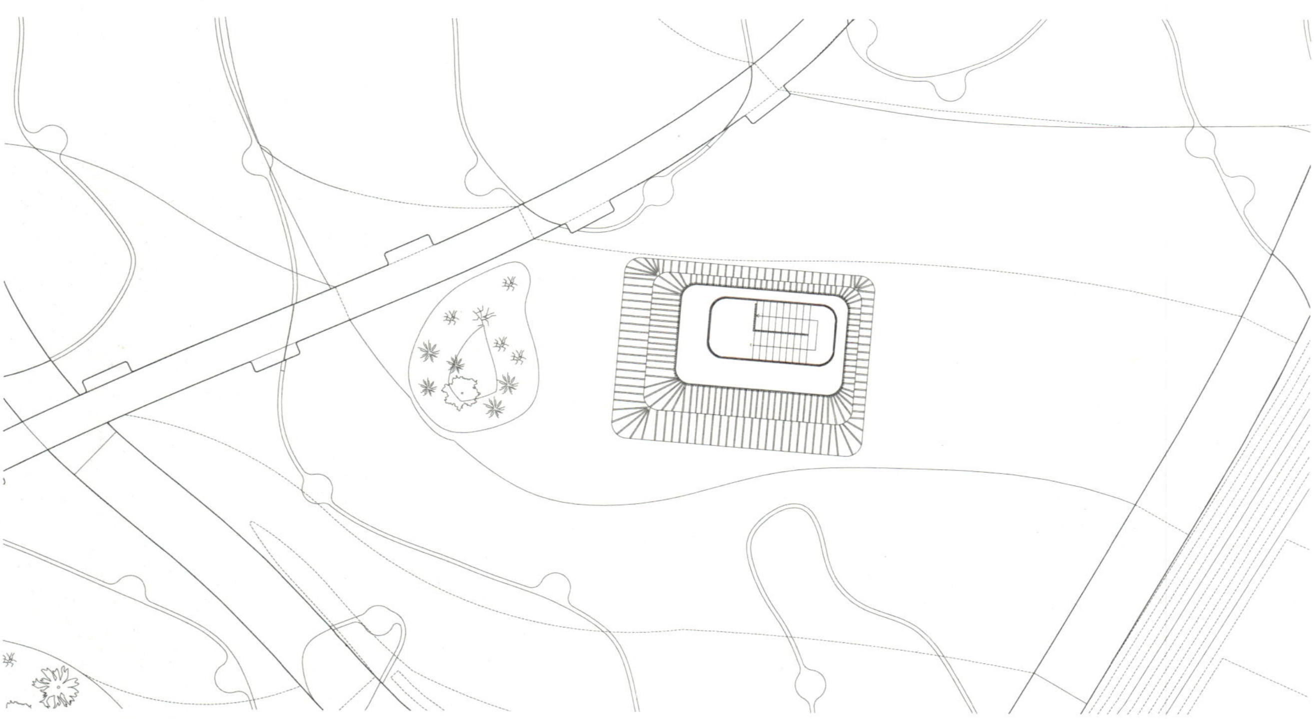

Floor plan {Fig. 3-438}

West View {Fig. 3-439}

Central Park, Taichung, Taiwan / Philippe Rahm architectes, mosbach paysagistes, Ricky Liu & Associates, 2011-2020

The pavilion is composed of 4 wooden structures nestled within each other to allow the refreshing wind to pass by convection but gradually reducing the light, allowing the visitor to get further and further into the shade to escape the heat of the sun's rays. Ascending, the visitor gradually loses visual contact with the outside and finds themselves increasingly in the shade in this dense structure before reaching the top of the tower which opens again to the sun and offers a panoramic view over the entire park.

III.IV. The Attenuated Exterior

Taichung, Taiwan, 2011-2020

The Attenuated Exterior

The 7,000 m² of photovoltaic panels of the Attenuated Exterior are located north of Central Park {Fig. 3-440}

The 7,000 m² of photovoltaic panels above the Attenuated Exterior {Fig.3-441}

Central Park, Taichung, Taiwan / Philippe Rahm architectes, mosbach paysagistes, Ricky Liu & Associates, 2011-2020

The Attenuated Exterior (Faded exterior) provides a place where the intense light and heat of the Taiwanese sun is softened, consisting of natural as well as artificial spaces, hills, parking lots, trees and buildings. The distribution of the photovoltaic panels follows a density gradient that generates a variety of different light intensities in this exterior whose burning sun is attenuated. The photovoltaic panels produce enough energy to supply the park with electricity at all times, both to buildings and for lighting. Beneath the panels are the gardens and grassy hills, parking lots, the maintenance center and toilets.

The Attenuated Exterior hosts buildings, landscape relief, parking facilities, toilets {Fig. 3-442} Photo: haixing0527

Beneath the Attenuated exterior of central park: a garden, a parking lot, the maintenance center. {Fig. 3-443} Photo: xuyuz0522

Faded Exterior

Master Plan

Faded Exterior

Ground Plan

Plans of the Attenuated exterior, Central Park, Taichung, Taiwan {Fig. 3-444}

III.V. Melatonin Café

Taichung, Taiwan, 2015-2020

Melatonin Café

Plan {Fig. 3-445}

Section and elevation {Fig. 3-446}

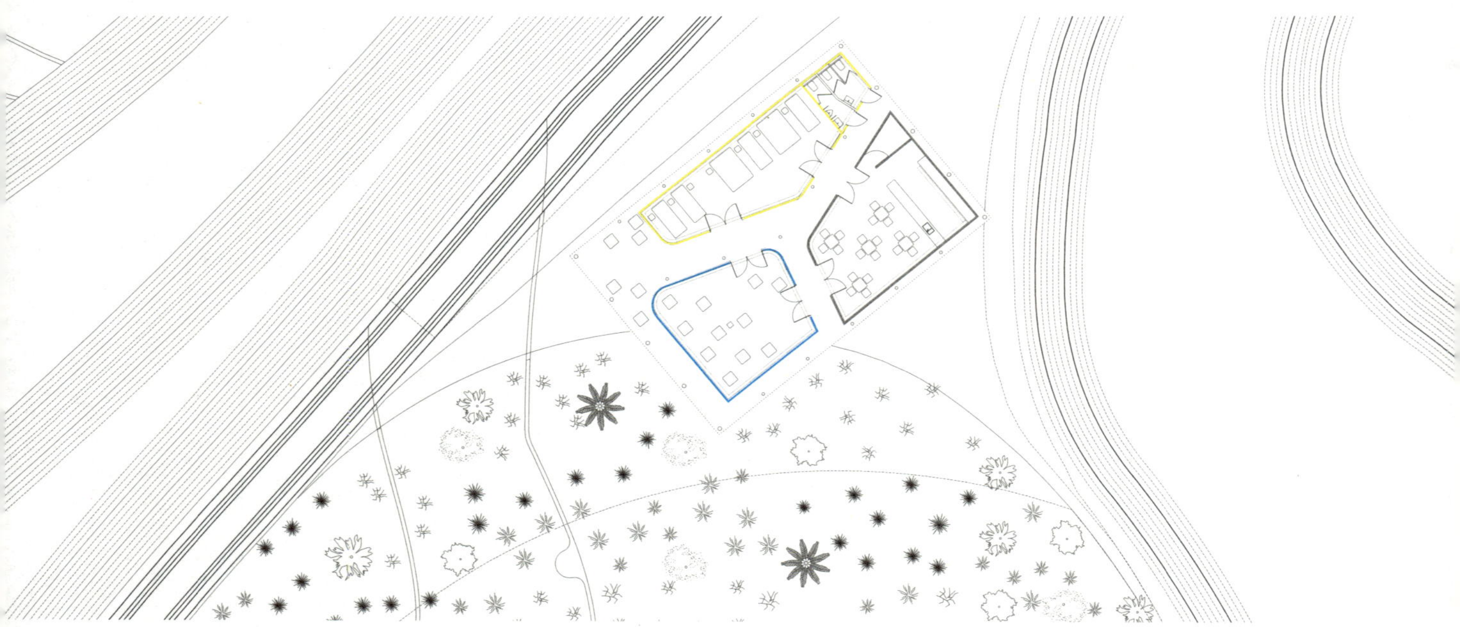

First floor plan {Fig. 3-447}

Melatonin Café in Taichung central park {Fig. 3-448}

Central Park, Taichung, Taiwan / Philippe Rahm architectes, mosbach paysagistes, Ricky Liu & Associates, 2011-2020

Melatonin café is a cafeteria composed of three glass spaces (blue, yellow and transparent) under a large roof. The different glasses create varied atmospheres as well as a perception of time related to the secretion of melatonin, the hormone that regulates the biological clock. The yellow glass space, located in the darker northern part, does not limit the production of melatonin, making it a pleasant space for evenings in the form of a lounge with benches where you can stretch out, like an endless night. The space with blue glass in the south-west creates a diurnal atmosphere even at night, by blocking the secretion of melatonin, as if perpetual daytime. The transparent space in the south-east hosts a cafeteria with tables, chairs and a kitchen, changing atmosphere according to the natural rhythm of the sun.

III.VI. Luminance
Arles, France, 2016

Luminance

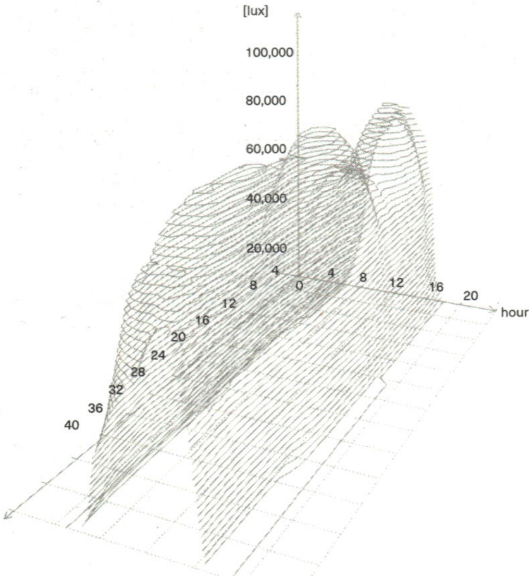

Solar radiation in Arles {Fig. 3-449}

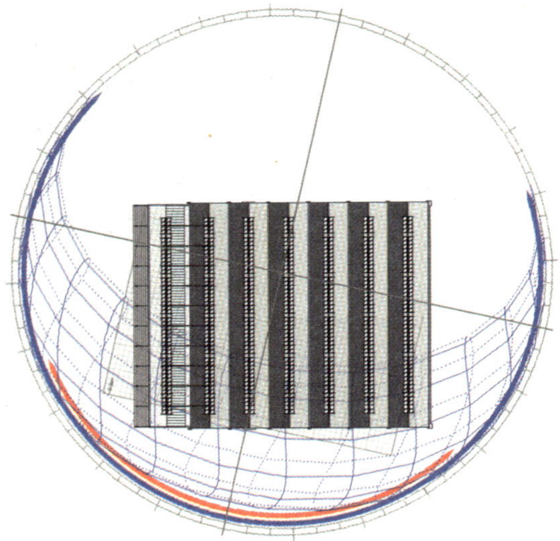

Path of the sun in Arles over the building {Fig. 3-450}

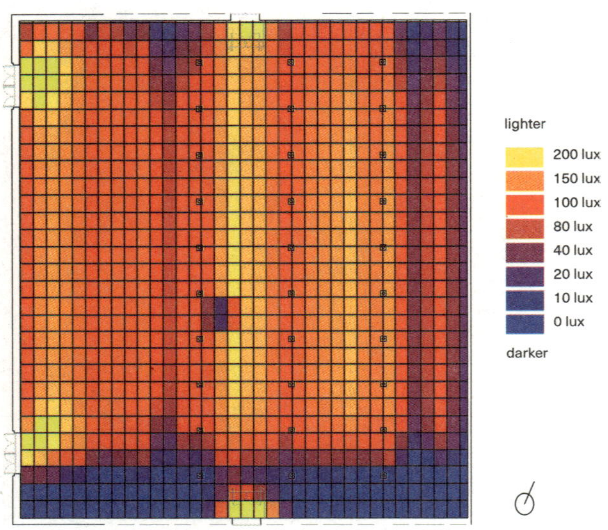

Computer simulation of mean solar radiation {Fig. 3-451}

Daily mean from July to September

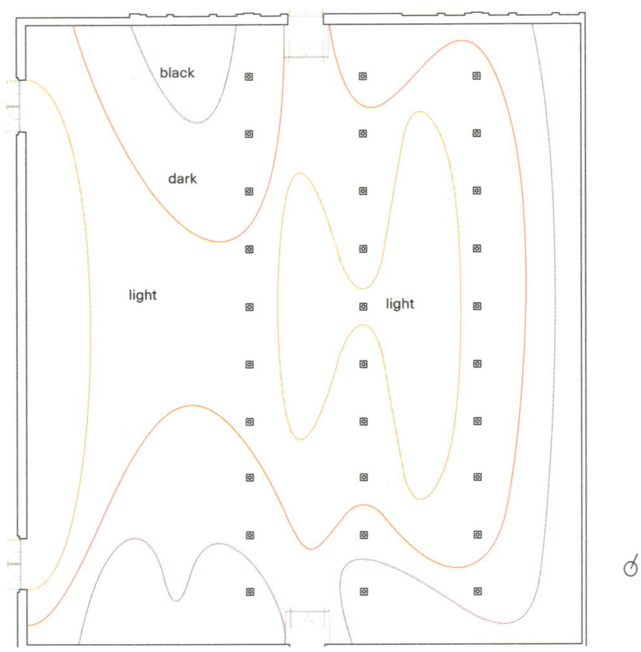

Map of solar radiation {Fig. 3-452}

Conversion of modeled solar gains into lines and surfaces

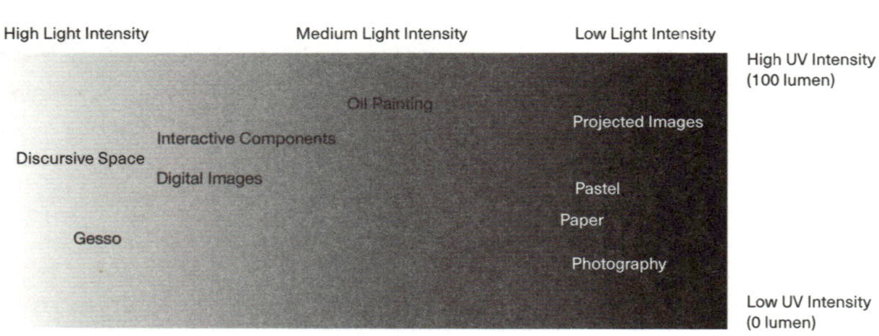

Recommended lights {Fig. 3-453}

Amount of light adapted according to the type of material of the work of art

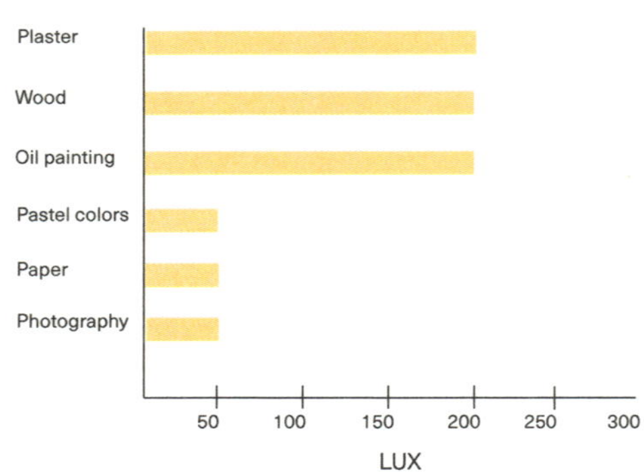

Recommended lights {Fig. 3-454}

Amount of natural light adapted according to the type of material of the work of art

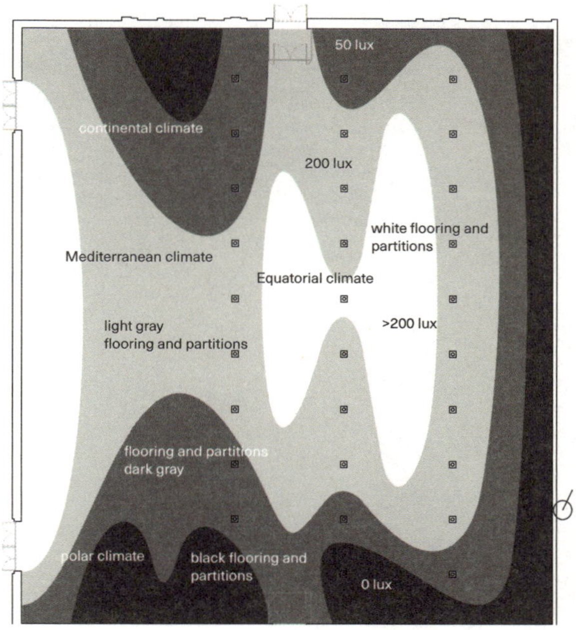

Distribution of the various reflecting surfaces {Fig. 3-455}
Analysis of the incident light and strengthening of the qualities by the treatment of the light reflection of the floors and the picture rails.

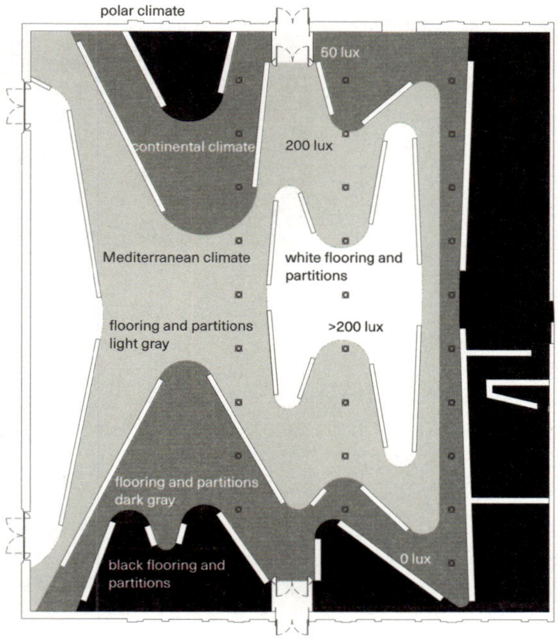

Plan with partitions {Fig. 3-456}
The picture rails are positioned according to solar insolation analysis

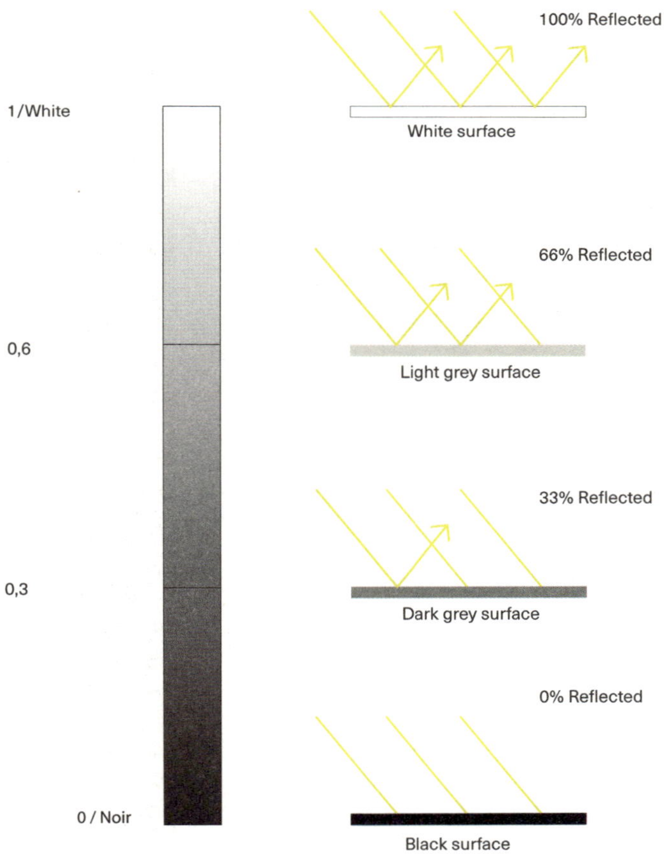

Incidence and reflectivity {Fig. 3-457}
The reflectivity (or albedo) of a material is the percentage of light reflected rather than absorbed.

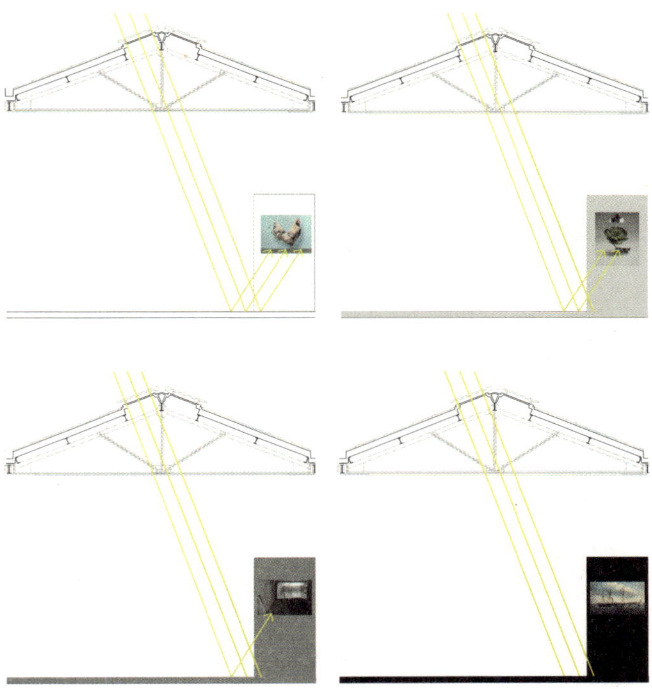

Incidence and reflectivity {Fig. 3-458}
Depending on the color of the floor and the picture rails, more or less light is reflected toward the works to suitably illuminate them.

Gradation of luminance
{Fig. 3-459}

Gradation of luminance
{Fig. 3-460}

Gradation of luminance {Fig. 3-461}

Fondation Luma, Arles, France/ Philippe Rahm architectes 2015
The "Luminance" project for the exhibition "Systematically Open?" at Fondation Luma in Arles proposes working on the quantity of light emitted and reflected, conceiving the background behind the works of art as a mediation of luminosity and a production of spaces with different degrees of light intensity, from the darkest to the lightest. Our proposal considers the display area as a natural landscape with varied qualities of luminosity suitable to accommodate works of art respecting lighting standards. Through its shape, its dimensions and its openings allowing natural light to penetrate, the building produces a certain interior solar landscape, with some sunnier, summery parts, and other slightly darker ones, like in early spring, and finally parts that are entirely in the shade, like in winter. The naturally present differences in brightness are then reinforced by surfaces capable of absorbing or reflecting the incident light thanks to variations between black, dark gray, light gray and white. Works of art take their places naturally in these luminous variations: videos are projected in darkness, old photographs are protected from excessive light and are exhibited in rather dim lighting, and contemporary photographic prints are displayed in the brightest areas. Rather than a scenography, we produce atmospheres that offer the same freedom of use and appropriation as a natural landscape, with its sunny meadows, shady forests and dark caves that suggest, characterize or enable certain actions and desires while respecting the standards for the conservation of works of art in terms of light intensity, according to the material the work is made of.

III.VII. Gradation of Interiority

Taichung, Taiwan, 2011-2020

View of the Maintenance Center of Taichung Central Park {Fig. 3-462}

The Maintenance Center of Taichung Central Park beneath photovoltaic panels {Fig. 3-463}

In the Maintenance center, beneath the photovoltaic panels {Fig. 3-464}

In the Maintenance center, beneath the photovoltaic panels {Fig. 3-465}

View of the Maintenance Center from the south {Fig. 3-466}

View of the Maintenance Center beneath the canopy of photovoltaic panels {Fig. 3-467}

Inside the Thermal Insulation of the Maintenance Center of Taichung Central Park, Taiwan {Fig. 3-468}

Under the rain protection of the Maintenance Center of Taichung Central Park, Taiwan {Fig. 3-469}

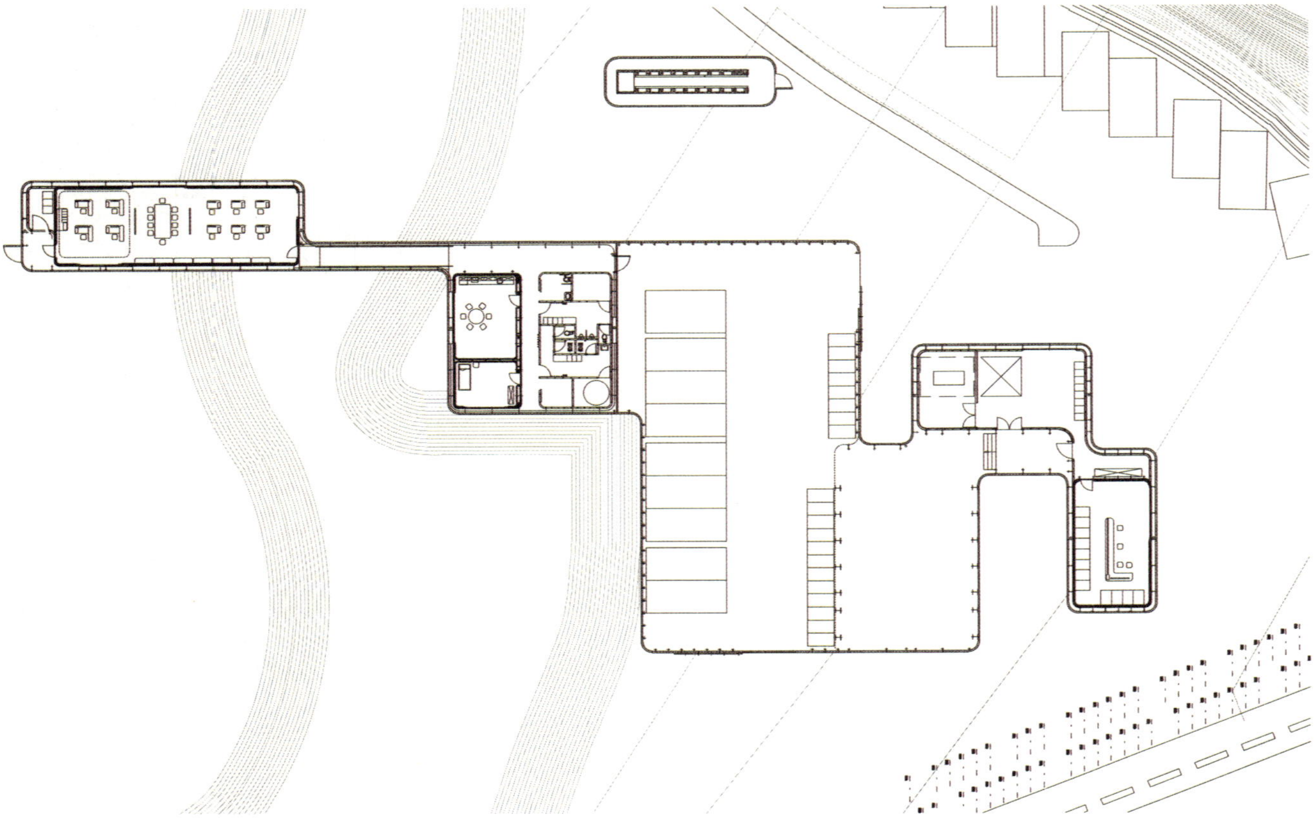

Plan of the Maintenance Center of Taichung Central Park {Fig. 3-470}

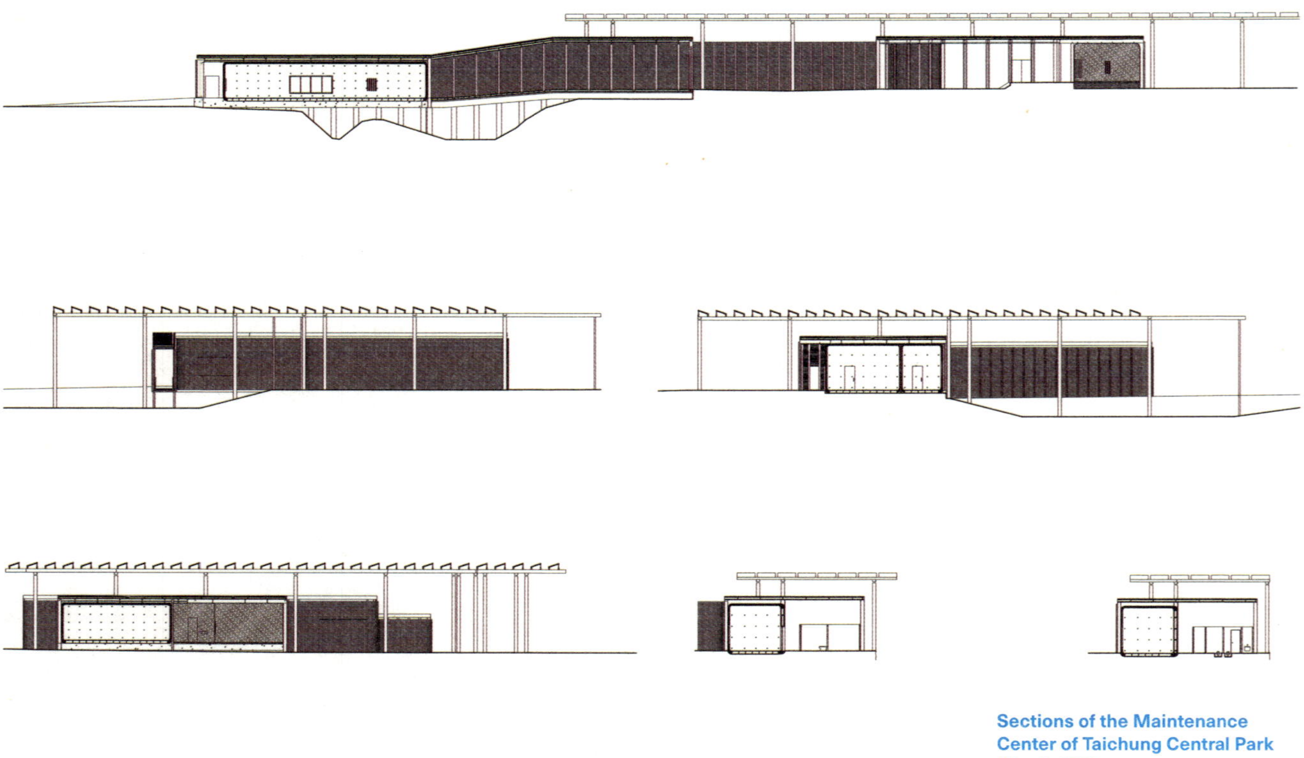

Sections of the Maintenance Center of Taichung Central Park {Fig. 3-471}

Maintenance Center

Elevations

South Elevation

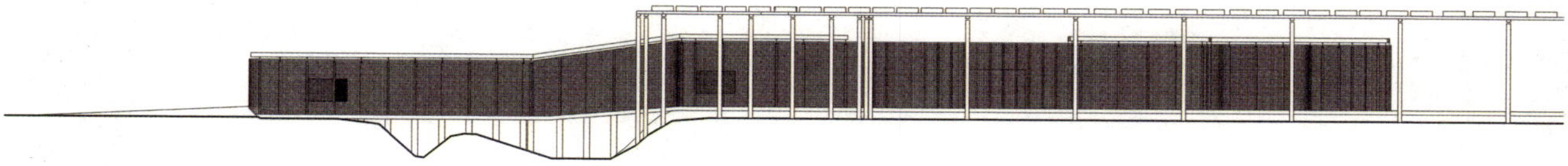

East Elevation

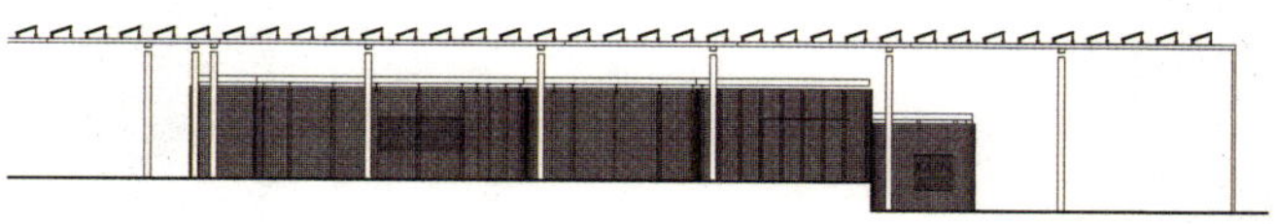

West Elevation

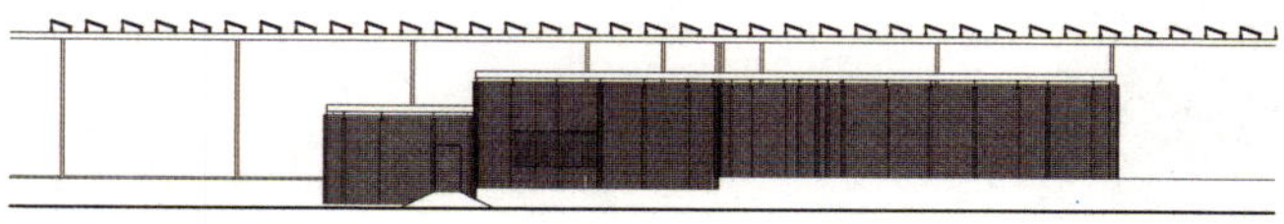

North Elevation

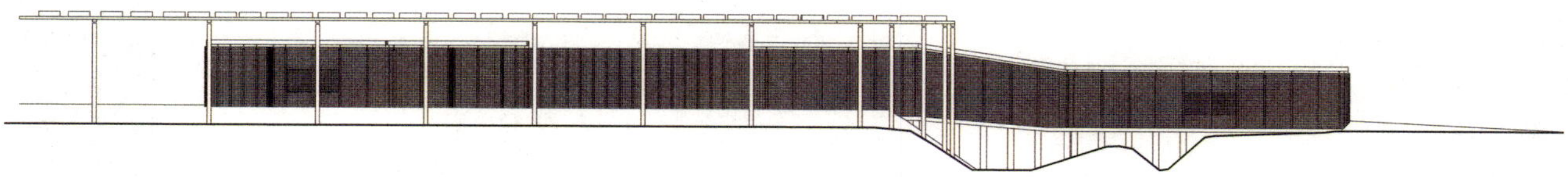

Physical Layer

200 x 200 x 200mm grid + 200 x 150mm structure

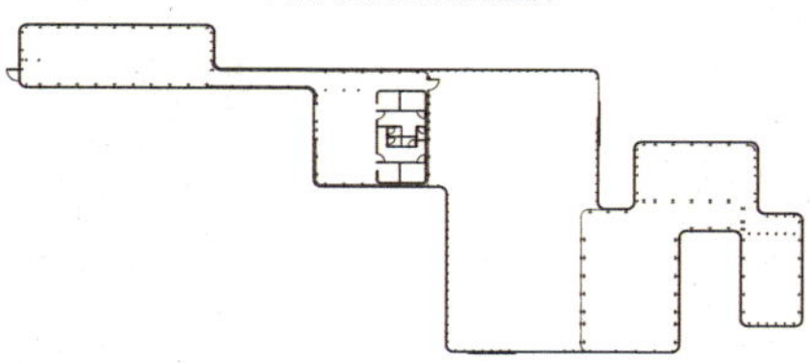

Waterproofing Layer

3mm waterproof membrane

Thermal Insulation Layer

160mm thermal insulation layer

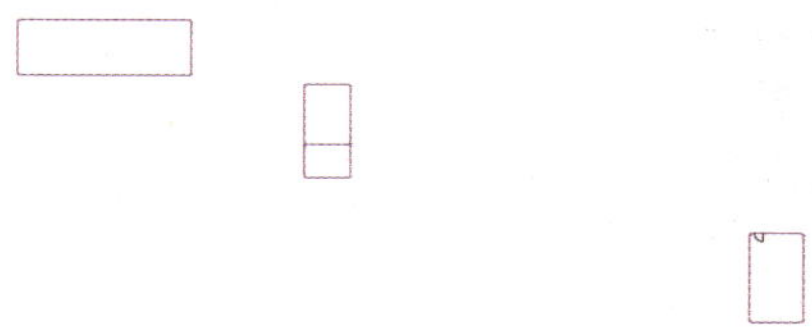

Layers of the façade of the Maintenance Center of Taichung Central Park {Fig. 3-472}

Central Park, Taichung, Taiwan / Philippe Rahm architectes, mosbach paysagistes, Ricky Liu & Associates, 2011-2020

The Maintenance Center (Gradation of interiority) consists of 3 layers (protection against intrusion, rain and heat), which shift with respect to each other to create different types of spaces. The intermediate space between the anti-intrusion layer (a 15 cm x 15 cm grille and the rain protection layer (composed of partial vertical impermeability and a horizontal roof) is used for parking the gardeners' tractors and vans. The intermediate space between the second and third layers (made of recycled PET and mineral insulation) houses storage areas for gardeners' equipment, the toilets and the showers. In the third layer, which protects against heat, are the offices as well as rest areas and computer equipment for park management. These latter spaces are the only ones that are air conditioned, as the others are naturally ventilated in order to limit energy expenditure. The whole construction is white, that is to say, with a high albedo, to preserve as low an overall temperature as possible by reflecting the sun's rays. All the electrical energy comes from renewable sources, thanks to the 7,000 m2 of photovoltaic panels covering the Maintenance Center.

III.VIII Meteorological Garden

Taichung, Taiwan, 2011–2020

The Anticyclone {Fig. 3-373}

The meteorological garden {Fig. 3-474}

The Desert Wind {Fig. 3-475} The Stratus Cloud {Fig. 3-476}

The Meteorological Garden

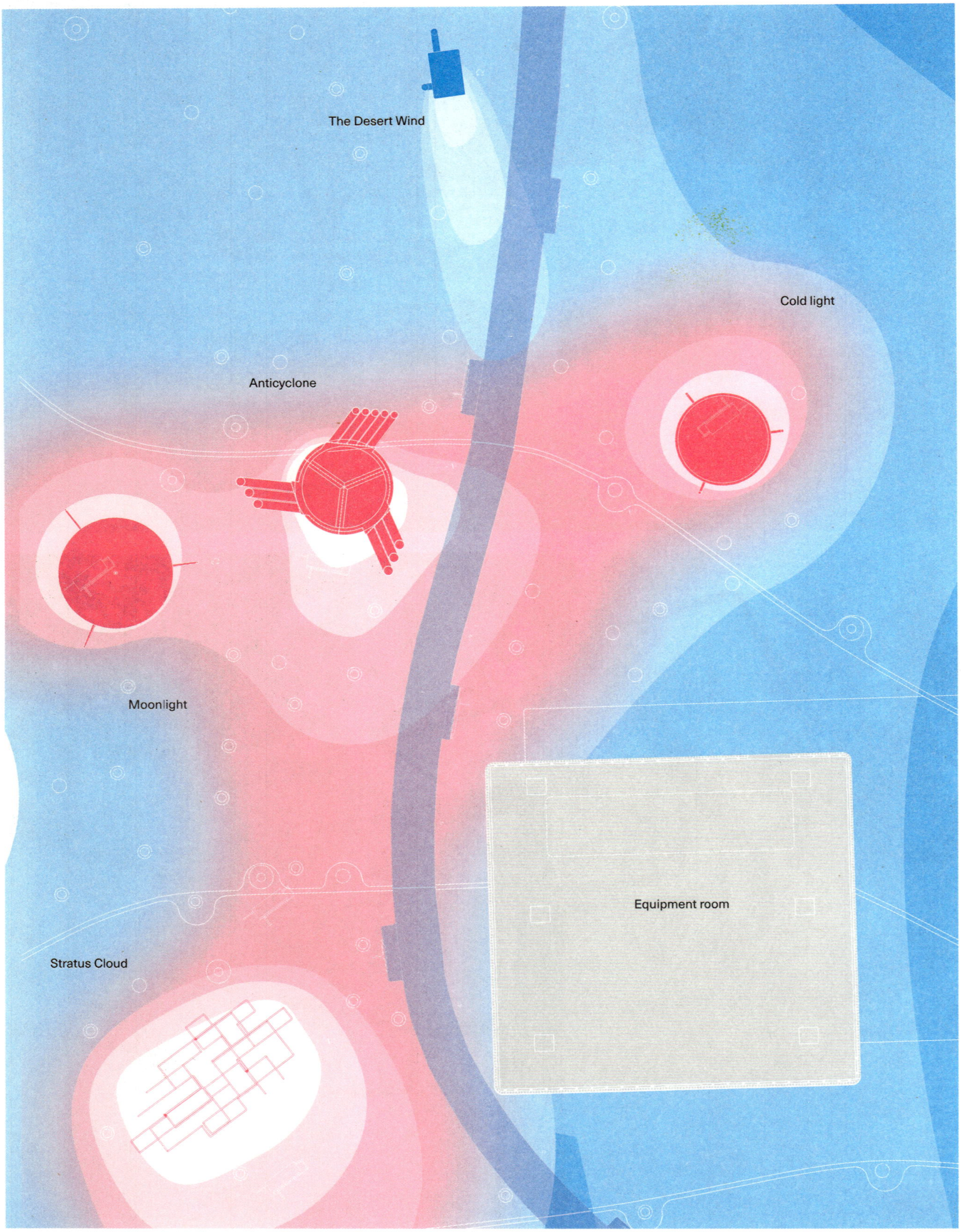

Plan of the Meteorological Garden
{Fig. 3-477}

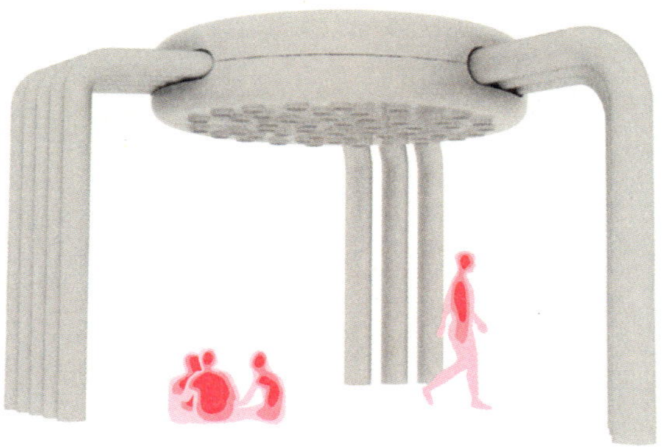

heat transfer from the hot skin to the cooler air through direct contact

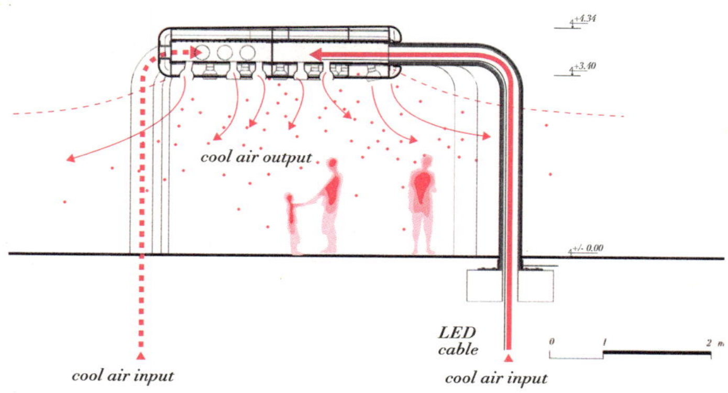

breeze facilitates convective heat loss from the skin

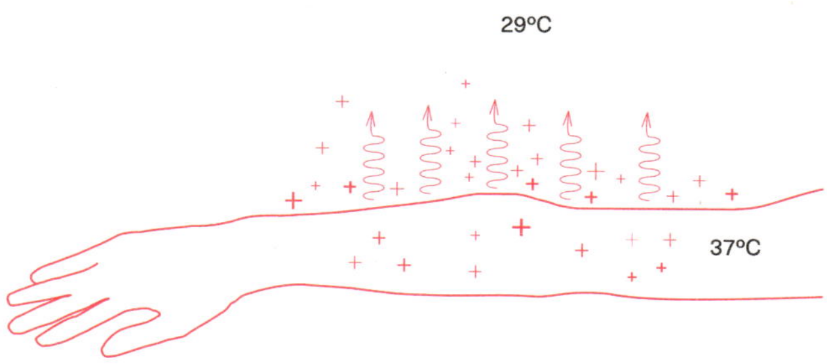

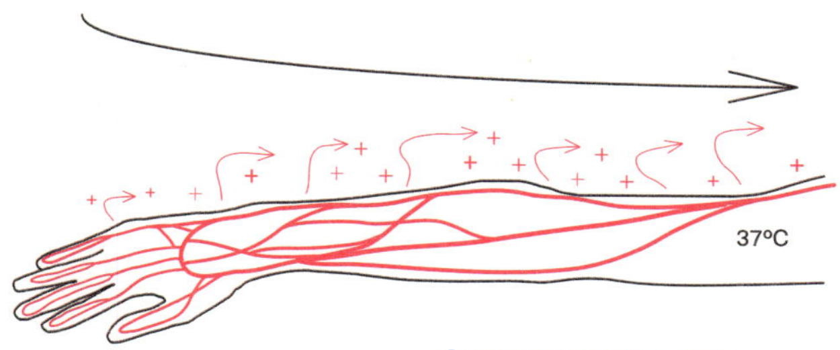

Convective cooling of the Anticyclone {Fig. 3-478}

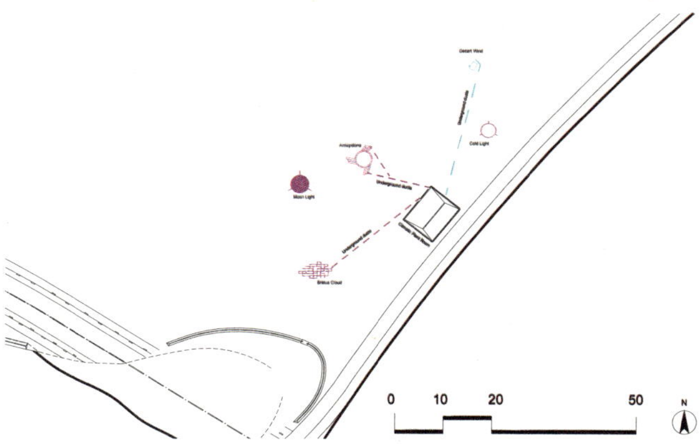

Underground ducts network

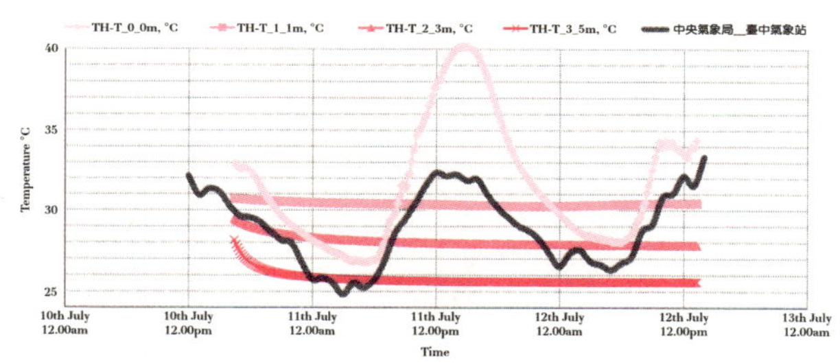

Soil temperature test on site

10th - 12th July 2013

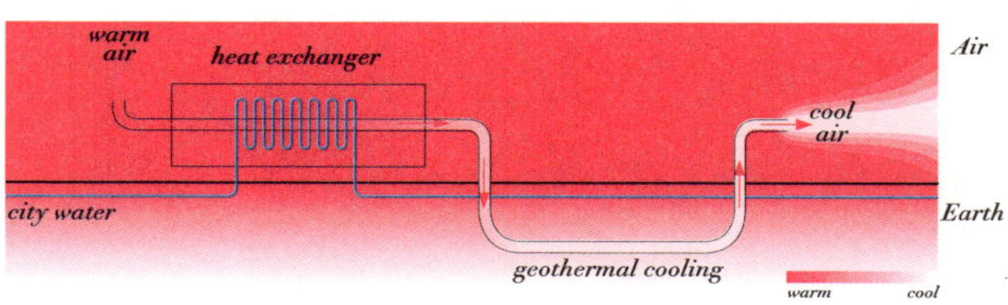

Geothermal cooling system

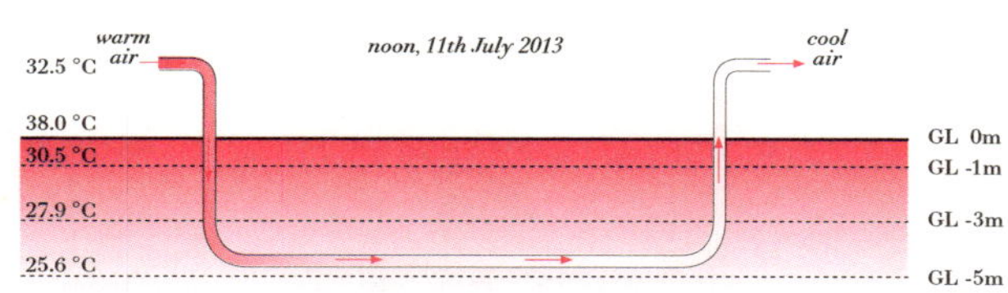

Geothermal cooling {Fig. 3-479}

The Meteors

Anticyclone

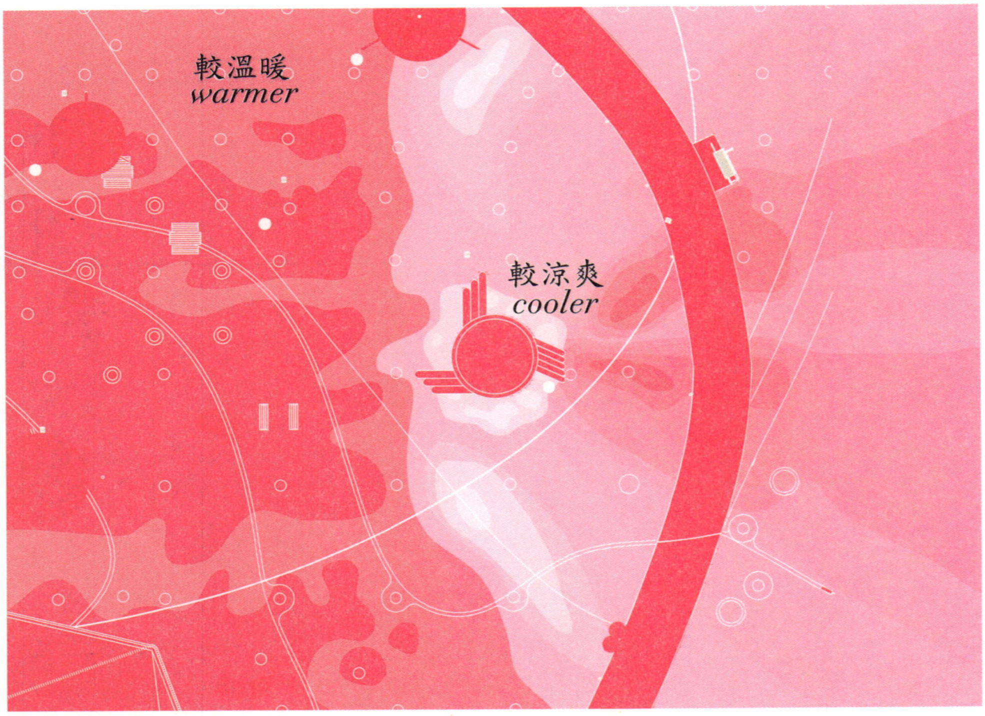

Plan

The Meteors

Anticyclone

Section

The Anticyclone {Fig. 3-480}

The Meteors

Anticyclone

Geothermal cooling

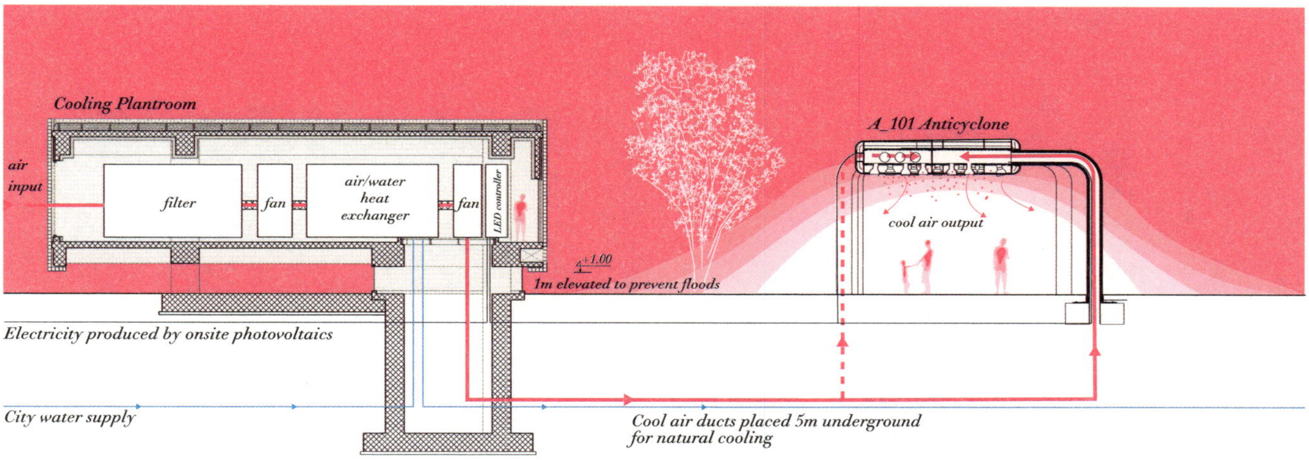

The Meteors

Anticyclone

Detail Design

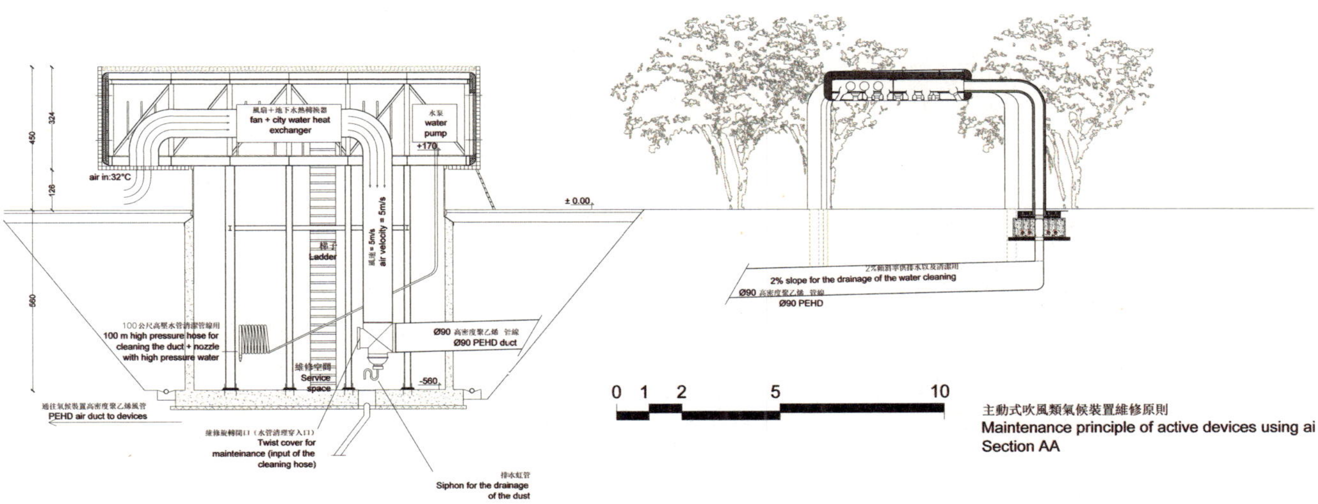

Section of the Anticyclone [Fig. 3-481}

The Meteors

Anticyclone

CFD analyses

15th June 1pm
Air Temperature: 35.4 °C
Relative Humidity: 50 %
Humidity: 18.3 g/kg
Air Velocity: 0.5 m/s

21st July 4pm
Air Temperature: 28.9°C
Humidity: 20.7 g/kg
Air Velocity: 1.5 m/s

Air Temperature

Air Velocity

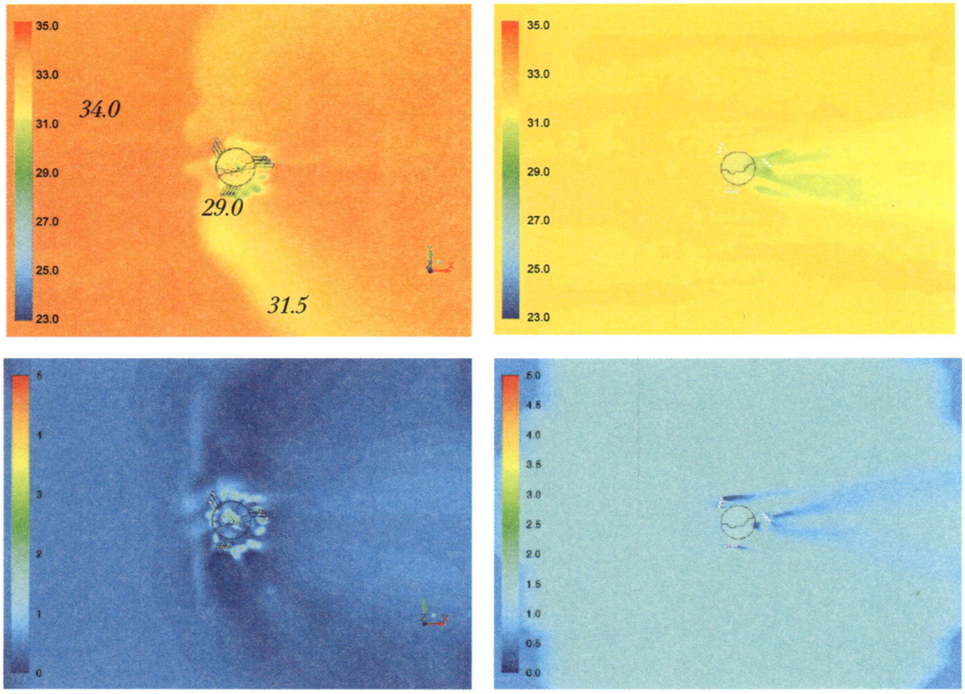

at 1pm 15th June while the ourdoor temperature is at 35°C
people can enjoy cooled air with temperature of 29°C underneath Anticyclone

Air volume 9000m3/h

The Meteors

Anticyclone

CFD analyses

Computational fluid dynamics (CFD) simulations {Fig. 3-482}
Simulations: Transsolar

The Meteors

Anticyclone

CFD analyses

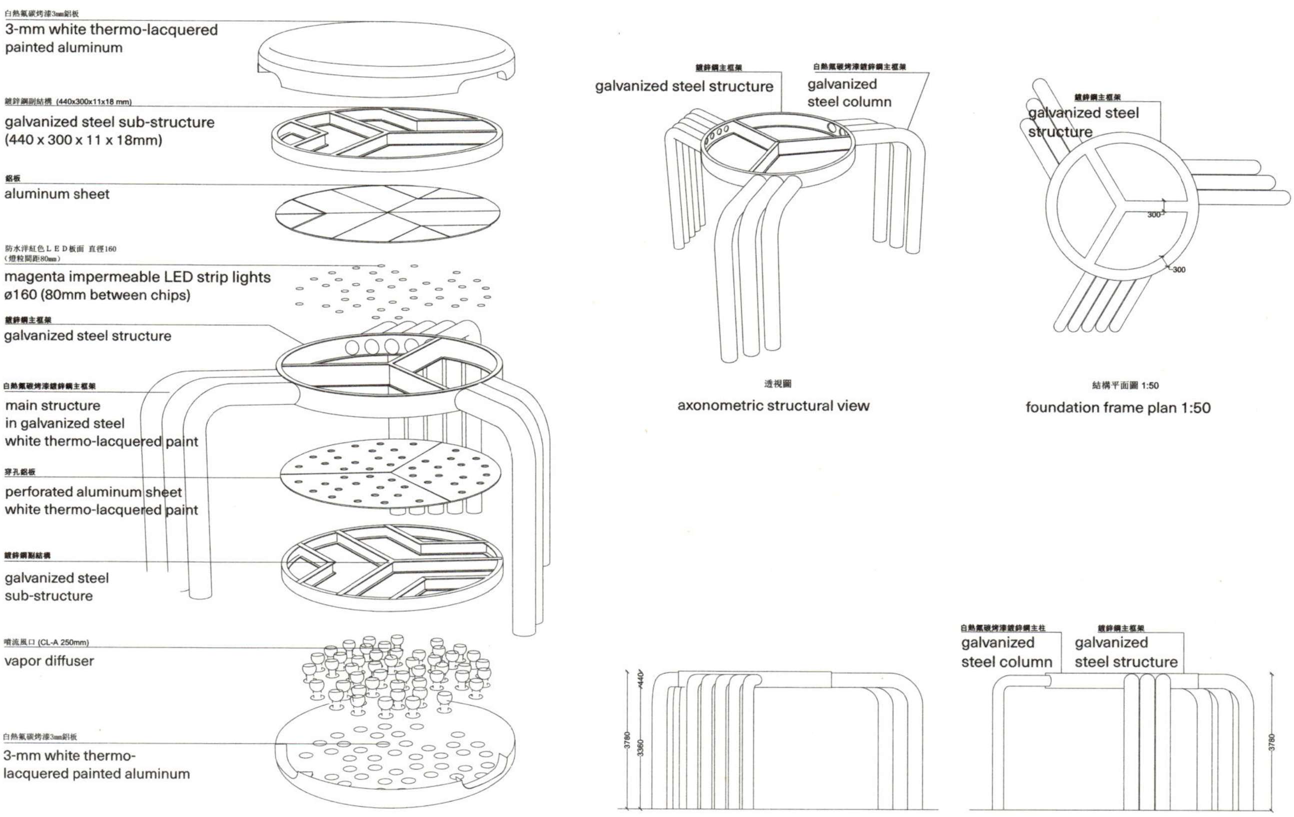

The Meteors

Anticyclone

Detail Design

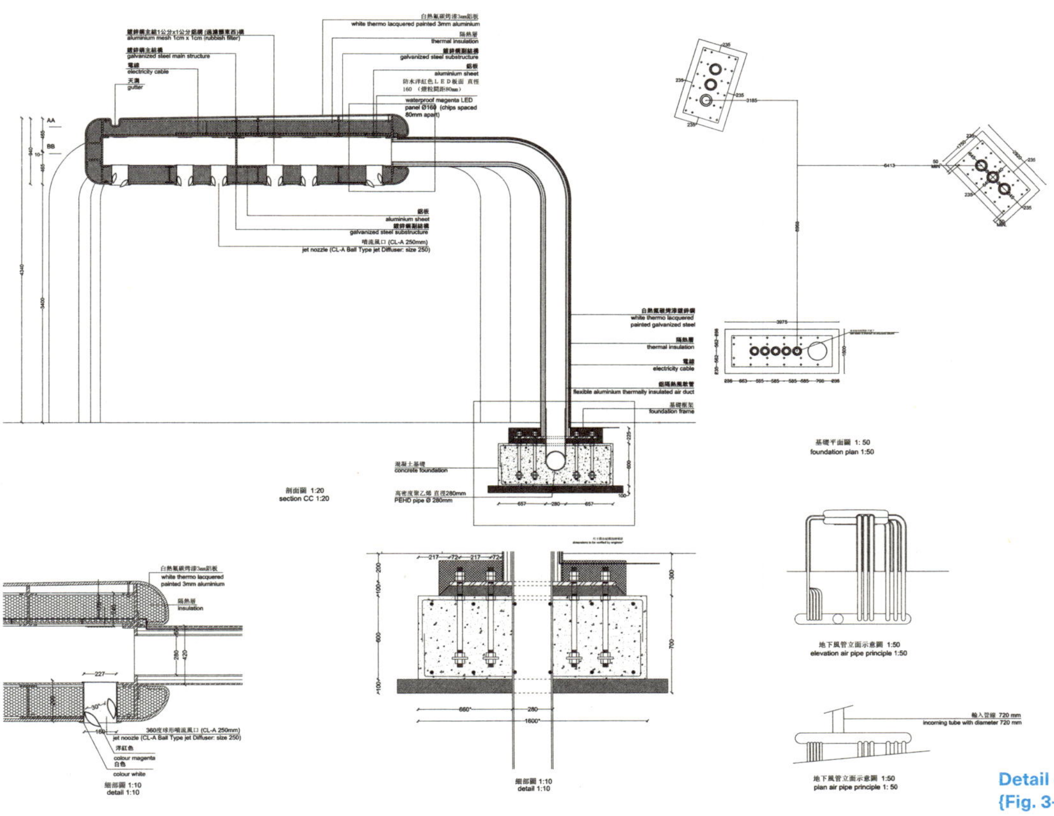

Detail of the Anticyclone {Fig. 3-483}

The Meteors

Stratus Cloud

Cooling by Evaporation

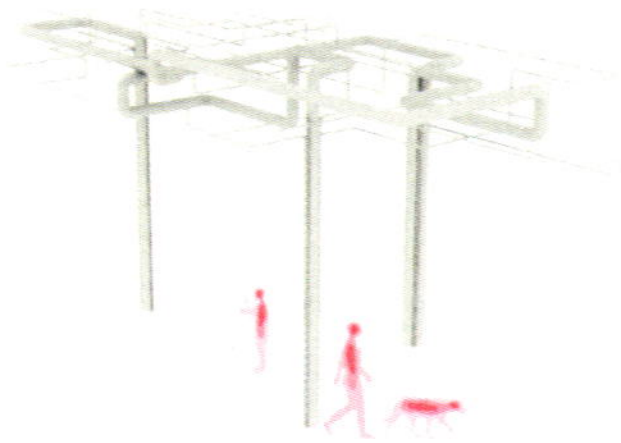

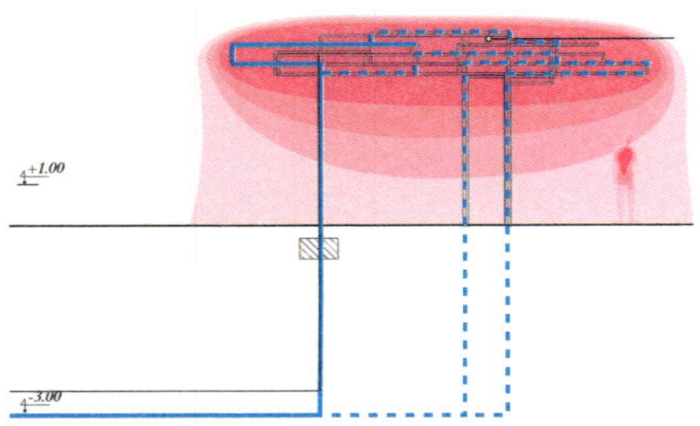

Water Evaporation Cooling Influence

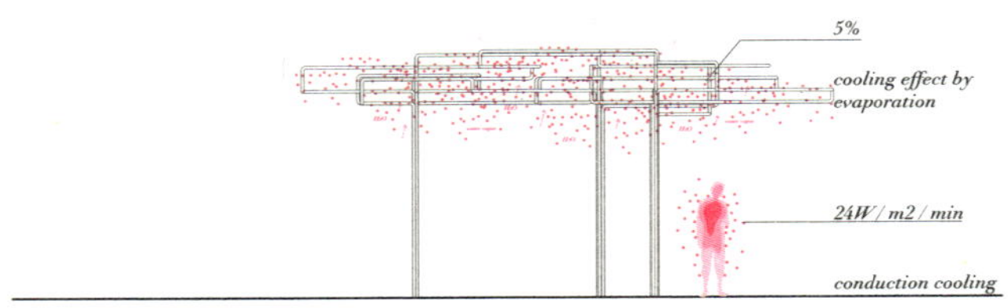

Mist production

The Stratus Cloud {Fig. 3-484}

The Meteors

Stratus Cloud

Detail Design

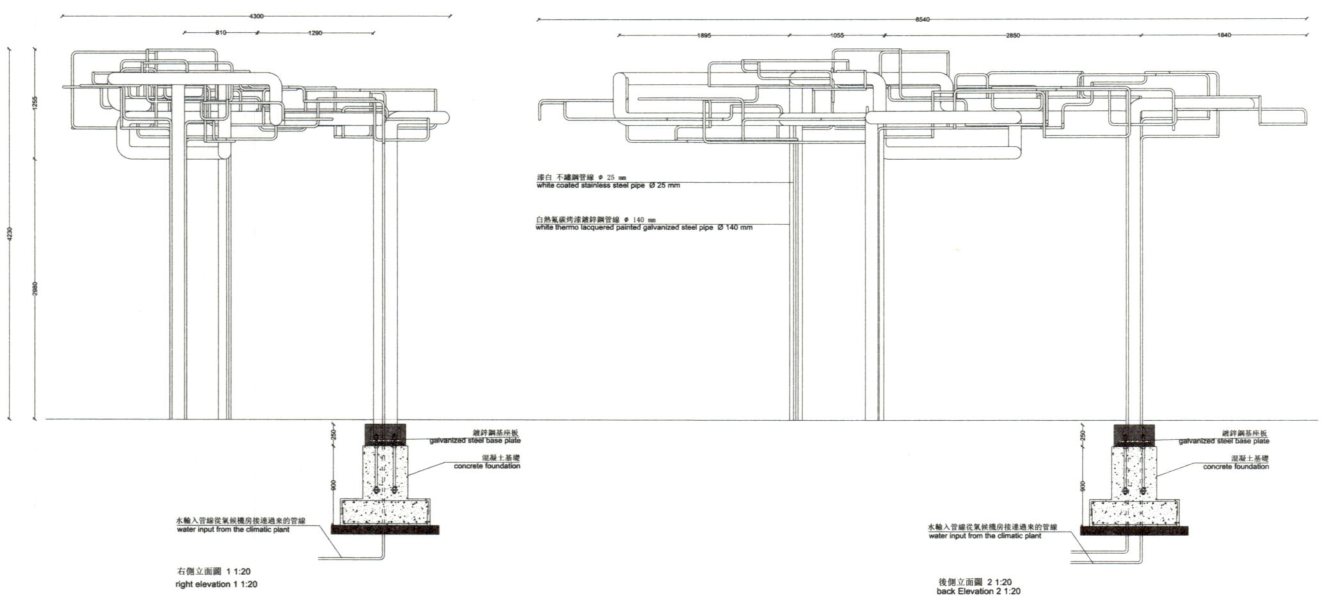

Details of the Stratus Cloud {Fig. 3-485}

The Meteors

Stratus Cloud

Detail Design

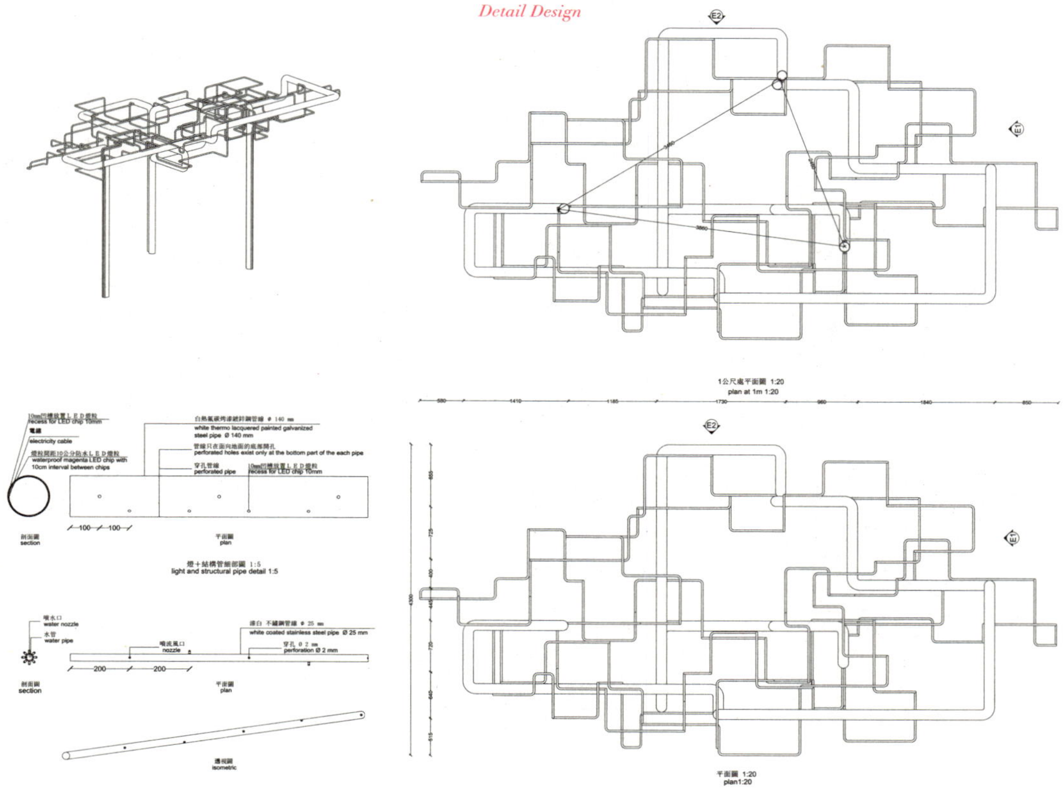

The Meteors

Stratus Cloud

Detail Design

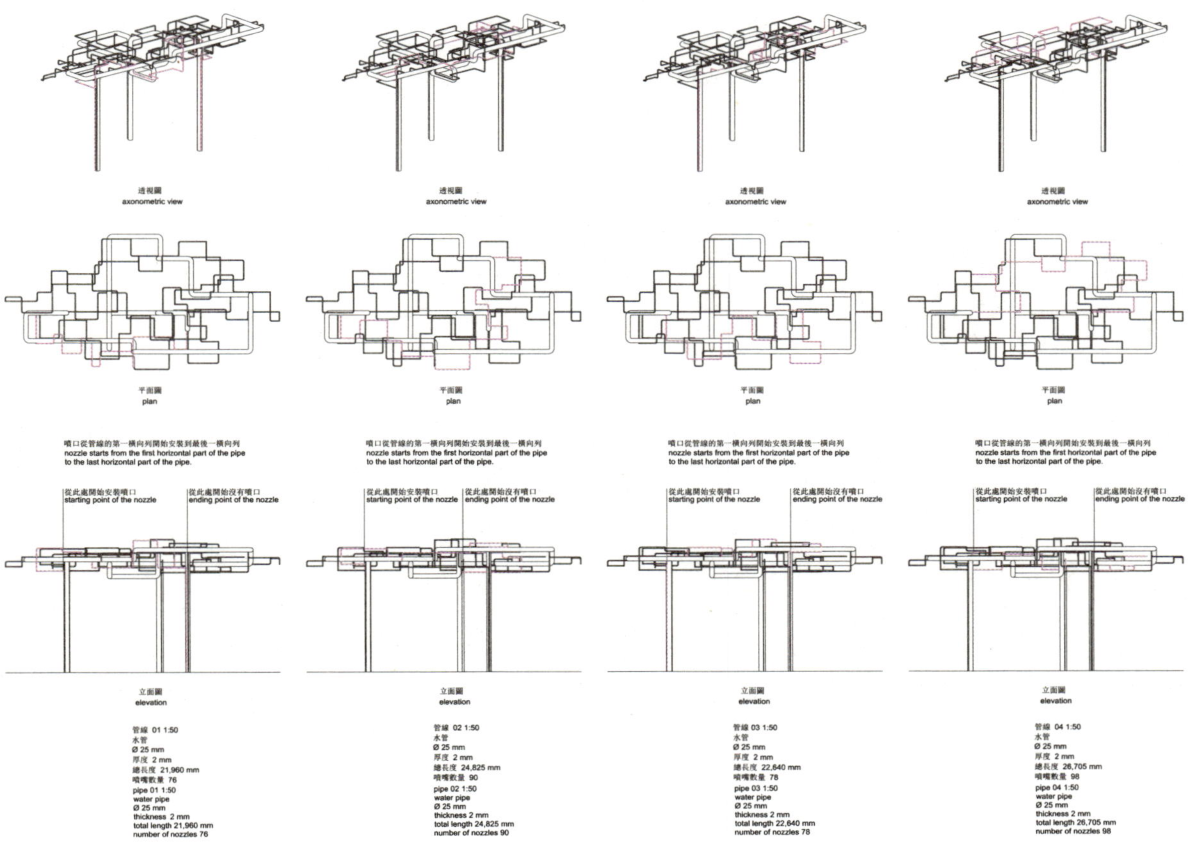

The Stratus Cloud {Fig. 3-486}

The Stratus Cloud and the Meteorological Garden {Fig. 3-487}

The Stratus Cloud and the Meteorological Garden {Fig. 3-488}

The Anticyclone and the Moonlight {Fig. 3-489}

Central Park, Taichung, Taiwan / Philippe Rahm architectes, mosbach paysagistes, Ricky Liu & Associates, 2011-2020
Located in the north-east of Taichung Central Park on Taiwan Island, the Meteorological Garden is an oasis of freshness in a warm, humid, subtropical climate. Spread over half a hectare, the garden is dotted with trees and Meteors, half-pavilion half-machine tree-sized artefacts.

During the day, the stomata of the trees, the small orifices beneath their leaves, open to release water, which, by evaporation, creates a pressure difference between the top and the bottom of the plant, activating the movement of sap in the xylem and allowing the water and mineral salts to be drawn by the roots for distribution throughout the plant. The water that escapes from the stomata evaporates when it comes into contact with the air. During this change of physical state, from liquid to gaseous state, energy present in the air is consumed, thus lowering the temperature of the air in the vicinity.
The trees' evapotranspiration thus refreshes the air of the Meteorological Garden during the day. With growth, the tree canopy rises and spreads. Composed of thousands of leaves, the canopies form thousands of small screens that gradually block the direct penetration of the sun's rays onto the ground and toward the people below, protecting them with their shade from this radiant energy source and its heat. The trees thus cool the air and humans via two types of energy transfer, evaporation (transpiration) and radiation (creating shade).

The Meteors, meanwhile, also cool the air, ground and humans via two thermodynamic modes of heat transfer. The first Meteor we encounter in the south is the Stratus Cloud. The water circulating in the pipes is vaporized in the air by nebulization into fine droplets that form a cloud. By evaporating almost instantaneously, they lower the temperature of the air in the vicinity. In addition, the droplets of water floating in the air under the effect of the wind form a cloud that protects the ground and humans from the rays of the sun and its heat.

The second Meteor is the Cold Light which cools by limiting heat transfer by radiation between the skin and the sun. It consists of a glass disc placed at a height that blocks all the sunlight except for violet light, which corresponds to the wavelength of the visible spectrum least charged with energy and therefore heats the least. The sun's rays appear white to us due to the additive synthesis of all the colored waves from violet to red. By precisely analyzing the visible light spectrum, that is to say the wavelengths between 380 and 800 nm, we can see that the part least charged with energy is violet, with a wavelength of around 400 nm and an intensity of about 500 W/m2 µm, while this intensity rises to 1,200 W/m2 µm for green rays and is slightly lower for red ones, at 1,000 W/m2 µm. Letting violet pass through and blocking the other colors of the visible spectrum therefore means choosing to expose the humans below to the light least charged with solar energy, the coldest part of visible sunlight.

The Meteor called Moonlight cools, limiting the heat transmitted by the sun's rays. It is a vault, black below and white above, a flat dome, its thickness containing 20 cm of thermal insulation whose purpose is to prevent heat from being transmitted by conduction between the top and the bottom of the vault. To stop the dome from overheating, it is painted white to approach an albedo of 1, reflecting the sun's incident rays and heat into space. Above, the vault is as black as the night sky, creating shade without any solar radiation gain except for a small crescent moon slit that occasionally allows a ray of sun through;

The last two Meteors, the Anticyclone and the Desert Wind, cool by convection and by conduction. The principle of convective cooling is what takes place when we blow on a spoonful of soup to cool it down. As it passes over the surface of the liquid, the wind carries away the heat and cools it.
The same principle is put in place with the Anticyclone and the Desert Wind, which blow air onto the humans in front of or below them, carrying away the heat from the skin and refreshing it. The wind also removes perspiration, increasing the air evaporation cooling rate near the skin. The speed of the blown air is 8m/s (29km/h), which is a force 5 wind, described as a good breeze. In addition, the blown air is cooled beforehand to 25°C by geothermal energy circulating at a depth of 5 m, where the temperature of the ground remains at a constant 25°C. Thus, by conduction, the blown air will cool the skin on contact. The energy consumed by the Meteors is entirely renewable.

The electricity supplied to the fans of the Anticyclone and Desert Wind is produced entirely by the photovoltaic panels located in the north of the park. The Meteors get to work when the sun provides energy. The air is cooled by geothermal energy. There is no consumption of fossil or nuclear energy to blow or cool the air.

Together, the trees and Meteors create a variety of sensations and pleasures, offering a host of ways to refresh space and body, naturally and artificially.

III.IX Structural Densification

Taichung, Taiwan, 2015-2020

Structural Densification

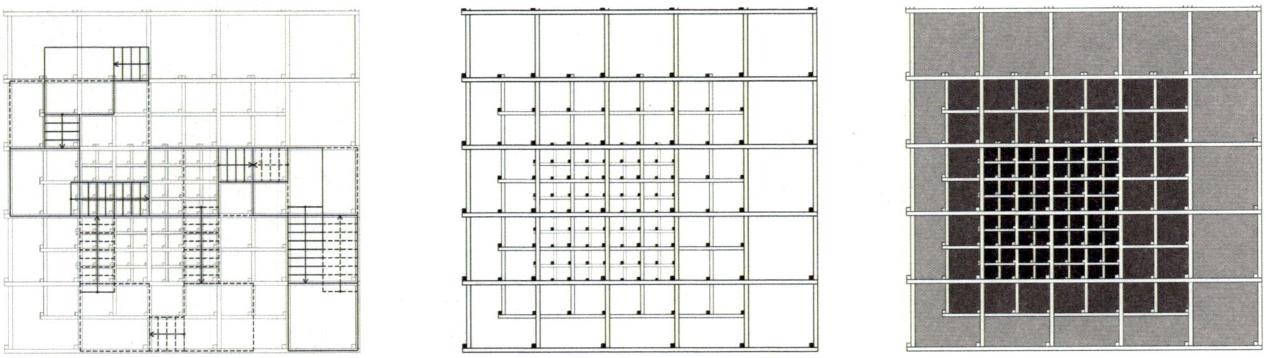

Plans {Fig. 3-490}

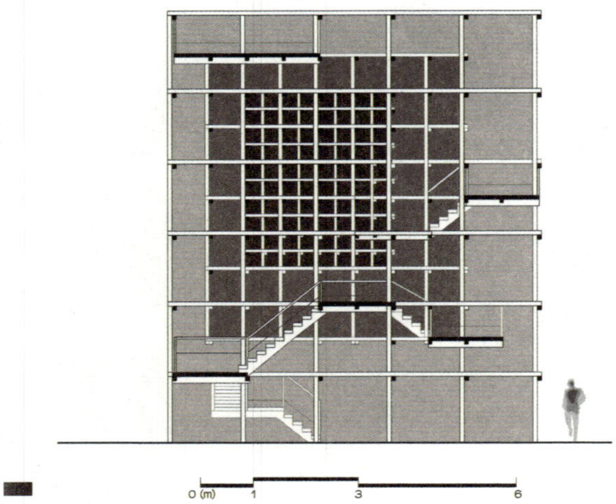

Section {Fig. 3-491}

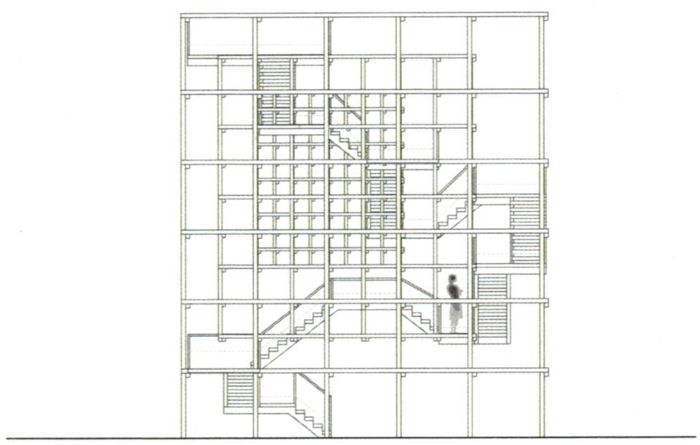

Elevation {Fig. 3-492}

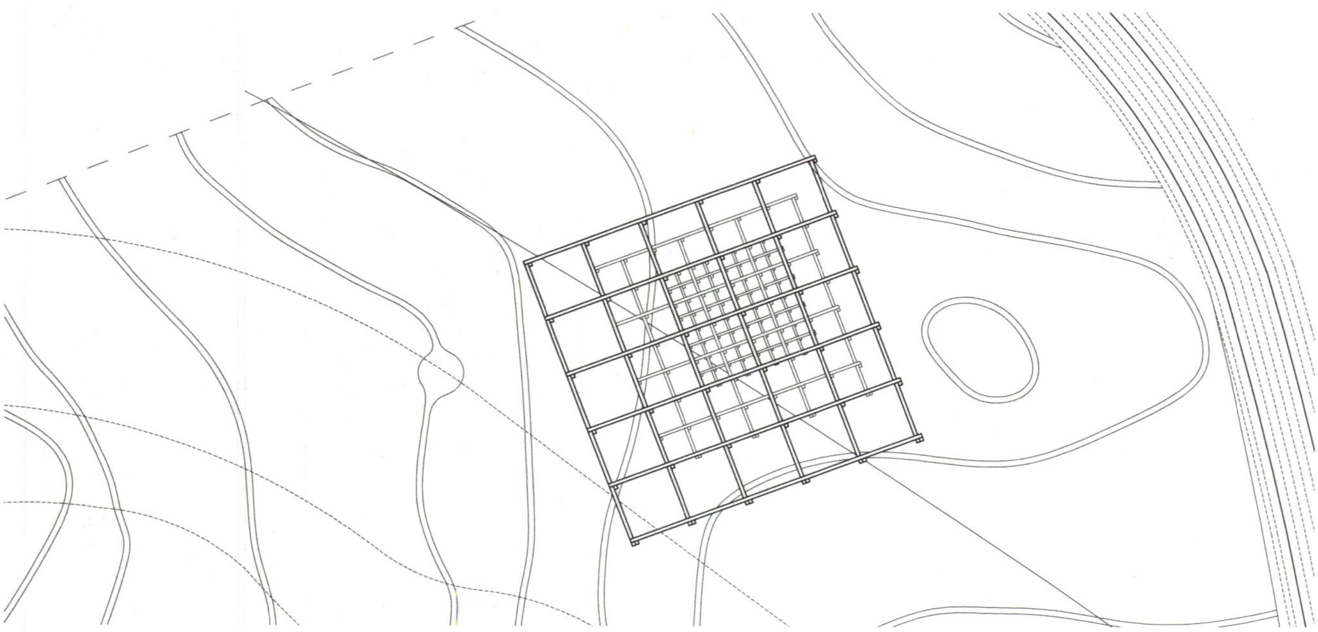

Plot plan {Fig. 3-493}

The Densification pavilion of Taichung Central Park {Fig. 3-494}

Central Park, Taichung, Taiwan / Philippe Rahm architectes, mosbach paysagistes, Ricky Liu & Associates, 2011-2020

Densification is a superimposition of 3 structural spatial grids of 3 different dimensions (2 meters, 1 meter and 0.5 meters) nestled in each other, creating a gradation from the outside to the inside, from high density to the low density of wood, from the heart to the edge of the structure, from the top to the bottom, from darkness to light, from the hidden to the most overt. It offers a variety of sensations of movement, with numerous breaks on platforms at different heights. At the top, visitors can enjoy a panoramic view over the park.

III.X. The Convective Cafeteria

Taichung, Taiwan, 2015-2020

The Convective Cafeteria

Plans {Fig. 3-495}

Section and elevation {Fig. 3-496}

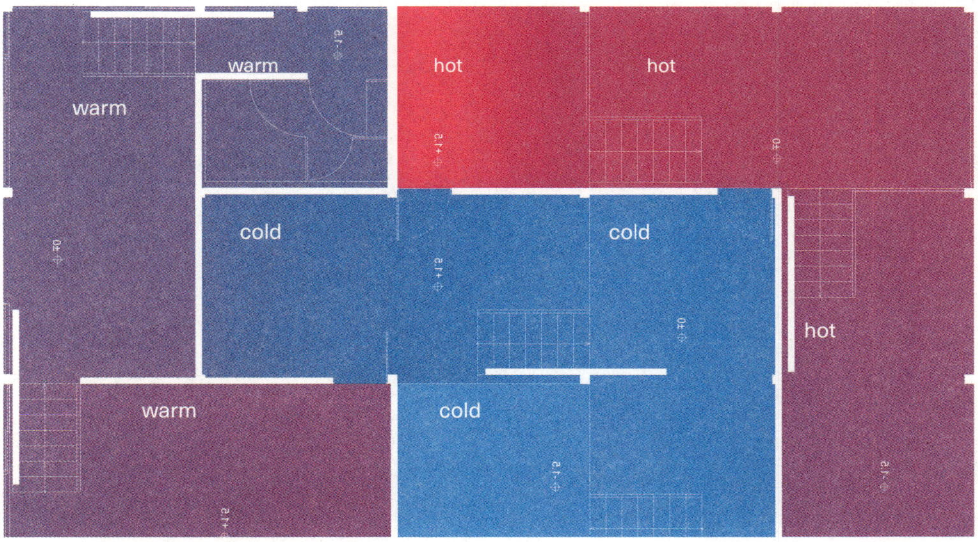

Plan of thermal depth {Fig. 3-497}

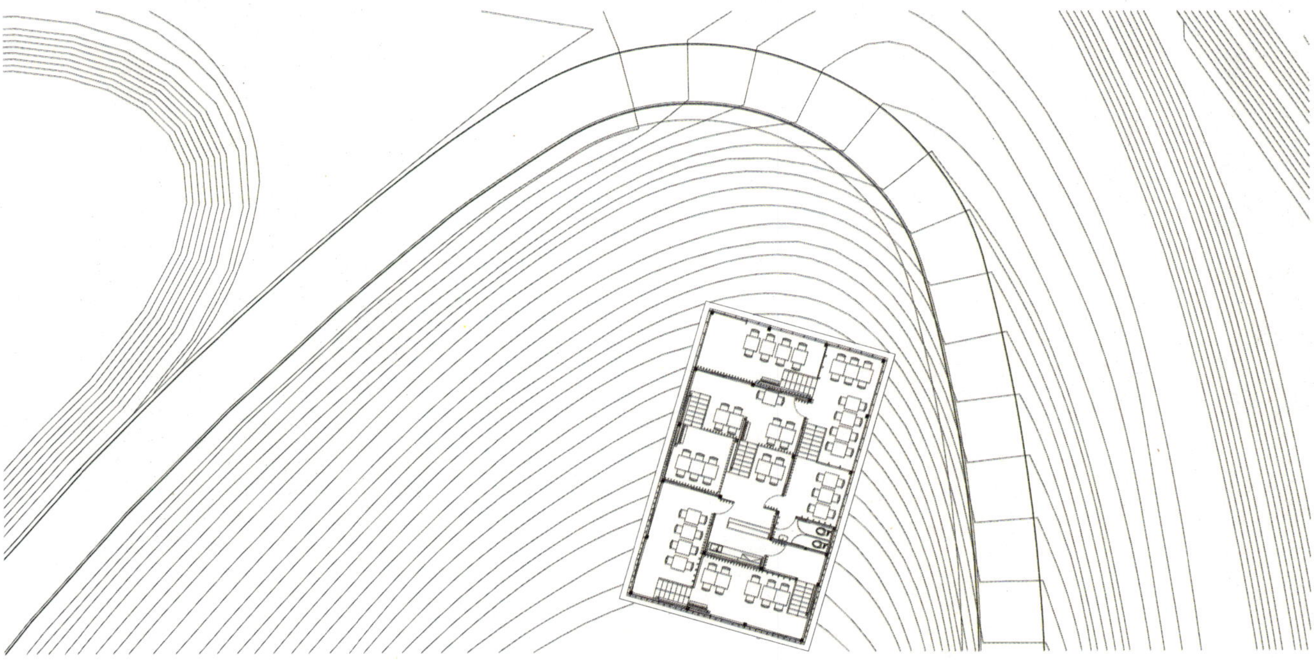

First floor plan of the cafeteria {Fig. 3-498}

Central Park, Taichung, Taiwan / Philippe Rahm architectes, mosbach paysagistes, Ricky Liu & Associates, 2011-2020

The Convective Cafeteria is composed of a succession of nine rooms organized in 3 series of 3 rooms according to a general gradation of light (from most open to most closed), and an internal gradation from warmest to coldest by convection according to 3 floor heights. Thus, the first series is completely open to the outside. The second is closed by windows and is naturally ventilated. The third is even more closed and cooled by air-conditioning. The different floor heights in each series provide three different air temperature experiences, from the coolest at the bottom to the warmest at the top, as the cold air descends and the warm air rises, allowing visitors to freely choose where to rest by electing one of these 9 climatic rooms, depending on time of day, season, light and temperature.

III.XI. Gradation of Privacy

Taichung, Taiwan, 2015-2020

Central Park toilets, Taichung
{Fig. 3-499}

Central Park toilets, Taichung
{Fig. 3-500}

Central Park toilets, Taichung
{Fig. 3-501}

Central Park toilets, Taichung
{Fig. 3-502}

Gradation of Privacy

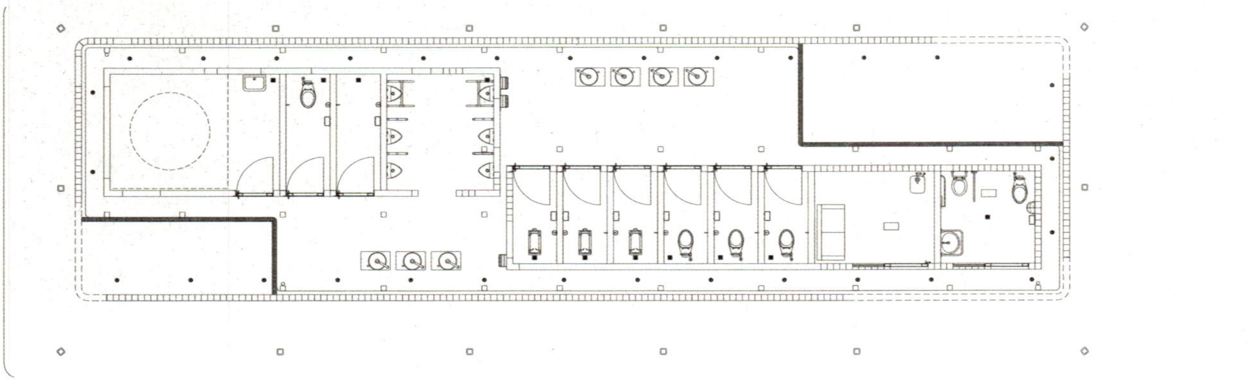

Plan {Fig. 3-503}

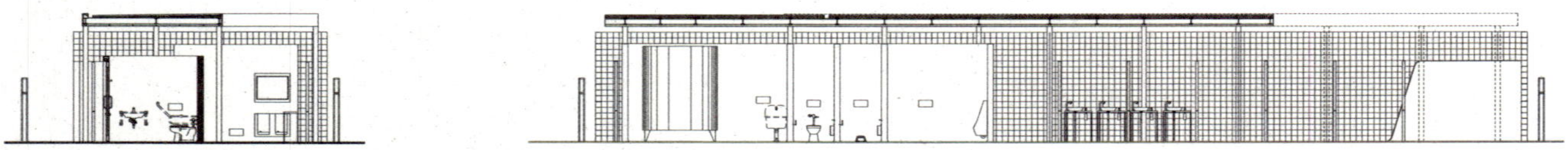

Section {Fig. 3-504}

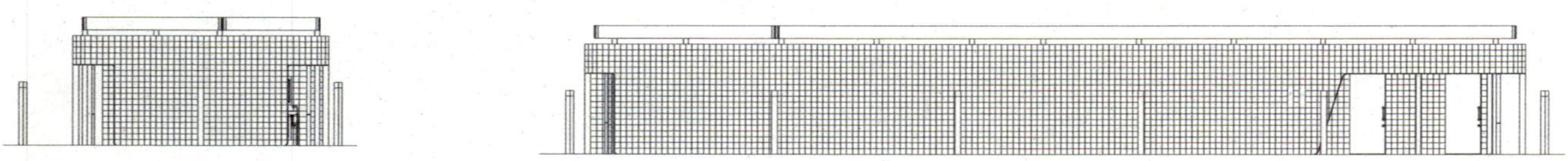

Elevation {Fig. 3-505}

Floor

Floor + Sound

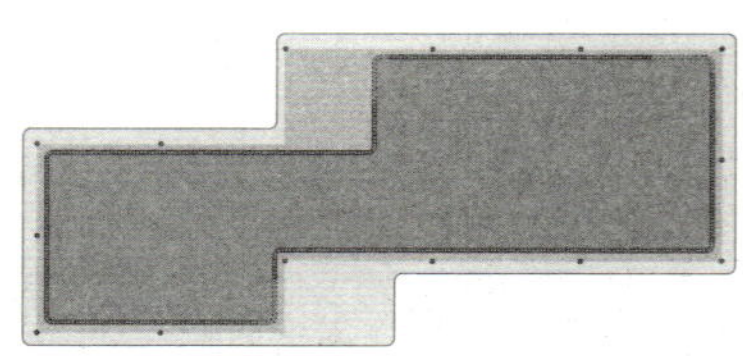

Floor + Sound + Wall

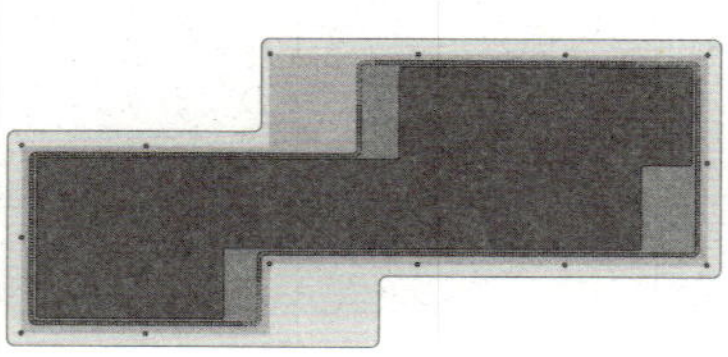

Floor + Sound + Wall + Roof

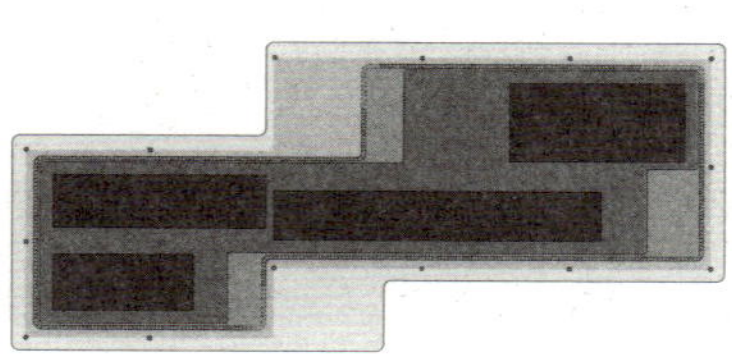

Floor + Sound + Wall + Roof + Wall

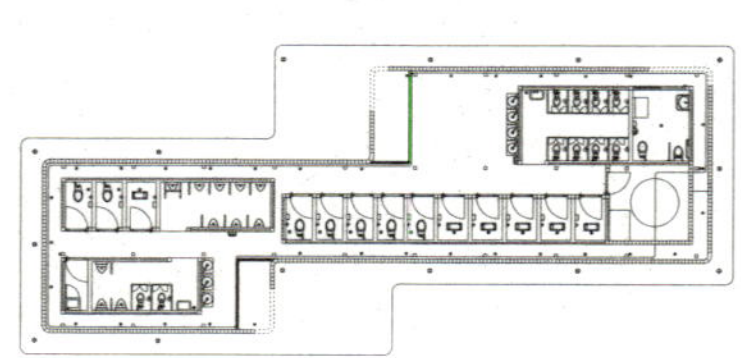

Floor + Sound

Central Park toilets, Taichung
{Fig. 3-506}

Gradation of privacy
Central Park, Taichung, Taiwan / Philippe Rahm architectes, mosbach paysagistes, Ricky Liu & Associates, 2011-2020

The concept of these toilets is based on a dissociation of horizontal and vertical layers of protection (mud, rain, light) and perception (noise, view), which jointly form the building, thus creating a concentric gradation of boundaries that gradually progresses from the most public, outside, to the most private, inside. First of all, a concrete surface defines a flat, practicable floor, in contrast to the natural environment of earth and grass. The next innermost layer is bounded by loudspeakers that play music inspired by the sounds of water, acoustically masking the sound of the toilets. The third layer is delimited by a vertical closure giving a little privacy to the common areas of the toilets. The fourth layer is one of artificial light in the form of vertical masts. The fifth layer is delimited by the roof which protects from the rain. Finally, the sixth and final layer that accommodates the toilets is the most private, made of a perforated partition that completely blocks the view without preventing the air from passing, without touching the roof.

MC
VB
2008

III.XII White Geology
Grand Palais, Paris, France, 2009

White Geology

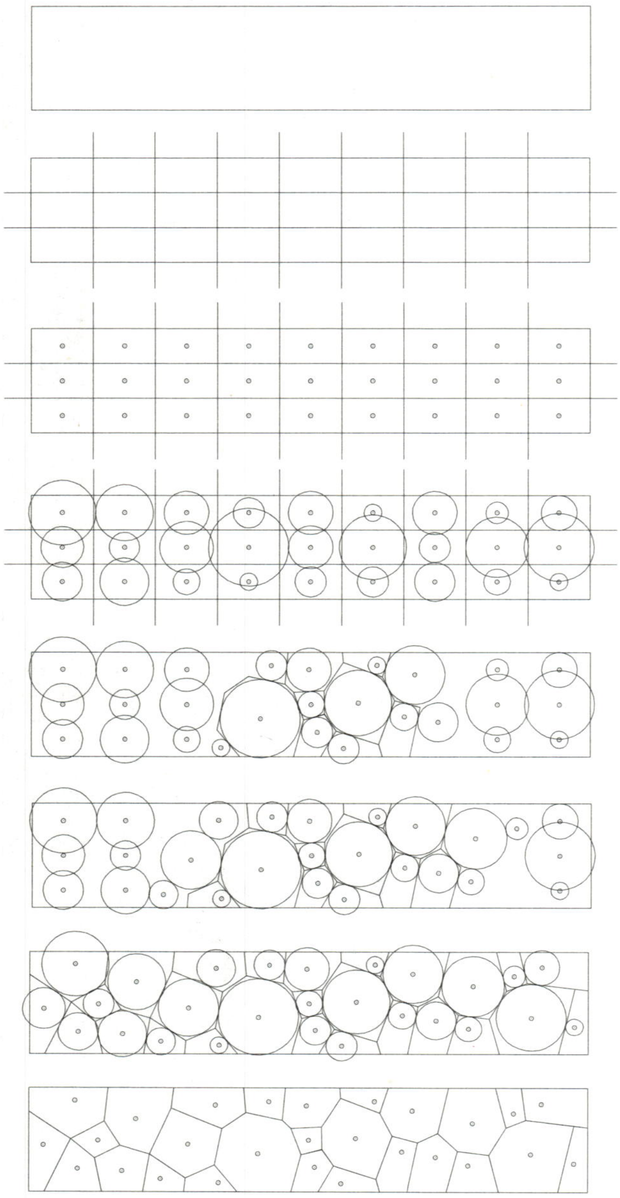

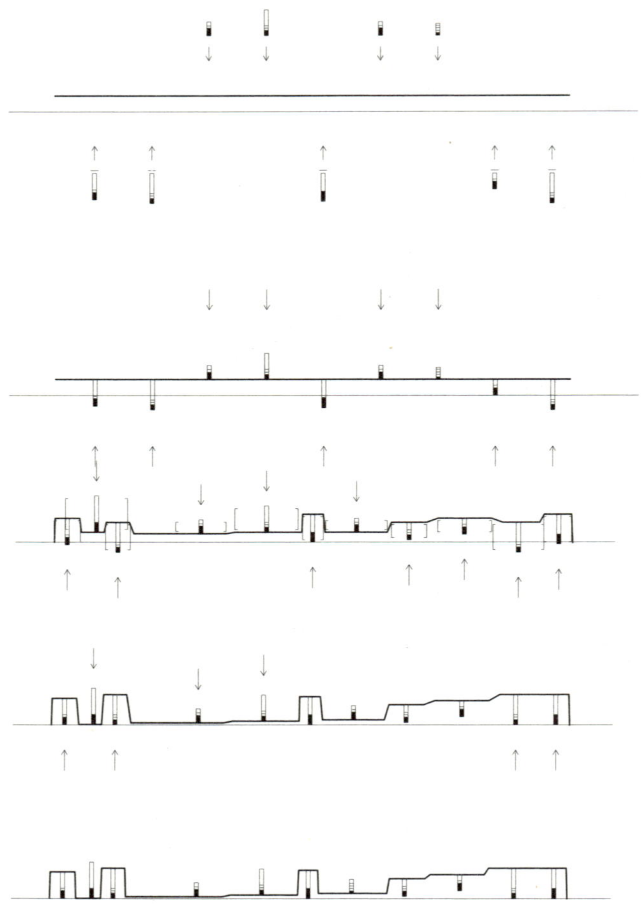

Plan formation process {Fig. 3-507}

The site is divided by the number of artists. Each cell is then sized according to the dimension of the artist's work. The clash of all the specific needs of the spaces renders the plane a tectonic geological movement.

Section formation process {Fig. 3-508}

Since the floor of the Grand Palais can only withstand 500kg/m2, heavy works of art must rest on panels to distribute their load. Thus, the final floor of heavier artworks is higher than lighter ones, thus creating an undulating terrain. Some works of art also require darkness, leading to the creation of hollow hills.

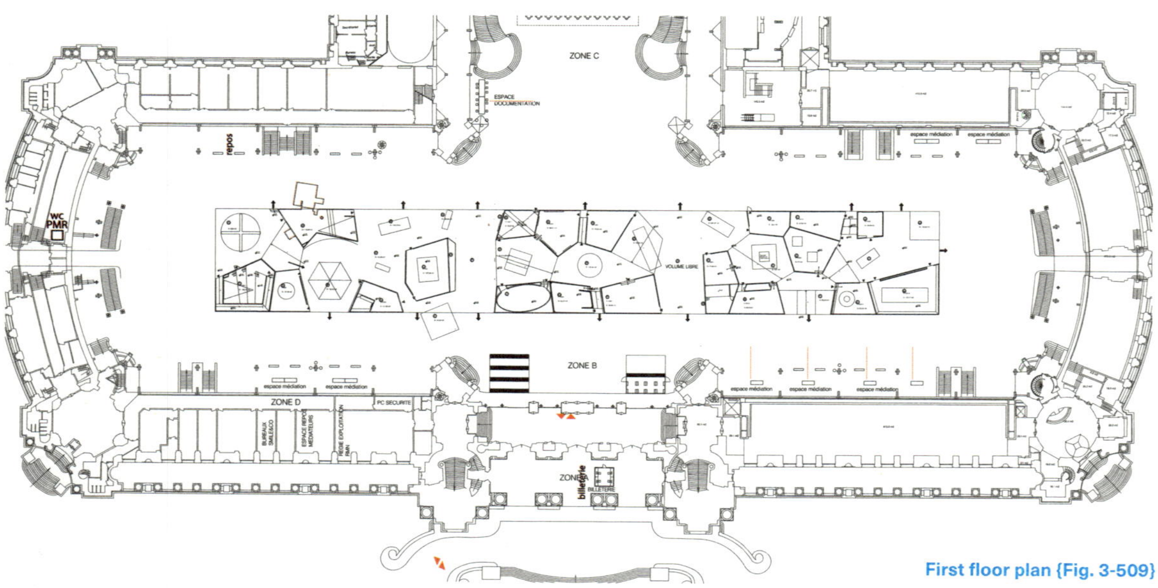

First floor plan {Fig. 3-509}

Views of White Geology in the Grand-Palais {Fig. 3-510}

La Force de l'Art 02, Grand Palais, Paris/Philippe Rahm architectes, 2009

White Geology, designed for Force de l 'Art 02, the second triennial of contemporary art at the Grand Palais, is an exploration of the reflective light of a background and the ability of a white surface to emit light and create a landscape. The material used to make this background serves only to ensure a high degree of reflection (80%) and to diffract and diffuse light to enhance the works of art. It is a white landscape, like an unfolded "white cube" that serves as a base for works of art that fit in like natural objects placed on top or in front. Their shapes and colors stand out clearly on this neutral white background, which blurs the distinction between vertical and horizontal. It is a white simplification of a geological form. We propose a geological process that is activated by the actual physical properties of the works of art. The original flat white rectangle measures 125 meters by 24 meters and buckles, hollows out and curves under the influence of geological the forces caused by the dimensions of the works and their weight, which deform the initial landscape: tectonic shifts, distortion, pressures and vacuums, folds. Here, the movements and distortions of the territory are caused by the abstract forces of the art works. At first, the same surface area and the same volume are assigned to each work. Depending on the size of the work and the distance from the viewer, they start to collide, as with tectonic plate movements. Their weight and the amount of light they require will determine how much they deform the surface, cutting into or swelling it as a result, creating valleys and hills.

III.XIII Spectral Light

Artemide, 2015

Spectral Light

Array of physiological spectra {Fig. 3-511}

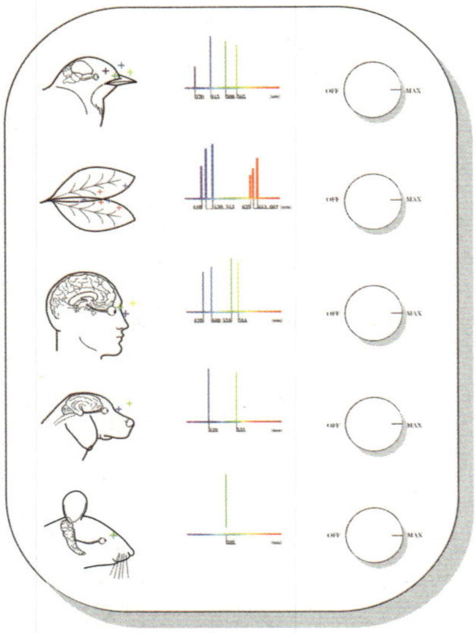

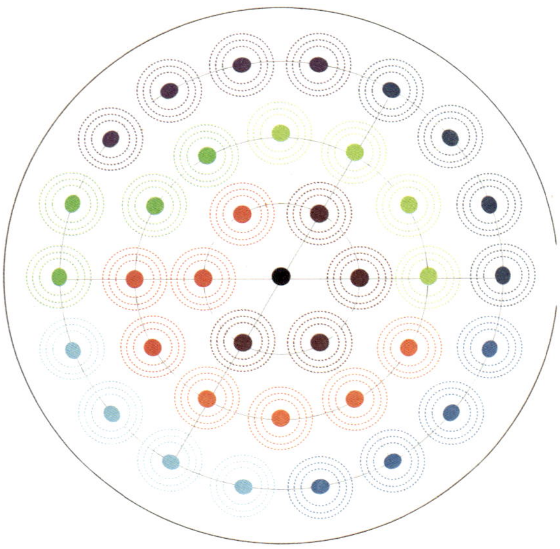

Switch {Fig. 3-512}

Artemide, Milan, Italy / Philippe Rahm architectes, 2015

Spectral Light proposes the creation of the most economical and healthiest indoor artificial lighting possible, by adding only the necessary wavelengths precisely chosen for economic and health reasons according to human, animal and plant physiology. We proceed according to a technique similar to that of Georges Seurat's pointillism by adding only the precise and distinct wavelengths actually perceived by the photoreceptors of the eye of each living species, which enables not spending energy and money on wavelengths invisible to these species. The type of wavelengths emitted is chosen by the user according to the type of occupants of the house (humans, cats, dogs, hamsters, birds, plants). For example for humans, it will involve assembling the specific wavelengths that are biologically active on humans during the day, namely 423 nm, 533 nm, 564 nm according to the 3 photoreceptors (the cones) around the blue, green and yellow, plus the 460 nm wavelength which blocks melatonin. To these specifically human wavelengths will be added the ones perceived by the other animal or plant species inhabiting the house.

For example, when you leave the house, you can switch off the 4 wavelengths specific to humans while leaving the 2 wavelengths for the dog staying at home switched on. Through addition and convergence, all these wavelengths ultimately reproduce a light close to white, but without the unnecessary wavelengths that are not perceived by humans, animals and plants, allowing energy savings while offering the best quality and the healthiest lighting, both for humans and for other animal or plant species living under the same roof.

Spectral Light Chandelier {Fig. 3-513}
Photos: Miro Zagnoli

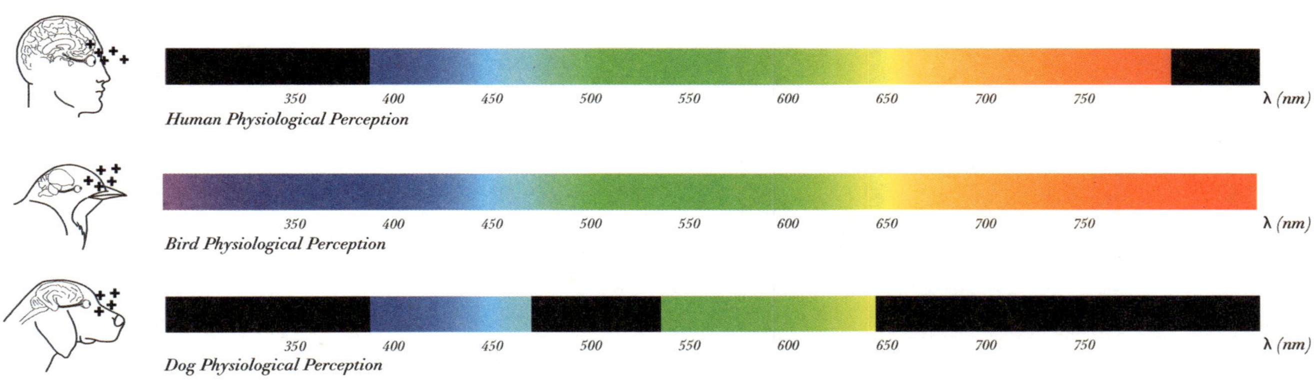

Spectrum of physiological perceptions of light {Fig. 3-514}

III.XIV. Evaporated Rooms

Lyon, France, 2012

View of the kitchen (below) and of the bathroom (above) {Fig. 3-515}

View of the bathroom grille, which does not block the passage of air and convective movements {Fig. 3-516} Photos: Nicolas Pauly

Evaporated Rooms

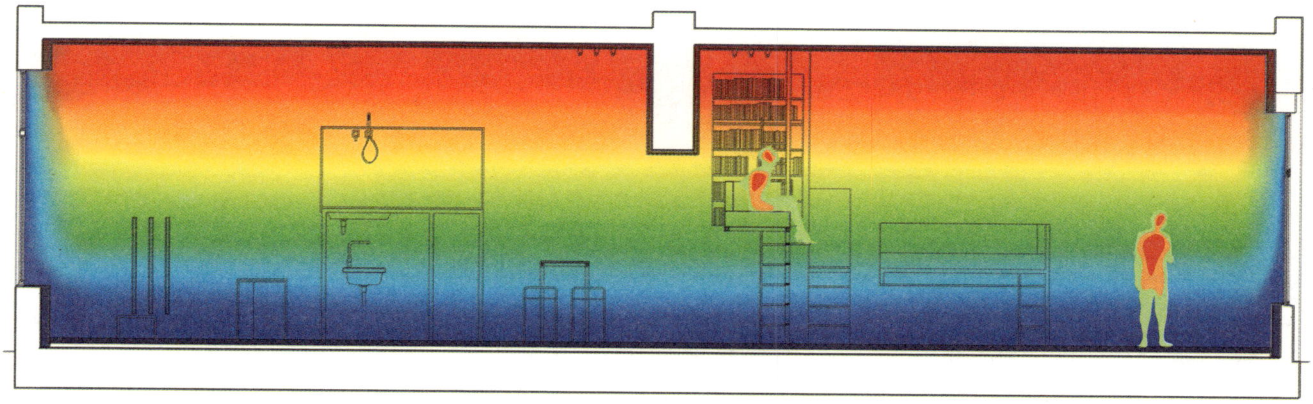

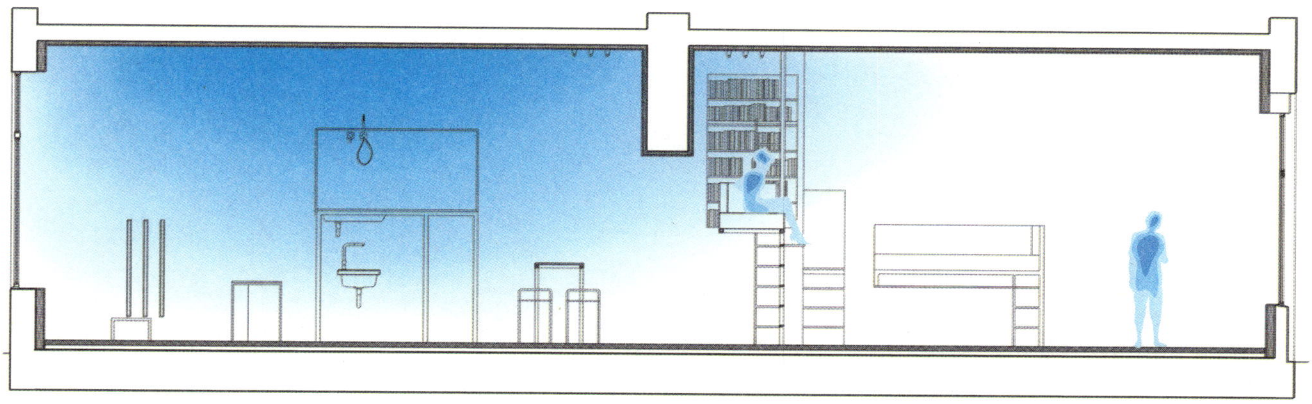

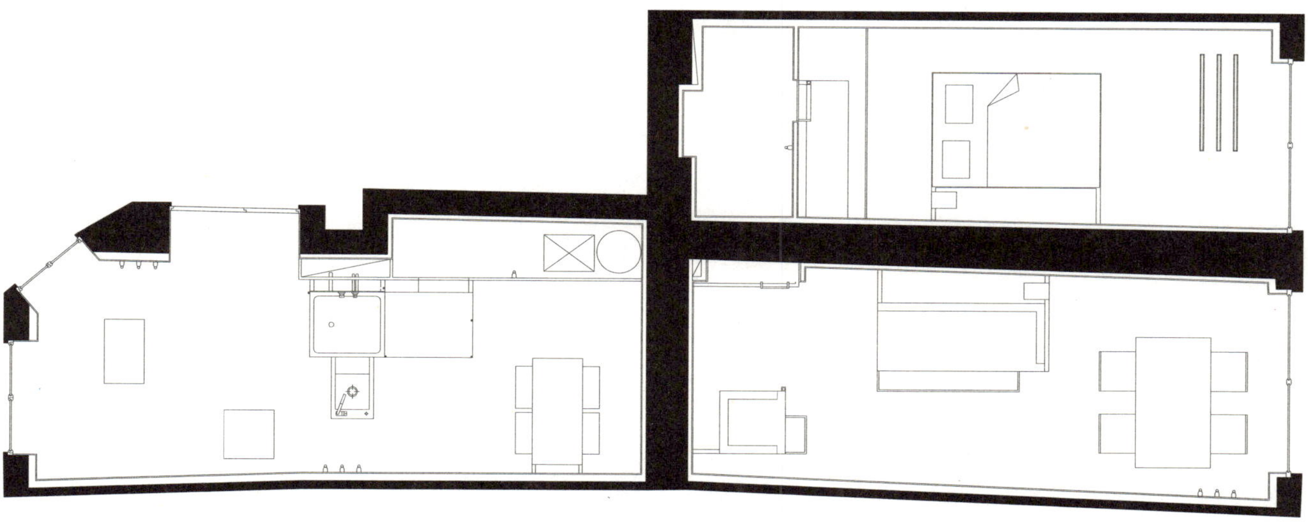

Thermal section {Fig. 3-517}

Hot air rises and cold air sinks, creating a stratified atmosphere with different temperatures. Functions naturally take place according to height: lower down, the bed, where you are covered by a blanket, upstairs, the bathroom where you are nude, and halfway up, the living room.

Vapor section {Fig. 3-518}

The house is designed plan-wise according to the movements of moisture caused by the movement of air, controlled by a mechanical ventilation system. Thus, all the functions that produce water vapor (shower and kitchen) are grouped together at the end of the path of the air, while the drier functions (living room, bedroom, dressing room) take place near the fresh air inlet.

Plan of the apartment for a young physician {Fig. 3-519}

View of the kitchen (downstairs, cooler) and of the bathroom (upstairs, warmer)

View of the bedroom toward the closet {Fig. 3-520}

Apartment, Lyon, France / Philippe Rahm architectes, 2011-2012

For this apartment in Lyon, we have based ourselves on recommendations concerning minimum domestic indoor temperatures in order to reduce energy consumption in the building sector (18°C for the bedroom, 19°C for the kitchen, 20°C for the living room, 21°C for the bathroom). The floor plan and section are composed according to the atmospheric variations of the space. We work in cross-section on one of air's intrinsic physical principles, convection, which causes air to rise when it is hot and sink when it is cooler. We can thus begin to arrange rooms and spaces in the vertical plane according to the required temperatures, or more precisely, to arrange usages in the atmosphere, without using walls to delimit them. We resort solely to the spatial arrangement of temperatures in the air to distribute uses. The aim is no longer to conceive a geometric design but rather an atmospheric one, with its climatic gradations in which the occupant moves to find a certain temperature, humidity (or air renewal rate) or light. So the chair where you just sit is the highest point and therefore the hottest, as is the shower, where you are naked. The couch, where several people can share their human warmth, will be slightly lower down, since the temperature tends to increase when several individuals are seated there. The kitchen floor is placed even lower down, at a lower temperature, while the bed is located in the very lowest part of the dwelling, on the ground floor, where it is coolest.

CHAPTER IV.

ATMOSPHERIC FRONTS

IV.I. Indoor Meteorology

Alain Robbe-Grillet

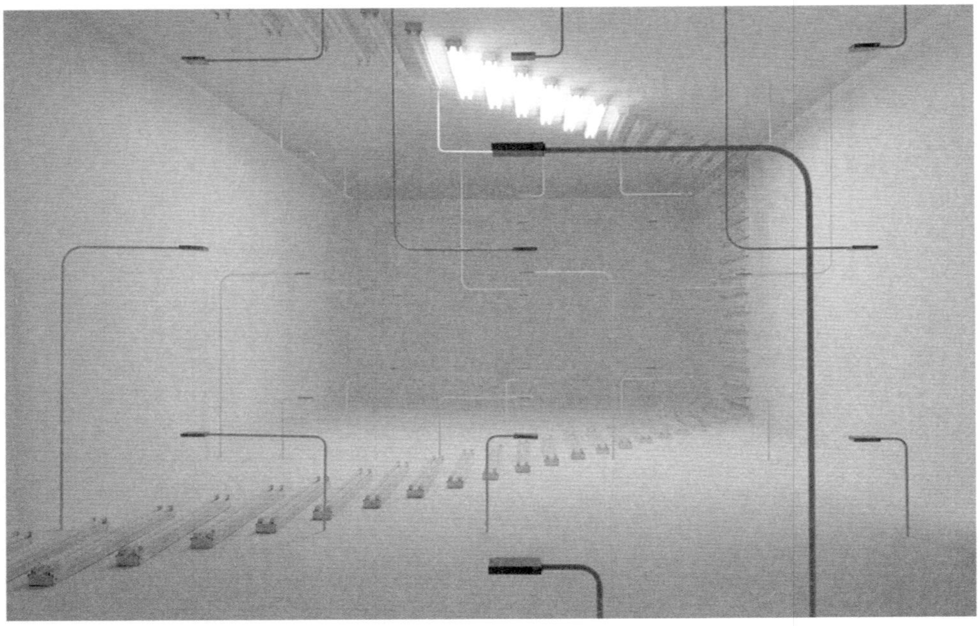

At first glance, the place where I find myself is neutral, white, let's say; not brilliant white, rather an indecisive, misleading, ephemeral hone, also rather absent. If there were anything to see before me, we would see it without any problem under this uniform lighting which is neither excessive nor sparse, devoid in the final analysis of any adjectivity. In such a space, unconvincingly affirming its indifference, it is neither hot nor cold.

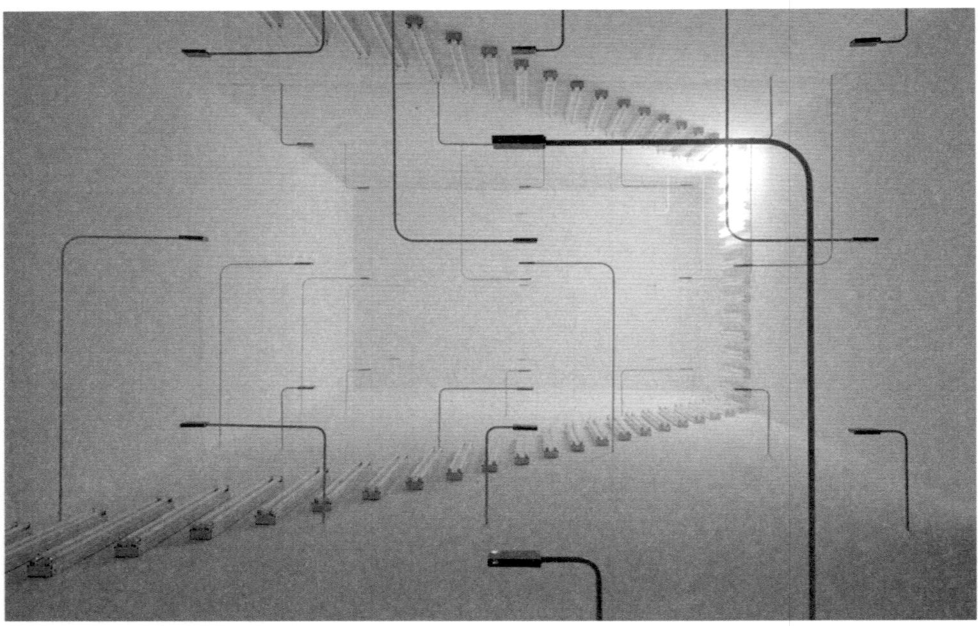

The room seems cubic, with no visible window or door, no furniture or decoration. I am motionless, lying on my back, legs stretched out, arms resting along my body, my chest slightly raised due to an inclination of about twenty degrees of the (metallic?) chassis of what must be a very low bed base, that may be adjusted and raised higher than normal, articulated like the ones for patients in hospitals. So, would I be resuscitating in some surgical or other clinic? The idea crosses my mind that it could actually be a morgue to which my lifeless body has been transported following an accident...

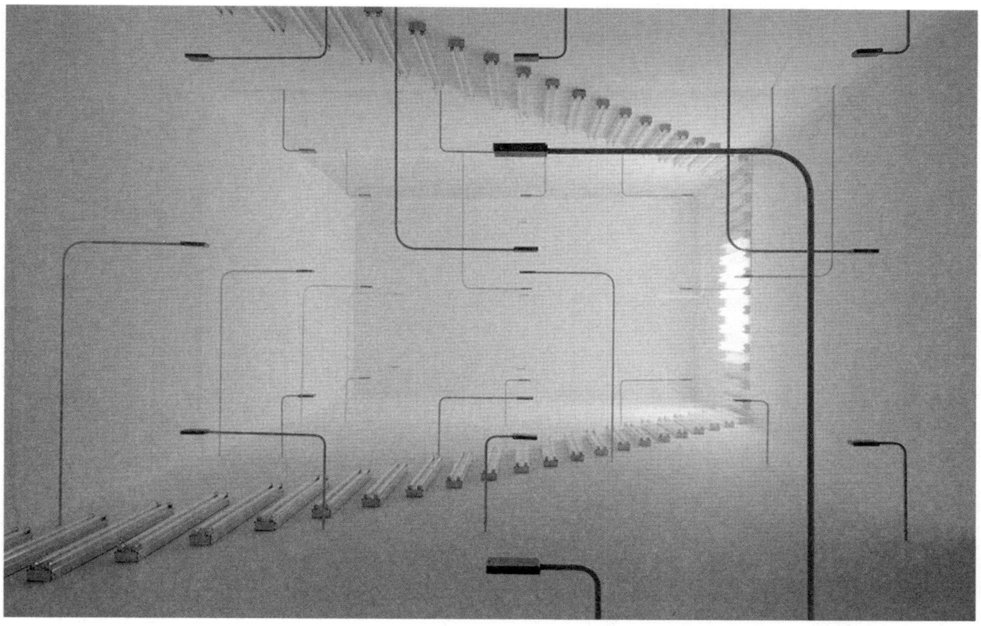

Something, however, immediately prevents me from giving credit to such a hypothesis: if I were dead, and especially in this exhibition, the icy atmosphere of a funeral chamber, I would feel the cold gradually gaining on me. While I have the opposite impression of the warmth of an alcove that soon even increases with heat exhaling from the tropical forest, whose moist, heavy puffs besiege, disorient and invade me. In my torpor, I think I see the diffused light from the walls around me move as if the sun, filtered by the foliage of massive trees, populated at the top by a hushed murmur, were reaching the ground (and me) in the form of a scrambling of particles without precise contours, without direction, aimless.

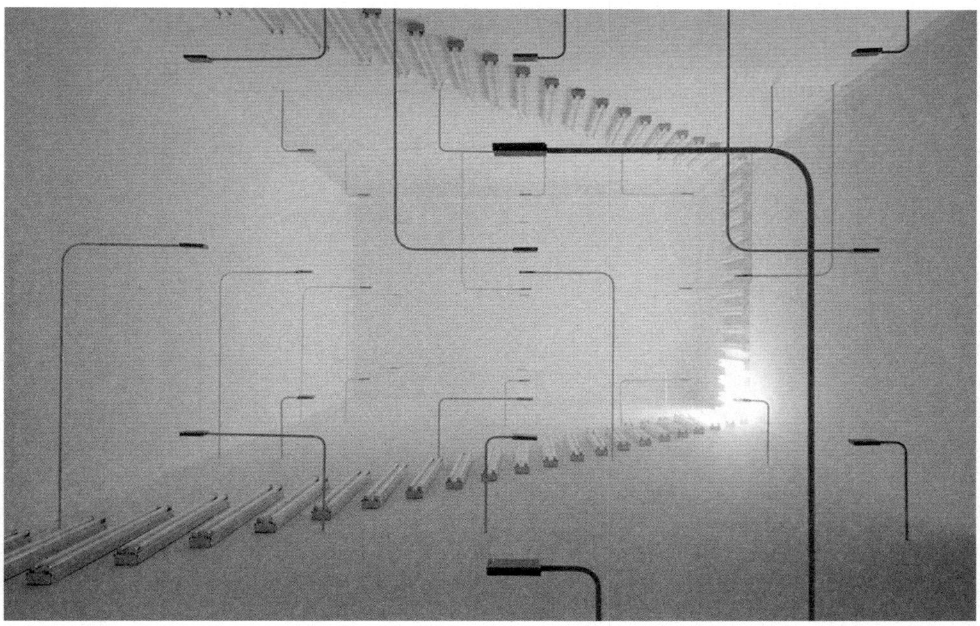

Toward the back wall, the one on which my languid eyes wander most easily, in the foreground of a drawing the evidence of which is quickly confirmed, I make out a forest in perspective with vertical, rectilinear trunks, a kind of pool of such clear water that it becomes almost immaterial, an oblong expanse of a lucid source as deep as or even deeper than a bathtub, set among gray round-shaped rocks, soft to the touch, welcoming.

Original text written by Alain Robbe-Grillet for the exhibition "Gilles Clément & Philippe Rahm: Environnement: Manières d'agir pour demain" at the Centre Canadien d'Architecture de Montréal, Canada, 2006

Photos: Michel Legendre / CCA

IV.II. Digestible Gulf Stream

Piero Macola

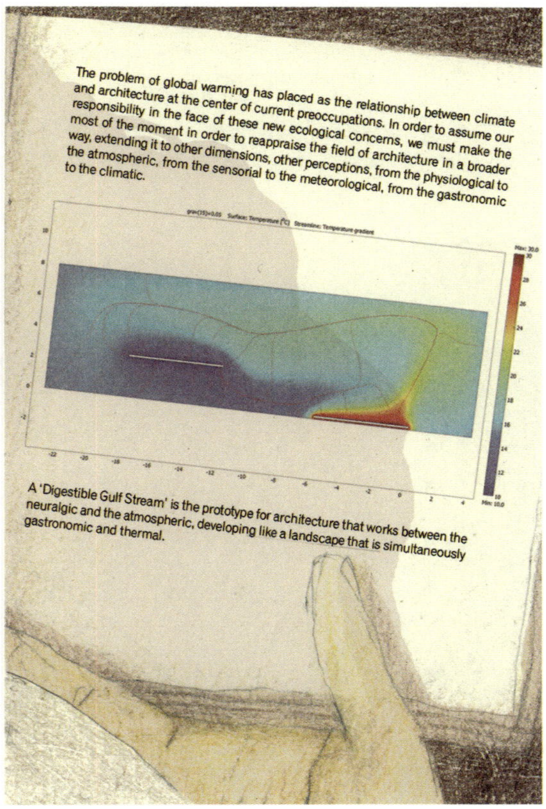
The problem of global warming has placed as the relationship between climate and architecture at the center of current preoccupations. In order to assume our responsibility in the face of these new ecological concerns, we must make the most of the moment in order to reappraise the field of architecture in a broader way, extending it to other dimensions, other perceptions, from the physiological to the atmospheric, from the sensorial to the meteorological, from the gastronomic to the climatic.
Surface: Temperature (°C) Streamline: Temperature gradient
Max: 30.0
Min: 10.0
A 'Digestible Gulf Stream' is the prototype for architecture that works between the neuralgic and the atmospheric, developing like a landscape that is simultaneously gastronomic and thermal.

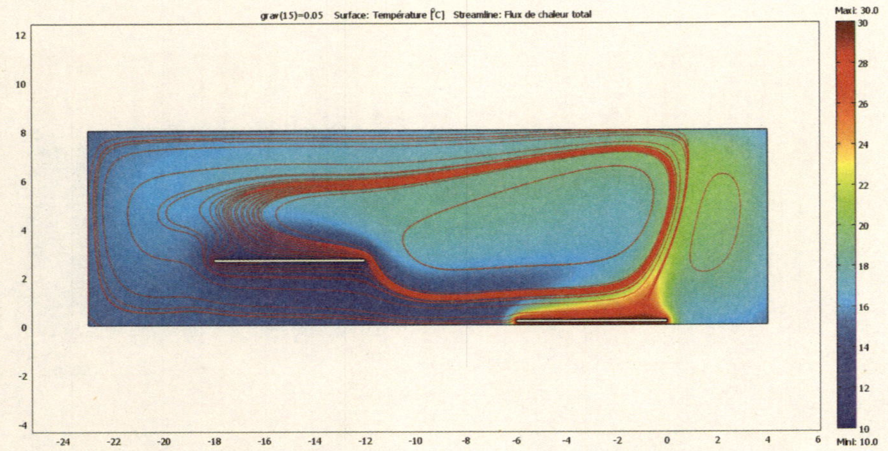

Architecture as Meteorology

Architecture should no longer builds spaces, but rather create temperatures and atmospheres. Here, two horizontal metal planes are extended at different heights. The lower plane is heated to 28°C, the upper one is cooled to 12°C. Like a miniature Gulf Stream, their position creates a movement of air using the natural phenomenon of convection, in which rising hot air cools on contact with the upper cool sheet and, falling, is then reheated on contact with the hot sheet, thus creating a constant thermal flow, akin to an invisible landscape. What interests us here is not the creation of homogeneous, established spaces, but of a plastic, climatic dynamic, the activation of forces and polarities that generate a landscape of heat. In this case the architecture is literally structured on a current of air, opening up a fluid, airy, atmospheric space. This architecture is based on the construction of meteorology. The inhabitant may move around in this invisible landscape between 12°C and 28°C, temperatures at the two extremities of the concept of comfort, and freely choose a climate according to his or her activity, clothing, dietary, sporting or social wishes.

The concept of thermal comfort depends not only on external temperature, but also on clothing, the physical activity of those who live in the space and their diet. For example, when we feel too hot, we have five ways of cooling down, which act on different scales:

1. reducing the air temperature in the room, for example by air conditioning (atmospheric solution);
2. drinking (physiological solution);
3. taking off clothes (social solution);
4. resting (physical solution);
5. stimulating a sense of coolness with the mind (neurological solution).

Each one of these solutions is architecture. Architecture is a thermodynamic mediation between the macroscopic and the microscopic, between the body and space, between the visible and the invisible, between meteorological and physiological functions.

Architecture as Gastronomy

We then propose to add two culinary preparations to the two plates that directly stimulate the sensory receptors of hot and cold at the cerebral level and that can be eaten or applied to the body.tThe first preparation, on the upper cold plate, contains mint, which has molecules of crystalline origin known as menthol that cause the same sensation in the brain as the coolness perceptible at a temperature of 15°C. The menthol activates the TRPM8 (transient receptor potential) molecular sensory receptors on the skin and in the mouth that stimulate the group of peripheral sensorial neurons known as cold-sensitive units. The second composition, on the lower hot plate, contains chilli, in which one of the molecules, capsaicin, activates the neuro-receptor TRPV1, which is sensitive to temperatures over 44°C.
The traditional field of architecture thus expands, operating on both the atmospheric and gastronomic scales, breaking down the barriers between internal and external, body and space, neurology and physiology.

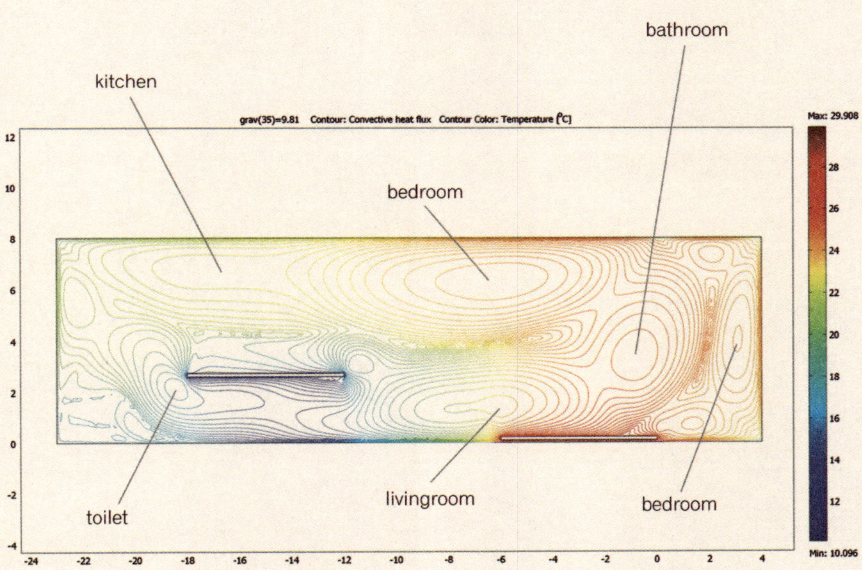

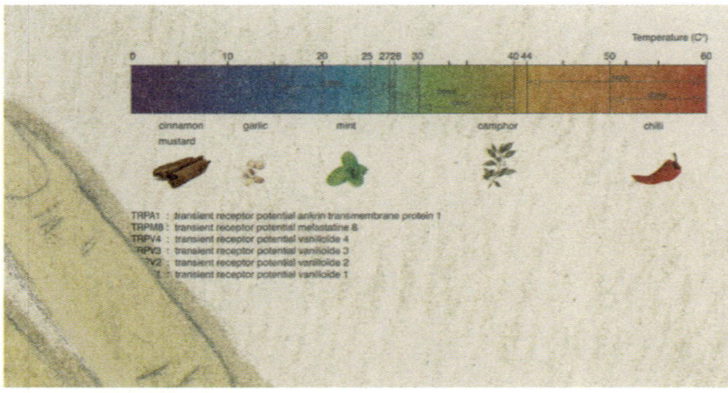

The sensations of hot and cold may be perceived as much inside the body (diet) as outside (atmosphere). So the question of diet also comes into the field of architecture, on a par with its climatic dimension.

Our architecture extends between the microscopic and the macroscopic, the gastronomic and the atmospheric, the visible and the invisible. Architecture becomes a 'Gulf Stream' that polarises the contrasts on different scales (hot/cold, low/high, clothed/unclothed, internal/external, rest/activity) to give rise to architecture as a convective movement of air, creating a place like geography, designingspace like climate, atmosphere and gastronomy.

Original drawings by Piero Macola for the international architectural exhibition "Out there", at the Venice Biennale of Architecture in 2008

CHAPTER V.

CLIMATE RESEARCH

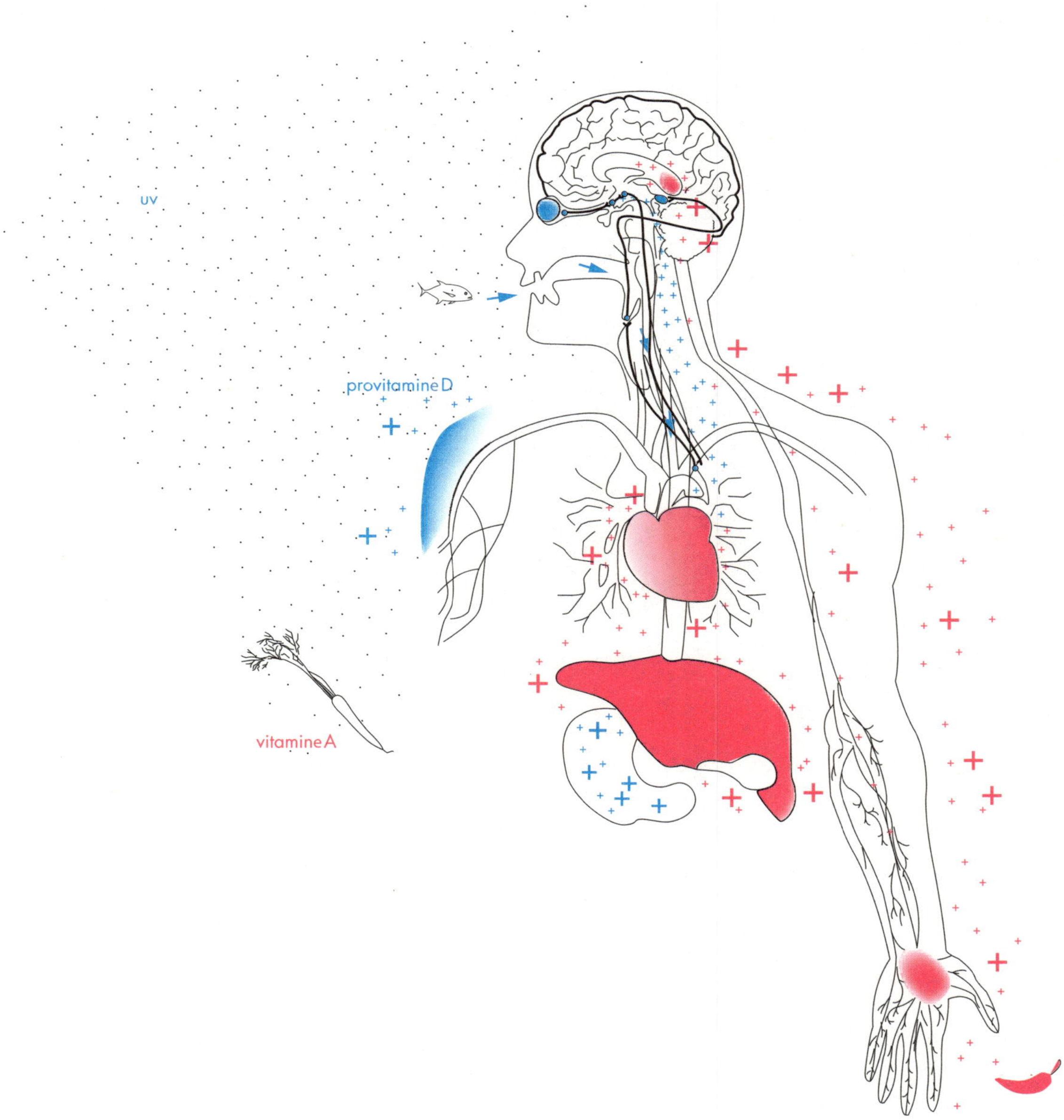

Deterritorialized Terroirs
Carte blanche, VIA, Paris, 2009
Photo: A. Dupuis/ VIA

V.I. Research into air

Climate Research

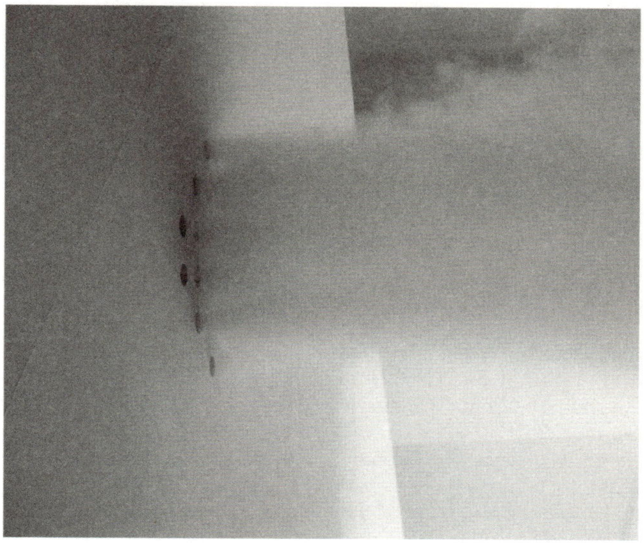

Paradise now!

Paradise now! is the project of an ineffable space. Without matter or any other physical limit than smell, the project develops in space, or more precisely develops a space as a meteorological modification of the atmosphere, which takes its real forms in the proteins of the ciliary membrane, in the abstraction of the nervous system. Musk, aloe, rayhan, milk, honey, wine as fragrances of the paradise of Islam, with touches of Christian paradise: incense, Judean balm. The International 3, Manchester, UK, 2005

Fœhn - Föhn

"Foehn" questions the nature of indoor ventilation by proposing the creation by mechanical ventilation of a 60 km/h air current, with an upper temperature of 8°C and a decrease in relative humidity of 17% as an indoor simulation of the "foehn", the famous wind "that drives people crazy". Insomnia, headaches, restlessness, the foehn is said to create tension, make one impulsive, darker, exacerbate libido and drive one to insanity. What really happens is that the foehn appears when the warm air masses from the south meet the southern face of the Italian Alps, cooling as they rise. The air caresses the granite mountains, reaches the ridge at the top, then as it descends toward Switzerland, it is compressed before heating again, quickly moving away from dew point. This makes the air very dry and clear, very powerful, charged with positive ions. After several days, there is great relief when the foehn finally calms, when the humidity of the air rises and the air cools and becomes electrically negative again, allowing one to breathe more easily, to release the tension, so Alpine mythology has it. Frac Lorraine, Metz, France, 2005

Absinth'Air

Absinth 'Air is a climate system that connects to the inlet of a ventilation or air-conditioning system. At the Swiss Cultural Center, it operates according to the principle of the hookah, forcing fresh air coming from the outside through a liquid volume of absinthe, harvested and distilled in the Val-de-Travers, to fill it with perfume and alcohol. This absinthe-conditioned air is then fed into the center's exhibition halls. The visitors perceive this intervention through the vapor, by breathing in the perfume, the aroma, with each breath, through the alcohol, testifying to ancient cultural relations between Switzerland and France, where the distillation of Swiss absinthe plants intoxicated Parisian artists of the bohemian era and cursed poets. Swiss Cultural Center, Paris, France, 2003 (Décosterd & Rahm, associés)

Placebo Paint

It is scientifically conceivable that a room painted in a soporific placebo, a paint containing a homeopathic dose of orange blossom, for example, would induce a person to rest, whether they are consciously or unconsciously convinced it secretes an endo-hypnotic. Similarly, a room decorated in an aphrodisiac placebo, such as ginger, would improve sexual performance by increasing the secretion of endorphins, precisely the substances of ecstasy. By incorporating an infinitely small quantity of ginger or orange blossom into industrial white paint, we seek to designate the use made of a room in a way other than by visual or plastic representation. Our goal is to act at a pre-sensory level, in order to create a kind of sub-functionalist architecture, generating its forms within the neurological and endocrine area of the brain.
Musée d'Art Moderne de la Ville de Paris, France, 2001 (Décosterd & Rahm, associés, with the group AIR - J-B Dunckel - N. Godin)

Pulmonary Space

Can wind instruments transmit the influenza virus? Can we feel the moisture of a singer's breath, which rises into the air when the choir sings louder? Do we inhale the breath emitted by the trumpeter? What part of music is chemical? Would music, with its waves, its physical impressions on the air and the distillation of vapor, not begin to take on an architectural form? An invisible yet perfectly audible form. What if we accept that music is not abstract but actually produces a volumetric space? Could we not finally accept, for once and for all, that music is not an immaterial void but rather a material void, however light and impalpable it may be, but above all moist, dense, physical and chemical? Music becomes viscous, fetid, moist, exhaled from the lungs and in turn inhaled. The spaces take on a form of architectural breathing in which the audience is immersed. Music is no longer simply heard but breathed and inhaled, constructing its own space.
Barbican Art Gallery, London, UK, 2009

Deterritorialized Terroirs

The aim of the project is to recreate a specific environment (the terroir) in an interior by using contemporary energy-saving techniques, dual-flow ventilation with heat recovery, optimized heating and energy-saving lights. The chemical, electromagnetic, physical, atmospheric and geological properties of a specific external microregion are reconstructed within an interior, reproducing a specific type of soil, a climate, exposure to the sun, smells, specific air movements, making a de-territorialized, a-territorial, domestic terroir. The heating involves different types of limestone, quarried in the Paris basin. The ventilation is rendered by a precise, homothetically reduced composition of plants and trees present in the Paris basin, oak and chestnut, but also pine and beech, which will produce the air fragrances, while glass, polyethylene and stainless steel are chemically neutral components in the architectural composition of the living space.
Carte blanche, VIA, Paris, 2009. Photo: A. Dupuis/ VIA

Polarized house

The electrical polarization of air, measured according to ionic equilibrium, is one of the real invisible dimensions that lead us to question the impact of spatial perception. Differences in ion intensities are imperceptible; humans have no sensory organ to detect ions. Scientific research is ongoing, however, increasing the implications and fantasies concerning the influence of the invisible on our mood and health. Dr. Michael Terman and his team at Columbia University in New York are working on the psychological/ psychiatric effects of the air ion balance. They came to the conclusion that the direction of the electrical polarity of ambient air influences our mood. Science - which is gradually penetrating further into the infinitely small - unveils new ground every day. Through new spatial dimensions, our Polarized House project seeks to create an environment of well-being that could turn into a nightmare at any time. By electrifying the air, it is possible - physically, chemically - to build contemporary fantasies as well as fears.
Kunsthaus Graz, Austria, 2006

The Hot Death

The performance is like an entropic process, passing from the multitude of energies, of individualities, to finally, gradually, homogenize, stabilize, become immobilized in an evolutionary impossibility. Passing from life to entropic death, that of a warm, gray universe, where all differences have disappeared. Heat and cold level out in warmth, white and black become gray, with no possible way back, a tragedy. At the beginning, the bodies on stage behave individually, each with its own movements, independent of the others, like a multitude of energies. Then gradually, the temperature, the humidity of the room increase to match that of the human body. Movements become slower, heavier, gravity increases to the point of bringing everything to the ground, motionless, with no more space in between, no more movement possible.
Fall Festival, Paris, France, 2004 (with Gilles Jobin, choreographer)
Photo: Gilles Jobin

A Winter Beach

The winter beach reproduces a summer beach indoors by diffusing a seaside aerosol, derived from marine spray, a cloud of iodine produced indoors, which is perceived upon breathing and develops its real form in the body itself. The winter beach is suggested as a new form of indoor public space with a bright, olfactory, thermal and gustatory atmosphere, somewhere between the pool and the exotic restaurant. What we take from the beach are just certain chemical and electromagnetic phenomena such as the sun and the sea spray, tanning and iodine: a certain radiation with its high ultraviolet component and a most particular angle of incidence, an aerosol with a specific chemical composition. It is these atmospheric phenomena specific to seaside coasts that are recreated indoors in the middle of winter.
Le Lift, Saint-Nazaire, France, 2008. Photo: Marc Domage

ND Cult

ND CULT is an artificial space defined by its atmosphere, at an altitude of 10,000 meters, in which the air's oxygen component is as low as 6%. It proposes a space on the verge of death, where our perception and our consciousness are modified with an intensity in all likelihood verging on mystical states. This space is a spatial reproduction of what the body experiences at the time of death, those few seconds during which the brain is deprived of oxygen during cardiac arrest. The reduced oxygenation of the brain at the time of death can be compared to an extremely rapid rise in altitude. The lack of oxygen, known as hypoxia, creates highly intense physiological responses. Hallucinations are common among mountaineers who climb to altitudes above 6,000 meters without oxygen. These symptoms of severe hypoxia occurring at very high altitudes are very similar to those reported by people who have experienced near death. Migros Museum Für Gegenwartskunst, Zürich, Switzerland, 2003 (Décosterd & Rahm, associés)

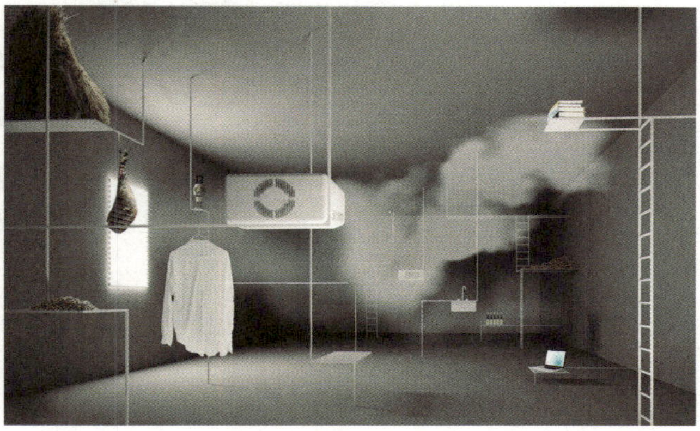

Relocated Cellar

The layout is arranged between two climatic devices on either side of the room, at medium height. The first device produces an ultrasonic mist that reproduces the relative humidity of the air in a cellar. On the other side, the second unit is a dehumidifier to dry out the surrounding air, creating an atmosphere close to that of an attic granary. In this new free plan based on relative humidity variations, different foods take their place according to their hygrometric needs in terms of conservation, as well as today's common furniture, fittings and accessories; shower, shelf, computer, books, clothes, which are also placed on the plan according to the water vapor they produce or contain. As for humans, here they can adopt a more sensitive and tactile way of life, between the moister, mistier air of the shower, of the kitchen, of potatoes and wine, and the drier, transparent air of the areas for resting, chatting or working, where seeds and cured meat are also kept to dry.
San Francisco Museum of Modern Art, USA, 2005

Parisian spring light
Carte blanche, VIA, Paris, 2009
Photo: A. Dupuis/ VIA

V.II. Research Into Light

Climate Research

The medium is the message

Information, most often cultural and visual, uses the media only to convey information. As such, the medium is usually not important: the quality of the paper in a book, for example, does not play a role in the meaning of the text; screen light technology does not influence the movie. Our project here proposes being critical of the lack of reality of the medium itself, and to accept that the medium is not neutral but has physical signs. In this space of fabrication, we accept the hormonal and invisible information transmitted by the medium, as the spectrum of light emitted and its intensity are chosen specifically to block the secretion of the sleep hormone, melatonin. Here, the medium really is the message.
Dysfashional, Luxembourg, 2007

Eternal spring

In the courtyard of the Swiss Cultural Center in Paris, we recreate a beautiful mid-May day, in the middle of March. This May 15 is artificially constructed, with its light variations, the time of its sunsets and sunrises, their intensity. In the courtyard, the sun sets at 9:24 pm, while outside, it is already dark, as for example on March 12, when the sun naturally sets at 6:50 pm in Paris. In the courtyard, architecture recreates the same day every day, that of May 15, which is the day on which we usually stop heating the house. It is the ideal, comfortable spring that we propose to extend and repeat endlessly throughout the year, by actually acting on the electromagnetic spectrum of space and light intensity, on photoperiodism, by artificially compensating for the decrease in winter sunlight by an inversely proportional increase in artificial light, by acting on the plant hormones cyclokinins, by allowing the trees to blossom in the courtyard 2 months early.
Swiss Cultura Center, Paris, France, 2005. Photo: Marc Domage

Second summer

Here, the light of the summer solstice, on June 21, with its intensity and variations, is perpetuated throughout the year. Here, the sun will rise every day at 5:03 am and set at 11:51 pm. In the clearing, the days will always have a duration of 15 hours and 53 minutes. The vegetation of the clearing will remain in this summer, and will not really change color. The junctions of the branches may never break, the leaves will not fall completely, because these seasonal phenomena are induced by the shortening of the length of the day. An artificial light whose electromagnetic spectrum reproduces that of sunlight will compensate each day for the shortening of the day, extending photosynthesis beyond autumn, beyond winter. It is a kind of second summer, lost in the Austrian winter, a climate in suspension. Here, summer will last indefinitely, it will stretch along the year, it will pass autumn by, and will cross winter until the following spring. Eibesfeldt Island, Austria, 2005

Upside-down

We propose to transform, or more precisely to "transtemporalize" a passage of the city of Winterthur, by creating a desynchronized place in the historical tradition of the climatic constructions of the 19th century such as glasshouses, winter gardens or covered ways, ambiguous spaces between outdoors and indoors, day and night, summer and winter, tropical and polar climates, where some, but not all climatic parameters are modified, where it does not rain when it rains outside, where it is night during the day, where it is day at night, where it is warm when it is cold outside. Our project highlights the ambiguity and divergence present in this passage to suggest reversing time. By working on the electromagnetic spectra, we produce night when it is daytime by emitting electromagnetic waves beyond 550 nm, and conversely daytime when it is night time by emitting electromagnetic waves below 550 nm. Lichttage, Winterthur, Switzerland, 2010

Cosmogony

All movements, all forms, the possibility of life itself are linked to a temperature difference, to the presence in a space of temperature differences. Energy, life, movement, is in tension between two poles, one hot, the other cold. Our cosmogony is based on a thermal difference: one hot, polluting source, and another, cold and ecological one. The hot source is an outdated, polluting 20th century sun, made of incandescent light, in which 90% of the energy is converted into heat instead of light. In contrast, we propose a cold, ecological moon, a frozen desert star, powered by non-polluting photovoltaic panels, shining with its electroluminescent light, consuming very little energy. The two stars produce a cosmogony due to their temperature and climatic differences, they generate meteorological phenomena and wind, creating an interior cosmogony. King Abdulaziz Center for World Culture, SaudiArabia, 2019

Glare

For the exhibition "Alpine Light", at the Art Museum of St. Gall, at an altitude of 669 meters, we propose producing the cold, dazzling light of the mountains, more intense and bluer, that of St. Moritz, at an altitude of 1,822 meters. To do so, we propose doubling the amount of incident light, as snow does, by creating a white floor on the existing brown parquet. The shape of this white floor is precisely sized and adjusted to reflect the sun's rays entering the exhibition room through the windows during the days of the fall. This passive, inert first device therefore plays on the material's property of physical reflection. The floor will be made of wood, painted white and varnished to increase the reflection of the white, dazzling the visitor entering the room. A second device consists of artificially adding a tiny amount of blue to the light, because the higher the altitude, the greater the layer of atmosphere through which the sun rays pass, the more the light loses its blue wavelength by diffusivity, becoming yellower. Kunstmuseum, St. Gall, Switzerland, 2019. Photo: Stefan Rohner

Sublimated Music

Our architectural work proceeds from a quasi-chemical separation of space into elementary particles: wavelengths, moisture content, light intensity, heat transfer coefficient, and sound frequencies. We are interested in forms that are no longer the result of a composition seeking to create a "whole" as proposed by the Gestalt, but in those that are dissociated, spectralized into fragments of reality, sensitive particles. For the Maxxi, we disrupt the beginning of Claude Debussy's piano music "Bells through the leaves" by breaking down the temporal linearity of the work into single notes (8 for the left hand and 15 for the right) played independently in space. The project consists of dissolving the linearity of the sequence of notes, abandoning the temporality of the music in favor of space and a static musical atmosphere. Debussy's work will no longer be performed in the linear time of a concert but immobilized in the spatiality of an exhibition space. Maxxi, Rome, Italy, 2014 (with Sebastian Rivas)

Uchronic Climate

What we once recognized as the natural climate - that has always been in the exterior, outside buildings - is no longer natural today. The classic architectural dichotomy between interior and exterior no longer exists. Since the industrial era, human activities and their consumption of CO2 have changed the global atmosphere of the planet, manipulating it like a product. The paradox is this: if we want to experience the "natural" climate we must now reproduce it "artificially" because it no longer exists. The "Uchronic Climate" is the construction of a false sun, the physical projection of a uchronic universe, the world as it would have been if global warming had not existed. Based on a computer model of the atmosphere that cancels out the effects of anthropogenic greenhouse gases since 1850, a computer can generate this uchronic climate for each minute of the duration of Manifesta 7 by varying light intensity and relative humidity, as calculated by the software. The "Uchronic Climate" offers the possibility of inhabiting the natural world as we will never know it again. Manifesta 7, Fortezza, Italy, 2008

White Geography

The aim here is to create a neutral background, the fundamental work of architecture for the visual arts: that of building a simple, neutral, white background in front of which the figure, that is to say the artist's work, takes its place. This is of course the "White Cube" specific to contemporary art galleries and museums where architecture remains in the background so as not to disturb the reading of the work of art, in the foreground. But in this uniform, white neutrality, our project proposes variations in light intensities, allowing a gradation of neutralities, combining the neutrality of the background and intensity value, and forming the space in a gradation of neutrality, ranging from paler to more intense light. Chapelle du Genêteil, Château Gontier, France, 2009 (with Didier Faustino)

Built shift

The current global warming in Europe is spatially represented by a geographical displacement from north to south of an estimated distance of about 200 km. This shift is slowly transforming the nature of the climate and therefore the nature of agricultural crops. Two physical dimensions are called into play, one outside the body, climatic, and the other inside, physiological. The first relates to sunlight, whose intensity increases by a factor of 0.75% when we move 2° in latitude, or about 200km southward. The second, internal dimension of the project concerns nutrition.

We simply offer the possibility of drinking mineral water imported from a region 200 km further south. Our project is a gustatory and environmental modification of our perception of global warming, combining dietary, geographical, ethical and esthetic considerations. Musée Zadkine, Les Arques, France, 2007

Chlorophyllian Remanence

Our proposal is to create a new type of garden furniture that produces the shade of a tree throughout the year, even when the leaves have fallen. To do so, we analyzed pigment by pigment the wavelengths absorbed during photosynthesis by the leaves of the trees in the spring and those that are reflected, subsequently touching the ground. The reflected light is the light that produces the tinge of spring beneath the plant cover of trees. The absorption ratio between the different pigments Chlorophyll A, Chlorophyll B, betacyanins, anthocyanins, lycopene, carotene and lutein is reconstructed according to their percentage presence in a spring leaf. The sunlight will thus be filtered by the device, and projected onto the ground throughout the year, recreating the particularly green light created by trees in spring. The visible spectrum of light is divided into different wavelengths from which we remove fourteen of them, corresponding precisely to the plant's pigment absorption spectrum. FIAC, Galerie Dominique Fiat, Paris, France, 2011

The New Olduvai Gorges

The exhibition investigates the most immediate, microscopic, digestible, electromagnetic forms of architecture, somewhere between dietetics, thermogenics and esthetics, just before deployment in space. In answer to the major reasons for the emergence of architecture (feeling too cold, too hot, not having enough light, etc.), we propose multiple endogenous, infra-architectural solutions, at the limit of physiology. Like adding ultraviolet or cod liver oil to compensate for vitamin D deficiency caused by the weak winter light. Like offering mint, peppers, a draft of air, to cool off in an overly hot climate. Like supplying proteins, an accelerated heartbeat to warm up. Here, all these measures bring back the original climate of the cradle of humanity, the eternal temperate spring, neither too hot nor too cold, and architecture appears as the art of rebuilding this so comfortable a climate for human beings, that of the Olduvai Gorge, throughout the earth. The Royal Danish Academy of Fine Arts, Copenhagen, Denmark, 2009

Gradation

The layout of our monographic exhibition at the Architecture Gallery Berlin uses the basic colors of our graphic representations: magenta to represent the heat or the hot part of the space, cyan to represent the humidity of the air or the cold parts. Since cyan and magenta are the basic pigments constituting printing inks in the CMYK system, their color and brightness are not altered when printed pure on paper, causing a strong visual impact. Since freedom of use is always at the heart of our architectural spaces, the distinction between cyan and magenta is made by a subtle gradation of colors, passing gradually from one atmosphere to another, from one thermal value to another, allowing the coexistence of all kinds of nuances and opening up to a variety of atmospheres and desires. Architektur Galerie Berlin, Germany, 2013

The New Olduvai Gorges

This space increases the amount of light during the period when the intensity of the sun becomes weaker and on winter's days in northern latitudes, far from the equator. The architecture provides endogenous solutions to the deficiency of vitamin D, vitamin A and melatonin caused by the absence of sunlight. It comprises:
-a visible field of carrots, which we can eat to improve our night vision and ingest vitamin A.
-a horizon of ultraviolet rays, to allow synthesizing vitamin D, missing due to Denmark's short winter days.
-an electromagnetic beam at 470 nanometers to block the secretion of melatonin caused by weak daytime light.
-a rectangular table (polyethylene) covered with raw oily fish such as salmon, trout, halibut, herring, sprat and mackerel presented in the form of sashimi, which can be eaten to obtain vitamin D, normally provided by the sun. The Royal Danish Academy of Fine Arts, Copenhagen, Denmark, 2009

Melatonin Chamber

The Melatonin Chamber is a physiological piece of architecture that acts on the space itself by modifying its electromagnetic composition. This room is defined as a space for hormonal stimulation. Two rooms are built next to each other. The first is defined by the emission of electromagnetic rays of less than 520 nm with an intensity of 5,000 lux, which suppresses the production of melatonin by the pineal gland. When secreted, this hormone provides information related to fatigue and sleep. The space becomes physically motivating and chemically exciting. However, yellow rays are emitted in the second room which, in contrast, encourage the endogenous production of melatonin. The Melatonin Chamber is a non-representational space that reduces the medium between the transmitter and the receiver to a minimum. Circulo de Bellas Artes, Tenerife, 2006 (Décosterd & Rahm, associés)

A Winter Beach

The winter beach recreates solar radiation indoors using rays similar to those of the sea, in summer, when the rays come from the sky but are also doubled up by the rays reflected by the foam of the sea. It is for this reason, because we are undressed, that we tan quicker on the beach or in the snow than in the city or in the country, where the sun's rays reaching the ground are absorbed and not reflected as they are by water or snow. What we reproduce here is an electromagnetic landscape, a horizon of ultraviolet rays, a certain amount of UV-A present on the beach in summer, which here we perceive no longer as something visible, but cutaneously, through the transformation of the skin, in the form of winter tanning. It is also this angle of incidence that is developed between the ground, the horizon and our body. Le Lift, Saint-Nazaire, France, 2008. Photo: Marc Domage

Three Parallel Spectralities

The scenography for our solo exhibition at the Frac Centre presents our projects in three spaces according to three phases. Each space is defined volumetrically by the 12 edges of a parallelepiped of variable volume (27 m3, 36 m3 and 60 m3), though empty, associated with a particular color (cyan, magenta and ultraviolet, respectively). Each phase is defined by the emission, during a given period, of light of a certain color (cyan, magenta and ultraviolet) associated with a video projection. 100 fluorescent tubes of 3 colors light up alternately over a 30-minute cycle. Depending on the weather and the color, the visibility of one space will take over, while the other volumes will disappear into the gray background that predominates in the Museum. Frac Centre, Orléans, France, 2005. Photo: Marc Domage

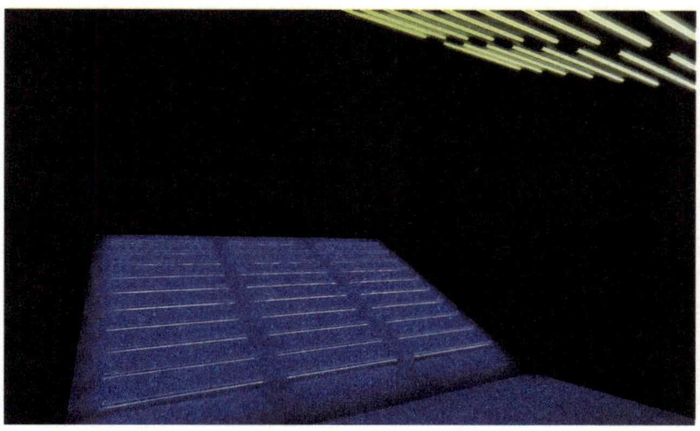

Simultaneous Day and Night

This installation builds both day and night on the basis of the specific emissions of these two times of day which are perceptible to the human body thanks to the retina, affecting the pineal gland and the secretion, or not, of melatonin. Geometry then provides a landscape of blue daylight that confronts the eyes of the human body when standing in a precise place and, conversely, a landscape of night time yellow light when the body is supine. The idea here is to build spaces that are no longer perceived only by our 5 senses but also by our endocrine system. In a way, it is a question of studying the forms and intensity of space in its subcorporeal qualities, creating endocrine landscapes, building a hormonal architecture. Dynamo, Grand Palais, Paris, 2013

Winter Beach

Our winter beach is as much a temporal shift between summer and winter as a shift of scale, from the macroscopic scale of an outdoor seascape, complete with sun and sea, to the physiological microscopic scale of the indoors, complete with iodine and ultraviolet rays. It is a composition of elements whose meaning and use are open-ended and can be freely interpreted by individuals and groups alike, as the shoreline has been down the ages, a space between rejection and desire. The winter beach is presented as a new form of indoor public place with atmospheres comprising light, smell, heat and taste - a quirky form, somewhere between the swimming pool and the exotic restaurant.
Casino Luxembourg, 2008. Photo: Axel Heise

Electromagnetic Chamber

Walls, partitions, curtains, shutters, blinds: the privacy of the house is traditionally defined by an infinite number of gauges and materials, which separate public space from private space visually and acoustically. The development of new means of communication has completely freed us from these physical limits. The Electromagnetic Chamber is the project for a new privacy in which the electric field is zero. The envelope is made of two conductive copper grilles forming a Faraday cage. With a first mesh of 30 cm, FM-AM television and radio waves are blocked out. A second mesh of 2 cm blocks out telephone waves. These two grilles can work together or separately. Against electromagnetic fields, away from public information flows, the inner space is electrically isolated from the public sphere. Bienale de Valencia, Spain, 2003 (Décosterd & Rahm, associés)

Parisian spring light

This low-energy LED lamp automatically or manually reproduces the luminous intensity and color temperature variation of May 15, 1832 in Paris. An automatic timer generates the day and night cycle of May 15, with sunrise at 6:08 am and sunset at 9:24 pm. The light varies according to an east-west movement of light, a transformation of the color temperature, hot in the morning and until noon, then colder until night. By means of a switch, the lamp also enables freely choosing the time of day, recreating the light of any time of that beautiful spring day of May 15 in Paris at any time of the year. Carte blanche, VIA, Paris, 2009. Photo: A. Dupuis/ VIA

Diurnism

The introduction of street lighting in cities during the 19th century was the reason for one of the major social and political revolutions of urban planning and of the form of the city. The goal was demiurgic: to create daytime at night, such is the program of noctambulism. Since then, within the city we have been living in an endless artificial day. Conversely, diurnism proposes creating night during the day. The aim is to reinvent a form of night in the artificial continuous daytime of modernity, to produce night, physically, during the day. It is an inverted response to the eternal daylight created by modernity, the Internet, and contemporary globalization, like a kind of second, or re-reversal. Following on from "Noctambulism", we would like to invent "Diurnism", using orange-yellow wavelengths between 570 nm and 800 nm, which are perceived physiologically by the body as night (following the melatonin rhythm). The room then becomes a paradox between the visible and the invisible, the visual and the physiological, a night that feels like a sunny day. National Museum of Modern Art, Centre Pompidou, Paris. Photo: Adam Rzepka, Centre Pompidou

Diurnes for piano

The first city street lighting was installed in London in 1812. And the first nocturne (slow, short and atmospheric musical compositions evoking the night) was invented by the composer John Field in 1814. The nocturne for piano could therefore be interpreted as the first modern musical figure that sought to create night time during the day. Our "18 Diurnes for piano" are a spectral inversion of John Field's "18 Nocturnes for piano". By reversing the form of these scores, we would like to introduce the work of art as a double artificialization of artistic representation, creating day within the night that was created in the day. The Diurnes for piano are a cultural transcription of the diurnism physiological project. Galleria Laurin, Zürich, Switzerland, 2006

Melatonin Weather

In the gallery, a wall of light crosses the room from south to north. The light is entirely generated by "i-weather.org" software, a circadian rhythm of 25 hours and 36 minutes oscillating between very intense green light and very weak violet light. I-weather.org produces a deterritorialized circadian rhythm that responds precisely to human physiological metabolism outside of any natural geography. It offers an artificial alternative, synchronized with the endogenous period of the human endocrine system.
Swiss Cultural Center, Paris, France, 2005. Photo: Marc Domage

Phantom Housing

Phantom housing (Ghost Flat) is a housing project whose rooms do not take shape spatially in three dimensions but through the wavelengths of the visible spectrum. Simultaneously occupying the same space, their mass and volume intertwined, each part of the program (bedroom, living room, bathroom) spans a specific fraction of the electromagnetic spectrum. The bedroom appears between 400 and 500 nm, the living room between 600 and 800 nm, and the bathroom, located in the ultraviolet, between 350 and 400 nm.
CCA Kitakyushu, Japan, 2004 (Décosterd & Rahm, associés in association with Marie Darrieussecq)

I-weather.org

Accessible everywhere and to all thanks to the Internet, I-Weather.org produces an artificial circadian rhythm synchronized to match the cycle of the human hormonal and endocrine system. In the absence of the natural terrestrial day and night cycle, it becomes apparent that this cycle in fact lasts around 25 hours, and that body temperature, alternation between sleep and wakefulness, and the accumulation and secretion of substances such as cortisone and oligopeptides, all depend on this cycle. I-weather.org has thus introduced the first climate specific to humans. I-weather 1.0b operates solely on the basis of fluctuations in the rate of melatonin, which in is turn influenced by variations in the intensity of light received by the retina. I-weather.org 1.0b acts as a kind of personal artificial sun, whose cycle runs for 25 hours and 36 minutes with a maximum light intensity of 509 nm and a minimum close to ultraviolet.
Museum für Gestaltung, Zürich, Switzerland, 2001. (Décosterd & Rahm, associés in association with fabric.ch). Photo: fabric.ch

Heat Square

The quality of the space provided for the Evento festival that we propose is one of a warm public space for visiting in winter. Indeed, the value of public spaces is historically acknowledged and they have always been linked to thermal values. In winter, we seek the warmth of the fireplace and shared human warmth when we drink a coffee. In summer, we seek the shade of a tree to protect ourselves from the sun. Here, we propose the construction of the first geometry produced by a floor of infrared lamps heating the individuals gathered in this space by radiation. The proposed spatial quality is above all thermal, reinstating the original value of public spaces, offering warmth in winter and cool in summer.
Evento, Bordeaux, France, 2009

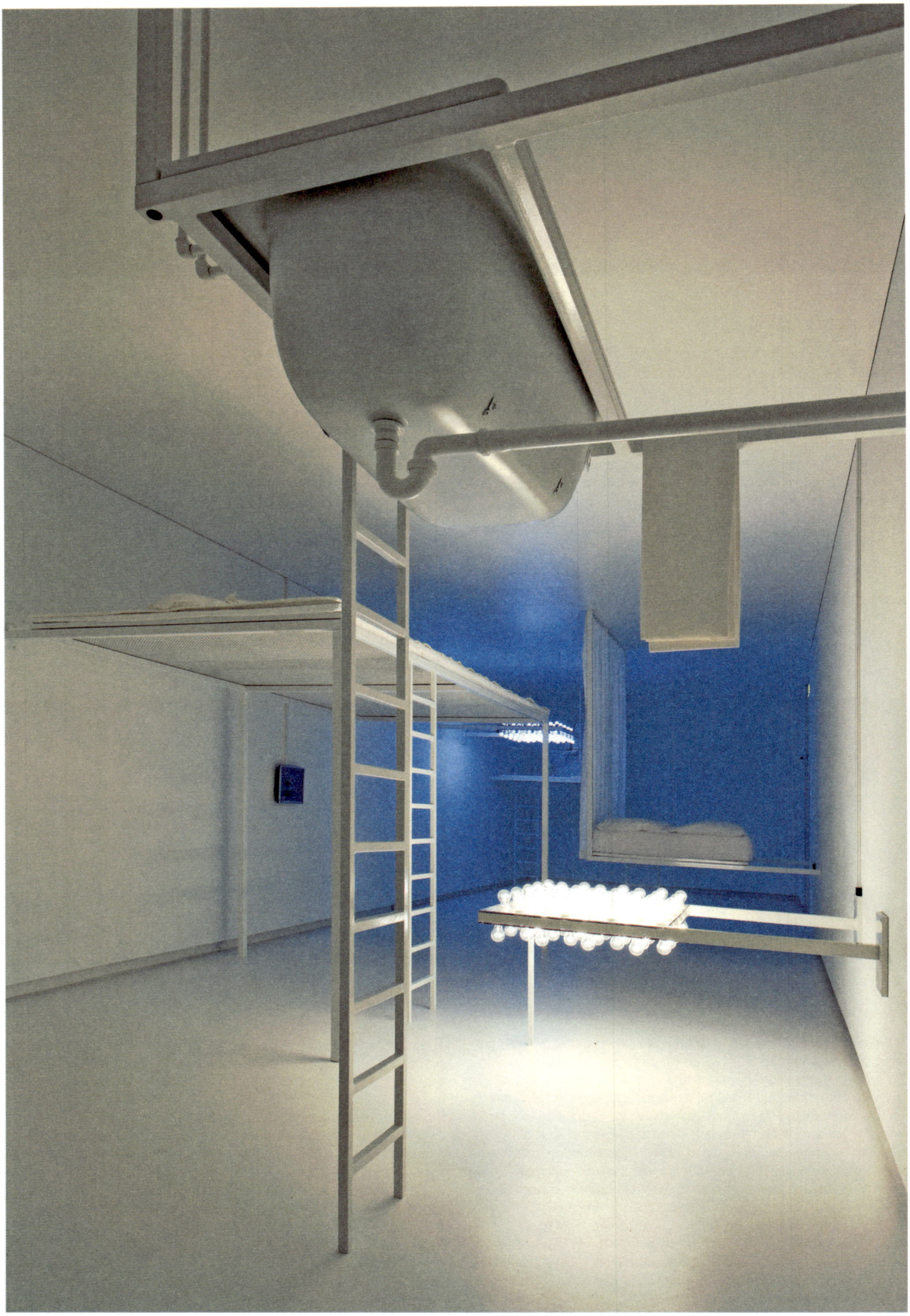

Domestic Astronomy
Louisiana Museum, Denmark, 2009
Photo: Brøndum & Co.

V.III. Research Into Heat
Climate Research

Geological looring

The aim here is to go beyond symbolic post-modernism to achieve a chemical post-modernism, moving from a visual and narrative memory to a material, olfactory and tactile memory. We propose a floor heating system that reproduces the natural geological qualities of the Parisian basins indoors. Thus, we propose a mineral quality for our floor heating, made of different types of limestone assembled in the right proportions to create a heated floor. The aim of the project is to recreate a specific environment (the terroir) in an interior by using environmentally friendly techniques such as dual-flow ventilation with heat recovery, optimized heating and energy-saving LED lights. The specific external chemical, electromagnetic, physical, atmospheric and geological properties of the Parisian terroir are used to operate in an interior, in a domestic reconfiguration of the Parisian terroir: a "deterritorialized" or "aterritorialized" terroir, reproducing a specific type of soil, climate, exposure to the sun, odors and air movements. Carte blanche, VIA, Paris, 2009. Photo: A. Dupuis/ VIA

The Effusivity Parquet

The effusivity parquet is an interior design proposal based on the climatic properties of materials. In a very obvious gradation, we put forward four materials with different intrinsic effusive properties whose characteristics include accelerating or slowing down body heat loss by conduction upon contact with the material. The progression from the coolest to the warmest is thus established between a woolen floor, then a wooden floor, then a stone floor, and finally an aluminum floor. Contact with low-effusivity wool will create a sensation of heat which will be less notable when in contact with the wooden floor. Stone, whose effusivity is higher, will cause a sensation of freshness, and finally aluminum with its very high effusivity will provide coldness to the touch. The goal here is to rethink the fashions of choice of materials in architecture, by moving away from metaphorical or narrative criteria in favor of real thermal values, ultimately leading to reduced energy consumption through heating in winter or air-conditioning in summer. Istituto Svizzero, Milan, Italy, 2018. **Photo: Giulio Boem**

Refreshing Equipment

This exhibition offers a variety of equipment to refresh visitors to the bienniale during the summer of 2019. Refreshing systems make use of different climatic and meteorological phenomena. Evaporation through the production of an ultrasonic mist creating a cloud of water that evaporates and cools the air. Convection too, thanks to a fan that removes the layer of heat accumulating on the skin and cools it. In addition, a Japanese chime, a glass bell called a Furin, rings pleasantly in the wind, signaling the refreshing breeze and providing psychological comfort. Two trees are also installed, a fir and a birch, chosen for their special capacity to absorb fine particulate pollutants (PM 2.5) from the air. Bap! Biennale d'architecture et du paysage Versailles, France, 2019

The Emissive Façade

The Emissive Façade proposes rethinking the materials of walls according to their thermodynamic performance. We offer two coverings, one for summer (or for warm climates) and the other for winter (or cold climates). The summer coating is made of aluminum, a material that will not re-emit radiant heat to people in the vicinity. To reinforce the refreshing property of the material, we propose adding a layer of violet paint since, at around 350 nm, this wavelength heats water least, which makes up some 70% of our body. Conversely, the winter coating is made of rubber, a high emissivity material that will strongly absorb heat from the sun's rays and re-emit it to nearby individuals. To reinforce this material's heating property, we propose painting it in red paint, because this 800 nm wavelength is one that most heats the 70% of water of which our body is composed. Chicago Architecture Biennal, USA, 2017

The Anthropocene Style

Thinking about wall coverings in terms of emissivity enables reconsidering thermal issues leading to the choice of a material. Emissivity is the physical property of walls or a tapestry to emit more or less heat by radiation to the bodies of individuals who are in their vicinity. Some materials have low emissivity, which means they do not heat by radiation, remaining cold under the sun. Others have high emissivity, and produce a significant amount of radiation that is sent to the bodies nearby. As for flooring, materials can be chosen according to their effusivity value. A low effusivity mat, made of wool or wood, will feel warm to the touch, as it exchanges little heat. Conversely, a high effusivity mat, made of metal or stone, will seem cold to the touch, because it quickly absorbs human heat, cooling of the body. The Anthropocene Style involves resetting the criteria for judging the decorative elements of architecture in a practical and thermal mode of evaluation, in order to fight against today's current global warming, mainly caused by the construction sector. San Francisco Art Institute, USA, 2018

The Low-Emissivity Base

The Low-Emissivity Base is a proposal for the lower outer part of the façade of buildings in warm climates. In order not to heat passers-by, we propose that the base of buildings, up to human height, should be made of aluminum, a very low emissivity material that does not re-emit heat by radiation to the people nearby. The surface is polished because this also lowers its emissivity. In order to reinforce the refreshing property of the material, it is painted in violet-colored aluminum paint, with a wavelength of about 350 nm. Because shorter wavelengths heat water less, it is possible to protect our skin and our body from radiant heating since human tissues are composed 70% of water. Seoul Biennale of Architecture, South Korea, 2017

The New Olduvai Gorges

To increase heat in winter, we propose different solutions that stimulate endogenous thermogenesis, which gradually develops outside the body without ever becoming clothing or house: a platform beneath which a field of peppers is growing, illuminated by photosynthesis-enhancing lights. Peppers stimulate the ion channel and the TRPV1 receptor, which detects temperatures over 44°C, also activated in the presence of "capsaicin", a molecule that processes intense heat stimuli. There is also a device for making hot chili tea for drinking, a tea dispenser, a professional cooking device for boiling meat and ingesting proteins that help the body produce heat and maintain a temperature of 37°C, a bench for five people, to huddle together and share their human warmth. The Royal Danish Academy of Fine Arts, Copenhagen, Denmark, 2009

Black Day

Black Day is a negative urban lamp, producing night during the day, physically. It is an inverted response to the perpetual day created by modernity, the Internet, and contemporary globalization. Following on from the "White Night" we would like to invent the "Black Day". In the wake of noctambulism, we propose diurnism. To create night in the streets during the day, we are working on creating a precise spectrum of electromagnetic radiation emissions, as for a traditional lamp. But if a common lamp, like a mini-sun, emits a portion of visible light (artificially reproducing the electromagnetic waves of the bright, hot sun), our outdoor lamp, to the contrary, emits cold, invisible electromagnetic radiation, like that of the night sky. Our lighting consists of a mini night sky, cooled by a flow of glycol water to a temperature close to 0°C. As on an icy night, our body, at 37°C, will lose energy by infrared radiation toward the celestial vault. Gdansk, Poland, 2005

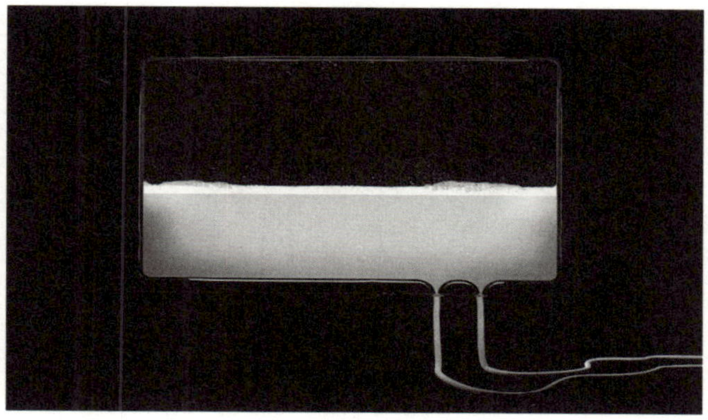

Fermented Films

Watching a movie also means losing energy through our bodies, even if we are seated in an armchair. This loss of physical energy, in the form of the emission of heat at 37°C, takes place to ensure the basic functioning of the metabolism - heartbeats, respiratory movements, hormone synthesis. We can quantify the energy lost over a period of 2 hours of movie at 120 kcal. We propose a cinema screen as a thin translucent cavity that will be filled with 9 liters of milk at the beginning of the screening. During the movie's two hours, the milk will heat up by absorbing the heat from the images and enter a forced fermentation process. At the end of the screening, the screen will be emptied of its 9 liters of milk to fill yoghurt machines in a nearby room. Then, with these 9 liters of milk we will make 60 yoghurts of 150 ml each containing 120 kcal. A yoghurt will be given to each viewer at the end of the movie, allowing them to recover the 120 kcal lost through metabolism during the screening. Kunst-Werke Berlin, Germany, 2009

Reversed Cinema

The Reversed Cinema proposes losing rather than receiving information. The goal is to reverse the roles between the broadcaster and the viewer, and make cinema, television or billboards absorb information in the form of electromagnetic projections rather than broadcast it. From being receivers, our bodies become emitters, emitting infrared rays that will be absorbed by the cold, black cinema screen. All bodies emit energy in the form of electromagnetic radiation. Our skin, at a temperature between 33°C and 36°C, emits infrared rays. If a nearby object is colder than our skin, the heat of our body is evacuated in the form of radiation toward that object. Our body, a projector of invisible information, thus placed in front of a screen cooled by a cold water system, will transmit its heat to this screen in the form of radiation, becoming an emitter, while the screen becomes the receiver. Centro Andaluz de Arte Contemporáneo, Seville, Spain, 2004 (Décosterd & Rahm, associés)

Digestive Gulf Stream

Two horizontal metal planes are placed at different heights. The lower one is heated to 28°C and the upper one is cooled to 12°C. Like a miniature Gulf Stream, their positions create a movement of air by means of the convection phenomenon in which the rising warm air will cool on reaching the cold pole and descend, and so on, creating a continuous flow of air, an interior wind like an invisible landscape. The architecture here is literally structured by a current of air, the design of the invisible, an atmospheric space. This architecture is a meteorological construction. The occupant can then move around in an invisible thermal landscape ranging from 12°C to 28°C, temperatures at both extremes of human comfort, in a space where it is possible to freely choose the climate according to one's activity, clothing, diet or social desires. Venice Biennale of Architecture, Italy, 2008. Photo: Noboru Kawagishi.

Convective Screen

The Convective Screen is a screen made of materials with the highest air impermeability. Placed in front of the door, it stops the air currents that would cool the skin of the human body by convection, allowing the heat generated by the human body to remain on the surface of the skin. Without any movement of the air, heat lost through the skin forms a layer of warmer air that stagnates around the skin, heating it. This layer can be lost if there is a movement of air, which gives a physiologically very real impression of cooling. By convection, the air carries this layer of heat around the skin away, bringing it down to a lower temperature. For this reason, the temperature felt depends on the wind. The less wind there is, the warmer it feels. By installing the most effective screen in terms of air impermeability at a human height, the movement of air around the body can be prevented and the layer of heat around the skin is preserved, creating a more comfortable environment.
25th Biennale Interieur, Kortrijk, Belgium, 2017

The Anthropocene Style

Designed for the low temperatures of continental regions, Anthropocene Style furniture and decorative items are made to improve indoor comfort by providing more warmth, while following precise physical properties. The "low-effusivity rug" is a new type of rug made of very low thermal effusivity materials, which do not transmit the cold of the ground when in contact with the human body. The "low-conductivity tapestry" is a tapestry made of materials with a high thermal insulation capacity that prevents the cold of the walls from being transmitted to the indoor air. The "convective chair" is a high chair that reaches the highest and warmest layers of air, according Archimedes' law that states that hot will rise and cold air will sink. These seven pieces of decorative art sketch out an "Anthropocene Style," where materials and shapes are chosen for their high capacity to fulfill a thermal role that reduces energy consumption and CO2 emissions.
25th Biennale Interieur, Kortrijk, Belgium, 2017

Low emissivity curtain

The "low emissivity curtain" is a curtain made of a very low thermal emissivity material, which therefore does not allow the cold radiation of the outer walls to reach the inner space, sending it back on itself. This prevents cold rays from reaching the human body located in the vicinity of the curtain, protecting it from radiative cooling. External walls normally have high emissivity, and because their temperature is permanently cooled in winter by the cold outside, they capture heat from any hotter body placed nearby, inducing a feeling of cold for the occupants, a very real sensation because the human body actually loses heat by radiation, and this thermal energy is absorbed by high-emissivity walls. By placing low-emissivity curtains, which are not capable of absorbing or emitting heat by radiation, against the walls, the heat exchange between the hot human body and the cold wall is then blocked by this interface, thus protecting the human body from cooling.
25th Biennale Interieur, Kortrijk, Belgium, 2017

Low-emissivity mirror

The "low emissivity mirror" is a mirror that reflects only the infrared waves of light, absorbing the visible wavelengths. It thus returns the thermal rays emitted by the human body toward the body itself. Indeed, the human body constantly loses heat by radiation in the form of infrared rays if it is in an environment whose temperature is lower than that of its skin, namely 34°C. By forming a mirror using a layer of chemical materials whose properties reflect the infrared, the specific part of the electromagnetic rays emitted by the human body will be returned to the emitter, warming the body and causing a sensation of warmth, or at least allowing the body to retain its own heat. 25th Biennale Interieur, Kortrijk, Belgium, 2017

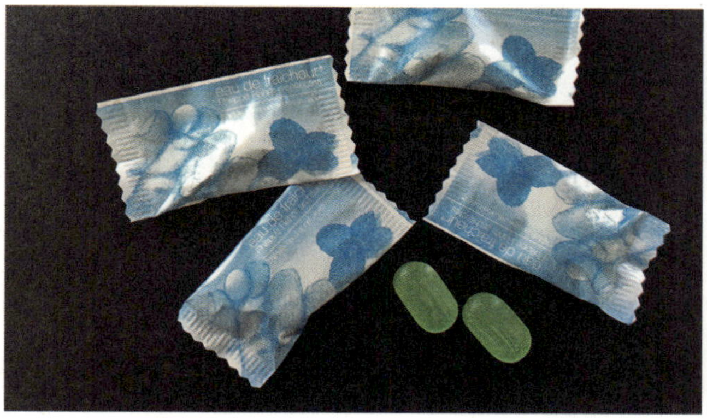

Digestible Cold

The temperature of our environment is perceived by neurons through their projections on the skin. Ion channels of the transient potential receptor (TRP) family participate in the mechanisms whereby we feel hot or cold, and these channels have a direct or indirect influence on our thermoregulation. TRP8 can be activated by cold temperatures but also by menthol when it comes into contact with the skin or other mucous membranes such as those of the oral cavity. The threshold for activation by cold is between 15°C and 25°C. Mint, meanwhile, has molecules of crystalline origin known as menthol that cause the same sensation in the brain as the freshness perceived at a temperature of 15°C. Thus mint has traditionally been used in excessively hot climates to trick the brain, providing a false, unreal feeling of freshness.
Venice Biennale of Architecture, Italy, 2008

The New Olduvai Gorges

As usable flooring for the exhibition, we propose planting chilis which contain molecules, capsaicins, that activate the TRPV1 neuroreceptor, which is sensitive to temperatures above 44°C. Thus, when you eat chili pepper or rub your skin with one, you will feel very strong heat (even pain), similar to that of very hot water, which will trigger a whole series of physiological mechanisms of thermolysis (vasodilation, perspiration) allowing to evacuate the heat more forcefully through the human body. Thus, eating spicy food in hot climates will distort the understanding of the temperature around us, making the brain believe that it is much higher than it actually is, accelerating the process of thermolysis to refresh the body, ultimately producing a real cooling effect.
The Royal Danish Academy of Fine Arts, Copenhagen, Denmark, 2009

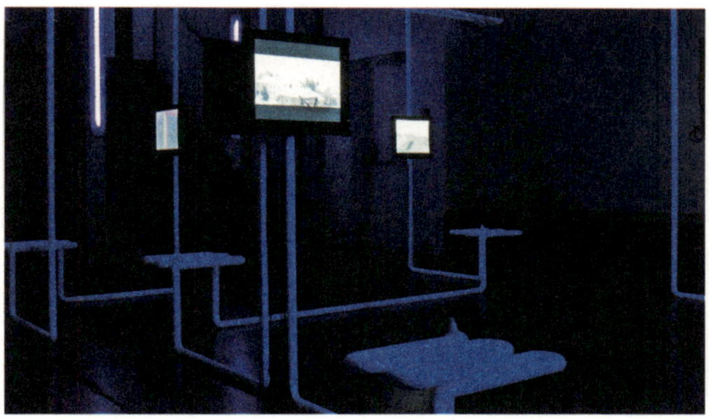

Thermal Chair

Architecture can be divided into three layers: "immobilier" (real estate), "emmobilier" (trimmings and fittings), and "mobilier" (furniture). "Emmobilier" is the halfway layer that contains all the intermediate responses to climatic problems - thermal insulation, vapor barrier, waterproofing, upholstery, carpets - as well as all the technicalities of the building - heating, ventilation, electricity, lighting, etc. It is precisely this intermediate layer that we would like to remove from the first and integrate into the second. We propose combining the hot water pipe of the radiator system with chairs, bringing the heat as close as possible to the body.
Swiss Cultural Center, Paris, France, 2005

Domestic Astronomy

"Domestic Astronomy" is a prototype for an apartment where it is no longer a surface but an atmosphere that is occupied. Its functions and furniture leave the floor and rise: they diffuse and evaporate into the atmosphere of the apartment, stabilizing where the temperatures correspond to certain activities, the body or clothing. Our proposal is to enable these physical differences in the distribution of temperature in space and to exploit them by changing lifestyles: replacing a horizontal lifestyle with a vertical one, where we occupy different heat zones, different layers, different heights. Thus we create a global ecosystem similar to an astronomy of the home, where temperature, light, time and space are redefined. Louisiana Museum, Denmark, 2009

CHAPTER VI.

WEATHER REPORTS

VI.I. Meteorological Architecture

2005-2009

INTRODUCTION

The concept of climate as a theme of architecture has gained in importance in view of the challenges of sustainable development in response to the problem of global warming. In architecture, we understand climate as the management of heating and cooling inside a building to achieve moderate comfort. It is of course one of the essential missions of architecture to make sure that, in the cold of winter, it is warmer indoors or, conversely, we are protected in summer from the sun and extreme heat by retreating to the cool of the indoors. Paradoxically, it is by heating or cooling the indoor climate that fossil fuels are burned, releasing gases that cause the greenhouse effect and global warming.

Today, we must ask ourselves about this climate-related mission of architecture, no longer accepting it merely as a goal but also as a means. Climate must be integrated upstream of the architectural project, in its very language. It must modify the elements, the structure, the modes of composition, and finally the esthetic criteria. Architecture, like climate, extends its scope of action to other sensitive, thermal, plastic, and material dimensions. It opens up to other olfactory, cutaneous and hormonal perceptions. It is now necessary to reassess the field of architecture, from the physiological to the spheric, to redefine its forms and typologies, undertaking our responsibilities in view of the new ecological concerns regarding climate.

If usually the form of a building and the quality of a space are approached in terms of surface and volume, here we would like to propose a meteorological and atmospheric architecture. The reason for this paradigm shift, from the surface to the meteorological, from the volume to the atmosphere, is driven by our conviction that the current challenges related to climate, protecting it against global warming and the depletion of resources, should cease to be considered problematic or worrying, but on the contrary, should bring about the emergence of new means and solutions. Architects must generously and enthusiastically latch on to these new climate data rather than view them as additional technical constraints. It is not solely a question of accepting them, but more deeply of making the physical elements of the climate, i.e., pressure, vacuum, temperature, relative humidity, become the actual materials of architecture, the new tools for architectural design. Approaching architecture in terms of climate means projecting into another spatiality, a more sensual relationship with space, inhabiting interior space as an atmosphere, with its diversity of climates, its meteorological variations and its gradients.

After decades devoted to the visible, in which a subjective approach and "storytelling" shamelessly replaced the progressive and moral programs of modernity, we are now in a new and extremely interesting period. A slippage of the real from the visible toward the invisible is taking place, and a shift of architecture toward the microscopic and the atmospheric, the biological and the meteorological. The tremendous breakthroughs in the life sciences now resonate with the climate and the concern for its warming. Between these two extremes, the field of the visible, hitherto saturated with symbols, morals, narratives and individual interests, is in the process of deflating, emptying, diffracting, lightening, deforming, de-programming. Deployed somewhere between the physiological and the climatic, between determinism and freedom, this now open, floating, undecided in-between is the space for a new humanistic landscape.

We wish to re-found the language of architecture as a result of this shift towards the invisible; to stretch architecture between the infinitely small and the infinitely large, between the physiological and the meteorological. We seek to get back to the essence of the architectural elements, subsequent to this collapse of the visible. The means of architecture must become invisible and light, produce places as free, open landscapes, new geographies, other meteorologies. They must renew the idea of form and use between sensation and phenomenon, between the neurological and the meteorological, between the physiological and the atmospheric. We are seeking meaningless, narrativeless, interpretable spaces, in which boundaries evaporate and structures dissolve. It is no longer a question of building images and functions, but of opening up climates and interpretations. Working on the void, on the air and its movements, on the phenomena of conduction, transpiration, convection as so many transient weather conditions as new paradigms of contemporary architecture. Moving from metric to thermal composition, from structural to climatic thinking, from narrative to meteorological thinking. Composition of the hourly air renewal, plan of relative humidity levels, inhabitable convection, thermal design, the path of air movements, pressures and vacuums, temperature stratification as new means of architectural composition and then reveal their programmatic, plastic and sensual potentials. Space, hitherto empty and abstract, materializes in an electromagnetic, chemical, sensory atmosphere, in which we are immersed and which, by inhabiting, we in turn compose by breathing, perspiring, through the thermal radiation of our body, our physical and hormonal activity, according to our movements and the clothes we are wearing. Between the infinitely small of the biological and the infinitely large of the meteorological, architecture must construct sensual exchanges between the body and space, the senses, the skin, breathing and climate, temperature, variations in humidity and light.

To date, the processes for building cities and buildings have produced petrified narratives, icy forms of social, political and moral conventions. They formed frozen cultural landscapes that once opposed the fluctuating and unreasonable nature of the countryside and climate. Overwhelmed by recent breakthroughs and failures in biology and air pollution, this dichotomy no longer exists. Then, in turn, we can appropriate the tools of physical geography in order to generate cities and buildings devoid of their narrative, their functionalism and their determinism; buildings and cities that then reveal themselves as pure presences. Fluctuating atmospheres, open, objective landscapes, "devoid of adjectivity", which we inhabit thus interpreting them. Architecture as a new atmosphere and a second meteorology is no longer the "subjective", closed place of the representation of social and political relationships, but rather becomes the "objective" and open place where new social and political relations can be invented.

ENERGY

Until the twentieth century, the notion of symmetry was the basis of architectural composition. Vitruvius announced this from the outset in the first tome of De Architectura: "The purpose of architecture is order, arrangement, eurythmy, symmetry, propriety and economy." He goes on to state that "symmetry is a proper agreement between the members of the work itself, and relation between the different parts and the whole general scheme, in accordance with a certain part selected as standard". In the 15th century, Alberti raised the notion of symmetry above others. For him, beauty was a kind of harmony and agreement between all the parts which form a whole, constructed according to a fixed number, a certain ratio, a certain order as required by the principle of symmetry, which is the highest and most perfect law of nature, (Book IX, chapter V). And Claude Perrault set out the reasons for it in the seventeenth century in his book Ordonnance for the Five Kinds of Columns: "The Ancients rightly believed that the proportional rules that give buildings their beauty were based on the proportions of the human body." That the human body has served as a model for architectural composition corresponds to this Hegelian idea of art as a means for man to project himself to the outside world to domesticate it. The classical artist in his image transforms the wildness of nature. He corrects its asymmetry and disorder by projecting the symmetry and proportions of his body and the mathematical rules of his thought.

Paradoxically, this project was carried out in classical era only in regard to the esthetics. The notions of balance, order and symmetry were applied to the design of façades, to the arrangement of the bodies of buildings but not at all in the invisible where disorder and disharmony continued to reign: insufficient ventilation, low, poorly distributed interior temperatures, darkness. In classical architecture, the external harmony of façades, ordered by symmetry, is unintentionally counterbalanced in the invisible by an imbalance and an asymmetry of the thermal qualities of the interior space. In the visible, the art of composition was perfectly articulated and mastered after Vitruvius. In the invisible realm, however, the poor heating techniques – which only started to be mastered at the end of the 19th century – induced uncontrolled and uncomfortable, asymmetrical, unbalanced and unharmonic thermal phenomena. The fireplace, lit only intermittently, performed poorly. Walls were insufficiently insulated.

For Hegel, the art of architecture was to order, arrange and reconstruct the initially messy lines of nature through symmetry. Although this program was perfectly executed in the visible after Vitruvius, it was not until the end of the 19th century that it began to be rendered in the invisible, thanks to the improvement of air-conditioning techniques. Central heating and radiators replaced fireplace and stove. Gas and coal replaced wood. Thermal equilibrium and homogeneity within the building were thus achieved for the first time. Surprisingly, this victory of equilibrium in the invisible was accompanied by an assumed return to formal disorder in the visible, as if this new mastery of the climate in the invisible enabled freeing oneself of the rules of harmony and equilibrium in the visible. At the same time that, in the visible, Le Corbusier rejected symmetry in favor of the regulatory forms, he paradoxically sought in the invisible to homogenize heating, to normalize the interior climate by elaborating the eloquent concepts of "neutralizing wall" or "correct breathing". Likewise, the tremendous asymmetries and plastic imbalances of the houses in Frank Lloyd Wright's high meadow are accompanied, as consideration, by the first modern implementations of heated floors that generate a perfectly balanced, standardized, silent indoor climate. We are witnessing a pendulum phenomenon between the visible and the invisible. There is a reversal to modernity, changing the values of balance and symmetry from the visible to the invisible and those of disorder and asymmetry from the invisible to the visible.

The triumph of thermal equilibrium in the invisible at the end of the 19th century paved the way to formal asymmetry and imbalance in the visible. While during the 20th century the forms of buildings moved visually further from the equilibrium, an inverse process took place in the invisible which tended toward the normalization, neutralization
and trivialization of indoor climatic forms. From Frank Lloyd Wright to the buildings presented by Phillip Johnson and Mark Wigley during the "Deconstructivist Architecture" exhibition at MoMa in 1988, formal asymmetry developed and was theorized to steal the esthetic place that was previously reserved for symmetry. Greg Lynn so stated in the article "The Renewed Novelty of Symmetry": "Symmetry breaking is not a loss

but an increase in organization within an open, flexible, and adaptive system....Symmetry is not a sign of underlying order but an indication of a lack of order due to an absence of interaction with larger external forces and environments." This is a paradox: if recent years have seen the emergence of extraordinarily visually unbalanced buildings, it is on the contrary a depletion of climatic diversity that has developed in the invisible where new heating and ventilation techniques, such as outdoor insulation or air-conditioning, have normalized and trivialized interior temperatures and climates.

This value reversal between symmetry and asymmetry, visible and invisible, formal and climatic, certainly began with the rise of thermodynamics in the 19th century which changed the esthetic notion of imbalance, from ugly to beautiful. The study of steam engines has shown that the imbalance of a system is the prerequisite for any movement and for life itself. To produce energy, a system must be open, unbalanced, and have two separate heat sources, one hot and one cold. Conversely, equilibrium does not allow the production of energy, it irremediably annihilates any possibility of movement and leads to death. The second law of thermodynamics, the principle of entropy, was established in the 19th century. It showed that in a closed system, energy is constant and that the world then inexorably evolves towards a maximum entropy, a collapse of its structures, a degradation of order, a decrease in potential energy, leading to thermal death, in a definitive final equilibrium where life and movement become impossible. Thermodynamics reverses the value criteria: equilibrium equals death, while imbalance opens up life in the notion of negative entropy: neguentropy. Invented in 1944 by the physicist Erwin Schrödinger in "What is Life?", the concept of neguentropy explains the phenomenon of life by an increase in order thanks to the metabolism of living beings. Made popular in France by the biologist Henri Laborit during the 1970s, it has become a key concept of ecology today. It is this phenomenon, formalized in the twentieth century, that allows animals to grow, to increase their structure by drawing their energy from food and, further upstream, plants to grow and develop through photosynthesis by harnessing the energy emitted by the sun. The thermal imbalance between the sun and the earth is the very basis of life. With his theory of dissipative structures, the chemist Ilya Prigogine showed that a system, far from equilibrium, can suddenly become ordered. The imbalance of a system is in itself responsible for the origin of life, that is, an open system can spontaneously structure itself and reverse entropy. Beings would be alive because their relationship with the environment is unbalanced. Through metabolism, i.e., breathing and nutrition, where the living being increases its order, its structure, its energy potential by drawing them from a "hot" source to reject them, disordered and degraded, in the "cold" source. Order, life, and more generally a form, only appear due to the imbalance and asymmetry that occur in a system. The void, for example, is shapeless, because it is symmetrical and invariant in whichever translational or rotational movement. Only an imbalance can cause a necessary phenomenon of asymmetry to break its isotropy and its homogeneity and allow the appearance of a form. The imbalance of a system is asymmetrical, since it produces irreversible phenomena where the forms are transformed in relation to time. In chemistry, the reduction in symmetries which, for example, takes place during crystallization when a material cools, corresponds to an increase in order. The 2008 Nobel Prizewinners for Physics, Makoto Kobayashi, Toshihide Maskawa and Yoichiro Nambu, have shown that the world we see exists because it is asymmetrical. A rupture in the symmetry of the Universe occurred during the Big Bang when the symmetrical cancellation of matter and antimatter left an excess of matter corresponding to what we can see of the Universe, the galaxies, the planets, the earth, life. The appearance of a form is the result of a break in symmetry, itself a consequence of an imbalance in a system.

Produced in the 20th century, this new scientific knowledge reversed the order of values between symmetry and asymmetry, equilibrium and disequilibrium, life and death, beautiful and ugly. They literally challenged the reasons invoked by Vitruvius for choosing the notions of symmetry, balance or homogeneity as criteria of beauty in architecture. Hence, during the twentieth century, there was a complete reversal of values in the face of the new definitions of the Universe and of life arising from thermodynamics and the breach of the symmetry of the Universe, for which asymmetry was the yardstick of beauty. Because without asymmetry, there would be nothing but death and emptiness. And to Heidegger's question, which is the basis of metaphysics: "Why is there something rather than nothing?" Today we can answer: "Because the Universe is asymmetrical. "

Our aim today is to fully accept this new esthetic field. The architectural projects presented below are based on thermal imbalance and climatic asymmetry. They explore plastic, formal, programmatic, ecological and esthetic potential. They transform the interior of the house into a living, asymmetrical, unbalanced atmosphere, with its cold poles and tropical equators, its positive and negative pressures, its variations in humidity and light where architecture becomes meteorology.

Interior Gulf Stream

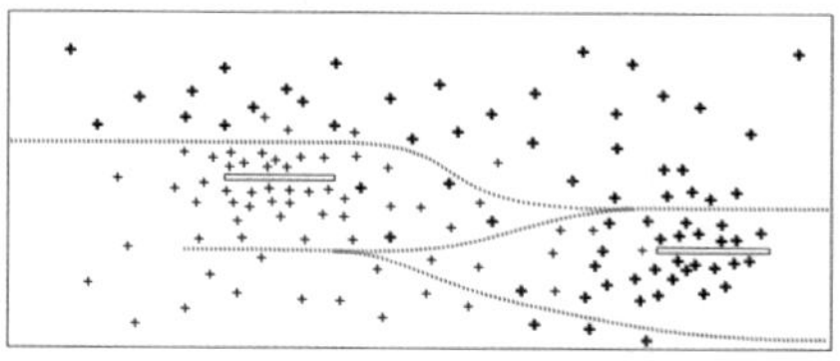

The thermodynamic phenomenon of the Gulf Stream is one of the most fascinating models today for conceiving architecture. It opens up a way of meteorological thought that enables escaping the normalization and homogenization of modern space. This climatic phenomenon is generated by the polarization in space of two different heat sources: a cold source, at the top, and a warm source, at the bottom. In space, this thermal polarization causes a convective movement of air (or water in the case of the Gulf Stream), sketching a complex, rich, invisible thermal landscape, defined according to multiple zones of different temperatures as climates, sensitivities, and terroirs. In the process of our proposed architectural project for the house and atelier commissioned in 2008 by the artist Dominique Gonzalez-Foerster, it is first of all an atmosphere that is created, before the program, before the spatial forms. An indoor atmosphere is initially generated in an unbalanced and dynamic manner, unlike the modern atmosphere that is standardized, conditioned and homogenized throughout the year. The atmosphere produced is moving, variable, like real meteorology, with its hot or temperate zones, its cold zones, its rising hot air currents, its high and low pressures, its sinking masses of cold air. It is only then that the program is placed in the space, in a movement of interior transhumance seeking sensual thermal conventions that intersect with the criteria of localized ambient temperatures, clothing, physical activity and pure subjectivity.

Modernity has led to standard, uniform spaces in which the temperature is regulated at around 21 degrees. We seek to provide the body with a rich, sensitive rapport with space and its temperature, to allow movement within the house, migrations from downstairs to upstairs, from cold to warm, winter and summer, dressed and undressed. Today, with the desire to save energy resources, the requirement, depending on the function of the room, is to install a precisely calculated thermal power in order to spend only what is strictly necessary on energy. The Swiss building standard SIA 384/2 thus sets the following ambient temperature values for each room of the house: bathroom at 22°C, living room at 20°C, kitchen between 18 and 20°C, bedroom between 16 and 18°C, corridors and toilets between 15 and 18°C.

These recommended temperatures obviously take account of clothing and activity, between the nudity of the bathroom, the thermal protection of bed covers, the light clothes worn in the living room.

Instead of spatially separating each room and heating each one specifically to a certain temperature according to the above recommendations, we propose thinking of the entire house as a global atmosphere. In the volume of the house there are two different sources of heat. Two radiators in which water circulates heated to two different temperatures; 15°C for the upper radiator, and 22°C for the lower radiator. These two thermal poles, at the two extremes of comfort, produce thermodynamic tension throughout the house. The cold pole, at 15°C, at the lower level of domestic comfort, is situated in the lower part of the house. The hot pole, at 22°C, however, is located in the upper part of the house. Like a miniature Gulf Stream, their asymmetrical positions generate a movement of air by causing a circular, dynamic convective phenomenon produced by the temperature difference between the two poles. In contact with the lower, hot plate, the air expands, heats, its density decreases, rising a few meters until it reaches the upper, cold plate where it will cool down. This contact causes it to sink to the ground until it again touches the hot plate where it reheats, and so on, creating a continuous heat flow sketching an invisible landscape.

With the help of a thermal modelling program, we analyze the spatial distribution of the air and the temperature variations caused in space. We thus discover the most suitable zones for certain activities according to their temperature, as recommended by standard SIA 384/2. The project process is thus reversed: it is first a climate that is built and then the functions are placed in it freely, according to temperature, clothing, activity and personal preference. An overall saving is made on heating the house, whose average temperature has been reduced to 18°C. The floors of the house are thus designed and deformed in order to seek out the right temperatures in the space according to the form the air takes in the entire volume of the house, according to its vertical and horizontal movements, according to the temperature and the functions that the atmosphere suggests. The floors and partitions are perforated so as never to stop air movements. What interests us here is not to create homogeneous, established climates, but rather to create an aerial plastic dynamics, to implement forces and a polarity that give rise to a thermal landscape. Between 15°C and 22°C, the occupant can then move within this interior geography and freely choose a climate according to their preferences regarding clothing, nutrition, sport, social and other activities, and mood. The architecture is literally structured around an air current, deploying a fluid, aerial, atmospheric and dynamic spatiality. It becomes the construction of meteorologies. Architecture no longer builds spaces but atmospheres, it designs temperature fields.

Digestible Gulf Stream

The notion of thermal comfort is not only quantifiable in terms of outside temperature. It is also dependent on clothing, on the physical activity of the occupant, on diet. It results from a relationship between different dimensions, between the macroscopic and the microscopic, the climatic and the nutritional, the atmospheric and the physiological. Architecture is then the medium that organizes these various scales, that amplifies one to allow the other to shrink, which transforms one sensation by working on another. There are four ways to cool down if you are too hot. They act on different scales: Reducing the air temperature of the space by cooling it with the help, for example, of air-conditioning (atmospheric solution). Drinking (physiological solution). Removing clothes (social solution). Resting (physical solution). Each of these solutions is related to architecture that becomes

the art of creating polarizations on different scales: atmospheric, physiological, social, physical, between hot and cold, dressed and undressed, movement and stagnation, meteorology and diet. Architecture is a thermodynamic mediation between the macroscopic and the microscopic, the body and space, the visible and the invisible, the meteorological and physiological functions. Digestible Gulf Stream for the Venice Biennale of Architecture in 2008 is a manifest installation that is suggested as a concrete exploration of architectural language as a thermodynamic imbalance. It seeks to increase the field of action of space towards the microscopic and the macroscopic, the edible and the atmospheric, the invisible and the visible.
The project builds a convective movement similar to the previous project as a sort of inner "Gulf Stream" in the space of the Corderia of the Arsenal. It acts on scales, polarizing their differences (hot/cold, low/high, dressed/undressed, indoor/outdoor, rest/activity) to give rise to an architecture as a dynamic convective movement, to create a place as a physical geography, to design space as a climate, to open architecture as a design of the atmosphere and as gastronomy.

Two horizontal metal plates are placed at different heights in space. Their dimensions are similar; 2.40 m by 6 m. The lower plate is heated to 28°C, the upper one is cooled to 12 °C. This architectural device polarizes the space between two temperatures.
The thermodynamic difference generates an atmosphere and causes a meteorology, with its air movements and its thermal variations. The device, in a simplified form, reproduces the terrestrial climate induced by a polarization of space and time between pole and equator, cold and hot, winter and summer, night and day. These astronomical asymmetries are responsible for the variety of climates and the wealth of landscapes. Suspended at a height of 2 m, the cold plate radiates at 12°C, and in this spatial thermal equation acquires the role of a boreal pole, something like a perpetual indoor winter or a domestic Arctic.
In contrast, the warm plate, slightly raised above the ground, at a height of 25 cm, forms a warm, comfortable mat, which offers a surface at the upper level of the thermal comfort zone, at 28°C, just before being "over warm". In space, it defines a kind of miniature summer or perpetual tropics, on which the occupants can sit, lie, warm themselves in the Venetian fall. A thermal oasis, this warm plate conjures up an whole mythological tradition of the temperature of 28°C, when house and clothing become superfluous. This is the temperature at which the air is heated in swimming pools so that bathers do not catch cold when leaving the water.
At this temperature, architecture can disappear, the external environment being perfectly comfortable and temperate, totally suited to the human metabolism, like paradise where no dwelling or clothing are necessary. In this thermal instant, the need for shelter disappears and architecture becomes superfluous.

A mythical tradition has been built around legendary or utopian places where it is neither too hot nor too cold, where one can live naked and without shelter, where the body and space become one. It has taken different forms over the centuries and mythologies: "Perpetual spring", "lost paradise", "garden of Eden", "golden age", "Ogygia", "Shangri-la". While architecture has always tried, and more or less succeeded, to reconstruct these paradises indoors at 28°C, thanks to heating and artificial light, it is in modern times that the reconstruction of paradise lost is imagined not just indoors, but also outdoors, where the borders between interior and exterior, artificial and natural, indoors and outdoors, architecture and furniture would disappear to achieve a shared and globalized 28°C. In Monte Verità, Switzerland, at the beginning of the 20th century, it was the political invention of the outdoors, the liberation of the body that undressed and opened freely to the outside air and sunlight. Yves Klein, in the 1950s, called for the disappearance of hard, solid architecture in favor of an architecture of air and fire where a pure dispenser of heat in space would suffice to make the world habitable. Buckminster Fuller dreamed of a Manhattan under a bell where eventually the whole planet would become a great interior in which one would no longer produce houses and real estate but only furniture and objects like the continuous monument of the radical Italian architects of the late 1960s. So many projections of spaces around 28°C, where the atmosphere of the outside world is at body temperature, where the need for architecture disappears and through which the hippies sought a new harmony between man and nature, interior and exterior.

This is the most ideal expression of architecture, one of a transformation of the entire outside world into a comfortable space acclimatized to human physiology, at 28°C: the transformation of the planet as a vast interior. Due to its asymmetry and imbalance, our architecture as meteorology refuses this climatic homogenization of space as it was envisaged in the 20th century. Digestible Gulf Stream produces a space at 28°C, but also a space at 12°C, and the quality of the project arises from the tension between these two contradictory polarities. It is only in this imbalance that the interest of an architecture as the construction of climates lies.

Convection

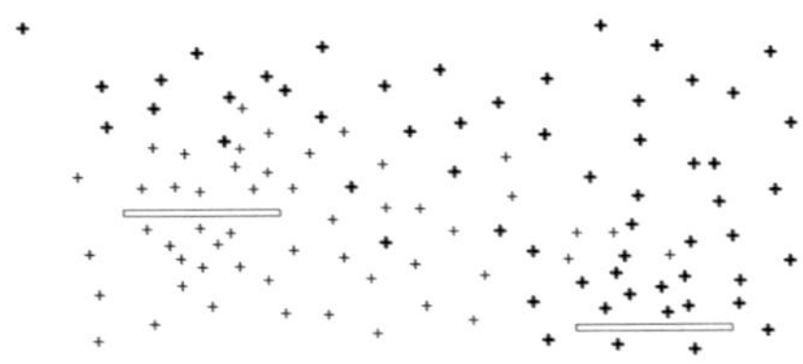

For the Museum of Contemporary Art in Wroclaw, Poland, we use this thermodynamic project method by integrating the climatic constraints specific to running a museum, between the conservation of the works and visits by the public. The project is developed in the same way as Digestible Gulf Stream, reversing the project timeline between climate and function.
The project process begins with the installation of two heat-generating poles, two radiators, one colder at 16°C and the other warmer at 22°C. We choose these two temperatures because they set the lower and upper limits of the temperatures required for a museum program, 16°C corresponding to storage or little used premises, to maximum office temperatures of 22°C. These two poles are arranged diametrically opposite on the plan, the cold pole being placed higher than the hot pole in order to create a thermodynamic imbalance. Thus, for the museum as a whole they generate an atmosphere composed of a multitude of microclimates and flows between 16 and 22°C, different temperature zones in which one can freely seek a temperature in relation to a programmatic requirement. When air heats, its weight decreases and it rises. At some point during its movement, this warmer air will encounter the colder pole, cool down upon contact and sink. And so on. This polarity of the two sources of different heats generates a circular flow of air throughout the museum whose average air speed remains at less than 0.15 m/s so as not to create any discomfort. What interests us is to first set up this movement of air which, throughout the entire volume of the building, will create multiple temperature zones, as different climates, atmospheres, spatial qualities where different activities and particular desires can be embodied. In this thermal landscape, special climatic situations emerge. The computer model of the climate inside the museum thus allows us to position the program by looking, on plan and in section, for temperatures in accordance with the uses and museographic recommendations, between storage at 16°C and offices at 22°C. Between 21 and 22°C we will locate the premises in which normally dressed people are at rest for some length of time, such as offices. Between 21 and 22°C we will locate the premises in which normally dressed people perform very light physical activity, such as in meeting or conference rooms, auditoriums, restaurants. Between 18 and 20°C, the premises in which normally dressed people perform light physical activities, such as in exhibition rooms, workshops, kitchens. Between 17 and 18°C, we will place the premises that only serve as a passage for normally dressed people, such as corridors, stairwells, toilets and the foyer. And finally between 16 and 17°C the premises that are not regularly occupied, such as archives, storage: 16 or 17°C. The air must circulate freely throughout the entire museum, infiltrating into all spaces. For this, walls, floors and ceilings are perforated with a varying full-to-empty percentage ratio according to the needs of the program. The architecture here arises from a modern uniform formal and programmatic determination to open up to a climatic indeterminacy in which the occupant no longer moves only in space, but also in an atmosphere, according to heat latitudes and altitudes. Their movements become transhumance and migration, the architecture of the building establishing a true geographical mapping of meteors.

Synchronized seasons

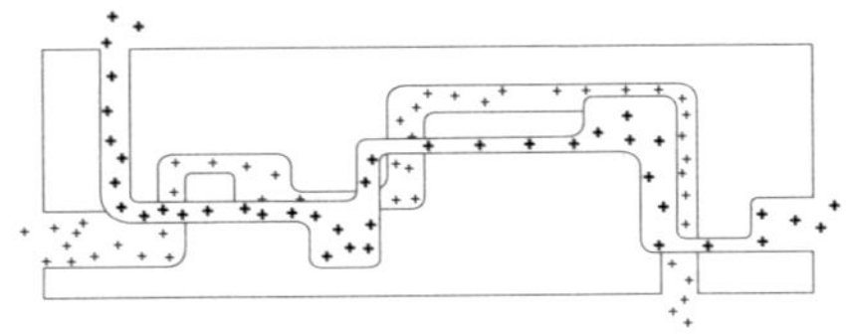

VEGA 3 is an urban development of 28,000 m3 in Marghera, a district opposite Venice, Italy, which involves a project to transform an industrial warehouse into offices and shops. This project thematizes the need for air renewal with heat recovery in order to make it the origin of the project's architectural form. This project revolves around two ventilation ducts whose dimensions are increased to become climatic landscapes, like two synchronized inhabitable seasons, which crisscross the entire building. The first duct, oriented to the north, physically accentuates the coldness of its interior climate, by precisely using a high-inertia mineral material, due to the whiteness of its high-albedo surfaces, which reflect the incident light to the maximum without transforming it into heat. The second duct, with opposite physical qualities to the first, is black, metallic, thermally conductive, oriented to the south, thus throughout the year seeking to intensify the heat of its indoor climate by means of its low albedo. These two climates intervene inside the building, inflate and deflate as they pass through it, twist and expand throughout the entire volume to distribute fresh air to all of its floors. There are two interior landscapes whose opposite thermal qualities are reinforced, right in the heart of the building, to the point of constituting a thermodynamic polarization between a kind of winter of continuous synthesis and a constituent perpetual summer.
The thermal qualities of these two geographical poles are extended to spatial, programmatic, social, esthetic and functional qualities. These two climates form two climatic shifts from Venice, northward for the white duct, and southward for the black duct. At the same time as being air reserves, interior landscapes, public spaces, the two ducts compose two gardens: an alpine garden with a Swiss interior climate, and a Mediterranean garden with its Pantellerian microclimate. The two ducts crisscross the building, widen to form public spaces such as interior squares, thoroughfares, landscape paintings or contract to again become ventilation ducts. Throughout the year, the interior functioning of the building evolves according to the seasons and the outside temperatures. The gardens serve as a thermal storage space for the building's thermodynamics. During summer, the cold garden produces freshness and during the winter, the warm garden produces warmth in order to best provide users with thermal comfort.

In summer, the cold garden is open to the outside during the night. Air at night time temperature runs through the entire duct and cools the materials. Throughout the night, heat exchange takes place with the high-inertia mineral material of the white duct. During the night, the temperature of the duct decreases accumulating the cool of night in its mass. During the day, the openings to the exterior are closed to allow only the required amount of fresh air to pass through the duct. The mineral matter of the duct conveys the freshness accumulated at night to the incoming air, which on contact, cools by a few degrees, supplying office spaces with cooler fresh air that is more pleasant in summer. During the winter, in the daytime the air inside the black duct is heated by a greenhouse effect which traps the sunlight inside the duct and converts it into heat in contact with the black metal surface, which, with no inertia, immediately redistributes this heat. Glazing exposed to the south creates solar inputs that heat the black garden. Unlike the white duct, the materials of the black duct have no inertia, immediately returning heat to the offices. The amount of heat absorbed by a surface depends on its color. The color of a surface determines the transformation of the energy of light wavelengths into heat. A white or light-colored surface (which is white because it re-emits practically all the wavelengths of white sunlight) absorbs about 30% of the radiation which is then transformed into heat, whereas dark or black surfaces absorb about 90%, so only 10% of the light is reflected.

Unlike conventional natural passive HVAC systems that alternate year-round between systems, our project implements both systems simultaneously and year-round, transforming a technical solution into landscapes thatcan be enjoyed year-round.

The building becomes a geography, between a cold north pole, the white duct, and a warm equator, the black duct, between which, in a comfortable temperate environment, the offices and shops are placed. The indoor/outdoor, hot/cold cleavage is overlapped by other geographies, other climates, other temporalities that organize the interior of the building like a meteorology, between a winter and a summer, a cold climate and a warm climate, perpetual and intermediate between the interior and the exterior of the building.

Expansion

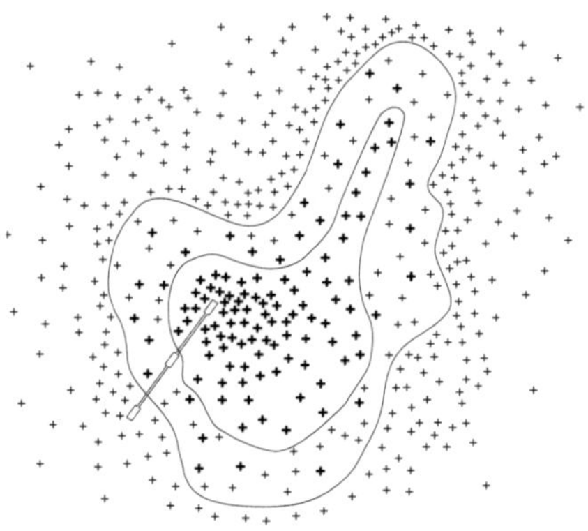

The various functions that make up a whole program are usually brought together in the same building, "under the same roof". Thus, the rooms that comprise the program of a dwelling can be found in a single block (the house). Bedrooms, kitchen, bathrooms, living room, are all bits of space, with as many functions, so is a question of agglomerating, assembling, compressing one against the other, between four walls until forming a more or less compact, more or less high, whole, depending on the architecture chosen. The house has a center and sides, to the north, south, east or west, toward which the functions are oriented. The bedroom is best located to the north of the house, because there is not much need for light as you only sleep there. The living-room is best located on the south or south-west side, to enjoy the afternoon sun during free time. The concept of view also determines the layout of the functions: the living room looks out over the beautiful view, the bedroom is happy to overlook the yard. The house therefore adapts its plan according to the external qualities of the site. It prioritizes its functions, composes a figure that opens to at least one side and to all four at best.

Our task for the project of an artist residency center in Grizedale in the heart of the countryside in the north of England extends over the whole site. It expands functions in different places, different climates, different qualities of light, temperature and humidity, to be defined according to the times of the day or the seasons. In reality, here architecture is stripped of its last layer of skin, of the last layer of its outer envelope, and it is the environment that takes on this role, becoming this last envelope, filtering more or less light, containing more or less moisture, heating or cooling according to the place and time of day. From an initially compact form, the construction expands to reach three geographical situations (in the meadow, on the forest edge, in the forest) which are each determined by specific climatic qualities (light/light-dark/dark, humid/medium/dry, hot/warm/cold), which vary according to the time of day, the amount of sunshine and the season. It is above all the trees that alter the climatic quality of the place, their foliage protecting in summer from too much heat while during photosynthesis they produce water vapor that lowers the surrounding temperature. However, during the night, the foliage protects from radiation from the night sky which sucks the heat of the sun that the earth has received during the day skyward. The night is therefore warmer in the forest in summer than in the meadow. Depending on the time of day, the time of year, the indoor climate quality of the rooms will vary and, depending on the activity, a certain type of climate will be sought, the heat of the night forest, the heat of the meadow in winter during the day, the cool of the forest edge in spring. Expanding the house on the site, scattering the rooms and the program over the entire site, choosing as many different places as individual climates. The house is no longer a compact, closed object: it explodes, its program and its spaces spreading throughout the site, multiplying views, situations, locations, atmospheres, climates. It is often reasons related to orientation that determine the choice of a café on one or the other side of the street depending on the time of day. In winter in Paris, a café on the west side of the street will be chosen in the morning and one on the east side in the afternoon, to always enjoy the sun.

In "Espèces d'espaces", Georges Perec evokes this possibility which we quote from memory as follows: "Having his living room in the Latin Quarter, his study close to the Champs-Elysées, his bedroom in Montmartre and his bathroom in Ile de la Cité." It is the idea that an apartment can be spread throughout the city of Paris, that one finds its rooms in different places, according to the hour of day, according to the atmosphere of the neighborhood, according to the whims of the moment and the season. The thresholds between two rooms, these doorways normally a few centimeters thick, extend to over hundreds of meters. The apartment is no longer 50 m2 but 1 km2. And the occupant inhabits this expanded space on a city scale, on the scale of a territory.

Heat

The principles of heating, that is to say the transmission of heat between one body and another, are based on three physical phenomena. The first is conduction, the direct transmission of heat from one body upon touching another. It occurs, for example, when your feet come into contact with a warm floor that then conducts its heat to the human body. It is also by conduction that air is heated by brushing against the multiple folds of a radiator. And then it's our turn to brush against the air to warm up. The second means of heat transmission is radiation, especially infrared. This mode of heating, which relates to electromagnetism, does not require direct contact in order to transmit heat from one body to another; here, heat is transmitted by means of waves that can cross space indefinitely as long as they do not encounter any solid surface. In solar electromagnetic radiation, for example, the waves emitted by the sun diffuse through the Universe, some passes through the Earth's atmosphere and finally transforms into heat when it comes into contact with a hard body, the skin for example. This is the phenomenon we experience in the mountains in winter when, even though the outside air is at a temperature below 0°C, we feel the heat of the sun on the side of the face exposed to it. Indoors, we use radiation for certain radiant ceiling systems in office fixtures or infrared lamps for bathrooms or, more recently, on the winter terraces of Parisian bistros.
The last means of heat transmission, which here takes an inverted thermal form, is that of convection causing a cooling sensation when the body comes into contact with the moving air. This natural phenomenon occurs with some cool summer winds that, like the breeze, cool the atmosphere when they blow and lower the feeling of bodily warmth. Convection is reproduced indoors by propeller fans.

These different heat transmission phenomena are all well known, yet Western modernity has concentrated on heating interior spaces exclusively on the basis of conduction. Western modernity has trivialized the climate of interior space by reducing means of heating to conduction only. The air temperature is raised uniformly and continuously to 21°C and the interior space becomes a kind of hot air bath where you drink from the cup with each breath. The modern performance goal is to achieve the most comfortable, stable and uniform indoor climate. The temperature of the interior space is raised to 21°C throughout, around our body, against our body and in our body with each breath. The air against our skin is at 21°C just as much as the air that penetrates our lungs. This trivialization of the interior space due to the homogeneity and consistency of temperature is symptomatic of the modern project. Based initially on an objectification and a normalization of the notion of comfort, it developed such concepts Le Corbusier's "correct breathing", which is based on a rational autonomy and precision of the artificial indoor climate that is totally separated from the irrational imponderables and variants of the outdoor natural climate. Le Corbusier thus radicalizes the artificialization of interior space, pushing the disjunction between the interior climate and the exterior climate to the limit. Vitruvius proposed this artificialization, but only as a means of adjustment. "Thus we may amend by art what nature, if left to herself, would mar", without going as far as the total rupture achieved by modernity. The Japanese space presents numerous distortions compared to the modern trivialization of the indoor climate, notably in a dissociation of the different modes of production and perceptions of heat, on different scales, from weather to food passing through bodily care or sound production. The Japanese interior, unlike the homogeneous and constant Western interior, is a complex, variable, diversified place of different climatic zones and moments where several phenomena are used distinctly in the same space to specifically heat certain parts of the body. It offers a great wealth of phenomena and sensations and interesting reversals of scale and dissociations of the senses in the responses to the problems of cooling or heating. This is the case with certain dissociations between radiation and conduction, clothing and nudity, heat and diet, which, in the same space, cause variations in temperatures and sensations. Most concretely, Japanese space provides many examples of significant dissociations between the temperature you breathe and the temperature you touch. Some toilets are not heated and the air inside the room is at outdoor ambient temperature, slightly attenuated by the thin walls of the house. If you are apprehensive in winter of undressing to sit on the seat, you are pleasantly surprised to find that it is heated. The heat is transmitted by conduction to the most sensitive part of the body at this precise moment as it is naked, while the rest of the space will remain at a lower temperature. Immediate thermal rebalancing comes into play, between the temperature of the air, the thermally protective clothing, the nudity of the skin of the buttocks, the heat of the toilet seat. This dissociation between the temperature of the air you breathe and the temperature of the object you touch is also found in the Japanese living room or bedroom. The latter can provide an air temperature close to 12°C while the bed, heated from the inside by an electric mattress, will directly impart to the body a gentle heat of 22°C. This creates a pleasant contrast between the cool temperature of the air you breathe and the warm temperature against your skin. In contrast to modern Western uniformization where the entire temperature of the room (air, surfaces, objects) is heated univocally to 21°C, this interior composition, which diversifies thermal emission modes, their sources and their intensity, the places of their sensitive reception, produces a spatial wealth of sensations, thermal contrasts, sensual qualities. The warm bath we take in Japan in winter is a way to warm our bodies by immediately immersing ourselves in heat; heat that we then keep for a few hours, beneath our clothes, in the cold of winter. In summer, small glass bells are placed on the terraces and, being most sensitive, at the slightest breath of air they emit a small crystalline sound that pleasantly signals a moment of possible freshness, sonorously. Radiation, convection, evaporation but also smell, sound and taste, combine in the same space in a subtle act of balance and sensory contrast, between the hot and the cold, the atmospheric and the dietary, the dressed and the undressed. This contrasting, complementary, dissociating work is reminiscent of certain processes of nouvelle cuisine that works on differences in temperature and textures as well as odors, looks and, of course, tastes and breaks down the barriers between salty and sweet or hot and cold.

Contemporary space, inherited from Western modernity and the precepts of climate rationalization and homogenization, has focused on its visual expression by reducing and normalizing the presence of other physical phenomena that may be perceived by other senses. It is not so much the diversity of sensations that interests us as the possible interaction between the different modes of perception. We would like to consider space as the place of synesthesia, of the multiplicity of phenomena and sensations, that is to say as a place of sensory encounters around the body but also on the body and in the body. We seek physical balances and imbalances, spatial compositions and sensitive contrasts between the heat of a cup of tea, for example, the heat of the air, and infrared electromagnetic radiation. Being multiphenomenal and multi-sensory, architecture must be built on the multiplicity of singularities that define space (volume, temperature, light, color, humidity, smell, diet) and on the diversity of bodily perceptions of it (touch, breathing, ingestion, radiation, perspiration, physical activity, clothing, hormonal perception).

Food

Food is a neglected part of the field of architecture. Its organic or neurological action helps warm or cool the body by producing real or simulated heat. It is therefore one of the environmental factors to be taken into account in the general equation of comfort, alongside clothing, physical activity, air temperature, wind speed, and electromagnetic radiation. The case of the St. Bernard, the rescue dog from the Swiss Alps, is a radical example of the possible reduction of the macroscopic space of a warm room, to a portable, microscopic and liquid scale, by a small barrel of alcohol that is given to travelers lost in the coldness of the winter mountain to drink, to warm them. Here there is a compensatory movement between the macroscopic and the microscopic, between the coldness of the external environment and the heat of the internal alcohol, where the deficiency of the one is counterbalanced by the increase in the quality of the other. This equilibrium is real, since man, being homeothermal, must maintain a constant body temperature of around 37°C. To do so, he must burn more nutrients when the outside temperature is cold in order to produce heat from the inside to stabilize his body temperature through thermoregulation. It is thus possible to imagine a less warm room where one would eat more, or, on the contrary, a warmer room where one would eat less. Food is a digestible form of furniture, a whole field in itself of the architectural project, where the poor thermal insulation of the walls, for example, can be compensated by ingestible calories. Food can really contribute to the climate comfort of indoor space and its esthetic quality. "Do we know the moral effects of foods? Is there a philosophy of nutrition?", asked Friedrich Nietzsche. Can we imagine shapes that are no longer really shaped in space, but in the body: an architecture that shrinks to become sound, smell, taste?

The organic reality of food in relation to environmental temperature also has a neurological correspondence because we perceive the temperature of the external environment thanks to the projection of neurons in the skin and in the oral mucous membranes. Information concerning heat is sent from the skin to the brain through neurological sensory channels through which ions flow. These ion channels are peripheral extensions of the heat-sensitive sensory neurons linked to the hypothalamus. Known as Transient Receptor Potential ion Channels, they contribute to heat perception (thermosensation) and thermoregulation in humans. The ion channels of the Transient Receptor Potential (TRP) family thus collaborate in the mechanisms that allow us to sense and differentiate cold or hot temperatures and contribute directly or indirectly to the thermoregulation of the human body. A physiological particularity is that these channels can be activated as much by temperatures produced outside the body (air temperature or radiation) as on or in the body by means of certain substances contained in plants that we might apply directly to the skin or to the oral mucous membranes. There is a neurological correspondence between an invisible physical value and a tangible organic temperature value. The immediate neurological stimulation of a feeling of coolness or warmth that does not really exist has been known about for a long time. This is what happens, for example, in hot countries upon ingesting mint (Moroccan mint tea), a plant that contains molecules of crystalline origin called menthol that lead the brain to feel the same sensation of freshness as a temperature of 15°C when it is actually 30°C. Recently, a calcium channel of the TRP-melastatin family (TRPM8) has been identified in neurons of dorsal root ganglia which reacts to cold from 15°C as well as to menthol substances. Indeed, menthol activates the TRPM8 molecular sensory receptors present on the skin and in the mouth that stimulate the group of peripheral sensory neurons called sensitive cold units. TRPM8s belong to the TRP channel family and are sensitive to temperatures between 15°C and 28°C. At these temperatures or in the presence of menthol, they activate calcium ions in the sensory neurons, transmit information to the brain through the spinal cord, and provide a cold sensation. Current research shows that there is another family of TRP including TRPA1 that would respond to thermal stimulation below 15°C. TRPA1 would also be stimulated by chemical molecules such as cinnamaldehyde (present in cinnamon, for example), allyl-isothiocyanate (present in mustard, wasabi or black radish), or allicin, a compound found in garlic, when it comes into contact with the skin or mucous membranes such as those of the mouth.

The opposite phenomenon occurs with the TRPV1 neuroreceptor which is sensitive to temperatures above 44°C but which also is also activated in the presence of capsaicin, a molecule which is found in red chili for example. Capsaicin and temperatures above 44°C stimulate TRPV1 which triggers a sensation of very high heat, like a burn. At lower temperatures, TRPV3 is sensitive to temperatures between 30 and 39°C and to a molecule, diphenyl, present in camphor or irritant extracts found in oregano or clove when it comes into contact with the skin.

For our Digestible Gulf Stream project presented atthe Venice Biennale of Architecture in 2008 and at the Biennale Internationale Design Saint-Étienne in the same year, we worked on the forms that could be taken by an architecture that would be developed at both atmospheric and neurological scale. With each metallic plate, we associated a culinary product for the cold and a dermatological one for the hot, as a kind of synchronization between the temperature one feels outside the body (coming from the hot or cold plate) and that caused directly on the skin or on the mucous membranes (produced synthetically in food and dermatological form). These products are a miniature resonance of the temperature of the plates, in a kind of synesthesia between the macroscopic and the microscopic, the meteorological and the dermatological, the atmospheric and the edible. The cold plate, cooled by the circulation inside of water at 12°C, is associated with a candy producing a sensation of freshness based on a precise assembly of the active molecules contained in mint, garlic and cinnamon, which stimulate exactly the same neurological sensory receptors of the brain (TRPM8 and TRPA1), around 12°C. In contrast, on the hot plate, a dermatological cream, applicable to the skin, is made from a precise dose of active molecules extracted from chili pepper and camphor, stimulating precisely the neurological sensory receptors, at the same temperature as that of the metal plate, at 28°C.

The sensation of cold and warmth can therefore be felt both inside the body (food) and outside the body (atmosphere). The traditional field of architecture extends by working on these different scales, blurring the boundaries between interior and exterior, body and space, neurology and physiology. The concept of food comes into the field of architecture at the same time as the climate dimension. Our architecture lies between the microscopic and the macroscopic, the gastronomic and the atmospheric, the visible and the invisible. By integrating this dimension of food, architecture is constructed as a thermodynamic imbalance between the temperature inside the body and the temperature outside the body, between conduction and radiation. It becomes the plastic expression of an inner Gulf Stream, working at different scales, polarizing contrasts (hot/cold, low/high, dressed/undressed, indoor/outdoor, rest/activity).

By becoming a convective movement, being constructed as a thermodynamic imbalance, the architecture abandons modern climate stability and its thermal trivialization to become meteorology. The space, like a physical geography and a dynamic climate, therefore evolves between balance and imbalance, macroscopic and microscopic, atmospheric and gastronomic.

ATMOSPHERE

Climate and meteorology have recently entered the field of architecture, first and foremost as concerns. Pertaining to the neglected category of the void, of the immaterial, of the invisible, the impalpable, they have long been excluded from the language of architecture based on the notion of the solid, immediately perceptible, divisible, composable and manipulable. There is, we know, a kind of misunderstanding in the history of architecture since its origins. It has been written through the history of the solid and has never, since Hippocrates, managed to establish a valid theory of the void. The predominance of the solid in the history of architecture increased yet further in the 20th century, when all the theories associated with the void were systematically dismantled. The latter were for a long time based on the empirically developed Hippocratic theory of humors, which began to be challenged in the 18th century. Yet the void is the subject of architecture. The solid is secondary.

The very mission of architecture is to define this hollow that is space, to describe a vacuum that contains a quantity of air from which we subtract certain properties from which we wish to protect ourselves such as the rain, the cold and the wind. The history of architecture is paradoxically that of the full and its associated classifications, between forms, structures and materials, because we managed to qualify the void, not by defining it itself, but by defining its visible limits, walls, ceilings and floors. For a long time without means to quantify the void itself, architecture has historically retreated to the solid. Formal classification, that of the decorative molding of its elements, the composition of the masses and their proportions, between classicism and deconstruction, balance and imbalance, order and disorder, connecting rooms.

Structural classification too, that of the design of the descent of the loads, the Romanesque vault or the Gothic arch, a small stone span and wide steel cantilevers. Ultimately, material classification between a pure tectonic expression of its materials and postmodern narrative. The history of the void, however, follows stranger, more obscure paths, sometimes superstitious or magical, flirting with the irrational or the utopian. Until the 18th century, the void remained a unique and indivisible element, that of the air, in Aristotle's theory of the four elements of the Universe. Alongside water, fire and earth, air is one of the four basic elements of an empirical and imprecise chemistry. It was not until the second half of the 18th century that we began to understand the real nature of the void, classifying it no longer as a single element but as a lighter, more transparent fluid than water, composed of different gases that English chemists began to decompose, to catalogue. This chemical revolution which, a century later, followed that of the discovery of the physical properties of air, of its elasticity, was followed in the second half of the 19th century by Louis Pasteur's biological revolution. The French scientist demonstrated that air is composed of suspended microscopic elements: dust, microorganisms, germs, microbes. Even today, we continue to explore the void and discover other properties, be they electromagnetic or electrical. This until very recent ignorance of the physical, chemical and biological reality of the void certainly explains its absence in the field of the history of architecture. The invisibility of the air, the impossibility to define the void, in which we bathe and which we nevertheless ingest with each breath we take, have generated more fiction or fantasies than architectural treatises or construction manuals. Unlike the fullness that is in before us and outside of us – frank, heavy, hard, with fixed and defined contours, whose shape and influence can be measured precisely – the void tends toward the indefinition of its shape and contours, to vagueness, to the invisible. Vaporous, dilated, permeable, it penetrates us at the same time as we penetrate it, it infiltrates our body, our lungs, enters our blood and our brain, mixes with our body. This carnal relationship with the air certainly explains the physiological and medical, even humoral value, which very quickly projected itself on the void, sometimes tinged with common sense, sometimes shrouded in nightmares.

In the fifth century BC, in the treatises of Hippocrates on winds or places, cause-effect links between climate and mood emerged, in which architecture and urban planning played a paramount role. For example, the orientation of a house with respect to the wind, the sun or humidity influences the temperament of its inhabitants. Hippocrates drew up a sort of urban planning treatise based on a physiological link between meteors, architecture and mood. Cities exposed to cold winds, for example, engender inhabitants with bilious temperaments. Faced with easterly winds, men have a lively character and an incisive intellect. This link between climate and temperament stems from his theory of humors where the qualities (hot, cold, humid and dry) of the four elements of the Universe (air, earth, water and fire) apply to the four humors of the body (blood, yellow bile, phlegm and black bile) and determine man's four temperaments: sanguine, phlegmatic, melancholic and choleric. This erroneous conception of the body and space, which modern science later overturned, continued until the sixteenth century. The sixth book of De Architectura by the

(1st century BC) Roman architect, Marcus Vitruvius Pollio, bases its chapter I, "On climate as determining the style of the house", on this humoral philosophy by affirming the links between the climate of a place and the character of its inhabitants. For example: "It is even to the subtlety of the air, to the warmth of the climate, that the peoples of the south are indebted for this activity in the design of their projects. Northern ones, on the contrary, drowsy due to the density of the air, cooled by the humidity of the atmosphere, have numbness in the mind... It is therefore not surprising that heat gives vitality to the spirit of man; the cold, on the contrary, gravity." While this climatic determinism of human nature proved to be inaccurate, it nevertheless proposed common-sense solutions such as these programmatic recommendations related to the course of the sun, to humidity and to the winds where "the location of houses must be appropriate to the nature of the place and the difference of climates." These theories of humor can be found in the 16th century with the English clergyman Robert Burton, who states in his Anatomy of Melancholy that melancholy can also be caused by an unhealthy air. And quoting Liévin Lemmens, Burton advocates a great deal of attention to the choice of window arrangement and ambient air management:"A pure air exhilarates the spirits, invigorates the brain; a thick, black, misty, stormy air contracts and harms it. The sun scatters the horror of a heavy spirit; for it is true that it is not the bodies but the souls that are thus altered by the heavens and the winds, and the health is affected differently by a cloudy sky or a clear sky."

While modernity has rightly eclipsed the humorous sense of climate and relegated it to a poetic form, its architecture and urban planning have been developed under the organic influences of the climate. The typical elements of modernity, length-wise windows, bay windows, balconies or solariums are devices derived immediately from architectural typologies for therapeutic purposes, developed for sanatoriums at the end of the 19th century. Ventilation is a priority. It requires designing the layout of apartments creating a current of air, eliminating nooks and crannies to allow the air to flow from one side of the house to the other and to expel germs. Good ventilation requires healthy air around the house, which involves working on city plans through zoning. Dwellings must be moved away from places of production polluting the air and buildings placed in large green parks that regenerate the air by absorbing CO2 and retain dust in the fluffy leaves of the trees. The fight against tuberculosis led to flooding light into interior space. The white color of the whitewash that covers walls and ceilings, is both disinfectant and reverberant, reflecting natural light throughout the dwellings. The cleanliness of the space also requires the removal of decorative elements, which clutter the surfaces.
Moldings that create shadows, paintings on walls that absorb light, carpets that accumulate shadow and dust are all rejected. This modern hygienist program, required after the urban health tragedies of the 19th century, profoundly changed the shape of architecture and cities in the 20th century. Yet it turned out to be as approximative as the programs of Hippocrates or of Richard Burton and its need was lost upon the discovery of antibiotics, which enabled curing tuberculosis with simple pills. With a profound and humanistic approach, modern architecture, uprooted from its therapeutic raisons d'être, only survived this medical discovery, after the 1930s, as a decorative style that one could henceforth mock, or accuse of the worst evils of contemporary society. From the 17th century onwards, the birth of analytical science gradually deflated the humoral fantasies linked to the void, yet the void remained the space where mysteries, worries and fantasies were projected; a phenomenon that is reinforced today by the shift of technologies towards the invisible and the nanometric.

Some of our projects have explored the restless field of the invisible by reintroducing a share of irrationality and superstition in modern techniques for controlling the environment. They thus lend themselves as a critical exploration of modern indoor air, conditioned and homogenized by working on a magical re-enchantment and an analysis of the plastic potential of invisible architectural elements.

Interior wind

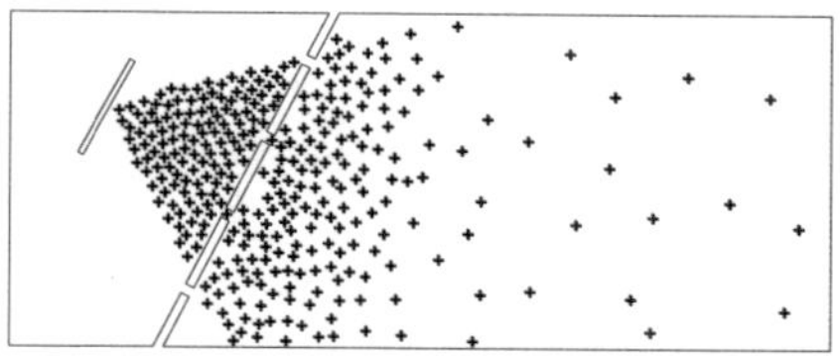

Normally, the air in modern houses is almost still. No winds, no tornadoes inside buildings, but stagnant warm air masses at 21°C barely move, at about 0.00000008 km/h, gently from one room to another. Heavily, slowly, they take up CO2, moisture, the breath rejected by our organism living in the space. Outside, the air moves, stirs irregularly, rages and calms, cools or heats In modern houses, on the contrary, the air is still, "neutralized" as Le Corbusier puts it. This movement of the air outside or its absence inside is a discreet yet fundamental difference between interior space and exterior space, between the climate of houses and that of the outdoors. It reveals the temptation to control, that of conditioned air that we would have rid of the imponderables of meteors and their irrationality. The unexpected turbulence of positive and negative pressures, sudden differences in temperature, air movements and winds give way inside houses to warmth, even pressures, the absence of draughts as the climatic construction of a moderate, stable, anthropic comfort.

With the "Foehn" project presented for an exhibition at the Frac Lorraine in Metz in 2005, we propose an irrational indoor climate construction: enchanted air-conditioning. This dysfunctionalist project re-engages in the domestic space the instability of the external natural climate as a possible opening to atmospheric influences on man, reactivating an uncontrolled relationship between the inhabitant and their environment, between fantasy, the humorist theory of the 17th century and popular imagination.
In an interior space of the Frac Lorraine, through controlled ventilation we cause a movement of air at a speed of 60 km/h from south to north. The temperature of the air is increased by 8°C compared to other rooms and the relative humidity is reduced by 17%. This wind blows through a wall composed of concrete with granitic aggregates, against which the air brushes before exiting into the room. Granitic gneiss, quartzite, Bergell granite, Grimsel granodiorite, green shale, micaceous shale, are the crystalline aggregates of the granitic Alps that we incorporate into the concrete. These physical, geological and climatic parameters, reconstructed indoors, reproduce the characteristics of the foehn, the dry, warm wind that blows from the south into the valleys of the Swiss mountains. Irrational, scientifically unfounded folk mythology is associated with this wind nicknamed "the wind that drives people insane", which causes insomnia, headaches, restlessness. The foehn is a warm wind that caresses the Alps on its way from Italy, climbing the southern slope of the granite mountains, passing the summit ridge to descend towards Switzerland as it compresses, and warming up as it moves rapidly away from dew point. The air then becomes dry and clear, blowing violently from the south, electrically charged with positive ions. After a few days, the foehn ceases, the humidity level rises, the air becomes cooler again, we then try to breathe better and the tension drops as the electric charge of the air reverses.

Ionized air

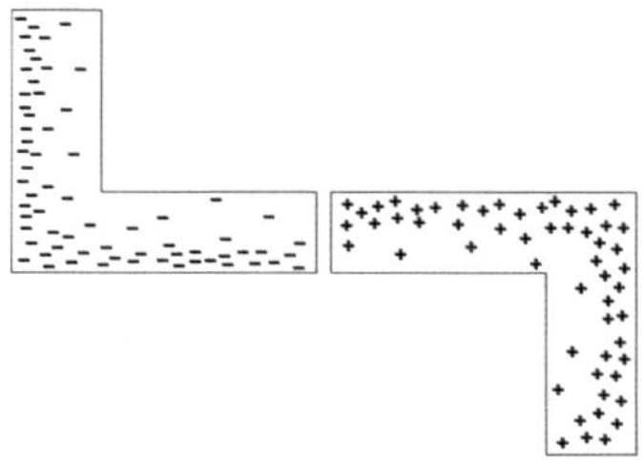

Can a modern house also be haunted? Where are the ghosts hiding if the house no longer has a cellar or an attic? Where do they hide if the thickness and humidity of the walls of ancient castles have given way to thin plaster and glass walls? What shapes do these specters take?
What is the nature of the ghosts of modernity? Fear is always of the order of the immaterial, it is formed in the invisible, in fields imperceptible to our senses. Science, because it penetrates ever deeper into the infinitely small, reveals more and more new fields of dread each day. Our project for the MAK Schindler House in Los Angeles in 2006 is to haunt a modern house, to physically and chemically render the appearance of modern and contemporary ghosts possible. The field of architecture is increasingly deployed in the invisible and it is less in terms of space than in terms of climate that the architectural form is expressed today. Displacement is actually an increase in the sensitivity of architecture. The visible, surfaces and square meters, is superimposed by other chemical, physical, electromagnetic dimensions, those of air, light, temperature, humidity rate, electromagnetic fields, in which the occupant is immersed just as much as it immerses them, through respiration, the epidermis, the retina. The perception of architecture today concerns all the senses, but also other means of the perception of space, via the endocrine system, by ingestion, by inhalation. It is at these levels, close to the invisible, that scientific research is evolving today, exploring the microscopic dimensions of space, at chemical, molecular, genetic and nanometric levels. And in this exploration of unknown fields of space, science produces uncertainty, new forms of fear and ghosts. Air polarization, measured in terms of ions, is one of these new invisible dimensions that are now being explored for their ability to determine certain qualities of space. "The differences in ion density are imperceptible and the organ that perceives ions remains unknown at present." Research is underway, and with it, so is the production of fantasies and concerns.
Dr. Michael Terman and his team at the Department of Psychiatry at Columbia University in New York are working on the questions concerning the effects of ions on the psychic, and their current findings point to an influence of ion polarization on mood. "The presence of negative ions in the air tends to be higher in summer than in winter and in humid environments rather than in dry environments. The effects on mood have long been disputed, although until recently no placebo studies have been commissioned. In response to exposure to negative ions, subjects reported reduced irritability, decreased stress and mood, and increased feelings of calm and relaxation. On the other hand, higher levels of positive ions in the air – such as in artificial, heated and air-conditioned environments at home or at work – have been associated with increased irritability, bad mood and tenseness. On the occasion of the "Gen(H)ome" exhibition at the MAK Schindler House, we propose to exacerbate the spatial duplication of the house and transform it into schizophrenia. The Schindler House was designed to allow two families to live separately under the same roof by differently orienting the nevertheless similar spaces of the two families, to the east for the first, to the west for the second. The Schindler House is a Janus house, with two faces, whose duality we would like to amplify by inversely polarizing the ions of the air. In the east, the air ions will be negative, in the west they will be positive. One calming side, one irritable side, similar visually, different in the invisible, open to subjectivity and interpretations, fantasies and ghosts.

Iodized spray

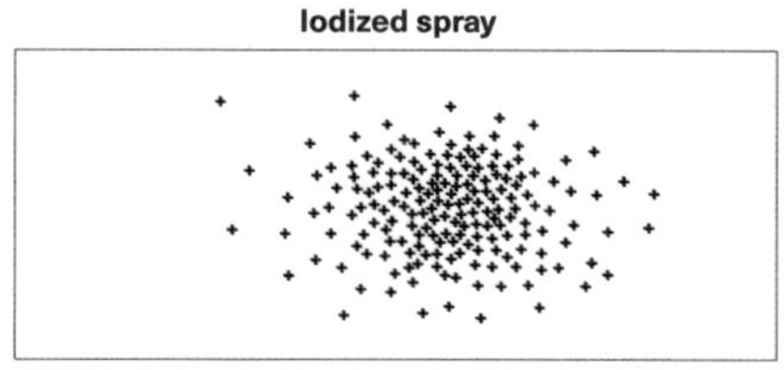

In 1857, the following description appeared in the flyers praising the bathing establishment of the village of Saxon in the heart of the Swiss Alps: "There is no valley in Switzerland of such a grand size. From Martigny to Sion, the vegetation is luxuriant and the climate, in summer, reminds one of that of Provence, in winter the warm softness of the beautiful sky of Nice.
That same year, the Paris Academy of Medicine clearly established the presence of iodine in the waters of the Saxon spa, about twenty years after a Swiss doctor, Maurice Claivaz, had praised the therapeutic properties of the waters of Saxon. We now know from Russell's research in the early 19th century that iodine is an essential trace element in the production of thyroid

hormones. Its absence is the cause of goitre and of this form of so-called endemic cretinism, which began to be mentioned in the middle of the eighteenth century to characterize a form of stupidity encountered among some inhabitants of the Alps. Iodine is absent from the soils of Alpine countries because the diluvial erosions that took place at the end of the last ice age of the Quaternary era impoverished the soil. The therapeutic properties of iodine-blessed Saxon water have been shown to be quite effective in controlling diseases in a population with chronic iodine deficiency. Iodine is supplied mainly through a diet of marine origin, but it can also be inhaled by means of sea spray. Iodine is indeed contained in large quantities in seawater, in all products derived from the sea such as algae, fish or crustaceans. And it is therefore naturally, at the same time, at the sea, during the first half of the nineteenth century, as opposed to the mountains, that these pharmacological virtues of sea water confirmed the therapeutic practice of bathing in the sea and partook in the invention of the beach and the development of seaside practices that we still know today.

It is particularly troubling to note this kind of geographical synesthesia perceived in the sentence quoted in the introduction, above taken from the prospectus praising the establishments of the baths of Saxon. As chemical element specific to the sea as is iodine could profoundly transform first the body of its inhabitants but also the geography itself turning the Alpine climate and mountains into a Mediterranean landscape, bathed in the light and softness of the sea. Is it this same synesthesia to which the Swiss government resorted when it decided in 1922 to add 3 mg of iodine per kg of salt to the cooking salt commonly sold in Switzerland to combat goitre and cretinism? This measure, which was followed by the United States in 1925 and subsequently by other countries cut off from the sea, took as support, by a twist of poetic fortune, a product of the sea itself, salt, as a microscopic "Mediterraneanization" of the mountain landscapes, a dietary "oceanization" of the inhabitants of the Alps. The sea, absent from Switzerland, is thus reintroduced in miniature into the diet, thus participating in this characteristic ubiquity of modernity where the seasons drift in the year until they overlap in a sort of perpetual spring, where the nights and days blend in a white luminosity as much daylight as night light, where distances shorten until they overlap in the immediateness of globalization. It is ultimately a synesthesia of the same order that we seek to produce at the LiFE event in Saint-Nazaire in the project of a winter beach presented in December 2008: a winter that becomes a form derived from summer, that of a place that is inhabited at the same time, both summer and winter, in a perception that is both atmospheric and physiological. A metamorphosis that is less geographic than temporal, that of a beach that shifts in winter, that contracts in a certain electromagnetic radiation, in an aerosol, like a "summerization" of the St. Nazaire winter. We retain only a few phenomena, sun and spray, tanning and iodine: a certain radiation and its angle of incidence, an aerosol and a certain chemical composition, a few summer and seaside phenomena that are reformed indoors, in the middle of winter. Inside the underwater base of LiFE in Saint-Nazaire a marine aerosol is produced as a form derived from marine spray: a cloud of iodine that is perceived when breathing and develops its real forms in the body itself. The space is warm, close to 28°C, a temperature found in summer in Saint-Nazaire, which is also the temperature to which the air is heated in swimming pools in winter, the temperature at which clothes become dispensable. In the foreground, natural mineral water from Saxon, the village up in the Swiss mountains, is available in bottles, stored in refrigerators.

Horizon of ultraviolets

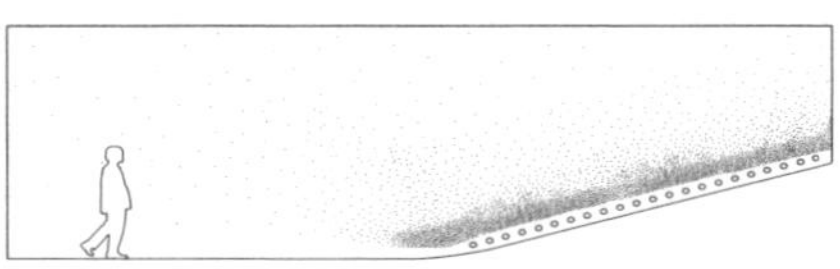

The second element reconstructed indoors in the Winter Beach at LiFE in Saint-Nazaire is the solar radiation we encounter by the sea, on the beach in summer, which reaches us from the sky but is also reflected on the water and thus reaches us on the beach, at twice the intensity. This is why we tan quicker on the beach than in the city or in the countryside, where the solar radiation reaching the ground is absorbed and not reflected as is the case by the sea or on the snow in the mountains. What we reproduce here is an electromagnetic landscape, a horizon of ultraviolet rays, a certain amount of UV-A present on the beach in summer, which we perceive here no longer as something visible, but cutaneously, through the transformation of the skin, winter tanning. This invisible seaside light is reproduced at a low angle of incidence that takes place between the ground, the horizon and our body. Our Winter beach is as much a temporal shift between summer and winter as a shift of scale, from the macroscopic scale of an outdoor seascape, complete with sun and sea, to the physiological microscopic scale of the indoors, with iodine and ultraviolet rays. It is a composition of elements whose meaning and use are open-ended and freely interpreted, as the shoreline has been down the ages, a space between rejection and desire. The winter beach is suggested as a quirky form of indoor public space with bright, olfactory, thermal and synthesized gustatory atmospheres, somewhere between the pool and the exotic restaurant.

Meteorology of mood

Today, it is not so much the strangeness of a form or of a matter that can bring about a feeling of anxiety, but rather the chemical composition of the air, of electromagnetic radiation, of an atmosphere, of a climate. Imperceptible yet terribly effective environments. Like these photographs of beautiful summer countryscapes, taken fifteen years ago in Chernobyl. The representation of the visible is an indirect means of transmitting information: you can at best show the symptoms, never the cause, or under the microscope, using a Geiger counter or a device to measure electromagnetic fields. Fear is no longer symbolic, it becomes physiological: it thickens, it concerns the space itself. It no longer only shows itself to the eye, but penetrates our skin. It goes beyond the retina, it invades our lungs with each breath, makes the air vibrate, conditions it, conditions us. Modernity has developed artificial physical environments. It was built on the mastery of space and its climate, both urban and domestic. Artificial gas lighting emerged in the second half of the 19th century, central heating of dwellings imposed itself in the first half of the 20th century, air-conditioning of office buildings became widespread in the 1960s.

There is, in the modern invisible conditioning of space, a search for certain effects. There is also an appearance of side effects, partly reality and partly fantasy. Because the invisible offers areas of shade, risk and the unknown that can cause anxiety. Modernity has gradually reduced thicknesses and materials. Heavy stone walls have given way to thin glass curtains, the smell of wood burning in the chimney disappeared in favor of the silent radiation of radiators. The whims of natural light and the flickering of candles have been replaced by the uniform and constant brightness of electricity. While we usually give lighting the mission of allowing us to read or not bump into objects when we move, we are now beginning to find out about some of its influences on the hormonal system, especially on the hormone melatonin. The poor quality of artificial lighting, its lack of intensity and its depleted electromagnetic spectrum would cause mood disorders, a degree of sadness and a decreased libido. This is known as Seasonal Affective Disorder (SAD) and is now being treated with light therapy.

Another invisible form of space is the chemical quality of indoor air.

In the past, the air in buildings was renewed irregularly by opening the window. With mechanical air-conditioning, whose application has been increasing since the 1970s, for economic and ecological reasons, we can now see the emergence of new conditions related to air confinement. The World Health Organization lists them as Building Related Illness (BRI) and Sick Building Syndrome (SBS). They manifest themselves in the form of headaches, allergies, respiratory problems, fatigue, stress, etc.

Recent developments in information technology and communications networks have accelerated the de-materialization of space. Computer screens, mobile telephony, multiple electromagnetic waves in turn change the climate. Invisible and artificial, non-ionizing radiation contributes to concerns about the nature of contemporary space. Previously electrically neutral space is charged and totally envelops us. The fear is that invisible networks will generate nuisances, ranging from simple discomfort to the generation of genetic mutations. Electrosmog and "electromagnetic pollution" are new terms that appear to designate the risks to which our body and perhaps also our mood might be exposed. In view of this electrical modification of the climate, scientific studies are under way to evaluate the organic and psychological implications of electromagnetic radiation on the body. No effects have yet been demonstrated. In the face of uncertainty, precautionary measures are being taken today in Europe, and on July 12, 1999 the European Union adopted a recommendation to reduce the average exposure of the public to non-ionizing electromagnetic radiation to the lowest possible level.

Right after the worry, there is the fear of being manipulated, physiologically, psychologically, genetically. This fear recurs regularly. It reveals itself through often disturbing fiction, somewhere between the prophetic and the naive. Both utopian and dystopian, these fictions accompany periods of strong technological development, revealing their risks, possibilities and uncertainties. Jules Verne, in Doctor Ox's Experiment in 1872, tells of the establishment of a certain Doctor Ox in a small town in Flanders, who has a project for new kind of modern gas-fired street lighting: "Oxyhydric". The works begin, and with them the moral transformation of the inhabitants of the city. Usually placid, they suddenly become quarrelsome, revelers, licentious, and then end up causing a war. The reason for this change in mood is purely chemical. It is caused by the science of Dr. Ox, who, using the network of gas he installs in the town, but also in theaters and other public places, insidiously emits a surplus of oxygen into space. "Well", resumed Dr. Ox, "was I not right? See to what not only the physical developments of an entire nation, but its morality, its dignity, its talents, its political sense have come! It's only a question of molecules... To conclude: Are virtue, courage, talent, wit, imagination,—are all these qualities or faculties only a question of oxygen?" (Jules Verne)

Two years later, in 1874, in "La Machine à Gloire" from the Cruel tales, the French writer Villiers de l'Isle-Adam extended the possibilities of Dr. Ox targeting the chemical manipulation of human emotions and the possibilities of artificially manufacturing glory. The theater becomes a machine, with the release of laughing or tear gas, a mechanical slap system that stimulates applause, sounds, in order to manipulate the spectators into granting glory to the author of the play performed, irrespective of all plastic or dramatic qualities of their work. In 1884, J.-K. Huysmans, despite his denunciation of positivism and his aspiration to the suprasensitive and the magical, found paradoxically in his novel Against Nature material reasons for the decline of faith and religious thought during his century. If one no longer believes in God today, it is, he asserts, because of the air breathed in the churches, which has been physically corrupted by modernity and economic concerns. A low-materialistic vision of spirituality "For years, holy oils were adulterated by poultry fat; wax by calcined bones; incense by vulgar resin and old benzoin." No image, no form, no free will: here, it is the smell of the church that brings faith, there it is the surplus of oxygen that causes war, elsewhere, it is laughing gas that allows man to know glory. There is, in these three nineteenth century French authors, the enunciation and the denunciation of the modern spatial conditioning behind behavior, our aspirations and faith, a questioning both of pre-scientific thought and of free will. This disenchantment with human nature, whose moods and aspirations would depend on the climate, architecture and urban planning, are cathartic representations of 19th century materialism, the physiology of Claude Bernard or the positivism of Auguste Comte, but whose disenchanted premises are found in the philosophy of the Enlightenment, in the 18th century, in writers such as La Mettrie whose book titles speak for themselves: The Natural History of the Soul or Man a Machine.

In the 1950s, we see the theme of manipulation that Guy Debord and the situationists borrowed as a revolutionary means of rethinking urban planning updated. "We can now launch human reactions with a sure hand in directions determined in advance." On the basis of this phrase by Serge Tchakhonine, in 1958 Internationale Situationniste proposed that artists should appropriate from the police the methods of modern conditioning of brainwashing with the goal of producing "exciting and liberating environments". Subliminal images are evoked, sensitive to the retina but remaining below a conscious perception, and subliminal advertising diffused by infrasound, or even temperature variations and the intervention of a light ray. "It is the entire humanist, artistic and juridical conception of the inviolable, unvarying personality that is being condemned. We shed no tears over its passing." By accepting the possibility of conditioning, the situationists seek to clear it of all repressive intent. In the authoritarian and moral visions of Jules Verne or Villiers de l'Isle-Adam, the situationists call for revolution and play. On September 20, 1955, the Lettrist International decided to establish by plans and mock-ups the model of a "haunted house". According to this objective, Jacques Fillon later suggested transforming the churches, maintaining their current ambience and accentuating their unsettling effects. Similarly, the Lettrists proposed opening the metro, at night, whose passageways and platforms should be poorly lit with dim, blinking lights. Or even leaving public squares open at night, unlit. So many climatically determined situations, likely to lead to the anguish of the inhabitants who have become players in a game. A formalization of the situationist city, the New Babylon project, which the Dutch artist Constant began to design as of 1960, is a universe where all climatic conditions (lighting, temperature, hygrometry, ventilation, olfactory and sound atmosphere) are under control in order to allow everyone to create climates that can be modified at will and thus invent situations that can relate as much to sexuality as to fear, in a permanent and each time renewed game.

Recent developments in molecular biology, genetics and nanotechnology have reactivated fears of the invisible manipulation of space. And through space, our human nature. What is dreaded is to be able to conceive spatialized forms of current developments in neurological and biological knowledge, evoked by the neurobiologists Jean-Pierre Changeux or Antonio R. Damasio in Neuronal Man or Descartes' Error. Shapes tinged with the concerns of the philosophers Peter Sloterdijk or Giorgio Agamben on human future, concerning his free will, his moods, his faith.

While there are currently uncertainties about the harmfulness of electromagnetic fields, much of the history of the void is based on misconceptions arising from fantasy and concern, certainly caused by its invisibility, its impalpability, and our inability to grasp it. Our purpose today is to invent a new history of the void that is dynamic, open, plastic, meteorological, interpretable, released from fear in order to produce new climates, a new form of beauty as sensory atmospheres in which the inhabitant might immerse himself sensually.

Dry air

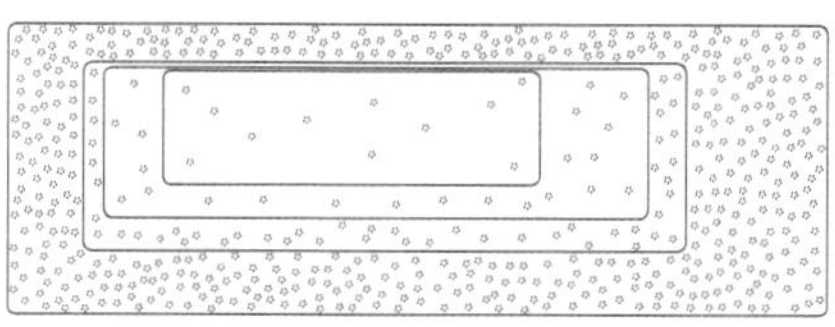

Our architecture lends itself to working in the very matter of space, by qualifying the thermal, hygrothermal balances, by literally modelling the air. But it is no longer in terms of function or square meter that we qualify space, but in terms of relative humidity and temperature, in order to create different atmospheric ambiances, different temperatures, whose use we freely interpret. Our goal is to penetrate into space itself, to decompose its atmosphere as physical quantities and qualities, and to make some of these specificities inhabitable. It is therefore the air itself that we work with and that we inhabit, that we modulate until we create varied atmospheres that are moister or drier, warmer or cooler, under positive or negative pressure. Working with the void, on the void, in the void implies a change in the hierarchies in developing the architectural project. The quality of a space can thus be defined in terms of air temperature, wall radiation, air speed, relative humidity, light intensity, chemical composition, electromagnetic field. Architecture, as the construction of atmospheres, engages physiological relations between the occupant and space, through their clothing, their metabolism, exchanges of evaporation, sweating, radiation, ingestion or convection.

Here, relative humidity, temperature gradient, light intensity and spectrality are the elements of architectural language that are studied in the production of the architectural project. We seek to produce spaces and times below the significant by working in the materiality of architecture itself. It is therefore no longer a predefined use or practice of space that determines the form of the building, but, on the contrary, it is the form that freely gives rise to different practical interpretations.

Our 2005 project for a new museum in Estonia refuses to be the embodiment of a cultural use or practice of museum space. Neither is it the symbolic expression or metaphor of a previously conceived sign or image. In reality, here, architecture is nothing more than a gradual lowering of certain ambient climatological values such as moisture content, the quantity of ultraviolet rays, light intensity. The aim is to respond to the museum's need to preserve the materials of works of art over time by shielding them from certain natural chemical and physical conditions that can deteriorate them. Depending on their organic or mineral nature, works of art require a specific climate for their conservation. Thus, metal must be kept in an environment whose degree of humidity must be very low, between 15 and 30%, in order to prevent rusting. Organic materials, however, require a higher level of relative humidity, up to 60%, for their preservation, in order to prevent them from dehydrating. However, it must not exceed 75% to prevent the risk of mold. Similarly, each material requires an acceptable amount of light. Light causes changes at molecular level and leads to the photochemical degradation of the material by shorter wavelengths such as ultraviolet, as well as heating of the material by longer wavelengths near the infrared. Thus, paper must be kept at a maximum intensity of around 20 lux while wood or metal withstand greater brightness. The museum is thus organized as a degradation of the natural climate, progressing rigorously and rhythmically from the outside to the inside, from the moistest to the driest, from the lightest to the darkest, from the strongest in ultraviolet to the weakest. It is conceived as a succession of selective filters that progressively reduce the chemical and physical parameters that deteriorate the materials according to their specific physical and chemical characteristics.

The layout of the museum is organized according to a series of concentric glass envelopes, which are gradually crossed one after the other to go from the more corrosive natural environment to one that is more chemically neutral. From the outside to the heart of the museum, five climates with different humidities succeed each other. It is first of all the humidity content of the air that gradually reduces from layer to layer, from 76% to 60%, 55%, 35%, 30% and finally to 20%. At the same time, the natural light intensity decreases as you plunge into the museum, gradually descending from 5,000 lux to 10 lux in its core. The premises are positioned in one or another of these climates according to the chemical conservation requirements of the works.

The quality of the architecture emerges from this pathway through the invisible. The principle behind the formal composition of the museum, its atmospheric conservation requirements, leads to the practice of the space. The museum visit takes place following this progression, by sinking into the chemical and physical data of the conservation of the works. The exhibition halls form a continuum that unfolds between the storage spaces that one can freely travel through, migrating through climates. The visitor is given a cultural but also a sensory visit. It is no longer only by sight that one can contemplate the works but also through the physiology of the space and of the body. It is a descent into darkness, the perception of increasingly dry air. It is the invention of a new way of visiting a museum, in the physical phenomenon itself of the conservation of the works, the physiological grasping of a climate increasingly close to a disappearance of time where chemical degradation is gradually erased and where the visitor enters an architectural form of durability.

Air density

For a pavilion project in Nantes designed in 2006, we were asked to reflect on the forms that a new school of fine arts might take. Our reflection examined the spatial concept of an atelier and its form in relation to the practice of art. Knowing, for example, the form and orientation to the light of a 19th century painter's workshop, then what would be the forms of an artist's workshop today and tomorrow? Rather than providing an inevitably false answer, our purpose here is to free the architecture from any programmatic predetermination of form and function. What we seek to do is to work on a new organization of space where function, but also form, could arise spontaneously from architecture as a climate. What matters to us today is to work on the matter of space itself, on the density of its air and the intensity of its light, to offer an architecture like geography and meteorology, open and variable, made of different climates, of different atmospheric qualities, that one inhabits freely according to the activity, according to the time of day, according to the season, according to one's desires.

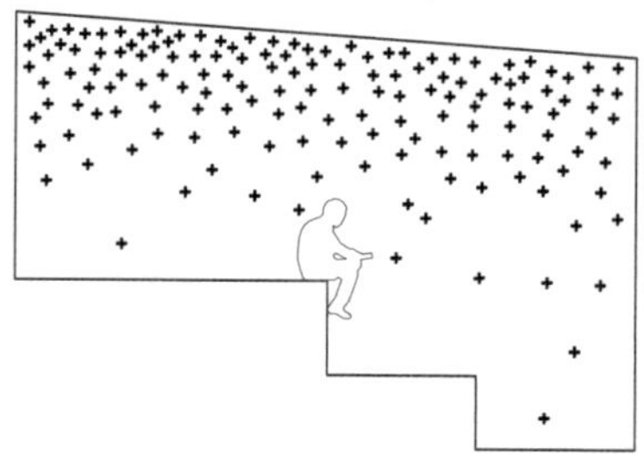

Called Open Climate, our pavilion works on two essential parameters of the development of space, that of light and that of temperature, according to two dimensions, the horizontal and the vertical. Horizontally, the plan is organized in the manner of a set of Russian nesting dolls, from the periphery towards the center, between the lightest and the darkest, in a succession of glass walls that each filter 33% of the brightness. Vertically, the space is designed from bottom to top, from the coolest to the warmest according to the principle of natural ascension of air masses according to their density and their temperature. The architecture finally takes shape here as a combination of three luminosities with three temperatures yielding nine climates, varying according to the time of day that can be enjoyed by everyone according to their activity, their work tools, their wishes.

Nine rooms on the same surface but like as many different climates, which unfold on the plane and in altitude. The choice is then made freely between the cooler places where one can be more active, weaker, more uniform lighting like for the 19th century painter, complete darkness in the center to work with video images, for example; as many behaviors and atmospheres to discover, experience or enjoy.

TIME

In "Building, Dwelling, Thinking", the philosopher Martin Heidegger extends his critique of modern technology to architectural and urban space. "Mortals dwell in that they receive the sky as sky. They leave to the sun and the moon their journey, to the stars their courses, to the seasons their blessing and their inclemency; they do not turn night into day nor day into a harassed unrest." This text, through its denunciation, reveals one of the very missions of architectural modernity: that of distorting the daily and seasonal rhythms, of provoking purely human spaces and times, outside the natural cycles, where night is day, where winter and summer are tempered to become spring. This alteration of the temporalities began in the 19th century, when public gas lighting lifted the city out of darkness, removed it from the natural daily rhythms, distorted the alternation of day and night. By rereading the writers of the era, we take the measure of an upheaval whose nature is truly demiurgic. Jules Verne in the Black Indies of 1877 mercilessly exposes the uprooting of the modern city from its natural context by inventing "Coal-city", an underground city with a completely artificial reconfigured climate. "Electromagnetic machines" produce heat and light, electric discs suspended from the vaults of the cave replace the solar disc and the stars of the night. Yet more radical, in his novel of 1889, Topsy-turvy, the French writer imagines the industrial project to rectify the Earth's rotational axis of 23° and 40 min in order to make it absolutely perpendicular to the orbital plane it follows around the Sun. There would then be no

winter or summer, but a perpetual spring in which the Earth would always enjoy a temperate climate like the golden age sung by Ovid. Jules Verne's book echoes the utopian futurology of Charles Fourrier who, with similar means, called for a "second genesis of the world".

The nineteenth century imagined a new electric and public golden era, on the city- and even planet-wide scale. But it was under individual and liberal configurations that it became a reality during the 20th century. In Ecumes (Schäume, Sphären III), the philosopher Peter Sloterdijk takes stock of the question of modern air-conditioning with the central idea that it is the main architectural innovation since ancient times and that with it "the construction of houses becomes an explicit climatic construction". Today, the axis of the planet has not been straightened, neither have cities been conditioned under a cloche. But in a more selective, private and unequal way, multitudes of fixed and mobile air-conditioned units of space – apartments, cars, offices, trains, planes – are connected uninterruptedly, are linked endlessly through private networks, throughout the planet.
A mesh of conditioned spaces draws a permanent temperate climate continuum throughout the year as the construction of an expanded perpetual spring. Safe, warm and dry, you can roam throughout, always, and never leave this electronic golden age uprooted from the natural world and repeated indefinitely.

The very origin of architecture lies in geographical and temporal distortions: the primary mission of architecture is to create pockets of spring climate, to locally deregulate natural climates to make them inhabitable: to create shade in full sunlight, places without rain or wind, warmer or drier, brighter or darker places. Anywhere, anytime, architecture transforms the irrational and brutal natural space into a comfortable ideal climate, around a 21°C, adapted to human metabolism. Geographically, the technique of heating houses in winter is a kind of motionless spatial displacement. It causes multiple small domestic shifts in latitude, in real time. As we walked through the door of his house on January 2, 2009 in Lausanne, we passed, in a single step, from the cold 0°C of the Swiss mountains, to the clemency of a climate of 18°C in the south of Spain. Returning inside the house thus amounts to an instant shift on the globe, from north to south, passing in an instant from Switzerland to Spain, from 46° to 37° latitude north. Architecture subconsciously causes a journey frozen in space, constructed from localized spatial contractions.

The technique of heating in winter is also a journey in time. It teleports you instantly from a today to a tomorrow. Crossing the threshold of the house in winter is an immediate leap, from the cool of January 2, 2009 to the mildness of July 2, 2009. The house is a machine to travel in time, to produce temporal and seasonal shifts. In one step, you go from -2°C in winter to 21°C in summer, to take off your coat and get comfortable.

With modernity, this phenomenon of geographical and temporal distortions has become global in scope. Today's cities are artificial environments where the old time correspondence between astronomical day-night cycles with physiological wake-sleep cycles and the rhythms of work-rest activities is blurred. The artificial lighting of cities and houses is a geographical misappropriation. It is an astronomical disturbance, where day is created at night, dawn at midnight, continuous days, an endless afternoon where twilight never comes. After this suppression of the night in the 19th century, winter was made to disappear during the first half of the 20th century by the widespread generalization of central heating. Today, it is summer that is reduced, due to air-conditioning enabling throughout the year establishing one long spring, at the same temperature, somewhere around 21°C. In these geographical and temporal displacements caused by heating and artificial lighting, we are clearly evolving in this "inspection" denounced by Heidegger, of modern technology that removes man from his natural climatic, geographical, temporal and astronomical conditions. With modernity and globalization, planning the climate has become one of the main reasons for the emergence of the architectural form. Today, it could be argued that global warming, the depletion of resources and the ensuing measures to combat these phenomena through reducing energy consumption, perfectly thermally insulating buildings and better territorial organization – could reduce this dissociation between natural and artificial space. In reality, the opposite phenomenon occurs. Increasing the thickness of insulation, the use of double or even triple glazing, the impossibility of opening windows in the controlled air renewal system: the most effective measures today to combat the release of greenhouse gases further dissociate internal space from the external environment, isolating the house from its context.

One of the key issues at the start of the 21st century is the matter of climate management. The challenge is posed not only on a global but also on an architectural scale, where both the causes and the consequences of global warming are found. Causes, because the energy consumed by the building sector accounts for 50% of the total energy consumed by European countries responsible for greenhouse gases. Consequences too, because architecture's primary mission is to provide places of comfort, protected from variations in the natural environment and therefore to increasingly respond with an artificial climate more tempered to the extremes of the natural climate. Global warming thus generates a vicious circle that drives toward increasingly air-conditioning the interiors of buildings, thereby accentuating the causes of this global warming. Today, there is a real abyss between the natural outdoor climate and the artificial indoor climate, and this is expected to increase in the coming years. Managing this disjunction between the uniform, stable interior and the indeterminate and unstable exterior has become central in architectural practice to combat unintentional heat exchanges and consequent energy wastage. It has become crucial to manage the boundary between the inside and the outside, to precisely control the thermal, chemical and electromagnetic exchanges between these two distinct universes. Current construction techniques bring about a perfect physical and chemical separation between the inside of the house and the outside.

The improvement of the thermal coefficient of the materials used, the maximum, continuous insulation of the building, the airtightness of its envelope cause and amplify this separation. Managing water vapor is a collateral problem of the building's impermeability where mechanical means must be put in place to evacuate the excess water vapor produced in the interior space by the breathing of the occupants and by the use of water. The impermeability of the building envelope requires the installation of an artificial system to renew the indoor air, formerly provided by cracks, construction defects, imprecise joins of the materials which unintentionally provided air passages between the inside and the outside of the house. In order to meet the demand for a reduction in energy consumption in buildings, we are now faced with a series of new building problems. They initially focus on the building envelope, its composition in layers: interior cladding/vapor barrier/load-bearing wall/thermal insulation/ventilation space/exterior cladding. Unlike the monolithic constructions of the past, where a single material solved all the structural and thermal problems, such as constructions built entirely of stone or entirely of wood, the contemporary construction proceeds by subdividing the constraints of the building which are solved separately in an optimal, economical fashion. The thermal issue is thus solved independently of the rest of the constructional problems by a material such as glass wool, for example, which has no other purpose than that of thermally insulating the building. The building's sealing against water vapor is regulated by a continuous plastic sheet which chemically and uninterruptedly encloses the interior spaces on themselves. Air renewal and the evacuation of excess water vapor take place in very specific parts of the façade, controlled, determined in relation to the internal ventilation air flows. The need to reduce energy expenditure and thus reduce greenhouse gas emissions is played out here by regulating the heating and cooling of buildings, by a reduction in the consumption of hot water and electricity.

All these measures produce an absolutely hermetic interior space isolated from the outside, neutral, free from random variations of the weather and the imponderables of the seasons, completely removed from the exterior nature. The interior space is then a manipulated, entirely artificial climate, whose parameters must be determined autonomously, calculated in accordance with its use, with the desired luminous and thermal environment. Without any possibility of returning to the old alliance between natural space and artificial space, the current technical measures insulate the interior space from the natural climate, deterritorialize it more strongly, abandoning it to the free choice of an artificial, autonomous climatic and temporal construction. By turning night into day, winter into spring, the contemporary architectural space today is floating outside of the astronomical rhythms. It is stagnating on a midday in May, in an endless pause in the Earth's rotational movement.

The two most important changes in this early twenty-first century, globalization and climate change, further accentuate the drift of modern space in autonomous spatialities and times, beyond astronomical cycles and biological rhythms, and are bringing about a profound transformation of the notion of space and time. Everywhere, light becomes the same, the temperature remains average and the degree of humidity constant. Architecture is deployed in a space that has become universal, ceaselessly projecting an eternal, continuous, invariable, homogeneous May midday, always present. It is a spatiality and a time beyond biological cycles, with no sleep, with no seasons, outside astronomical and climatic rhythms, with no night or winter, with no rain or cold: an uninterrupted global climatic network. Here and now, but also there and tomorrow, all meteorological variables have been stabilized around a shared standard of comfort. Somewhere in the region of 21°C, with relative humidity of 50%, a light intensity of 1,000 lux, like a beautiful spring day in Paris that it has been decided to repeat endlessly, all over the world. A Parisian May 15, for example, the day when buildings' central heating is usually switched off, until the fall. The climate continuum today is planet-wide: one of a global air-conditioning system where we find, everywhere and at all times, the same temperature, the same light intensity, implemented continuously, between housing, transport, offices and supermarkets, air-conditioned automobiles, airports, airplanes, and other cities, on all sides of the Earth, day and night, summer and winter. Without interruption from Paris to Neuchâtel, from Melbourne to Minneapolis, from Kyoto to Venice, the same spring day on which one dresses light, repeats itself eternally, winter, summer, day, night, the world over. It is the perpetual spring of the mythical Ogygia, the island of Calypso, the mild and invariant climate of San Diego, California that has now been extended to the rest of the world as a perpetual equinox, a nightless day, beyond the seasons. Distances are veiled with globalization and the Internet. The journey is instantaneous, from one climate to another, from one geographical location to another. We encounter the north of Africa in real time when opening the door of his house in winter in Vendée. Melbourne's morning light is instantly brought into the Parisian apartment when the electric light is switched on at midnight. The natural climate is no longer in real time but delayed. The season stretches, the day extends. By switching on a lamp at midnight, the light of the late afternoon is summoned abruptly. By starting the heating in winter, we are immediately projected into the coming spring. Contemporary space is the permanent artificial construction of an Indian summer, an American night or a Tahitian winter.

Abandoning public spaces in favor of private and commercial spaces, the design of this continuum amplifies economic and social disparities.

An unequal geography is established between south and north, the center and the periphery, where the cooling of one space is conditional upon the warming of another. In this widespread air-conditioning system, however, no architecture has yet been really defined other than by simple engineering whose sole inflection today is that of sustainable development and its cost. Faced with the depletion of resources and global warming, the

new ecological recommendations and standards, HQE in Europe or Minergie in Switzerland, are the standards in the making with which architects and planners must comply before studying them.

Against this standardization, architecture must become the critical instrument that enables articulating this continuum, to create faults and breaks in it, to engage it in sustainable development, to create balances between north and the south, the center and the periphery, to snap its global homogenization. Today's architecture must seize the project design means of this shift from the spatial to the temporal but refuse the end, as it is usually defined today. New scenarios are appearing: to occasionally or momentarily inflate certain climates, to naturalize a context, or on the contrary to distance it, to create temporalities, to generate meteorologies, to project seasons and days, to spatialize functions, to shorten distances or on the contrary to increase them, to reduce the lengths of the day or to create an endless night, here and there. These are times projected by architecture , these are entire days that are built. Here, architecture generates no sign or function, but moments, seasons, days, for inhabiting physically. It distorts, shifts, adjourns, prolongs or provokes suspended, supernatural temporalities, outside of astronomical rhythms, which one can penetrate freely. It is no longer a question of controlling the climate but of projecting time frames. Projecting eastern nights, anticipating springs, superimposing nights over days, creating twenty-five-hour days, living two summers in the same year can today become new themes of architecture. Architecture builds journeys in time, motionless paths, displacements through the day and the seasons. Going back, from winter to summer, working by climatic disturbance, by desynchronization, bringing out and modifying the form by seasonal shift, projecting by thermoperiodism, by dormancy or vernalization. Going back a few hours, a few months, a season, regaining that moment of comfort that we lost as the year progressed, returning from winter to fall, from evening to afternoon. Today's architecture is thus less a matter of space than of time. It does not represent. The times it builds are not allegorical. They are built in the very matter of space and the void, in their physical density, according to electromagnetic vibrations, chemical compositions. Astronomical time, physiological time, the persistence of a past light, circadian distortion, future day ahead, time shift. Architecture as built temporalities.

The contemporary issues of globalization and the immediateness of communications revive the idea of eternal return and, even more, the notion of a continuous present. Beginning in the early 19th century, a constant deterioration of the notions of rhythms and cycles took place to favor a continuous balancing. And it is within this new astronomy, distorted by modern technology, that some of our projects are conceived.

Black Day

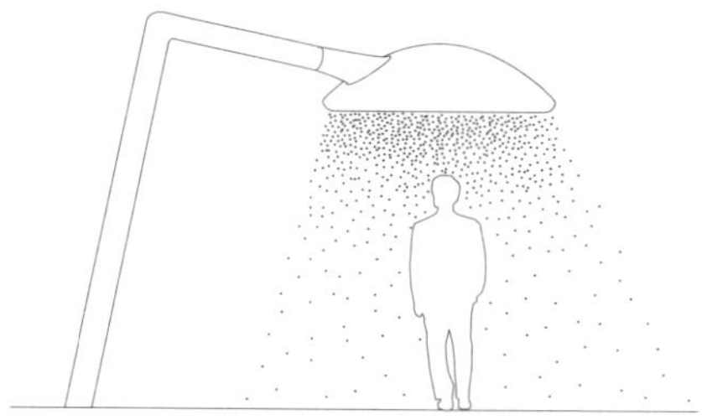

In Poland, in Gdansk, "Black Day" consists of fitting out a street, with its furniture and lighting, causing a false night to appear in the continuous day of modernity. This project invents an artificial night by diffusing, during the daytime, cold, black night time electromagnetic radiation like a mise-en-abyme of the urban night turned into a false day with street lighting. The proposed new urban infrastructure becomes the possible venue for inventing new social behaviors. This project is modeled, but negatively, on the technical and social revolution brought about in the nineteenth century with the introduction to public spaces of night lighting. In the past, the goal, demiurgic, was to create daytime at night, to prolong the day and its activities of wakefulness into the darkness of the evening. This project reverses this phenomenon. It renders the night day, draws the darkness of the night and its sleep into the light of the day. The introduction of street lighting in the 19th century revolutionized the perception of the city. New urban typologies emerged, such as the boulevard. A similar goal is proposed here, but flipped, by perverting the globalization of the uniform temporality of the contemporary city. The 19th century invented the idea of noctambulism. Gas street lights allowed using the city at night with its entertainments, strolls along the boulevards and night working. The conquest of the space of the night is now complete. Television and radio operate uninterruptedly, the Internet broadcasts a globalized time, permanently, services operate 24 hours a day, commercial late-night opening is becoming more widespread, and cultural late-night opening is increasingly commonplace. Today, on the contrary, it is the space of the day, though according to other modes of occupation, that interests us, where we would like, in turn, to deploy new temporalities. The "Black Day" project is a temptation to invent "diambulism", like the conquest, in the middle of the day, of the space and time of the night and of sleep. What we are pursuing is a disruption of behaviors, the possibility of a new use of urban space through the distortion of the astronomical cycles and an even greater freeing of human activities from natural rhythms. The man slept at night. The nineteenth century allowed (or forced) staying awake at night, strolling through the streets at night, working or cultivating, having fun under the light of the street lamps. This phenomenon of the colonization of the night continued in the twentieth century and today we are heading toward a continuous day where the ancient alternation between activity and rest, day and night, is shattered. This domestication of the urban night and the social behaviors it has brought about, has been associated with the production of a new conception of the public space and its furniture. The boulevard was one of its first manifestations. Lights, the illumination of monuments, neon lights, multicolored commercial signs, LED screens built a nightscape and a night time geography that differed from the daytime form of the city. The idea behind the "Black Day" project is to provoke night time activities in the middle of the day. It defines the urban space of "diambulism", its geometric structure, its furniture, its physical and electromagnetic definitions. Sleeping in the street in the afternoon, lying on a bed in the middle of the day and in the middle of the street, listening in the morning to nocturnes for the piano, this night-evoking music played in the daytime, feeling on your skin the coldness of the night sky during the day, plunging your gaze into the darkness of the night in the middle of the afternoon. To create night time during the day, we work on the electromagnetic radiation emitted, as the traditional street lamp does, but in an inverted way. Like a mini sun, the classic urban street light emits a visible, warm electromagnetic radiation of light at night, similar, in miniature, to that emitted by the sun. We invent a new kind of street lamp, to be lit during the day, which diffuses the negative light of the night composed of invisible and cold electromagnetic radiation, similar to that produced by the night sky, in miniature. These street lights that turn day into night consist of a concave diffuser in the form of a mini night time celestial vault, cooled by the flow of glycol water at a temperature of 1°C. The black diffuser absorbs the entire spectrum of light according to the same physical rules as the universe at night. The human body, at a temperature of 37°C, loses energy, by infrared radiation, in the direction of this celestial, nocturnal and icy mini-vault. In a phenomenon of thermodynamic equilibrium, our warm body loses its heat to this cold body, as it does at night under the night sky, whose temperature is estimated at -22°C. Aligned along the street, following the traditional arrangement of street lighting, these new outdoor lamps thus produce during the day cold, black light reproducing the electromagnetic spectrum of the night sky in miniature. They operate during the day, more strongly at noon, in a temporal inversion of night and day.

Yellow night

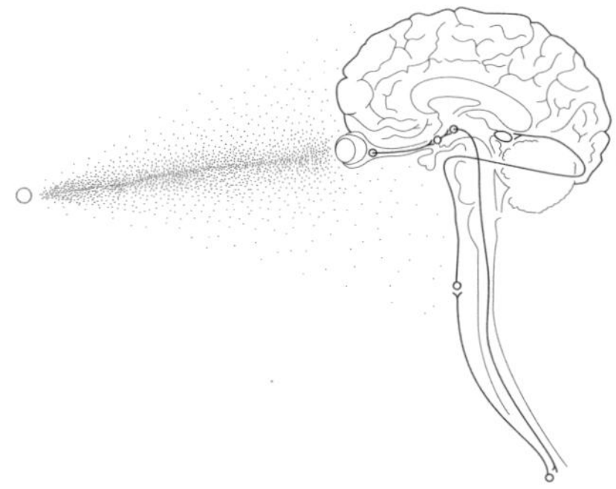

Presented at the Centre Georges Pompidou in 2007, "Diurnism" is an extension of "Black Day". But while "Black Day" recreates night during the day, or, more precisely produces a false night during the natural day, "Diurnism" makes a false night during the false day produced during the natural night. In a doubly perverse move to "make day during the night", we progress to "make night during the day made during the night", perverting the modern perversion which is to create daytime at night. Our project is to reinvent night time during the continuous artificial day, electromagnetically to produce a real physiological night during the false day. It is a perverse response to the eternal day created by modernity, the Internet, and contemporary globalization. After "noctambulism", we are inventing "diurnism". We work on an intense yellow light whose wavelengths, above 570 nm, are perceived by the body through the hormonal rhythm of melatonin as during true night. Melatonin is a hormone produced by the pineal gland located in the brain. This gland reacts to the light information received by the retina. Melatonin secretion gives the body and brain information related to circadian rhythms and biological clocks, sleep and fatigue in its presence, wakefulness in its absence. Melatonin secretions normally occur at night because they are activated by the absence of light. During the day, melatonin is only scarcely secreted. In 1980, A. Lewy, T.A. Wehr, and F.K. Goodwin showed that exposing the eyes to intense light inhibits the human production of melatonin (Sciences No. 210, 1980). In 1998, G.C. Brainard established that the maximum suppression of melatonin is achieved with green light at close to 509 nm wavelengths. This inhibition has since been known to be even stronger in the blue. In contrast, wavelengths above 570 nm have no effect on the secretion of melatonin, which can then spread freely in the body. At the 2005 World Congress of the Society for Light Treatment and Biological Rhythms in Eindhoven, a Canadian team showed that above 570 nm, in yellow, orange and red, light, even very intense, had absolutely no influence on melatonin variation, the body perceiving these wavelengths as night time. "Diurnism" thus creates a concocted night, visually resembling the day but physiologically similar to the night. Illuminated by a very strong yellow light, the Centre Pompidou room becomes a paradoxical place between the visible and the invisible. The very powerful electric light emitted in the room is comparable in intensity to that of the outdoors, at more than 7,000 lux. But it is emitted only in wavelengths above 570 nm, in the yellow. Hence the mismatch between the cultural image we have of the night and its physiological reality.

"Diurnism" builds a temporality, causes the emergence of the night in the false modern day. This work does not use any symbolic or analogous narrative process. The architecture lies in temporal desynchronization, at the heart of physical, physiological, electromagnetic and biological phenomena.

Simultaneous time frames

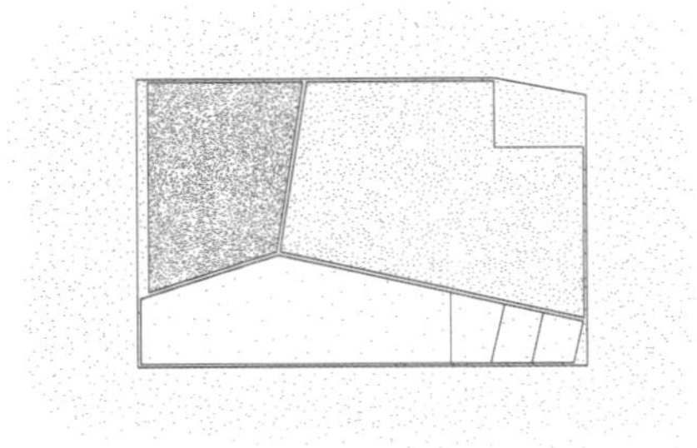

"Split Times" is a project for a pavilion commissioned in 2007 for implementation near Graz in Austria. This project explores a spatiality derived from an architectural language as a construction of temporalities. It works on a division of the present time, into three parallel times: a natural astronomical cycle, a perpetual night, a perpetual day. In this project, these three times are present simultaneously.
The architecture opens up the possibility of inhabiting three time frames at the same time: one is natural, the other two are artificial but physiologically real. The project is thus a machine to travel in time, from day to night, from natural to artificial, and architecture here is literally the construction of temporalities, thus offering a variety of uses and atmospheres, to choose freely, passing instantly, in one stride, from night to day and vice versa. The first temporality, enveloped in clear glass, is, in real time, that of the natural, changing, circadian, seasonal solar cycle. The furniture is that of a traditional café, with tables and chairs. The second temporality is delimited by a yellow-colored glass envelope, blocking the wavelengths of light responsible for the decrease in melatonin in the body. It reproduces a true physiological night while being luminous. The furniture here is similar to that of a lounge bar, tending towards sofas and couches. The third temporality is defined by a blue glass envelope, letting through the wavelengths of light that block the secretion of melatonin in the body. This last room builds a permanent day. The associated furniture is that of a bar, with tables and chairs.

Continuous Present

Like an unconscious design, modern techniques of central heating and electric lighting aim to extract space out of the astronomical natural variations of the seasons and day-night cycles to attain a sort of perpetual spring, beyond what is natural. Deeply attached to any architectural production is a desire for stability, balance and eternity that is expressed both through space and in time. The statics of constructions, the balance of masses, are accompanied by the durability of structures and the standardization of climates. Architecture generates immobility and heritage. The notion of heritage is to time what air-conditioning is to space: an extraction of man's environment outside of time and astronomical and climatic movements. To the breaking of day and night, to the variations of the meteors, to the alternations of the seasons, architecture opens immobile and stable spaces, it generates a constant temperature and intensity of light, atmospheres that repeat, day after day, year after year, until forming an artificial landscape and artificial weather. Architectural spaces are traps in the course of time. These are moments that are snatched, to be repeated endlessly and without escape, to infinity, in the same space. The production of a temporal heritage, the constitution of a space as an eternal recurrence, the moment in winter one extends incessantly until the following spring, eternal noon, continuous spring, architecture deploys an abstract temporality outside the world that turns into heritage, generating cultural geographies that overlap on the natural ones of landscapes in movement and times.

Continuity, perpetuity, eternity and repetition appear as possible states of the reality of contemporary housing and modern times. These themes find their philosophical origin in Nietzsche in "the death of God" and in his vision of "the eternal return of the same" that the German philosopher experienced during his stay in Switzerland, in Sils-Maria, in 1881. This concept certainly comes from his interest in thermodynamics which dominated the late nineteenth century.
"The law of the conservation of energy demands eternal recurrence", Friedrich Nietzsche wrote in 1887. In 1934, Jorge Luis Borges recalled the content of this concept in his essay on "The Doctrine of Cycles" published in "The History of Eternity": "The number of all the atoms that compose the world is immense but finite, and as such only capable of a finite (though also immense) number of permutations. In an infinite stretch of time, the number of possible permutations must be run through, and the universe has to repeat itself. Once again you will be born from a belly, once again your skeleton will grow, once again this same page will reach your identical hands, once again you will follow the course of all the hours of your life until that of your incredible death."

The idea of an "eternal return" is fundamental for Nietzsche, because it gives him the means to invent a moral philosophy without God, marked only by "the theory of return as the greatest weight".
"He who does not believe in the circular process of the universe must pin his faith to an arbitrary God – thus my doctrine becomes necessary as opposed to all that has been said hitherto in matters of Theism!"
The eternal return, therefore, is the invention of a metaphysics without God, which makes us responsible for our actions apart from the threat of a divine chastisement. In this modern morality, each of our actions, each of our decisions, whatever it is, will be repeated an infinite number of times, for all eternity. In the History of Eternity, Borges contests Nietzsche's explanation of circular times by arguing that he relies only on the first law of thermodynamics ("Nothing is created, nothing is lost") and that he forgets the second, that of entropy, certain processes being irreversible. But it is not so much the idea of the eternal return that Borges contests as the way that Nietzsche states it. Borges would prefer circular but gradually deteriorating time. He combines the two principles of thermodynamics in one theory. In this regard, Schopenhauer said: "The form in which the will appears is only the present, not the past or future; these exist only conceptually and for the enchainment of consciousness, submitted to the principle of reason. No one has lived in the past, no one will live in the future: the present is the form of all life."

Perpetual present and temporal distortion are some of Borges' fictional devices. It is for example the Secret Miracle where we see a present of one second dilated over a year. In the analysis of Herbert Quain's "fictional" work April March, based on the inverse world of a certain Bradley, in which death precedes birth, Borges describes the novel as regressive and ramified. Borges places the origin of the idea of inverted world with Plato who, in Politicus, describes a regression of the same kind: "That of the Sons of the Earth or Autochthons, who, subjected to the influence of an inverse rotation of the cosmos, passed from old age to maturity, from maturity to childhood, from childhood to disappearance and nothingness. A friend of Borges', Adolfo Bioy Casares, was the author in 1940 of Morel's Invention. The aim of this book is the repetition of the same present fixed at a single moment in time that no longer evolves. It describes a machine for recording the present and replaying it. The book draws on the visions of the 19th century, at the time of the invention of cinema, where it was imagined that it would be possible, without difficulty, in the near future, to record and to rebroadcast, in addition to image and sound, volumes, smells, temperatures, humidity levels, light intensities, that is to say, the entirety of a climate, of a moment. In Morel's Invention, as in Jules Verne's Castle of the Carpathians, the cinematographic technique grows to allow the recording and rebroadcasting of living beings, half-opening the possibility of a form of eternity. In Morel's Invention, it is a summer's week that is recorded and repeated tirelessly, for all eternity, at the site where it was recorded.

The thematic similarity of Alain Robbe-Grillet's Last Year at Marienbad with Morel's Invention is obvious. There are similar moments like this fantastic scene where the characters of the island of Morel are bathing as if it were the middle of summer in a swimming pool full of dead leaves, beneath two suns, simultaneously summery and autumnal. But for Robbe-Grillet, the temporal disturbances are generated by cinema as a form and not by a mere fictional effect. In his work, the content and the structure participate in this work of physical influence. The specificities of cinematographic or literary language, the material that makes up the movie or novel, influence the narrative. The novel emerges in the present of the writing, the story is invented gradually, in the dark, without the continuation being known. Writing continuously, without a plan, without going back, following the trace of the pen on the page, from right to left, line after line, which is Robbe-Grillet's modus operandi, brings about the emergence of a kind of extraordinary temporality in which the reader loses their way. In the seeming strangeness of the images of the movie Last Year at Marienbad, directed by Alain Resnais based on a screenplay by Robbe-Grillet, the latter sees only a "concern for realism", as referenced by Gérard Genette in his very beautiful essay Vertige fixé. This realism is that of cinema as a medium whose images are always in the present. Cinema can only be in the present for Robbe-Grillet, because the images captured by a camera are those of the present. No camera can film the past or the future, unlike writing where the same verb can be conjugated in the present, the past or the future. Last Year at Marienbad is therefore a movie in always the present: all the images we see on the screen, whether they tell the past, the present or the future, collide in a single time, in the present. The same image sometimes combines several temporalities. In the park scene, the motionless characters cast long late afternoon shadows on the gravel, while the trees in the gardens have none, as if under the midday sun. All temporalities intersect in the fixed frame of the image, the different years intermingle, amalgamate, dissolve in the same shot. The characters of the movie, lost in this continuous present, no longer know if it was last year at Marienbad or next year, if they have already met or if they have already experienced the same situation. The medium influences the course of fiction: the characters in the movie and their story are prisoners of the instrument that stages them – a camera that films always in the present.

The projects that follow relate to the reality of today's space: ubiquitous, supermodern, supernatural. They work on perpetual presents and time distortions, not through subjective intention, but because architecture's unconscious mission is to create interior spaces where time stands still, where the same light of a May 15 is artificially repeated, where heating perpetuates a beautiful spring day, beyond the night, beyond the seasons. The frozen temporalities of architecture are exacerbated, globalized, they reactivate the myths of the golden age, provoking spaces and times when it is never too hot or too cold. These projects theme time as architecture and thereby reveal its artificial content. In reality, this artificiality has been part of the very program of modernity since the 19th century and the phenomenon of global warming accentuates it. It is no longer just the interior of houses that is transformed into an artifice but the atmosphere of the entire planet as well. Increasingly early springs, trees budding as early as February, summer extending until the end of fall, these are phenomena we are all familiar with and are starting to experience. Architecture as a construction of temporality presented here is not a modernist project, but a critical project of a modern project that has already been carried out.

Second summer

This is a duplication of the present, where two seasons overlap, where the present stretches beyond itself, which is at work in out project of 2005 for the development of an island in Austria, in Eibesfeldt. Here, a second summer, a false summer season drifts through the fall into winter. On this island, the ground temperature is artificially kept below 0°C throughout the year. Artificial lighting reproduces the length of the day of the summer solstice providing 15-hour summer days throughout the fall and in winter. The project takes place on a marshy island, in the dead arm of a river. It is another time that we seek to produce, outside of astronomical rhythms, a drifting time, a shifting season, perhaps a summer, which first becomes an Indian summer to stretch even further into the year until December. A kind of second summer, like a suspended climate, a seasonal alteration, an even more vigorous Indian summer, strays into the Austrian winter. In the heart of the island, behind a thick belt of tall deciduous trees, we choose to define a clearing as a summer, as an uncertain and supernatural space. Here, the summer lasts indefinitely. It stretches along the year, it passes fall by, and crosses winter until the following spring.

In the clearing, we reconstruct the light and temperature of June 21, whose physical characteristics are prolonged and repeated every day of the year, in the fall and in winter, thanks to electronically controlled lamps and a buried circuit of hot water that prevents the ground from freezing completely. It is the summer solstice that is reproduced and repeated. June 21 is the longest day of the year after which the days begin to get shorter, signaling to the plants to drop their leaves. Computer programs provide the exact amounts of sunlight (in lux) every minute of the day at this precise place on the planet, on June 21. These data are then entered into a computer controlling the lamps, their intensity and their illumination time, causing artificial sunsets and sunrises every day. In the clearing, the sun will rise every day at 5:03 am and set at 11:51 pm. Days in the clearing always last 15 hours and 53 minutes, while on December 21, for example, beyond the clearing there are only 8 hours and 30 minutes of daylight. The vegetation of the clearing remains in this summer, and will not really change color. The junctions of the branches may never break, the leaves do not fall completely, because these seasonal phenomena are induced by the shortening of the length of the day and its perception by means of a plant hormone. The electromagnetic spectrum of the artificial light illuminating the clearing reproduces that of the sun on June 21, compensates each day for the shortening of the day, extending photosynthesis beyond the fall, beyond winter. Powered by an energy-independent water-water heat pump, a water circuit, at a temperature of 35°C, is installed in the ground of the clearing, at a depth of 25 cm. The heat radiates into the ground, spreading to the surface, throughout the year. During winter, a new population of plants and shrubs grows there, in a staggered climate and time, lost in this unlikely summer. Rather than determining a place with the help of spatial tools, it is through temporal tools that we delimit a space in the vegetation of the island, as a sort of temporal clearing. "Second Summer" is a drifting season, a staggered climate, a summer that extends too far into the year, that eclipses the natural cycle of the seasons. Architecture here is the construction of a season, and we are finally close to the art of gardens, with the idea of greenhouses, exotic gardens or winter gardens, that seek to build not only shapes and colors, but also climates and other geographical latitudes. This is the attraction of these motionless journeys, those of orangeries, exotic follies or jardins d'acclimatation. But here, it is not a journey in space, but a journey in time, through the seasons: living a few moments of summer in winter, living dawn at night, feeling the intensity of the summer sun beneath the winter sun. In this clearing, the vegetation does not behave exactly as in the surroundings: the leaves fall later or not at all, it snows less. It is a microclimate that recalls Shangri-La in Frank Capra's movie Lost Horizon, the myth of the Golden Age and Arcadia, legendary, seasonally offset places. "Second Summer" is a drifting season, an overly long summer that, slightly veiled, overlaps winter.

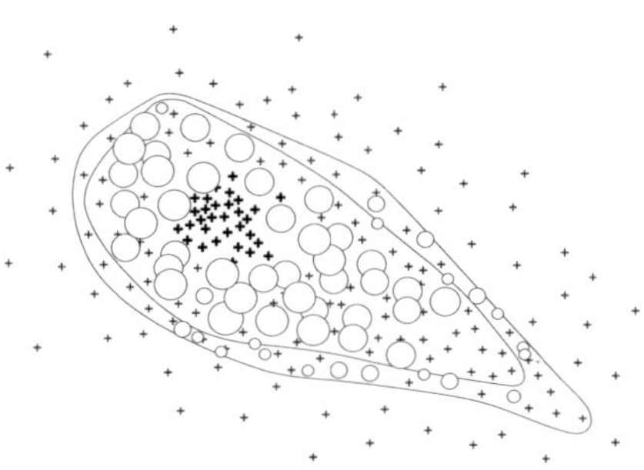

Perpetual spring

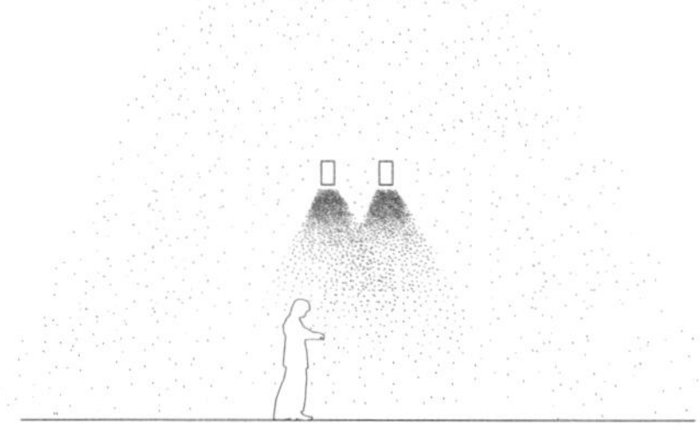

For the exhibition "Invisible Architecture" in 2005 at the Swiss Cultural Center in Paris, the means are similar: the physiological construction of a false spring that overlaps the natural winter, as if two temporalities, two seasons, two days straddled each other. It is the ideal, comfortable spring that has been recreated in winter by acting on the electromagnetic spectrum of space and its intensity, on photoperiodism, on dormancy. In the courtyard of the Swiss Cultural Center, it is a beautiful mid-May day that we recreate on March 12. It is a May 15 whose light intensity and variations we anticipate every day. From the opening of the exhibition, on March 15, in real time a powerful lighting system injects brightness into the courtyard to create "May 15" every day until disappearing on the "natural" May 15. Because of the position of the Earth's axis in relation to the sun and its annual movement, sunlight is less intense in winter than in spring, just as the days are shorter. The lamps thus compensate each day for the decrease in the intensity of the sun and the length of the day by artificially creating sunsets similar to those of May 15 at 9:27 pm (whereas on March 15, for example, the sun sets at 6:56 pm) and sunrises at 6:08 am (instead of 7:04 am). In the courtyard, every day the lamps reproduce the electromagnetic spectrum of the sunlight of May 15, its intensity at each minute as well as the sun hours of this spring day. Reproducing the spectrum of daylight, these lamps form a luminous sheet, a supernatural duplication of the light of the sky, which reacts precisely to variations in natural light intensity, compensating whenever necessary for the lack of sunshine in order to build this spring day of a May 15 on every day of the exhibition. Programmed by an electronic clock, reacting to the variation of the ambient luminosity in real time by means of photosensitive cells, the lamps light up at the end of the afternoon to compensate for the decrease in sunshine hours and unvaryingly maintain a day of 15 h and 18 min in length. Plant flowering, budding or leaf fall cycles are totally related to day length through plant cyclokinin hormones. Offset by a few months, spring will come earlier in the courtyard of the Swiss Cultural Center where fruit trees bud as early as March.

Uchronia

The comparative study of computer models of climate and its actual evolution very clearly relates the anthropogenic influence (i.e., caused by human activities) with the current phenomenon of global warming.

It is by artificially introducing carbon dioxide (CO_2) and chlorofluorocarbons (CFCs) into simulated models of the Earth's natural climate that a change in temperature of the order of 0.2°C per decade is generated. The phenomenon is today clearly identified: gases derived from carbon, mainly CO_2, are produced by the burning of fossil fuels such as oil, coal and gas released by human activities: heating buildings, transport, electricity production. For a long time, there was an ecological balance between the release of CO_2 into the air produced by breathing, burning or the chemical decomposition of living bodies and the absorption of this CO_2 by plants during photosynthesis. This balance meant that there was about as much CO_2 gas in the atmosphere as there was solid CO_2 in plants and other living beings. But by starting to burn fossil forms of carbon, in other words oil, underground natural gas or coal that had been "sidelined" from ecological exchanges for millions of years, fossilized, and thus inert underground, humans introduced a surplus of carbon in gaseous form into the atmosphere that the existing terrestrial biomass has not been capable of recycling. This results in an ecological imbalance that translates into the accumulation of these gases in the atmosphere, which form a chemical layer at altitude. This layer prevents some of the infrared radiation emitted by the earth from dissipating into the universe, beyond the planet. Just like in a greenhouse, infrared heat is trapped in the Earth's atmosphere and finally heats up the climate of the entire planet: this is what is known as the greenhouse effect.

Mathematical climate models that began to be developed in the 1960s very clearly show the influence of this release of CO_2 on the increase in the Earth's temperature. Computer models of climate consist of a set of equations derived from fluid mechanics, the water cycle, and the geography influencing the absorption or reflection of solar radiation. Thus, by introducing or modifying certain parameters (so-called "forcing"), these models enable understanding climate change. It thus becomes very clear that the anthropogenic release of CO_2, since the beginning of the industrial era with a very pronounced acceleration as of the 1950s, is directly responsible for global warming. Indeed, depending on whether this anthropogenic forcing of CO_2 is input into the equations, we will yield different curves of evolution of the temperature of the Earth's atmosphere.

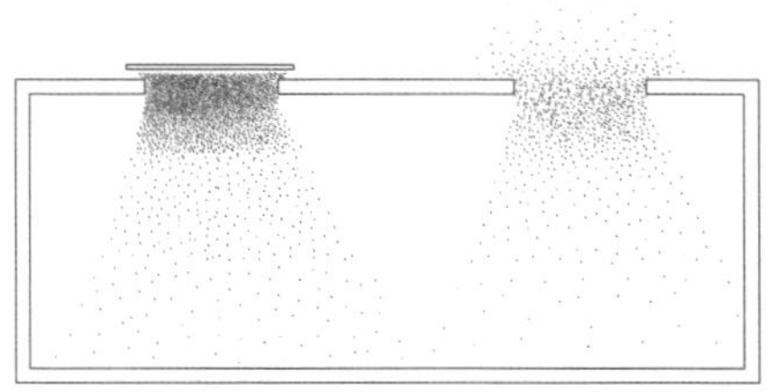

These computer simulations, which demonstrate the link between global temperature warming due to the release of CO_2, were initiated more than 20 years ago and their results have been found to correspond exactly to the measured real change in the earth's climate over the past 20 years, which seems to prove their validity. It was on the basis of these data that, in 1990, the Intergovernmental Panel on Climate Change (IPCC) published its first report showing global warming of a few degrees (between 2 and 5°C) for a two-fold increase in CO_2 emitted into the atmosphere. This report was the basis for the second Earth Summit, held in Rio, Brazil. This meeting alerted the political authorities to the phenomenon. Today, all computer models confirm the observations, and governments are beginning to take serious action against the emission of greenhouse gases. While computer models of climates have therefore proved effective, they are not intended to be used for short-term weather forecasts. Meteorological predictions are dependent on microphenomena like the famous wing flapping of the butterfly that will trigger a hurricane a few days later. These models work on time scales of the order of a year, decades or thousands of years to prove effective.

The real phenomena of global warming are transforming our perception of what is natural and what is artificial. What we once recognized as the natural climate, which was outside the building, is no longer natural today. Since 1850, the industrialization of human activities and the release of CO_2 have transformed the planet's atmosphere, manipulated it like a product, and warmed it. The comparative study of temperature trends between computer climate models and actual climate measurements demonstrates very clearly the anthropogenic influence on the phenomenon of global warming caused by the emission of greenhouse gases. The classic architectural dichotomy between interior climate, that was artificial, and exterior climate, that was natural, no longer exists. The paradox is this: if we want to experience the "natural" climate, that may never again exist, we must now reproduce it "artificially". Presented for Manifesta 7, the European Biennial of Contemporary Art which took place in Fortezza, Italy in 2008, Climate uchronia is the construction of a false sun, the physical projection of a uchronian universe: the world as it would have been if global warming had not existed. To define this climate, we recalculate the climate of summer 2008 by thermochemical digital modeling of the atmosphere from which we remove the variables of gas emissions caused by humans that has been responsible for the greenhouse effect since 1850. The computer then calculates the climatic data (temperature, cloud cover, humidity) in Fortezza for the summer of 2008 and generates precise and specific climatic values for each minute of the three months of the summer of 2008. We transcribe into light intensity the variations calculated by the software of the moisture content of the sky (cloud, fog) which causes solar light intensity to vary; a cloudless sky obviously being brighter than a misty sky. These data are entered into a computer that controls the variations of an electric light placed in false windows that illuminates the interior of the space and generates light variations during every day, with its artificial sunrises and sunsets, corresponding to what the weather would have been like in Fortezza if global warming had never existed. Coexisting within the exhibition space, through the real and the false windows, these two luminosities, these two temporalities, without one knowing any longer what is natural and what is artificial, what is live and/or what is recalculated.

Built shift

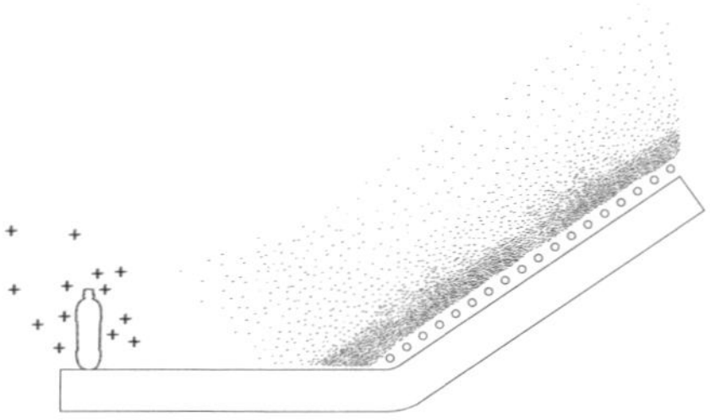

The current global warming in Europe is represented spatially by a geographical displacement from north to south of a distance estimated at about 200 km by the year 2050. This shift is measured by a rise in temperature of a few degrees and a change in relative humidity level.

It is slowly transforming the nature of the climate and therefore the nature of agricultural crops. There is therefore, at the same time as a transformation of the external, meteorological climate, which we perceive outside our body, with our senses, by sight, by the heat on our skin, a physiological transformation, inside the body, intimate, in nutritional terms, which manifests itself through taste, through digestion. Global warming is bringing about a macroscopic change in nature, on a territorial scale, in the order of a latitudinal subsidence of about 200 km southward and a microscopic change in nature, in the order of a flavor, of a chemical and gustatory quality. It is then also the notion of terroir that is transformed and warmed by sliding on the terrestrial globe southward, modifying the ancestral habits of cultures, both those of their lands and of their inhabitants.

The foreshadowing of this displacement and its interpretation become the formal reasons for the project to develop one of the interior spaces of the Zadkine Museum in Les Arques. Two physical dimensions, one climatic, outside the body, the other physiological, inside the body, are called into play. The first relates to sunlight, whose intensity increases by a factor of 0.75% with each degree of latitude less. By descending on the terrestrial globe some 200 km south, i.e., 2° in latitude, the sunlight increases by a factor of 1.5%. If the light intensity is 108,732 lux at noon on July 1 in Les Arques, at a latitude of 44°36'11" north, it will be 110,347 lux 2° further south, at 42°36'11". The light intensity of the sun is therefore stronger in the Pyrenees. A difference of 1,615 lux whose intensity we reproduce in the museum by using 18 daylight-type 58 W fluorescent tubes. The second project dimension, the internal one, concerns food. We propose to offer drinking water in the museum brought from a region 200 km south of Les Arques in the central Pyrenees. Thus, our project is both a gustatory and an environmental modification of our perception, combining dietary, geographical, ethical and esthetic considerations.

GEOLOGY

Architecture is organic. It is part of the ecological cycles. It is one of the links in mineral, chemical, biological and also food chains that form and deform materials and substances through time and space. Architecture is entropic at first, because construction is irretrievably subjected to the dissipation of its energy, to the corruption and degradation of its forms and materials. First of all, there is gravity, which tends to make heavy, pull down, undo all the light movements that the architect directs upwards to the point of ruin. Gravity also played a major role in the status of architecture in the work of the German philosopher Hegel. Because of its formal dependence on this natural force, precisely because of its heaviness, Hegel demotes architecture's position among the Fine Arts, relegating it to the lowest level of art. Whereas Le Corbusier spoke of architecture as "a pure creation of the mind" in a typical aspiration of the modern esthetics for dematerialization, weightlessness and the disavowing of the connection to the ground, Hegel, on the contrary, described architecture as "an imperfect symbol of the mind" because gravity dictates architecture's physical laws, threatening at every instant to make its forms which were so painfully erected toward the sky fall to the earth. In the system of the arts conceived by Hegel, architecture is placed at the lowest level, as the poorest art as it is the least free to represent ideas. Architecture is described as the first of the arts, primary, because it is too strongly constrained by the external forces of nature which impose its lines and its forms. If walls are straight, if the column is vertical, it is above all because the force of gravity is precisely vertical, and not because the architect decided it should be so. For Hegel, the forms of architecture are above all the outcomes of the perceivable world, of weightiness, of the physical properties of the materials used, regardless of human will. The architect, with no room for maneuver, can only order these forces, arrange them according to rules of symmetries, proportions, regularities, as classical architecture did. Hegel's system of the arts is based on a hierarchical system of value, simultaneously historical and artistic, between a lowly material art, architecture, and the highest art, intelligible and pure; poetry. The purpose of art is clearly articulated by Hegel as the transformation of the exterior world perceivable to man into a representation of himself. Through art, man projects the forms of his mind beyond himself, in the manner of the 17th century French gardens where the mathematical human forces of composition, symmetry and regularity come, through pruning, alignment, and perspective, to contradict the natural irregularity and instability of the plant forms and materials generated by the climatic and geological forces upon which they depend. At the lowest level of Hegel's system of the arts, therefore, is architecture, because it is dependent upon weightiness and material constraints. Above architecture, because it is formally freer, comes sculpture. Above sculpture is painting, where matter progressively becomes less resistant and more malleable, like the oil colors of classical painting which one mixes, which one masters until matter and pigment disappear to reveal only color and light and an image. In this relentless progression toward the pure mind, by the historical overcoming of its carnal condition, art gradually loses its materiality to become sounds and waves in music, then just words in poetry, before dying, replaced by scientific prose and religious thought.

The development of the life sciences and of ecology in the 20th century invalidates Hegel's hierarchical system, founded on an ontological separation of man from his environment. Biological, chemical, and electromagnetic exchanges tie man to his environment through ecological, physiological, and neurological processes, through breathing, eating, through conduction and radiation. Similarly, the extension of the domain of knowledge to the invisible, at microscopic or nanometric scale, transforms the validity of an esthetic judgment which is happy to distinguish only the phenomena which can be perceived by our five senses. The field of art and artifice is today deployed in invisible dimensions, imperceptible to the eye, which emerge, for example, from genetics or electromagnetics. To take the case of the French garden, a garden could appear disorderly, irregular and natural and yet be a pure construction of the mind by consisting of genetically modified plants. The principle of the dematerialization of art, a sign of the elevation of the mind according to Hegel, is no longer valid today when the very notions of thought and even morality are embodied henceforth as much in the baseness of material, neurological, and physiological processes. Similarly, the ecological sciences have made visible the invisible chemical and biological mechanisms required for life and for our own survival. Accordingly, the landscape architect Gilles Clément, through his praise of brownfields and of weeds, renewed the esthetics of the garden by shifting the reading grids from the visual toward the biological: weeds acquired a positive value as their presence and growth attests to a neguentropic dynamism of the living and to the augmentation of the richness of the ecological exchanges of a place.

This shift from the visible to the invisible is also at work in a second entropic line of thought of architecture that concerns the degradation of its materials. In the same way that the form of architecture is subjected to natural physical forces such as weightiness, its materials are subjected to a slew of exchanges with the environment and modifications to their substances, the most obvious of which, chemical, physical or biological, are erosion, corrosion, oxidation or rotting. If Hegel classifies architecture at the lowest level of the arts, it is also because the materiality of architecture is subject to the corruption of its material state according to uncontrollable and involuntary processes of pollution, contamination and alterations. If for Hegel music and poetry belong to the highest of the arts, it is because their material dimension is reduced to almost nothing and to the almost pure: a sound, a word, a wave.

What we can keep of the Hegelian system of the arts today is not its moral and philosophical dimension but the precise designation of the scope of action of each of the arts. Architecture is characterized by its submission to gravity and the corruption of its materials, its material baseness, its heavy and tangible incarnation in the perceptible world, its chemical, biological, electromagnetic links with the environment. What belongs to the esthetically poor and the negative according to Hegel, the impure and the base, can today be interpreted in an opposite and positive manner by way of an ecological and physiological reading grid. Rather than tending toward a false dematerialization of architecture, as a certain form of modernity has striven since El Lissitzky, we remain jubilantly, and in contradiction to Hegel, among the lowly, the material and the impure. The natural forces rejected by the field of classical art according to Hegel such as gravity, photosynthesis, or oxidation, then become the efficient forces of the architectural project.

From the modern struggle to overcome gravity to the erasure of physical materiality in favor of the narrative sign in post-modernism, architecture can today reverse its movement which tends toward a dematerialization of its appearance, in order to reengage in the physical world, to regain thickness, sink itself into the responsive ground, in the obscurity and the humidity, a certain lukewarmness, a rate of emissivity of matter. Architecture then becomes geological and mineral, but of a reformulated geology. A second geology, a second minerality, which we build by taking certain mineral substances from the natural soil, certain ecological or physiological sequences, sometimes impoverished, reduced to a few elements but nevertheless efficient. From an entropic status, architecture becomes neguentropic. By accepting to no longer be at the end of an ecological chain as the only moment of consumption and unproductive expenditure, it becomes a link, it acquires a nutritive, physiological, and intermediary status in a series of movements and processes of transformation of materials and of forms in time and in space.

Brown geology

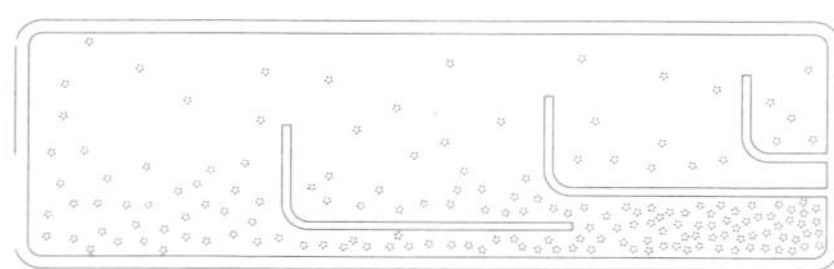

From the multiplicity of physical phenomena specific to the quality of a soil and the constitution of a geology, for the project of houses in Vassivière in Limousin in 2005 we retain only the notion of inertia due to the thickness of the rocks and their physical constitution:

the low conductivity of the lithosphere which stabilizes the temperature beneath the earth. The inertia of a basement becomes an architectural theme: the constitution of a certain thermal quality, constant year round, warm, that is to say cooler in summer, hotter in winter. The project develops according to this idea, as an entrenchment in the granite soil. The houses are implemented horizontally in two superimposed volumes. The first is above the earth. The second, lower, buried one, envelops a certain amount of air in the earth's thickness and constitutes a stock of warm, fresh air, at a constant temperature and humidity. It is an underground landscape, a granitic climate, offset from the atmospheric seasons, in the region of 12°C which has traditionally been used as a cellar or a grotto, to store wine, preserves or fresh food. But it is air that is stored here in the inert subsoil, serving to renew, to ventilate the air of the house. The fresh air, harnessed above ground level, is fed underground in the manner of a Provençal or a Canadian well, to then be stored in the depth of the Limousin granite subsoil, where it cools in summer or, conversely, warms in winter. What interests us is this possibility of the air literally becoming a product of the terroir, preheated by its time spent in the lower layers of the earth, warmed by the geothermal differences of the subsoil. Can the air of this house acquire an olfactory quality specific to the place? Can it transmit to the house an earthy flavor, a slightly brownish tonality? In fact, what the air acquires from the ground is nothing but a thermal quality, a degree of warmth. But the goal of the project is to build this deep physiological and sensual bond with the place and its subsoil where the quality of the architecture, through the air and breathing, takes on a geological dimension.

Domestic geology

Today, the living space of contemporary people is clearly confronted with the disintegration, not to say complete reversal, of the traditional organization of its categories and its balances between interior and exterior, artificial and natural, enclosed and open. The "natural" as opposed to the "artificial" (insofar as the term "natural" qualifies that part of the real world that has not been modified by humankind), literally no longer exists. As a consequence of humanity's technical progress and its meteoric development as of the second half of the 19th century, it is today the planet itself, its atmosphere and its geology which are transformed by human activity into artifices through the phenomena of pollution and global warming. The German philosopher Martin Heidegger put the origin of these phenomena in the project of modern technology, which proceeds from what he termed "provocation" or "critical ordering", that is to say by subjecting nature to reason, taming it. If, prior to this, nature had neither a goal nor a duty, the aim of modern technology was to impose on nature a single purpose, an unequivocal application, a single use, thus depriving it of its indetermination, open-endedness and irrationality. Technology is the tool for this transformation of the natural into the artificial. And architecture is part of this, since its aim, locally and temporally, is to extract climate and geography from their state of nature to render them domestic and reasoned.

Ideally, we may argue that at one time there was a simple, clear-cut dichotomy between the interior of the house, its architecture, the city (artificial space), and the exterior of the house, its environment and atmosphere (natural space). In the interior, architecture was the technically modified space in which an artificial climate and a reconfigured geography were implemented, smoothing out the conditions, providing convenience and comfort, a more temperate environment all year round, somehow constructing domestic geologies and humanist meteorologies, a place of symmetry, balance and homogeneity. Outside, the environment was the domain of the natural climate and primitive geography, unpredictable and intemperate, wild as the forest or the mountains, and of the wind, the cold and the rain: asymmetrical, unbalanced, heterogeneous. This dichotomy no longer exists today. Change began at local level with agricultural exploitation and pollution to then spread worldwide, to the scale of the entire planet, leading to the present-day phenomenon of global warming. It is no malapropism to say that the entire globe and its climate have today become products of human activity: although we previously heated only the interiors of our houses, today we also heat the exterior of our houses, in other words the planetary climate, buy several degrees, due to the phenomenon of global warming. There was a time when the external environment, the terrestrial atmosphere and the countryside, belonged to the realm of the natural, but from now on they too belong to the category of the artificial.

The reaction to climate change today is to implement a policy known as "sustainable development". In construction, this policy translates into fairly basic technical solutions, foremost involving the reduction of energy used for heating. Such techniques rely mainly on good thermal insulation, good weatherproofing, and the need for the renewal of filtered, controlled air.

Ironically, these measures conspire to produce an interior environment that is even more removed from the exterior than was previously the case, precluding all direct sense relationship with the external local context, other than that of a comprehensive interface: insulated, separated, controlled. This is a paradox: the technical measures put forward by architecture to achieve sustainable development accentuate the modern-day chasm between external environment and interior environment.

Our living space has been upended by these transformations of hierarchies and of the dialectical rapport between the natural and the artificial, the exterior and the interior. If the external environment, outside the building, is no longer natural, could we not posit the hypothesis of a "naturalization" of interior space? Could the interior environment of buildings not be reconfigured through sustainable development and its techniques no longer as the place of the artifice but also as that of nature, that is to say, of what is asymmetrical, unbalanced and heterogeneous? A second nature as a domestic geology and an interior meteorology. If nature, which by definition has always been excluded from the interior of the dwelling, is now also expelled from the exterior, could we not imagine a reversal of this situation, by reintroducing nature into the interior, by making the interior of a building more natural than the exterior, which it no longer is? This implies another upending of scale, hierarchy and classification: what once belonged to the macroscopic, to the realm of asymmetry, to the exterior of the house, to the natural and to the atmospheric worlds, suddenly find themselves inside the house, on a microscopic scale. Nature would then no longer be outside the house, but inside.

In some way, this upending has existed since the 19th century, just as it was at work in the program of modernity, while remaining an involuntary consequence of it. One example is the drinking water of Paris, which still today comes from the Seine, after filtering and purification, which means that it is ultimately cleaner and more natural in our water pipes than it is in the river, polluted as the latter is. Here is the reversal. On the microscopic scale of the water in the pipes, the interior of houses becomes the place where the resurgence of the natural, the geological and the non-polluted takes place, while on the macroscopic scale of the water in the river, the exterior becomes the domain of pollution, of the artificial, the accidental, unwanted by-product of human activity. From Gallo-Roman times to the mid-19th century, the water drunk in Paris was drawn straight from the Seine, hitherto clean, natural and potable. River pollution rose sharply during the 19th century, thanks to urban development, until the tragic cholera epidemic of 1832. In response to this crisis, a water management policy was implemented to control water supply and drainage by the prefect Haussmann as of 1854. Subsequently, and up until 1901, Paris drinking water no longer came from the river but from the countless natural springs of the Paris basin, whose water was brought to the city in closed aqueducts. Not until 1901, with the construction at Ivry-sur-Seine, upriver from Paris, of the plant for slow-filtering over gravel and sand, did river water once again become fit for human consumption. But no longer drawing straight from the Seine, outside: but by bringing it indoors, as if recreating a second Seine, flowing through the pipes, and, paradoxically, more natural than the Seine itself. Today, a second pre-modern Seine flows along the city's water supply network: a miniature re-formation of natural hydrology, unpolluted water, a chemical reproduction of a state of nature prior to 1832.

Terroir

The principle of our work for the 2009 Carte Blanche of the VIA (Valorisation de l'Innovation dans l'Ameublement - Valorization of Innovation in Furnishing) precisely concerns this reversal, an upending of the interior to the exterior. It aims to encapsulate nature in architecture, in order to qualify interior space sensually, chemically and spatially, and confer on it visual qualities that can also be appreciated by our senses of smell, touch and taste. Our project is implemented within the framework of sustainable development applied to construction design according to the goal of significantly curbing energy consumption and a consequent reduction in the release of greenhouse gases. In our project, the elementary issues of air and temperature generate new interior landscapes, like a second nature, where geological, plant and atmospheric conditions reappear in all their asymmetry, re-shaped to scale inside architecture by means of technical systems that strive to fulfill the goals of sustainable development.

Our proposal chemically and mechanically reconstitutes the geology and atmosphere of Paris prior to the appearance of widespread pollution in the 19th century, like a natural reality that is filtered and regenerated at the same time. It thus represents a process that is both nostalgic and prospective, a miniature reconstitution of the Parisian geology and of its atmosphere in pre-modern times, with its limestone strata and semi-Atlantic climate, its prevailing winds blowing mainly from the ocean to the west, eroding the limestone strata of Normandy and the white chalk bedrock of the Loire valley, becoming impregnated with their minerality and the scents of oak and chestnut forests typical of the soils of these regions. It is also a movement of air, a wind direction that we reproduce homothetically on the scale of the house; a south-west to north-east vector, like the invisible passage of the Paris air. Here, our architecture reconstitutes these limestone soils within the building, using construction techniques associated with sustainable development. It renders visible the filtered and recomposed reality of the contemporary dwelling by increasing the spectrum to include geological, plant and atmospheric values.

In this respect, our project integrates certain practices common to the arts of cuisine and wine-making, the composition of fragrances and flavors, brought about by the precise choice of ingredients as if following a recipe or creating a wine. It is not so much a question of the visual aspect or the colors of the materials that interest us as their chemical nature in terms of smell and taste, and how they are assembled. Our furnishings are composed of elements that are qualified on the corporal and physical level, and others that are chemically neutral. This process might resemble the notion of aging as conceived by wine experts, and is applied here to the air and the interior living space, with the aim of conferring certain objectives of wine to air: to rid the air of its impurities, to enhance its aromas, to complete the structure of the air by adding external tannins. Certain materials remain chemically neutral, with neither smell nor taste, and do not influence the tang of the air, just as stainless steel used in kitchen pans or cellar vats does not modify the taste of food or wine. Or polyethylene, for example, a neutral, non-acid, non-alkaline, pH 7 material that can be recycled; as such it does not interfere chemically with other materials, hence it is often used in the conservation of works of art, or for packaging foodstuffs. This project aims at recomposing a specific terroir in an interior, applying construction techniques, air management, aeration with heat recovery, heating and ventilation. There will then be chemical, electromagnetic and physical terroirs indoors, deterritorialized terroirs, like post-terroirs, which chemically evoke a type of soil, a climate, exposure to the sun, a perfume, a certain tang in the air. Different kinds of limestone, quarried from the Parisian basin, steer the air through the interior. A precise homothetically reduced composition of the proportions of plant species and trees common to the Parisian basin, mainly oak and chestnut, but beech and pine too, which impart fragrances to the air, while glass, polyethylene and stainless steel constitute chemically neutral components in the architectural composition of the scented living space.

Our proposal primarily addresses the immaterial, aerial, and invisible because it is precisely in building design, heating and ventilation that the challenge is posed to reduce greenhouse effect emissions caused by the construction industry. Today, it is not so much the choice of building materials that is responsible for climate change as the quantity of energy expended on everyday living in buildings due to heating and cooling. We are working on this level, on the movement of air, on its thermal, olfactory and electromagnetic qualities because it is here, in this invisible part of the living environment, that some of the major issues of this new millennium are at stake. Our proposal relies on three elements: gentle aeration by means of dual-flow air renewal, an asymmetric heating system by means of split radiation, and a low-energy lighting. Each of these elements comes in response to three objectives of enhancing the interior space: an ecological goal, a physiological goal and a third goal, mentioned above, that of naturalizing the interior space, reformulating a terroir within the building. And it is the atmosphere of Parisian limestone terroirs that we seek to reproduce indoors: "Parisian air" and pre-modern "Parisian light". And so we seek to recover a certain age before the anthropogenic discharge into the atmosphere of greenhouse gases, circa 1831, that we start to repeat perpetually. The interior environment of the house becomes a kind of phantom temporally displaced from the exterior environment, a kind of uchronia of a Paris "where global warming would not have existed"; a distortion of both time and scale, indoors, of a time, of an air, of a light of Paris that is repeated to infinity, in the house, by applying building techniques. For its inscription in history, the reference to localism, this work revives some project themes of the post-modernism of the 1970s-1980s, in its symbolic, imaginary and narrative dimension. But it would then be a kind of chemical, almost culinary, postmodernism at the very heart of matter and its phenomena.

Greenery

Today, gentle, automatic ventilation is the most energy-efficient and economical method of renewing the air inside a well-insulated and sealed building in order to bring in fresh air for breathing and to evacuate pollutants and moisture released when breathing. In the past, aeration took place either accidentally through the cracks and poorly closed joints of the façade or windows, or routinely opening the windows to air. Since nowadays buildings must be perfectly insulated to reduce energy consumption by heating or cooling, these spontaneous solutions are no longer acceptable because they are too random and unintentionally dissipate too much heat. The principle of gentle aeration involves ventilating the interior of the house, with the help of two small fans, by continuously renewing the minimum amount of air required according to the number of occupants, thus allowing a four-fold reduction in the energy consumed on heating or cooling. The dual-flow principle consists of preheating the cold outside air entering the building by bringing it into contact with the hot inside air leaving the building. This method recovers up to 80% of the calories usually lost. This system is presented here in the form of a compact ventilation device for the entire house, with a system of pulsed air and extracted air, while recovering the heat contained in the discharged air by means of a plate exchanger. The air is thus preheated in winter and cooled in summer by the exhaust air. This exchange, which is purely thermal and not chemical, thus allows not to waste the temperature of the evacuated indoor air by recycling it in the incoming air which is preheated by the outgoing air. Our proposal is to integrate this new technique not by hiding it in the cellar or in equipment rooms, but by accepting the compact dual-flow device as a new object of the domestic living space, a kind of new chimney whose narrative and symbolic potential we anticipate for the future.

The renewal of indoor air, also known as ventilation, is necessary above all to remove the excess moisture that we confer to the air upon breathing. It is also used to bring new oxygen to the metabolism which is consumed through muscular activity. It is also necessary to evacuate metabolic waste products, such as H+, CO2 and lactates. The renewal of air inside a building takes place at a certain rate according to its use, number of occupants and their physical activity: a low rate of occupation or activity will require a lower renewal rate while a space occupied by many people or by people engaged in strenuous physical activity or sport requires a much higher rate. The wood present inside the apparatus helps regulate the atmospheric humidity by absorbing water vapor when the air is moist (wood can hold up to 11% water content when the relative humidity of the air is high). Conversely, it will slowly release moisture when the air is dry (descending to a humidity of 6% of its weight).

Before reaching Paris, the air brought by the wind brushes against the landscape of the Parisian basin, sweeping through the typical forests of the limestone soils surrounding Paris, becoming charged with the scent of the branches and leaves. It is this Parisian plant terroir that we seek to reproduce in order to confer a particular quality to the air. The wind rose indicates the origins of the winds and their respective force. We superimpose the wind rose on the map of the distribution of the arboreal flora of the Paris region to choose, proportionally, the different species of wood for incorporating inside the dual-flow apparatus. The air leaving the apparatus is a miniature reproduction of a kind of plant synthesis of the air of Paris that would have reached the city before the industrial era, charged with natural particles giving flavor to the air. And, finally, we choose only to use wood whose growth rings date from before 1832, the year of the establishment of the first iron foundry and the start of the industrial era in France. All other materials and elements making up the monobloc dual-flow ventilator, such as screws, hood, gaskets, and so on, are specifically chosen for their chemical neutrality; glass, polyethylene and stainless steel, so as not to influence the contact of the air with the wood and its subtle beech and chestnut scents, typical of limestone soils.

Light

The average amount of electricity consumed by a family on lighting is about 400 KWh per year. This figure can be halved by using new light sources which, yielding the same luminescence, consume less energy. With the incandescent light bulb, only 7% of the electricity consumed is turned into light, the remaining 93% being lost through heat. We propose working with a lamp using light emitting diodes (LEDs) as a light source. LEDs use very little electricity thanks to their excellent performance as the light they emit is practically all in the visible range of the spectrum, so there is little dissipation by heat loss. If modernity has deterritorialized and dephased light by making it available everywhere and at all times, we now know, thanks to the studies by Alfred Lewy in the early 1980s, that certain circadian hormonal cycles are in phase with the daily variations in the luminosity of natural light, between day and night. The modern break of this natural physiological relationship between the human body and the astronomical rhythm of sunlight has been the subject of studies ever since, which seem to reveal both organic and psychological disorders as a result of this. In this context, we propose to restore the temporal dimension to electric light by resynchronizing it with the solar cycle. We propose a lamp that can be used in two modes. The first is an automatic mode that controls a daily light cycle corresponding to a spring day, May 15 in Paris. This mode automatically varies the intensity of the electric light over 24 hours, with artificial sunset at 9:26 pm and sunrise at 6:07 am. This default mode is satisfying from a physiological point of view as it enables synchronizing the body clock to the cycle of natural luminescense. The second mode is manual with 24 positions available via a switch, thus allowing the user to choose the desired light intensity of any given moment during May 15, the lamp then diffusing the programmed intensity. The device, which can be adjusted to requirements, is a kind of physiological clock whose timetable responds to light intensity. Using archived meteorological data and digital models of climate, we capture the solar cycle of May 15, 1831, from morning to night, from midnight to midnight, over twenty-four hours. We choose the year 1831 as it is the year before the commissioning of the first coal-fired plant in France which marks the beginning of the ecological upset of the carbon cycle caused by the vaporization into the atmosphere of fossil carbon, responsible today for the greenhouse effect and global warming. We therefore propose a "day before" to experience indoors, like a beautiful spring day, in Paris, in 1831.

Vertical thermal air stratification

In this new indoor geological and climatic environment, we propose another type of furniture whose horizontal mobile function we accept but we amplify vertically, by allowing inhabiting different heights, different air strata, different temperatures. The height of the seats, between floor and chair top, between table and ceiling are unusual, and this enables conceiving the interior space not only in terms of surface area, but as a volume, an atmosphere allowing interior climatic migrations, domestic thermal transhumance.

Limestone

The 2007 project for the Parisian branch of Voies Navigables de France (Navigable Waterways of France) involves a similar upending to the VIA project, where nature is found within the architecture and sensually, chemically, spatially qualifies the interior space, conferring its plastic, olfactory and gustatory qualities. The basic issues of air and water shape interior landscapes, like a second nature, geological, hydraulic, atmospheric, reformed, scaled down, at the heart of architecture and its technical systems in response to the goals of the Kyoto Protocol.

Contemporary building techniques developed to meet the sustainable development goals are themes whose architectural potential we expand to the point of producing new types of buildings, new forms, new functional organizations, new spatial poetics.

The building's air management generates the spatial form and poetics of the Parisian branch of VNF. The project composition is based on the requirements governing air management (gentle, automatic ventilation with heat recovery/Minergie, Switzerland) in order to reduce the energy consumed in the building by a factor of 4. The MINERGIE® seal is awarded in Switzerland to buildings whose design enables reducing consumption by at least one factor, two compared to current legal requirements. Reinforced thermal insulation, an airtight casing, controlled air renewal and high performance glazing are the requirements for achieving these objectives. Our project transforms the question of automatic gentle aeration into an architectural theme that literally gives body to architecture. Rather than considering this technical issue as secondary in the architectural composition, the project makes it a major point, the very root of the form and program of the building and the poetic qualities of the space. Another facility is proposed. The project consists of a compact, dual-flow ventilation unit for the entire building, with a forced and extracted air system, and recovery of the heat contained in the evacuated air by means of a plate exchanger. Automatic gentle ventilation is the most environmentally and economically efficient method today for renewing the air inside the building in order to bring in fresh air for breathing and to evacuate pollutants and moisture released when breathing. It combines comfort, cleanliness, air quality, low energy consumption and performance. The fresh air input in each room is provided at any time and separately. Pollutants are removed continuously. There is no excess moisture, therefore no surface condensation, no mold or uncomfortable drafts. This air renewal technique requires a network of fresh air injected into each of the rooms, which is then recovered by an air extraction network which returns it to the compact air renewal apparatus incorporating a heat exchanger.

Our project themes this supply of forced air in the building by transforming a technical problem of an internal duct and the routing of air in the rooms into an architectural and spatial principle. The size of the forced air ventilation duct is augmented to become the living space itself, flowing, deforming, folding and unfolding, compressing and relaxing throughout the entire length of the building. Rather than being made of galvanized steel sheet, this duct is made of solid load-bearing limestone, from west to east encompassing the length of the geological course of the Seine, from the sea to the Paris basin, from Le Havre to Paris, reproduced in miniature. The architectural quality we provide is that of an atmosphere, an odor, a perfume, an air, an air of Paris, coming from the sea, from the west which, in the heart of the building, will fill with the fine scent of limestone. A few oaks are planted along this course, their presence adding a touch of vegetation to this Parisian fragrance. Due to its continuous, rounded shape, the mineral ventilation duct sends the air to each room noise- and turbulence-free. Inhabiting a ventilation duct is only possible for uses that do not themselves require significant air renewal in order to preserve the quality of the air distributed to the offices. This is the case of the lobby and corridors, which only require air renewal twice per hour as they are used only temporarily, while in the offices, as they are in constant use, the air requires renewing ten times per hour.

White geology

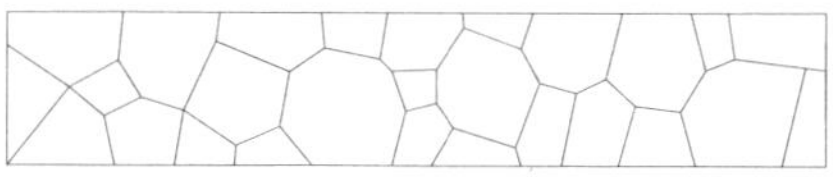

"White Geology" for Force de l'Art 2009, the second triennial of contemporary art at the Grand Palais in Paris, in fact deals with nothing more than the reflection of a background, on the emissivity of white, a certain optical reflection rate, set back. The material for this background is of no further importance than to provide a high reflection rate, 80%, where the light diffracts, evens out and homogenizes. It is a white landscape, an unfolded museum White Cube upon and before which, in the foreground, works of art are placed, set like natural objects. The particular shapes and colors of the works of art stand out and are enhanced on this uniform, neutral, white background, which blurs the distinction between horizontality and verticality.
It is a white geological form reduced to just a few elements. Rather than an architectural project, we propose an impoverished geological process, generated by the strength of the works of art themselves. It is first and foremost a surface in space, a rectangle measuring 125 m by 24 m which begins to deform, hollow out, swell according to a series of forces specific to the geological language of the formation of the landscape by tectonic shifts, deformations, pressure and vacuum, folds. Here, the plastic movements and distortions of the territory are caused by the abstract forces of the art works themselves. Each of the works is given the same space and the same volume at the outset. Then, depending on the size of the work and the distance from the viewer, they start to collide, as in tectonic plate movements. Their weight and the amount of light they require will determine how much they deform the surface, cutting into or swelling it, creating valleys and hills.
The museographic device is therefore reversed: it is not the work of art that adapts to the architecture, but the architecture that folds and deforms according to the requirements of the work of art. In a game of mutual movements of reciprocal and simultaneous thrusts and equilibria, the works will, in unison, bring about the emergence of a landscape. A landscape deforms, hollows out and swells, housing spaces beneath extracted from the natural light of the Grand Palais and above, open to natural variations of sunlight. Superficial, objective, shallow and unmysterious, the surface of the geology has a similar albedo to snow, 80%, in which shadows and decorative moldings are obliterated.

CLIMATE

**Working on the atmosphere, breaking down the boundaries between interior and exterior, between body and space, physiology and meteorology, atmosphere and sensuality.
Working on the void, on the air and its movements, pressures and vacuums, conduction, transpiration, convection. Architecture as the dispersion of borders, the vaporization of structures, the evaporation of boundaries. Invisible, light, clear, architecture becomes the design of air movements, the composition of temperatures, humidity levels, pressures and vacuums.**

Our architecture develops as a plastic and poetic consequence of sustainable development that induces a broadening of the climatic techniques of the building with the purpose of reducing the emission of greenhouse gases that cause global warming. In this mission, architectural hierarchies shift and transform. Emphasis is placed on controlling the heat transfer coefficient of materials, on insulation, on controlling the airtightness of the casing, on studying air movements, on gentle aeration with heat recovery, on managing the path of water vapor, convective movements and temperatures. A shift takes place from the structural to the atmospheric, from the solid to the void, from the terrestrial to the aerial. And the forms of architecture emerge as new atmospheric relationships. Unprecedented typologies emerge as poetic and spatial extensions of these techniques.

Our interest in sustainable development applied to construction is twofold. First, because it induces a renewal of the architectural elements. Second, because new modes of architectural compositions appear. There is a shift in hierarchy insofar as the elements and means of architecture that slide from the visible toward the invisible, from metric to thermal composition, from a structural to a climatic thought process. There is, therefore, a typological reformulation of the architectural composition that throws the modern criteria of form and program into crisis. In a process of clarification, vaporization, simplification, the meteorological and physiological essence of architecture is reformulated: on the one side that of the phenomenon, and of the sensual on the other, and of their exchanges. Space becomes a physical and a chemical atmosphere in which one is immersed and which one transforms by inhabiting it, through breathing, perspiring, through our body's heat radiation. At the architectural level, the goal of sustainable development is to reduce the amount of greenhouse gases released by the building during its operation, that is to say to reduce the energy consumed in the building, mainly through heating and the production of hot water. In architecture, this goal is expressed less at a formal than at a climatic level, which essentially requires rethinking the matter of building techniques, those of heating, ventilation, insulation of the envelope, moisture management as enhanced elements of architectural language and its design. Thermal composition, hourly air renewal structure or the hierarchy of relative humidity levels can thus become the new esthetic criteria of the architectural composition. These invisible compositions of space are then transfigured to reveal plastic and sensory qualities where a sensual relationship is invented between the body and space. On the one hand, the senses, the skin, breathing. And on the other hand, the climate, the temperature, variations in humidity and light. The programs, functions and uses will then appear as poetic deployments through the variations of spaces produced by these novel techniques.

Historically, we have seen that the introduction of a new technique often modifies the spatial and functional organization of the building and consequently lifestyles. It causes a typological renewal, the abandonment of forms that have become inadequate. Central heating, which distributes heat evenly throughout the house, has produced very different living arrangements from those generated by spot heating, using stoves or chimneys. In the country houses of yesteryear, the single fireplace served both for heating and as a place for cooking, that is, of course, when its fumes was not being used for smoking meat to preserve it. The fireplace organized the space thermally, distributing the floor plan and the section in relation to the radiating diffusion of temperatures in the air. Activities that are today separated spatially would all be carried out In the same place. The fireside, in equal measure and at the same time, was given over to the kitchen, the living area and sleeping quarters for the elderly. The floor plan and the section of the house were organized according to thermal qualities, allowing seasonal variations of uses and thus migrations through the house. Central heating, meanwhile, deterritorializes the house in relation to its external context, it eliminates the geographical and climatic relationship between the interior space and the immediate external environment in favor of a homogeneous continuity of the climate throughout the house, however it may be oriented.

**There is, therefore, an archaeology of architectural typology, strewn with ancient programs, outdated architectural organizations, rooms whose name and use have been lost. While we have been working since the 19th century with certain well-defined spatial elements such as the "bedroom", the "living room", the "bathroom" or the "kitchen", to name but the most common ones, along the way we have abandoned many other rooms whose meaning and function have been forgotten. Bourne, chauffoir, burons, vivoir, caouhade, voûte, zadrouga, rafraîchoir, ramonétages, curralѕ, these unspoken communities are all rooms, spaces or places that have disappeared since the introduction of modern heating. When in the 17th century the Italian Baroque architect Le Bernin presented his plans to King Louis XIV to transform the Palais du Louvre, the minister of State, Jean-Baptiste Colbert, was worried about where the king would sleep. Bernini replied that this type of question was not of interest to architects but only to stewards.
At that time, the architects to the court did not take care of the function of the rooms, thus working upstream of the notion of program, the king being able to sleep freely wherever he pleased, depending on the season, according to his mood. In a reaction against modern functionalism, Louis Kahn updated this principle during the 1960s by proposing to rethink the composition of the architectural plan in more abstract terms of spatial hierarchy, served and servant space, and no longer in terms of program and function, according to the idea that the structural form of a space could be deployed in temporal permanence while its function could change over time.**

The spatial design and organization of the city and of housing have thus varied greatly throughout history and geography. Programmatic elements, certain types of rooms have disappeared over the centuries.

In many villages of France, there was once a room called "voûte" or vault, which served as a living room, a stable and a dung heap, where people would gather in winter to enjoy both the shared animal heat and that released by fermentation. The poor insulation of olden day houses and the insufficient thermal performance of their heating drove the inhabitants to congregate in the same confined room, in order to share their body heat. A heating problem gave rise to a social practice: the "veillé" or evening gathering. This principle of communal thermal assemblies is found in the Aleutian Islands, where during the winter, up to three hundred people gathered in the same community house sharing a single fireplace. And we could cite many more examples of these unusual hybrid programs that are socially inconceivable today, such as the "Tué", the central room of the Jura farm which was at the same time a hearth, a living room and a butchery since up to three whole oxen were preserved there in the smoke of the fire.

But it is not so much the evocation of different lifestyles that interests us here but rather the way in which a problem or a typically architectural solution (fighting the cold, enduring a humid climate) have led to the emergence of other unpredictable and unimaginable lifestyles. In the end, it is not clear whether it was the wine that invented the cellar or whether it was the cellar that invented the wine. But here we go along with the second hypothesis. By reversing the hierarchical order between function and climate, we hypothesize an architecture whose functionality would emerge "as if by chance" from climate-related problems or solutions. Architecture would have the capacity to be no longer functional. It might not respond to a pre-established use, but be open, interpretable, suggest, enable, through answers to climatic or technical problems, the emergence, almost in spite of itself, of other functions. Certain programs, certain social practices appear by chance or by necessity in the management of the climatic forms of architecture. Working in this way with new climate techniques, freely accepting their formal and programmatic potential, would make it possible to move the built space out of its unambiguous functional determination to render it freer, more interpretable and more sensual.

Architectural responses to sustainable development should not be a mere technical layer that would be superficially superimposed on the project. They must become the driving

force behind a fundamental renewal of the urban and architectural form and, consequently, enable the invention of new functions and other uses. It is a radical work that questions, at the root, the essence of architecture to redefine its elements. The "form and function follow climate" formula condenses this viewpoint by putting forward the idea that architectural form and function could be not upstream but downstream of the architectural project. Form and function would then emerge as formal and programmatic consequences of environmental and climatic factors, the latter becoming the efficient forces of the project. The introduction of climatic parameters, in the city and the building will lead to the emergence of new forms that will call for a redefinition of their social, cultural or political interpretations. This formula reflects the theses of the professor of geography Jared Diamond of UCLA University in the United States. By confronting the major events of social, political, religious or cultural history with geographical and climatic upheavals, Jared Diamond reverses the understanding of human history. He establishes sequences in which human history is no longer separated from an environmental history, but on the contrary, it is completely intertwined, sometimes to the point of being not the engine but the consequence. "...environmental geography and biogeography influenced societal development ... History followed different courses for different peoples because of differences among peoples' environments, not because of biological differences among peoples themselves", Jared Diamond writes.

The monofunctional programming of the rooms in which we live today, such as the bathroom or the bedroom, originated at the beginning of the 19th century, with the introduction of the corridor into the middle-class home. Thanks to this typological finding, each room, now independent, then takes on a specific function: children's room, library, boudoir. Other rooms were invented later and have undeniably enjoyed great success until today, such as the famous Frankfurt kitchen, a functionalist archetype of the kitchen, created in 1926 by Margarete Schütte-Lihotzky, whose minimum dimensions are based on a precise description of the movements of the arm and its length to grab a pan on one side of the room and put it on the stove on the other.

Ultimately, we live in a rather recent interior typological landscape which is a mixture of a bourgeois organization of the house dating from the 19th century and the modern free plan, dating from the early 20th century. Some functions are starting to decline, such as the dining room. We can see a kind of evolution of architectural typology throughout history, with its past genres, its endangered spaces, its nascent programs. Our work is at this last level, in the desire to bring out new typologies of housing according to the modification of the building's air-conditioning modes related to sustainable development. But our goal is not to produce new functions for rooms, but rather to take advantage to break with the unbending and unambiguous designation of the rooms. We would like to free the space from the function by considering the whole house as a climate atmosphere. The inner space becomes a physical geography, in which the functional interpretation of the space is left free. Architecture becomes meteorological. Wall, floor, ceiling, column are no longer the main elements of architecture, now replaced by vapor, heat, air and light. And if we still talk about distribution or composition, it is no longer that of rooms, corridors or spaces but of convection, conduction, radiation or condensation.

The emergence of new forms in the history of architecture is often not the result of a process of imagination, but rather the consequence of an intelligent and critical consideration, but also sometimes accidental, thanks to emerging techniques or technologies. Faced with global warming, today we would like to posit the hypothesis of a modification of architectural language and of a typological change as a consequence of a disruption of architectural hierarchies in connection with sustainable development. We wager that the plastic and structural potential of new air-conditioning and insulation techniques is capable of renewing the field of architecture and urban planning with equal intensity to what happened with the appearance of reinforced concrete at the start of the 20th century or the arrival of new digital technologies at the end of the 20th century. The urban history of Venice provides a perfect example of this thesis where an urban form and the uses it enables are unexpected consequences of a physiological necessity and a geographical reality: that of drinking in a place without drinking water. Venice owes its urban form and the labyrinthine multiplication of its squares, the campi, solely to the absence of available fresh water. Built in the brackish waters of the lagoon, Venice is not fed by any underground source of drinking water or freshwater river. If each of us can now enjoy the picturesque charm of the campi, the restaurant terraces that have been set up there, and admire the tremendous social attraction of these little squares, we must remember that they were not originally designed for such social activities. The campo is above all the site of an ingenious rainwater catchment system, the only source of drinking water in Venice. It was harnessed within a certain perimeter, sent to the edges of the square, cleaned in sand beds isolated from the lagoon by a waterproof clay cone, to be finally stored in a well, in the center of the square, ready for drinking. The shape of these squares, their dimensions are the result of a match between the amount of rainwater that can be recovered according to a certain floor area and the number of inhabitants occupying the surrounding houses. The number of occupants is therefore directly related to the amount of rainwater recovered. The latter then sets building density and height, an occupancy rate not according to legislative, cultural or political rules, but according to geographical and physiological equations. If each inhabitant needs 5 liters of water a day to live and the well of a campo allows recovering and storing 200 liters, then ultimately a population of 40 people will be able to live around the square. Venice is the most beautiful city in the world, even though its urban structure is, in reality, just a contingent response to the lack of drinking water. This real and necessary balance between physiological necessity and social life, as expressed through the Venetian campo, may explain the failed attempts to create public squares, ex nihilo, during the 1980s. These last attempts, void of the physiological necessities of the Venetian model, then produce superficial and dispensable urban forms, incapable of actually establishing a use on a form.

The radical decomposition of new climatic techniques according to their geographical and physiological dimensions could give rise to a new method of meteorological composition. Unimagined forms and unexplored uses could arise from this new architectural language as a construction of atmospheres and trigger a new form of beauty. Today, reducing the release of greenhouse gases by a building during its operation involves three main measures: high performance thermal insulation of the building envelope, perfect airtightness, and controlled dual-flow air renewal. This program, related to air and energy, is an opportunity we wish to seize to transform architecture by re-founding it on other hierarchies, to allow the emergence of a new, atmospheric, sensual language as an overture to a meteorological architecture.

The following projects are an immediate exploration of this language. With no formal or functional presupposition, they allow themselves to be surprised by the emergence of possible spatialities not established at the outset. These projects are the result of a reversal between use and form. Upstream in these projects are the problems of building techniques related to sustainable development. Rather than adapting these techniques to an architectural project that has been preconceived according to a symbolism or a use, here they bring about the actual form of the architecture. The question of the degree of humidity in the air, its management in the dwelling, the distribution of the air through the space according to the density and its temperature define, through their physical and sensory quality, the organization of the layout and section of the building. Unexpected new types of housing then emerge. They are not based on modern housing planning, with its agreed private/public social distributions, but on sensual and physiological constituents emerging from the treatment of building techniques, as climates for living in.

Writing about Alain Robbe-Grillet's "Objective literature", Roland Barthes points to a determination to strip objects of their metaphoric or analogic associations and thus to destroy "the singular adjective that ties up all its metaphysical threads". "The qualification is only spatial, situational, absolutely not analogical". Denouncing the "tyranny of significations", Robbe-Grillet criticized the use of these adjectives (austere landscape, huddled village, etc.) modifying nouns to impose a psychological reading of the space and steer the reader's interpretation in their perception of the outside world. This is often what architecture is all about and what we dispute: that the project is based on a pre-existing analog meaning, on a predefined form, a pre-standardized function. Quite the opposite, we would like to allow space to be free of meaning, open to interpretations, to changes in behavior, to the invention of new ways of living, to the unexpected, to the unknown. We seek an architecture that can be "there before being something" as Robbe-Grillet stated in 1956 concerning the gestures and objects of the novel of the future. Architecture must precede significations, that is, it must produce meaning but within its own language, through space and time. It must create hot, cold, moist, dry, high, low zones as so many climatic potentials for interpreting, for inhabiting. Architecture must not illustrate but give substance for illustration. It does not represent, but presents physical, climatic, geographic, physiological spaces and times. Thus we seek to redeploy space in a freedom of signification to allow the emergence of original practices. This architecture replaces functional and symbolic constraints with freedom of use and interpretation. For this is what fascinates us in Robbe-Grillet's work. Beyond all theoretical dogmatism or morality, it is this formidable capacity to make spaces and times exist in the present tense, to draw with the movement of the sentence these extraordinary labyrinths where the real is reinvented with every word, where "places, objects and situations combine, transform, fuse, overturn, multiply, before our eyes", as Gérard Genette describes.

The French literary critic Jean Ricardou said about the work of Robbe-Grillet and more generally of the "Nouveau roman" that it was a question of reversing the dogma of representativeness (the pre-eminence of fiction) in favor of the dominant producer of writing (the pre-eminence of textuality) and called for abandoning the "adventure novel" for the "adventure of the novel". Our goal is of the same order: an architecture that emerges starting not from an image or a habit of use, but from its technical constraints, from its physical and physiological reality until reinventing its forms and uses, in an unexpected way, sometimes unusually at first, but always necessarily and essentially. Architecture is a language. Our work questions this language and reformulates its elements. An architectural project is not just about applying an elsewhere predefined architectural language in order to decline it according to a variety of contexts and programs.

To imagine an architectural project is also to imagine the project of the architectural project, or to conceive the "architecture of the project", that is to say question the language of architecture at its root, agreeing to change its paradigms, redefine its foundations and invent new compositions.

Invent new plans, formulate new typologies between meteorology and physiology. Articulate air movements, water vapor displacements, air renewal rates, define pressures, temperatures, work on breathing, perspiration, nutrition. The proposals below are of this order; the invention of plans as a literal, unexpected and jubilant expression of current building techniques.

Air flow

Reducing the question of architecture to ventilation and aeration is at stake in two projects that we proposed for a sports hall in Martigny, Switzerland in 2007 and in Lyon in 2008. The architectural project is based literally on the need to bring oxygen into the building and to evacuate the waste air produced by breathing. We thus seek a way to reconnect the question of form and program to the physiological essence, to free them from the transient influences of representation.

What is the deep link between the space and sports activity? It is undoubtedly the question of the supply of oxygen required for the athlete's muscles, the consumption of which is multiplied by 500 during significant muscle activity compared to being at rest. The renewal of air inside a building takes place at a certain rate according to its use, number of occupants and their physical activity: a low rate of occupation or activity will require a lower renewal rate while a space occupied by many people or by people engaged in strenuous physical activity or sport requires a much higher rate. Air renewal is necessary to provide the metabolism with fresh oxygen whose consumption is multiplied 500 times during heavy muscle exercise. It is also necessary to evacuate metabolic waste products, such as H+, CO2 and lactates. From about 7.5 l/min at rest, the ventilatory flow rate rises with corporal activity up to 120 l/min. This increase is possible thanks to an increase in the respiratory rate which then requires a quicker exchange between fresh air and stale air. There is, therefore, a precise relationship between the form of the space and the amount of air to be supplied and renewed, again posing the question of the ventilation of the building as the main efficient strength of the architectural form.

Here, the architecture is movement of air, with its tributaries that increase the amount of air brought to the space. The plan is drawn gradually, in cascade, between spaces with a low hourly air renewal rate (less busy and/or with gentle muscular activity) until spaces with a high hourly air renewal rate (sports activity) by adding injections of air to obtain the following air renewal rates:

foyer: 4 times per hour;
store rooms: 10 times;
toilets: 15 times;
sports changing rooms: 20 times;
sports halls: 25 times;
showers: 30 times.

The project is formed precisely from the management of air in the building, along of its path, its speed, its volume, its renewal rate (calculated in m3 air/hour).
The architecture is nothing more than a ventilation system where spaces literally take on the soft, rounded forms of ventilation ducts into which low-pressure air flows smoothly and noise-free. The building is shaped as a continuous movement of air, with its tributaries that progressively increase the amount of air. Gradually the architecture is composed, linearly, between spaces with a low rate of air renewal to spaces with a high air renewal rate. The plan first deals with the spaces where one spends the least amount of time, where the body is little active in terms of burning muscle, where the least amount of water vapor is emitted, then with those where one breathes more quickly and sweats more. The building is a wind whose flow gradually increases according to the increase in the respiratory frequency of its occupants according to their activities and the amount of water vapor released. This movement of air, whose volume of fresh air is increased at each stage, begins in the foyer and passes into the changing rooms. It then swells in the gym, accelerates in the showers where the air renewal rate is highest, to pour, at the end of the trajectory, into dense vegetation, chlorophytum-type plants that transform the CO2 of the air produced by the athletes into oxygen in photosynthesis. Before exiting the building, the air passes through a dual-flow heat exchanger to restore the heat acquired during its passage within the building to the incoming fresh air.

This project is a habitable air flow, the design of an air movement. The plan involves wind whose qualities are expressed in terms of air speed, volume, forcing and extraction as a poetic and spatial transfiguration of the question of ventilation. The composition of our architecture is based on ecological requirements of air management (gentle, automatic ventilation with heat recovery) in order to reduce the energy consumed in the building by a factor of 8. It thematizes the subject of hourly air renewal rate.

Thermal convection

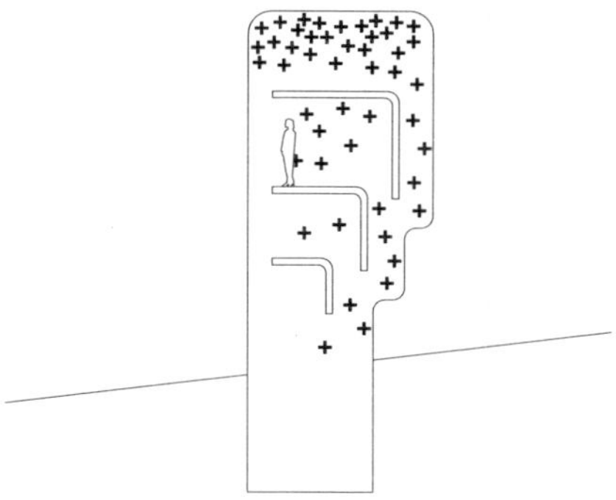

Archimedes houses designed during the urban study of the region of Vassivière in 2005 seek to establish an intrinsic relationship between the house and the air, by spatializing the functions of the dwelling – sleeping, living, washing – in the air itself, in its density, its temperature, its movements. The house is organized according to the physiological needs of the occupant in relation to their bodily activity and degree of nudity. The architecture is deployed vertically and is structured according to the specific heat requirement of each space. It is a matter of ensuring the thermal comfort of the dwelling, of architecturally projecting places where the indoor temperature is adapted to the activity and the clothing of the occupants. Modernity has led to uniform, consistent spaces in which the temperature is regulated at around 21°C. Here, we seek to provide diversity in the rapport between the body and space, with its temperature, to allow movements within the house, migrations from downstairs to upstairs, from cold to warm, winter and summer, dressed and undressed.

The plan and section of the house are then drawn according to the form that the air takes throughout the height of the house, according to its vertical movements depending on its temperature and the functions it may suggest. Warm air will tend to rise In a heated room, it is common to find temperature differences between the bottom and the top, 21°C at one meter from the floor, 27°C at ceiling level. Obviously, this results in wasted energy with respect to the desired temperature, 21°C. The principle of the ascendancy of warm air is linked to the principle of density. As the hot air is less dense, it rises in, in accordance with Archimedes' principle. Temperature differences are stratified in relation to the height of the house, with a first floor at 16°C, a second floor at 18°C, a third floor at 20°C and a top floor at 22°C. The functions then evidently take place vertically in the section according to the horizontal thermal strata, from the coolest at the bottom to the warmest at the top, following the Swiss SIA 384/2 construction recommendations – bathroom at 22°C, living room at 20°C, kitchen at 18-20°C, bedroom at 16-18°C, corridor/toilet at 15-18°C. The sanitary facilities and the laundry room are located on the first floor, the bedroom on the second floor, the living room and the kitchen on the third floor, the bathroom on the fourth floor. Without prejudices, the design of the section reveals new hierarchies in the organization of the spaces of the house, unknown proximities and other connections between the rooms.

Our architecture is climatic. It does not presuppose any custom or agreed spatial practices, it is developed only in its own subject matter. By not illustrating any programmatic use or narrative interpretation, it provides for new ways of inhabiting space that emerge from the very language of architecture, in time and in space.

Water vapor

The Mollier houses project of 2005 in Vassivière in the Limousin reveals and qualifies an invisible but nevertheless essential relationship between interior space and humidity. It seeks to transform a problem of building physics into a question of architecture, to the extent that it becomes the effective cause of the form. It introduces novel sensual and physiological relations between the inhabitant and the space into the constraints of the technical equipment of the building. It also seeks to engage closer ties with the lake landscape of Vassivière, physical and chemical ties, as it is set in the material character of the territory itself, in its humidity.

An occupant of an indoor space produces water vapor, not always the same amount, but according to the function for which a room is primarily intended. The natural breathing of the occupants and the use of hot water cause the presence of water vapor in the air and potential risks of condensation and damage to the construction. Hitherto, the means to prevent excessive water vapor indoors has been purely technical, in the form of an everyday ventilation system. Our project intends to transform the technical response into an architectural response by formalizing the space according to the water vapor itself, as an opening to a deep and complex relationship between the inhabitant, their body and the void, according to its physical and chemical characteristics. Our architecture designs the variations in relative humidity, formalizes the spaces of the dwelling, from the driest to the most moist, from 20% to 100% relative humidity. It establishes a spatial stratification of humidity levels. A person asleep produces some 40 g of water vapor per hour (bedroom), while someone who is awake produces up to 150 g/h (living room). The use of a bathroom releases up to 800 g in 20 min and a kitchen 1,500 g/h. Like a set of Russian nesting dolls, the house is designed according to the air renewal path through the house, from the driest to the most humid, from the freshest to the stalest, from the bedroom to the bathroom. The first room has a relative humidity of between 0 and 30% and can be used as a drying room or sauna. The relative humidity of the second room is between 30 and 60% and can accommodate a bedroom or a study. The third room, whose relative humidity is at between 60 and 90%, can be used as a bathroom, living room or/and kitchen. And finally, the last room that falls into the water of the lake, as a living room and/or a swimming pool with a moisture content of between 90 and 100%. But we reject functionally programming the space according to specific activities. It is not so much rooms with their univocal function that are created, but rather spaces as more or less dry, more or less humid atmospheres, to occupy freely, to appropriate according to the weather and the seasons. The quality of the architecture takes shape in the water vapor content and causes a real and carnal immersion of the occupants' bodies in the humid and variable body of space. The plan of the house is a spatial representation of the Mollier diagram which represents the variations in relative humidity according to temperature. The project creates new programmatic correspondences in which the same space can, a priori, accommodate different functions. From a landscape standpoint, the project amplifies hygrometric stratification in relation to the context, integrating the physical presence of the water of the lake and the natural outdoor moisture as one of the rooms of the house, at 100%

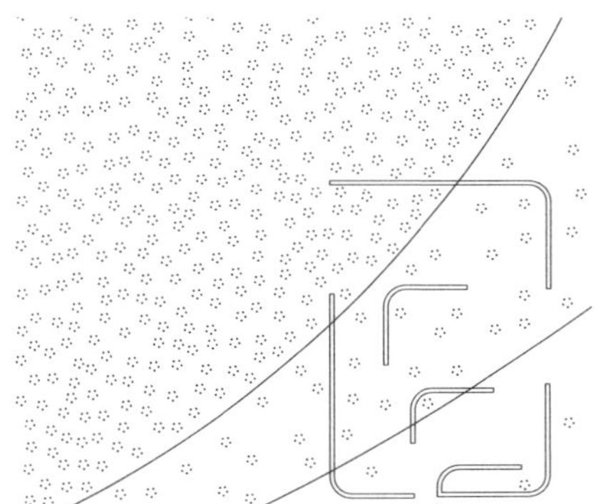

For the Lyon water sports center project, it is the variation of water vapor released by athletes' perspiration and the management of relative humidity levels induced by sustainable development that are transfigured to invent a new typology in which architecture is formed as a horizontal linear variation, in layout, of humidity levels, between the driest in the north and the most humid in the south. The technique of gentle aeration with dual-flow heat recovery creates a continuous, constant movement between a pulse of fresh air in the driest areas in the north and the extraction of stale air from the moistest areas in the south. This technique saves up to eight times the energy spent in the building because it renews the indoor air while minimizing heat losses. Vertically, temperature variations are stratified in section being constructed according to Archimedes' principle, between the cooler layers at the bottom and the warmer layers at the top. The project is an architectural geography, an atmosphere where water vapor forms and concentrates progressively, with its relative humidity longitudes, which deform to accommodate a program for a spa, a sauna, a hammam. The body is deployed sensually in space, due to variations in moisture, breathing, differences in temperature perceived by the epidermis, between the driest and the wettest places, between the sauna in the north and the hammam in the south. Thus, it is not only a program that is being implemented in space, but literally a geography that is being deployed, with its arid areas and wetlands, its deserts, its temperate zones, and its tropical climates. Thermal conductivity Our 2007 project for an elementary school in La Neuveville, Switzerland, is an explicit formal roll-out of the issues of reducing energy consumption in buildings. Our architecture meets the goals of the Swiss Minergie-P program, in other words, the most stringent requirements in terms of thermal control and energy efficiency. While the ecological target contents us as citizens, it is the plastic dimension that the building can take through the application of these thermal measures that interests us as architects. We seek to turn technical requirements into architectural qualities, to transform solutions into space that, a priori, are only technical in order to make them habitable. We want to infiltrate these new building techniques and discover their new architectural forms and uses. The Minergie-P standard is a building construction concept that allows making considerable energy savings by relying on two main maximized solutions: highly efficient insulation of the building façade, and an air renewal system including heat recovery. While fulfilling the second solution, our project more specifically focuses on the first solution, that of the building's insulation, taking advantage of its plastic, practical and spatial potential, which is today largely unexplored.

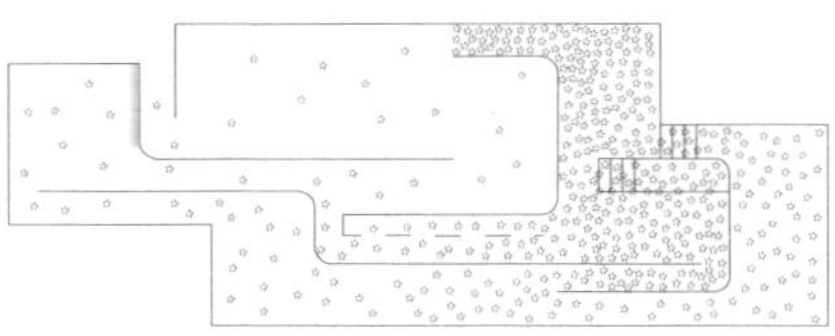

According to this principle, the purpose of a building's excellent thermal insulation is expressed in figures by a thermal transmission coefficient, U, of the façade of less than 0.15 W/m2K. The U-value is the measurement that indicates the heat loss per square meter of a building element. A low U-value is a sign of good thermal protection. The definition of the U-coefficient is the quotient of the heat flow per unit area of a building element, under steady-state conditions, by the temperature difference between the contiguous atmospheres of the element. Depending on the nature of the material, its thickness, the U-coefficient will be more or less low and consequently more or less effective. In other words, it means having extremely thick peripheral insulation. The traditional thermal requirement according to the SIA standard is 0.3 W/m2K, which corresponds to insulation of a thickness of some 12 cm. Here, we require a U-value of 0.15, or even 0.1 W/m2K, corresponding to insulation of at least 25 cm and up to 40 cm thick. Such building principles require the addition of layers of insulation, 10 cm plus 20 cm, plus 10 cm, to reach the goals. Quite literally, the envelope of a building resembles layers of clothing that are superimposed on one's own clothes to feel less cold. The addition of layers of insulation in the Minergie-P standard is exactly the same as the accumulation of clothes on the body according to the outside temperature, but also depending on the activity being carried out: just a T-shirt when running, a light sweater when walking, a thicker sweater if inactive. The Swiss KBOB recommendations of January 1, 2000 thus provide very clear recommendations concerning the relationship between ambient temperature, clothing and physical activity. They advocate heating a sports hall to 16°C because the athletes themselves produce heat when doing their physical exercise, or heating corridors to only 15°C because they are only places of passage and therefore the user does not spend long enough there to get cold. Conversely, they recommend heating changing rooms to 22°C because the user is naked and inactive. The rooms can therefore be heated to different temperatures depending on their uses, the clothing worn and the activity carried out there. And these precise temperature differences become important in order to save energy and therefore not overheat a space that does not so require. The elementary school is stratified like a thermal onion according to the temperatures required in relation to the activity conducted. The device renders the various layers of thermal insulation into which one physically penetrates, infiltrates, visible and sensitive. Our project is thus defined as a series of insulating layers that gradually reduce the U-coefficient of heat transmission, from 0.3 W/m2K in the first stratum to 0.1 in the third stratum. More sensitive to external variations, the first stratum, on the periphery of the building, is the one where people spend the least time. It houses the toilets and equipment rooms. The second layer, slightly further indoors and more insulated, accommodates the intermediate spaces, the passageways and the assembly room. The last, innermost layer, protected by 40 cm of insulation, hosts the classrooms and, with a thermal coefficient of 0.1, provides the children with maximum comfort.

Thermal expansion

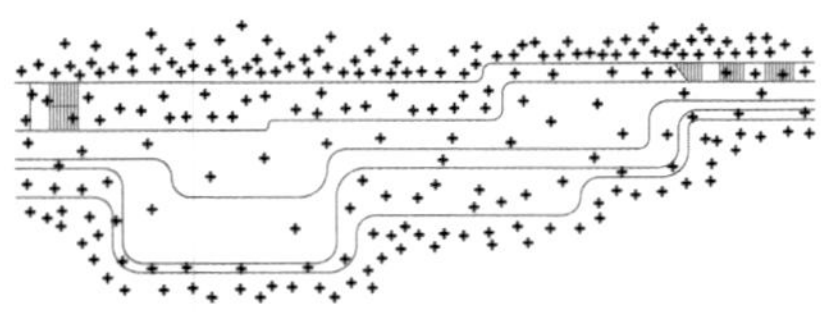

The 2006 project for a museum in Poland in Krakow for the artist Tadeusz Kantor also works on an expansion of the thermal coefficients of the building envelope by literally proposing to live between the panes of the double glazing by increasing the small spacing of a few millimeters between the panes of a double- or triple-glazed window, until it becomes a living space of a several meters. In the precise manner of contemporary windows that add layers of glass (single/double/triple glazing) to improve insulation by decreasing the thermal transmission coefficient k (W/m2K) (single glazing: k = 5.6; double glazing: k = 3; triple glazing: k = 2), this museum adds strata to gradually improve the thermal coefficient, layer by layer, thus offering a variety of temperatures and brightnesses. In contrast to the homogeneity of the modern climate, we propose a variety of ambiances, lights and temperatures that invite one to move freely. Depending on the type of activity and the season, we will choose one layer over another, warmer or further north, with a more constant temperature or greater brightness. Architecture, in its language, even in the invisible language of light and temperature, and, in the energy needs linked to sustainable development, finds the means to conjure up interpretable, extraordinary spaces, freely open to future behaviors and functions. Here there are no clear boundaries between the inner space and the outer space, no more precise delimitations, rather there is a succession of layers which, one after the other, progressively filter a certain quantity of temperature and of light. The design of the project involves a window composed of five panes whose dimensions are enlarged to the scale of a building allowing inhabiting the spaces in between the double glazing which are normally a few centimeters wide but we have increased here to several meters.

Intermediate climate

The transformation of an old house in the Ticino region of Switzerland also questions the values of the thermal conductivity U-coefficient of the façade according to the thickness of the thermal insulation of the building envelope. Through a critical approach to these values, the project aims to develop and, almost by chance, invent new types of spaces and functions. The current house is a typical regional construction made of dry granitic stone. Despite its 60 cm-thick walls, the thermal conductivity coefficient of the envelope remains too high, at about 0.7 W/m2K, and does not reach the current target for good insulation, which is set at 0.2 W/ m2K. To reach this figure, a layer of insulating material must be added. Our work as architects starts here, with the critical questioning of the addition of this new layer of insulation. Should we line the entire stone wall with this new insulation? Should we completely transform the indoor climate of the house in a homogeneous and regulated manner?

One of the qualities of this house is at the same time its defect: its rooms each have very distinct and well established climates. Thus, the lower rooms are very dark, humid and always fresh because in the north they are half dug into the mountain from which runoff water seeps. They have nevertheless found their use: they store not only equipment, but also wine and apples that need this cool, moist climate to keep properly. The sunny, temperate intermediate upstairs rooms are used as living spaces. At the back, now off the ground, are dark, warm but drier rooms that used to fulfill the role of refrigerators, where milk and produce were kept fresh. Above, beneath the roof, grass and fodder were stored in warm, dry spaces. These different degrees of humidity, temperature and luminosity, which vary throughout the year according to the seasons, are the opposite of the normative thinking and climate homogenization stance of architecture today. Without meeting today's standards of comfort, we nevertheless see a rich thermal composition of the space, a diversity of atmospheres that may suggest uses, such as cooling down in summer in the dark and humid freshness of the lower floors, or warming up in winter in the warm upper layers of air.

We propose to line the stone wall from the inside using 10 cm-thick insulation. But rather than adhering it throughout the existing stone wall, we dissociate it and keep it away from certain places in order to open up intermediate spaces that are thermally less efficient but more sensitive to the quality of the immediate climatic context, re-engaging architecture in a sensual dialogue with the seasons, the meteorological movements, exposure to the sun and the physical nature of the soil. A secret and irrational passage in the thermal organization of the renovation as a whole, a kind of climatic chimney, less controlled but nevertheless temperate, zigzags from bottom to top, freely developing a series of climatic situations, between the coldest and the most humid in the lower part of the house to the warmest and driest at the top. These spaces store garden produce, preserved according to their nature at a certain humidity level and at a certain temperature. Mint that dries in the upper part of the house, vegetable preserves protected from the light, syrup, broth, apples. This intermediate space is a new atmospheric landscape that is neither fully interior nor fully exterior, no longer totally natural, yet not fully artificial, in which the controlled and perfectly isolated interior spaces open up. New inverted windows look onto this indoor climate landscape and its intermediate climate.

These projects are, therefore, not only a technical response to the challenges of sustainable development. What we are pursuing is to bring out new, more sensual, almost dermatological spatialities between space and the body through the new needs for energy reductions. Unlike a standard architecture that normalizes and uniformly conditions the entire interior space to a single high-performance thermal coefficient, our aim is to create a variety of atmospheres, temperatures, and ambiances in and through which we can migrate, move, and feel through the skin, through the body. Here, architecture becomes the construction of atmospheres and climates.

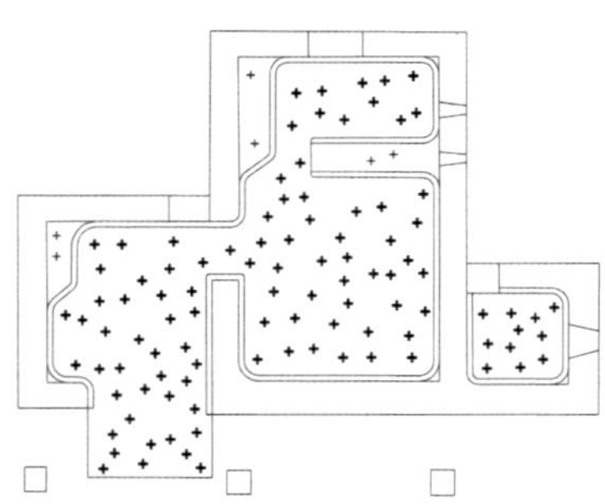

VI.II. Climatic Urbanism

2010-2023

Climatic park

Central Park is a new 67-hectare public park built on the abandoned site of an old airport to the north-west of the center of Taichung, a city of 3 million inhabitants in the center of Taiwan. The park is at the heart of a new development covering 256 hectares. It is a hive of nature, planted with nearly 12,000 trees. Given its topography, its rolling hills acting as holding tanks, it also plays the role of a sponge for all the rainwater of the new district, preventing flooding and overloading the network.

Taichung has a warm, humid and polluted tropical climate. The goal of the park is to offer cooler, drier and less polluted spaces. The formal compositional strategy of the masterplan is based on the climate, on a mapping and a design of the variations of heat, humidity and air pollution, as three independent layers initially which, by superimposing each other, invent a multitude of microclimates and as many variations in character and atmospheres, giving the park its esthetic richness and multiple uses. The choice of the location of the trees and their essence, the nature of the soils, the program, are steered by the climate. Thus, in the coolest places, naturally the most exposed to the cold wind from the north, we extend the woodland cover to create shade, thus increasing their freshness. The soils are rich in evaporating or high albedo grasses to increase freshness. In places furthest from moisture sources we plant trees with floating roots to reduce their moisture and the soils are chosen for their drainage capacity. In the areas farthest from the roads, we increase the number of trees with fluffy leaves to reduce air pollution. The program follows the climate: the less polluted areas host children's games, the drier areas are used for sports activities, and the cooler areas for various leisure pursuits.

The microclimates created provide the places with their identity. Thus, the coolest places are called Coolias, the driest Dryias, and the least polluted, Clearias, which are distinguished by the predominance of a species of tree.

The planting of trees follows a gradual order of composition, from open field to dense woodland, slowly amplifying the effects of cooling, de-watering and de-pollution.

The paths follow the climate like isothermal lines. The Cool Path connects all the coolest parts of the park, the Dry Path connects the driest parts, while the Clear Path connects the least polluted.

Sensors measuring heat, humidity and air pollution are placed every 50 meters. The data are then transcribed in the form of a real-time chart, accessible via a smartphone, indicating the variations in climatic values throughout the park, thus orienting the walkers.

10,000 m2 of photovoltaic panels placed like canopies over the park to the north and to the south, produce all the energy used in the park, which is thus totally carbon neutral.

Night lighting gradually decreases as one moves away from the roads, changing from white to orange in order to preserve animals' circadian cycles.

Our intention is to provide the residents of Taichung with a comfortable nature area, encouraging outdoor activities, and the diversity of microclimates allows use during the 12 months of the year and at any time of the day, between very hot and cooler parts, to suit every individual's needs and desires.

Attenuated exterior

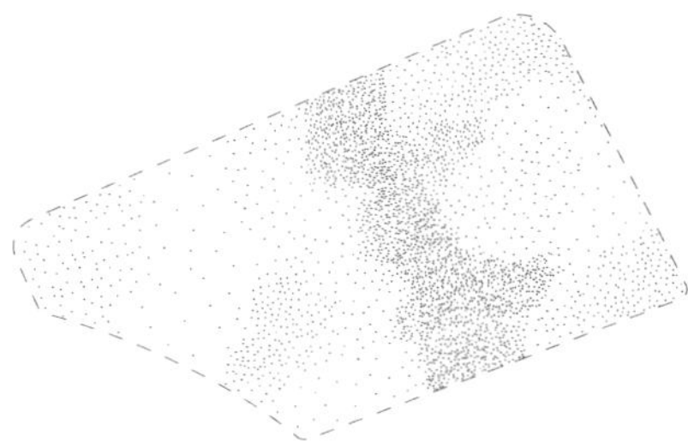

Located just north of Taichung Central Park, the Attenuated exterior is a 10-hectare site that hosts different programs such as a garden, a lake, a pavilion, parking lots for buses, cars, two-wheeled vehicles, the maintenance center for park gardeners, toilets, a flower kiosk, extensive grasslands left to nature, as well as holding ponds, large natural grassy and wooded basins that can temporarily collect water in case of heavy rain and prevent flooding. At the heart of this site are 7,000 m2 of photovoltaic panels raised to 7 meters in height above a hilly and wooded park, also covering part of the parking lot and various buildings that are installed beneath, sliding between the poles, spanning the topographic rifts. This artificial canopy, composed of thousands of solar panels, spaced more or less close to each other, each casting its shadow on the ground, creates an area where the heat of the Taiwanese sun is softened to a greater or lesser extent, composing a kind of sweetened exterior where the warm climate, the intense heat of the sun's rays, the fiery light of the Tropic of Cancer, are attenuated.

Taichung has a warm, humid subtropical climate. Relative humidity averages 83%. The average annual temperature is 22.2°C. It generally varies between 33°C in summer and exceptionally 12°C in winter. By way of comparison, the relative humidity of Paris is 79.2% with an average temperature of 12.4°C which ranges between 25°C in summer and 3°C in winter. The solar radiation in Taichung is intense due to the city's relative proximity to the equator. While Paris is located 5,432 km from the equator, at a northern latitude of 48.8', Taichung is located only 2,685 km from the equator, at a latitude of 24.8' north. The ferocity of the sun's rays, its radiant heat, intensifies as we move closer to the equator where it reaches a maximum because the sun's rays are perpendicular to the earth's surface, their energy concentrated on a minimum surface area, also passing through a thinner layer of atmosphere. The further you move away from the equator towards the poles, the greater the angle between the sun's rays and the Earth's surface, decreasing the intensity of the light and that of the infrared radiant heat, since the rays of the sun then spread over a larger area of the earth, thus diminishing the energy reaching the ground. Moreover, the further you go from the equator, the thicker the atmosphere crossed by the sun's rays, thus weakening the radiant light and heat, as the rays are partially intercepted and diffused by the atmospheric gases, oxygen and water vapor in particular. The amount of solar energy (solar radiation or insolation) reaching Taichung varies between 6.66 kWh/m2/day in summer and 2.6 kWh/ m2/ day in winter, while in Paris, by comparison, the energy received will vary between 5.41 kWh/m2/day in summer and 0.81 kWh/m2/day in winter. We can thus understand that the sun's rays, throughout the year, are warmer, fierier in Taichung than in Paris because they are more charged with energy. Getting into the shade, under a sunshade, under a roof, under a hat (Chinese hat), an umbrella or a parasol thus becomes a necessity to avoid the excessive radiant heat, in addition to the conductive heat of the air in these latitudes closer to the equator as in Taichung, during much of the year. This shade provided by the photovoltaic panels sieves the incident light and the radiant solar energy descending to the ground, attenuating the heat and making the landscape below more comfortable, more habitable for humans. At the same time, this canopy enables recovering surplus energy, catching it in the photovoltaic panels to use this solar energy (visible and infrared light) beyond the bounds of human comfort, to produce electricity to be made available elsewhere, for purposes other than immediately heating the earth and the people present in the landscape below the canopy, dephasing it for use at night. The excess solar energy is thus used for a purpose other than heating the already overheated ground and people, and at other times. This excess energy is recovered to produce light at night or cold in the heat of the day, for night lighting, or for air-conditioning inside the buildings by day. The climate of this exterior is thus mitigated, temperate, and enables occupying the outdoors comfortably during the day, in the radiant coolness of the shade of the photovoltaic panels.

These panels are laid out in staggered fashion, varying in density, thus generating below a diversity and a richness of landscapes with multiple light intensities and degrees of shade. Trees and hills, pavilions and buildings; both natural and artificial forms inhabit this climate-softened exterior. The Attenuated exterior allows for more outdoor activities, to enjoy it in greater comfort in a subdued light, parking there in a relative coolness. Modulated by this pointillism of shadows and lights generated by the panels, the hues and biology of the natural landscape and the plant species that grow there differ very slightly from those of the surroundings.

The 7,000 m2 of photovoltaic panels in this northern part of the park, together with the 3,000 m2 of photovoltaic panels located in the south of the park, produce all the necessary energy used there day and night, supplying all buildings, meteors or night lighting with electricity. Thus, no energy responsible for global warming, namely fossil and carbon fuels, natural gas, oil, fuel oil, coal, wood or biomass, nor waste-generating energy, such as nuclear, are used in the park. The park's only energy sources are renewable: solar for electricity and geothermal for cooling. The entire park, its buildings, its lighting, therefore emit no greenhouse gases. The park can therefore be classified as carbon neutral.

Atmospheric urban planning

To establish the urban principles of the new district of Farini in Milan, we respond to the three contemporary challenges (global warming, urban heat islands, air pollution) that affect the city today by proposing a cool, unpolluted city in summer, minimizing its impact on global warming, that is to say rendering it carbon neutral.

Our project is based on two environmental devices: one creating a green environmental zone, the other blue that we call Limpidariums, that is to say environmental urban devices that clean, clear and regenerate the air and water.

The green environmental feature, the Aria Limpidarium, is a large linear forest along the front of the entire new district of Farini that is capable of cooling the warm winds coming from the south-west and purifying the air of fine PM2.5 particles. Its name means "render [the air] clear". The new district of Farini is based on the environmental substance of the Limpidarium, to the south-west. Its role, besides being a park for leisure pursuits, is foremost to constitute a filter against the excess heat and pollution brought by the prevailing wind that blows over Milan from the south-west. It then serves to produce shade and freshness, thanks to the trees and fountains whose water evaporation lowers the air temperature. By convection, this new, cooler, less polluted air will blow into the new neighborhood, driving away heat and pollution. The new buildings are placed parallel to this wind to allow it to pass freely through the new district from south to north.

The second environmental device, the Limpidarium d'Acqua, or water Limpidarium, is blue. It is a long linear system in San Cristoforo to purify the groundwater coming from the north, creating a landscape that passes from a pond attracting wildlife to a natural pool that can be used by people. Its name means "render [the water] clear". Our project for San Cristoforo involves a park that welcomes both humans and non-humans, but whose function also consists of acting as a filter to enable de-polluting the groundwater coming from the north-west flowing south- east. This park naturally cleans underground runoff water by first mechanically filtering it and then biologically regenerating it until it constitutes the water of a natural swimming pool where you can bathe. The water then continues its course towards the south-east, thus cleansed and returned to the territory in a clean state.

Socialism

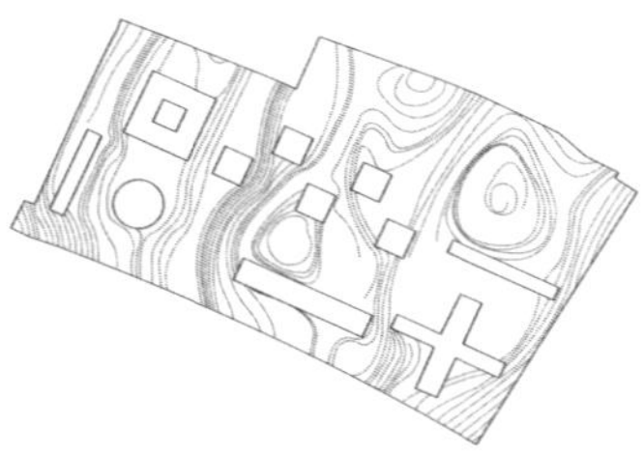

Social life, living together, public space, these shared community values are not abstract concepts. They were formed in the material, physical world, following the essential needs of gathering: to warm up in winter by sharing a fire, to cool down in summer by sharing the shade of a tree and the water of a fountain. And it is in these fundamental gatherings of people around a fire, in the shade of a tree, around the well, that politics, cohabitation and social life began.

Our proposal for the Health Campus is to create outdoors, in the Campus park, these two fundamental places of human gathering and social life, that provide warmth in winter, and refreshment in summer.

To do so we work on two scales, that of the neighborhood, that of the materialization, whose optimal and efficient place and materials we will define to get warm in winter and cool in summer.

At the neighborhood level, we use computer modelling to analyze first of all the amount of sunshine on the site in winter, the shade projected by the future buildings during the day of the winter solstice, on December 21. We identify the sunniest places in the new neighborhood that will be a candidate for hosting our place for the winter, as they will be the places that will receive the most sunshine and will be warmed by solar radiation.

At the neighborhood level, we then analyze the movements of the winter wind between the new buildings. We identify the areas most protected from the wind that might host our place for winter, as they will be the places where students and residents of the neighborhood will be least cooled by the wind by convection.

By crossing these two models, we identify the most naturally warm place in winter, exposed to the maximum sun, protected to the maximum from the wind. This is where we will set up our place for the winter.

In order to reinforce the warmth of this winter place, we build it using materials whose physical characteristics we choose precisely: albedo, emissivity, effusivity.

In order for the winter place to heat up as much as possible under the visible rays of the sun, we choose dark or black materials, so that they absorb and transform the energy of the sunlight into heat upon contact, according to the albedo principle.

A social place to meet in summer: the Coldarium

At the neighborhood level, we use computer modeling to analyze first of all the amount of sunshine on the site, the shade projected by the future buildings during the day. We identify the least sunny places in the new neighborhood that might host our place for the summer, as they will be the places that will be the most shaded during summer days and will not warm up due to solar radiation.

At the neighborhood level, we then analyze the movements of the summer wind between the new buildings. We identify the areas most exposed to the wind that might host our place for summer, as they will be the places where students and residents of the neighborhood will be most cooled by the wind by convection, as if under the blades of a fan.

By crossing these two models, we identify the most naturally cold place in winter, exposed to minimum sun, exposed to maximum wind. This is where we will set up our place for the summer.

In order to reinforce the coldness of this summer place, we build it with materials whose physical characteristics we choose precisely: albedo, emissivity, effusivity.

In order for the summer place to heat up as little as possible under the visible rays of the sun, we choose light-colored or white materials, so that they reflect the energy of the sunlight and thus do not heat up, according to the albedo principle.

Then, and so that the material reflects the infrared rays of the sun as much as possible and thus does not rise in temperature, we choose a low-emissivity stone, namely granite, whose emissivity, on a scale of 0 to 1, is 0.45.

Finally, to ensure a cool contact of the human posterior with the bench, we propose aluminum for the seat, which has a high effusivity of 23,377 J/m2K.s, that is to say that the coldness of the bench in summer (which will be at ambient air temperature, always lower than that of the human body, and therefore always refreshing) will be transferred rapidly towards the human body, thus giving a feeling of coolness as if touching the bench in winter.

We propose to boost the freshness and the collective and public value of the summer place by the possibility of drinking water, making it possible to replace the excess loss of fluid by perspiration during the hot summer days.

This winter place we call Coldarium.

Physiological urban planning

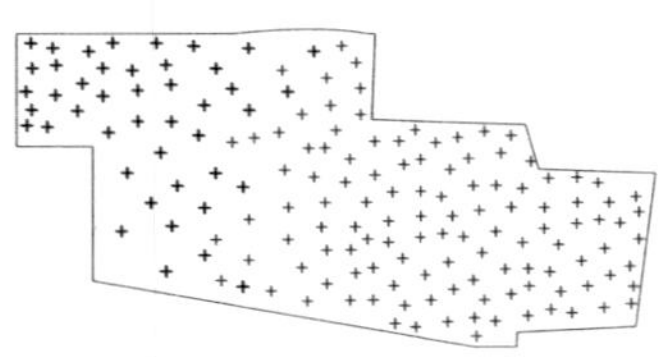

It is important to understand the direction of the arrow regarding the mechanisms of causality in the formation of a city and a microscopic analysis often reverses the order: what is taken as being a cause on the macroscopic scale can prove to be a consequence on the microscopic scale. The nutritional precedes the esthetic, the physiological allows the symbolic, the microscopic forms the macroscopic. While urban planning since the 1970s has not been a complete success, it may be because we have built forms without cause, with only the consequences of urban actions remaining. The medieval square was a spatial extension of the fountain to allow the surrounding inhabitants to come and fetch water. And it was precisely here, due to the physiological need to drink water, that social and political life started to be practiced. The shape of the square, its social and political function, are therefore not the cause of the invention of the place but rather the consequence of the need to drink, at a specific geographical point, that of the source where the water resurfaces and at the intersection of several roads leading to the different fields providing protein, carbohydrates and lipids. The social, political, esthetic or symbolic purpose is not upstream of the invention of the square. It is downstream.

If we want to define an urban and territorial strategy for the future, we must then analyze what today are the real effective causes at work in the transformation of the territory. In this analysis, parameters of climate and energy, which are very closely linked from an architectural and an urban point of view, appear as the main vectors which act and will act on the renewal of spatial planning. The term thermodynamic urban planning could cover all the active criteria that can be mobilized in a renewal of the urban project.

Rather than conceiving urban planning in terms of social and political functions, we propose to analyze and design the energy-related and physiological causes. Eating, heating, lighting and drinking, these necessities must correspond to energy sources provided by the territory: food, sun, river in natural or artificial form. Our project proposes to design the urban plan of the new center of Kiruna as an ecological formulation of these needs, to first spatialize the energy and food requirements to then draw the social and political consequences. Thus, the street where we stroll in the evening will be aligned with the rays of the setting sun to enjoy its light and warmth. Thus, the design of the heating and water networks will not be consistent with the morphological plan of the city, but will actually fabricate the plan itself. We will take advantage of new energy sources, such as electronic servers, geothermal and solar energy, to orient, to draw the shape of the city. We will design the physiological pathway of water and food, from its production to fermentation that will form a new, more economical, more ecological urban plastic, capable of generating real social and political links.

Thermal public space

During the 20th century, the modernity with which the concepts of geographical globalization, temporal desynchronizations and the industrialization of technology were associated had led us to lose the fundamental sense of the urban form, which then no longer met clearly identified and immediate needs or local resource management, as was the case at the start of the formation of European cities.
Everything was possible in recent years and the principles of urban planning took as many pretexts for developing, ranging from a pure semiology of signs as advocated by a certain form of post-modernism to the simple speculative expansion of built surfaces. It is today difficult to know how to extend the urban forms of the European city, which are the result of a prolonged and even secretion over the centuries, but whose causes were brutally rendered obsolete by industrialization and globalization from the 19th century onwards.
Since the European city built since the Middle Ages is a political, social and esthetic consequence of a mastery of production means and local resources, stone, water and land. The materials, the structures and the forms whose harmony of the colors of the façades is extolled today, the gray of Freiburg, the yellow of Neuchâtel, the beige of Paris, we immediately grasp their material causes, the city appearing simply as an obvious, practical and inexpensive manufactured form of the geological resources on which it stands. The town of Molasse, yellow stone, limestone with terracotta, slate or stone roofs. But when it becomes more profitable to build with a material produced 500 km away, when the very material for the construction of the façades passes from the geological to the plastic, to the metallic or to fiber cement, then no natural reason can guarantee the harmony of the colors of the town other than some totally arbitrary legislative regulation. Similarly, the quantity of water immediately available on site defined the size of towns and villages, some hamlets not being able to exceed a certain size because with more than nine houses for example, the source of the village no longer made provided enough drinking water for all its inhabitants. The agricultural quality of the land has also been a determining factor in the growth of a town or city. Modernity had sought to get rid of it with its scandalous but honest and sincere blank slate. Post-modernism and Italian tendenza, on the contrary, have tried to preserve it, but by reducing it to a mere role as a visual vector of collective identification and memory, thus rendering

their irrelevance too patent. And today it is the tourist economy that has taken its best pieces to amplify postcard picturesqueness to kitsch.

With the concept of sustainable development and the urgency of responding to the problem of global warming, we are witnessing the return of a certain form of necessity, where a thinking of resource management, an immediate geographical recontextualization of the built environment, a new form of localism, both ecological and climatic are being reformed, in which a renewed alliance between indoor private space and outdoor public space can be forged. We would like to take advantage of this state of affairs to re-found the rules governing urban planning and rediscover a form of necessity in the relations between interior and exterior spaces, architectural environment and urban environment. Our project seeks to establish an inalienable link between public space and indoor air quality by proposing to rebuild this broken physiological link that once existed between outdoor and indoor, production and consumption. For example, the square and its fountain provided much-needed drinking water to the surrounding residents. The square thus took on its true sense of public space, building a social bond around this physiological necessity of the need to drink. The fountain generated the sense of the void and organized the solid around it. It then aroused a social bond through this daily activity of fetching water outside. Reintroducing a physiological necessity to the principles of urban planning seems to us to be a responsible and fascinating attitude today when the management of resources and the preservation of ecosystems become paramount. Without nostalgia, we are passing from one of the urban reasons for the public space of yesteryear, the water supply, to a contemporary reason: air supply. That the public space becomes the place where the air is shaped, its quality, its temperature, hot in winter or cold in summer, in relation to a climate, an albedo, vegetation, its de-pollution, that the air organizes the form, the layout and the materials of the public square, seems to us today to be necessary for updating the meaning of the public space, a re-qualification of the outside in a certain physiological necessity of the inside from which only a social life and exchanges between the inhabitants can be born. The project is implemented between two thermally differentiated poles, two public spaces built at opposite ends of the site. One is to the north-west, the other to the south-east, one is hot, the other cold, one mineral, the other vegetable, one in the light, the other in the shade, one absorbing the heat, the other reflecting it. From the first space, in the north-west, the warmer, mineral, light and black, we will draw air during winter to supply dwellings with preheated air, in the manner of the Canadian well. From the second place in the south-east, plant covered, filled with trees, shady and cooler, the space of which we virtually extend to the cemetery park, we will draw cooler air in summer, in the manner of the Provençal well, to supply all the dwellings in a dual-flow ventilation system. For each of these two poles, we carefully choose the thermal qualities of the materials of which they are made, their albedo, their emissivity, their effusivity, their revegetation, in order to augment the heat of the air in the north-west or to decrease it in the forest in the south-east. These two public spaces, hot and cold, are the sources of air to which the dual-flow ventilation system of residential buildings and the school is connected. The urban agenda of our project is to construct two squares at opposite ends of the site and to build between them. The buildings are implemented between these two public spaces. The distribution corridors of the buildings become ventilation ducts augmented to human scale, the ends of which, like nozzles connect to the public squares to draw the ambient air whose thermal quality will have been raised at the mineral square to the north-west or lowered in the forest to the south-east. The district is polarized between two main public squares whose development is based on an initiative to enhance the air, its structuring, where housing is implemented between two poles of air, one hot and one cold. Between these two principal areas, 7 strips of landscape unfold stretching from north to south, which gradually, from west to east, take you from the mineral square in the north-west to the forest in the south-east, from hot to cold, from mineral to vegetable, with alternating strips of different geological and plant qualities, ranging from brownfield land, vegetable gardens, shrubs, fruit trees and forest with which uses are associated that take advantage of the landscape qualities and atmospheres of each of the strips: beach games, gardens, sport on mineral surfaces, games, jogging, relaxation, walks in the meadows, tree houses

A kilometer of shade

Basra has a hot, dry climate, with an annual average temperature of 22.7°C and intense sunlight. Shade and freshness are the qualities that we will seek most of the time, and it is these fundamental qualities that we wish to confer to the new public space overlooking the Chatt el Areb river, where the Tigris and the Euphrates meet. Thus, offered to the population, transitioning to the north-east between the river and a new city district overlooking Dinar Street, a kilometer of shade is created by a pergola 7 meters in height, a kilometer in length and 60 meters wide, composed of hundreds of photovoltaic panels. This refreshing walk, in the shade, pleasant all day, allows joining up all the programs of housing, work, dining (the terraces of outlets extend beneath the canopy and take advantage of the shade provided) and leisure by this vast walk re-engaging the link with the river. The kilometer of shade, Basra's new landmark, also becomes a social and public gathering place, where the new inhabitants of the neighborhood meet, cross paths, interact. Punctuating the kilometer of shade, clearings like oases, climatic squares create areas of attraction, landmarks and meeting points, adding other modes of refreshment, by convection, conduction, evaporation and radiation to the shade of the photovoltaic panels that cast their shadow. They are all equipped with benches and chairs, allowing visitors to rest in the freshness enjoying the originality of each of the squares. Inertia Square is a kind of cold, dark cave during the day, with 150 cm-thick stone walls, whose ceiling opens at its center to evacuate the warm air. The thick walls of Inertia Square absorb the coolness of the night during the small hours, and distribute it indoors during the day, according to the principle of thermal phase shift (inertia), which, by conduction, offsets the temperatures between the outside and the inside by 12 hours. It will be cooler inside than outside during the day, warmer inside than outside during the night. The Dark Square is like an onion consisting of three layers of moucharaby which, by adding one after the other, gradually remove the square from the sunlight until the very dark, fresh core, while letting the air pass. The layers, like four rectangles embedded in each other, are made of moucharaby filtering the light and its radiative heat, surrounding both the vertical faces of the walls and the horizontal face of the roof, superimposed to gradually attenuate the light passing from layer to layer. With its slightly labyrinthine composition, the Dark Square leads to a nucleus where you come across a few benches to rest and chat. River Breeze Square is a slatted floor surface that lets the view and air pass from the basement to part of the ground in the shade. Here, in the basement, water taken from the river flows through a pipe, like a slight arm that returns to the river after being channeled beneath the kilometer of shade. The pipeline is half filled with moving water driven by the flow of the river, and above, by the air taken from the river level, which will cool down following the pipeline and will reappear in the middle of River Breeze Square to bring a cool, refreshing breeze to passers-by who can also stop on the slatted benches to cool off by convection. Fountains Square is a series of fifteen drinking fonts mounted on vertical tubes whose different sizes allow people of all ages and heights to drink, as well as animals, dogs and cats. Fountains Square thus offers the possibility of rehydrating, which then allows better perspiration, which is one of the major means of cooling the body in warm climates. Chairs scattered among the fountains allow you to pause for a few moments. Sunset Square is an elevated platform above the canopy of photovoltaic panels, offering a view to the west, towards the last rays of the setting sun. The platform is accessed by two double helix staircases, also supporting it structurally, allowing people to climb up and sit there to admire the setting sun and enjoy the refreshing evening breezes. Solar Chimney Square is a room that can be entered and is extended in height by a white metal thermal chimney whose end, 15 meters high, is painted black in order to heat up under the sun's rays, thus naturally sucking the air inside the duct to create a wind that will sweep the base of the room, providing visitors with convective freshness, like that of a fan or when blowing on a hot spoon of soup to cool it down. Tempered Garden is a small garden in the shade of the canopy. Because of the reduced brightness, caused by the photovoltaic panel roof, more northern types of vegetation can be planted here, northern Mediterranean-type flora, such as olive trees, fig trees, thyme and lavender. Tables with chairs allow you to pause, have a picnic or play there. Oasis Garden is a large garden that houses a palm grove. Because of the trees' great height, the canopy is interrupted to allow the trees to grow above 7 meters in height. Tables with chairs allow you to pause, have a picnic or play there. Cold Conduction Hall is a public space which you can enter, sit in, climb, nestle in, made of stone and concrete whose interior and exterior are refreshed by coils of water cooled by exchanging heat with the cooler river water. By touching the stone with your body, sitting down, lying down, nestling within or against it, you can benefit from the coolness transmitted by the stone by conduction. Dense Forest is a garden composed of a very high density of palm trees creating a large amount of shade to amplify radiative refreshment. In the heart of the forest two clearings open up which are fitted with benches.

Climate gradient

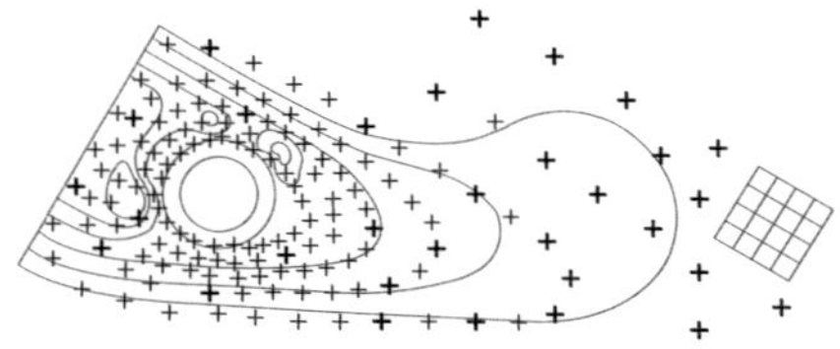

A famous Russian tale entitled "The Twelve Months" tells that a little girl saw the twelve months of the year, as as many legendary characters, gathered in the same place, at the same time, in the center of a clearing in the middle of winter. In the middle of January, she finds snowdrops that normally do not appear until March, strawberries from June, cucumbers from July, apples and pears from August, and hazelnuts from September, which she takes home through the snow-covered forest. The tale ends with a final paragraph that expresses the ideal of human habitat, architecture's primordial task, the fundamental purpose of landscaping: to render a place habitable for humans by tempering a space that is extracted from the uninhabitable vigor of the seasons and the wrath of the climates: "It is told that around her house was the most marvelous garden in the world. The flowers bloomed, the strawberries ripened, apples and pears swelled with juice earlier than everywhere else. There, coolness reigned during hot summers and calm during winter storms. The good people said then — and who knows? Perhaps they were right: "Happy is the house that sees the twelve months together!" There is in this fable the raison d'être of both the house - offering warmth in winter when it is too cold - and the garden, offering freshness in summer when it is too hot. There is also a goal we set ourselves in response to the design of Tuchkov Buyan Park, which is to offer the people of St. Petersburg a park that is open and active throughout the year, each month providing a new vegetal event. Our project for Buyan Park proposes to offer the inhabitants of St. Petersburg twelve months together in a park, in a gradient of atmospheres built according to real physical parameters, passive climatic strategies providing all the comfort, pleasantness and pleasures of a park throughout the year, but with economic and ecological intelligence and thriftiness. To be able to shelter from the rain, snow, cold winds and enjoy the warmth of the few rays of sun in winter. To be able to shelter from the sun, to be exposed to fresh breezes in summer, this is what we propose as a general plan for composition of the park. The park thus stretches between two opposite poles, two opposite temperatures, the winter pole and the summer pole. They constitute the park's two largest public architectural spaces. To the west of the park, the circular "Winter Pole" (зимний столб) incorporates the ancient technique of ice caves commonly found in 18th-century parks. Under a slight hill, in the existing excavation of the Plinth, an underground ice cave is implemented to enjoy a constant cold temperature all year round according to the principles of seasonal inertia (Seasonal Thermal Energy Storage - STES). The technique is the same as for natural ice caves like the ones found in the Jura, for example, or artificial, like in the parks of the past. We propose using the cleared hollow of the foundations to establish an ice cave in which during winter we store snow that will be kept all year long like a glacier, cooling this underground site that is turned into an amusement park offering a natural ice rink whose use, by thermal inertia, can extend beyond spring. During summer heat waves that are likely to become increasingly commonplace thanks to global warming, this Winter Palace, of ice and snow, is an attractive pole for winter games, an underground ice rink, which is suitable for cooling down and which transforms into a cool lake during the hottest days. The Winter Pole, in the basement, is naturally illuminated by a huge circle of more than 40 meters in diameter, allowing natural light to enter. The Summer Pole (летний лагерь) is a glass greenhouse, measuring 47 meters by 47 meters, which rises above the ground to a height of 25 meters. It is composed of multiple floating platforms, spaced apart to allow the movements of air, heat, the rays of the sun and gazes to pass through. Some platforms are made of a simple slab to accommodate cafés, restaurants, shops. Others, 2 meters thick, are filled with soil allowing for the planting of trees and vegetation. Following a natural vertical convective stratification of the air going from least hot at the bottom to warmest at the top, the plants follow this stratification, between continental plants at the bottom, followed by Mediterranean plants and then tropical plants at the top. The greenhouse is a kind of winter palace that welcomes the public in winter who can warm up, spend some time, have a coffee, visit a few shops, admire the plants, rest there sheltering from the cold, the wind and the snow. To the north-east, 4 black-colored blocks, of 60 cm-thick stone and concrete, absorbs the excess heat and by inertia, constitutes a store of heat that can be redistributed in the evening. These 4 blocks accommodate the greenhouse services and brace the entire white steel supporting structure of the greenhouse. In summer, the first floor can open completely to let the air in and out through the roof, thus avoiding overheating. Between these two poles, from west to east, there is a gradient of atmospheres, from coldest to warmest, which allows offering a variety of situations, climates, surfaces, activities, games throughout the year, starting with a cool, shady forest to the west around the winter pole, suitable for summer, to an open, sunny meadow in the east, conducive to winter. To implement this gradation, we vary actual physical parameters, types of trees and their density, the nature of the soils, furniture, according to albedo, emissivity, effusivity, convection and conduction criteria.

The forest is composed according to a gradient of tree density, high around the winter pole to the west, dispersing and spacing out heading eastwards. The species of tree follow this gradient. They are essentially coniferous to the west, comprising a winter forest, and are gradually replaced by deciduous trees toward the east, more spaced out letting more light pass in winter upon shedding their leaves, offering maximum sunshine around the Summer Pole. On this climatic gradient, we propose 12 seasonal stations, 12 clearings, each corresponding to a month of the year, to host as many events that will each be activated during a specific month of the year, with the blossoming or the budding of the trees and the ripening of their fruits, with the blooming of flowers, creating a different event each month in the park, a destination for strolling that gives the park renewed interest throughout the year. These 12 events take place in "The glades of the 12 months" (12-месячные освобоцения), which are broken down as follows: For example, the December clearing (Декабрьская расчистка), the April clearing (Апрельские Поляны), the July clearing (июльская поляна), etc. For example, the April clearing will be composed of cherry trees that will blossom at this specific time of the year, that of March by the cultivation of snowdrops, that of September by a chestnut tree that will yield chestnuts during the months of the fall. The January glade will be embodied by the "Winter Pole"("зимний столб"). The July glade by the "Summer Pole" (летний лагерь). Associated with these floral and fruity events, we offer small pavilions (like the April Pavilion - Апрельский павильон-, or the October Pavilion - октябрьский павильон-, etc.) that will enhance the atmosphere of the clearing by offering a unique sensual and meteorological experience. These pavilions will sometimes also be associated with a specific leisure program accordingly.

Evaporium – Ombrium – Ventorium – Inertium

Originally, the first public spaces and facilities fulfilled essential roles, meeting vital needs: drinking, eating, cooling down, warming up. It was the role of the fountain, the market place, the basilica, the thermal spas, the church, the swimming pool or the Biergarten, the thermopolium, the *veillée* evening gathering, the café, the bistro. Social and political life was created around this need to go and get something to drink, to eat, to cool down or to warm up. One would meet there, talk there, love each other, argue.

During the second half of the twentieth century, with cheap energy, all these public spaces largely lost their sense, replaced by individualistic and private responses. No need to go to the fountain, the water comes through the faucet straight into the kitchen. No need to go to the market, the shopping can be bought on the Internet. No need to go to the church for refreshment, air-conditioning is installed in each apartment. And no more having to go to the café after work to warm up, we have electric radiators at home. The old public squares have become traffic circles for automobiles, the cafés have closed one after the other.

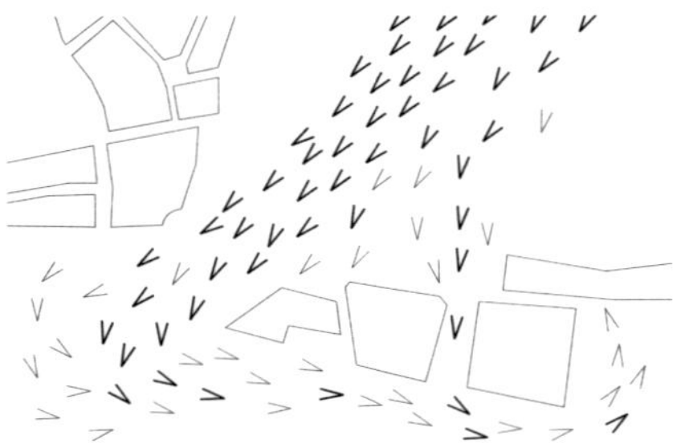

Today, there is a depletion in the old models of public spaces, little renewed, which, moreover, have been relieved by modernity of their primary reason for being. Secondly, because the individualistic and energy-intensive model of the second half of the twentieth century no longer works at a time when the fight against global warming has become a priority.

Our proposal is to invent and propose new types of public space or public facilities for the 21st century that will take their place beside the church, pool or library, in a square, in a street, in a neighborhood. For the time being we propose four: the Evaporium, the Ombrium, the Ventorium and the Inertium each of which offers a fresh public space, where the inhabitants can come and take refuge during heat waves, meet, talk, love each other, argue. The Evaporium offers a space where water evaporates to lower the air temperature, a public fog, reminiscent of the function of the Roman Natatio or the frigidarium. The Ombrium is a covered square, resembling the model of the Roman basilica or the cathedral, where you are protected from the sun. The Ventorium offers a room where air blows to cool by convection. Finally the Inertium invites you to cool down by taking advantage of the cool of the Munich subsoils, where the temperature is at 10°C year round. All the energy consumed is renewable, geothermal and solar. All materials are chosen wisely according to their thermal and sanitary, durable and non-toxic values.

We propose equipping each of our new urban poles with a new type of public facility designed for the coming years, placed at the heart of the new neighborhoods. These are new values, new roles, new types of urban public spaces that we propose in Munich for the 21st century.

ATMOSPHERIC ARCHITECTURE

Evaporated rooms

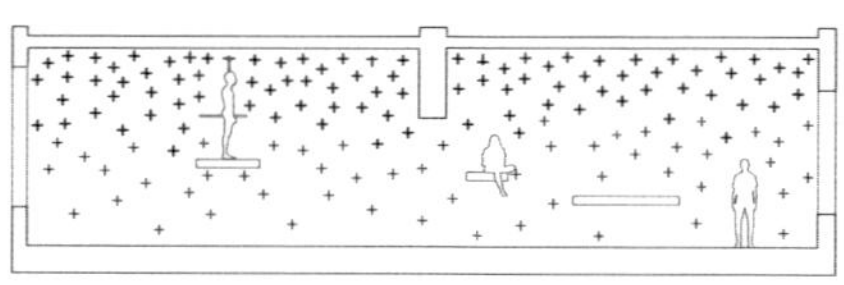

For an apartment for a young physician in Lyon, we rely on the new thermal recommendations for reducing the energy consumed in the building to steer architectural design based on a compositional mode by plotting on the plan to a compositional mode by sectional atmospheric gradients. The new guidelines for reducing the energy consumed by a building recommend heating spaces such as the corridor (to 16°C), the kitchen (18°C), or the bedroom (16°C) less, because you are either physically active or well clothed there. Spaces where you are inactive or undressed, such as the living room (20°C) or the bathroom (22°C) will instead be heated more. These energy reduction targets make it necessary, if they are strictly adhered to, to physically separate the different rooms and their specific functions, by walls and closed doors so that their air at different temperatures does not mix and homogenize as hot water and cold water do if they are allowed to meet, so becoming lukewarm. Following these recommendations to the letter would mean having to give up the free plan and the spatial continuity acquired during modern times to return to plans as they were drawn in the 19th century when each room was separated from the other by walls and doors. We can avoid this regression by working on the intrinsic physical behaviors of the air that rises when it is hot and sinks when it is cold. We can start to compose rooms and chambers, or more precisely begin to distribute functions in space, no longer using walls that would delimit the contours of the rooms, but only the spatial distribution of temperature and luminosity in the air. It is no longer a question of drawing the plan, but rather of designing an atmosphere, with its different weather gradients in which one moves to find a certain temperature, a certain light. A bit like in a natural landscape, where we take refuge under a tree when it rains, where we seek the shade of a rock when it is too hot, we propose to no longer design rooms and spaces, but to stabilize the furniture and uses at certain heights, at certain temperatures, at certain light intensities. Thus, the shower where you are naked, is the highest up, in the hottest strata of the air. Just below floats the armchair, where you sit alone, while the sofa, where you are joined by several other people, seated side by side, will be a little lower, in the more temperate temperatures, since the ambient temperature there will increase due to the human warmth generated when several people are seated there together. Further down, cooler, the kitchen and, on the lowest level, the bed, plunge into the coolest layers of the air of the house.

Domestic astronomy

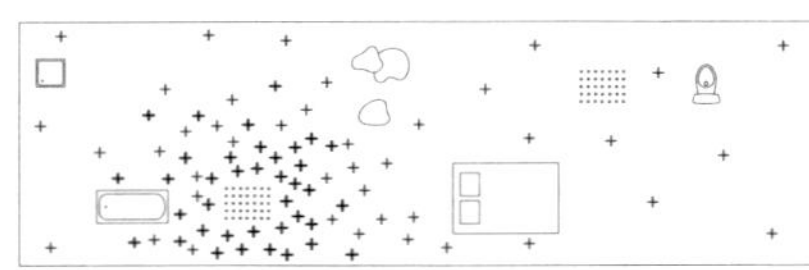

Domestic astronomy is a prototype for an apartment where it is no longer a surface but an atmosphere that is occupied. The functions and furniture rise from the floor up: they spread and evaporate in the atmosphere of the apartment, and they stabilize at certain temperatures determined by the body, the clothes worn and the activity being pursued. Our proposal is to take these physical differences in the distribution of temperature in space into account and to exploit them by changing the lifestyle: replacing a horizontal with a vertical lifestyle, where we can occupy different heat zones, different layers, different heights. And so we create a global ecosystem as a kind of astronomy in the home, where the combinations of temperature, light, time and place are reconfigured.

Interior transhumance

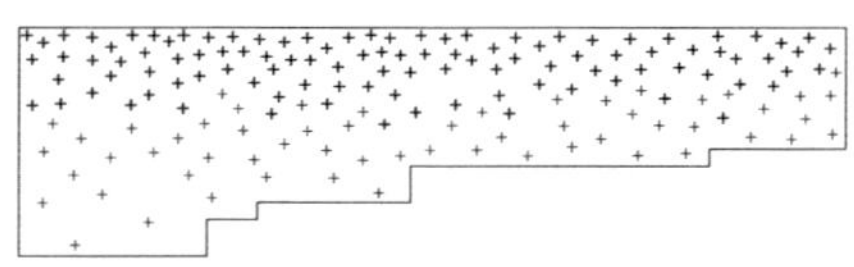

The design of this jointly owned property in Hamburg is based on Archimedes' natural law which states that hot air rises and cold air sinks. Very often, in an apartment, an actual temperature difference can be measured between the floor and the ceiling, which can sometimes reach 10°C. Depending on our physical activities and the thickness of the clothes we are wearing, the temperature of the rooms in an apartment need not be the same throughout. As we are protected by the bed covers, the temperature of the bedroom can be as low as 16°C. In the kitchen, because we are dressed and physically active, we could have a temperature of 18°C. The living room is often heated to 20°C because we are dressed but immobile, sitting on the sofa. The bathroom is the warmest space in the apartment because we are naked there. By keeping these precise temperatures in these specific areas, a great deal of energy could be saved by reducing the temperature level to the exact requirements. In connection with these physical and thermal figures, we propose to adapt the apartment at different depths and at different heights; the space where we sleep will be lower down while the bathroom will be higher up. The apartment becomes a thermal landscape, with its different altitudes of heat, where the occupant can walk freely as if in a natural landscape, in search of specific thermal qualities related to the seasons or the time of day. By deforming the horizontal floor slabs, different heights of the spaces are created with different temperatures. The deformation of the slabs also gives the building its exterior appearance.

In physics and thermodynamics, heat is the process of transferring energy from a body or system by thermal contact, which in turn is defined as transferring energy to a body in a way other than due to work performed by the body. Heat energy transfer can occur between objects by radiation, conduction and convection. Convection is generally the dominant form of heat transfer in gases. It is a term used to characterize the combined effects of fluid conduction and flow. In convection, enthalpy transfer occurs by the movement of hot or cold parts of the fluid as well as by conduction heat transfer. Generally, an increase in temperature causes a reduction in density. Thus, when air is heated, it rises, displacing the denser, cooler air that sinks. In this free convection, the forces of gravity and buoyancy cause fluid to move. Buoyant convection is due to the effects of gravity, and therefore does not occur in microgravity environments.

The public space of the building extends from the basement to the roof as a continuous movement of air from lower to upper levels. Since hot air is lighter than cold air, the upper level will be warmer than the lower level of the public space. The common spaces will be placed in this vertical heat geography. Spaces to dry laundry after washing will be located at the upper level of the building, while we could use the cooler, lower floors of the public spaces to store fresh food or objects without specific heat needs such as bicycles, buggies or detergent. But some places in this public space have no function, becoming free spaces whose social uses can be invented, children's games, between a cold interior garden on the first floor and a warm garden in the attic.

If the design process follows the new energy reduction targets linked to sustainable development recommendations, above all, through these new constraints, it offers new forms and new ways of life.

Indoor Clouds

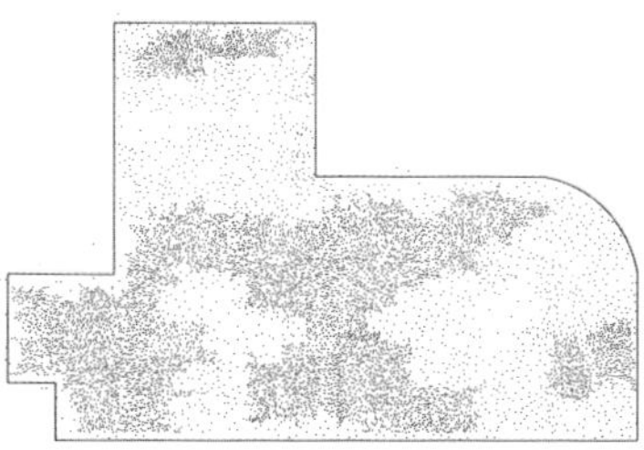

Our proposal for the development of the new Palais de Tokyo is, literally and figuratively, to build a meteorology; variable and fluctuating climates, lights and spaces that we can always make and un-make and re-make, according to new configurations, unexpected desires, requirements or demands that we do not yet know today. As part of this reconfigurable atmosphere, the historic building looms in the background as a mineral landscape of limestone and concrete on which the program of the new Palais de Tokyo floats like a fluctuating atmosphere whose ephemeral forms and invisible matter, air, light, heat, sound are worked on. And if there are forms, they will be only transient, fleeting, the length of an exhibition, the duration of an economic context, like the form of a cloud, which suddenly takes the shape of an elephant or a face, before transforming into the shape of a butterfly, for example, or disappearing altogether. The same will apply to the shapes and layout of the exhibition rooms and the concessions which, as they can continuously be transformed, modified and improved, will take adopt a very classic form during an exhibition, for example, by matching the existing volumes and the canopies to reform for the following exhibition, if required, in a completely different configuration. Our architectural project is based on an antinomy between a historical background, that of the existing building, as a light gray, solid, uniform landscape, and picture rails, objects, mobiles, like meteors, fluctuating and variable.

Evaporated building

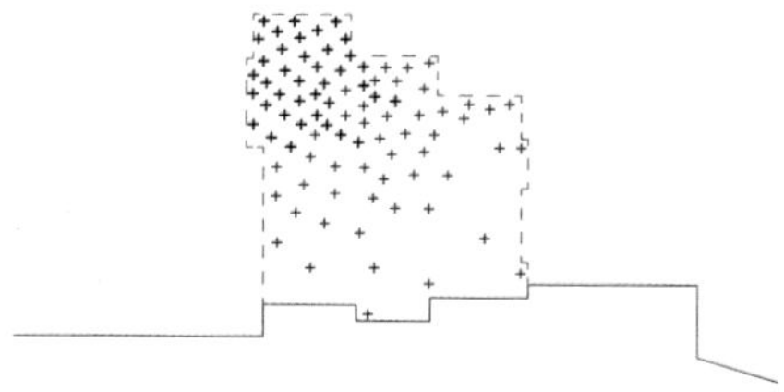

A relationship is usually established between a building and its surroundings that results from contrast, rupture and opposition: a solid versus the void, the opacity of the walls versus the sunlight, the weight of the constructions versus the lightness of the air, a finite and fixed form of a building versus the infinite and always renewed form of the atmosphere. It is paradoxically toward the qualities of the latter that we wish to head for this office building in La Défense, near Paris, toward an architecture that evaporates, disintegrates and dissolves. We wish to build our construction with climatic and meteorological qualities, evaporating heavy structures and fraying opacities. For we believe that halting global warming and the depletion of energy resources must not remain as a mere goal that must be achieved, rather they can constitute a new formal vocabulary, provide architecture with new means, become new tools. Climatic phenomena such as convection, vacuum and evaporation might become the new principles of architectural composition. And meteorological elements such as light, heat or water vapor can become the new building materials of contemporary architecture, its new bricks. Today, architecture can become meteorological, working with the climate, responding to the challenges of sustainable development until it becomes climatic, atmospheric, meteorological. And this is not an allegory, but constitutes the very reality of the building's construction and heating and ventilation techniques in order to improve its ecological efficiency and offer a new, healthier, more sensual and more pleasant quality of life. Whereas it is usually through the fullness and opacity of the walls, floors and ceilings that we define the void of the interior spaces we inhabit, our goal today is to define these voids no longer by means of the solid but also through void. To evaporate structures, to sublimate the walls and the ceilings, to go from solid to gaseous state, to create an architecture of air and light, transparent, open, airy, fragmented. A building in a gaseous state, a disintegrated architecture, an evaporated building. Beyond the metaphor, it by basing ourselves on our physical and ecological reality that we must accomplish this phase shift, on the scale of the city and of architecture, but also of the user, for their well-being and comfort. For an imposing volume, we propose a volume that crumbles, from the inside and the outside, giving the new building a certain ecological and physical permeability and transparency to the air and to the light. A building that aerates, dissolves, expands and spreads to allow the natural movements of the environment, the winds and the sun's rays to take place. It is a response on an urban scale, but also a social response: the building should not be a mask or an obstacle but a link, not a solid, opaque screen, but be distended, disintegrated, hollowed out and opened up to allow the winds, the light, gazes and strollers to pass, beyond it, to the Cité des Provinces Françaises. Thus the warm winds from the south do not come crashing into and bouncing off the southern façade of the new building, but cross it, penetrate it to emerge on the other side, to the north, thus bringing to the neighborhood of Les Provinces Françaises the warmth of the south in winter. Thus the rays of the sun no are longer blocked by our building, but pass through it, letting the rays of the low winter sun through the building, illuminating the offices located to the north of the building, until lighting up the neighborhood of Les Provinces Françaises. We therefore model the shape of the building following the principles of aeraulics to allow the wind from the south-west to cross the building in several places to emerge from the north-west. The building's initially compact and regular shape is hollowed out, swollen in some parts, retracted or diffracted to cause the wind to slip inside it and leave via the other side. The building's aeraulic deformations then also meet the specifications of the area to offer a certain variety of depths on the façade. Likewise, the interior rooms dissolve and separate, from south to north, following the lines of the sun's rays at certain times of the day. The building then appears as a sort of inverted sundial, letting the sun's rays pass through its thickness, uninterrupted, as far as Les Provinces Françaises quarter. Views are also provided from the regional express train station to La Défense, visual crossings, between Les Provinces Françaises and Les Terraces. It is ultimately a veritable public route that allows crossing the building from the unusual walk to descend to Les Terraces. It is thus an evaporation of architecture itself. While an office building usually consists of numerous compartments separated and arranged by opaque walls and thick ceilings filled with ventilation ducts, we now want to expand these partitions, open them out and empty them until they evaporate and become hollow and transparent. We seek to distend the ventilation ducts to make them disappear, until they evaporate to become the building itself, or more precisely so that the building will expand, evaporate, diffract to become cloud, breeze, light. Evoking the idea of a building that evaporates goes beyond the metaphor. It is a real implementation of a process of lightening the structures and getting rid of the heavy techniques of the building like the air-conditioning and ventilation systems. Our intention is to truly eliminate the ventilation ducts, to reduce the electrical supply for lighting, by emptying the building, decomposing it, disintegrating it to allow natural light to pass throughout, so that it floods with a new atmosphere of clean, fresh air. We suggest architecture in its gaseous state, traversed by air and light. We distend the thicknesses of the walls and of the floors, which we vacate until they vaporize letting the air and the light flow freely throughout the building, into the tiniest corner, as far as the spaces usually relegated into darkness, the most northerly and the lowermost. And we will re-encounter simple, personal gestures like closing a window if we are too cold, opening the curtains when it is gray outside. This evaporation process is developed with a view to minimizing the energy consumed by the building and the release of greenhouse gases. The building consists of two atmospheres. The first, more interior, consisting of the workspaces themselves, heated in winter or cooled in summer, embedded in the second, more exterior atmosphere, formed by the voids and circulations that appear with the expansion of the walls and of the floors. The building's external double glazed envelope constitutes its thermal insulation. It holds this second atmosphere composed of fresh air which serves as a reservoir to ventilate all the offices. All the workspaces that are separated from this atmosphere by single glazing are floating in this immense pool of fresh air, heated only by the immediate natural solar input and its supply by means of a Canadian well. These offices, which are the only heated spaces, are therefore thermally insulated from the outside by these three layers of glazing and two layers of air. This second atmosphere consists of the voids between the offices and the horizontal and vertical circulation areas and the entrance halls. Here, air can circulate throughout the building, from top to bottom, from south to north. The floor of the corridors is glazed and perforated to allow the air to move freely throughout the building. In this second atmosphere floating at different heights are the office surfaces like flying carpets, or more precisely decomposed floating boxes whose adjoining walls and ceilings are spaced apart to allow a continuity of the void of the second atmosphere through which the fresh air and the light pass. The offices are manually ventilated. Everyone can open the window to air the offices. This is done on 53% of the days of the year (the fine days) by opening the building's exterior windows. On the other 47% of days, during winter and on hot summer days, the windows will be opened in the intermediate space to take in fresh, more temperate air, in order to save energy to normalize the air temperature to a comfort threshold. The stale air is evacuated from the offices directly via the façade, but through a series of metal plates that act as heat exchangers when the stale air passes through the second atmosphere, between the offices and the outside of the building. This system enables recovering the heat (or freshness in summer) by bringing it into direct contact with the intermediate atmosphere. This second, intermediate atmosphere is a semi-natural, semi-artificial climate, where we take advantage of the extra heat input in the south to bring it back to the north in winter, or on the contrary, cool the building by sending the cooler air from the north to the south. Ascending currents of cooler air in summer or, in winter, descending movements of warm air, reproduce a form of meteorology in the heart of the building, with its interior Gulf Streams, its depressions, its formation of clouds through the vaporization of water to the south at the top of the building in summer, in order to lower the temperature and trigger air movements. The aim, just as the Gulf Stream does in Europe, is to mitigate the differences between north and south, bottom and top, to homogenize indoor temperatures, to heat the fresh air in this atmosphere in winter or to cool it in summer so that it enters office spaces at a more moderate temperature than that of the outside and so that the energy expended through heating (in winter) or cooling in summer is reduced. It is the design of these diagonal movements of indoor air from top to bottom, from south to north, that lead these rooms to vibrate with respect to each other, these slight deviations in plan and in section that ultimately give the building a generally vaporous, almost cloudy form. The expansion of the interstices of the rooms is therefore carried out in such a way as to expand or contract the air flows in order to cause upward or downward movements of air. This interior expansion of the interstices (between two of the partitions, between two of the ceilings) generates different thicknesses of void, the effect of which is felt on the façade, on the roof and on the underside of the building, which deforms more significantly where interior expansion is greatest.

The shape of a building is usually initially compact, closed, opaque, generating its own shadows. If the rooms in the north of a building do not receive sunlight, it is not really due to a lack of sun but rather because the rooms to the south shade the rooms to the north. If the southern rooms did not exist, the northern rooms would be in the sun. The architecture here poses its own drawbacks. It is against this common situation that we propose to expand the interstices between the rooms, to separate them, both in plan and in section, in order to allow the light from the south to pass beyond the first row of rooms to naturally illuminate the second and third rows of rooms. In our project, the walls decompose and break up, allowing the sun's rays to penetrate further, as far as the rooms located to the north of the building, bringing these occupants the happiness of beautiful direct sunlight. This fragmentation of the thickness of the building follows the position of the sun, cutting at certain angles corresponding to the paths of the sun's rays at the hours when they are at their lowest. By expanding along these angles of incidence of the sun's rays, the voids which appear then push back the rooms to allow light to pass through, thus deforming the linearity of the volume of the building. The quality of the architecture lies in the visible and the macroscopic, but also in the invisible and the microscopic, in what could be called the chemistry of the atmosphere. To live is also to breathe and see effortlessly. Improving air quality, the quality of light, the quality of the materials and even the quality of sound becomes a major theme of our proposal. The project takes into account this sensual dimension, linked to the well-being and comfort of its occupants. If we propose a vaporized architecture, disintegrating the building, it is above all to improve the comfort of its occupants by allowing the rays of sun and the fresh air to pass through the building, to reach its innermost depths. By breaking down the solid to make it diaphanous and vaporous, we seek to improve visual comfort and air quality and offer natural light and fresh air to the heart of the building. The nature of the materials helps to give a certain tone, a certain almost terroir-like flavor to the light and to the air. The slightly beige and soft tone of the natural stone also confers a certain visual and tactile comfort to the occupants. Whilst most materials and paints are chosen for being chemically neutral, we suggest forming a contextual and local link with the Seine valley by proposing different limestones from the geological soils of the Seine valley and Grand Paris to Le Havre as floor coverings. The different limestones are grouped by zone according to their color, albedo and reflection rate, by placing lighter, whiter limestone further north in order to improve and intensify the light intensity by allowing greater reflection of the surfaces, while in the south, we will tend towards very slightly darker, more ochre tones. And yet more chemically, we propose using concrete whose aggregates also consist of locally-sourced limestone.

River of air

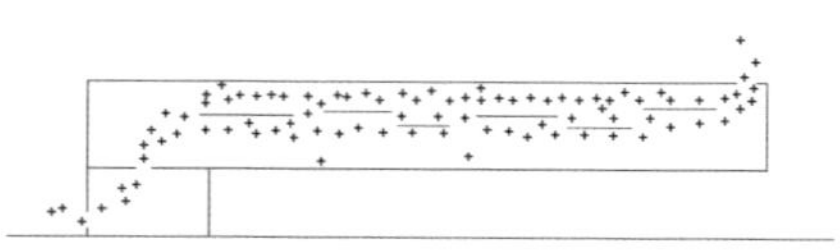

It is also possible to conceive space according to formal meteorological rather than geometric principles according to the design of the air flows that are inevitably found in buildings due to the implementation of ventilation required to evacuate moisture from the air and renew oxygen input. Thus, one can agree to design the building around this air flow, to structure the space around a current of air, by studying its flow and its variations in speed and all of the spaces one can occupy differently. So, we must ventilate a work room occupied by many people who will each breathe and give off water vapor far more than a library occupied by a lot of books that do not breathe and is frequented by fewer people. Each time we ventilate, we must warm the cold air taken from the outside in winter to raise its temperature to that of human thermal comfort, at around 21°C, which results, each time, in new energy expenditure and therefore CO_2 emissions. We therefore understand that to fight global warming we must try to ventilate as little as possible, or more precisely we must change very exact quantities of the air of a room according to the activity of the people inside (the more intense their physical activity, the more quickly we will have to renew the air) and the number of people present. Thus, if we have to place people and books in the same room, we can do so according to the air current, by designing the space and the building like a river, with faster moving points (like a torrent) and slower ones (like a lake), a strong central current and a sidestream, where the air stagnates. By designing this airflow, we can locate work spaces where the airflow is faster while we can place the books where the air tends to languish. Thus, the air of the libraries will not be renewed whatsoever, thus saving energy. By relying on fluid mechanics rather than on Euclidean geometry, we can thus solve the problems of space and the fight against global warming together, inventing forms according to the logic of meteorology rather than of plasticity. These few examples seek to explain how to design architecture based on the void rather than the solid, how to project a building taking space as the subject, rather than walls. Here, the architecture is based on meteorology rather than geometry, taking advantage of climatic phenomena to reduce energy consumption while discovering new forms of void and spaces, new ways of living.

Thus, for a project for a library at the University of Nancy, we design the building as a river of air, with parts where the air flows faster (where the users who make the air stale by breathing will be placed) and the parts where the air flows more slowly (where we will house the storage of books, which needless to say do not foul the air). To define its form and spatial organization, the architecture of the Library quite literally follows the climatic recommendations for sustainability in order to reduce energy consumption in buildings. In a volume fully thermally insulated with 35 cm-thick insulating wool, meeting today's most stringent thermal requirements, the library takes shape in layout and in section along the air ventilation path and on the natural ascent of hot air according Archimedes' law. The aim is to minimize heating and air renewal (because a high rate increases energy costs for heating and produces CO_2, which is responsible for global warming) by heating only the areas that so require to precise temperatures and by renewing the air only in the areas that so require, to save energy expenditure in the building.

Air flow

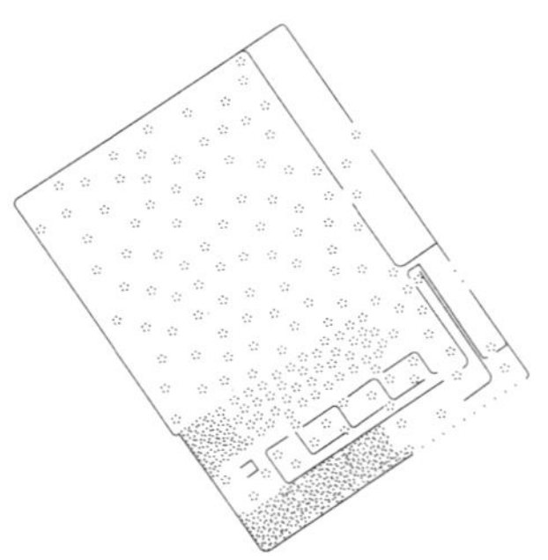

To condense the question of the architecture of a sports hall to one of ventilation and aeration, that is to say to re-found the architectural project literally on the need to bring oxygen into the building and to evacuate the discharged air produced by breathing. It might seem like a provocation, but on the contrary, we see it as a way to reconnect the question of form and program to its essence today, to free them from transient influences and representational automatisms. The French literary critic Jean Ricardou said about the work of the French author Alain Robbe-Grillet and more generally of the "Nouveau roman" that it was a question of reversing the dogma of representativeness (the pre-eminence of fiction) in favor of the dominant producer of writing (the pre-eminence of textuality). Meanwhile, Robbe-Grillet wrote: "Be there before being something". That is what we are seeking: an architecture that emerges starting not from an image or a habit of use, but from its technical constraints, from its physical and physiological reality reinventing its unexpected and perhaps also sometimes incongruous but always innovatory forms and uses.

What is the deep link between the space and sports activity? It is undoubtedly the question of the supply of oxygen required for the athlete's muscles, the consumption of which is multiplied by 500 during significant muscle work compared to muscle rest.
At the same time, the organism must evacuate metabolic waste products, such as H^+, CO_2 and lactates.
During physical activity, the ventilatory flow rate rises from about 7.5 liters per minute at rest to 120 liters per minute. This increase is possible thanks to an increase in the respiratory rate which then requires a quicker exchange between fresh air and stale air. There is, therefore, a precise relationship between the form of the space and the amount of air to be supplied and renewed, again posing the question of the ventilation of the building as the main efficient strength of the architectural form.

Our project consists of, is ordered, is indeed formed precisely from the management of air in the building, by means of its path, its speed, its volume, its renewal rate. The architecture thus becomes a mere ventilation system and the spaces literally take the form of ducts in which air flows at low pressure. The building becomes a continuous movement of air, with its tributaries that increase the amount of air brought to the space. The architecture is gradually composed, linearly, between spaces with a low rate of air renewal to spaces with a high hourly air renewal rate. So, first the spaces where one spends the least amount of time, where the body is little active in terms of burning muscle, where the least amount of water vapor is emitted, then gradually, everything increases.

Our project, therefore, is gradually composed like a stream of air which, joined by others, gradually turns into a tributary, then into a river, then into a sea. The architecture we propose is thus a habitable flow of air, a wind that is designed and augments according to the average hourly rates of fresh air renewal depending on the type of room.

Thus the sports hall is defined by the path of the air according to the hourly air renewal rate:

Store:	3 to 10 times / hour
Hall:	2 to 4 times / hour
Classrooms:	5 to 10 times / hour
Toilets	10 to 15 times / hour
Sports changing rooms	10 to 20 times / hour
Sports halls	10 to 25 times / hour
Showers	20 to 30 times / hour

The school itself is defined as follows:

Hall:	2 to 4 times / hour
Classrooms:	5 to 10 times / hour
Toilets	10 to 15 times / hour

The architecture of our project is organized like a wind, a movement of air whose flow gradually increases according to the increase in the respiratory frequency of its occupants according to their activities and the amount of water vapor released.

This path of air, whose volume of fresh air is increased at each stage, begins in the foyer and passes into the changing rooms. It then swells in the sports hall to finally concentrate its highest renewal rate in the showers to pour out into dense vegetation, chlorophytum-type plants that rapidly remove the CO_2 of the air produced by the athletes, recomposing oxygen. This air finally passes through a dual-flow heat exchanger to convey its heat to the incoming fresh air, before exiting. The building meets the requirements of the Minergie standard thanks to its high-performance insulation, compact shape and dual-flow controlled air renewal.

Habitable duct

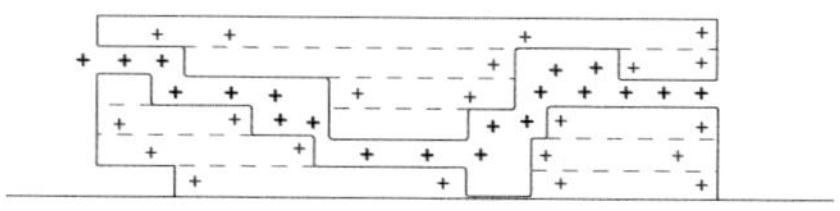

The buildings are each arranged around an enlarged ventilation duct to become an interior street that runs continuously and smoothly through the buildings linking up the hot square in the north-west to the cold square in the south-east. This duct collects cooler or warmer air depending on the season. An automated system, depending on the season and the temperature outside, enables opening the duct at the hot square or alternatively the cold square, in order to draw the air at the most suitable temperature and thus gain a few degrees. In winter, it will be the air from the northern mineral square we will breathe. In summer, it will be fresh air drawn from the shade of the trees of the south that we will breathe. At each end of the duct is a dual-flow air exchanger which preheats / cools the incoming air using the stale recovered air exiting through another channel. The ventilation duct, which also serves as a corridor, is made of thick clay in order to store the heat or the cool, respectively, in the inertia of its mass while regulating the degree of humidity. This duct distributes the building's air horizontally in order to strengthen social ties while expanding in certain places in order to accommodate common and collective programs and some indoor gardens composed of cleansing plants. The design of this duct can be deformed according to the different types of apartments, thus enabling a strong and individualized characterization of its route.

The buildings' supporting structure is composed of structural columns allowing great variations in the layout and modifications of the apartments over the years. Our goal, for each apartment, is to offer a unique spatial quality by working at different heights, different depths, mostly cross-cutting, between very classic typologies on one level to more ambitious typologies spreading over several levels, to be chosen according to age category or lifestyle. The layout of the apartments is according to a calculation of the temperature variations dependent on the entry of fresh air into the apartment via the duct. By means of a physical thermographic study, we arrange the various functions in relation to the movements of internal temperatures according to the recommendations of the SIA 384/2 standard by placing the functions which require the coolest temperatures nearer the duct (bedrooms, toilet, kitchen) and those that require higher temperatures, according to Archimedes' principle, by following the upward movement of heat. The entire construction is made of wood. Toilet and garden water comes from harvested rainwater. The domestic hot water is heated by solar panels installed on the roofs. The basements are concentrated beneath the school and in the western part of the site in order to facilitate construction in stages according to the different investors but also to allow the east to enjoy real vegetation, in the open ground, where real trees can grow. In the north, the new buildings close the urban islet by being aligned along the street. In the south, they open up toward the cemetery park, accentuating the opening and sliding of the street front toward the park and nature.

Summer wind

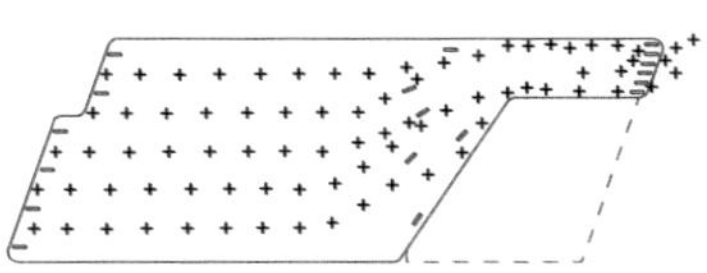

Our project for a new swimming pool considers environmental issues as the end and the means to its architecture. The climate and its protection are a goal of civic responsibility for sustainable development. But the climate and its meteorological phenomena are also the architectural means, the plastic and technical tools that we use to conceive and design this new swimming pool. Our architectural forms and solution thus follow the climate. They are designed and arranged, taking meteorological phenomena into account, air and wind movements, water vapor displacements, the conduction of materials, temperature variations.

Our building is set on the north-western edge of the plot. Its longest façade opens to the south to the outdoor areas of the pool, the areas for relaxation and playing, the restaurant terrace, protected in the heart of this new island. But its shape is mainly due to the prevailing summer winds of the region, from the south-west, whose natural force is harnessed to ventilate the pool during the summer. Our architecture slips into the path of the wind, gently interfering with its flow, bringing it in throughout the length of the building, taking advantage of this natural source of ventilation in summer. Our building is constructed around the wind, it follows its forms, it is designed along its path, it accompanies it for a few dozen meters, harnesses then releases it, contracts then expands it and allows it to follow its course. The shape of the building and its supporting structure are defined by following the guiding principles of wind direction, from south-west to north-east, accompanying it, deflecting it slightly, without interrupting it, conducting it along the length of the building. The structure supporting the roof is made of glue-laminated timber. It is made of long longitudinal beams that channel the wind from south-west to north-east by gradually lowering it toward the changing rooms in order to evacuate the stale air in summer and, in winter, to input heat. In summer, the building, placed in the direction of the wind, is naturally ventilated allowing the wind to pass throughout its entire length. Five openings are placed partly on the south-west façade, between the supporting columns. They harness the wind and bring it into the building. The roof, high up above the pool, descends towards the changing rooms. This enables channeling the wind downward and to ventilate the changing rooms. The wind finally emerges from the north-east façade of the building. The beams, whose load direction is longitudinal, in the direction of the wind, do not interrupt the movement of the wind and steer it throughout the building. In winter, a dual-flow mechanical ventilation system with a heat exchanger controls the path of the wind in the building. The fresh air is thus guided between the beams, then it is sent down toward the changing rooms due to the shape of the roof, which enables driving to the floor the heat of the hot air above the pool that has tended to rise and accumulate beneath the ceiling. So we use the heat of the pool's ambient air to heat the changing rooms. But to avoid introducing moisture to places where people get changed, we "chemically" separate them from the pool space, connecting their air supply to the entrance hall. The heating of these changing areas, therefore, is by conduction from outside the cubicle. The hot air coming from the pool "kisses" the highly heat-conducting aluminum walls and therefore, by conduction, sends the heat inside the cubicles. Conversely, ventilation is rendered from the entrance hall to avoid inputting the water vapor coming from the pool. The heat contained in the stale air is removed by a dual-flow heat exchanger before exiting the building. The glazing is triple and the roof is perfectly insulated.

Windtrap

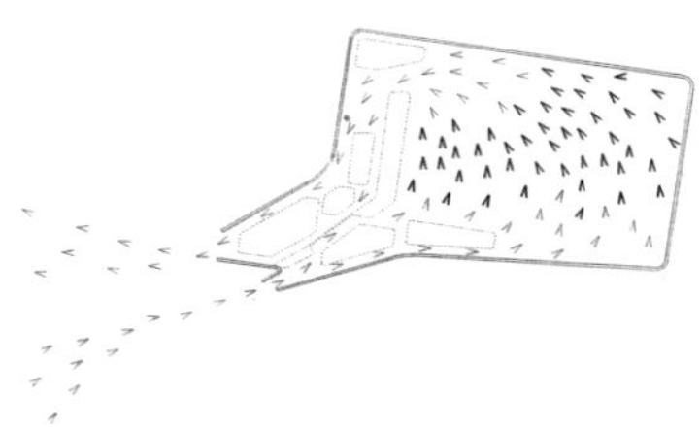

The fight against global warming has become one of the main challenges of contemporary architecture. Better insulation, controlled air renewal with heat recovery, and a compact building form are today the main measures to be taken to reduce a building's energy expenditure. As can be seen here, the main architectural elements concerned by these measures relate to the building's heating and ventilation. If preserving the global climate has become the challenge facing architecture, today we would like the climate to also become the means of architecture. Architecture can become the construction of climate and we can now conceive architecture as meteorology, working mainly on the air and its movements, on heat and its transfers. Our project for a new sports hall in Menges, Slovenia, concerns meteorological architecture, settled in nature in the form of a wooden object slipped into the path of the wind. It captures warm air from the south and later discharges the stale air further north, after removing its heat. Our sports hall emerges as a slight inflection in the natural movement of the wind. Technically, this perfectly insulated building involves the dual-flow air renewal system, which preheats the fresh incoming air with the heat of the outgoing stale air, forcing them to meet between large conductive metal surfaces, in winter allowing the outgoing hot air to confer its heat to the incoming cold air. Our project expands the dual-flow air renewal system, which is usually the size of a machine, to the scale of the building. Thus, an entire part of the building hosting the sports facility's ancillary programs, are involved in this heat exchange, according to the temperatures recommended depending on their functions: from the hottest at 22°C for showers and changing rooms to the coldest at 12°C for equipment storage rooms. The WCs and corridors form the chemically sealed boundary between the incoming fresh air and the outgoing hot air, where heat is exchanged on the metal walls between the hot, more to the north of the building and the cold, more to the south. The main structure and the façade are made entirely of wood. The adjoining rooms, allowing heat exchange between the incoming and the outgoing air, are made of metal in order to allow rapid heat transfer. The sports hall is entirely lit at the top. A walkway upstairs connects the new sports hall to the school, allowing the passage of people, a view and wind on the first floor.

Relocated barn

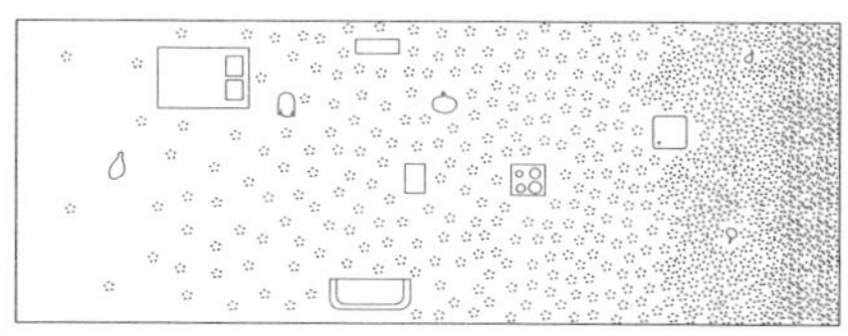

The urban, social and political crisis of modernity that began in the 1960s with books like "The Death and Life of Great American Cities", by Jane Jacobs, or "The Architecture of the City", by Aldo Rossi, was accompanied by an irrational debilitation of modernity that could be described as poetic and even "psychoanalytical". What has been criticized of modern architecture is a kind of dehumanization, with its destruction of the ideas of collective and individual memory, its lack of symbolism, the loss of the "spirit of place", of historical and cultural contexts, and a break with all notions of peculiarities and differences within tradition and the land. It was said that if modern architecture had indeed failed, it was because it had killed the symbolic part of the house by eliminating the cellar and the attic, two highly emotional and poetic spaces.

To a certain extent, the postulate of this critique of modernity is based on the work of the French writer Gaston Bachelard. In his book of 1957, "La poétique de l'espace", through literary and subjective desires he seeks to re-establish the possibility of a poetic imagination of the notion of space, room and house. Bachelard first proposes that the range between the cellar and the attic is the essence of the house, its "raison d'être". "Verticality is ensured by the polarity of the cellar and the attic, the marks of which are so deep that, in a way, they open up two very different perspectives for a phenomenology of the imagination. Indeed, it is possible almost without commentary, to oppose the rationality of the roof to the irrationality of the cellar". Bachelard then discusses in detail how a house is structured between these two poles, which appear as the two symbolic extremes: below, the basement, the darkness, the fear, the crime, the heavy air, the black and dirty water; above, the attic, the smell of the grapes that dry on the grid, during the day. And between these two, one or two floors of a calm and soft atmosphere. Postmodern criticism claims that by emptying the cellar of its modernity, the house has lost its roots and its cause for its occupants: "Home has become mere horizontality,... the relationship between house and space becomes an artificial one", Bachelard writes. But beyond the allegory, what the house really lost with the advent of modernity is a rich diversity of climatic spaces, between the moisture of the cellar and the dryness of the attic, cold and warm, dark and light. In addition, the areas where food was stored have disappeared. Until the twentieth century, the house was equally a place to protect, store and preserve food, as well as the space for its consumption. Several preservation methods had prevailed until modern

times: fermentation, refrigeration, preserving, smoking, salting, drying and desiccation, each corresponding to an region with a certain temperature and humidity. The types of rooms in the house were far more diverse and far more hybrid than today. Besides the dining room, bedroom or living room, there were other rooms and spaces where food was kept. There were cold and moist areas, such as the cooler, the larder, the bottlery, the place for the butter, the *komara*, the *cassinette*, the *fleurines*, the *pantry* (where bread was kept), the fruitery, and the cellar. As well as hot and dry spaces like the smoking chamber, the *dispensa* (store room), the *lardarium* (where the dried meat was kept), the locus ad torrendas annonas, the smokehouse, the herbarium, the *horreros*, the *cencalli*, the *luova*, the *poignards*, the *vignaux* and the attic. But there were sometimes rooms with mixed functions, that is, the identity of a room is then defined by its particular climate in which we can keep the goods as well as consume them at that time. The Diolas de Casamance kept their rice harvest in the room where they slept, the people of Las Hurdes in Spain kept chestnuts, their main source of nutrition, in the living room itself. Modernity has got rid of spaces for conserving while it has removed the sloping roof to allow terraces and the basement to allow piles. Or more precisely, it has contracted these spaces to the point of making furniture of them. The cellar, where dried pork, apples, potatoes, cabbage, celery and carrots were kept, became the refrigerator. The attic, where we kept cereals, flour, nuts and salt, was reduced to the scale of a cabinet. Our proposal is now to transcend modernity, by seeking a synthesis between conservation and consumption, between stores and lounges, between modern rationality and poetic imagination, by reorganizing the space of the home by mapping a climate. From the refrigerator and the cabinet, we propose to extract Bachelard's two polarities of the cellar and the attic and to redeploy them spatially as living spaces. What we propose is to reshape the modern horizontal space from one end to the other with a source of water vapor on one side and an absorbent sink on the other. In addressing the barn, a historical element of American identity, our work also contributes to the relocation of irrationality, myths and folklore in contemporary architecture in the context of globalization and sustainability. In this new free plan based on variations in relative humidity, different food, as well as everyday objects such as a shower, a stove, computers, books, clothes, are located according to their conservation needs, participating in this water vapor landscape. Occupants encounter a more sensual, tactile and poetic way of living between moist, misty areas - where they take a shower or cook - and where they sleep, talk and work, in drier, brighter areas.
The plan consists of two devices fixed on each side of the room at medium height. The first device is an ultrasonic device, which produces a fine water vapor mist, a sort of cellar air. On the other side of the room, the second device is a dehumidifier that dries the surrounding air, providing an attic-like atmosphere. Forming an ecological cycle, the water passes continuously from one side to the other, from gaseous to liquid phase, in the form of vapor between the atomizer and the dehumidifier, and in the form of a liquid between the dehumidifier and the atomizer. Our target would then perhaps be to re-situate contemporary housing into a poetry of space, to re-expand today's space from the cellar to the attic, from the atomizer to the dehumidifier.

Interior tropics

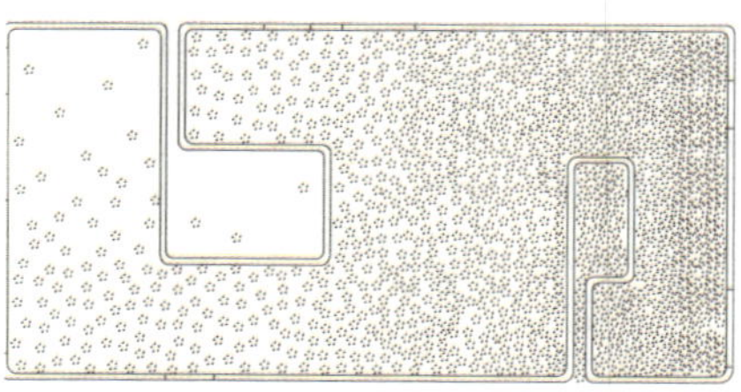

The design of this building is based on the path followed by vapor in the house. Indeed, the building's new dual-flow ventilation system for air renewal (one of the most efficient ecological solutions today for reducing the energy consumed in the building) requires a precise one-way flow of air in the house. Air renewal starts in the driest part of the house and ends in the most humid part of the house as we do not want to bring moisture into the dry part of the house. There is therefore a slow flow of air, which circulates throughout the building, evacuating moisture from the air in one direction, starting from the dry part (the bedroom) and becoming ever more humid along its way through the house towards the moist part (the bathroom). This path of the air is determined by the use of space and the vapor produced by the body according to physical activity. A person produces 40 grams of vapor per hour when asleep, 150 grams of vapor when awake, 1,500 grams when cooking, 2,400 grams of vapor when taking a shower.

Water vapor is the gaseous phase of water at a temperature at which the same substance may also exist in liquid or solid state, below its critical temperature. In the atmosphere at ordinary temperatures, gaseous water is therefore called water vapor and it condenses into a liquid if its partial pressure is sufficiently increased. Under normal atmospheric conditions, water vapor is continuously generated by evaporation and eliminated by condensation. Atmospheric water vapor is found near the earth's surface and may condense into small droplets of liquid and form meteorological phenomena such as fog or mist. Humidity is the amount of water vapor present in the air.

The main objective of this project is to compose the plan of the apartment around the mapping of the interior humidity to create a more sensual relationship with the space. This spatial solution will allow reintroducing a diversity of climatic zones into the modern apartment since it will be drier in one area and moister in another. Instead of simply responding with a technical solution to the problem of air humidity, we wish to transform this problem poetically into a sensual landscape. The apartment will be closer to a natural landscape, with some kind of variation in the latitudes of relative humidity. The apartment becomes an indoor geography, stretched between a miniature desert and an indoor tropical jungle, between a dry zone at 30% relative humidity and a humid zone at 90% relative humidity through which the occupants can wander freely.

The public space of the building extends from the basement to the roof as a continuous movement of air from lower to upper levels. Since moist air is lighter than dry air, the upper level of the public space will be moister than the lower one. The common spaces will be placed on this vertical vapor geography. The areas for drying clothes after washing will be placed at a higher level of the building, while the lower, moister floors of the public areas can be used to store potatoes, apples or wine. Like in olden-day houses, with their dry cellars and humid basements, used for specific storage depending on the type of food or article, potatoes and preserves in the basement, cereals and dried fruit under the roof, our public space will contain not only the functional distribution, stairs and elevator, but also a variety of common spaces where social life can take place.
The different humidities of these public spaces will be used as a kind of collective refrigerator or shared pantry, scaled up to the size of the building.

Different thermal conductivities

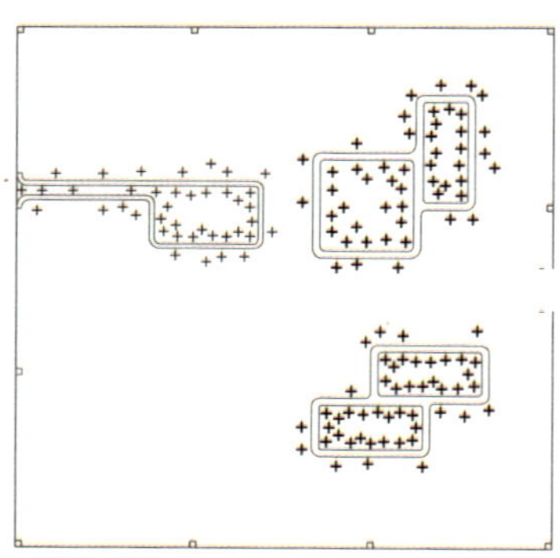

The design of this building is based on thermal conductivity. The main building materials have a very low coefficient of thermal conductivity (less than 1 W/ (m.k); wood and hollow filling fiber insulation), with the exception of a few parts, some made of concrete and others of steel. These two different materials, with different coefficients of thermal conductivity, will provide from the environment through the insulated glass of the façade heat through the sun's radiation or the coolness of the dark of the night.

In physics, thermal conductivity, k, indicates a material's ability to conduct heat. It appears mainly in Fourier's law of thermal conduction, the transfer of thermal energy between neighboring molecules in a substance due to a temperature gradient. Thermal conductance is the amount of heat that passes in units of time through a plate with a particular surface and thickness when its opposed faces differ in temperature by one Kelvin. Thermal conductivity depends on many properties of a material, in particular its structure and its temperature.

The main purpose of this project is to reintroduce different climatic qualities to the modern apartment in different places and at different times, as was done in the old house depending on the building material and its exposure to the sun. With this in mind, it could become most comfortable in summer to sit near or inside the concrete rooms, which retain the coolness of the night during the day. Or conversely, it could become very pleasant in winter to sit near the steel parts heated by the sun's rays on the façade.

The building's collective interior space, which houses its common program of stairs, elevator, laundry, storage, drying room, etc. Some spaces are made of steel, others of concrete. The program will be implemented according to the climatic qualities of the two materials. Inside the concrete-cement spaces is the common wine cellar, the laundry room, the bicycle room, for example. Inside the steel rooms, we will find clothes dryers, corn storage and strollers, for example. Instead of keeping all our items and provisions in our individual refrigerator and cupboard, we propose to use the natural climatic conditions of the public spaces to store them.

Thermal interlocking

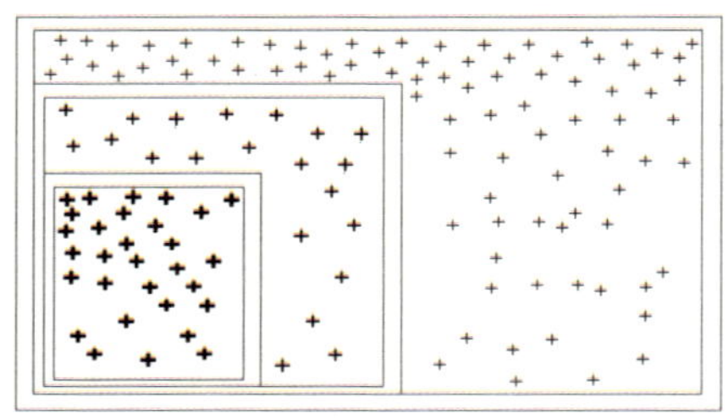

The project proposes to completely redefine access to the school site. To do so, our idea is to create a new single entrance to the three differentiated programs: the new sports hall (from which we also separate the program of the cafeteria, which can be opened independently of the sports hall), the old sports hall, and the school itself. A single foyer unites and distributes these three programs, each of which can operate independently of the other. This foyer opens onto a pedestrian access to the south-west which is extended by a covered area, created under the extension we propose from the west wing of the existing school building. The changing rooms of the existing sports hall and their independent entrance to the west are preserved as they are. However, they are rendered directly accessible from the new entrance by a covered way, which runs along the new sports hall to the south, behind the existing school building.

Our new sports hall project draws on the building's current thermal insulation requirements according to the Minergie standard, on which it focuses in structural and economic terms by increasing the thickness of the insulation from the MINERGIE construction standard on the periphery to the even more stringent MINERGIE-P standard at the heart of the building. Our building consists of different thermal and programmatic "bubbles", nestled in each other, from the most thermally protected, innermost, to the outermost. Each bubble consists of a load-bearing wall and 16 cm-thick thermal insulation. These 16-cm layers of insulation are therefore added bubble by bubble the further you go toward the building's interior, by the provision of a first 16-cm layer for the periphery, then of 32 cm for the second layer, and finally of 48 cm in the heart of the building. The first layer, the peripheral bubble, has 16 cm-thick insulation, that is, a maximum U-value of 0.23W/m2K. In this first bubble, we find the spaces that require less heating, such as supply depots, the entrance hall, the corridors as well as the spaces where the body is highly active physically and produces its own heat, that is to say the sports hall itself, which can remain in winter at a temperature of 14°C. Within this first bubble, there is a second bubble which contains the spaces requiring a higher temperature, at 18°C or 19°C, like the cafeteria, the reception hall, the toilets. This second layer forms a thermal insulation envelope with a thickness of 16 cm which, added to the first, therefore attains at a total thickness of 32 cm of insulation, that is, an even more efficient U of 0.12 W/m2K. Within this second bubble is the third bubble, hotter still, at 22°C, whose thermal insulation of 16 cm is also added to the two previous insulations, to reach at a total thickness of 48 cm, that is, a maximum U of less than 0.1W/m2K, thus meeting the requirements of the Minergie-P Standard. This last "bubble", the hottest and the most protected from the outside, is suited to changing rooms and showers, where people are undressed.

Our project renders these three thermal layers, embedded in each other like a set of three Russian nesting dolls, structural, visible and habitable, its volumes rationalized for the sake of economy and thermal precision. This principle of thermal composition becomes the principle of architectural composition whose perception is reinforced by slightly offsetting the layers and their openings relative to one another. Some openings, like the ones that lead to the heart of the building, in the most insulated layer, housing the showers and changing rooms, thus appear as extrusions of one layer through the other two until reaching the outside. This offsetting of the layers relative to one another blurs the perception of the outside, since the second layer will appear as the outside of the first layer, while the third layer will appear as the outside of the second layer and the outside of the building will be the outside of the third layer.

The choice of materials contributes to the poetization and to the perception of these three thermal bubbles through the use of three different concretes and three different insulation materials. Thus, in the building's core, for the layer of insulation around the places where people are undressed, we provide sheep's wool-type insulation and a softer concrete containing sedimentary aggregates. In the intermediate layer, we propose a more solid concrete with eruptive, granitic aggregates and cellulose-type insulation of vegetable origin.
The last layer, on the periphery, forming the façade, will be made of concrete with highly resistant aggregates of metamorphic origin and rock wool insulation.

Form and function follow climate

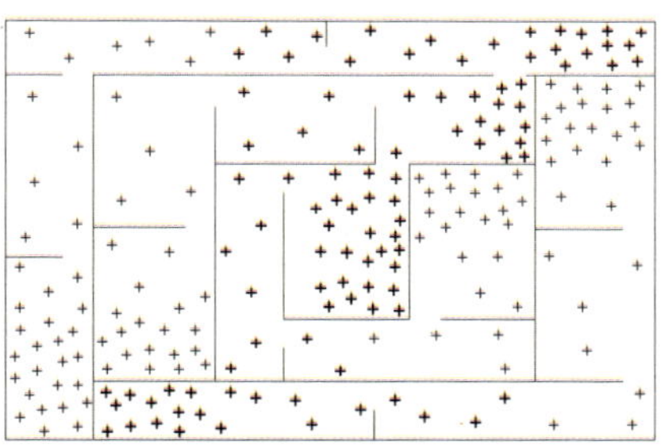

With modern times, two major theories clash concerning the relationship between program and architectural form. The first, that of heroic modernity, states that form must follow function, according to the famous phrase by the American architect, Louis Sullivan: "Form follows function". What was asserted then was the utilitarian role of architecture as a spatial response to a program, in terms of surface areas, spatial and organizational relationships, and constructive techniques. This first position was affirmed at the beginning of modernity in contrast to the symbolic and decorative status of 19th century architecture. Rationalist, functionalist and universalist, this architecture tends to give a "just form" with clearly defined functions and practices, from the social, technical and ergonomic points of view. What we see then is nothing other than the spatial expression of the program itself, with no semantic or affective excesses. The architecture here is indeed the spatial response to a program developed upstream, for immediate inhabitation.

It was from the 1960s onwards that there was opposition to this modern ethos with Louis Kahn's "Function follows form". Here, it is explicitly a criticism of the program as an univocal response to needs or activities. Because these needs can change, because the history of architecture shows the permanence of the architectural form in contrast to the program and function that can change over time, or even be totally transformed. Taking this hypothesis as his basis, Louis Kahn developed a new method of composition, where architecture is no longer the expression of a program but of hierarchies. The latter, in a more abstract manner and without really establishing functions, define a set of spaces that establish a kind of society of spaces, where each space is qualified in relation to the other spaces, establishing a structurally rigid but programmatically open framework.

Our purpose here is to prolong this release from all predetermined form and function. What we seek to do is to work on a new organization of space where function, but also form, could arise spontaneously from architecture as a climate. What matters to us today is to work on the matter of space itself, on the density of its air and the intensity of its light, to offer an architecture like geography, or, rather as an open, moving meteorology comprising different climates, different atmospheric qualities that one inhabits according to the activity, according to the time of day, according to the season, freely. Our architecture works on two essential parameters of the development of space, that of light and that of temperature, according to two dimensions, the horizontal and the vertical. Horizontally, the plan is organized in the manner of a set of Russian nesting dolls, from the periphery towards the center, between the lightest and the darkest, in a succession of glass walls that each filters out 25% of the brightness. Vertically, the space is designed from bottom to top, from the coolest to the warmest following the principle of natural ascension of air masses according to their density and temperature. The architecture finally takes shape here as a combination of four luminosities with four temperatures yielding sixteen climates, varying according to the time of day that can be enjoyed by everyone according to their activity, their work tools, their wishes. Sixteen rooms on the same surface but with as many different climates, which unfold horizontally and vertically. The choice is then made freely between the cooler places where one can be more active, with weaker, more uniform lighting like for the 19th century painter, complete darkness in the center to work with video images: as many behaviors and atmospheres to discover, experience or enjoy.

Impressionist Greenhouse

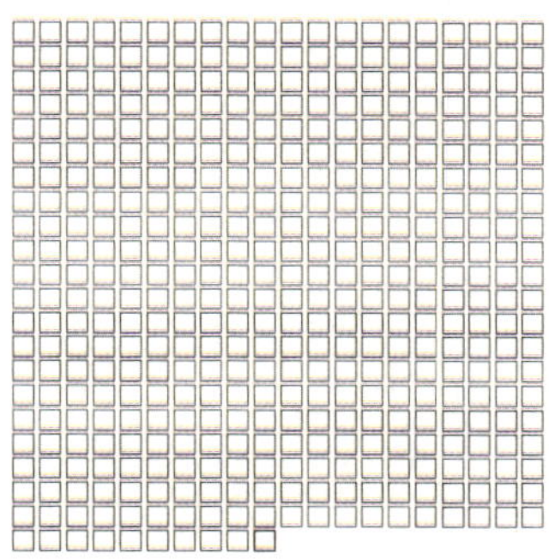

One of the keys to modern and contemporary art is the notion of dissociation, decomposition, analysis (in the chemical sense of the term) and synthesis (artificial), all methods and processes derived from scientific culture. Because with the breathtaking scientific progress initiated in the 19th century in the fields of chemistry, physics or biology, the status of the work of art has changed. It is no longer a question of art representing the real, of imitating it, of composing with it, but rather of decomposing it, of disintegrating it, of pulverizing it, of vaporizing it into a multitude of microscopic particles of colors, of sounds, of words disunited from the whole, atomized. The real is no longer a visible, tactile macroscopic block, it expands, diffracts, opens up to the infinitely small and to the invisible. It is this method that is initially found behind impressionism in painting, with Monet, in music, with Debussy, or in literature, with Mallarmé. Each time, they seek to break down the "whole" into basic particles, to pulverize the blocks of reality to then vaporize them into strokes of colors, fragments of sounds and words. In the 1950s, the "Nouveau Roman" by Alain Robbe-Grillet or Nathalie Sarraute also arose from a deconstruction of literary language in which narrative is emptied of all psychological intention to become pure sensations, literal and objective descriptions of the world with no further subjective, globalizing and unifying intention. Closer to us, the spectral music of Gérard Grisey or Tristan Murail no longer com-poses, but de-composes or, simply, sets the sound by disintegrating the instrumental sounds, reducing them to their essential components, then, ultimately, recomposing them, or rather using these elements to synthesize new aggregates. Dissociating the real, breaking down the common places to recompose them otherwise, in a different order, are compulsory moments of the reformation and evolution of forms at the same time as those of society and of techniques.

Our work as architects is part of this "tradition", in a quasi-chemical dissociation of space into elementary particles, into wavelengths, humidity levels, light intensities and heat transmission coefficients, for example, but also into hormone secretion rates, kilocalories and nanometers. The broadening of the field of the real produced by increased scientific knowledge modifies the field of art, which shifts into new dimensions, slides onto other phenomena, arouses other perceptions. We are now interested in forms that are no longer composed together to form a whole, but dissociated, exploded into fragments of the real, into particles of the sensitive. Our project for the Greenhouse of South Bristol Skills Academy thus proceeds by decomposing the real, a real no longer "perceived according to human physiology" that we can experience via our five senses, on the visual spectrum between 380 nanometers and 780 nanometers, but real commensurate with the plant, its physiology, reduced to the electromagnetic spectra of photosynthesis, at their different wavelengths. We thus very literally accept the idea of a greenhouse as an environment that does not respond to human physiology, but only to plant physiology, with the goal of accelerating, conditioning and modifying the existence and growth of certain plants. Our project therefore departs from a visual world disaggregated into just the wavelengths absorbed by certain plant pigments allowing photosynthesis: chlorophyll a, chlorophyll b, anthocyanin, betacyanin, carotene, lutein, and lycopene. The visible spectrum of white light is therefore broken down into wavelengths from which only the seventeen, corresponding precisely to the light absorption range of plant pigments are retained. The space of the greenhouse is thus a dissociated real according to these plant-specific wavelengths and then vaporized on the site according to certain diffraction factors, certain orientations relating to the sun's path and to the different tones of light according to the time of day.

Luminance

The "Luminance" exhibition architecture at the Luma Arles Mécanique Générale thus proposes to work on emitted and reflected light intensity values, to approach the background of the works of art as the mediation of luminosities, the production of spaces with differentiated light intensity values, between the darkest and the brightest. We propose to accommodate the indoor exhibition space as an existing natural landscape.
Due to its shape, its dimensions, its openings to natural light, the architecture produces a given interior solar landscape, with its always sunnier zones, like the summer tropics, its slightly darker zones, like temperate spring climates, and parts that are always in the shade, like polar winters. The works of art will naturally find their position in these light variations, between video screenings in darkness, old photographs, to be protected from excess light in the slightly lighter regions, installations or contemporary photographic prints in the more illuminated parts. More than a scenography, we produce atmospheres, offering the same freedom of use and appropriation as a natural landscape, with its sunny meadows, its shady forests, its dark caves that suggest, characterize or enable certain actions, certain desires.

Using solar insolation simulation software, according to the actual path of the sun in Arles between July and September, we map the exhibition space according to the average natural light intensities produced by the sunlight penetrating the building through the windows and skylights. We thus obtain a standard map of the variations in natural sunlight in the exhibition space during the day, according to gradients of light intensity ranging from the brightest, in the center and north of the space, to the darkest, in the periphery and in the south.

Our proposed scenography is based on this objectivity of incident natural light. It first maps the space in terms of light intensity, solar insolation, to define 4 major zones, as 4 different light climates, naturally existing, according to a gradient ranging from the brightest, at 100% incident solar insolation (summer or equatorial climate), to less bright, at 66% incident solar insolation (spring or Mediterranean climate), to less dark, at 33% incident solar insolation (autumn or continental climate), to the darkest, at 0% incident solar insolation (winter or polar climate).

The distribution of natural light is doubled up in parallel by electric lighting in the evening, whose number of projectors is maximized in the equatorial climate (3 projectors per beam), and gradually attenuated (2 projectors, 1 projector) then no projectors in the polar climate.

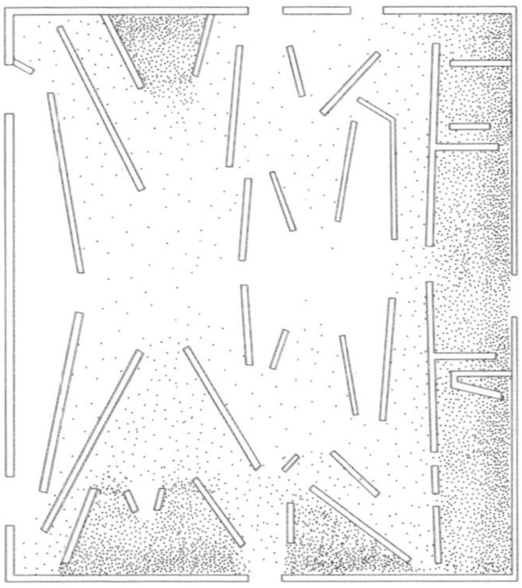

To better distribute the light in space and attenuate shadowing and vertical dramatic effects, we propose to reflect or absorb the incident light coming through the skylights on the ground and the picture rails, whose color values that will cover them will range from white in the bright areas (equatorial climate), to light gray in the slightly less bright areas (Mediterranean climate), to dark gray in the dimmer areas (continental climate) to black in the very dark areas (polar climate). To do so, we propose to cover the ground, following the lines of natural insolation, of different values (white for the equatorial climate, light gray for the Mediterranean climate, dark gray for the continental climate, and black for the polar climate), either with white, light gray, dark gray or black carpets, or by self-adhesives of different sizes and values, in order to accentuate the light in the equatorial climate by completely reflecting it (100%) by means of the white flooring, partially reflecting it (66%) in the Mediterranean climates by means of a light gray colored floor, very slightly reflecting it (33%) using a dark gray floor cover in the continental climates, and not reflecting it at all, using a black floor cover, in the polar climates. The picture rails of each climate will be painted with the same values, white in equatorial climates, light gray in Mediterranean climates, dark gray in continental climates. Their height will vary according to the climates, lower for the equatorial climate to let in more light, rising progressively higher in the climates becoming darker to reduce the incident light.

By following the cartography of natural light in the exhibition area, our scenography proposes to construct four climates, as four different light levels: one extremely bright, white space, one light space, one dim space and one black space. These spaces lend themselves as natural landscapes that can be freely appropriated by curators, by placing the works of art according to their curatorial preferences and according to the requirements for preserving the works. Thus, the white space of the equatorial climate can accommodate installations, reproducible photographic prints or events, for instance. The Mediterranean climate can host digital images. The more dimly lit spaces, in the continental climates, will display antique photographs that must be shown under low light to preserve them. The very dark spaces, the polar climates, can offer video screenings.

The primary mission of a museography or scenography intended to exhibit art is to form a neutral and objective background, a backdrop on which characterized and subjective works of art can stand out in the foreground and be enhanced as figures. In simpler terms, architecture must retreat leaving the work of art at the front of the stage. Therefore, the scenography must produce a background landscape, having no particular form or materiality, so as not to disturb, distort or distract from the reading of the form and the materiality of the work of art as a foreground figure that stands out clearly and precisely. The modern "white cube" is the quintessence of this as it dematerializes a rather formless background into a homogeneous and uniform, almost abstract whiteness which in reality becomes a high, single light reflection value, an imageless mirror that diffuses and diffracts, directing all the light toward the work of art, without absorbing any wavelength. By means of continuous illumination, with no dramaturgic effect, the "white cube" produces a homogeneous, general white light, with no shadow, no specificity, allowing the work of art to express only itself, its own colors, light intensities, textures, its own forms, its own materiality, its own image, without confusion, with no disturbance from the background. So, a museography or an art scenography must be first and foremost the design of a luminosity, without form, without texture, like a bath of homogeneous light that reveals the intrinsic characteristics of the works of art, their forms, their material, their color, their meaning. The work of the museographer or the scenographer is, therefore, specifically to design this light, an ambient luminosity. It is first of all a question of characterizing the light emission from natural or artificial sources, the illuminance, the density of the photon flux, the radiation, the irradiance, the degree of insolation, the electromagnetic spectrum emitted, the wavelengths and their respective intensities. It is then necessary to qualify the reflection of the incident light on the ground, on the walls and the picture rails that is to say to characterize their values of reflection, emissivity, absorption, their albedo, diffraction, their spectral and diffuse reflectance. This design of the emission and reflection of light is carried out according to two main criteria: one of perception, visual apprehension of the work by the visitors, and one of conservation, of the preservation of the work of art against the physical and chemical damage that can be caused by light. First, the perception of a work, its presentation to the visitors, will depend on the type of work, on a perceptual and/or curatorial and subjective choice: a sculpture can be on display in full light or in the dark, a photograph in the light or in the penumbra to bring out its own brightness. In this register, habits and certain automatic reflexes are in place, such as exposing contemporary art under bright lighting, homogenized in space, whereas ancient art is entitled to medium lighting centering on the work, while a video projection will tend to be screened in the darkness of a "black box". Secondly, the physical and chemical preservation of a work of art has an important role in defining the light spectrum and its intensity, or even the length of time it is on display. In fact, the light incident upon a work of art damages the materiality of the artwork, its colors, due to the chemical reaction of light when it touches the work of art. Thus, depending on its medium, as of a certain light intensity and according to certain wavelengths, a work of art will deteriorate, lose its colors, undergo the disintegration of its materials. Thus, a medium of organic origin, like paper or linen, as well as a photographic print and its colors, pastel paints or watercolors, for example, will begin to deteriorate as soon as the intensity of the short wavelengths, in the ultraviolet in particular, exceeds 30 lumens as well as at an intensity exceeding 50 lux in the visible wavelengths, between 400 and 760 nm. Conversely, an oil painting can be exposed to more intense lighting, up to 200 lux in the visible and up to 80 lumens in the ultraviolet. Other inorganic, metal, stone or glass works of art can be subjected to higher light intensities.

A museum or an art scenography can therefore be understood and designed solely according to these values of light emission and reflection, in order to produce a variety of spaces, in the form of differentiated "baths" of light, between 30 and 200 lux, between 30 and 80 lumens for reasons of preservation, between 0 and 10,000 lux for reasons of their presentation.

Different light temperatures

The architecture of a contemporary art exhibition space must above all provide a neutral background, a homogeneous background, without form, color, shadow or texture. It must be unremarkable, barring that it must produce some ambient light to highlight the works of art that then stand out as unique figures in the foreground.

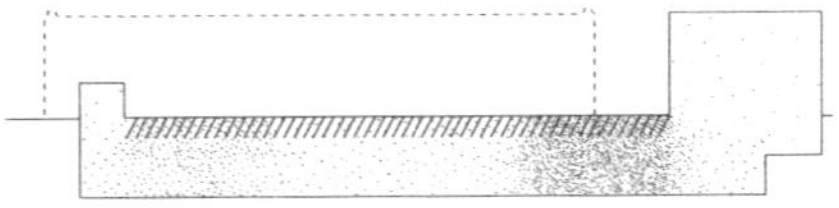

Avoiding shadows, avoiding backlight, avoiding glare, avoiding the drama of the horizontal and the vertical, reducing the ambient brightness and allowing the light to achieve a homogeneous bath of light without a source or any reflection is the work of the museum architect. Our project proposes to combine two sources of continuous and indirect light; one is the continuous and homogeneous natural light of the ceiling, while the other emanates from the ground, creating a soft LED light to erase shadows on the ground and enhance the work of art. For a museum, specific light intensities are desired, with values of between 150 and 250 lux. The light should be white, with a temperature of between 4,000 and 8,000°K. Our project proposes working with these subtle variations in the intensity and temperature of light to create changes in light found naturally in Peru, from the neutral light of the coast, through the cold light of the Andes, and finally to the warm light of the Amazon. The light in the gallery is white. But should this light be slightly more orange and less intense like at the seaside or in areas far from the equator, or slightly bluer and more intense, like in the mountains or in latitudes close to the equator? Sunlight initially has a spectrum that contains more short wavelengths (violet and blue between 450 and 480 nm) than the light that eventually reaches the earth. The oxygen and nitrogen molecules in the atmosphere gradually block the shorter wavelengths of light by optical diffusion, giving the sky its blue color. When light is forced through a greater thickness of atmosphere, it will appear less blue and more orange as it approaches sea level. In reality, white light varies slightly between red and blue, between warm white and cold white, which photographers describe in terms of "color temperature", between 2,000°K (warmer and red) and 10,000°K (cooler and blue). Sunlight loses its blue wavelengths as it passes through the atmosphere. Therefore, high up in the mountains, at noon or in equatorial countries, the light will be colder (bluer) at around 7,000°K. In these locations, the light passes through the atmosphere at its thinnest point. However, in locations where the atmospheric layer is thicker, like at the seaside, in Nordic countries, or late in the afternoon, it will be more orange, at around 4,000°K. From the point of view of an art curator, we understand that a work of art presented in the natural light of Lima in Peru seems more neutral than in the pinker light of the Amazon. And it will appear bluer if it is presented at the altitude of Cusco. It is by taking these fine variations in color and brightness into account that we propose to build for Lima Art Museum, rather than simply working on the shape of the museum in the style of early 21st century architecture. Due to its relief and geography, Peru enjoys an exceptional diversity of climates and light: the intense cold light of the mountains, the orange light in the Amazon, and the more neutral light of Lima, uniformly filtered by clouds. In the linearity of the exhibition space, we propose to organize a journey through Peru's luminous geography, from west to east, starting from the Pacific coast, passing through the high altitudes of the Andes, descending slightly in the Altiplano, and finally lower still, into Amazonia. We transcribe this movement in the grand gallery through variations in light intensity and color temperature. We enter the southern gallery bathed in the more neutral light of Lima, with a color temperature of 5,000°K, and a low light intensity of 150 lux. Then, moving northward, we reproduce a climb from sea level to the summit of the Andes Cordillera, where we find a colder color temperature of 8,000°K and a maximum luminous intensity of 250 lux. Continuing north, we then descend to the Altiplano, a slightly cooler color temperature, at 7,000°K and slightly less intense, at 210 lux. We then descend further, into Amazonia, moving to the end of the gallery, in the north, where the color temperature will be the warmest, at 4,000°K and with a higher average light intensity, at 170 lux.

Climatorium

The Climatorium, a building of 3,500 m2, is the visitor center of Taiwan's Taichung Central Park. It is located in the middle of the park, on its eastern boundary. The program consists of a cafeteria, an information desk, toilets and educational spaces and exhibitions on climate change. These include a public reception room opening onto 3 climatic rooms each of some 200 m2 to which

3 themes are attached: heat, humidity and pollution, themes that also prevail in the general make-up of the park. The actual architecture of the Climatorium, its constructive mode, assumes and reveals architecture's primary climatic role and mission. It thus consists of 4 physical layers placed one inside the other, 4 filtering climatic planes arranged in a concentric climatic gradient, from the outermost and most natural to the innermost and most artificial, from the most variable and uncomfortable to the most permanent and comfortable. Each layer tackles a particular mission in terms of climate mediation, the debilitation, the waning of the natural climate.

The climate of Taichung (24°8'48.84"N, 120°41'2.04"E.) is indeed hot and humid, classified Cwa (humid subtropical climate) according to the Köppen classification. In December, the coldest month in Taiwan, the average daytime temperature is around 24°C, with peaks of up to 31°C in recent years. The relative humidity of the air is on average quite high throughout the year. November is the driest month, with relative air humidity at 60%. In summer, the temperature is highest in August, with a daily average of 33°C and peaks of up to 38°C in recent years. The average relative humidity of the air is 80% in summer.

Because humans are homeothermal animals, we must maintain our body temperature at 37°C because the enzymes necessary for the biochemical reactions of human metabolism, present in their billions in our body, can function optimally only at a temperature of between 35.5 and 37.6°C. According to some speculations by chronobiologists, the sub-Saharan origin of humans, 2 million years ago, on the highlands of East Africa where the average temperature of the air remains constant throughout the year around 22°C, would explain the suitability of the human body to an outdoor temperature of 22°C, colder than the body, necessary to lose by conduction and convection on and through the skin's surface the excess heat that it produces continuously. This temperature of 22°C establishes the upper limit of the thermal neutrality zone which is situated at between 20 and 22°C. Any variation in the outdoor temperature below or above 20 to 22°C will increase the basic energy expenditure to combat the risk to the body's core temperature of falling below 35.5°C or rising above 37.6°C, which we call thermoregulation. And of course, the greater the difference, the greater the body's effort to warm it up or cool it down. When the temperature of the air outside drops below 20°C, then the body loses too much heat to the outside air, and must fight against such losses by thermogenesis, by producing more internal heat, by shivering, through vasoconstriction, by reducing the diameter of the veins to limit the exchange of heat between the blood and the cold air via the skin, which we call thermogenesis. Conversely, when the outside air temperature exceeds 26°C, the body overheats because it can no longer evacuate its excess heat to the outside air. The body then reacts according to a physiological process called thermolysis which makes it possible to reduce this internal heat, by increasing perspiration (the forehead and the armpits sweat abundantly), by vasodilation, that is to say by exposing more vein surface to the outside air in order to increase blood heat loss, which also leads to an increase in heart rate. Perspiration, a constant means of heat dissipation, is prevented if the air is already saturated by water vapor, which is the case in a tropical or subtropical humid climate like Taiwan's. This is why dry heat is more comfortable than humid heat, because in dry heat, the human body can make full use of the phenomenon of perspiration to cool down. On the contrary, in humid heat, the air cannot accept any additional water vapor, which prevents the perspiration by the human body from becoming vapor, from passing from liquid to gaseous phase, cooling by evaporation, which should lower the temperature of the air and of the skin during the physical change from liquid to gaseous phase, which removes energy from the air resulting in a decrease in body temperature.

It is therefore because of this homeothermal condition of the human body and its suitability to a sub-Saharan climate with an outside air temperature around 22°C that humans require an outside air temperature at between 20°C and 26°C allowing us to maintain our body temperature between 35.5°C and 37.6°C. When the outside air temperature drops below 20°C or increases above 26°C, humans combine the means available inside their body, the various thermoregulatory mechanisms, and the means outside the body including clothing, shelter and migration. Architecture is therefore not autonomous and cultural. It actually falls within the range of physiological means to keep our body at a steady 37°C. It is one of the responses, along with food, clothing and migration, to a large drop or increase in body temperature, alongside the internal mechanisms of vasodilation, sweating, muscle contractions or catecholamine secretion. To keep our body temperature constant, various natural or cultural, microscopic or macroscopic, biochemical or meteorological, dietary or urban planning actions are implemented. In this original thermoregulatory mission, architecture thus appears as a larger form of vasoconstriction, or conversely, digestion appears as a slightly smaller variant of architecture. Because ultimately, architecture is nothing more than an augmented form of bodily thermoregulatory mechanisms, an augmented, exogenous and artificial form of thermogenesis and thermolysis.

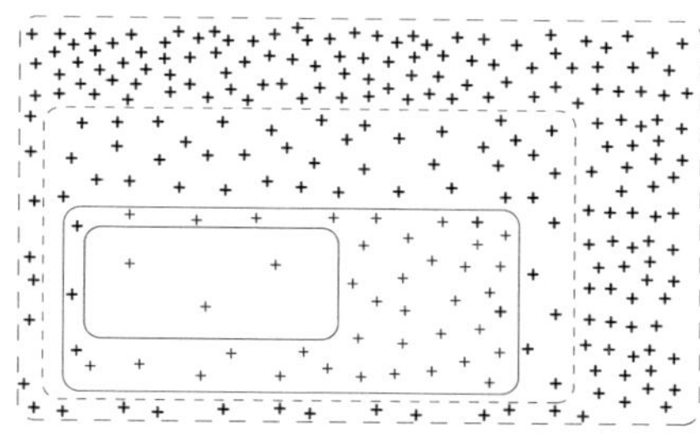

It is thus understood that the climate of Taichung is most often uncomfortable because it is too hot for human metabolism, because the air temperature varies on average during the year between 24°C and 33°C with regular highs of between 31°C in winter and 38°C in summer, and a high degree of humidity. In Taiwan, more often than not, the air temperature is too high and it is too humid for the human body which is naturally made for a cooler environment, that of the cool altitudes of the African highlands of the Olduvai Gorge, at an altitude of 1,400 meters. In Taichung, at an altitude of 100 meters above sea level in a warm and humid climate, the human body is constantly under stress to evacuate the excess heat it produces that the already saturated outside air cannot receive. It must then avoid hot flashes, sun exposure, fatigue, thermal collapse due to a drop in pressure when all modes of thermolysis are exhausted. It should seek the shade, avoid moving too much to prevent muscle overheating, try to take advantage of a draft of air to dissipate more quickly the heat emerging from the blood toward the skin, drink even more to sweat even more and increase the sweat that evaporates on the skin to lower its temperature, get undressed. Air-conditioning that was to arrive en masse as of the 1990s was adopted unanimously as the panacea and its huge success is understandable given this need to reduce thermal stress and, once put in place, given the thermal relief that air-conditioning provides. Today, most homes, offices, trains and automobiles are air-conditioned, making it possible both to lower the temperature of the ambient air to bring it back to the thermal neutrality zone, at around 22°C, and to lower the relative humidity of the air, thus improving the body's evacuation of heat through perspiration. In 2017 Taiwan had 1,090,000 air-conditioning appliances for a population of 23.58 million inhabitants, or 1 appliance for every 23 inhabitants, while in France, by way of comparison, there were 534,000 air-conditioning appliances for a population of 67.12 million inhabitants, or 1 appliance for every 125 inhabitants, a figure that has doubled in France in just a few years.

In Taiwan, the role of a building is to first passively reduce excess heat from the exterior climate, by lowering the input of radiant heat first, by means of a roof and walls blocking out or limiting the sun's rays, preventing rain and runoff water from entering the interior through waterproofing, and removing heat from the outside air through air-tight sealing. Next, the building allows creating a spatial pocket that is thermally insulated from the outside, whose air can then be conditioned, that is to say its temperature lowered to the thermal neutrality zone at 22°C and also to benefit from decreasing its moisture content. All these actions to reduce the excess conductive and radiant heat as well as rain are distinguished by different layers, different materials, different thicknesses, precisely chosen to fulfill a mission: to block out light, block out water and block out heat. Thus, each filtering layer comprising the façade of the Climatorium, each part remains isolated, detached, individualized from each other and from everything: the banishment of water, the banishment of heat, the banishment of the sun, the banishment of public access, each is treated independently, forming its own geometry. The façade is not a simple line, but becomes thick, it thickens, expands, shrinks, distends, contracts, divides, splits into multiple independent lines.

The first layer of the façade, the outermost, is a white aluminum grid whose purpose is to provide a physical but visually transparent barrier against the sun with its large frame made of cubes open at two ends measuring 20 cm by 20 cm. It establishes the perimeter of the Climatorium and renders an initial layer of privacy, allowing opening the complex to visitors during the day and closing it at night thanks to two sliding gates. Between these first two layers there are covered and open, mineral and plant spaces, a courtyard that can be used to meet up or to set out the outdoor terrace of the cafeteria, another covered space as well as a rear service yard. This first physical boundary provides an initial degree of interiority, gradually progressing from the outside to the inside, from the largest to the smallest. This initial layer still allows the rain, the sun, the heat and the wind to permeate but establishes a physical boundary of the property, which is opened and closed according to the opening hours of the Climatorium. Inside this first layer, a second layer unfolds to seal against water, sometimes adhered to the first layer, the grid, and sometimes detached, thus generating external intermediate voids, that is, the courtyard, the terrace, the café and the rear yard. This second layer blocks out the moisture of the soil, the rain, it filters out the sun and air with a waterproof polymer painted white to achieve a high albedo to reflect the incident sunlight and prevent it from turning into heat upon contact with the material. This second waterproof layer, inside, sometimes adheres to the first outer layer or is dissociated from it, it expands, it retracts toward the inside to open up the spaces where it can rain, where the wind can pass, where you can stand pleasantly outside when it is not too hot. Inside this second layer is the third layer, the thermal insulation consisting of recycled PET wool covered in a textile finish. This third layer is sometimes detached from the second, the sealing, to open up spaces between the two, protected from the light and the rain, but still not thermally insulated, without air-conditioning, only ventilated naturally, perhaps sometimes a little too hot but which is acceptable since the functions in place there are of minor importance as regards the time spent there or the use made of it. These spaces, slightly further away from the outside, yet still not completely interior, are given over to the corridors, access points, toilets, storage, equipment rooms: all places of passage, where one never remains long enough to feel uncomfortable due to excess heat. Here, in the shade, the floor, ceiling and walls are entirely covered in aluminum and steel, high effusivity and low emissivity materials that are therefore cold to the touch and do not radiate the heat of the material, whose temperature is the same as that of the air, often too hot, on the visitors. Inside the third, thermal insulation layer, we find the most comfortable spaces, completely insulated from the outside, which are the only ones whose air is conditioned in order to lower the temperature there to around 22 degrees Celsius. This is where you will find the park's information center and climate change interpretation center, coffee shop and offices. Finally, in the heart of the building, inside the three layers, there is the fourth and final layer, the innermost one, made of concrete ensuring the structural role of the building, bracing its entirety. Inside this fourth layer, there are three Climatoriums, three rooms each of which reproduces a particular climate: colder for the first (the Coolium), drier for the second (the Dryium), less polluted for the third (the Clearium), like three ideal climates considering Taichung's excess heat, humidity and air pollution. These three spaces are places of sensual experience, built climates in which you can take a moment's refuge away from the heat, the ambient humidity and the air pollution. They are constructions of ambiances, built atmospheres in which the visitor is immersed in body and senses. Amazing as these three built climates may seem, they actually reflect the techniques and issues commonly put in place since the origins of architecture, whose purpose is merely to modify the natural climate to make it habitable and comfortable for humankind, cooler and less humid in the tropics, less polluted today. The roof (against the rain, against the sun), the walls (against the wind, the cold or the heat), the fireside (against the cold), the candle (against the night), then radiators, electric lighting, air-conditioning, are but different elements established to render the same climatic result, to fulfill the same goal, that of making the natural climate more comfortable: creating shade where the sun is too hot, warmth where the air is too cold, light when it is night, cool when it is too hot.

Gradation of privacy

Nine public toilets, small buildings, are scattered in Taiwan's Taichung Central Park. The construction of these toilets is based on a dissociation of layers of protection and perception, which each slide in respect to the other, from the outside to the inside, creating a concentric gradient of boundaries from the most public to the most private, the most exposed to the most intimate, the most open to the most closed. An initial concrete area is first defined, to shape the flat, practicable flooring, contrasting with the natural soil and grass environment. Further inside, a second area consists of a series of loudspeakers that play music inspired by the sounds of water, acoustically masking the sound of the toilets. A third area is then built vertically, made of a grille bringing a little privacy to the common parts of the toilets. The fourth layer is one of artificial light for evening and night time, comprising vertical mast lights. The fifth area is the roof, which protects from the rain. The sixth area, where the toilets themselves are located, is the most private and consists of a second grille that completely blocks one's view, without obstructing the passage of air. Each of these layers has its own construction system, its own supporting structure, independent of the others.

Inhabiting technical facilities

The mechanization of the building with the advent of modern times in the 20th century separated the notions of space and technology, the latter thus becoming uninhabitable, reduced to ducts, false ceilings, false floors, basement premises, all uninhabitable space; constrained to their functionality. This was not the case before modern times where cooling, heating and ventilation spaces, equipment rooms, found their own qualities, even to the point of being architecture itself. Thus, the dome of the Pantheon in Rome, or the cupola of Palladio's Villa Rotonda, are the spatiality itself of ventilation, habitable ventilation ducts, so to speak, where the rising hot air, according to Archimedes' law, is expelled through the roof, actually shaping the building itself. It is to this reconciliation that we appeal, to make technical installations habitable to the point that they become shared, public places, the communal areas of the new social sciences and humanities building. Thus, the fresh air ventilation duct forms a huge virtual cone in the heart of the building, allowing the fresh air to rise from the basement to the roof (in winter or on very hot days) and, thanks to its conical shape, distribute the fresh air laterally to each floor to renew the air. This broad ventilation duct, which could also be a hollowed out atrium, we propose to cover on each floor with a fireproof concrete grating (concrete is chosen here for its fire resistance and for its thermal inertia), in order to extend the continuity of the entire space of the floor, its free use, interrupted from one side to the other of the building, and to constitute, as a clearance space, a shared space, a public meeting place for all the users of the building, like a thermal agora, a combined ventilation duct and public place, thus regaining certain qualities of pre-carbon architecture. The building's single, wide staircase serving its entire height, is not enclosed (the fire escape stairs are in the two wind towers outside the building) and so fully open, it crosses, in a panoptic position conducive to exchanges, the entire height of the building, on each floor allowing multiple opportunities for encounter, path- crossing, glances and the chance to pause on each story on the slatted flooring of the thermal agora. For reasons of health and comfort, the five floors of the faculty offices where people will spend most time on a daily basis, are built entirely from natural materials (wood, stone, natural wool) free of glue or any chemical binder, without glued laminate, (therefore without releasing volatile organic compounds (VOC) or formaldehyde, both for the supporting structure (timber frame) and for the secondary structures (stone, wood and natural wool flooring, vegetable fiber thermal insulation, glass partitioning, raw earth, wood and recycled wood). Thus, from the 3nd floor, the study spaces, then for all five floors dedicated to the faculties, as well as for the restaurant at the very top, the structural load-bearing system is composed vertically of raw wood poles spaced every 6 meters, with a section of 50 cm × 50 cm (on the 3nd floor), whose section tapers vertically upward, changing to 45 cm × 45 cm (on the 4th floor), 42 cm (on the 5th floor), 38 cm (on the 6th floor), 35 cm (on the 6th floor), then 30 cm × 30 cm for the following floors. The horizontal load-bearing structure is composed of 65 cm-high square timber beams, running every 6 meters, to which 35 cm-high joists are attached every 50 cm. On top, a 12 cm-thick wooden floor supports sound insulation in which the floor heating coils are placed. On top of this, the coating is chosen depending on the material's thermal capacities in terms of effusivity and inertia. Thus, where the rays of the direct sun fall, 6 cm-thick stone, will be placed in order to capture the solar energy, store it, and slightly offset the reversion of heat in winter (avoiding overheating at noon and enjoying it at the end of the afternoon);
while in summer, the stone, protected during the day from the sun's direct rays by balconies, roses and blinds, will conversely feed back the coolness of the night during the day.

THERMAL DECORATION

Thermal tapestry

We propose for the call for the creation of the Cité internationale de la tapisserie in Aubusson a tapestry that evokes the historical use of tapestries. Beyond its esthetics and luxury, over the course of cultural, technical and esthetic evolution, the tapestry has replaced the animal skins used to insulate primitive houses. Tapestries hung on walls in the stone constructions of the Middle Ages and the Renaissance served to protect from the cold coming from the outside, even through thick walls. We therefore propose an updated tapestry in accordance with the modern dwelling, redesigned with contemporary thermal problems in mind. Temperature distribution in a room is not homogeneous but according to a physical principle, that of Archimedes, which means that the hot air will tend to rise because it is lighter, and accumulate under the ceiling while cold air will tend to sink and then stagnate at ground level. These differences in temperature within the same room can be substantial, sometimes of the order of almost ten degrees, resulting in an unnecessary expenditure of heat that will often not even be taken advantage of since it is unattainable beneath the ceiling. This results in stagnant air masses at, for example, 18°C at floor and foot level, 21°C at the level of the torso, 25°C at head height, and 28°C below the ceiling. What happens is that people are sometimes tempted to raise the temperature of the radiators because they feel cold at the feet whereas the problem is only a question of the distribution of heat in the room. We propose a tapestry that balances the temperature differences that might exist in a space. We use the gold or silver thread weaving technique already practiced in the weaving tradition of Aubusson. We replace the gold and silver threads with copper threads to obtain a thermal distribution effect. We weave with copper wire wool, a highly powerful thermal conductor of the order of 348 to 390 Watt meter-1 Kelvin-1. The copper wire weave is designed to be denser at the top of the tapestry so as to capture maximum heat. Then this design is refined as far as the bottom third of the tapestry in order to redistribute the heat evenly. The precision of the weaving will enable precisely quantifying the proportion of copper wire required and consequently the density of the weave. Some of the weave will consist mainly of copper wire to capture more hot air. We offer two versions of the tapestry.

Fermented cinema

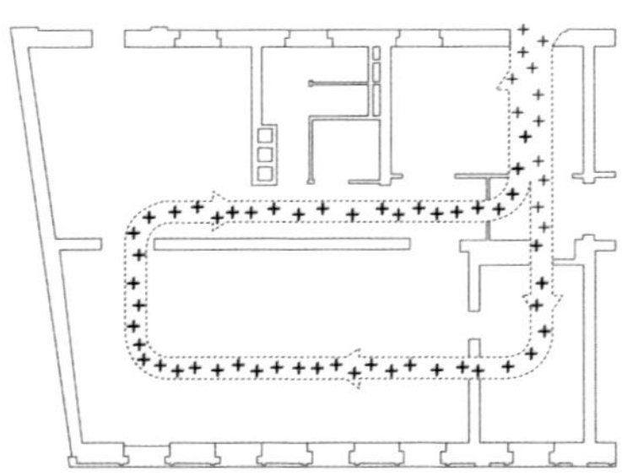

The new environmental concerns linked to our awareness of our responsibility with regard to global warming are changing our relationship with the abstraction and dematerialization of contemporary technologies and information. If modern techniques, such as electricity, electromagnetism or digital technology, and their derived uses (artificial light, the cinema, mobile telephony, internet, etc.) have hitherto appeared to us to be relatively neutral, dematerialized, light and inconsequent, we are now witnessing their thickening and obfuscation. Dematerialization was only visible and we discover that even a simple search on Google consumes energy and releases 0.2 grams of CO2 into the atmosphere. In reality, today everything seems to be gaining weight, nothing escapes the physicality of the sentient or virtual world where our avatar, like our body, also produces CO2. Even wireless, with its non-ionizing radiation could affect health. Carbon and energy footprints are now taken into account at all levels of the world. Chemical, physical and biological exchanges are described between the body, the planet's atmosphere and the most dematerialized technologies. This early twenty-first century is witnessing the interconnection of all things of the world, in which no form any longer exists independently of others. The living and the lethargic are now chained to each other, in infinite ecological cycles of exchanges of matter and energy, and merely switching on a light bulb in one's kitchen now has an impact on the global climate.

Our cinema proposal recovers this new energy and material consciousness. Our aim is to accept today that physical links, transfers of materials and energy, exist between things that we had so far presented separately. We want to consciously build exchanges of energy, matter, food, to abandon the autonomy of the elements. Our project creates chemical and biological links, between things that are usually separate. For what we seek to do is to organize the transformation of energy from one form to another, materialize the invisible flows, reuse wasted energy, organize the interaction between the organic and the inorganic world, between space and body, computing and physiology, culture and food.

Thus, a 35 mm movie screening is no longer the disembodied place of a pure distribution of immaterial images. The images come to the light and we discover them made from light and so the light appears as if made by heat. Moving images consume energy and produce greenhouse gases. For a 2-hour movie, more than 2,000 grams of CO2 are released into the atmosphere, produced by the 35 mm projector which consumes 4 KW/hour of electricity which it converts almost entirely into heat. Because cinema images are electromagnetic brilliance issuing from a projector, they are waves of heat and light that cross the room to burst onto the white screen, on it communicating its colors and intensities, crashing into it to become heat. The projection beam is an electromagnetic field composed of visible and invisible wavelengths, light and heat. Upon contact with the screen, this energy is partly reflected, partly absorbed transforming into heat.

If the transmission of information in the screening room is not energetically neutral, neither is its reception. Watching a movie also involves losing energy through our bodies, even if we are just sitting in an armchair. This loss of physical energy, in the form of the emission of heat at 37°C, takes place to ensure basic metabolic functioning - heartbeats, respiratory movements, hormone synthesis. We can quantify the energy lost during a 2-hour movie at some 120 kcal. When we sleep, this energy expenditure falls to 50 kcal. If we do sport it increases to 2,000 kcal. It is this energy lost during the day that we compensate by ingesting at least 1,500 kcal. Here, what interests us are these hidden energy links between a movie screening, the loss of calories as a viewer, and then food. It is a cycle we wish to highlight, where esthetics and physiology, culture and food overlap. It is also a typological questioning about the notion of program, where a cinema is today coupled with a bar selling popcorn and soda pop. To realize the energy, as in a scientific experiment where a dye is added to a substance to trace and distinguish it from others, we suggest giving energy a white color. We suggest realizing this plastic characteristic using milk, whose different forms punctuate the transformation of energy during the movie session. White is the color of the screen, required to reflect the colors and bright shapes. We propose a screen as a thin translucent cavity that we will fill with 9 liters of milk at the start of the screening. During the two hours of the movie, the milk will be heated up by absorbing the heat from the projected images and start a forced fermentation process. At the end of the screening, we will empty the screen of these 9 liters of milk to fill the yoghurt makers in the adjacent room.
Then, these 9 liters of milk will be used to make 60 yoghurts of 150 ml each containing 120 kcal. A yoghurt will be given to each spectator at the end of the movie, so that they can recover the 120 kcal lost through metabolism during the spectacle.

In reality, this project is very close to current movie practices, where the screening is accompanied by the consumption of popcorn or candy bars, where the projectionist is also a cook, where culture and food mix in the same energy cycle. In this process, we find certain pre-modern forms of thought, where morality and physiology intersect, where esthetics and biology, synesthesia and irrationality construct a strange but nonetheless premonitory, empirical, ecological system of current exchanges of energy. While milk is chosen because its white color becomes the screen, its nutritional value also interests us. It is also, perhaps, even a certain irrational echo of wet nurse nourishment in the early nineteenth century when it was thought that the morality of the wet nurse could influence the quality of her milk and be transmitted to the infant and influence its morality... In this case, sixty yoghurts that will have fermented during the screening of a movie might become a fermented form of this movie, so that we can taste it afterwards.

The 60 yoghurts are fermented by yoghurt manufacturers in the room adjacent to the exhibition hall. These yoghurts are either immediately consumed or stored in refrigerators. A very simple pump cleaning system, as used in dairies, using water and an alkaline sodium hydroxide cleaner like Ecolab P3-mip, ensure the hygiene of the system and the cleanliness of the screen and pipes.

The projection room is completely matt black with the exception of the milk on the screen, the other room is totally gloss white, partially rendered in silicone and glass.

The project is set in the context of sustainable development. A dual-flow air renewal system is installed to reduce the energy spent on heating by a factor of four. The flow of the renewed air follows the passage of energy from one room to another. It allows linking the different moments of the session. So, the fresh air coming from the outside is preheated in winter in the cinema and then passes into the cafeteria and then into the toilets before exiting, preheating the incoming air in a heat exchanger.

Artificial spring

If we want to discover the essence of architecture, it is finally to our "endothermic" condition that we must return, to this necessity we have to maintain our body temperature at 37°C. The very existence of architecture is due to the enzymes required for the biochemical reactions of human metabolism. These molecules, which are present in our bodies by the billion, can only function optimally at a temperature of between 35 and 37.6°C. Humans must, therefore, keep their body temperature constant, regardless of the outside temperature. To do so, they combine internal bodily means, the various physiological thermoregulatory mechanisms, and means outside the body, which include clothing or constructed shelter. Architecture is therefore not autonomous. It actually belongs to the range of means to keep our body at around 37°C. It is one of the responses to a large drop or increase in body temperature, alongside vasodilation, sweating, thirst, and muscle contractions, for instance. These responses are applied alone or in combination. They develop from the natural to the artificial, from the microscopic to the macroscopic, from the biochemical to the meteorological, from food to urban planning, between physiological determinism and pure cultural freedom. In this mission, architecture appears as a larger form of vasoconstriction, or conversely, food appears as a slightly smaller variant of architecture. Because architecture is nothing more than an augmented form of bodily thermoregulatory mechanisms, an augmented, exogenous and artificial form of thermogenesis and thermolysis.

From an anthropic point of view, when we say we are too cold or too hot, we find the cause outside of us, in an inadequate external climate, at atmospheric level. And we try to make this outdoor climate habitable and comfortable by correcting it, which is the origin and the intrinsic mission of architecture. In reality, the first signs of architecture are physiological and totally interior and autonomous, sweating, if it is too hot, or shivering, if it is too cold. These are the first responses to an increase or decrease in body temperature due to an unfavorable thermal environment. Then, in the simplest way, straight afterwards, come the most rudimentary gestures towards the exterior, of drinking if it is too hot to lower body temperature through evaporation, or eating if it is too cold to start the process of burning nutrients that will produce heat in the body.

After these endogenous corrective measures, if the body nevertheless fails to compensate for the excess cold or heat of the external environment, the range of geographical corrections comes into play. The first corrective action is a movement; migration or transhumance, in other words moving, changing place, passing from the cold to the warm, positioning oneself in the sun or in the shade. The second action is to get dressed or undressed, to wear white that reflects heat or, on the contrary, to wear thick clothes that insulate from the cold. The third action is to artificially build shade and freshness or, on the contrary, shelter, with no air movements and heated. These exogenous measures, which are taken in the outdoor world, are but an out-of-body projection of the phenomenon of thermogenesis when it is too cold, or of thermolysis when it is too hot. To paraphrase Vitruvius, the architecture of cold countries and winter appears as an augmented, exogenous thermogenesis, outside the body. And the architecture of warm countries and summer presents itself as an exteriorized thermolysis, artificially correcting what nature has rendered uncomfortable.

Thermogenesis is the increased production of heat in the body in the event of cold. It is consistent with doing sports activity or the ingestion of food. It consumes a great amount of energy. Thermogenesis triggers shivering, it stimulates metabolism and various hormones, which has the effect of warming our body from the inside to maintain it at around 37°C. The basis of thermogenesis is food, mainly proteins, that is to say meat or fish, as well as sugar. Most of the food we consume serves this purpose. The digestion of proteins, their combustion in the Krebs cycle in particular, causes a great deal of heat which raises body temperature.

Eating is therefore an inner form of the same process of thermal reactions that drives us to build houses in cold climates. Likewise, the typical constructions of hot countries, riads, porticos, thick walls, are externalized forms of thermolysis, this bodily function that serves to dissipate excess heat through vasodilation or sweating.

It may seem surprising to wish to return in such depth to the understanding of the reasons for and means of architecture, but the problem of global warming has suddenly brought architecture's climate mission to the fore: to help manage energy and resources. Today's architects have a duty to explore the scope of architectural means that point toward sustainable development. They need to understand how they can restrict energy consumption and the production of greenhouse gases and, as we now know, this is primarily a question of reducing the energy used in buildings for heating or cooling.

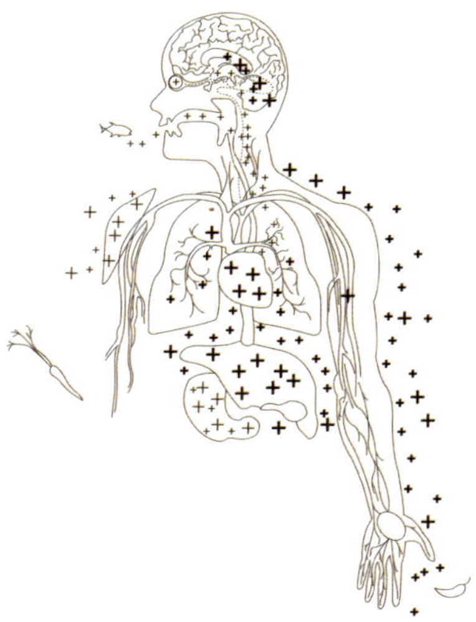

Today, we know the measures to be taken in the field of urban planning, where the main emphasis is to densify cities and concentrate their functions in order to curb energy wasted through travel. Then there are the measures to be taken with regard to building where the use of renewable energy and a considerable improvement in the thermal insulation of the envelope of the buildings coupled with controlled air renewal are advocated. But does a kind of architecture that would act at a more modest, finer, smaller, more economical level exist? Architecture that is less burdensome, less present, a diluted, almost homeopathic sort of architecture, whose aim of correcting the climate would be respected? Today we would like to go a little further down the scale of measures to rectify the climate, to explore sensitive areas closer to the body, on the surface of the skin, to reach the point where architecture dissolves and becomes pure thermogenesis. But let us not be misguided! While this research obviously consists of saving the energy expended in the building and fighting against global warming, it also involves discovering new ways of housing and spatial compositions, developing new design and beauty strategies, where scales mix, where architecture becomes as much construction and structure as food and sweating.

The exhibition at the Royal Danish Academy of Fine Arts School of Architecture is part of such a strategy. It brings about the emergence of new architectural solutions by no longer acting on the delimitation from the outside of a comfortable climate where the temperature of the body can be maintained effortlessly at 37°C, but in the emergence, from the inside, of endogenous architectural solutions. It explores the microscopic, digestible and electromagnetic forms of architecture, somewhere between dietetics, thermogenics and esthetics.

So, in the northern winter, we propose to switch off the heating, to turn off the electric light, to open the windows of the exhibition room and to let in the night and the cold throughout the duration of the exhibition, from November 13 to December 13, 2009. Our work begins then, by recomposing the space based on bodily needs, by alleviating shortcomings and deficits one by one, by developing an endogenous thermogenesis that gradually develops outside the body without ever becoming clothing or house. The architecture is here a juxtaposition of elements that each responds to a shortcoming, a insufficiency, caused by the coldness of the Danish winter and the decrease in the sunshine and the length of the day at these latitudes so distant from the equator, plunged into winter. Architecture closer to the body that, element by element, provides responses by means of vitamins D and A, melatonin, heat, nutrients

-Electromagnetic radiation at 470 nanometers to block the excessive secretion of melatonin caused by the weak or absent daylight.

-A rectangular table (polyethylene) covered with raw oily fish such as salmon, trout, halibut, herring, sprat and mackerel presented in the form of sashimi, ready to eat to immediately supply the body with vitamin D, normally provided by the sun.

-A square, matt black, polyethylene platform, to capture insofar as possible the electromagnetic radiation present in the room, harness it and transform it into heat.

-A horizon of ultraviolet rays, to allow synthesizing vitamin D, missing due to Denmark's short winter days.

-A slatted platform beneath which is a field of peppers (ground heating) illuminated on the periphery by photosynthetic lights. Peppers stimulate the ion channel and TRPV1 neuroreceptors which detect temperatures above 44°C and is also activated, in the presence of "capsaicin", a molecule found in red peppers. Capsaicin and temperatures above 44°C stimulate TRPV1 which triggers a sensation of very high heat.

-A hot chili tea production device to drink the heat of the chili peppers

-A cup dispenser for the tea

-A professional cooking appliance for boiled meat to be able to ingest proteins that constitute the fuel for the heat produced by the body to maintain its temperature at 37°C.
-Bourguignonne-style fondue forks or disposable wooden chopsticks

-A visible field of carrots, which can be eaten to improve night vision by providing vitamin A

-Techno music played via directional loudspeakers at a 120 beats per minute to help the heart rate increase from 60 to 120 BPM, thus increasing metabolism and muscle activity.

-A compact bench for 5 people to share human warmth

Gradient of effusivity

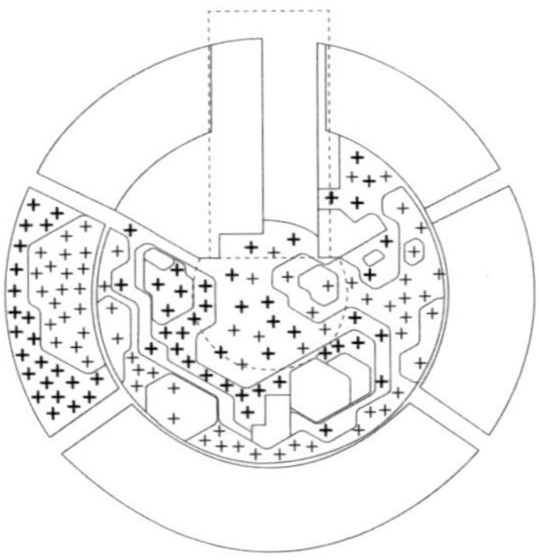

Through computer simulation of the sunlight of the Maison de la Radio, we map the mean annual distribution of the sun in the Agora to establish two extreme poles of a gradient of intensity, from the sunniest to the shadiest, from warm to cold, from soft to resonant, from winter to summer. In the shadiest regions of the Agora (which will certainly be preferred in summer), as revealed by computer analysis, we boost its freshness solely by using high effusivity materials, in other words, materials that will appear cold to the touch: polished granite, metallic tin. More resonant, these areas will host the bar and a more distracted and public regime of speech. The places heated by the direct rays of the sun will constitute the wintriest region of the Agora, where you can warm up with a hot drink, in the soft atmosphere of wool and then fir, low effusivity materials, warm to the touch, absorbing sound and conducive to whispering. Because going for a drink at the bar is originally about going to cool down in the summer by seeking the cool of the shade, to quench the thirst with a cold drink. On the contrary, enjoying a hot drink in winter means getting warm, warming up with a steaming drink, a cup of coffee or hot chocolate. It is these two fundamental polarities for the cool and the heat, for the shade and the sun, for cool drinks and hot drinks, that we map and spatialize throughout the space of the Agora, in an acoustic and thermal progression, gradually between the high thermal effusivity sound echoer and the low thermal effusivity sound absorber.

Reverberant carpet

The light in the mountains is stronger and bluer than the light on the plain. This is due to the fact that the higher you climb, the thinner the atmosphere becomes. Because the gases that make up the atmosphere block certain wavelengths of the sun's electromagnetic radiation, which then diffuse, transforming into heat. Thus, the greater the thickness of the atmosphere traversed by the sun's rays (as in winter, in the morning or in the evening, at sea level or at latitudes distant from the equator), the lower the intensity of light coming from the sun and the redder the light becomes. Indeed, nitrogen molecules block and diffuse blue wavelengths, which gives the sky its blue color. And the other molecules, of oxygen and ozone, as well as water vapor and water in liquid form, that is to say in cloud form, attenuates the intensity of the light. The more you descend in altitude the further you head toward the poles, or in the evening and in the morning, the light intensity decreases proportionally, while the thickness of the atmosphere through which the light passes increases. Thus, the higher up you go in altitude, the higher you climb up the mountains, the stronger and bluer the light will be.

For the exhibition La Luce Alpina, at the Museum of Saint Gall, at 669 meters in altitude, we propose to reproduce this dazzling, cold mountain light that is more intense and bluer than that of Saint Moritz, at an altitude of 1822 meters.

To do so, we propose doubling the amount of incident light, as snow does, by creating a white floor on the existing brown flooring. The shape of this white floor is precisely sized and adjusted to reflect the sun's rays directly entering the exhibition room throughout the day. This passive, inert first device therefore plays on the physical reflective property of the materials. The floor will be made of wood, painted white and varnished to increase the reflection of the white, dazzling the visitor entering the room.

To slightly tinge the light of the room with the blue missing from the natural light of Saint Gall having disappeared in the layer of 1153-meter thicker atmosphere that separates Saint Moritz from Saint Gall, we propose to replace the bulbs of the room's artificial lights with blue bulbs that will tint the natural yellower light of Saint Gall blue, to make it resemble that of Saint Moritz.

Finally, we propose to house in our room the painting by Giovanni Segantini, Mezzogiorno sulle Alpi (o Meriggio), of 1891 whose pictorial technique, that of multiplying, by slipping between the patterns of colors, tiny dots of pure white paint to increase the brightness of the painting, is close to our architectural standpoint, which is to drastically increase the brightness of the room by incorporating this white floor.

The Anthropocene Style

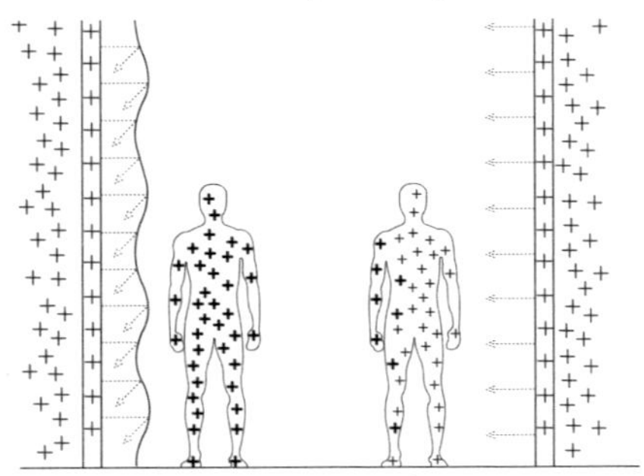

Our goal is to propose a new style in the History of the Decorative Arts, which after Regency, Louis XV, Empire and Louis-Philippe, would be coined "Anthropocene Style", a style we aim to be elegant, innovative and contemporary, updating the art of interior design and decoration to meet the contemporary challenges of sustainable development, the reduction of energy consumption and the emission of greenhouse gases. Redesigning the lines, the patterns and the geometries of walls, ceilings, floors, woodwork, parquet flooring and moldings according to the optical behavior of the sun's rays, to multiply the natural sunshine, to reduce by conduction the excess heat accumulated beneath the ceiling, to increase the thermal insulation coefficient of the walls and avoid (cold) thermal bridges. Rethinking the intelligence of the materials used according to a precise choice of physical, optical, thermal and acoustic properties, their porosity to water vapor, to air, their conductivity, their diffusivity or their effusivity. Rethinking the materials, colors and textures of interior design materials and innovating. Rethinking materials that in some places reflect infrared, absorb them in others, let it pass elsewhere, while at the same time they reflect the short wavelengths of visible white light or absorb it to still take advantage of their heat.

For the 25th Biennale "Interieur" in Kortrijk, we propose to design 7 pieces of decorative art in Anthropocene style. Designed for temperate and continental regions, our rooms come in response to the cold exterior climate by improving indoor comfort and warmth according to precise physical properties. The "low effusivity rug" is a new type of carpet composed of materials with the lowest thermal effusivity so as not to take up the heat of the human body in contact with it. The "low emissivity curtain" is a curtain made of a material with the lowest thermal emissivity so as not to allow the radiant coldness of the outer walls to emit inwards by reflecting it back on itself, thus not cooling the human body placed before it. The "high reflectance mirror" is a mirror reflecting not in the visible, but in the infrared, reflecting the radiant heat emitted in the infrared by the human body on itself. The "low thermal conductivity tapestry" is a tapestry made of higher thermal insulation materials so as not to transmit by conduction the coldness of the exterior walls to the indoor air. The "convective screen" is a screen made of the most airtight material available today. Placed in front of the entrance door, it prevents draughts which cool the skin by convection, preventing the heat produced by the human body from remaining on the surface of the skin and warming it. The "Spectral light" is a lamp whose light spectrum is reduced to wavelengths corresponding to human vision, thus making it possible to reduce energy consumption by the same amount. The "convective chair" is a chair whose elevated seat enables reaching warmer layers of air, which, according to Archimedes' law, is vertically stratified causing cold air to sink and warm air to rise. The materials of these 7 items of "Anthropocene Style" decorative art are chosen for their great ability to respond to the physical behaviors appropriate to their task.

Spectral light

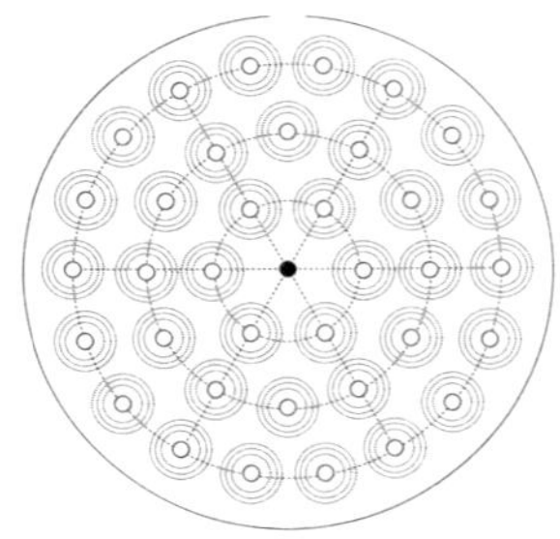

Over the past twenty years or so, the lighting industry has rightly begun to focus more on the composition of the artificial electromagnetic light spectrum, the wavelengths of which it is composed. There are three reasons for this: the first concerns health and biology, the second, energy saving and the fight against global warming, and the third and last reason is intrinsic to the technical development of artificial lighting: the physical features of electroluminescent lighting (LED) to compose light by additive synthesis of monochromatic wavelengths; each of the semiconductors emitting a single wavelength. Until now, the spectrum emitted by electric lights was more or less an unintended physical consequence of the use of a certain material, a certain technique. The incandescence of the tungsten filament produced a poor light spectrum in the blue spectrum but extremely rich light in the red and infrared without anyone really deciding that this should be so. Fluorescence produced not a complete and continuous, but a discontinuous, fragmentary electromagnetic spectrum, limited to a few peaks in a few visible wavelengths, omitting some colors. This was primarily for health reasons thanks to the development of scientific and medical knowledge carried out by health research institutes, particularly in the field of chronobiology. What was observed was that the poverty of the spectra of fluorescent light, in particular, could lead people working in artificially lit environments to suffer headaches or problems of insomnia that did not occur in people working in natural light. The causes of these health problems were twofold. The first cause is the poor spectrum of light emitted by fluorescence, which is highly partial and discontinuous compared to the broad, linear and complete spectrum of natural sunlight, which covers all visible wavelengths. Made up of a few solitary peaks in bad wavelengths compared to those perceived by the human eye, fluorescent light forced the human eye to invent for itself the missing colors that it did not see, to reconstruct them by interpreting them, which caused eye fatigue and headaches. The second reason is the discovery of the influence of light, the role of its intensity and certain wavelengths in the synchronization of daily human biological rhythms known as circadian cycles. Indeed, from the 1980s, thanks to the input of Alfred Lewy (A. Lewy, T.A. Wehr, F.K. Goodwin, et al., "Light Suppresses Melatonin, Secretion in Humans." Science 210:1267-1269, 1980), it is understood that light variations between day and night have a direct effect on the variations of certain hormones in the human body, mainly melatonin, but also growth hormone or cortisol. What is demonstrated, therefore, is that intense light, like daylight, blocks the secretion of these hormones, while darkness, like at night time, activates their secretion and participates in the arousal of sleep. This explains why some people working in environments under electric lighting sometimes present a state of permanent fatigue and a risk of developing what has been called SAD (Seasonal Affective Disorder), that is to say a kind of nervous depression caused by the excessive presence of melatonin in the blood due to the lack of light stimulation. The intensity of artificial light sources as emitted by fluorescent or incandescent lamps is not strong enough to block the secretion of melatonin. While light intensity was the first criterion, it was as of the new millennium that the light spectrum began to be precisely decomposed to highlight the effect of certain wavelengths either blocking or not blocking melatonin secretion. It was noted that wavelengths of around 460 nanometers, in the blue spectrum, even at a low intensity, can block melatonin when they reach the retina. However, wavelengths beyond 550 nanometers, beyond the yellow, even at high intensity, play no role in melatonin suppression. Halogen or fluorescent light have very poor spectra with respect to these criteria since they emit practically no electromagnetic waves at 460 nanometers. As they do not block melatonin, these lamps are responsible for increased fatigue and sleep disorders. Armed with this knowledge, the lighting industry timidly began to readjust the electromagnetic spectra emitted by artificial sources.

Along with health, it is then reasons of economics and sustainability that have prevailed to increase the interest in an enhanced definition of the artificial electromagnetic spectra by appreciating the scale of the considerable energy loss of incandescent or halogen lamps, the emission of unnecessary wavelengths in light beyond 800 nanometers, in the infrared. Indeed, an incandescent bulb would convert only 5% of the energy (and therefore of the electricity consumed) into light, that is to say in the visible spectrum between 400 and 800 nanometers. The rest of the energy would be wasted by the emission of infrared, invisible heat, which represented a disproportionate loss of energy. Halogen had the same defect of emitting few visible waves compared to invisible ones, in the infrared, at the same ratio of about 7% light to 93% heat. By observing the electromagnetic spectrum of their respective light, the two spectra can be seen to continuously cover all wavelengths but go far beyond the visible to spread incommensurably in the infrared. The increased awareness of the responsibility of energy consumption in the light of global warming has aroused criticism of these light sources primarily in favor of the fluorescent light. The advantage of fluorescent or fluorocompact light from the point of view of energy savings is, on the one hand, that it does not emit in the infrared.
On the other hand, it does not emit a continuous spectrum, unlike incandescent or halogen lamps, rather a series of peaks in some visible wavelengths, which enables considerable energy savings due to not emitting in the entire visible spectrum. However, while this reduction in the fluorescence emission spectrum is welcome from the point of view of reducing energy expenditure, it poses a problem from a health point of view as it emits precisely none of the wavelengths perceived by the human eye, thus leading to eye fatigue.

So, today biological and health-related concerns, the fight against global warming and to reduce energy consumption that are driving the design of light, less concerned with the material and the container than with the immaterial and the content, that is to say a design of the light spectrum itself, of its electromagnetic composition, somewhere between pointillism, divisionism and additive color synthesis. The new technique of illumination by light-emitting diodes further provokes this design of the spectrum by the addition and juxtaposition of monochromatic light waves because each LED emits independently in a precise wavelength according to the semiconductor material used.
The design of light by electroluminescence is therefore a spectral composition that involves a precise choice of certain semiconductor materials emitting certain wavelengths which are then added. If the spectral design of electroluminescent lamps has so far been carried out according to psychological or analog criteria (reproducing the comforting spectrum of incandescent light or natural sunlight), we propose instead to work according to physiological criteria by adding, wavelength by wavelength, the ones considered healthy, necessary and economical, which we then delicately assemble, in the manner of a perfumier with their fragrant essences and molecules or a cook with their ingredients, to reconstitute a novel light, with a biological and an ecological value.

Our project proposes to work on the light spectra of indoor life, adding only the necessary but vital wavelengths, for the sake of economy. We proceed according to a technique similar to the impressionist Claude Monet or Georges Seurat's pointillism, by adding particular, separate wavelengths, which, through addition and convergence, will reproduce near-white light, but lacking the wavelengths that are not perceived by human, animal and plant physiology. The precise composition of the wavelengths of our lamp will be at the discretion of the user who can choose the spectral composition according to their wishes for their home, according to the use made of the room, at their pleasure. Thus, first only the precise wavelengths are assembled that are biologically active in humans during the day: 437 nm, 533 nm and 564 nm, according to the three photoreceptor cones around the blue, green and yellow, and 460 nm to block melatonin. We then propose a second light, a night light, the light of the full moon, lower in intensity according to two wavelengths: 498 nm for the reception by the rods of human scotopic vision, 662 nm as a non-stimulating wavelength of the melatonin hormone. But we generously open up a specifically human universe to the other inhabitants of the house, to domestic animals, dogs, cats and birds, to green plants, whose perceived wavelengths we include, to photo reception, to growth, to photosynthesis that each individual can assemble freely in the lamp to compose its spectral make-up like an indoor sun, economical and healthy. The lamp we propose consists on the one hand of a transparent glass globe whose lower part is opaline in order to mix and diffuse together the different emitted wavelengths whose different colors will blend toward white through additive color synthesis. The upper part of the globe is peppered with 36 electrical connectors into which the light tubes will be inserted and electrically connected. The convex shape of the upper part of the globe enables directing each of the 36 light tubes towards the lower opaline surface of the globe in order to force their light to converge so that it mixes and homogenizes with the others in order, through addition and synthesis, to produce white light close to the spectrum of natural sunlight.

The second part of the lamp is composed of light tubes that we call "Artificial rays". Each of these artificial rays has one and only one wavelength. There are 18 different types of artificial rays with 18 different wavelengths, corresponding to the 18 peak wavelengths perceived by the living species we have chosen. Six wavelengths for humans (4 during the day, 2 during the night), 6 for plants, 4 for birds, 2 for dogs, and 1 for hamsters. Depending on the natural intensity of the sunlight, some artificial rays must be doubled or tripled in number (especially around the blue) to respect the natural proportion of solar light intensity according to the wavelength. Each individual can then freely assemble in the lamp the number of rays they desire, in order to produce a lamp for humans and plants, for example, or humans, dogs and plants, and so forth. The 36-hole lamp exhibited at the Artemide Corso Monforte Showroom for the Salone del Mobile 2015, presents all wavelengths together, for humans, plants, birds, dogs and hamsters. There are 11 rays for plants, 9 rays for humans (we will not include the night rays), 8 rays for birds, 5 rays for dogs, 3 rays for hamsters, 36 rays in all. All the lights of these rays converge towards the lower, opaline part of the lamp to jointly recompose a white light close to that of the sun, but more economical and more biologically true. Our work as architects and designers involves a quasi-chemical dissociation of space into elementary particles, into wavelengths, humidity levels, light intensities and heat transmission coefficients, for example, but also into hormone secretion rates, kilocalories and nanometers to then be synthesized into a new, more essential form responding to the contemporary issues of health and sustainable development. The broadening of the field of the real produced by increased scientific knowledge modifies the field of art, which shifts into new dimensions, slides onto other phenomena, arouses other perceptions. We are now interested in forms that are no longer composed together to form a whole, but dissociated, exploded into fragments of the real, into particles of the sensitive. We explore the infinitely small, we analyze the invisible, optical or sound spectra, we decompose the real into electromagnetic, chemical or biological particles, which we then recompose, but with only some, not all of its elements. There is a light, the rational clarity of the Enlightenment, a whiteness of writing, an almost chemical objectivity, the absence of fiction and narration, but from which finally emerges a sensation of wonder, a "disturbing unreality", which would be "a further realism more than a deliberate fiction", as Gérard Genette rightly said about Robbe-Grillet. But our interest in decomposing reality and the synthesis of two or three chemical and electromagnetic elements that compose it is not just an esthetic project. More deeply, it seems to us to be a necessary process in the reassessment of the underlying, often hidden fundamental historical reasons that generated the human landscape, triggered a certain architecture, a kind of urban planning, a way of designing furniture. It is then a method that allows rethinking architecture and industrial design outside of any common place, cliché and picturesque until reaching a certain form of truth, economy and beauty.

INDEX

List of projects

Projects presented in Chapter I. III. V. VI

Conduction

Three thermal bubbles
Program: New Sports Hall
Authors: Philippe Rahm architectes
Client: Municipality of Vétroz
Location: Vétroz, Switzerland
Date: 2011

Conductive Condominium
Program: Housing
Authors: Philippe Rahm architectes
Client: IBA-Hamburg
Location: Hamburg, Germany
Date: 2010

Thermal Conductivity
Program: School
Authors Philippe Rahm architectes
Location: La Neuveville, Switzerland
Date: 2007

Conductive Layers
Program: Competition for the Kantor Museum
Authors: Philippe Rahm architectes
Location: Krakow, Poland
Date: 2006

Convection

The park as a climatic gradient
Program: Tuchkov Buyan Park
Authors: Philippe Rahm architectes, Agence Ter
Client: City of Saint Petersburg
Location: Saint Petersburg, Russia
Date: 2020

Convective House
Program: Project for a block of Four Apartments
Authors: Philippe Rahm architectes
Client: IBA-Hamburg
Location: City of Hamburg, Germany
Date: 2010

Domestic Astronomy
Program: Exhibition: "Green Architecture for the future"
Authors: Philippe Rahm architectes
Collaboration: fabric.ch, Amy O'Neill
Client: Louisiana Museum of Modern Art
Location: Humlebæk, Denmark
Date: 2009

Meteorological Museum
Program: Competition for a Contemporary Art Museum
Authors: Philippe Rahm architects
Location: Wroclaw, Poland
Date: 2008

Interior Gulf Stream
Program: House and Atelier
Authors: Philippe Rahm architectes
Client: Dominique Gonzalez-Foerster
Location: Countryside near Paris, France
Date: 2008

Digestible Gulf Stream
Program: Exhibition:
Venice Biennale - 11th Architecture Exhibition
Authors: Philippe Rahm architectes
Collaboration:
Piero Macola, Syd Matters, Comsol
Date: 2008

The Archimedes Houses
Program: Holiday Homes
Authors: Philippe Rahm architectes
Client: SYMIVA,
Location: Île de Vassivière, France
Date: 2005

Emissivity

The Anthropocene Style
Program: Renovation of Hôtel de Coulanges
Authors: Philippe Rahm architectes
Client: URBEM, City of Paris
Location: Paris, France
Date: 2014

Climatic Gradient
Program: Competition for the Storage of Works of Art
Authors: Philippe Rahm architectes
Client: Musée National de Suisse
Location: Affoltern am albis, Switzerland
Date: 2013-2014

Limpidarium d'Aria
Program: New district in Milan covering 44 hectares
Authors: Philippe Rahm architectes, OMA, Laboratorio Permanente
Client: COIMA, City of Milan, trenitalia
Location: Milan, Italy
Date: 2019

Effusivity

The Effusive Carpet
Program: Exhibition
Authors: Philippe Rahm architectes
Client: Office KGDVS, Kortrijk Biennale
Location: Kortrijk Biennale
Date: 2018

The Effusive Pool
Program: Exhibition
Authors: Philippe Rahm architectes
Client: Istituto Svizzero
Location: Istituto Svizzero, Milan, Italy
Date: 2018

Gradation of Effusivity
Program: Public Space (cafeteria, book store, stage)
Authors: Philippe Rahm architectes, Nicolas Dorval-Bory Architectes
Client: Radio France
Location:
Maison de la Radio, Paris, France
Date: 2017-

Radiation

Open Climate
Program:
Conceptual Pavilion for an Art School
Authors: Philippe Rahm architectes
Client: École Nationale Supérieure des Beaux-Arts de Nantes
Location: Nantes, France
Date: 2006

Expanded house
Program: Artist Residency
Authors: Philippe Rahm architectes
Client: Grizedale Arts
Location: Ambleside Cumbria, England
Date: 2006

Black Day / Toward Diurnism
Program: Urban Furniture
Authors: Philippe Rahm architectes
Client: City of Gdansk
Location: Gdansk, Poland
Date: 2005

The Various Lights of Peru
Program: Competition for the Contemporary Art Museum
Authors: Philippe Rahm architectes
Client: Lima Art Museum
Location: Lima, Peru
Date: 2016

Climatheque
Program: Competition for the Human Sciences building
Authors: Philippe Rahm architectes
Client: University of Lausanne, Vaud Canton
Location: Dorigny, Lausanne, Switzerland
Date: 2020

One Kilometer of Shade
Program: Masterplan for a new district of Basra, Irak, on the side of the former oil port
Authors: Philippe Rahm architectes, Technital, Roberto D'Agostino
Client: Republic of Irak, City of Basra
Surface: 80 hectares
Date: 2019

Evaporation

Relocated Cellar
Program: Exhibition
Authors: Philippe Rahm architectes
Client: SF-Moma
Location: San Francisco, USA
Date: 2010

The Evaporating House
Program: Housing
Authors: Philippe Rahm architectes
Client: IBA-Hamburg
Location: Hamburg, Germany
Date: 2010

Descending climates
Program: Competition, New National Museum
Authors: Philippe Rahm architectes
Location: Tarfu, Estonia
Date: 2005

Mollier Houses
Program: Holiday Homes
Authors: Philippe Rahm architectes
Client: SYMIVA
Location: Vassivière en Limousin, France
Date: 2005

An Adiabatic Quarter
Program: Art Museum
Authors: Philippe Rahm architectes
Client: Sharjah Art Foundation
Location: Sharjah, United Arab Emirates
Date: 2019

Pressure

Interior Wind
Program: Covered Pool
Authors: Philippe Rahm architectes
Location: Lieu dit Montoly, Switzerland
Date: 2011

Filtered Realities / Paris Air
Program: Floating Office Building
Client: Voies Navigables de France (VNF).
Authors: Philippe Rahm architectes Client: VNF - Voies Navigables de France
Location: Paris, France
Date: 2008

Airscape
Program: Sports Center
Authors: Philippe Rahm architectes
Client: Rhône-Saône Développement
Location: Les Docks, Quai Rambaud, Lyon, France
Date: 2008

Windtrap
Program: Competition for a Sports Hall
Authors: Philippe Rahm architectes
Location: Slovenia
Date: 2008

Airflux
Program: Sports hall
Authors: Philippe Rahm architectes
Client: Town of Martigny
Location: Martigny, Switzerland
Date: 2007

Indoor Seasonal Shelters
Program: Office building and shopping center
Authors: Philippe Rahm architectes
Client: Bureau Vallée
Location: Les Clayes-sous-Bois
Date: 2022

Built Projects

Taichung Central Park
Program: Urban park and buildings
Authors: Philippe Rahm architectes, Mosbach paysagistes, Ricky Liu and Associates
Client: Government of Taichung city, Taiwan
Location: Taichung, Taiwan
Surface: 67 hectares
Design: April 2012 - December 2013
Construction: January 2014 - December 2020

Luminance
Program: Exhibition architecture
Authors: Philippe Rahm architectes
Client: Luma Foundation
Location: Arles, France
Surface: 2700 m2
Date: 2015

Spectral Light
Program: Lighting Design
Authors: Philippe Rahm architectes
Client: Artemide, Milan, Italy
Date: 2015

Evaporated Rooms
Program: Apartment for a Young Physician
Authors: Philippe Rahm architectes
Client: Louis M.
Location: Perrache, Lyon, France
Surface: 70 m2
Design: January 2011 - June 2011
Construction: September 2011 - January 2012

White Geology
Authors: Philippe Rahm architectes
Program: Architecture for the "la force de l'art 2009" Triennial of Contemporary Art
Client: Centre National des arts Plastiques, Réunion des Musées Nationaux, Ministère de la Culture
Location: Grand Palais, Paris, France
Date: 2009

Bibliography (in order of importancy)

Vitruvius, Cesare Cesariano, and Carol Herselle Krinsky. 1969. *Vitruvius De architectura.* München: Fink.
Tacitus, Cornelius, and John Jackson. *The Annals of Tacitus*, Book V. Cambridge: Hardvard University Press, 1937.
Alberti, Leon Battista. *On the Art of Building in Ten Books.* Cambridge, Mass: MIT Press, 1988.
Intergovernmental Panel on Climate Change. "Climate Change 2014 Synthesis Report Summary for Policymakers", p.4. IPCC, 2014.
Global Alliance for Building and Construction. "2018 Global Status Report, Towards a Zero-emission, Efficient and Resilient Buildings and construction Sector", p.11. GlobalABC, 2018.

Image credits

The physical values presented in this book come from multiple cross-sources, such as https://www.engineeringtoolbox.com, sciencedirect.com, or thermworks.com

Acknowledgments
Philippe Rahm

Collaborators at Philippe Rahm architectes

Andrej Bernik, Pei-Yao Wu, Gabriel Cuellar, Cara Liberatore, Muriel Maggiol, Valentina Leodori, Liam Doyle, Brice Pannetier, Cyrille Berger, Quentin Vaulot, Jérôme Jacqmin, Mustapha Majid, Mathieu Bujnowskyj, Joao Miguel Bento Asper Banha, Jessica Leung, Enrico Prenna, Aleksandra Duczmal, Tina Tengxiao Gao, Patrick Harvey, Steven Ischkin, Nastya Kalmykova, Beomki Lee, Nichola Czyz, Mark Lien, Ryu Matsuzaki, Dmitry Stolbovoy, Josh Van Zak, Victor Wichrowski, Malgorzata Wylupek, David Colombini, Michael Liu, Hung Yunya, Vicky Chuang, Yifan CAI, Eddie Hsieh, Chihchen Huang, Juliette Valat, Thomas Flores, Lior Ramon, Caroline Spielvogel, Jeanne Guerin, William Solis, Sofia Boarino, Sofia Gozzi, Lea Mosconi, Rena Gienseck, Yifain Cai, Marion Valli, Cyril Assaad, Mayumi Iitsuka, Steven Ischkin, Aryan Ofeany, Timothée Boitouzet, Anastasia Kalmykova, Victor Wichrowski, Agnieszka Nycz, Charlotte Schmidt, Elie Moutel, Ariane Merle D'Aubigné, Yoann Jacquon, Beomki Lee, Alexandra Cammas, Marc Eychenne, Renaud Pinet, Marina Huguet i Blasi, Patrick Harvey, Rocco Vitali, Olimpia Starzycka, Isabella Vecchini Ferraz, Simone Iannucci, Enrico Prenna, Pauline Alexandrou, Fanny Betfert, Sung Di, Etienne Gilly, Nicola Fiodor, Lucien Chartier, Maelle Saunois Alexandre, Killian Dumolin, Sampah Pediredla, Alessia Zambon, Maxime Leclerc, Audrey Tseng Fischer, Xiaoyu Pan, Maria Paulina Chaparro Garzon, Miachael Liu, Pedro Alcantara, Szymon Łapajm, Yun-Chih, Slah Ben Chaabane, Konrad Chmielewski, Yu Yang Huang, Emilie Kang, Cheng Peng, Camille Lacadee, Min Sun, Chih-Heng Chuang, Nicolas Souchko, Alejandro Falcon, Aleksandra Wolffgram, Chia Chieh Lee, Amir Halabi, Yifain Cai, Mou Yuyangguang, Aurore Chartier, Mégane Sorres, Victoria Ciprien, Charles Guinant, Evan Ribadeau Dumas, Isabela Ferrari, Yann Schwaller, Bertha Chen, Ambre Lahlou, Filippo Montagna.

Acknowledgements

Lucien Rahm, Louise Rahm, Jean-Max Colard, Isabelle Mical, Dominic Thomas, Francis Rambert, Antoine Picon, Dominique Gonzalez-Foerster, Bruno Latour, Philippe Potié, Liz Diller, Inaki Abalos, Philippe Parenno, Maja Hoffmann, Emmanuel Caille, Werner Rahm, Anne-Françoise Rahm, Jean de Martini, Roberto D'Agostino, Laura Bunagenti, Lorenzo Benedetti, Jenny Sabin, Hou Hanru, Marcello Smarelli, Olivier Marguerit, Marc-Olivier Wahler, Sébatian Rivas, Thomas Lévy-Lasne, Guy Lelong, Ippolito Pestellini Laparelli, Ryoko Sekiguchi, Sana Frini, Aurelien Gillier, Didier Rittener, Benoit Peeters, Kim Hou, Patrick Keller, Christophe Guignard, Benjamin Lafore, Sébastien Martinez Barat, Paulo Dam Mazzi, Giacinto Cerviere, Andrea Simitch, Luc Debraine, Massimiliano Scuderi, Sascha Marko Roesler, Evelyne Jouanno, Nicolas Dorval-Bory, Shantel Blakely, Bart Lootsma, David Gissen, Jimenez Lai, Ulrich Müller, Bénédicte Dorighel, Javier Fernandez Contreras, Henri Bava, Andrés Jaque, Irene D'Agostino, Beatrice Galilee, Sean Lally, Marco De Michelis, Ellen Wahler, Emanuele Quinz, Matthieu Poirier, Manou Farine, Maddalena Terragni, Annie Ratti, Jérôme Mauche, Alexandre Labasse, Marianne Carrega, Emeric Lambert, Jean-Christophe Quinton, Marie Darrieussecq, Alain Robbe-Grillet, Piero Macola.

This book Philippe Rahm architectes Climatic Architecture has received the support of Istituto Svizzero (Swiss Institute in Rome), the Graham Foundation, USA, Swissnex, San Francisco, USA, the MAAT, Museum of Art, Architecture and Technology (Fundacao EDP), Lisbon, Portugal, San Francisco Art Institute, USA, and École Nationale Supérieure d'Architecture de Versailles

Istituto Svizzero fundação edp

Graham Foundation SFAI San Francisco Art Institute

swissnex maat

École nationale
supérieure d'architecture
Versailles

Biographies
Principal and Office

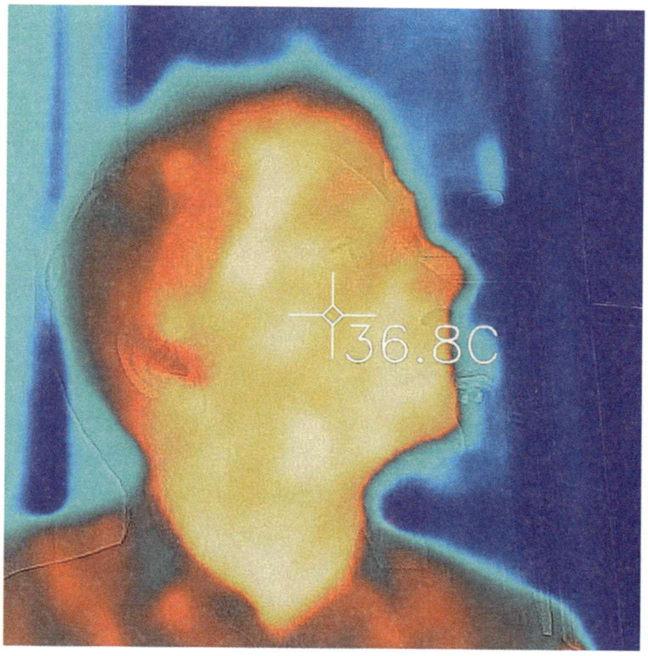

Thermography: Philippe Rahm

Philippe Rahm

Philippe Rahm (born 1967) is a Swiss architect who graduated from the École Polytechnique Fédérale de Lausanne in 1993 and received his doctoral degree in Architecture from the University of Paris-Saclay in 2019. His thesis, "Histoire naturelle de l'architecture" (Natural History of Architecture) is published in 2020 by the Éditions du Pavillon de l'Arsenal. In 2019, he was awarded the Silver Medal from the Académie Française d 'Architecture. He is a knight of the Order of Cultural Merit of Monaco. He has given numerous lectures on his work, including at Yale, Beijing Forum, ETH Zurich and UCLA. Philippe Rahm was a resident of Villa Medicis in Rome in 2000. He has been a visiting professor at the AA School in London (2005-2006), at the Mendrisio Academy of Architecture in Switzerland (2004-2005), at the EPFL (Ecole Polytechnique Fédérale de Lausanne) in 2006-2007, at the School of Architecture at the Royal Danish Academy of Fine Arts in Copenhagen (2009-2010), at the AHO School of Architecture in Oslo, Norway, in 2010, at Princeton University in the USA (2010-2012), at Harvard GSD from 2014 to 2016, at Columbia University in 2016 and 2023, and at Cornell in 2019 and 2020. He is a full professor at the Ecole Nationale d'Architecture de Versailles and an associate professor at the HEAD (HES-SO) in Geneva.

Philippe Rahm architectes

The architectural firm Philippe Rahm architectes has been established in Paris since 2008. Its work, which stretches the field of architecture between physiology and meteorology, has acquired an international audience in the context of sustainability. In 2011, Philippe Rahm architectes together with Catherine Mosbach & Ricky Liu won the international competition for Central Park, a new, 57-hectare urban park in Taichung, Taiwan and its buildings, that opened in December 2020. In 2017, together with Nicolas Dorval-Bory, he won the Agora architecture competition at La Maison de la Radio (Radio-France) in Paris, currently under study. In 2019, with OMA, he won the urban redevelopment project for the 62-hectare Farini district and a 14-hectare park in San Cristoforo in Milan, Italy. In 2008, he was one of the twenty international architects selected for the 11th Venice Architecture Biennale. In 2017 he took part in the Seoul and Chicago Architecture Biennales, the biennales of Sharjah in 2019 and Tallinn in 2022. In 2002, he represented Switzerland at the 8th Venice Architecture Biennale with Décosterd & Rahm, associés. In 2007, a solo exhibition Philippe Rahm architectes (together with Gilles Clément) was dedicated to his work at the Canadian Centre for Architecture in Montreal. The books *Architecture météorologique*, published in 2009 in France by Archibooks, *Constructed Atmospheres* in Italy, in 2014, *Le jardin météorologique* and *Écrits climatiques* published by Éditions B2 in 2019 and 2020, *Arquitectura meteorológica* published by Éditions Arquine in Mexico in 2020 and *The Anthropocene Style* in 2023 (HEAD Publishing), are devoted to the work of Philippe Rahm architectes

CLIMATIC ARCHITECTURE

Published by
Actar Publishers
New York, Barcelona
www.actar.com

Author
Philippe Rahm

Book Conception
Philippe Rahm / Philippe Rahm architectes

Editorial Development
Marta Bugés / Actar Publishers

Book Design
Philippe Rahm architectes
Sofía Sandoval, Ramon Prat / Actar Publishers

Graphics and Illustrations
Philippe Rahm architectes

Texts
Philippe Rahm

Translations
Andy Clarke, Shantel Blakely

Printing and Binding
Arlequin & Pierrot, Barcelona

Indexing
ISBN: 978-1-63840-039-4
LCCN: 2022942869

Distributed by
Actar D, Inc.

New York
440 Park Avenue South, 17 Fl.
New York, NY 10016, USA
+12129662207
salesnewyork@actar-d.com

Barcelona
Roca i Batlle, 2-4
08023 Barcelona, Spain
+34 933 282 183
eurosales@actar-d.com

Printed in Europe
Publication date: September 2023